HYPOCRETINS

Integrators of Physiological Functions

HYPOCRETINS

Integrators of Physiological Functions

Edited by

Luis de Lecea
J. Gregor Sutcliffe
The Scripps Research Institute
La Jolla, California

Springer

Luis de Lecea
The Scripps Research Institute
La Jolla, CA 92037
USA

J. Gregor Sutcliffe
The Scripps Research Institute
La Jolla, CA 92037
USA

ISBN-10: 1-4614-9806-6
ISBN-13: 978-1-4614-9806-3

9 8 7 6 5 4 3 2 1

springeronline.com

FOREWORD

The first report that rapid eye movements occur in sleep in humans was published in 1953. The research journey from this point to the realization that sleep consists of two entirely independent states of being (eventually labeled REM sleep and non-REM sleep) was convoluted, but by 1960 the fundamental duality of sleep was well established including the description of REM sleep in cats associated with "wide awake" EEG patterns and EMG suppression. The first report linking REM sleep to a pathology occurred in 1961 and a clear association of sleep onset REM periods, cataplexy, hypnagogic hallucinations and sleep paralysis was fully established by 1966.

When a naïve individual happens to observe a full-blown cataplexy attack, it is both dramatic and unnerving. Usually the observer assumes that the loss of muscle tone represents syncope or seizure. In order to educate health professionals and the general public, Christian Guilleminault and I made movies of full-blown cataplectic episodes (not an easy task). We showed these movies of cataplexy attacks to a number of professional audiences, and were eventually rewarded with the report of a similar abrupt loss of muscle tone in a dog. We were able to bring the dog to Stanford University and with this as the trigger, we were able to develop the Stanford Canine Narcolepsy Colony. Breeding studies revealed the genetic determinants of canine narcolepsy, an autosomal recessive gene we termed *canarc1*. Emmanuel Mignot took over the colony in 1986 and began sequencing DNA, finally isolating *canarc1* in 1999.

As the leading instigator of the early efforts, I am content that the considerable outlay of funds to house and feed a large colony of narcoleptic canines for twenty years has paid off, and paid off handsomely I might add. This book is mainly about what has happened and is happening after the isolation of *CANARC1*. For individuals interested in sleep disorders, circadian rhythms, sleep regulatory processes, and other brain mechanisms, this book is a must read.

William C. Dement, M.D.
Stanford University,
Sleep Disorders and Research Center,
Palo Alto, CA

PREFACE

1. HOW IT STARTED

Although our publication of the discovery of the two peptides we named the hypocretins did not occur until 1998, the road to their discovery began in the spring of 1979. One of us (JGS) was completing postdoctoral studies with Richard Lerner at the Scripps Clinic and Research Foundation (now The Scripps Research Institute) following doctoral work performed under Wally Gilbert at Harvard, where the main thesis efforts had been in scaling up the DNA sequencing procedures developed by Gilbert and Allan Maxam. Those studies, which were described in a *Reflections* piece in *TIBS*,[1] represented the first time DNA analysis was used to determine the sequence of a protein in the absence of peptide sequence information from the protein itself. Thus, one learned first hand about the superior accuracy and rapidity of carefully collected DNA-based sequence information and the possibility that one could use the methods to determine the sequences of previously unrecognized genes for which neither proteins nor RNAs were known. This was a quarter of a century before mammalian genome sequences began to appear.

In Lerner's laboratory, JGS was working with Tom Shinnick on determining the first retrovirus genome sequence, that for Moloney murine leukemia virus.[2] Lerner knew of his long-term interest in neurobiology. The eminent neurobiologist Floyd Bloom, then a professor at the neighboring Salk Institute, asked Lerner if the technologies JGS had brought to Scripps could be used to study the brain. Bloom had just read the report describing the cDNA cloning of proopiomelanocortin (POMC),[3] and was particularly interested in finding out about undiscovered peptide neurotransmitters. Lerner passed alog the question.

JGS answered that, since so little was known about the molecular operation of the brain, a rational approach would be to construct cDNA libraries from brain mRNA. Individual cDNA clones could then be isolated from such libraries and their nucleotide sequences determined, thus allowing the amino acid sequence of the protein encoded by the corresponding brain mRNA to be conceptually translated. cDNA cloning had recently been developed, and it was the route by which clones were obtained for mRNAs encoding particular, already known proteins, such as POMC. cDNA libraries represent all of the mRNAs expressed in the tissue from which the sample was isolated and, thus, such libraries could inform us about the complete protein set, including those proteins that were not yet identified. But that led to an obstacle: how could we use the conceptual protein sequences translated from the cDNA sequences to learn about the putative proteins themselves? The proteins would not have been seen previously, and there were few protein sequences to which to compare the new sequences. The solution was to prepare synthetic peptide fragments of the proteins and use these to elicit antibodies that would react with both the peptides and the novel protein itself, thus facilitating its biochemical and anatomical characterization. The first opportunity to apply this approach on a putative protein in the Moloney virus sequence.[4]

Bloom, his then-postdoctoral fellow Rob Milner and JGS began a collaboration to test these ideas. The first effort was to use northern blot hybridization to characterize the mRNAs corresponding to clones in a brain cDNA library. By analyzing the size, abundance and tissue distributions of the mRNAs corresponding to nearly 200 clones isolated randomly from a rat brain cDNA library,[5] the team calculated that the 10^8 to $2x10^8$ nucleotides of mRNA complexity expressed by brain corresponded to 20,000 to 40,000 distinct mRNAs, numbers that compare favorably with modern estimates since entire genome sequences have been solved. Of these, approximately 65% were enriched in the brain compared to peripheral tissues. Most were of low abundance, on the order of one part in 10^5. The team raised antisera directed against synthetic peptides corresponding to one of the first partial putative brain protein sequences determined, and used these to detect the protein in brain extracts and to conduct a preliminary anatomical description of the protein later shown to be myelin-associated glycoprotein.[6]

These early studies represent the beginning of what have since come to be known as open-system approaches to mRNA expression analysis: mRNAs are detected because of their property of being expressed in the tissue sample isolated for study. Refinement of this approach led to the discovery of the hypocretins. The notion that the tools of molecular biology could be used to address fundamental questions about the operation of the mammalian brain was controversial at the time, but is one that no one today would argue. From the sequences of brain cDNAs, we learn about the nature of brain proteins. The cDNA clones also allow the study of the brain genes themselves and, with the advent of transgenic and knockout technologies, allow the power of genetic analysis to be brought to bear on the central nervous system, permitting a forceful molecular dissection of CNS physiology.

2. REFINEMENTS

The team was inspired by the success of its initial collaborative studies, which were among the first applications of what would later become known as genomics to neurobiology. Bloom, Milner and John Morrison moved from Salk to Scripps, and together with JGS initiated a program project aimed at expanding the effort. The Sutcliffe laboratory's role in the program was that of discovery of novel brain-specific proteins.

Which of the thousands of brain-specific mRNAs to characterize? The early studies demonstrated that many neuronal mRNAs exhibited differential distributions within the CNS; however, their expression was generally not restricted to a few discrete loci that could be attributed to specialized functions, but rather was variegated across the CNS. The studies therefore evolved to focus upon the identification of mRNAs that show a high degree of regional enrichment within the CNS. The logic that motivated this focus was that mRNA molecules with restricted expression were likely to encode proteins that are singularly associated with the unique functions of the cells that contain them and, perhaps, might be preferentially associated with particular physiological or behavioral processes compared with molecules with more general patterns of expression. Hence, their functional roles might be more transparent to investigation. Furthermore, such molecules might, in the future, provide highly specific targets for pharmaceuticals that would act only at the restricted site of target expression.

In order to enrich our libraries for such mRNAs, and the team turned to subtractive hybridization. Subtractive hybridization refers to a series of methodologies that compare cDNA sequences from one RNA sample, the target, with those from a second sample, the driver. Nowadays, the two complementary, antiparallel strands of cDNA can be produced in either of their single-stranded orientations, sense or antisense, using modern cloning and enzymological procedures. When sense strand from the driver is supplied in great excess over antisense strand from the target and these reagents are coincubated under conditions that favor the formation of double-strand hybrids, most of the mass of those sequences that the target and driver have in common becomes double stranded, whereas sequences from the target population that are absent from the driver population remain single stranded. The single-stranded material is isolated and used to create cDNA libraries enriched for target-specific mRNAs or enriched, radioactive probes for screening libraries. This methodology, originally developed by Timberlake[7] for studies on gene expression in fungi, has been progressively improved in the ensuing decades to a degree that it has allowed identification of mRNAs selectively expressed within complex mammalian nervous tissue.

Gabe Travis joined the group and introduced mixed-phase methods for increasing the apparent concentrations of the target and driver nucleic acids, thus vastly increasing the extent of their hybridization and hence enhancing enrichment for target-specific sequences.[8] He and Miles Brennan applied the improved subtraction method to isolate mouse retinal photoreceptor-specific mRNAs, including that corresponding to the product of the *retinal degeneration slow* gene, whose human homologue accounts for a considerable portion of heritable late-onset blindness.[9] Joe Watson had success in the identification of forebrain-enriched mRNAs, including those RC3/neurogranin[10] and G protein γ7.[11] Nevertheless, the method, although occasionally effective, was cumbersome and inconsistent.

These shortcomings were overcome with the advent of PCR. Hiroshi Usui, with design input from Mark Erlander, developed a method (simplified here) in which the target cDNA is cloned into a vector that introduces PCR primer binding sites on both sides of the cDNA insert. After hybridization with the driver, the single-stranded target is PCR amplified and cloned. The refined method, called directional tag PCR subtractive hybridization,[12] was used to prepare cDNA libraries enriched for clones of mRNAs specific to the striatum. Screening the large number of clones produced required high-throughput in situ hybridization. LdL developed a free-floating section method and brought anatomical expertise to the group.

As a result of implementing the refined subtraction method and high throughput in situ hybridization, we identified cortistatin, a neuropeptide of the somatostatin family expressed in the neocortex and hippocampus.[13]

3. ON TO THE HYPOTHALAMUS

With this powerful arsenal of new techniques and success record, we were prepared to search for new peptide neurotransmitters, and turned to the hypothalamus as a likely place to find them. We were joined in the search by Kaare and Vigdis Gautvik, who performed the subtractive hybridization studies that led to finding a clones for what later became known as the hypocretins.[14,15] Those studies are described in Chapter 1. Patria Danielson and Pam Foye conducted the bulk of the blotting and sequencing experiments; Tom Kilduff and Cristelle Peyron most of the neuroanatomical characterizations.

These studies were not big biology, but were so cross disciplinary as to demand extensive collaborations. The trail that the team embarked upon in 1979 has been consistently productive, but finding the hypocretins has been particularly gratifying, in part because finding such proteins was a large part of its original impetus.

4. REFERENCES

1. J. G. Sutcliffe, pBR322 and the advent of rapid DNA sequencing, *TIBS* **20**, 87-90 (1995).
2. T. M. Shinnick, R. A. Lerner and J.G. Sutcliffe, Nucleotide sequence of Moloney murine leukemia virus, *Nature* **293**, 543-548 (1981).
3. S. Nakanishi, A. Inoue, T. Kita, M. Nakamura, A. C. Chang, S. N. Cohen and S. Numa, Nucleotide sequence of cloned cDNA for bovine corticotropin-beta-lipotropin precursor, *Nature* **278**, 423-427 (1979).
4. J. G. Sutcliffe, T. M. Shinnick, N. Green, F.-T. Liu, H. L. Niman and R. A. Lerner, Chemical synthesis of a polypeptide predicted from nucleotide sequence allows detection of a new retroviral gene product, *Nature* **287**, 801-805 (1980).
5 R. J. Milner. and J. G. Sutcliffe, Gene expression in rat brain, *Nucleic Acids Research* **11**, 5497-5520 (1983).
6. J. G. Sutcliffe, R. J. Milner, T. M. Shinnick and F. E. Bloom , Identifying the protein products of brain specific genes with antibodies to chemically synthesized peptides, *Cell* 33, 671-682 (1983).
7. W. E. Timberlake, Developmental gene regulation in Aspergillus nidulans, *Dev. Biol.* **78**, 497-510 (1980).
8. G. H. Travis and J. G. Sutcliffe, Phenol emulsion-enhanced DNA-driven subtractive cDNA cloning: Isolation of low abundance monkey cortex-specific mRNAs, *Proc. Natl. Acad. Sci. USA* **85**,1696-1700 (1988).
9. G. H. Travis, M. B. Brennan, P. E. Danielson, C. A. Kozak and J. G. Sutcliffe, Identification of a photoreceptor-specific mRNA encoded by the gene responsible for retinal degeneration slow (rds), *Nature* **338,** 70-73 (1989).
10. J. B. Watson, E. F. Battenberg, K. K. Wong, F. E. Bloom and J. G. Sutcliffe, Subtractive cDNA cloning of RC3, a rodent cortex-enriched mRNA encoding a novel 78 residue protein, *J. Neurosci. Res*. **26**, 397-408 (1990).
11. J. B. Watson, P. M. Coulter II, J. E. Margulies, L. de Lecea, P. E. Danielson, M. G. Erlander and J.G. rat Sutcliffe, G-protein γ-7 subunit is selectively expressed in medium-sized neurons and dendrites of the neostriatum, *J. Neurosci. Res*. **39,** 108-116 (1994).
12. H. Usui, J. Falk, A. Dopazo, L. de Lecea, M. G. Erlander and J. G. Sutcliffe, Isolation of clones of rat striatum-specific mRNAs by directional tag PCR subtraction, *J. Neurosci.* **14,**4915-4926 (1994).
13. L. de Lecea, J. R. Criado, O. Prospero-Garcia, K. M. Gautvik, P. Schweitzer, P. E. Danielson, C. L. Dunlop, G. R. Siggins, S. J. Henriksen and J. G. Sutcliffe, A cortical neuropeptide with neuronal depressant and sleep-modulating properties, *Nature*. **381**, 242-245 (1996).
14. K. M. Gautvik, L. de Lecea, V. T. Gautvik, P. E. Danielson, P. Tranque, A. Dopazo, F. E. Bloom and J. G. Sutcliffe, Overview of the most prevalent hypothalamus-specific mRNAs, as identified by directional Tag PCR subtraction, *Proc. Natl. Acad. Sci. USA* **93,** 8733-8738 (1996).
15. L. de Lecea, T. S. Kilduff, C. Peyron, X.-B. Gao, P. E. Foye, P. E. Danielson, C. Fukuhara, E. L. F. Battenberg, V. T. Gautvik, F. S. Bartlett, W. N. Frankel, A. N. van den Pol, F. E. Bloom, K. M. Gautvik and J. G. Sutcliffe, The hypocretins: Hypothalamus-specific peptides with neuroexcitatory activity, *Proc Natl Acad Sci USA* **95,** 322-327 (1998).

J. Gregor Sutcliffe and Luis de Lecea
Department of Molecular Biology
The Scripps Research Institute
La Jolla, CA 92037

CONTENTS

DISCOVERY OF THE HYPOCRETINS/OREXINS AND THEIR RECEPTORS

1. THE DISCOVERY OF THE HYPOCRETINS: New Hypothalamic Peptides 3

Luis de Lecea and J. Gregor Sutcliffe

1. CLONES OF HYPOTHALAMUS-ENRICHED mRNAS 3
2. THE CLONE 35 SEQUENCE 6
3. DETECTING THE PROTEIN 7
4. ARE THE PEPTIDES NEUROTRANSMITTERS? 8
5. GOING PUBLIC: A VOTE ON NOMENCLATURE 9
6. INDEPENDENT DISCOVERY 10
7. FUNCTIONS GALORE 10
8. REFERENCES 11

2. OREXIN AND OREXIN RECEPTORS 13

Takeshi Sakurai

1. INTRODUCTION 13
2. IDENTIFICATION OF HYPOCRETIN AND OREXIN 13
3. PREPRO-OREXIN GENE, STRUCTURE AND REGULATION OF EXPRESSION 15
4. STRUCTURES AND PHARMACOLOGY OF OREXIN RECEPTORS 16
5. GENETICS OF OREXIN RECEPTORS 17
6. HOW MANY OREXIN RECEPTOR GENES? 18
7. SIGNAL TRANSDUCTION SYSTEMS OF OREXIN RECEPTORS 18
8. DISTRIBUTION OF OREXIN RECEPTORS 21
9. STRUCTURE-ACTIVITY RELATIONSHIPS 21
10. REFERENCES 22

ANIMAL MODELS IN THE STUDY OF THE HYPOCRETINERGIC SYSTEM

3. RODENT MODELS OF HUMAN NARCOLEPSY-CATAPLEXY 27

Takeshi Sakurai, Michihiro Mieda, and Masashi Yanagisawa

1. DISCOVERY OF MOUSE NARCOLEPSY 27
2. REM SLEEP-RELATED SYMPTOMS 30
3. NON-REM SLEEP-RELATED SYMPTOMS 30
4. DIFFERENTIAL REGULATION OF SLEEP/WAKE STATES BY OX1R AND OX2R 33
5. RODENT MODELS OF PATHOPHYSIOLOGY OF HUMAN NARCOLEPSY 34
6. MORE THAN SLEEP/WAKE ABNORMALITIES 36
7. CONCLUSIONS 37
8. REFERENCES 37

4. THE CANINE MODEL OF NARCOLEPSY 39

Seiji Nishino

1. INTRODUCTION 39
2. SYMPTOMS OF CANINE NARCOLEPSY 40
3. INHERITANCE OF NARCOLEPSY IN CANINES 40
4. DOG LEUKOCYTE ANTIGEN (DLA) AND CANINE NARCOLEPSY 43
5. DISCOVERY OF CANINE NARCOLEPY GENE (*Canarc-1*) 43
6. HYPOCRETIN LIGAND DEFICIENT SPORADIC NARCOLEPTIC CANINES 45
7. PHARMACOLOGICAL CONTROL OF CATAPLEXY AND EDS 45
   7.1. REM Sleep/Cataplexy and Narcolepsy 45
   7.2. Monoaminergic and Cholinergic Interactions and Cataplexy 46
   7.3. Dopamineregic Transmission and EEG Arousal 49
8. HISTAMINERGIC SYSTEM AND NARCOLEPSY 51
9. HYPOCRETIN REPLACEMENT THERAPY 53
10. HYPOCRETIN DEFICIENCY AND NARCOLEPTIC PHENOTYPE 54
11. CONCLUSION 54
12. ACKNOWELDGEMENTS 54
13. REFERENCES 55

DETAILED ANATOMY OF THE HYPOCRETINERGIC SYSTEM AND RELATED HYPOTHALAMIC CIRCUITS

5. ANATOMY OF THE HYPOCRETIN SYSTEM 61

Teresa L. Steininger and Thomas S. Kilduff

1. INTRODUCTION 61

2. DISTRIBUTION AND MORPHOLOGY OF HCRT NEURONS 61
3. COLOCALIZATION OF NEUROCHEMICALS IN HCRT NEURONS 62
4. DEVELOPMENT OF HCRT NEURONS 63
5. COMPARATIVE STUDIES 63
6. DISTRIBUTION OF HCRT AND RECEPTORS IN NON-NEURAL TISSUES 64
7. DISTRIBUTION OF HCRT EFFERENT AXONS 65
8. LOCALIZATION OF HCRT RECEPTORS 67
9. AFFERENT CONNECTIONS OF HCRT NEURONS 69
10. CONCLUSIONS AND PERSPECTIVE 70
11. ACKNOWLEDGEMENTS 70
12. REFERENCES 70

6. THE ANATOMY OF HYPOCRETIN NEURONS 77

Tamas L. Horvath

1. INTRODUCTION 77
2. HCRT PERIKARYA 77
   2.1. Light Microscopy 77
   2.2. Electron Microscopy 79
3. AFFERENT INPUT TO THE HCRT NEURONS 81
   3.1. Light Microscopy 81
   3.2. Electron Microscopy 82
4. HCRT EFFERENTS 87
   4.1. Light Microscopy 87
   4.2. Electron Microscopy 89
5. SUMMARY 91
6. REFERENCES 91

7. TRANSMITTER-IDENTIFIED NEURONS AND AFFERENT INNERVATION OF THE LATERAL HYPOTHALAMIC AREA: Focus on Hypocretin and Melanin-concentrating Hormone 95

Christian Broberger and Tomas Hökfelt

1. INTRODUCTION 95
2. MELANIN-CONCENTRATING HORMONE AND HYPOCRETIN DEFINE TWO SEPARATE NEURONAL POPULATIONS IN THE LHA 96
3. COEXISTENCE WITH OTHER SIGNALLING MOLECULES 97
   3.1. Cocaine- and Amphetamine-Regulated Transcript 98
   3.2. Dynorphin 98
   3.3. Substance P 99
   3.4. Galanin 99
   3.5. Neurotensin 99
   3.6. Other Neuropeptides 99
   3.7. Nitric Oxide 100

3.8. Amino Acid Transmitters 100
4. LOCAL CONNECTIONS WITHIN THE LHA 102
4.1. Hcrt-MCH: Anatomical Interactions 102
4.2. Distribution of Hcrt and MCH Receptors in the LHA 102
4.3. Hcrt-MCH: Functional Interactions 102
4.4. Hcrt Autoregulation 103
5. AFFERENT INNERVATION OF THE LHA 103
6. INPUTS FROM THE ARCUATE NUCLEUS 104
6.1. Role in Feeding Behaviour 104
6.2. Parallel Pathways from the Arcuate Nucleus to the LHA 104
6.3. Functional Role Within the Feeding Circuitry 106
7. OTHER INPUTS FROM THE HYPOTHALAMUS 107
7.1. The Suprachiasmatic Nucleus 107
7.2. Ventrolateral Preoptic Area 108
7.3. Dorsomedial Hypothalamic Nucleus 108
7.4. Ventromedial Hypothalamic Nucleus 108
8. INPUTS FROM OTHER AROUSAL SYSTEMS 109
8.1. Noradrenergic and Adrenergic Innervation 109
8.2. Serotonergic Innervation 109
8.3. Cholinergic Innervation 109
8.4. Histaminergic Innervation 110
9. INPUTS FROM THE BASAL GANGLIA 110
10. CORTICAL INPUTS 111
11. CONCLUDING REMARKS 111
12. ACKNOWLEDGMENTS 111
13. REFERENCES 111

PHYSIOLOGICAL CONSEQUENCES OF HYPOCRETIN ACTIVATION

8. PHYSIOLOGICAL CHARACTERISTICS OF HYPOCRETIN/OREXIN NEURONS 123

Anthony N. van den Pol

1. INTRODUCTION 123
2. HYPOCRETIN NEURONS SHOW SPONTANEOUS REGULAR ACTION POTENTIALS 124
3. SYNAPTIC INPUT TO HYPOCRETIN NEURONS IS MEDIATED BY GLUTAMATE AND GABA 126
4. INHIBITORY RESPONSE TO NOREPINEPHRINE AND SEROTONIN 126
5. HYPOCRETIN ACTIVATES HYPOCRETIN NEURONS BY AN INDIRECT EXCITATION OF GLUTAMATERGIC INTERNEURONS 128
6. GROUP 3 METABOTROPIC GLUTAMATE RECEPTORS INHIBIT HYPOCRETIN NEURONS 128

7. GLUCAGON-LIKE PEPTIDE 1 EXCITES HYPOCRETIN NEURONS 132
8. METABOLIC SIGNALS MODULATE HYPOCRETIN NEURON ACTIVITY 132
9. HYPOCRETIN CELLS ARE EXCITATORY 132
10. ACKNOWLEDGEMENTS 133
11. REFERENCES 133

9. THE NE SYSTEM AS A TARGET FOR HYPOCRETIN NEURONS: IMPLICATIONS FOR REGULATION OF AROUSAL 137

Gary Aston-Jones, J. Patrick Card, Yan Zhu, Mónica González, and Elizabeth Haggerty

1. INTRODUCTION 137
2. EXCITATORY EFFECT OF HCRT ON LC NEURONS 139
3. HCRT INNERVATION OF THE LC 140
4. GABA INTERNEURONS IN THE PERI-LC: TARGET FOR HCRT INPUTS 142
5. A CIRCUIT FROM THE SCN TO THE LC AND CIRCADIAN REGULATION OF AROUSAL: A POSSIBLE ROLE FOR HCRT? 147
6. ACKNOWLEDGEMENTS 151
7. REFERENCES 151

10. HYPOCRETIN/OREXIN ACTIONS ON MESOPONTINE CHOLINERGIC SYSTEMS CONTROLING BEHAVIORAL STATE 153

Christopher S. Leonard, Christopher J. Tyler, Sophie Burlet, Shigeo Watanabe, and Kristi A. Kohlmeier

1. INTRODUCTION 153
2. EXPERIMENTAL 154
3. HYPOCRETIN/OREXIN STIMULATES THE FIRING OF LDT NEURONS 155
4. HYPOCRETIN/OREXIN STIMULATES CHOLINERGIC AND NON-CHOLINERGIC LDT NEURONS BY DIRECT AND INDIRECT MEANS 155
5. HYPOCRETIN/OREXIN STIMULATES EXCITATORY AFFERENTS TO LDT 157
6. HYPOCRETIN/OREXIN HAD INCONSISTENT ACTIONS ON INHIBITORY AFFERENTS TO LDT 159
7. HYPOCRETIN/OREXIN EVOKED A NOISY CATION CURRENT IN LDT NEURONS 159
8. HYPOCRETIN/OREXIN ELEVATES INTRACELLULAR CALCIUM IN LDT NEURONS 161
9. CONCLUSIONS 164
10. FUNCTIONAL IMPLICATIONS 164

11. ACKNOWLEDGEMENTS 165
12. REFERENCES 165

11. THE AMINERGIC SYSTEMS AND THE HYPOCRETINS 169

Oliver Selbach and Helmut L. Haas

1. INTRODUCTION 169
2. THE HYPOCRETIN SYSTEM 169
   2.1. Glutamate and GABA 170
3. THE AMINERGIC SYSTEMS 171
   3.1. Tuberomamillary Nucleus (Histamine) 172
   3.2. Dorsal Raphe (Serotonin) 175
   3.3. Ventral Tegmental Area / Substantia Nigra (Dopamine) 178
   3.4. Locus Coeruleus (Noradrenaline) 180
   3.5. Laterodorsal Tegmentum / Basal Forebrain (Acetylcholine) 180
4. HIPPOCAMPUS AND CORTEX 180
5. CONCLUSION 183
6. REFERENCES 183

12. EFFECTS OF HYPOCRETIN/OREXIN ON THE THALAMOCORTICAL ACTIVATING SYSTEM 191

Evelyn K. Lambe and George K. Aghajanian

1. INTRODUCTION 191
2. THALAMOCORTICAL ACTIVATING SYSTEM 191
3. SELECTIVE HYPOCRETIN PROJECTIONS 192
4. HYPOCRETIN EXCITES MIDLINE -INTRALAMINAR THALAMIC NEURONS 193
5. HYPOCRETIN EXCITES THALAMOCORTICAL TERMINALS IN PREFRONTAL CORTEX 195
   5.1. Pharmacology and Lesion Studies 195
   5.2. Two-Photon Calcium Imaging Studies 198
6. AROUSAL AND ATTENTION 199
7. CONCLUSIONS 200
8. REFERENCES 200

PHARMACOLOGY OF THE HYPOCRETINS AND DRUG DESIGN

13. *IN VIVO* PHARMACOLOGY OF OREXIN (HYPOCRETIN) RECEPTORS 205

Neil Upton

1. INTRODUCTION 205

2. PHARMACOLOGICAL TOOLS FOR CHARACTERIZING THE OREXIN PEPTIDE-RECEPTOR SYSTEM *IN VIVO* 206
2. 1. Orexin Receptor Agonists 206
2.2. Orexin Receptor Antagonists 207
3. *IN VIVO* PHARMACOLOGY OF OREXIN RECEPTORS 208
3.1. Feeding and Appetite 209
3.2. Arousal and Sleep 210
3.3. Pain Modulation 213
3.4. Other Actions 214
4. CONCLUSIONS AND THERAPEUTIC OPPORTUNITIES FOR THE FUTURE 216
5. REFERENCES 217

14. INTRACELLULAR SIGNAL PATHWAYS UTILIZED BY THE HYPOCRETIN/OREXIN RECEPTORS 221

Jyrki P. Kukkonen and Karl E. O. Åkerman

1. INTRODUCTION 221
2. CELLULAR SIGNALING PATHWAYS 221
2.1. G-proteins 221
2.2. Hypocretin Receptor Signaling in Neurons 223
2.3. Hypocretin Receptor Signaling in Endocrine Systems 225
2.4. Hypocretin Receptor Signaling in Heterologous Expression Systems 226
3. CONCENTRATION-RESPONSE RELATIONSHIPS 227
4. RECEPTOR SUBTYPE DIFFERENCES IN SIGNALING? 228
5. FUTURE PERSPECTIVES 228
6. REFERENCES 228

THE HYPOCRETINS IN NARCOLEPSY AND AROUSAL

15. THE HYPOCRETINS AND NARCOLEPSY: Pathophysiology and Diagnosis 235

Wynne Chen, Jamie M. Zeitzer, and Emmanuel Mignot

1. INTRODUCTION 235
2. CLINICAL ASPECTS OF NARCOLEPSY 236
3. GENETIC ASPECTS OF NARCOLEPSY 239
4. HYPOCRETIN DEFICIENCY IN NARCOLEPSY 242
5. ROLE OF CSF HYPOCRETIN IN THE DIAGNOSIS OF NARCOLEPSY . 244
5.1. Narcolepsy with Definite Cataplexy 246
5.2. Narcolepsy without Cataplexy or with Atypical Cataplexy 246
5.3. Narcolepsy Associated with a Known Physiological Condition 246
6. HYPOCRETIN DEFICIENCY AND PHARMACOLOGIC CORRELATES 247

7. FUTURE PROSPECTS 248
8. REFERENCES 250

16. AN APPROACH TO DETERMINING THE FUNCTIONS OF HYPOCRETIN (OREXIN) 255

Jerome M. Siegel

1. INTRODUCTION 255
2. ARE HYPOCRETIN CELLS HOMOGENEOUS? 256
3. REGULATION OF HYPOCRETIN/OREXIN RELEASE 256
4. ARE HYPOCRETINS ASSOCIATED WITH LOCOMOTOR ACTIVITY? 258
5. REFERENCES 260

17. HYPOCRETIN IN NEUROPSYCHIATRIC DISORDERS 263

Patrice Bourgin and Yves Dauvilliers

1. INTRODUCTION 263
2. HYPERSOMNIAS (EXCEPT TYPICAL NARCOLEPSY-CATAPLEXY) 263
   2.1. Atypical Narcolepsy 263
   2.2. Idiopathic Hypersomnia 264
   2.3. Post -Traumatic Hypersomnia 264
   2.4. Obstructive Sleep Apnea Syndrome 265
   2.5. Hypersomnias: Conclusion 265
3. IMMUNE DISORDERS 265
   3.1. Immune Polyneuropathies 265
   3.2. Encephalitis and Demyelinating Disorders 266
   3.3. Immune Disorders: Conclusion 266
4. OTHER NEUROLOGICAL DISORDERS 267
   4.1. Kleine-Levin Syndrome (KLS) 267
   4.2. Prader Willi Syndrome 267
   4.3. Niemann-Pick Disease 267
   4.4. Myotonic Dystrophy 267
   4.5. Hypothalamic Lesions 268
5. MOVEMENT DISORDERS 268
   5.1. Parkinson Disease 268
   5.2. Restless Legs Syndrome 269
6. DEMENTIA 269
7. PAIN DISORDERS 271
8. PSYCHIATRIC DISORDERS 271
   8.1. Depressive Syndrome 272
   8.2. Schizophrenia 273
9. LIMITATIONS AND PERSPECTIVES 273
10. CONCLUSION 274
11. REFERENCES 275

18. HYPOCRETIN/OREXIN AND SLEEP: Implications for the Pathophysiology of Human Narcolepsy 279

Gert Jan Lammers and Sebastiaan Overeem

1. INTRODUCTION 279
2. HYPOCRETIN DEFICIENCY IN HUMAN NARCOLEPSY 280
3. ANIMAL MODELS 282
4. THE CAUSE OF HUMAN NARCOLEPSY 283
5. ROLE OF HYPOCRETIN IN SLEEP REGULATION 285
6. ENDOCRINE RHYTHMS, AUTONOMIC TONE AND OBESITY 285
7. REFERENCES 287

19. MODULATION OF CORTICAL ACTIVITY AND SLEEP-WAKE STATES BY HYPOCRETIN/OREXIN 291

Barbara E. Jones and Michel Muhlethaler

1. INTRODUCTION 291
2. MODULATION AND ACTIVITY OF HCRT/ORX NEURONS 293
3. EXCITATORY INFLUENCE OF HCRT/ORX UPON THE DIFFUSE THALAMO-CORTICAL PROJECTION SYSTEM 294
4. EXCITATORY INFLUENCE OF HCRT/ORX UPON THE CHOLINERGIC BASALO-CORTICAL PROJECTION SYSTEM 296
5. EXCITATORY INFLUENCE OF HCRT/ORX UPON CORTICO-CORTICAL PROJECTION NEURONS 297
6. INDIRECT INFLUENCE OF HCRT/ORX UPON SLEEP-PROMOTING NEURONS 299
7. SUMMARY AND CONCLUSIONS 299
8. ACKNOWLEDGMENTS 300
9. REFERENCES 300

THE HYPOCRETINS IN FEEDING AND ENERGY BALANCE

20. REGULATION OF HYPOCRETIN BY METABOLIC SIGNALS 307

Katherine E. Wortley and Sarah F. Leibowitz

1. INTRODUCTION 307
2. REGULATION OF HYPOCRETIN/OREXIN SYSTEM BY METABOLIC SIGNALS RELATED TO NEGATIVE ENERGY BALANCE 308
3. REGULATION OF HYPOCRETIN/OREXIN SYSTEM BY METABOLIC SIGNALS RELATED TO POSITIVE ENERGY BALANCE 308
4. CONCLUSION 312
5. REFERENCES 312

THE HYPOCRETINS IN ADDICTION AND HYPERAROUSAL

21. HYPOCRETIN AND BRAIN REWARD FUNCTION 317

Benjamin Boutrel, Paul J. Kenny, Athina Markou, and George F. Koob

1. INTRODUCTION 317
2. LATERAL HYPOTHALAMIC SELF-STIMULATION AND THE HYPOCRETIN SYSTEM 318
   2.1. Lateral Hypothalamic Self-Stimulation Paradigm 318
   2.2. Lateral Hypothalamic Self-Stimulation May Activate Hypocretin Neurons 318
3. BRAIN REWARD CIRCUITRY AND HYPOCRETIN PROJECTIONS 319
4. HYPOCRETIN AND BRAIN REWARD MODULATION 320
5. HYPOCRETIN AND RELAPSE FOR DRUG-SEEKING 321
6. SUMMARY AND PERSPECTIVES 322
7. REFERENCES 323

22. OREXIN/HYPOCRETIN AND OPIOID DEPENDENCE 327

Ralph J. DiLeone

1. INTRODUCTION 327
2. EARLY STUDIES OF THE LH 327
3. THE VIEW OF THE LH IN THE MODERN HYPOTHALAMUS 328
4. NEW LH PEPTIDES 328
5. THE HYPOTHALAMUS AND ADDICTION 329
   5.1. The LH and Drugs of Abuse 329
6. ORX/HCRT NEURONS RESPOND TO CHRONIC MORPHINE AND MORPHINE WITHDRAWAL 329
7. THE ORX/HCRT GENE IS UPREGULATED AFTER PRECIPITATED MORPHINE WITHDRAWAL 331
8. ORX/HCRT NEURONS EXPRESS THE μ-OPIOID RECEPTOR 331
9. ORX/HCRT KNOCKOUT MICE SHOW ATTENUATED MORPHINE WITHDRAWAL 332
10. SUMMARY OF ORX/HCRT DATA 333
11. INTERPRETATION AND IMPLICATIONS FOR ADDICTION BIOLOGY 333
    11.1. Is ORX/HCRT Essential for the Development of Dependence and/or the Expression of Withdrawal? 333
12. ORX/HCRT RECEPTORS AND NEURAL CIRCUITS RELEVANT TO ADDICTION 333
    12.1. What Brain Regions are the Critical ORX/HCRT Targets? 335
13. REFERENCES 335

23. DOPAMINE - HYPOCRETIN/OREXIN INTERACTIONS: The Prefrontal Cortex and Schizophrenia 339

Ariel Y. Deutch, Jim Fadel, and Michael Bubser

1. INTRODUCTION 339
2. ANATOMICAL BASIS OF DOPAMINE-OREXIN INTERACTIONS 340
   2.1. Orexin Projections to Midbrain Dopamine Neurons 340
   2.2. Overlap of Forebrain Hcrt/Orexin and DA Axons 341
   2.3. The Dopaminergic Innervation of the Lateral Hypothalamus and Orexin 341
3. DOPAMINERGIC REGULATION OF OREXIN NEURONS 342
   3.1. Amphetamine and Apomorphine Effects on Fos Expression in Orexin Cells 343
   3.2. Activation of Orexin Neurons by D1 and D2 Agonists 343
   3.3. Mechanisms of Dopamine Agonist-Induced Activation of Orexin Neurons 344
4. ANTIPSYCHOTIC DRUGS AND OREXIN NEURONS 344
   4.1. Effects of Typical and Atypical Antipsychotic Drugs on Orexin Neurons 345
   4.2. How Do Both Anorexic and Orexigenic Drugs Both Active Orexin Cells? 345
5. DOPAMINE, OREXIN, AND SCHIZOPHRENIA 347
6. CONCLUSIONS 348
7. ACKNOWLEDGMENTS 348
8. REFERENCES 348

24. HYPOCRETIN/OREXIN IN STRESS AND AROUSAL 353

Craig W. Berridge and Rodrigo A. España

1. INTRODUCTION 353
2. THE NEUROBIOLOGY OF STRESS 353
3. HCRT AND WAKING 355
4. CIRCADIAN-INDEPENDENT ACTIONS OF HCRT 357
5. HCRT AND STRESS 360
   5.1. Stress-Like Physiological and Behavioral Actions of HCRT 360
   5.2. Effects of Stress on *c-Fos* Expression in HCRT-Synthesizing and HCRT-Receptive Neurons 360
   5.3. Activating Actions of HCRT on CRH Neurotransmission 362
6. APPETITIVE HIGH-AROUSAL STATES 362
7. REFERENCES 363

ROLE OF HYPOCRETINS ON PERIPHERAL SYSTEMS

25. THE HYPOCRETINS/OREXINS AND THE HYPOTHALAMO-PITUITARY-ADRENAL AXIS 371

Willis K. Samson, Meghan M. Taylor, and Alastair V. Ferguson

1. INTRODUCTION 371
2. THE ANATOMICAL FRAMEWORK 371
3. HYPOCRETIN AND THE CENTRAL ARM OF THE HPA AXIS 373
4. HYPOCRETIN ACTIONS AT THE LEVEL OF THE PITUITARY GLAND 375
5. HYPOCRETIN ACTIONS IN THE ADRENAL GLAND 377
6. AN INTEGRATIVE VIEW OF THE ACTIONS OF THE HYPOCRETINS IN THE HYPOTHALAMO-PITUITARY-ADRENAL AXIS 378
7. CONCLUDING REMARKS 380
8. REFERENCES 380

26. OREXINS (HYPOCRETINS) IN THE GUT 385

Annette L. Kirchgessner and Erik Näslund

1. INTRODUCTION 385
2. DISTRIBUTION OF HCRT/OREXIN IN THE GASTROINTESTINAL TRACT 386
   2.1. Rodent Gastrointestinal Tract 386
   2.2. Human Gastrointestinal Tract 387
3. HCRT/OREXIN EFFECTS ON VAGAL AFFERENT SIGNALING 387
4. FASTING SMALL BOWEL MOTILITY 388
5. GASTRIC EMPTYING 389
6. GASTRIC AND INTESTINAL SECRETIONS 390
7. SUMMARY 391
8. ACKNOWLEDGEMENTS 391
9. REFERENCES 391

27. HYPOCRETINS IN ENDOCRINE REGULATION 395

Miguel López, Manuel Tena-Sempere, Tomás García-Caballero, Rosa Señarís, and Carlos Diéguez

1. INTRODUCTION 395
2. NEUROENDOCRINE ANATOMY OF THE HYPOCRETIN SYSTEM 396
3. THE HYPOCRETIN SYSTEM IN PERIPHERAL ENDOCRINE TISSUES 397
4. ENDOCRINE ACTIONS OF HYPOCRETINS 398
   4.1. Hypocretins and Adrenal Axis 399

4.2. Hypocretins and Growth Hormone Axis 400
4.3. Hypocretins and Gonadal Axis 403
4.4. Hypocretins and Lactoprope Axis 406
4.5. Hypocretins and Thyroid Axis 407
4.6. Hypocretins and Pancreatic Function 407
4.7. Endocrine Actions of Hypocretins in the Gastrointestinal Tract 409
4.8. Hypocretins and Drinking Behaviour 409
5. THE ROLE OF HCRT2 410
6. SUMMARY 410
7. ACKNOWLEDGEMENTS 412
8. REFERENCES 412

28. THE HYPOCRETINS IN CARDIOVASCULAR REGULATION 425

Tetsuro Shirasaka and Hiroshi Kannan

1. INTRODUCTION 425
2. HYPOCRETINS AND CARDIOVASCULAR REGULATIONS *IN VIVO* ... 426
3. HYPOCRETINS AND CARDIOVASCULAR REGULATIONS *IN VITRO* 430
4. CONCLUSIONS 433
5. REFERENCES 434

INDEX 437

DISCOVERY OF THE HYPOCRETINS/OREXINS AND THEIR RECEPTORS

THE DISCOVERY OF THE HYPOCRETINS:

New hypothalamic peptides

Luis de Lecea and J. Gregor Sutcliffe*

1. CLONES OF HYPOTHALAMUS-ENRICHED mRNAS

The hypothalamus has long been recognized as a site for central regulation of homeostasis. In contrast to laminar cortical structures such as the cerebellum and hippocampus whose final functions rely on input from the thalamus and brain stem, the hypothalamus is organized as a collection of distinct, autonomously active nuclei with discrete functions. Ablation and electrical stimulation studies and medical malfunctions and misadventures have implicated several of these nuclei as central regulatory centers for major autonomic and endocrine homeostatic systems mediating processes such as reproduction, lactation, fluid balance, blood pressure, thermoregulation, metabolism, and aspects of behaviors, such as circadian rhythmicity, basic emotions, the sleep/wake cycle, feeding and drinking, mating activities, and responses to stress, as well as normal development of the immune system. Distinct hormones and releasing factors have been associated with some of these nuclei.

To illuminate additional molecules that contribute to the specialized functions of hypothalamic nuclei, we embarked on a systematic analysis of the mRNAs whose expression is restricted to or enriched in the hypothalamus. Our plan involved the following steps: Directional tag PCR subtractive hybridization[1] was used to enrich a cDNA library for clones of mRNA species selectively expressed in the hypothalamus. Candidate clones identified by their hybridization to a subtracted hypothalamus probe were validated in three stages. First, a high throughput cDNA library Southern blot was used to demonstrate that the candidate corresponded to a species enriched in the subtracted library. Second, candidate clones positive in the first assay were used as probes for Northern blots with RNA from several brain regions and peripheral tissues.

* Luis de Lecea and J. Gregor Sutcliffe, Department of Molecular Biology, The Scripps Research Institute, 10550 N. Torrey Pines Rd., La Jolla, CA 92037

Finally, candidate clones that were still positive were subjected to in situ hybridization analysis to detect the hypothalamic regions that express the corresponding mRNAs.

Kaare and Vigdis Gautvik of the University of Oslo visited the lab for a sabbatical year to learn the subtraction method. They were given the hypothalamus project and applied the directional tag PCR subtractive hybridization method[1] to construct a rat hypothalamic cDNA library from which, in two consecutive steps, cerebellar and hippocampal sequences had been depleted, enriching 20-to-30 fold for sequences expressed selectively in the hypothalamus. They made a radioactive probe from the library inserts and used it to screen 648 clones from the subtracted library that they had spotted into grid arrays. Approximately 70% of the colonies gave significant signals with the subtracted target probe compared to 50% with an unsubtracted target probe. Only 10% of the colonies gave signals with a mixed-driver probe. These data indicated that the enrichment by subtraction was substantial.

We studied a sample of 100 clones selected because of their enrichment in the subtracted library.[2] The sequences of 94 of these were and more than 90% of these suggested that they were clones of the 3' ends of bona fide mRNAs. These clones corresponded to 43 distinct mRNA species, about half of which were novel. They were analyzed to see whether they were enriched in the hypothalamus cDNA library compared to hippocampus or cerebellum libraries. Thirty eight of these 43 mRNAs (corresponding to 85 of the clones in the sample) exhibited enrichment in the hypothalamus; 23 were highly enriched. Among the clones showing the highest degree of hypothalamus enrichment were cDNAs for oxytocin, vasopressin, CART, melanin concentrating hormone, POMC, VAT-1, and a novel species called clone 35.

Northern blots were performed by for 15 of the species that showed hypothalamus-enriched or -specific distributions in the cDNA Southern blot assay. The blots included RNA samples from 6 grossly dissected regions of rat brain in addition to pituitary, liver, kidney and heart. For the clones of species that had been isolated two or more times, the correspondence with the cDNA library Southern blot assay was excellent. Thus, oxytocin and clone 35 each detected a band that was strong in the hypothalamus lanes, but only very faint or undetectable in the other lanes.[2] In samples of RNA from brains of developing rats, we detected the 700-nucleotide clone 35 mRNA at low concentrations as early as embryonic day 18, but its concentration increased dramatically after the third postnatal week.[3]

In situ hybridization on coronal sections of brain from adult male rats was performed, using the inserts from clones representing many of the RNAs, including clone 35 (Figure 1). The clone 35 mRNA displayed a striking pattern of bilaterally symmetric expression restricted to a few cells in the dorsolateral hypothalamic area, with no signals outside the hypothalamus.[2] Other novel mRNAs showed substantial enrichment in basal diencephalic structures, particularly the hypothalamus, without restriction to single hypothalamic nuclei.

Typically, subtractive hybridization protocols utilize a single target-driver dichotomy for enrichment of target-specific species. In the implementation reviewed above, we utilized a two-step subtraction, first depleting hypothalamus sequences with a cerebellum driver, and then with a hippocampus driver. We included the second subtraction because, in some previous studies using this methodology, we had been successful in finding clones of species enriched in a target compared to a single driver tissue, only to find considerable expression in yet other brain regions. The revised protocol was designed to provide a more stringent selection for clones of mRNAs with high selectivity for the target. The data suggest that this strategy was effective: 53 of the 94 clones studied were shown to correspond to mRNAs expressed in the hypothalamus at much higher concentrations than in either the hippocampus or cerebellum. An additional 32 of the clones were enriched in both hypothalamus and hippocampus over cerebellum, indicating that the first subtraction was more efficient, probably because the target concentration was higher in the hybridization reaction, thus a greater portion of the common species were driven into hybrids. Cumulatively, 85 of the 94 candidates were found to be enriched in the target hypothalamus compared to the cerebellum, a quite acceptable success rate.

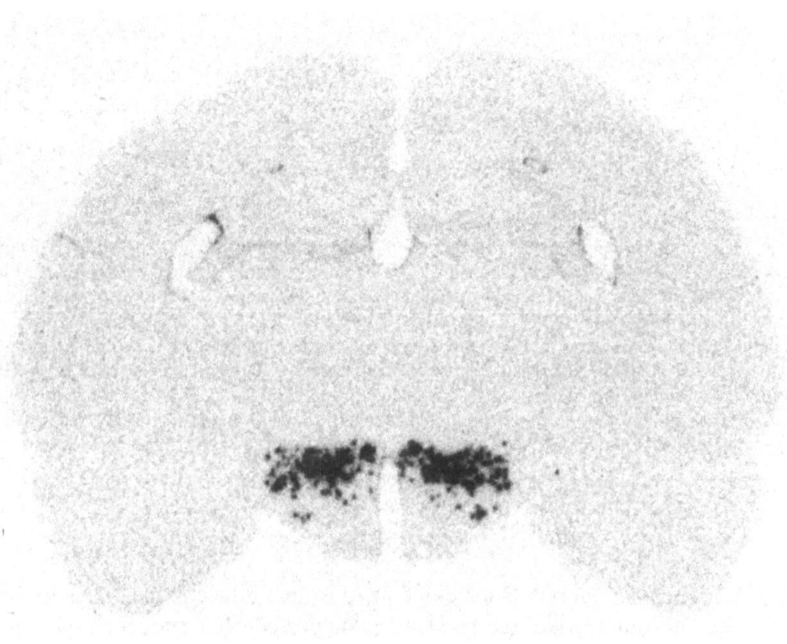

Figure 1. The first glimpse of the hypocretin system. In situ hybridization of rat coronal section with cDNA isolated in subtractive hybridization study detecting a few thousand neurons in the dorsal-lateral hypothalamus (from reference 2).

Among the novel species we identified, only clone 35 met our starting criterion in its strictest sense: the mRNA appeared to be restricted to a single nucleus in the dorsolateral area of the hypothalamus (the in situ hybridization image in reference 2 represents the first published description of the hypocretins). With Wayne Frankel at the Jackson Laboratories, we assigned the mouse homologue of the clone 35 gene to chromosome 11, a region that shows conserved synteny with human Chromosome 17q21-q24.

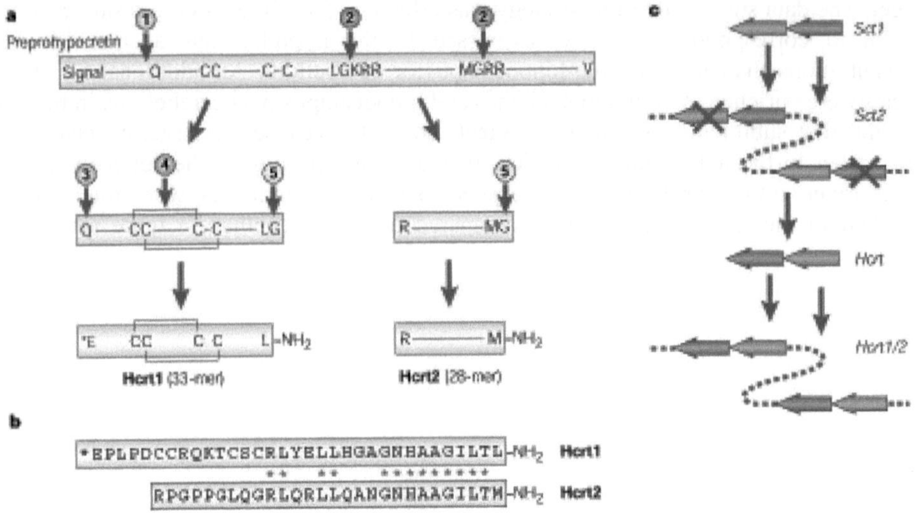

Fig.2. A. Maturation of preprohypocretin. Only amino acid residues key to the processing of the prepropeptide are shown. After removal of the secretion signal (1), the prohypocretin is cleaved at two pairs of tandem basic amino acids (KR, RR: 2). The genetically encoded glutamine (Q) is derivatived to form pyroglutamate (*E: 3), two intrachain disulfide bonds (C-C) are formed (4), and the C-terminal glycines (G) are modified by peptidylglycine alpha-amidating monooxygenase (5), leaving C-terminal amides on the resulting 33-mer and 28-mer peptides, Hcrt1 and Hcrt2, respectively. B. The primary amino acid sequences of rat Hcrt1 and Hcrt2. The *E at the N-terminus represents the pyroglutamate residue; the asterisks between the sequences indicate the positions of identity between the two peptides. The disulfide bonds are as in fig.2A. Notice the C-terminal amide groups (-NH_2). C. Model for the evolution of hypocretin from secretin (from reference 10).

2. THE CLONE 35 SEQUENCE

We used the original rat cDNA clone 35 to isolate full-length cDNAs for both rat and mouse. The 569-nucleotide rat sequence[3] suggested that the corresponding mRNA encoded a 130-residue putative secretory protein with four pairs of tandem basic residues for potential proteolytic processing. The mouse homologue nucleotide sequence differed relative to the rat in 19 positions within the putative protein-coding region, only 7 of which affected the encoded protein sequence: one amino acid difference at residue 3 was a neutral substitution in the apparent secretion signal sequence; the remaining 6 differences were near the C-terminus, one of which obliterated a potential proteolytic cleavage site. The absence of this site and the nature of the other differences made it unlikely that two of the four possible rat maturation products were functional. The two remaining putative peptides were absolutely preserved between rat and mouse (Figure 2). Both of these terminated with glycine residues, which in proteolytically processed secretory peptides typically are substrates for peptidylglycine alpha-amidating monooxygenase, leaving a C-terminal amide in the mature peptide. These features suggested that the product of the clone 35 hypothalamic mRNA served as a preprohormone for two C-terminally amidated, secreted peptides. One of these, which was later to be named hypocretin 2 (hcrt2), was, on the basis of the putative preprohormone amino-acid sequence, predicted to contain precisely 28 residues. The other (hcrt1) had a defined predicted amidated C terminus but, because of uncertainties as to how the amino terminus might be proteolytically processed, an undefined N-terminal extent.[3]

The C-terminal 19 residues of these two putative peptides shared 13 amino acid identities. This region of one of the peptides contained a 7/7 match with secretin, suggesting that the preprohormone gave rise to two peptide products that were structurally related closely to each other and distantly to secretin. We initially commented on the secretin similarity. Subsequently, Carlos Alvarez conducted phylogenic studies and detected genes that encode conserved homologues in pufferfish and frog species, suggesting that the hypocretin gene arose early in the chordate lineage.[4] Sequence similarities with various members of the incretin family, especially secretin, suggested that the gene was formed from the secretin gene by three genetic rearrangements: first, a duplication of the secretin gene; second, deletions of the N-terminal portion of the 5' duplicate and the C-terminal portion of the 3' duplicate to yield a secretin with its N- and C- termini leap-frogged (circularly permuted); and third, a further duplication of the permuted gene, followed by modifications, to form a secretin derivative that encoded two related hypocretin peptides (Figure 2).

3. DETECTING THE PROTEIN

We were now faced with the situation alluded to above: we had the sequence of a putative protein that had not been never before been seen. Tom Kilduff came to the lab on sabbatical just as the Gautviks departed for home. He was immediately interested in the clone 35 project and was given the task of preparing antibodies to the putative hypocretin protein and subsequently of using the antisera to map out the anatomical distribution of the protein. We decided to take two approaches and raised polyclonal antisera against chemically synthesized peptides corresponding to different regions within the putative protein sequence and to bacterially-expressed, histidine-tagged hypocretin protein. Tom was joined during his year in the lab by two postdoctoral fellows, Chiaki Fukuhara and Christelle Peyron, who participated in antibody generation and application.

In Western blots using these antisera and, as target extracts, bacteria expressing the fusion protein, we observed a single prominent immunoreactive band with a migration of approximately 19kDa with the hyperimmune sera, but not with the preimmune sera. Control extracts contained no immunoreactive targets, demonstrating the specificity of the antisera for the hypocretin.[3]

Immunohistochemical studies on sections from perfused adult male rats detected prominently granular immunoreactivity within widely spaced, large polymorphic neurons exclusively in the dorsalolateral hypothalamic area, coincident with the in situ hybridization-positive cells.[3] This coincident staining and its elimination when the sera were preincubated with their immunogens, together with the Western blot studies, provided strong evidence for the specificity of the antiserum for the hypocretin. A few thousand reactive cell bodies were observed between the fornix and the mammillothalamic tracts, 1 mm lateral to the midline, at the level of the median eminence. The neurons spanned the perifornical nucleus and the magnocellular nucleus of the lateral hypothalamus from the medial hypothalamus across the supra-fornical region at mid-to-posterior hypothalamic levels into the myelinated axons of the retrochiasmic optic radiation.

In addition to the hypothalamic neurons, the antisera detected a prominent network of axons located within the posterior hypothalamus and beyond. Fiber projections were observed in apparent terminal fields within septal nuclei in the basal forebrain, the preoptic area, the paraventricular nucleus of the thalamus, the central gray, and the locus coeruleus. Less prominent fiber projections were observed in apparent terminal fields within the colliculi, the laterodorsal tegmental nucleus, and the nucleus of the solitary tract. A thorough mapping of these extensive projections from a relatively small number of hypocretin-expressing neurons was published late in 1998.[5] The neuroanatomy of the hypocretins in detail is discussed in other chapters.

The deposition of the immunoreactivity along axons was clearly grainy. Elena Battenberg and Bloom conducted an electron microscopic examination, which revealed hypocretin immunoreactivity within the lateral hypothalamus on perikaryal rough endoplasmic reticulum, cytoplasmic large granular vesicles, and vesicles within myelinated axons and at presynaptic terminals opposite non-immunoreactive dendrites. Within the relatively dense terminal fields in the periaqueductal gray, immunoreactive boutons consistently made asymmetrical synaptic contacts with small-to-medium-sized dendritic shafts. Boutons contained 2-8 large, intensely immunoreactive granular vesicles in the plane of section, but in heavily reactive boutons a lighter peroxidase reaction was observed between the small agranular vesicles.[3] The accumulation of the no-longer-putative hypocretin peptides within dense core vesicles at axon terminals suggested that they might have intercellular signaling activity.

4. ARE THE PEPTIDES NEUROTRANSMITTERS?

The putative structures of the hypocretin peptides, their expression within the dorsolateral hypothalamus and accumulation within vesicles at axon terminals suggested that they might have neurotransmitter activity. To test this hypothesis, we collaborated with Tony van den Pol. We supplied a synthetic peptide corresponding to the amidated form of hcrt2 and Tony applied this to rat hypothalamic neurons that had been cultured for 10 days, and recorded postsynaptic currents under voltage clamp.[3] At 1μM the peptide evoked a substantial, but reversible, increase in the frequency of postsynaptic currents in 75% of the neurons tested, indicative of an excitatory effect (Tony's later studies suggest that approximately 33% of hypothalamic neurons respond to the hypocretins). The other 25% of the cells showed no response to hcrt2. There was little response by hypothalamic neurons that had been in culture for only 3-5 days, suggesting that a certain degree of synaptic maturity was required for the effect. Hcrt2 elicited no response in cultures of synaptically coupled hippocampal dentate granule neurons, which demonstrated target selectivity and suggested that specific receptors for hcrt2 may exist. The physiology of these peptide neurotransmitters is discussed in detail in the chapter by van den Pol.

5. GOING PUBLIC: A VOTE ON NOMENCLATURE

As we began to write the paper describing our discovery of the peptides via cDNA cloning, their immunohistochemical detection, their presence in dense core vesicles at synapses, and their neuroexcitatory properties on hypothalamic neurons, we realized that we needed a name other than clone 35. There were several possible functions for the peptides, but direct evidence for none. We came up with several non-pejorative possibilities, most of which were variations on syllables abstracted from *hypo*thalamic member of the *incretin* family. The most straightforward of these possibilities was "hypocretin". However, we were aware that this might be perceived as having negative connotations and, instead, initially settled on "hypoincretin". We submitted our manuscript to *Science*, referring to the peptides as hypoincretins. We then attended the 1997 Society for Neuroscience Meeting, at which JGS presented a poster describing the sequence of the new protein, the expression of its mRNA exclusively in a small number of neurons in the dorsolateral hypothalamus, the electron microscopic detection of immunoreactive vesicles in presynaptic boutons, and the neuroexcitatory properties of the amidated hcrt2 peptide.[6] At the adjacent poster, Cristelle Peyron and Tom Kilduff presented the data detecting hypocretin immunoreactivity in the dorsolateral hypothalamic neurons and immunoreactive fibers through the CNS.[7]

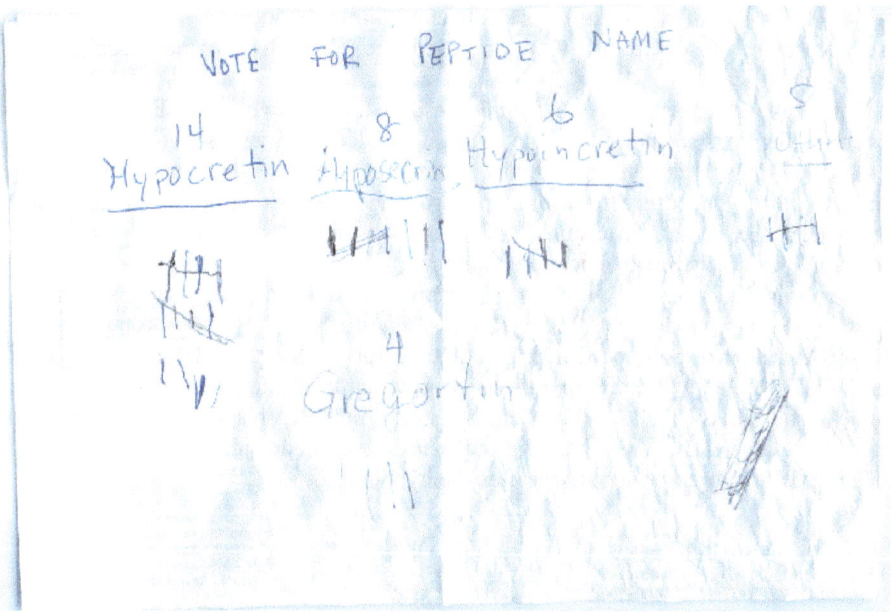

Figure 3 Ballot to name the new peptides by vote at the 1997 meeting of the Society for Neuroscience.

We were not yet comfortable with the name hypoincretin. Therefore, we posted a ballot listing several of the names under consideration and asked poster attendees to express their preference (Figure 3). The plurality of the votes were cast instead for hypocretin. The people spoke. Thus, this became the first democratically named neurotransmitter. Ironically, one of the names suggested as a write-in by an attendee (Karl Bauer) was "gregortin". While he probably meant this in jest, its adoption would have been prescient since the name Gregor is derived from the Greek *gregoros*, meaning watchful or alert, and the peptides have since been found to have their main function in maintaining arousal.

We came home from the meeting to find our manuscript, returned by *Science* unreviewed. We sent a reformatted text using the revised name hypocretin to the *PNAS*, where it was accepted and published on January 6, 1998. In the paper we noted that the existence of two hcrt peptides that differ in their amino acid sequences might indicate two Hcrt receptor subtypes.[3]

6. INDEPENDENT DISCOVERY

The February 20, 1998 issue of *Cell* carried an article[8] describing the identification of two endogenous peptides that stimulated calcium flux in cells transfected with an orphan G protein-coupled receptor. Chapter 2 traces this independent discovery of the hypocretins and their two receptors. Sakurai and colleagues demonstrated that intracerebroventricular administration of either hcrt1 or hcrt2 increased food consumption in rats. Furthermore, rats fasted for 48 hours increased the concentration of hypocretin mRNA and peptides. Based on these observations, they proposed the alternative name, orexins, for the hypocretin peptides.

This study demonstrated the actual presence of the two proteolytically processed hypocretin peptides within the brain, and elucidated by mass spectroscopy the exact structures of these endogenous peptides, which could not be deduced from nucleic acid sequences alone. The structure of orexin B was the same as that predicted from the hcrt2 cDNA sequence. The N-terminus of orexin A (hcrt1) was defined and found to correspond to a genetically encoded glutamine derivatized as pyroglutamate. The two intrachain disulfide bonds within hcrt1 were also defined. The exact structures and processing of these novel peptides are shown in Figure 2.

A commentary about new hypothalamic factors accompanied the *Cell* report.[9] It mentioned both the hypocretins and the orexins, without realizing that they were the same peptides. The omission was inadvertent but, unfortunately, this was the beginning of a dual nomenclature. The Yanagisawa group published an addendum in the March 6 *Cell* indicating that the orexins are the same as the hypocretins.

7. FUNCTIONS GALORE

The papers describing the independent discovery and characterization of the hypocretins/orexins have catalyzed a great body of work. There are now already over 800 papers concerning aspects of these neurotransmitters. The perifornical region of the hypothalamus has been associated with nutritional homeostasis, blood pressure and thermal regulation, neural control of endocrine secretion, and arousal. Thus, these activities ranked among those that might be affected by the peptides. The Chapters that follow show convincingly that the hypocretins/orexins are involved in all of these processes, although with special emphasis on arousal.[10]

8. REFERENCES

1. H. Usui, J. Falk, A. Dopazo, L. de Lecea, M. G. Erlander and J. G. Sutcliffe, Isolation of clones of rat striatum-specific mRNAs by directional tag PCR subtraction, *J. Neurosci.* **14,**4915-4926 (1994).
2. K. M. Gautvik, L. de Lecea, V. T. Gautvik, P. E. Danielson, P. Tranque, A. Dopazo, F. E. Bloom and J. G. Sutcliffe, Overview of the most prevalent hypothalamus-specific mRNAs, as identified by directional Tag PCR subtraction, *Proc. Natl. Acad. Sci. USA* **93,** 8733-8738 (1996).
3. L. de Lecea, T. S. Kilduff, C. Peyron, X.-B. Gao, P. E. Foye, P. E. Danielson, C. Fukuhara, E. L. F. Battenberg, V. T. Gautvik, F. S. Bartlett, W. N. Frankel, A. N. van den Pol, F. E. Bloom, K. M. Gautvik and J. G. Sutcliffe, The hypocretins: Hypothalamus-specific peptides with neuroexcitatory activity, *Proc Natl Acad Sci USA* **95,** 322-327 (1998).
4. C. E. Alvarez, and J. G. Sutcliffe, Hypocretin is an early member of the incretin gene family, *Neurosci Lett* **324,** 169-172 (2002).
5. C. Peyron, D. K. Tighe, A. N. van den Pol, L. de Lecea, H. C. Heller, J. G. Sutcliffe and T. S. Kilduff, Neurons containing hypocretin (orexin) project to multiple neuronal systems, *J. Neurosci.* **18,** 9996-10015 (1998).
6. J. G. Sutcliffe, K. M. Gautvik, T. S. Kilduff, T. Horn, P. E. Foye, P. E. Danielson, W. N. Frankel, F. E. Bloom, L. de Lecea, Two novel hypothalamic peptides related to secretin derived from a single neuropeptide precursor, *Soc. for Neurosci. Abstracts* **23**, 2032 (1997).
7. C. Peyron, D. K. Tighe, B. S. Lee, L. de Lecea, H. C. Heller, J. G. Sutcliffe, T. S. Kilduff, Distribution of immunoreactive neurons and fibers for a hypothalamic neuropeptide precursor related to secretin, *Soc. for Neurosci. Abstracts* **23**, 2032 (1997).
8. T. Sakurai, A. Amemiya, M. Ishii, I. Matsuzaki, R. M. Chemelli, H. Tanaka, S. C. Williams, J. A. Rickson, G. P. Kozlowski, S. Wilson, J. R. S. Arch, R. E. Buckingham, A. C. Haynes, S. A. Carr, R. S. Annan, D. E. McNulty, W.-S. Liu, J. A. Terrett, N. A. Elshourbagy, D. J. Bergsma, M. Yanagisawa, Orexins and orexin receptors: A family of hypothalamic neuropeptides and G protein-coupled receptors that regulate feeding behavior, *Cell* **92,** 573-585 (1998).
9. J. S. Flier and E. Maratos-Flier, Obesity and the hypothalamus: novel peptides for new pathways, *Cell* **92,** 437-40 (1998).
10. J. G. Sutcliffe and L. de Lecea, The hypocretins: setting the arousal threshold, *Nature Reviews Neuroscience* **3**, 339-349 (2002).

OREXIN AND OREXIN RECEPTORS

Takeshi Sakurai *

1. INTRODUCTION

"Reverse pharmacology", i.e., ligand identification using cell lines expressing orphan receptors, combined with genetic engineering techniques, has increased our understanding of novel signaling systems in the body.[1] Orexin is the first and most successful example of the factors to which such an approach was applied.[2] Our group initially identified orexin A and orexin B as endogenous peptide ligands for two orphan G-protein–coupled receptors found as human expressed sequence tags.[2] Living in post-genome era, the success of the orexin story is driving many researchers to dig up novel bioactive peptides and their receptors for another discovery of new phases in physiology and novel opportunities for clinical treatment. This chapter discusses structures and functions of orexin neuropeptides and their receptors.

2. IDENTIFICATION OF HYPOCRETIN AND OREXIN

Orexin/hypocretin peptides were identified by two independent groups utilizing completely different methodologies. de Lecea *et al.* utilized a subtractive-PCR technique to identify transcripts that are expressed specifically in the hypothalamus.[3] They had previously isolated a series of cDNA clones that are expressed in the hypothalamus but not in the cerebellum and the hippocampus. One of these was expressed exclusively by a bilaterally symmetric structure within the posterior lateral hypothalamus. They subsequently cloned cDNAs covering the entire coding region, which encodes a putative secretory protein of 130 amino acids. According to its primary sequence, they predicted that this protein gives rise to two novel peptide products that are structurally related to each other. They named them hypocretin-1 and hypocretin-2.

Around the same time with the report by de Lecea *et al.,* our group reported identification of orexins (orexin-A and orexin-B) by "reverse pharmacological" approach.

* T. Sakurai. Institute of Basic Medical Sciences, Departments of Pharmacology, University of Tsukuba, Ibaraki 305-8575, Japan and ERATO Yanagisawa Orphan Receptor Project, Japan Science and Technology Corporation, Tokyo 135-0064, Japan

Most neuropeptides work though GPCRs. There are numerous (approximately 100-150) "orphan" GPCR genes in the human genome; the cognate ligands for these receptor molecules have not been identified yet. We were expressing orphan GPCR genes in transfected cells and used them as a reporter system to detect endogenous ligands in tissue extracts that can activate signal transduction pathways in GPCR-expressing cell lines. In this process, we identified orexin A and orexin B as endogenous ligands for two orphan GPCRs found as human expressed sequence tags (ESTs) and now named orexin receptor 1 (OX1R) and orexin receptor 2 (OX2R).[2]

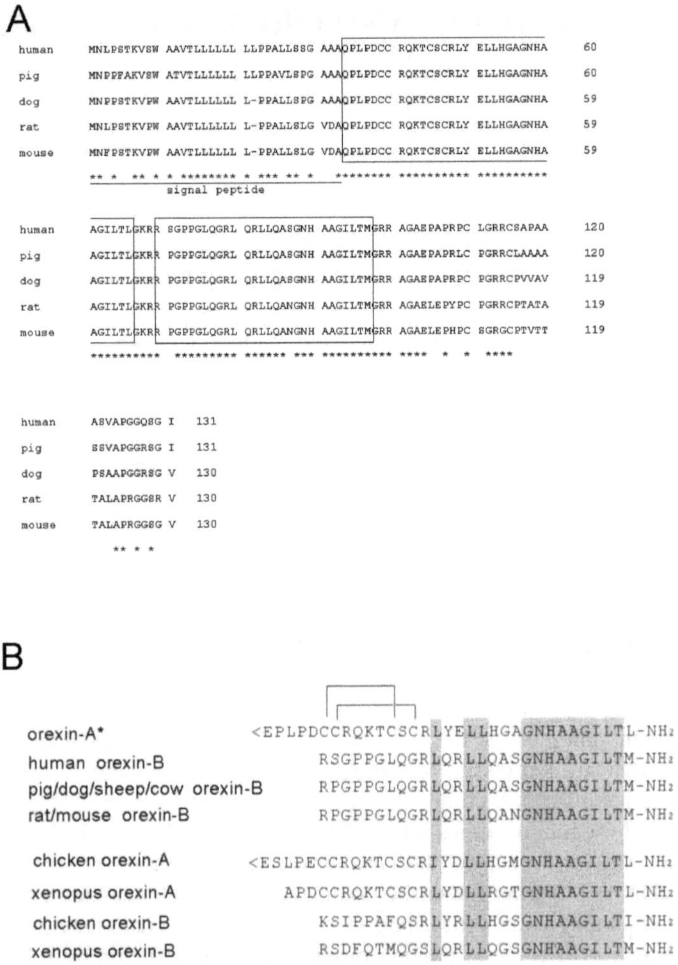

Figure 1 A. Aminoacid sequences of human and rat preproorexin. Asterisks indicate the identical aminoacids between human and rat sequences B. Structures of mature orexin-A and -B peptides. The topology of the two intrachain disulfide bonds in orexin-A is indicated above the sequence. Amino acid identities are indicated by boxes.

Structures of orexins were chemically determined by biochemical purification and sequence analysis by Edman sequencing and mass spectrometry.[2] Orexin A and orexin B constitute a novel peptide family, with no significant homology with any previously described peptides. Orexin A is a 33-amino-acid peptide of 3,562 Da, with an N-terminal pyroglutamyl residue and C-terminal amidation (Fig 1B). Molecular mass of the purified peptide as well as its sequencing analyses indicated that the four Cys residues of orexin A formed two sets of intrachain disulfide bonds. The topology of the disulfide bonds was chemically determined to be [Cys^{6}-Cys^{12}, Cys^{7}-Cys^{14}]. This structure is completely conserved among several mammalian species (human, rat, mouse, cow, sheep, dog and pig). On the other hand, rat orexin B is a 28-amino-acid, C-terminally amidated linear peptide of 2,937 Da, which was 46% (13/28) identical in sequence to orexin A (Fig 1B). The C-terminal half of orexin B is very similar to that of orexin A, while the N-terminal half is more variable. The mouse orexin B was predicted to be identical to rat orexin B. The human orexin B has two amino acid substitutions from the rodent sequences within the 28-residue stretch. Pig and dog orexin B have one amino acid substitution from the human or rodent sequences. Other than mammalian species, structures of xenopus and chicken orexin A and orexin B, which have also conserved structures as compared with mammalian sequences, was elucidated (Figure 1A).

The prepro-orexin cDNA sequences revealed that both orexins are produced from the same 130-residue (rodent) or 131-residue (human) polypeptide, prepro-orexin by proteolytic processing. The human and mouse prepro-orexin sequences are 83% and 95% identical to the rat counterpart, respectively (Fig 1A).[2] The majority of amino acid substitutions were found in the C-terminal part of the precursor, which appears unlikely to encode another bioactive peptide (Fig 1A).

It turned out later that prepro-orexin is identical to prepro-hypocretin and that orexin-A and –B correspond to hypocretin1 and 2, respectively. The original description of predicted structures of hypocretin 1 and 2 was incomplete as it did not predict the signal peptide proteolytic site, nor two intrachain disulfate bonds and N-terminal pyroglutamylation of hypocretin-1.[2,3] Nevertheless, hypocretins and orexins currently recognized as same peptides and used as synonyms.

We reported orexins initially as orexigenic peptides.[2] Subsequently they have been reported to have a variety of pharmacological actions (see other chapters). Especially, recent observations implicate orexins/hypocretins in sleep disorder narcolepsy and in the regulation of the normal sleep process. The biological activities of orexins are discussed in other chapters of this book.

3. PREPRO-OREXIN GENE, STRUCTURE AND REGULATION OF EXPRESSION

The human prepro-orexin gene, which is located on chromosome 17q21, consists of two exons and one intervening intron distributed over 1432 bp.4 The 143-bp exon 1 includes the 5'-untranslated region and the coding region that encodes the first seven residues of the secretory signal sequence. Intron 1, which is the only intron found in the human prepro-orexin gene, is 818-bp long. Exon 2 contains the remaining portion of the open reading frame and the 3'-untranslated region.

The human prepro-orexin gene fragment, which contains the 3149-bp 5'-flanking region and 122-bp 5'-non-coding region of exon 1, was reported to have an ability to express lacZ in orexin neurons without ectopic expression in transgenic mice, suggesting that this genomic fragment contains all of necessary elements for appropriate expression of the gene.[4] This promoter is useful to examine the consequences of expression of exogenous molecules in orexin neurons of transgenic mice, thereby manipulating the cellular environment in vivo.[4-6] For example, this promoter was used to establish several transgenic lines, including orexin neuron-ablated mice and rats, and mice in which orexin neurons specifically express green fluorescent protein.[5,6]

The mechanisms that regulate expression of the prepro-orexin gene still remain unclear. Prepro-orexin mRNA was shown to be upregulated under fasting conditions, indicating that these neurons somehow sense the animal's energy balance.[2] Several reports have shown that orexin neurons express leptin receptor- and STAT-3-like immunoreactivity, suggesting that orexin neurons are regulated by leptin.[7] Consistently, we found that continuous infusion of leptin into the third ventricle of mice for 2 weeks resulted in marked down-regulation of prepro-orexin mRNA level.[5] Therefore, reduced leptin signalling may be a possible factor that up-regulates expression of prepro-orexin mRNA during starvation. Prepro-orexin levels were also increased in hypoglycemic conditions, suggesting that expression of the prepro-orexin gene is also regulated by plasma glucose levels.[8] These observations are consistent with our electrophysiological study of GFP-expressing orexin neurons in transgenic mice showed that orexin neurons are regulated by extracellular glucose concentration and leptin.[5]

4. STRUCTURES AND PHARMACOLOGY OF OREXIN RECEPTORS

Initially, we used an orphan GPCR termed HFGAN72 (which now named as orexin-1 receptor, OX1R or hcrtrl to identify orexins. Since we found that orexin B has significaltly lower affinity for the human (OX1R), we sought other orexin receptors for which orexin B has high affinity. A BLAST search of the dbEST database with the OX1R amino acid sequence detected another orexin receptor, orexin-2 receptor OX2R, or hcrtr2.[2] The actions of orexins are mediated by these two receptors.2 Among various classes of G protein-coupled receptors, OX1R is structurally more similar to certain neuropeptide receptors, most notably to the Y2 Neuropeptide Y (NPY) receptor (26% similarity), followed by the thyrotropin-releasing hormone (TRH) receptor, cholesystokinin type-A receptor and NK2 neurokinin receptor (25 %, 23% and 20% similarity, respectively).

The amino acid identity between the deduced full-length human OX1R and OX2R sequences is 64%. Thus, these receptors are much more similar to each other than they are to other GPCRs. Amino acid identities between the human and rat homologues of each of these receptors are 94% for OX1R and 95% for OX2R, indicating that both receptor genes are highly conserved between the species. Competitive radioligand binding assays using CHO cells expressing OX1R suggested that orexin A is a high-affinity agonist for OX1R. The concentration of cold orexin A required to displace 50% of specific radioligand binding (IC_{50}) was 20 nM. Human orexin B also acted as a specific agonist on OX1R. However, human orexin B has significantly lower affinity compared to human OX1R: the calculated IC50 in competitive binding assay was 250 nM for human orexin B, indicating 2 orders of magnitude lower affinity as compared with orexin. A (Fig 2).

On the other hand, binding experiments using CHO cells expressing the human OX2R cDNA demonstrated that OX2R is a high affinity receptor for human orexin B with IC_{50} of 20 nM. Orexin A also had high affinity for this receptor with IC_{50} of 20 nM, which is similar to the value for orexin B, suggesting that OX2R is a non-selective receptor for both orexin A and orexin B (Fig 2).

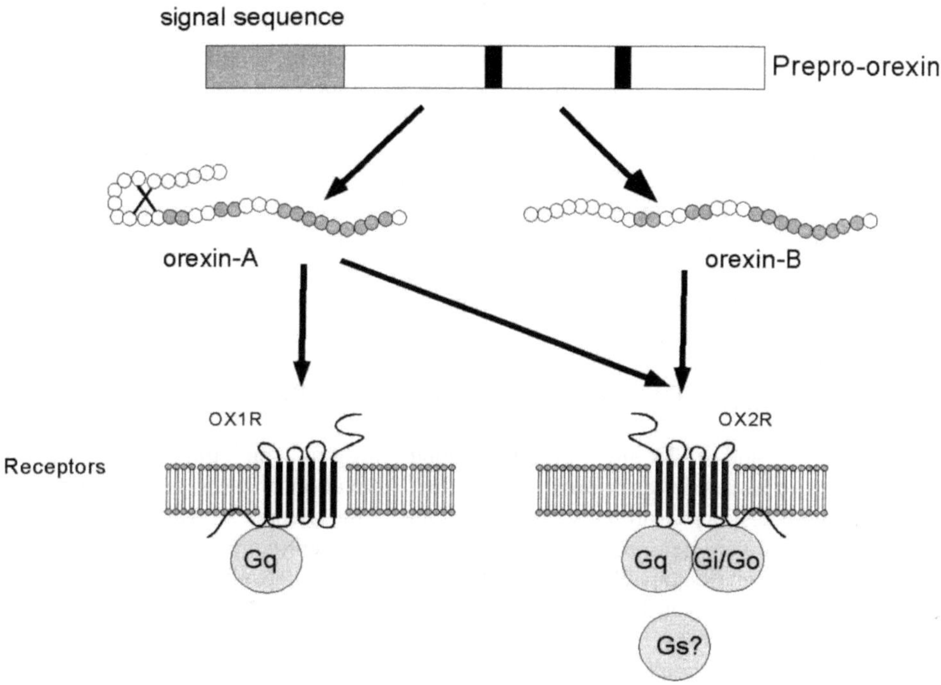

Figure 2. Schematic representation of orexin system. Orexin-A and orexin–B are derived from a common precursor peptide, prepro-orexin. The actions of orexins are mediated via two G protein-coupled receptors named orexin-1 (OX1R) and orexin-2 (OX2R) receptors. OX1R is selective for orexin-A, whereas OX2R is a nonselective receptor for both orexin-A and orexin-B.

5. GENETICS OF OREXIN RECEPTORS

Genetic studies revealed that dogs with a mutation of Hcrtr2 gene or OX2R-knockout mice displayed a narcolepsy-like phenotype,[9,10] while OX1R knockout mice did not reveal any obvious abnormality in the sleep/wake states.[10] These studies provide strong evidences for the roles of OX2R in regulating the vigilance state in human and animals. However, the behavioral and electroencephalographic phenotype of OX2R knockout mice is less severe than that found in prepro-orexin knockout mice.[9] OX2R knockouts are only mildly affected with cataplexy-like attacks of REM sleep, whereas orexin knockout mice are severely affected.[9] Double receptor knockout (OX1R- and OX2R-null) mice appear to have the same phenotype of prepro-orexin knockout mice, suggesting that OX1R also has additional effects on sleep/wakefulness. These findings suggest that loss of signaling through both receptor pathways is necessary for emerging all of narcoleptic characteristics.

The phenotypes of orexin receptor knockout mice are more precisely discussed in another chapter.

6. HOW MANY OREXIN RECEPTOR GENES?

Two genes for orexin receptors have been identified in mammalian species thus far. The phenotypes of OX1R and OX2R double-deficient mice were analyzed and shown to have sleep state abnormalities, which were indistinguishable from that of prepro-orexin gene-deficient mice. This observation suggests that only two receptors for orexins might exist in mammals. However, it is possible that there are other subtypes of receptors produced from OX1R or OX2R genes by alternative splicing. In fact, two alternative C-terminus splice variants of the murine OX2R, termed m OX2alphaR (443 aa) and m OX2betaR (460 aa) have been reported,[11] although orexin A and orexin B showed no difference in binding characteristics between the splice variants.

7. SIGNAL TRANSDUCTION SYSTEMS OF OREXIN RECEPTORS

Both OX1R and OX2R are G-protein coupled receptors, which transmit information into cells by activating heterotrimeric G proteins. Activation of the signaling pathways associated with distinct G-proteins may contribute to the diverse physiological roles of orexin in particular neurons. Although many G-protein-coupled neurotransmitter receptors are potentially capable of modulating both voltage-dependent calcium channels and G-protein-gated inwardly rectifier potassium channels (GIRKs), there might be a substantial degree of selectivity in the coupling to one or other of these channels in neurons (Fig. 3). The signal transduction pathways of orexin receptors were examined in cells transfected with OX1R or OX2R. In OX1R-expressing cells, forskolin-stimulated cAMP was not affected by orexin administration. In addition, PTX treatment did not show any effects on orexin-induced increases in $[Ca^{2+}]_i$. These results suggest that OX1R does not couple to PTX-sensitive $G_{i/o}$ proteins.[12] In contrast, orexin inhibited forskolin-stimulated cAMP production in a dose-dependent manner in OX2R-expressing cells. The effect was abolished by pretreatment with PTX. However, orexin-induced increases in [Ca2+]i was not affected by PTX treatment in OX2R-expressing cells. These results indicate that the OX2R couples to PTX-sensitive G proteins which were involved in the inhibition of adenylyl cyclase by orexin. These results suggest that OX1R couples exclusively to PTX-insensitive G proteins, and OX2R couples to both PTX-sensitive and -insensitive proteins (Fig 3). The relative contribution of these G-proteins in the regulation of neuronal activity remains unknown. Orexins have been shown to have an excitatory activity in many types of neurons in vivo. For instance, noradrenergic cells of the LC,[13] dopaminergic cells of the ventral tegmental area,[14] and histaminergic cells from the TMN[15] have been shown to be activated by orexins. Because, LC neurons exclusively express OX1R, while TMN neurons exclusively express OX2R, these observations suggest that both OX1R and OX2R signaling are basically excitatory on neurons.

However, these studies only examined effect of orexins on receptor-expressing cell bodies. There is a possibility that orexin receptors are also on presynaptic terminals, because Li *et al.* [16] reported that orexin increases local glutamate signalling by facilitation of glutamate release from presynaptic terminals. Therefore, it is possible that activation of PTX-sensitive G proteins in the downstream of OX2R might be involved in functions other than activation of neurons, such as in the tips of developing neurites and on presynaptic nerve terminals, leading to growth cone collapse and enhanced synaptic release of the transmitter. Alternatively, OX2R mediated activation of Gi might result in inhibition of some population of neurons. In fact, orexin was recently reported to inhibit proopio-melanocortin neurons in the arcuate nucleus in vitro.[17]

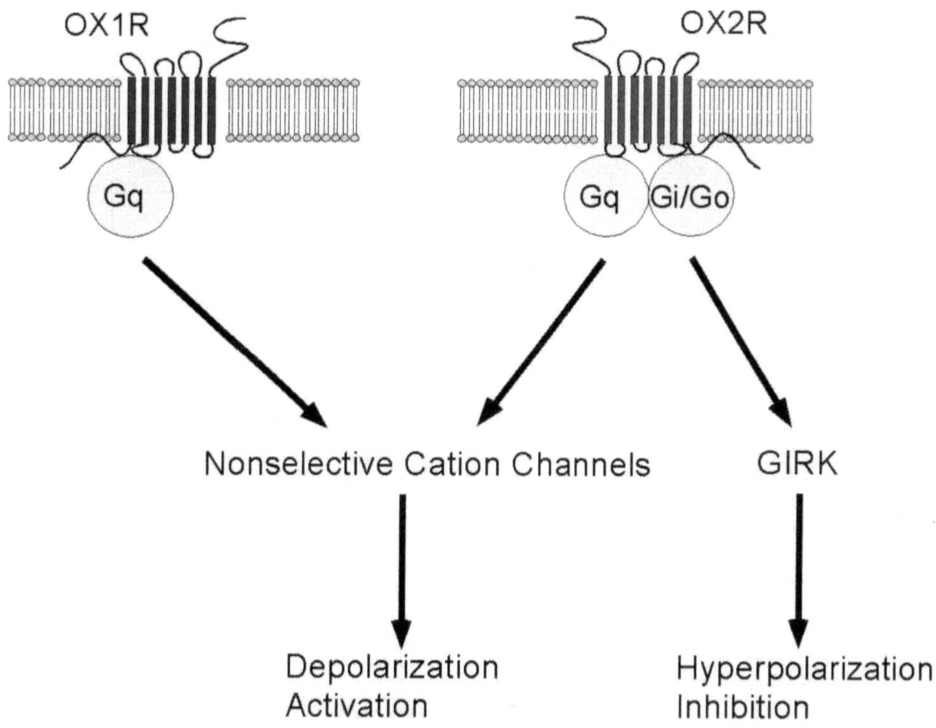

Figure 3. Schematic drawing of the intracellular signal transduction systems of orexin receptors in neurons. OX1R is coupled exclusively to the Gq subclass of heterotrimeric G proteins, whereas OX2R may couple to Gi/o, and/or Gq.

The other reports also suggested that in adrenocortical cells, orexins stimulate corticosterone secretion through the activation of the adenlylate cyclase-dependent signaling via an activation of OX2R.[18-20] The induction of cAMP in the adrenal cortex might be a cell-type dependent phenomenon, since in PC12 cells, orexins inhibited the PACAP-induced increase in the cAMP.[21] These differences in G-protein coupling might be influenced by receptor density and/or densities in each G-proteins.

Recently, several studies showed molecular interactions and cross-talks between orexin receptors and other receptors. Hilairet et al. showed that when the cannabinoid receptor (CB1) and OX1R are co-expressed, there is a CB1-dependent enhancement of the orexin A potency to activate the mitogen-activated protein kinase pathway. It was also shown that CB1 and OX1R are closely apposed at the plasma membrane to form

Table 1: Distribution of OX1R and OX2R in the rat brain. The relative density of labeling is classified as: absent (-); sparse (+); moderate (++); extensive (+++); very extensive (++++)

Region	Ox1 r ir	Ox1r mRNA	Ox2r mRNA
Telencephalon			
Olfactory system			
Anterior olfactory nu	+++	++	-
Piriform cortex	+++	-	++
Tenia tecta	++	+++	+
Cerebral cortex			
Agranular insular cortex	+	+	-
Neocortex layer 6	+	+	++
Neocortex layer 5	+	+	-
Neocortex layer 2	-	-	++
Claustrum	+	+	-
Cingulate/Retro splenial cortex	++	++	-
Basal Ganglia			
Caudate putamen	+	-	-
Globus palidus	++	-	+
Substantia nigra, pars compacta	++	+	+
Subthalamic Nucleus	++	+	++
Nucleus accumbens, rostral	-	-	+
Hippocampal formation			
CA1 region	+	+	-
CA2 region	++	++	-
CA3 region	+	-	+++
Dentate gyrus	+++	++	+
Amygdala			
Amygdaloid nuclei	++	++	++
Substantia innominata	++	+	+
Septal and basal magnocellular			
Bed nucleus of the stria terminals	+++	+++	++
Lateral septal nucleus, dorsal part	++	-	+
Medial septal nucleus	++		+++
Nucleus of the horizontal limb of the diagonal band	++	++	+++
Nucleus of the vertical limb of the diagonal band	+	+	++
Thalamus			
Antermedial thalamic nucleus	+++	++	+
Centrolateral thalamic nucleus	+++	+	++
Centromedial thalamic nucleus	+	+	++
Paracentral thalamic nucleus	++	-	+
Paraventricular thalamic nucleus	+	+++	++
Reticular thalamic nucleus	+++	++	-
Zona incerta	+++	++	++
Lateral & medial geniculate nuclei	++	-	-
Hypothalamic Preoptic Nuclei			
Anteroventral preoptic area		+	
Magnocellular preoptic area	+	+	++
Medial preoptic nu.	-	-	++
medial preoptic nucleus	++	++	++
supraoptic nucleus	+	+	+
ventrolateral preoptic area	+	+	+
ventromedial preoptic area	-	-	+
Hypothalamus			
Anterior hypothalamic area	+++	++	+
Arcuate hypothalamic nu	+++	-	+++
Dorsomedial hypothal. nu	+	+	+++
Lateral hypothal. area (LHA)	++	+	+++
Magnocellular preoptic nu	+	+	+
Medial mammillary nu	++	-	++
Paraventricular hypothalamic	+++	-	++
Posterior hypothal. area	++	++	++
Premammillary nu	++	++	+++
Supraoptic nu	+++	+	+
Suprachiasmatic nu	+++	-	-
Ventromedial hypothalamic nu	+++	++	-
Mesencephalon			
Dorsal tegmental nu	+++	++	+
Inferior colliculus	+	+	+
Interpeduncular nu	++	+	++
Periaqueductal grey	++	++	++

Principal sensory trigeminal nucleus	++	-	++
Raphe nuclei	++	++	++
Substantia nigra, pars compacta	++	++	++
Superior colliculus	+	+	+
Rhombencephalon			
Facial nucleus	+++	-	++
Locus coeruleus	++++	++++	-
Pontine reticular nucleus	++		++
Spinal trigeminal nucleus	++	-	++
Cerebellum			
Cerebellar cortex			
Deep cerebellar nuclei	++		
Spinal cord and dorsal root			
Spinal cord (grey matter, dorsal and ventral horn)	+++		
DRG	+++		

heterodimers.[22] It was also showed that OX1R and OX2R are capable of forming a homo- or heterodimer [23]. These observations suggest complex signaling cascade might exist in the downstream of orexin receptors, although we have to be careful about interpretation of these results from in vitro experiments.

8. DISTRIBUTION OF OREXIN RECEPTORS

Although orexin receptors are basically expressed in a pattern consistent with projections of orexin-producing neurons, mRNA for OX1R and OX2R were shown to be differentially distributed throughout the brain (Table 1). For instance, within the hypothalamus, a low level of OX1R mRNA expression is observed in the dorsomedial hypothalamus (DMH), while a higher level of OX2R mRNA expression is observed in this region. Other areas of OX2R expression in the hypothalamus are the arcuate nucleus, paraventricular nucleus (PVN), LHA, and most significantly, the tuberomammillary nucleus (TMN)[24]. In these regions, there is little or no OX1R signal. In the hypothalamus, OX1R mRNA is abundant in the anterior hypothalamic area and ventromedial hypothalamus (VMH).

Outside the hypothalamus, high levels of OX1R mRNA expression are detected in the tenia tecta, hippocampal formation, dorsal raphe nucleus, and most prominently, the locus coeruleus (LC). OX2R mRNA is abundantly expressed in the cerebral cortex, nucleus accumbens, subthalamic nucleus, paraventricular thalamic nuclei, anterior pretectal nucleus, and the raphe nuclei.

Within the brain, OX1R is most abundantly expressed in the LC, while OX2R is most abundantly expressed in the TMN, regions highly important for maintenance of arousal. The raphe nuclei contain both receptor mRNAs. These observations suggest strong interaction between orexin neurons and the monoaminergic systems. More precise description of the distribution of orexin receptors is discussed in other chapters.

9. STRUCTURE-ACTIVITY RELATIONSHIPS

Activities of synthetic orexin B analogs in cells transfected with either OX1R or OX2R were examined to define the structural requirements for activity of orexins on their receptors.[25] The ability of N- or C-terminally truncated analogs of orexin B to increase cytoplasmic Ca^{2+} levels in the cells showed that the absence of N-terminal residues had little or no effect on the biological activity and selectivity of both receptors. Truncation from the N-terminus to the middle part of orexin B resulted in moderate loss of activity, in the order of peptide length. In particular, deletion of the conserved sequence between orexin A and orexin B caused a profound loss of biological activity, and the C-terminally truncated peptides were also inactive for both receptors. These results suggest that the consensus region between orexin A and orexin B is important for the activity of both receptors.

10. REFERENCES

1. A. Wise, S. C. Jupe and S. Rees, The identification of ligands at orphan G-protein coupled receptors., *Annu Rev Pharmacol Toxicol.* **44**, 43-66 (2004).
2. T. Sakurai, A. Amemiya, M. Ishii, I. Matsuzaki, R. M. Chemelli, H. Tanaka, S. C. Williams, J. A. Richardson, G. P. Kozlowski, S. Wilson, J. R. Arch, R. E. Buckingham, A. C. Haynes, S. A. Carr, R. S. Annan, D. E. McNulty, W. S. Liu, J. A. Terrett, N. A. Elshourbagy, D. J. Bergsma and M. Yanagisawa, Orexins and orexin receptors: a family of hypothalamic neuropeptides and G protein-coupled receptors that regulate feeding behavior., *Cell.* **92**, 573-85 (1998).
3. L. de Lecea, T. S. Kilduff, C. Peyron, X. Gao, P. E. Foye, P. E. Danielson, C. Fukuhara, E. L. Battenberg, V. T. Gautvik, F. S. n. Bartlett, W. N. Frankel, A. N. van den Pol, F. E. Bloom, K. M. Gautvik and J. G. Sutcliffe, The hypocretins: hypothalamus-specific peptides with neuroexcitatory activity., *Proc Natl Acad Sci U S A.* **95**, 322-7. (1998).
4. T. Sakurai, T. Moriguchi, K. Furuya, N. Kajiwara, T. Nakamura, M. Yanagisawa and K. Goto, Structure and function of human prepro-orexin gene., *J Biol Chem.* **274**, 17771-6 (1999).
5. A. Yamanaka, C. T. Beuckmann, J. T. Willie, J. Hara, N. Tsujino, M. Mieda, M. Tominaga, K. Yagami, F. Sugiyama, K. Goto, M. Yanagisawa and T. Sakurai, Hypothalamic orexin neurons regulate arousal according to energy balance in mice., *Neuron.* **38**, 701-13 (2003).
6. J. Hara, C. T. Beuckmann, T. Nambu, J. T. Willie, R. M. Chemelli, C. M. Sinton, F. Sugiyama, K. Yagami, K. Goto, M. Yanagisawa and T. Sakurai, Genetic ablation of orexin neurons in mice results in narcolepsy, hypophagia, and obesity., *Neuron.* **30**, 345-54 (2001).
7. M. Hakansson, L. de Lecea, J. G. Sutcliffe, M. Yanagisawa and B. Meister, Leptin receptor- and STAT3-immunoreactivities in hypocretin/orexin neurones of the lateral hypothalamus., *J Neuroendocrinol.* **11**, 653-63 (1999).
8. B. Griffond, P. Y. Risold, C. Jacquemard, C. Colard and D. Fellmann, Insulin-induced hypoglycemia increases preprohypocretin (orexin) mRNA in the rat lateral hypothalamic area., *Neurosci Lett.* **262**, 77-80 (1999).
9. J. T. Willie, R. M. Chemelli, C. M. Sinton, S. Tokita, S. C. Williams, Y. Y. Kisanuki, J. N. Marcus, C. Lee, J. K. Elmquist, K. A. Kohlmeier, C. S. Leonard, J. A. Richardson, R. E. Hammer and M. Yanagisawa, Distinct narcolepsy syndromes in Orexin receptor-2 and Orexin null mice: molecular genetic dissection of Non-REM and REM sleep regulatory processes., *Neuron.* **38**, 715-30 (2003).
10. J. T. Willie, R. M. Chemelli, C. M. Sinton and Y. M., To eat or to sleep? Orexin in the regulation of feeding and wakefulness., *Annu Rev Neurosci.* **24**, 429-58 (2001).
11. J. Chen and H. S. Randeva, Genomic Organization of Mouse Orexin Receptors:Characterization of Two Novel Tissue-Specific SpliceVariants., *Mol Endocrinol.* (2004).
12. Y. Zhu, Y. Miwa, A. Yamanaka, T. Yada, M. Shibahara, Y. Abe, T. Sakurai and K. Goto, Orexin receptor type-1 couples exclusively to pertussis toxin-insensitive G-proteins, while orexin receptor type-2 couples to both pertussis toxin-sensitive and -insensitive G-proteins., *J Pharmacol Sci.* **92**, 259-66. (2003).
13. J. J. Hagan, R. A. Leslie, S. Patel, M. L. Evans, T. A. Wattam, S. Holmes, C. D. Benham, S. G. Taylor, C. Routledge, P. Hemmati, R. P. Munton, T. E. Ashmeade, A. S. Shah, J. P. Hatcher, P. D. Hatcher, D. N. Jones , M. I. Smith, D. C. Piper, A. J. Hunter, R. A. Porter and N. Upton, Orexin A activates locus coeruleus cell firing and increases arousal in the rat., *Proc Natl Acad Sci U S A.* **96**, 10911-6 (1999).
14. T. Nakamura, K. Uramura, T. Nambu, T. Yada, K. Goto, M. Yanagisawa and T. Sakurai, Orexin-induced hyperlocomotion and stereotypy are mediated by the dopaminergic system., *Brain Res.* **873**, 181-7 (2000).

15. A. Yamanaka, N. Tsujino, H. Funahashi, K. Honda, J. L. Guan, Q. P. Wang, M. Tominaga, K. Goto, S. Shioda and T. Sakurai, Orexins activate histaminergic neurons via the orexin 2 receptor., *Biochem Biophys Res Commun.* **290**, 1237-45 (2002).
16. Y. Li, X. B. Gao, T. Sakurai and A. N. van den Pol, Hypocretin/Orexin excites hypocretin neurons via a local glutamate neuron-A potential mechanism for orchestrating the hypothalamic arousal system., *Neuron.* **36**, 1169-81 (2002).
17. S. Muroya, H. Funahashi, A. Yamanaka, D. Kohno, K. Uramura, T. Nambu, M. Shibahara, M. Kuramochi, M. Takigawa, M. Yanagisawa, T. Sakurai, S. Shioda and T. Yada, Orexins (hypocretins) directly interact with neuropeptide Y, POMC and glucose-responsive neurons to regulate Ca 2+ signaling in a reciprocal manner to leptin: orexigenic neuronal pathways in the mediobasal hypothalamus., *Eur J Neurosci.* **19**, 1524-34 (2004).
18. L. K. Malendowicz, C. Tortorella and G. G. Nussdorfer, Orexins stimulate corticosterone secretion of rat adrenocortical cells, through the activation of the adenylate cyclase-dependent signaling cascade., *J Steroid Biochem Mol Biol.* **70**, 185-8. (1999).
19. G. Mazzocchi, L. K. Malendowicz, L. Gottardo, F. Aragona and G. G. Nussdorfer, Orexin A stimulates cortisol secretion from human adrenocortical cells through activation of the adenylate cyclase-dependent signaling cascade., *J Clin Endocrinol Metab.* **86**, 778-82 (2001).
20. H. S. Randeva, E. Karteris, D. Grammatopoulos and E. W. Hillhouse, Expression of orexin-A and functional orexin type 2 receptors in the human adult adrenals: implications for adrenal function and energy homeostasis., *J Clin Endocrinol Metab.* **86**, 4808-13 (2001).
21. T. Nanmoku, K. Isobe, T. Sakurai, A. Yamanaka, K. Takekoshi, Y. Kawakami, K. Ishii, K. Goto and T. Nakai, Orexins suppress catecholamine synthesis and secretion in cultured PC12 cells., *Biochem Biophys Res Commun.* **274**, 310-5 (2000).
22. S. Hilairet, M. Bouaboula, D. Carriere, G. Le Fur and P. Casellas, Hypersensitization of the Orexin 1 receptor by the CB1 receptor: evidence for cross-talk blocked by the specific CB1 antagonist, SR141716., *J Biol Chem.* **278**, 23731-7 (2003).
23. M. B. Dalrymple, K. M. Kroeger, R. M. Seeber and K. A. Edine, Use of BRET analysis to determine arrestin interactions, homo- and hetero-dimerisation between G ptotein-coupled receptors orexin type I and II., *85th Annual Endocreine Meeting.* P1-66 (2003).
24. J. N. Marcus, C. J. Aschkenasi, C. E. Lee, R. M. Chemelli, C. B. Saper, M. Yanagisawa and J. K. Elmquist, Differential expression of orexin receptors 1 and 2 in the rat brain., *J Comp Neurol.* **435**, 6-25. (2001).
25. S. Asahi, S. Egashira, M. Matsuda, H. Iwaasa, A. Kanatani, M. Ohkubo, M. Ihara, T. Sakurai and H. Morishima, Structure-activity regulation studies on the novel nuropeptide orexin., *Peptide Sci.*, 37-40 (1999).

ANIMAL MODELS IN THE STUDY OF THE HYPOCRETINERGIC SYSTEM

RODENT MODELS OF HUMAN NARCOLEPSY-CATAPLEXY

Takeshi Sakurai, Michihiro Mieda and Masashi Yanagisawa*

1. DISCOVERY OF MOUSE NARCOLEPSY

The rodent models of narcolepsy are summarized in table 1. The initial narcoleptic mouse model was the prepro-orexin gene knockout (*orexin*$^{-/-}$) mice. When orexin (hypocretin) knockout mice (*orexin*$^{-/-}$ mice) were created, intensive studies of their behavior using conventional methods failed to reveal any overt abnormalities during the light period, except mild decrease in their food intake.[1] To examine feeding behavior at night, when mice are normally most active, mice were recorded by infrared video-recordings.[2] This revealed frequent periods of obvious behavioral arrest in *orexin*$^{-/-}$ mice during the dark phase with full penetrance, which were totally unexpected (Fig. 1A).

Mouse	Phenotype	Ref.
Prepro-orexin knockout	Cataplexy (+), sleep attack (+) Sleep/wake fragmentation (severe)	2
OX1R knockout	Cataplexy (-), sleep attack (-) Sleep/wake fragmentation (mild)	1
OX2R knockout	Cataplexy (-), sleep attack (+) Sleep/wake fragmentation (severe)	5
Orexin/ataxin-3 mouse	Cataplexy (+), sleep attack (+) Sleep/wake fragmentation (severe)	6
Orexin/ataxin-3 rat	Cataplexy (+), sleep attack (+) Sleep/wake fragmentation (severe)	12

Table 1 Rodent narcolepsy models produced by genetic engineering

* T. Sakurai, Institute of Basic Medical Science, Departments of Pharmacology, University of Tsukuba, Ibaraki 305-8575, Japan; T. Sakurai and M. Yanagisawa, ERATO Yanagisawa Orphan Receptor Project, Japan Science and Technology Corporation, Tokyo 135-0064, Japan; M. Mieda, Department of Molecular Neuroscience, Medical Research Institute, Tokyo Medical and Dental University, 1-5-45 Yushima, Bunkyo-ku, Tokyo 113-8519, Japan.

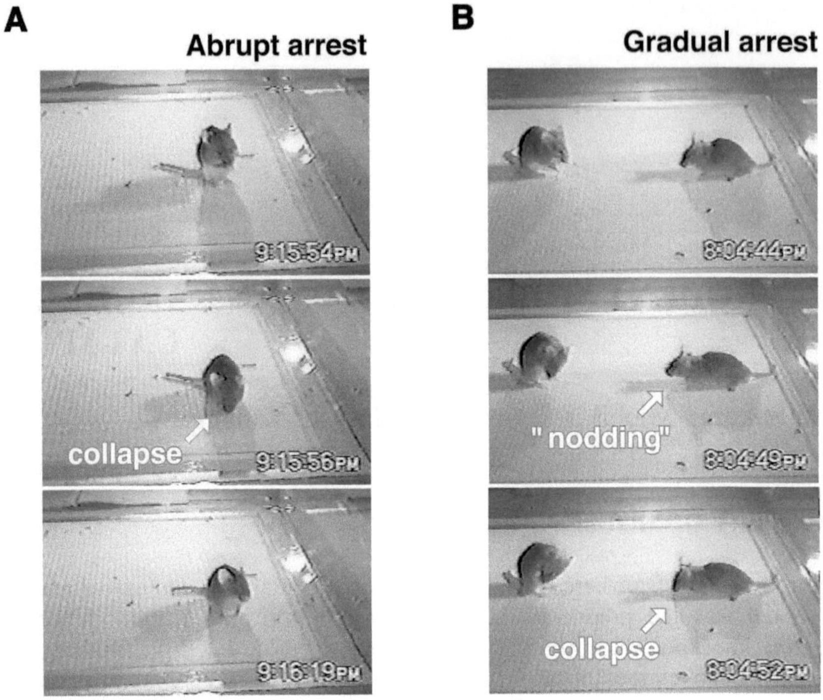

Figure 1. Comparison of behavioral arrests by infrared videophotography in knockout mice. (A) Time-lapse images portraying an abrupt arrest, rarely observed in an *OX2R* $^{-/-}$ mouse. Note the collapsed posture in the second panel. (B) Time-lapse images portraying a gradual arrest in an *OX2R*$^{-/-}$ mouse. "Nodding" behavior (second panel) occurs just prior to postural collapse (third panel). (Modified from ref. 5).

No sign of serum electrolyte imbalance or hypoglycemia was observed in *orexin*$^{-/-}$ mice. Bodily collapse associated with episodic rocking behavior initially suggested the possibility of a seizure disorder in the *orexin*$^{-/-}$ mice. However, electroencephalograph/electromyographic (EEG/EMG) recordings from *orexin*$^{-/-}$ mice showed no evidence of epileptic seizure. Rather, these EEG/EMG recordings revealed abnormal intrusions of REM sleep into wakefulness and fragmentation of sleep/wakefulness cycle (Figs. 2 and 3). Reduced latency to REM sleep and increase in REM sleep during the dark phase were also observed (Fig. 4). These behavioral and electrophysiological characteristics of *orexin*$^{-/-}$ mice were strikingly similar to characteristics human narcolepsy-cataplexy; Human narcolepsy-cataplexy is a debilitating neurological disease characterized by disorganization of behavioral states. This disorder affects approximately 1 in 2000 individuals in the United States.[3] Most cases of human narcolepsy usually start during adolescence. A cardinal symptom of the

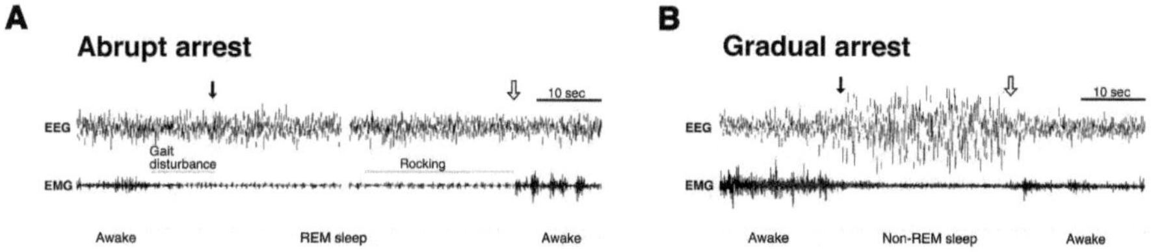

Figure 2. Typical EEG/EMG traces during behavioral arrests in *orexin* $^{-/-}$ and *OX2R* $^{-/-}$ mice. Solid and gray arrows demarcate onsets and terminations of arrests, respectively. Gray bars reflect the timing of gait disturbances and rocking behavior associated with arrests. Behavioral states are classified as awake, non-REM sleep, or REM sleep based on EEG/EMG features. (A) Abrupt arrest in *orexin*$^{-/-}$ mouse. Excited ambulation (high-amplitude nuchal EMG) accompanied by an EEG typical of normal active wakefulness (low-amplitude, mixed frequency activity) gives way to rapid onset of ataxic gait, reduced neck tone, and an EEG resembling REM sleep. Postural collapse is accompanied by neck atonia and continued REM sleep EEG. Rocking behavior from limb movement occurs exclusively during periods with EEG/EMG indistinguishable from REM sleep pattern. Residual low-amplitude noise remaining in the EMG during atonia consists primarily of electrocardiographic contamination. (B) Gradual arrest in *OX2R*$^{-/-}$ mouse. Feeding behavior with high-amplitude EMG and a waking EEG gives way to gradual onset of postural collapse with reduced but not atonic nuchal EMG and transition to an EEG indistinguishable from non-REM sleep. The mouse remains immobile until sudden recovery of waking EEG and purposeful behavior. (Modified from ref. 5).

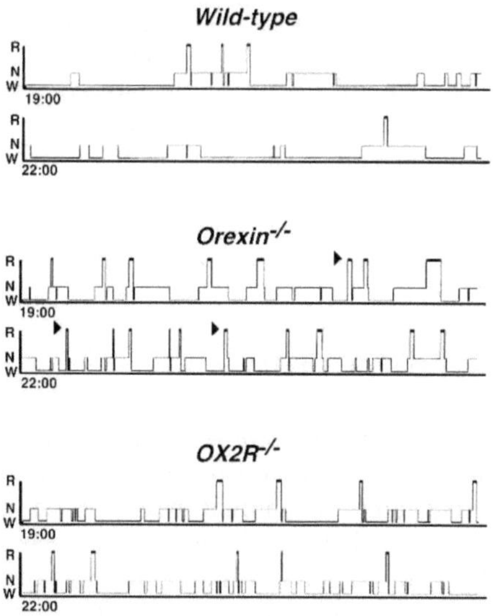

Figure 3. Representative hypnograms of wild-type, *orexin*$^{-/-}$, and *OX2R*$^{-/-}$ mice over the first 6 hr of the dark phase (19:00 to 01:00) obtained by concatenating 20 s epoch EEG/EMG stage scores. The height of the horizontal line above baseline indicates the vigilance state of the mouse at the time (min) from the beginning of the dark phase. Baseline, W, represents periods of wakefulness; N, non-REM sleep; R, REM sleep.Arrowheads highlight direct transitions from wakefulness to REM sleep. *Orexin*$^{-/-}$ and *OX2R*$^{-/-}$ mice show similar fragmentation of vigilance. (Modified from ref. 5).

disorder is excessive daytime sleepiness (an insurmountable urge to sleep), manifested particularly as attacks of somnolence at inappropriate times. Nocturnal sleep is also sometimes disturbed by hypnagogic hallucinations, vivid dreaming, and sleep paralysis. Narcolepsy patients often suffer from cataplexy, which is an attack characterized by sudden muscle weakness, which can range from jaw dropping and speech slurring to a complete bilateral collapse of the postural muscles. These attacks of muscle weakness are most often triggered by strong emotional stimuli. The latency for REM sleep is notably reduced in narcolepsy patients, and the presence of sleep-onset REM period is a diagnostic criterion for narcolepsy. It is generally accepted that the symptoms of narcolepsy can be considered as a pathological intrusion of factors of sleep, and especially REM sleep-related phenomena, into the state of wakefulness. Thus, narcolepsy can be viewed as a behavioral state boundary disorder.

An independent discovery around the same time that mutations in *OX2R* gene are responsible for familial narcolepsy-cataplexy in dogs, which is an accepted animal model of human narcolepsy, further supported the idea that deficiency in orexin-OX2R signaling might have an important role in pathophysiology of narcolepsy.[4]

Since mutations of *OX2R* gene are responsible for canine narcolepsy, it was expected that *OX2R*$^{-/-}$ mice exhibited cataplexy-like abrupt behavioral arrests (abrupt arrests), as observed in *orexin*$^{-/-}$ mice and narcoleptic dogs. In infrared videotaping studies in the dark phase, we indeed saw that kind of behavioral arrests in *OX2R*$^{-/-}$ mice. However, its frequency was much less than in *orexin*$^{-/-}$ mice (31-fold lower frequency in *OX2R*$^{-/-}$ mice as compared to *orexin*$^{-/-}$ mice).[5] Instead, *OX2R*$^{-/-}$ mice showed a distinct variety of behavioral arrests with onsets that were more gradual in nature (gradual arrests)(Fig. 1B), and it turned out that *orexin*$^{-/-}$ mice also exhibit gradual arrests with the frequency similar to *OX2R*$^{-/-}$ mice in addition to plenty of abrupt arrests. Detailed characterization of behavioral, pharmacological, and electrophysiological features of *orexin*$^{-/-}$ and *OX2R*$^{-/-}$ mice defined abrupt and gradual arrests as the presumptive mouse correlates of cataplexy and sleep attack in human narcolepsy-cataplexy, respectively. In following sections, we will describe narcoleptic phenotype of *orexin*$^{-/-}$ and *OX2R*$^{-/-}$ mice in detail and discuss in comparison with symptoms in human and canine narcolepsy.

2. REM SLEEP-RELATED SYMPTOMS

Characteristics of "abrupt arrests" were very different from those of quiet behavioral states with decreased overt activity, as well as from normal transitions into sleep. They were specifically recognized by the abrupt cessation of purposeful motor activity associated with sudden, sustained change in posture that was maintained throughout the episode, ending abruptly with complete resumption of purposeful motor activity (Fig. 1A).[2,5] Essentially, the episodes looked as if a behavioral switch had been turned "off" and then "on". Each episode lasted for a very short period, mostly less than a minute. Occasionally, gait disturbance lasting 1-3s was observed immediately before episodes. Also grossly observable motor activity causing side-to-side rocking, without change in overall posture, frequently occurred several seconds after the start of the arrest.

Detailed observations of behaviors along with EEG/EMG recordings revealed that "abrupt arrests" in *orexin*$^{-/-}$ mice occurred during direct transitions from wakefulness to REM sleep (Fig. 2A) or during pre-REM phase immediately after a waking period: the pre-REM phase shows EEG with high-amplitude spindle oscillations superimposed on non-REM sleep background, and these spindles are observed only during the transition phase immediately prior to REM sleep in wild-type mice. Direct or very rapid transitions from wakefulness to REM sleep are the most intriguing characteristic of EEG/EMG in *orexin*$^{-/-}$ mice or *orexin/ataxin-3* mice,[6] which were observed almost exclusively in their active phase, i.e. the dark phase, and were never exhibited by wild-type mice (Fig. 3).

Abrupt arrests in *orexin-/-* mice were suppressed by systemic administration of clomipramine, an anticataplectic agent used for treatment of human narcolepsy-cataplexy. Whereas, administration of caffeine, a psychostimulant used to treat excessive sleepiness in human narcolepsy, tended to produce a mild exacerbation of abrupt arrest frequency in mice.

These behavioral, electrophysiological, and pharmacological characteristics of abrupt arrests observed in *orexin*$^{-/-}$ or *orexin/ataxin-3* mice are very similar to cataplexy of human narcolepsy. An accepted definition of cataplexy in humans includes sudden bilateral weakening of muscles, usually induced by sudden strong emotions, lack of impairment of consciousness and memory, short duration (less than a few minutes), and responsiveness to treatment with clomipramine or imipramine.[7] Some studies performed during cataplectic attacks in humans as well as dogs reported REM sleep characteristics in the EEG, while other studies reported characteristics of wakefulness in the EEG during cataplexy.[8-11] In most cases, consciousness and memory is preserved in cataplexy in human.[7] Thus, we cannot definitely conclude definitively that abrupt behavioral arrests observed in *orexin*$^{-/-}$ mice are the equivalent of cataplexy in human until we confirm preservation of consciousness during the arrests in these mice. In relation to this discussion, we have occasionally observed apparent consciousness shortly after arrest onset in *orexin*$^{-/-}$ mice, and also wake-like EEG during abrupt behavioral arrests in transgenic rats lacking orexin neurons (*orexin/ataxin-3* transgenic rats, described below).[12]

Abrupt arrests in *orexin*$^{-/-}$ mice further share several features with cataplexy in humans and dogs. Cataplexy has been known to be triggered by strong emotions such as laughing, anger, fear, surprise, and excitement in human narcolepsy patients.[13] Consistently, in narcoleptic dogs, food and play with other dogs are the well-documented paradigms used to trigger cataplexy.[14] In *orexin*$^{-/-}$ mice, purposeful motor activity always precedes episodes of abrupt arrests; ambulation, grooming, burrowing, and climbing were most frequently observed prior to abrupt arrests. The dramatic increase in the number of abrupt arrests noted in the group-housed mice as compared with the individually-housed littermates suggests that social interaction may significantly enhance this phenotype. Chasing, tail biting, and social grooming were often observed to immediately precede abrupt arrests in the group-housed mice. The facts that abrupt arrests are observed mainly in their active phase and that novel environment enhances occurrence of abrupt arrests further supports emotional activation triggers these arrests in *orexin*$^{-/-}$ mice or *orexin/ataxin-3* mice.

Only approximately one-third of human patients experience full loss of muscle tone causing collapse to the floor with the majority having partial cataplexy evidenced by jaw sagging, head bobbing, arm dropping, ptosis, or dysarthria. Partial cataplexy in the dog is often evidenced by hindlimb buckling. Unambiguous full postural collapse was frequently observed in young *orexin-/-* mice, while adults tended to collapse onto their ventral surface at odd angles, suggesting some residual muscle tone.

3. NON-REM SLEEP-RELATED SYMPTOMS

Increased daytime sleepiness is well described in human patients.[7,13] Human patients often complain of involuntary or irresistible urge to sleep, which can occur while talking, standing, walking, eating, or driving. In addition to REM sleep-related abnormalities, another prominent feature of sleep/wake patterns in *orexin*$^{-/-}$ mice or *orexin/ataxin-3 mice* was shortened durations of both wakefulness and non-REM sleep in the dark phase, causing increased fragmentation of sleep/wake cycle (Fig. 3). *OX2R-/-* mice also showed sleep/wake fragmentation, while occurrence of REM sleep-related abnormalities was very rare as compared to *orexin*$^{-/-}$ mice (Fig. 3). This fragmentation was accompanied by statistically insignificant tendency toward reduced amounts of wakefulness and increased amounts of non-REM sleep during the dark phase (Fig. 4). Fragmentation of vigilance states suggests the inability to maintain sleep and wakefulness states. Presumptive excessive sleepiness in *orexin*$^{-/-}$ mice was further clarified by detailed analyses of "gradual arrests" in *orexin*$^{-/-}$ and *OX2R*$^{-/-}$ mice, which can be interpreted analogous to sleep attacks in human narcolepsy-cataplexy. In contrast to abrupt arrests, such arrests have their onsets that were more gradual in nature (Fig. 1B). The "gradual arrests" typically began during quiet wakefulness and could be easily distinguished from the normal onset of resting behavior by (i) the absence of stereotypic preparation for sleep (e.g., nesting and/or assumption of a curled or hunched posture, with limbs drawn under the body) and (ii) a characteristic ratchet-like "nodding" of the head over a period of several seconds, with a transition to a collapsed posture.[5]

Systemic administration of caffeine, which is a psychostimulant used for treating excessive sleepiness in human narcolepsy-cataplexy, dose-dependently suppressed gradual arrests, while administration of an anticataplectic agent clomipramine did not affect the frequency of gradual arrests in both *orexin-/-* and *OX2R-/-* mice

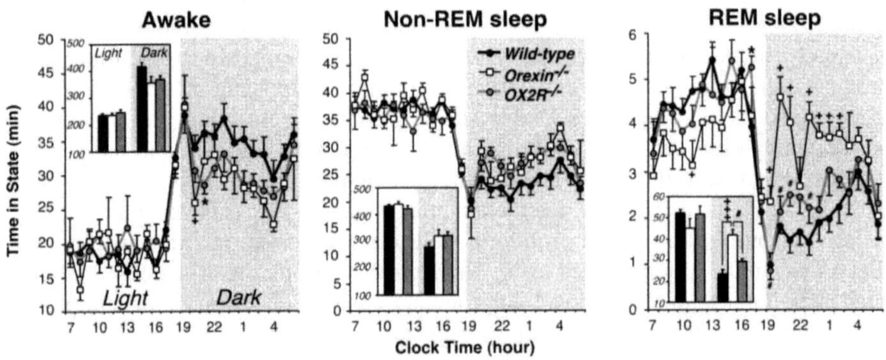

Figure 4. Hourly amounts of awake, non-REM sleep, and REM sleep states (means and SEM) plotted over the 24 hr day for each group. Data collapsed over 12 hr light and dark phases are displayed as graph insets. Data for the light and dark phases are displayed on white and light gray backgrounds, respectively. Wakefulness and non-REM sleep are disrupted to a similar degree in *orexin*$^{-/-}$ and *OX2R*$^{-/-}$ mice, especially during the dark phase. *Orexin*$^{-/-}$ mice consistently show significantly increased REM sleep times during the dark phase compared to normal mice; *OX2R*$^{-/-}$ mice do not. Significant differences between *OX2R*$^{-/-}$ and wild-type mice: *, $p < 0.05$; **, $p < 0.005$. Significant differences between *orexin-/-* and wild-type mice: +, $p < 0.05$; ++, $p < 0.005$; +++, $p < 0.0005$. Significant differences between *orexin-/-* and *OX2R*$^{-/-}$ mice: #, $p < 0.05$. (Modified from ref. 5).

EEG/EMG recordings concurrent with videotaping further differentiated gradual arrests from abrupt ones. As described above, abrupt arrests were accompanied either by direct transition from wakefulness to REM sleep or to the pre-REM stage with atonia in both *orexin*$^{-/-}$ and *OX2R*$^{-/-}$ mice (Fig. 2A). In contrast, EEG/EMG correlates of gradual arrests in both *orexin*$^{-/-}$ and *OX2R*$^{-/-}$ mice invariably revealed the onset of attenuated muscle tone, but not atonia, and an EEG transition from wakefulness to non-REM sleep (Fig. 2B). Gradual arrests were occasionally accompanied by apparent automatic behavior (continuation of semi-purposeful motor activity after the onset of light sleep such as stereotypic chewing of food) in both genotypes. The EEG features of these episodes conform unambiguously to the spectral features of non-REM sleep. In *orexin*$^{-/-}$ mice, a large proportion of sleep attacks transitioned prematualy from non-REM sleep to REM sleep, while such rapid transitions to REM sleep were observed only rarely in *OX2R*$^{-/-}$ mice.

Overall, gradual arrests in both *orexin*$^{-/-}$ and *OX2R*$^{-/-}$ mice resemble the sleep attacks in human narcolepsy. They might be associated with impaired but not with strong emotions or abrupt muscle weakness. The identification of sleep attack is based upon a behavioral context: patients report sudden irresistible sleepiness occurring during unusual circumstances (e.g., meals, conversations, driving), and such attacks may also be associated with automatic behavior. Sleep attacks are generally associated with the onset of early stages of non-REM sleep, reflect a compression of the normal process of entering sleep, and mimic the effects of sleep deprivation in normal humans. Unlike cataplexy, sleepiness is reduced by psychostimulants such as amphetamines, modafinil, and caffeine[7]. Similarly, canine narcolepsy exhibit sleepiness based on increased tendencies to fall asleep and more fragmented wakefulness patterns. Behavioral attacks of “drowsiness” associated with “stop motion” and nodding of the head and neck are observed in genetically narcoleptic Dobermans, although these have not been reliably distinguished from cataplexy by EEG/EMG, possibly due in part to the more staged EEG patterns of non-REM sleep in canines (S. Nishino, personal communication).

4. DIFFERENTIAL REGULATION OF SLEEP/WAKE STATES BY OX1R AND OX2R

As described above, *orexin*$^{-/-}$ and *OX2R*$^{-/-}$ mice are indistinguishable with respect to parameters of wakefulness and non-REM sleep. Both mice exhibited gradual arrests and fragmentation of sleep/wake cycle. In contrast, *orexin*$^{-/-}$ and *OX2R*$^{-/-}$ mice are distinct from each other in the expression of abnormalities of REM sleep. *Orexin*$^{-/-}$ displayed abrupt arrests and direct transitions from wakefulness to REM sleep without exception, the hallmark of rodent narcolepsy, while *OX2R*$^{-/-}$ mice exhibited no such transitions or did with far less frequency. *Orexin*$^{-/-}$ mice exhibited an increase in the amount of time in REM sleep over the entire dark phase, increased frequency of REM sleep in the dark phase, and reduced latency to REM sleep in both phases.[2] While in comparison to *orexin*$^{-/-}$ and wild-type mice, *OX2R*$^{-/-}$ mice tended to exhibit intermediate values for some of these same parameters of REM sleep, only a reduced REM sleep latency during the dark phase differed significantly from wild-type mice.[5] Overall, the general patterns of REM sleep of *OX2R*$^{-/-}$ mice more closely resemble those of wild-type mice than those of *orexin*$^{-/-}$ mice. OX2R-deficiency causes abnormal transitions from wakefulness to non-

REM sleep, while lack of orexin results in abnormal transitions from wakefulness not only to non-REM sleep but also to REM sleep.[5]

OX1R$^{-/-}$;OX2R$^{-/-}$ mice appeared to have the same phenotype with *orexin-/-* mice. However, *OX1R-/-* mice did not have overt behavioral abnormalities and exhibited only mildly increased fragmentation of sleep/wake states. Overall, it should be concluded that normal regulation of wake/non-REM sleep transitions depends critically on OX2R activation, while the profound dysregulation of REM sleep control unique to the narcolepsy syndrome emerges from loss of signaling through both OX1R- and OX2R-dependent pathways.

A complementary experiment we carried out in order to further confirm this conclusion was examination of arousal effect of i.c.v. orexin-A administration in each strain of knockout mice (unpublished observation). As previously reported in rodents, i.c.v. administration of orexin-A in wild-type mice increased wakefulness and suppressed both non-REM and REM sleep in dose-dependent manner.[15] Effects of orexin-A on sleep/wakefulness pattern were not significantly different in *OX1R$^{-/-}$* mice, while effects on wakefulness and non-REM sleep in *OX2R$^{-/-}$* mice were dramatically attenuated as compared to wild-type and *OX1R$^{-/-}$* mice. These data suggest that OX1R is dispensable for increase of wakefulness and suppression of non-REM sleep by orexin-A administration in the presence of OX2R, although stimulation of OX1R in the absence of OX2R is capable of promoting wakefulness and suppressing non-REM sleep with efficiency much lower than stimulation of OX2R. As for suppression of REM sleep by orexin-A administration, there was no significant difference among wild type, *OX1R$^{-/-}$*, *OX2R$^{-/-}$* mice, although suppression tended to be less efficient in *OX2R$^{-/-}$* mice, suggesting both OX1R and OX2R mediate REM sleep suppression. Furthermore, we obtained data suggesting that activation of OX1R directly suppresses REM sleep. Importantly, no significant effect of orexin-A administration on sleep/wakefulness pattern was observed in *OX1R$^{-/-}$;OX2R$^{-/-}$* mice, implicating arousal effect of orexin-A is mediated by both receptors. Overall, these results are consistent with the model described above derived from comparison of behavioral and baseline sleep/wake cycle characteristics of *orexin$^{-/-}$*, *OX1R$^{-/-}$*, and *OX2R$^{-/-}$*, and *OX1R$^{-/-}$;OX2R$^{-/-}$* mice.

5. RODENT MODELS OF PATHOPHYSIOLOGY OF HUMAN NARCOLEPSY

Electrophysiological abnormalities in *orexin$^{-/-}$* are strikingly similar to human narcolepsy. However, unlike *orexin$^{-/-}$* mice as well as *OX2R*-dificient narcoleptic dogs, familial transmission of human narcolepsy is rare; even in these rare cases penetrance is far less than 100%. No mutation has been found either in the *prepro-orexin* or orexin receptor genes of human narcolepsy-cataplexy patients except an unusually severe, early onset case, which is associated with mutation in the *prepro-orexin* gene that impairs peptide trafficking and processing[16]. In most cases of human narcolepsy-cataplexy, symptoms start to appear during adolescence. Nevertheless, dramatic reduction of orexin levels in the CSF and postmortem brains of narcolepsy-cataplexy patients has been reported.[17-19] Based on these observations, as well as a strong association of human narcolepsy with certain HLA alleles,[3] it has been speculated that narcolepsy-cataplexy may result from selective autoimmune degeneration of orexin neurons.

In order to mimic such pathophysiological condition of human narcolepsy-cataplexy, we also generated transgenic mice (*orexin/ataxin-3* transgenic mice) in which orexin neurons are ablated by orexinergic-specific expression of a truncated Machado-Joseph disease gene product (ataxin-3) with expanded polyglutamine stretch.[6] In these mice, number of orexin neurons gradually decreased after their birth, and at 12 weeks of age, over 99% of orexin neurons were lost (Fig. 5). *Orexin/ataxin-3* transgenic mice exhibited behavioral and electrophysiological defects that are essentially same as those displayed by *orexin-/-* mice: abrupt and gradual arrests, direct transitions from wakefulness to REM sleep, shortened latency to REM sleep, fragmentation of vigilance states, and increases in REM sleep time and duration in the dark phase. Only the exception is that the behavioral arrests seen in *orexin/ataxin-3* transgenic mice typically began about 6 weeks of age, while arrests were observed in some *orexin*$^{-/-}$ mice earlier than 3 weeks of age. Thus, *orexin/ataxin-3* transgenic mice have etiology and the course of disease similar to human narcolepsy, and *orexin/ataxin-3* transgenic mice may represent the most accurate pathophysiological model of narcolepsy available. Another conclusion derived from analyses of *orexin/ataxin-3* transgenic mice is that although orexin neurons produce other neurotransmitters such as glutamate and dynorphins,[20,21] it is orexin that is important for the regulation of the sleep/wakefulness states by these neurons. Indeed, we utilized this mouse model of narcolepsy to rescue its narcoleptic phenotype by genetic (overexpression of orexin peptides throughout the brain) and pharmacological (i.c.v. administration of orexin-A) means,[22] demonstrating that mice retain the ability to respond to orexin neuropeptides even if they lack endogenous orexin neurons and that a temporally regulated and spatially targeted secretion of orexins is not necessary to prevent narcoleptic symptoms. Thus, orexin receptor agonists would be of potential value for treating human narcolepsy.

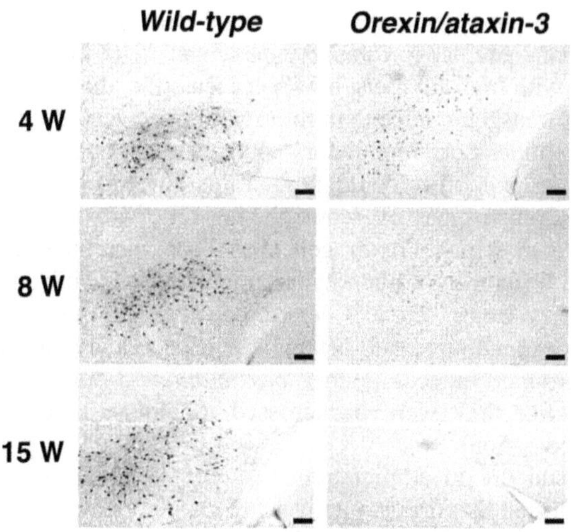

Figure 5. Ablation of orexin neurons in *orexin/ataxin-3* transgenic mice. Matched brain sections (bregma 22.1 mm) from 4-, 8-, and 15-week-old *orexin/ataxin-3* transgenic mice (right panels) and their transgene-negative littermates (left panels) were stained with anti-orexin antiserum. Bar, 100 μm. (Modified from ref. 6).

Recently we reported the transgenic rats that have the same transgene as *orexin/ataxin-3* transgenic mice (*orexin/ataxin-3* transgenic rats).[12] In these rats, number of orexin neurons was gradually reduced after their birth, and at 17 weeks of age hypothalamic orexin expression was undetectable. Again, *orexin/ataxin-3* transgenic rats showed a narcoleptic phenotype, with a decreased latency to REM sleep, increased REM sleep time, direct transitions from wakefulness to REM sleep, and a marked fragmentation of vigilance states. Brief episodes of muscle atonia and postural collapse resembling cataplexy were also observed while rats maintained the EEG characteristics of wakefulness, suggesting they were conscious during these episodes as human patients are during cataplexy. Since the rat has been more widely used for physiological and pharmacological studies in the field, *orexin/ataxin-3* transgenic rats are likely to be a valuable resource for the study of human narcolepsy-cataplexy.

6. MORE THAN SLEEP/WAKE ABNORMALITIES

The most profound symptoms of human narcolepsy are sleep-related disturbances. However, discovery of the orexin system and analysis of mice with deficiency in the orexin signaling pathway shed light on another aspect of human narcolepsy. Orexins were initially recognized as feeding peptides; i.c.v. administration of orexins increased food intake in rats.[23] Consistent with this observation, *orexin*$^{-/-}$ and *orexin/ataxin-3* transgenic mice were hypophagic.[1,6] Nevertheless, *orexin/ataxin-3* transgenic mice showed late-onset obesity.[6] This apparently contradictory phenomenon was likely to result from reduced energy expenditure in these mice since they were less active in the dark phase as compared to wild-type mice.[6] It has been reported that frequencies of obesity and non-insulin dependent (Type II) diabetes, as well as body mass index, are increased in human narcolepsy-cataplexy patients.[24,25] *Orexin*$^{-/-}$ mice also show mild obesity when fed with high-fat diet[1], however metabolic abnormality seems more severe in *orexin/ataxin-3* transgenic mice than in *orexin*$^{-/-}$ mice according to their initial reports. Loss of neuropeptides or modulatory factors expressed in orexin neurons in *orexin/ataxin-3* transgenic mice could explain this difference. Another possibility is that compensatory mechanisms may overcome the impact of orexin-deficiency on metabolism of *orexin-/-* mice, in which orexin gene has been disrupted from the beginning of development. The difference of phenotype might be also stem from genetic backgrounds used in these reports. Body weight is determined according to the balance between food intake and energy expenditure, and metabolic phenotypes are known to be very sensitive to genetic background. Indeed, neither *orexin/ataxin-3* transgenic nor *orexin*$^{-/-}$ mice showed obesity after they were backcrossed to almost pure C57BL/6J background (manuscript in preparation).

Pharmacological orexin administration studies, as well as anatomy of the orexin system, have suggested that orexin peptides are also involved in regulation of autonomic and endocrine systems.[26] Defects in these systems could result in disturbance of metabolism, leading to changes in energy expenditure. In *orexin*$^{-/-}$ mice, it has been reported that basal arterial blood pressure is significantly lower, and autonomic "fight or flight" response is attenuated as compared to wild-type mice.[27] We also demonstrated that orexin/ataxin-3 mice fail to respond to fasting with increased wakefulness and activity, suggesting that orexin neurons constitute central pathways regulating arousal and instinctual motor programs (such as food seeking) according to homeostatic need.[28]

7. CONCLUSIONS

Behavioral and electrophysiological analysis of *orexin*$^{-/-}$ mice, together with studies on narcoleptic OX2R-deficient dogs, made a big jump in our understanding on narcolepsy-cataplexy, as well as regulation of sleep/wake cycle. Mouse molecular genetics succeeded to dissect non-REM and REM sleep regulatory processes by comparing mutants of every component of the orexin signaling pathway. Furthermore, obesity in *orexin/ataxin-3* transgenic and *orexin*$^{-/-}$ mice directed researchers' interest toward metabolic abnormalities in narcolepsy patients. One of big advantages of using rodent models is easiness to prepare large number of animals for systematic studies. Detailed analyses of mouse and rat models of narcolepsy described here would not only deepen our understanding of disturbances of sleep regulation in narcolepsy but might also find out novel aspects of physiological abnormalities in human narcolepsy patients.

8. REFERENCES

1. J. T. Willie, R. M. Chemelli, C. M. Sinton and M. Yanagisawa, To eat or to sleep? Orexin in the regulation of feeding and wakefulness., *Annu Rev Neurosci.* 24, 429-58 (2001).
2. R. M. Chemelli, J. T. Willie, C. M. Sinton, J. K. Elmquist, T. E. Scammell, C. Lee, J. A. Richardson, S. C. Williams, Y. Xiong, Y. Kisanuki, T. E. Fitch, M. Nakazato, R. E. Hammer, C. B. Saper and Y. M., Narcolepsy in orexin knockout mice: molecular genetics of sleep regulation., *Cell.* 98, 437-51 (1999).
3. E. Mignot, Genetic and familial aspects of narcolepsy., *Neurology.* 50, S16-S22 (1998).
4. L. Lin, J. Faraco, R. Li, H. Kadotani, W. Rogers, X. Lin, X. Qiu, P. J. de Jong, S. Nishino and E. Mignot, The sleep disorder canine narcolepsy is caused by a mutation in the hypocretin (orexin) receptor 2 gene., *Cell.* 98, 365-76 (1999).
5. J. T. Willie, R. M. Chemelli, C. M. Sinton, S. Tokita, S. C. Williams, Y. Y. Kisanuki, J. N. Marcus, C. Lee, J. K. Elmquist, K. A. Kohlmeier, C. S. Leonard, J. A. Richardson, R. E. Hammer and M. Yanagisawa, Distinct narcolepsy syndromes in Orexin receptor-2 and Orexin null mice: molecular genetic dissection of Non-REM and REM sleep regulatory processes., *Neuron.* 38, 715-30 (2003).
6. J. Hara, C. T. Beuckmann, T. Nambu, J. T. Willie, R. M. Chemelli, C. M. Sinton, F. Sugiyama, K. Yagami, K. Goto, M. Yanagisawa and T. Sakurai, Genetic ablation of orexin neurons in mice results in narcolepsy, hypophagia, and obesity., *Neuron.* 30, 345-54 (2001).
7. M. S. Aldrich, Diagnostic aspects of narcolepsy., *Neurology.* 50, S2-7 (1998).
8. M. M. Mitler and W. C. Dement Sleep studies on canine narcolepsy: pattern and cycle comparisons between affected and normal dogs. Electroencephalogr., *Clin. Neurophysiol.* 43, 691-699 (1977).
9. M. E. Dyken, T. Yamada, D. C. Lin-Dyken, P. Seaba and M. Yeh, Diagnosing narcolepsy through the simultaneous clinical and electrophysiologic analysis of cataplexy. Arch. Neurol., 53, 456-460 (1996).
10. C. A. Kushida, T. L. Baker and W. C. Dement, Electroencephalographic correlates of cataplectic attacks in narcoleptic canines. Electroencephalogr., *Clin. Neurophysiol.* 61, 61-70 (1985).
11. C. Guilleminault, R. A. Wilson and W. C. Dement, A study on cataplexy., *Arch. Neurol.* 31, 255-261 (1974).
12. C. T. Beuckmann, C. M. Sinton, S. C. Williams, J. A. Richardson, R. E. Hammer, T. Sakurai and M. Yanagisawa, Expression of a poly-glutamine-ataxin-3 transgene in orexin neurons induces narcolepsy-cataplexy in the rat., *J Neurosci.* 24, 4469-77 (2004).
13. C. Bassetti and M. S. Aldrich, Narcolepsy, *Neurol. Clin.* 14, 545-571 (1996).
14. J. Riehl, S. Nishino, R. Cederberg, W. C. Dement and E. Mignot, Development of cataplexy in genetically narcoleptic Dobermans., *Exp. Neurol.* 152, 292-302 (1998).
15. J. J. Hagan, R. A. Leslie, S. Patel, M. L. Evans, T. A. Wattam, S. Holmes, C. D. Benham, S. G. Taylor, C. Routledge, P. Hemmati, R. P. Munton, T. E. Ashmeade, A. S. Shah, J. P. Hatcher, P. D. Hatcher, D. N. Jones , M. I. Smith, D. C. Piper, A. J. Hunter, R. A. Porter and N. Upton, Orexin A activates locus coeruleus cell firing and increases arousal in the rat., *Proc Natl Acad Sci U S A.* 96, 10911-6 (1999).
16. C. Peyron, J. Faraco, W. Rogers, B. Ripley, S. Overeem, Y. Charnay, S. Nevsimalova, M. Aldrich, D. Reynolds, R. Albin, R. Li, M. Hungs, M. Pedrazzoli, M. Padigaru, M. Kucherlapati, J. Fan, R. Maki, G. J.

Lammers, C. Bouras, R. Kucherlapati, S. Nishino and E. Mignot, A mutation in a case of early onset narcolepsy and a generalized absence of hypocretin peptides in human narcoleptic brains., *Nat. Med.* 9, 991-997. (2000).
17. S. Nishino, B. Ripley, S. Overeem, G. J. Lammers and E. Mignot, Hypocretin (orexin) deficiency in human narcolepsy., *Lancet.* 355, 39-40. (2000).
18. C. Peyron, D. K. Tighe, A. N. van den Pol, L. de Lecea, H. C. Heller, J. G. Sutcliffe and T. S. Kilduff, Neurons containing hypocretin (orexin) project to multiple neuronal systems., *J Neurosci.* 18, 9996-10015 (1998).
19. T. C. Thannickal, R. Y. Moore, R. Nienhuis, L. Ramanathan, S. Gulyani, M. Aldrich, M. Cornford and J. M. Siegel, Reduced number of hypocretin neurons in human narcolepsy., *Neuron.* 27, 469-474 (2000).
20. T. C. Chou, C. E. Lee, J. Lu, J. K. Elmquist, J. Hara, J. T. Willie, C. T. Beuckmann, R. M. Chemelli, T. Sakurai, M. Yanagisawa, C. B. Saper and T. E. Scammell, Orexin (hypocretin) neurons contain dynorphin., *J. Neurosci.* 21, RC168 (2001).
21. E. E. Abrahamson, R. K. Leak and R. Y. T. Moore, The suprachiasmatic nucleus projects to posterior hypothalamic arousal systems., *Neuroreport.* 12, 435-40 (2001).
22. M. Mieda, J. T. Willie, J. Hara, C. M. Sinton, T. Sakurai and M. Yanagisawa, Orexin peptides prevent cataplexy and improve wakefulness in an orexin neuron-ablated model of narcolepsy in mice., *Proc Natl Acad Sci U S A.* 101, 4649-54 (2004).
23. T. Sakurai, A. Amemiya, M. Ishii, I. Matsuzaki, R. M. Chemelli, H. Tanaka, S. C. Williams, J. A. Richardson, G. P. Kozlowski, S. Wilson, J. R. Arch, R. E. Buckingham, A. C. Haynes, S. A. Carr, R. S. Annan, D. E. McNulty, W. S. Liu, J. A. Terrett, N. A. Elshourbagy, D. J. Bergsma and M. Yanagisawa, Orexins and orexin receptors: a family of hypothalamic neuropeptides and G protein-coupled receptors that regulate feeding behavior., *Cell.* 92, 573-85 (1998).
24. Y. Honda, Y. Doi, R. Ninomiya and C. Ninomiya, Increased frequency of non-insulin-dependent diabetes mellitus among narcoleptic patients., *Sleep.* 9, , 254-259. (1986).
25. A. Schuld, J. Hebebrand, F. Geller and T. Pollmacher, Increased body-mass index in patients with narcolepsy., *Lancet.*, 1274-1275 (2000).
26. A. V. Ferguson and W. K. Samson, The orexin/hypocretin system: a critical regulator of neuroendocrine and autonomic function., *Front. Neuroendocrinol.* 24, 141-150 (2003).
27. Y. Kayaba, A. Nakamura, Y. Kasuya, T. Ohuchi, Y. M., I. Komuro, Y. Fukuda and T. Kuwaki, Attenuated defense response and low basal blood pressure in orexin knockout mice., *Am. J. Physiol.Regul. Integr. Comp. Physiol.* 285, R581-593 (2003).
28. A. Yamanaka, C. T. Beuckmann, J. T. Willie, J. Hara, N. Tsujino, M. Mieda, M. Tominaga, K. Yagami, F. Sugiyama, K. Goto, M. Yanagisawa and T. Sakurai, Hypothalamic orexin neurons regulate arousal according to energy balance in mice., *Neuron.* 38, 701-13 (2003).

THE CANINE MODEL OF NARCOLEPSY

Seiji Nishino*

1. INTRODUCTION

Human narcolepsy is a chronic sleep disorder that affects 1/2000 of the general population. One of the major research aims of the Stanford Sleep Research Center is to find the etiology of the disease and to develop better treatments for human narcolepsy. However, human narcolepsy is very complex and multifactorial, and is (1) genetically heterogeneous, (2) a polygenic trait and (3) environmentally influenced.[1] The development of a simpler animal model of narcolepsy was thus the high priority in the Center to pursuer this aim.

In 1972, the Stanford Sleep Research Center (directed by Dr. William C. Dement) had an educational exhibit about human narcolepsy at the annual convention of the American Medical Association in San Francisco. As part of the exhibit, a film clip of narcoleptic patients having cataplexy was shown, and one member of the audience (a faculty member from the school of veterinary medicine at UC-Davis), mentioned that he took care a dog that had the same attacks. Unfortunately, the dog had been euthanized since it was thought to have been affected with refractory epilepsy. However, film clips of cataplectic attacks from this dog remained. Dr. Dement started to use those film clips along with those of human cataplexy for his lectures, including the one at the annual meetings of American Academy of Neology in Boston in 1973. After that lecture, a neurologist talked to Dr. Dement and mentioned that one of his friends had a dog that showed similar attacks. The dog was a miniature French Poodle named Monique, and was donated to Stanford. Monique revealed that the emotion of eating triggered multiple flaccid paralyses, and using surface electrodes, the loss of electromyogram (EMG) activities was observed during the attacks. An electroencephalogram (EEG) was found to be normal, without any signs of an epileptic seizure, but showing patterns of typical of REM sleep and/or wakefulness.[2] This together with an independent report of a narcoleptic dachshund dog by Knecht *et al.*,[3] confirmed the existence of narcolepsy in dogs. In 1976, four narcoleptic Dobermans were obtained, and a breeding program for

* Seiji Nishino Stanford University. Center For Narcolepsy 701 Welch road B. Palo Alto CA 94304-5742

these animals was initiated.[4] Subsequently, it was demonstrated that narcolepsy in Dobermans (and Labradors) was transmitted with single autosomal recessive gene, named *canarc-1*, while the disease in other breeds were sporadic.[4] Twenty-three years later, the *canarc-1* was identified by positional cloning, and found to be a gene encoding one of the two hypocretin receptors (i.e., hypocretin receptor-2).[5] This, together with an independent discovery of narcolepsy phenotype in mice lacking hypocretin ligand (preprohypocretin gene knockout mice),[6] immediately lead to the discovery of the major pathophysiological mechanism of human narcolepsy (i.e., hypocretin ligand deficiency).[7-9] In this chapter, I will summarize the knowledge we have gained from these narcoleptic dogs over a 30-year period.

2. SYMPTOMS OF CANINE NARCOLEPSY

Affected dogs exhibit very pronounced attacks of cataplexy (which are mainly triggered by positive emotional experiences), such as being fed a favorite food or engaging in play. Cataplectic attacks in dogs often begin as a buckling of both hind legs, and this is often accompanied by a drooping of the neck (see Figure 1). The dog may collapse to the floor and remain motionless for a few seconds or several minutes. In contrast to the some forms of epilepsy, excess salivation or incontinence are not observed during cataplectic attacks. During long cataplectic attacks, rapid eye movements, muscle twitching and/or slow repetitive movement of the fore and hind limbs may occur. These phenomena are related to the active phase of REM sleep. The muscle is always flaccid and never stiff during cataplectic attacks, and this is also different from most forms of seizure attacks. Dogs usually remain conscious (especially at the beginning of attacks) with eyes open, and are capable of following moving objects with their eyes. If the attack lasts for an appreciable length of time (usually longer than 1 to 2 minutes), the dog may transit into sleep (often REM sleep). Dogs are often easily aroused out of an attack either by a loud noise or by being physically touched. Like narcoleptic humans, narcoleptic dogs are sleepier (fall asleep much more quickly) during the day. However, this was not noticeable at usual circumstances, because even normal dogs take multiple naps during the daytime. While hypnagogic hallucinations and sleep paralysis may also occur in these dogs, there is no objective way to determine this at the present time.

A series of polygraphic studies clearly demonstrated that the dog's sleep/wake pattern is very fragmented and their wake/sleep bouts much shorter than age, breed-matched dogs, and narcoleptic dogs change their sleep states more frequently than they change their vigilance states (Figure 2).[10-12] In other words, narcoleptic subjects could not maintain long bouts of wakefulness and sleep. By systematic polygraph assessments with multiple daytime day time nap test, it was demonstrated that narcoleptic Dobermans showed shortened sleep latency and reduced latency to REM sleep during (Figure. 2),[10] suggesting that these dogs have very similar phenotype to those in human narcolepsy.

3. INHERITANCE OF NARCOLEPSY IN CANINES

Unrelated narcoleptic poodles, beagles and dachshunds donated to the Stanford were bred, but none of the offspring from two primary crosses or a backcross were affected.[13] Thus, it appears that in these breeds, narcolepsy seems to be sporadic and an involvement of a high penetrant single major gene is unlikely. In humans, about 95% of narcoleptic subjects are sporadic, but familial occurrences were noted in about 5% of narcoleptic subjects.[14] In 1976, the Stanford Sleep Research Center received 4 affected Dobermans (including 2 littermates).[4] These dogs were also bred, and it was discovered that all offspring from the 2 affected parents developed cataplexy around 2 months old. In 1978, familial narcolepsy in a Labrador Retriever was also discovered.[4] Thereafter, more sporadic cases of canine narcolepsy were identified in Collies, Dachshunds, Beagles, Fox Terriers and in several mixed breeds [4].

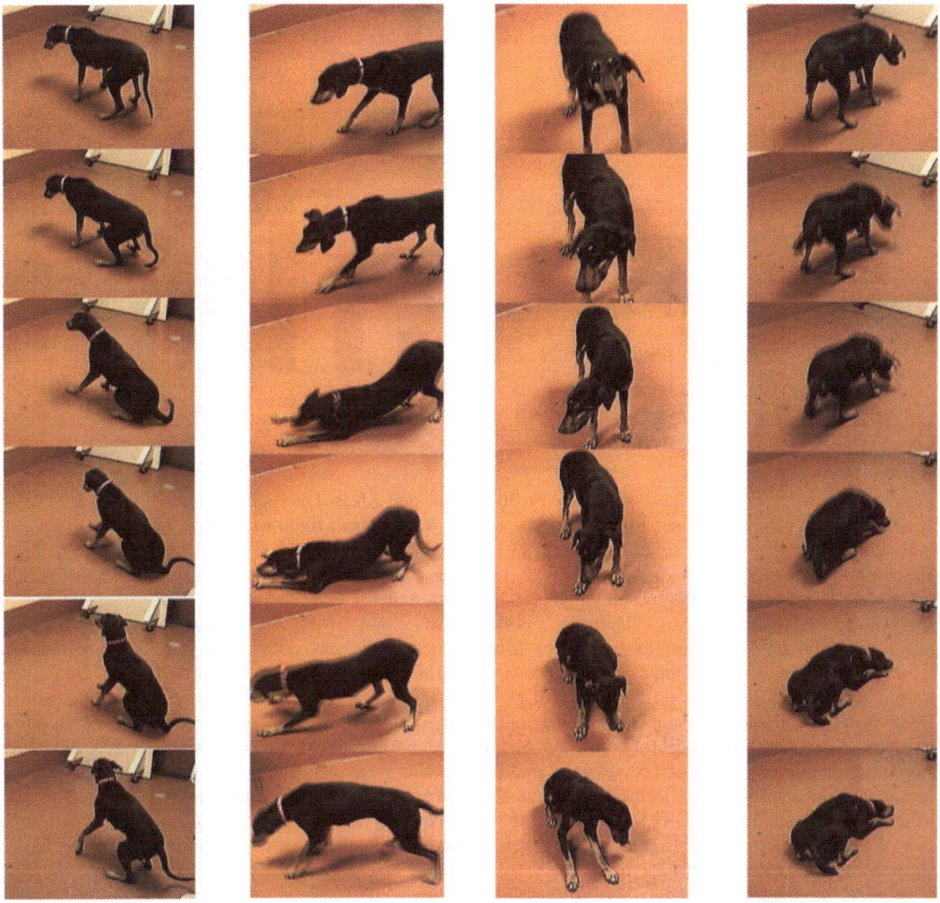

Figure 1. Cataplectic attacks in Doberman pinschers. Emotional excitations, appetizing food or playing easily elicit multiple cataplectic attacks in these animals. Most cataplexy attacks are bilateral (97.9%). Atonia initiated partially in the hind legs (79.8%), front legs (7.8%) neck/face (6.2%), or whole body/complete attacks (6.2%) Progression of attacks was also seen (49% of all attacks).[92]

Genetic transmission in Dobermans and Labradors has been well established as autosomal recessive with full penetrance.[4,15] Puppies born from narcoleptic Doberman

Pinscher-Labrador Retriever crosses are all affected. Thus, both breeds are likely to have mutations at the same locus, coined *canarc-1.*[4,15] As in human cases,[16] the disease onset in familial cases is earlier than in sporadic cases.[13,17] In sporadic cases, the disease (cataplexy onset) begins as early as 7 weeks and as late as 7 years old,[13] suggesting the

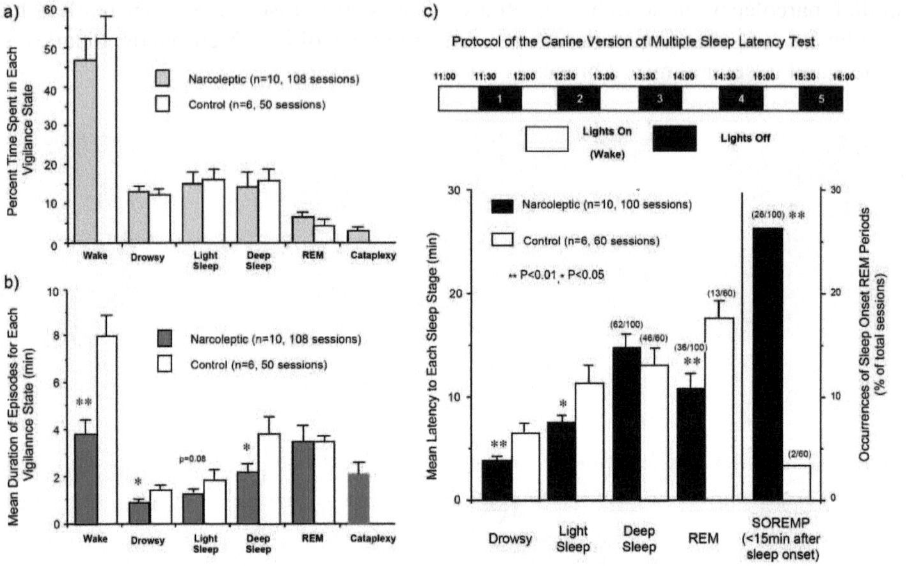

Figure 2. Percent of Time Spent in, Mean Frequency of, and Mean Duration for Each Vigilance State of Narcoleptic and Control Canines during Daytime 6-Hour Recordings (10:00 to 16:00). (a, b) No significant difference was found in percent of time spent in each vigilance state between narcoleptic and control dogs. However, the mean duration of wake, drowsy, and deep sleep episodes were significantly shorter in the narcoleptics, suggesting a fragmentation of the vigilance states (wake and sleep) in these animals. To compensate for the influence of cataplectic episodes on wake and drowsy those episodes interrupted by the occurrence of cataplexy were excluded. (c) Mean latency (min) to each sleep stage and occurrences (number/total sessions) of cataplexy and sleep onset REM periods (SOREMPs) during the multiple sleep latency test (MSLT) in narcoleptic and control Dobermans. Drowsy and light sleep occurred in all sessions. Deep sleep, REM sleep or cataplexy (for narcoleptic dogs) occurred in some sessions, and the number of sessions where each state occurred/total number of sessions are shown in parentheses. Narcoleptic dogs exhibited cataplexy in 9 out of 100 sessions, and these events were differentiated from REM sleep episodes. Narcoleptic dogs show a significantly shorter latency to drowsy and light sleep in overall sessions. Note that narcoleptic dogs exhibited SOREMPs (i.e., REM sleep occurring within 15 min of sleep onset) significantly more often than control animals, although both narcoleptic (36.0 % of total session) and control dogs (21.7 %) showed REM sleep during the MSLT.

acquired nature of the disease in these cases. In Dobermans, affected dogs display spontaneous complete cataplexy as early as 4 weeks, but almost always by 6 months.[4,17] Symptom severity increases until 5-6 months of age (with females being more affected during development), and it appears to decrease slowly and then remains stable through old age.[17,18]

Genetic canine narcolepsy was thought to be an invaluable model in searching for narcolepsy genes, since it is possible that the canine narcolepsy gene (its equivalent or genes with a functional relationship with the canine narcolepsy gene) may also be involved in some human cases.

4. DOG LEUKOCYTE ANTIGEN (DLA) AND CANINE NARCOLEPSY

Since human narcolepsy-cataplexy is specifically associated with the HLA gene HLA DQ*0602 (and DR15 [see human narcolepsy section]), 3 populations of dogs were tested to determine if a specific dog leukocyte antigen (DLA) allele was present in affected animals as in narcoleptic humans. These included genetically-narcoleptic Doberman Pinschers and Labrador Retrievers and small breed dogs with sporadic narcolepsy. Unlike humans, narcoleptic dogs tested do not share any single DLA locus reactivity, suggesting that a specific MHC class II haplotype is not a requirement for the disease.[19] In further experiments, a human HLA-DRb hybridization probe was used on DNA from narcoleptic dogs to determine whether there was an association between the DLA allele and susceptibility to narcolepsy.[20] This probe detected polymorphisms in both Doberman Pinschers and Labrador Retrievers. Results of this study also excluded the possibility of a tight linkage between DLA and the canarc-1 locus in these narcoleptic dogs.[20] However, it now appears that more extensive searches specifically in sporadic dogs (i.e., hypocretin ligand-deficient, see below), including other DLA regions are required to examine the involvement of the histocompatibility molecules/mechanisms in the development of narcolepsy in these animals.

5. DISCOVERY OF CANINE NARCOLEPY GENE (*Canarc-1*)

Screening of genetic markers, including mini satellite probes and functional candidate gene probes, revealed that canarc-1 cosegregates with a homolog of the switch region of the human immunoglobulin μ heavy-chain gene (Sμ).[15] The genuine Sμ segments are involved in a complex somatic recombination process, allowing individual B cells to switch immunoglobulin classes upon activation (see ref. 15). Fluorescence in situ hybridization (FISH) indicates that *canarc-1* is located on a different canine chromosome from the canine immunoglobulin switch loci.[21] Sequence analysis of the Sμ-like marker indicates that the Sμ-like marker has high homology to the true gene but is not a functional part of the immunoglobulin switch machinery.[22] Thus, positional cloning of the region where the Sμ-like marker is located was initiated.

After 10 years of work, *canarc-1* was finally identified, and narcolepsy in Dobermans and Labradors was found to be caused by a mutation in the hypocretin receptor 2 gene (*Hcrtr 2*) (Fig. 3). The mutations in Dobermans and Labradors were found in the same gene, but different locus, and both mutations cause exon skipping deletions in the *Hcrtr2* transcripts and the loss of function of Hcrtr 2, and thus impairs postsynaptic hypocretin neurotransmission. Therefore, it appears that these mutations occur independently in both breeds. Another mutation in Hcrtr 2[23] was also found in a new narcoleptic family of Dachshunds, but the reason that *Hcrtr 2* mutation often occurs in canines (no human case was yet identified) is not unknown.

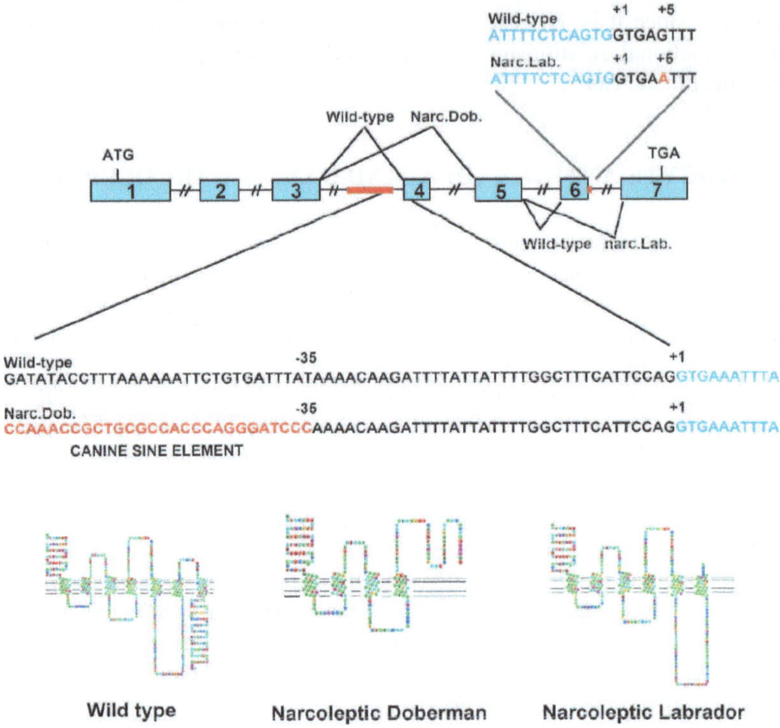

Figure. 3. Genomic organization of the canine Hcrt 2 receptor locus. The *Hcrtr 2* gene is encoded by 7 exons. Sequence of exon-intron boundary at the site for the deletion of the transcript revealed that the canine short interspersed nucleotide element (SINE) was inserted 35 bp upstream of the 5' splice donor site of the fourth encoded exon in narcoleptic Doberman pinschers. This insertion falls within the 5' flanking intronic region needed for pre-mRNA Lariat formation and proper splicing, causing exon 3 to be spliced directly to exon 5, and exon 4 to be omitted. This mRNA potentially encodes a non-functional protein with 38 amino acids deleted within the 5th transmembrane domain, followed by a frameshift and a premature stop codon at position 932 in the encoded RNA. In narcoleptic Labradors, the insertion was found 5 bp downstream of the 3' splice site of the fifth exon, and exon 5 is spliced directly to exon 7, omitting exon 6.

Almost simultaneously, along with the discovery of the canine narcolepsy gene, a report that preprohypocretin (preproorexin) knockout-mice also exhibited a narcolepsy-like phenotype, including sleep fragmentation and episodes of behavioral arrest similar to cataplexy in canine narcolepsy,[6] which was made by a group led by Dr. Yanagisawa. Considering how similar human and canine narcolepsy are at the phenotypic level, it was thought that abnormalities in the hypocretin system is likely to be involved in some human cases, either at the functional or the genetic level. Subsequent neurochemical screening revealed that the hypocretin ligand deficiency is indeed found in most human narcolepsy-cataplexy by CSF hypocretin measures and postmortem studies.[7-9]

6. HYPOCRETIN LIGAND DEFICIENT SPORADIC NARCOLEPTIC CANINES

In parallel with the progresses in human narcoleptic subjects, hypocretin contents in the brains and CSF in sporadic narcoleptic dogs (as well as in *Hcrtr 2*-mutated narcoleptic Dobermans) were also examined.[24] Hypocretin neurons and contents were found not to be altered in adult *Hcrtr 2*-mutated narcoleptic Dobermans.[24] A slight upregulation of ligand production was observed in affected young animals around the disease onset.[24] These results suggest that the hypocretin neurotransmission mediated by the Hcrtr 1 remains, but is not sufficient to prevent narcoleptic symptoms. We also examined the brains and/or CSF from 7 sporadic narcoleptic dogs, and found that the hypocretin contents in all 7 sporadic narcoleptic dogs were undetectably low.[24-27] Thus, sporadic narcoleptic dogs share the similar pathophysiological mechanisms to most human narcoleptic-cataplexy subjects.

7. PHARMACOLOGICAL CONTROL OF CATAPLEXY AND EDS

Beside the discovery of canarc-1, biomedical research in narcolepsy has been also greatly facilitated using narcoleptic canines.[2,13] Based on the data obtained from the canine model, current understanding of the neuropharmacological control of cataplexy and excessive sleepiness (as well as some prospects for the new treatment of narcolepsy) is discussed. For more details for the neuropharmacological results, please refer to a review article by Nishino and Mignot.[1]

7. 1. REM Sleep/Cataplexy and Narcolepsy

After the discovery of sleep onset REM periods in narcoleptic patients,[28] narcolepsy has often been regarded as a disorder of REM sleep generation. REM sleep usually appears 90 minutes after sleep onset and re-appears every 90 minutes in humans (30 minutes in dogs). Therefore, it was thought that in narcolepsy, REM sleep can intrude in active wake or at sleep onset, resulting in cataplexy, sleep paralysis and hypnagogic hallucinations, and these 3 symptoms are often categorized as "dissociated manifestations of REM sleep" (see [1]). Abnormal generation of REM sleep might therefore be central to narcolepsy, but this has not been previously demonstrated experimentally. We have therefore analyzed the REM sleep and cataplexy cyclicity in narcoleptic and control canines to observe whether the cyclicity at which REM sleep occurs is disturbed in narcoleptic canines.[10] Interval histograms for REM sleep episodes revealed that a clear 30-minute cyclicity exists in both narcoleptic and control animals, suggesting that the system controlling REM sleep generation is intact in narcoleptic dogs. In contrast to REM sleep, cataplexy can be elicited anytime upon emotional stimulation (i.e., no 30-minute cyclicity is observed).[10]

These results taken together with the results of extensive human study show that cataplexy is tightly associated with hypocretin deficient status (cataplexy appears now to be a unique pathological condition caused by a loss of hypocretin neurotransmission)[29] suggest that mechanisms for triggering of cataplexy and REM sleep are distinct. However, previous electrophysiological data has also demonstrated various similarities between REM sleep atonia and cataplexy.[30] Since H-reflex activity (one of the monosynaptic spinal electrically-induced reflexes) profoundly diminishes or disappears during both REM sleep and cataplexy, it is likely that the motor inhibitory components of

REM sleep are also responsible for the atonia during cataplexy.[30] Thus, the executive systems for the induction of muscle atonia during cataplexy and REM sleep are likely to be the same. This interpretation is also supported by the pharmacological findings that most compounds that significantly reduce or enhance REM sleep, reduce and enhance cataplexy respectively. However, some exceptions, such as discrepant effects of D2/D3 antagonists on REM sleep and cataplexy, also exist (see below for more detail).[31]

7.2. Monoaminergic and Cholinergic Interactions and Cataplexy

The importance of increased cholinergic activity in triggering REM sleep or REM sleep atonia is well established (see ref. 32). Similarly, activation of the cholinergic systems using the acetylcholinesterase inhibitor physostigmine also greatly exacerbates cataplexy [33]. This cholinergic effect is mediated via muscarinic receptors since muscarinic stimulation aggravates cataplexy, while its blockade suppresses it, and nicotinic stimulation or blockade has no effect.[33]

Monoaminergic transmission is also critical for the control of cataplexy. All therapeutic agents currently used to treat cataplexy (i.e., antidepressants or monoamine oxidase inhibitors [MAOIs]), are known to act on these systems. Furthermore, whereas a subset of cholinergic neurons are activated during REM sleep, the firing rate of monoaminergic neurons in the brainstem (such as in the locus coeruleus (LC) and the raphe magnus) are well known to be dramatically depressed during this sleep stage.[34,35] Using canine narcolepsy, it was recently demonstrated that adrenergic LC activity is also reduced during cataplexy.[36] In contrast, dopaminergic neurons in the ventral tegmental area (VTA) and substantia nigra (SN) do not significantly change their activity during natural sleep cycles.[37,38]

Since cataplexy in dogs can be easily elicited and quantified, the canine narcolepsy model has been intensively used to dissect the mode of action of currently used anticataplectic medications. The most commonly prescribed anticataplectic mediations in humans are tricyclic antidepressants. These compounds, however, have a complex pharmacological profile that includes monoamine uptake inhibition (dopamine [DA], epinephrine, norepinephrine [NE] and serotonin [5-HT]), anticholinergic, alpha-1 adrenergic antagonistic and antihistaminergic effects. It is thus difficult to conclude which one of these pharmacological properties is actually involved in their therapeutic effects.

We therefore first studied the effects of a large number (a total of 17 compounds) of uptake blockers/release enhancers specific for the adrenergic, serotonergic or dopaminergic system, and adrenergic uptake inhibition was found to be the key property involved in the anticataplectic effect.[39] Serotonergic uptake blockers were only marginally effective at high doses and the dopaminergic uptake blockers were completely ineffective. Interestingly, it was later found that these dopamine uptake inhibitors had potent alerting effects in canine narcolepsy.[1]

We also compared the effects of several antidepressants with those of their demethylated metabolites. Many antidepressant compounds (most typically tricyclics) are known to metabolize significantly by a hepatic first pass into their demethylated metabolites that have longer half-lives and higher affinities for adrenergic uptake sites.[40] During chronic drug administration, these demethylated metabolites accumulate[40] and can thus be involved in the drug's therapeutic action. The effects of 5 available antidepressants (amitriptyline, imipramine, clomipramine, zimelidine, and fluoxetine) were compared with those of their respective demethylated metabolites (nortriptyline, desipramine, desmethylclomipramine, norzimelidine and norfluoxetine).[41] In all cases, the demethylated metabolites were found to be more active on cataplexy than were the parent compounds. We also found that the active dose of all anticataplectic compounds tested positively, correlated with the *in vitro* potency of each compound to the adrenergic transporter but not with that of the serotonergic transporter.[41] In fact, the anticataplectic effects were negatively correlated with the *in vitro* potency for serotonergic uptake inhibition, but this may be a bias since potent adrenergic uptake inhibitors included in the study have a relatively low affinity to serotonergic uptake sites.

Although the results presented were obtained from inbred *Hcrtr-2* mutated narcoleptic Dobermans, similar findings have been obtained in more diverse cases of sporadic canine narcolepsy in various breeds donated to our colony (see ref. [33]). Protryptiline and desipramine (two compounds with no significant serotonergic uptake blocking properties), have also been shown to be very effective for the treatment of human cataplexy.[42-45] Thus preferential involvement of adrenergic system for the modes of action of anticataplectic effects of antidepressants are suggested regardless of the form of deficit in hypocretin neurontransmission (receptor mutation vs, hypocretin ligand deficiency).

In order to dissect receptor subtypes that significantly modify cataplexy, more than 200 compounds with various pharmacological properties (cholinergic, adrenergic, dopaminergic, serotonergic, prostaglandins, opioids, benzodiazepines, GABA-ergics and adenosinergics) have also been studied in the narcoleptic canine model (see[1] for details). Although many compounds (such as M2 antagonists, alpha-1 agonists, alpha-2 antagonists, dopaminergic D2/D3antagonists, 5HT1a agonists, TRH analogs, prostaglandin E2, and L type Ca2+ channel blockers reduce cataplexy), very few compounds significantly aggravate cataplexy [1]. Since cataplexy can be easily and non-specifically reduced such as by unpleasant drug side effects, we assume the cataplexy-aggravating effects are more specific. Stimulation of muscarinic M2 (non-M1) receptors significantly aggravates cataplexy. Among the monoaminergic receptors, stimulation of the postsynaptic adrenergic alpha-1b receptors[46] and presynaptic alpha-2 receptors[47] was also found to aggravate cataplexy, a result consistent with a primary adrenergic control of cataplexy. We also found that small doses of DA D2/D3 agonists significantly aggravated cataplexy and induced significant behavioral sedation/drowsy state in these animals.[48,49] These pharmacological findings parallel neurochemical abnormalities previously reported in canine narcolepsy, namely significant increases in alpha-2 receptors in the LC,[50] D2 receptors in the amygdala, nucleus accumbens[51] and M2 receptors in the pons.[52] To date, no other receptor ligands (i.e., adenosinergic, histaminergic or GABA-ergic) have been found to aggravate cataplexy, but thalidomide (an old hypnotic with an immunomodulatory property) significantly aggravates cataplexy, but the mechanisms involved in this effect are not known.[53]

The sites of action of D2/D3 agonists were also investigated by local drug perfusion experiments, and a series of experiments identified acting sites for these compounds. These include dopaminergic nuclei or cell groups, such as the VTA,[49] SN[54] and A11[55] (a diencephalic DA cell group that directly project to the spinal ventral horn), suggesting a direct involvement of DA cell groups and DA cell body autoreceptors for the regulation of cataplexy. The cataplexy-inducing effects of D2/D3agonists are, however, difficult to reconcile considering the fact that dopaminergic uptake blockers have absolutely no effect on cataplexy.[39] We believe that D2/D3 receptor mechanisms are more specifically involved in the control of sleep-related motor tonus (i.e., cataplexy or muscle atonia without phasic REM events) than those for active REM sleep. A recent finding in canine narcolepsy that sulpiride (a D2/D3 antagonist) significantly suppresses cataplexy but has no effect on REM sleep, also supports this notion.[31] It should be also noted that D2/3 agonists are clinically used for the treatment of human periodic leg movements during sleep (PLMS).[56] Involuntary leg movements during sleep are often associated with restless leg syndrome (RLS) and disturbed nighttime sleep. As reported in human,[57] narcoleptic Dobermans often exhibit PLMS-like movements.[58] It thus appears that there is an overlap in pathophysiological mechanisms between between cataplexy and PLMS, and dopaminergic system (i.e., D2/D3 receptor mechanisms) may be specifically involved both symptoms.

The effects on cataplexy by cholinergic stimulation in various brain regions were also examined in narcoleptic and in control canines. Local injection or perfusion of carbachol (a predominantly muscarinic agonist) into the pontine reticular formation (PRF) was found to aggravate canine cataplexy in a dose-dependent fashion.[59] The results obtained in the PRF with cholinergic agonists were somewhat expected considering the well established role of the pontine cholinergic systems in the regulation of REM sleep and REM sleep atonia. Surprisingly, however, we also found that the local injection/perfusion of carbachol unilaterally or bilaterally into the BF (rostral to the preoptic area, in the vertical or horizontal limbs of the diagonal band of Broca and medial septum) dose-dependently aggravated cataplexy and induced long-lasting episodes of muscle atonia accompanied by desynchronized EEG in narcoleptic canines.[60] Physostigmine was also found to aggravate cataplexy when injected into the same site, thus suggesting that fluctuations in endogenous levels of acetylcholine in this structure may be sufficient to induce cataplexy.[60] The carbachol injections did not induce cataplexy in normal animals, but rather induced wakefulness.

The BF is anatomically connected with the limbic system, which is regarded as a critical circuit for integrating emotions. Furthermore, BF neurons are known to respond to the arousing property of appetitive stimuli,[61] which we use to induce cataplexy in narcoleptic dogs. Considering the fact that emotional excitation is an alerting stimulus in normal animals but induces cataplexy in narcoleptic animals, the BF may be involved in triggering a paradoxical reaction to emotions-atonia rather than wakefulness, in narcoleptic animals. These results also suggest that more global brain structures (than those for REM sleep generation) are involved in the induction of cataplexy. Cataplexy is now demonstrated to be tightly associated with the loss of hypocretin neurotransmission. Global and persistent cholinergic/monoaminergic imbalance due to the loss of hypocretin neurotransmission may be required for the occurrence of cataplexy, and cataplexy could not be induced simply by an increase in REM sleep propensity and/or vigilance state instability that occurs in various disease conditions (such as depression) or in some physiological conditions (such as REM sleep deprivation). The findings that REM sleep abnormalities and sleep fragmentation are often seen in other sleep disorders such as narcolepsy without cataplexy, sleep apnea, and even in healthy subjects when their sleep patterns are disturbed, but these subjects who never develop cataplexy further supports this hypothesis. The mechanism for emotional triggering for cataplexy remains to be studied, but multiple brain sites and multiple functional and anatomical systems are likely to be involved.

7.3. Dopamineregic Transmission and EEG Arousal

Narcoleptic Dobermans were also used for a series of pharmacological experiments to dissect modes of action of wake-promoting compounds. Amphetamine-like CNS stimulants currently used clinically for the management of sleepiness in narcolepsy enhance monoaminergic transmission presynaptically.

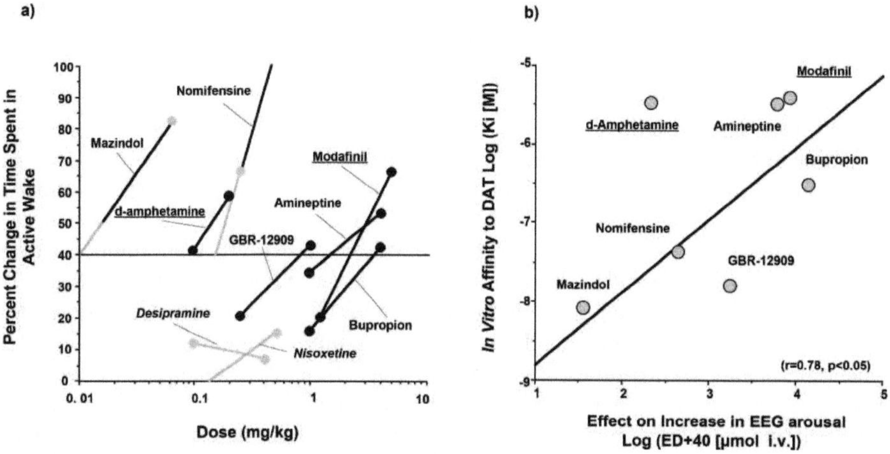

Figure 4. Effects of various DA and NE uptake inhibitors and amphetamine-like stimulants on the EEG arousal of narcoleptic dogs and correlation between in vivo EEG arousal effects and in vitro DAT binding affinities. The effects of compounds on daytime sleepiness was studied using 4 hrs daytime polygraphic recordings (10:00-14:00) in 4-5 narcoleptic animals. Two doses were studied for each compound. All DA uptake inhibitors and CNS stimulants dose-dependently increased EEG arousal and reduced slow wave sleep (SWS) when compared to vehicle treatment. In contrast, nisoxetine and desipramine, two potent NE uptake inhibitors had no significant effect on EEG arousal when doses which completely suppressed REM sleep were injected. Compounds with both adrenergic and dopaminergic effects (nomifensine, mazindol, D-amphetamine) were active on both EEG arousal (left panel) and REM sleep. The effects of the two doses studied for each stimulant was used to construct a rough dose-response curve (left panel). The drug dose which increased the time spent in wakefulness by 40% more than the baseline (vehicle session) was then estimated for each compound. The order of potency of the compounds obtained was: mazindol > (amphetamine) > nomifensine > GBR 12,909 > amineptine> (modafinil) > bupropion. In vitro DA transporter (DAT) binding was performed using [3H]-WIN 35,428 onto canine caudate membranes and demonstrated that the affinity of these DA uptake inhibitors varied widely between 6.5 nM and 3.3 mM. In addition, it was also found that both amphetamine and modafinil have a low, but significant affinity (same range as amineptine) for the DAT. A significant correlation between in vivo and in vitro effects was observed for all 5 DA uptake inhibitors and modafinil (right panel).

However, these compounds also lack pharmacological specificity. In order to study the mode of action of these wake-promoting compounds on daytime sleepiness, the stimulant properties of several dopaminergic and adrenergic uptake inhibitors were quantified and compared to the effects of amphetamine and modafinil using 4-hour daytime polygraphic recordings.[62] In spite of their lack of effect on cataplexy, all dopaminergic uptake inhibitors induced significant EEG arousal (Fig. 4). In contrast, nisoxetine and desipramine, two potent adrenergic uptake inhibitors, had little effect on EEG arousal but significantly suppressed REM sleep. Furthermore, the *in vivo* potency of DA uptake inhibitors on EEG arousal correlates well with their *in vitro* affinity to the DA transporter (DAT), but not to the norepinephrine transporter (NET). These results are consistent with the hypothesis that presynaptic modulation of dopaminergic transmission is a key property mediating the EEG arousal effects of these compounds. Interestingly, we also found that modafinil binds to the DAT site with low affinity,[62,63] similar to the affinity range for amineptine (a DA uptake inhibitor which also enhances EEG arousal in our model). Thus, the DAT binding property may also contribute to the stimulant properties of modafinil. We recently demonstrated that the wake-promoting effect of modafinil and amphetamine were completely absent in mice lacking DAT (i.e., DAT KO mice).[64] These results clearly indicate that DAT is required for the mediation of the wake-promoting effect of modafinil (and amphetamine), but other pharmacological properties may also involve in wake-promoting effects of modafinil.[65,66]

Amphetamine (which showed a potent EEG arousal property) has however, a relatively low DAT binding affinity,[67] suggesting that other mechanisms, such as monoamine release (exchange diffusion through the DAT) by amphetamine, are also involved in the mechanism of EEG arousal. Amphetamine is also reported to block the vascular monoamine transporter 2 (VMAT 2) and to induce reverse-transport of monoamines into the synaptic area.[68] To further assess the net effects of amphetamine on monoaminergic neurotransmission, we measured DA and NE efflux together with wake-promoting effects of amphetamine-analogs and isomers in the canine narcolepsy model.[69] Polygraphic recordings demonstrated that d-amphetamine was about twice as potent as l-amphetamine, and was 6-times more potent than l-methamphetamine in increasing wakefulness, while d-amphetamine and l-amphetamine were equipotent in reducing REM sleep and cataplexy, and l-methamphetamine was about half as potent as l- and d-amphetamine.[69] By measurements of extracellular levels of DA and NE, we found that d-amphetamine was more potent in increasing DA efflux than l-amphetamine, and l-methamphetamine had little effect. In contrast, there was no significant difference in the potencies of theses 3 derivatives on NE efflux. Thus, the potencies of amphetamine isomers/analogs on wakefulness correlated well with DA, but not NE efflux in the brain of narcoleptic dogs, and further exemplifies the importance of the DA system for the pharmacological control of EEG arousal.[69]

However, the involvement of the dopaminergic systems in the regulation of the sleep/wake process that has not been given much attention, mostly due to early electrophysiological findings demonstrating that dopaminergic neurons in the ventral tegmental area (VTA) and substantia nigra (SN) do not change their activity significantly during the sleep cycle,[38] in contrast to noradrenergic cells of the locus coeruleus (LC) or serotonergic cells of the raphe which increase firing in wake versus sleep. Although firing patterns of DA neurons during slow wave sleep are different from those during wake or REM sleep, DA release in the LC and amygdala measured with microdialysis experiments failed to demonstrate the state dependent change.[70] These experimental results led most investigators to believe that adrenergic tone was more important than dopaminergic transmission for the control of EEG arousal.

The dopaminergic system may thus be not so important for the normal sleep wake cycle regulation, but may play a key role for forced wakefulness by motivation and/or by stimulants, and the hypocretin-dopaminergic interaction may also involved in these alerting mechanisms. In clinical conditions, a disturbance of this mechanism may be more troublesome, since it may induce intolerable sleepiness at the situation required the high level of vigilance.

An involvement of dopaminergic system in intolerable sleepiness is also noted in some pathological conditions, such as Parkinson's disease. Frequent sleep attacks and associations with accidents are reported by patients with Parkinson's disease treated with DA D2/D3 agonists,[71,72] the classes of compounds that induce drowsy states and cataplexy in the canine model of narcolepsy.[48,49]

8. HISTAMINERGIC SYSTEM AND NARCOLEPSY

After the discoveries of the involvement of the hypocretin system in narcolepsy, we focused on how these deficits in hypocretin neurotransmission induce narcolepsy. One of the keys to solving this question is revealing the functional differences between Hcrtr 1 and Hcrtr 2, since it is evident that Hcrtr 2-mediated function plays a critical role in generating narcoleptic symptoms in canine model.[73] *In situ* hybridization experiments in rats demonstrated that Hcrtr 1 is enriched in the ventromedial hypothalamic nucleus, tenina tecta, the hippocampal formation, dorsal raphe, laterodorsal tegmental nucleus and LC.[74] In contrast, Hcrtr 2 is enriched in the paraventricular nucleus, magnocellular preoptic area, cerebral cortex, nucleus accumbens, VTA, SN and histaminergic tuberomammillary nucleus (TMN).[74] These findings are consistent with our pharmacological understanding of the regulation of cataplexy and EEG arousal discussed above, and the lack of Hcrtr 2 mediated function on the monoaminergic/cholinergic nucleus/receptive sites that are likely to result in cholinergic/monoaminergic imbalance.

Histamine is another important wake-promoting monoamine. Histamine neurons are typically wake-active neurons,[75] located exclusively in the TMN of the hypothalamus, from where they project to practically all brain regions, including areas important for vigilance control, such as the hypothalamus, BF, thalamus, cortex, and brainstem structures (see [76,77] for review). Since histaminergic TMN exclusively express Hcrtr 2, we initiated a study of the roles of the histaminergic system in the canine narcolepsy. As a first step, we measured histamine contents in the brain of *Hcrtr 2*-mutated and control Dobermans.[78] Contents of DA, NE and 5-HT and their metabolites were also measured. We found that the histamine content in cortex and thalamus was significantly lower in narcoleptic Dobermans compared to controls (Figure 5). In contrast to histamine content, DA, NE and 5HT content in familial narcolepsy was high in these structures. Since it was recently reported that hypocretins strongly excite TMN histaminergic neurons *in vitro,* through Hcrtr 2 stimulation,[79-82] the decrease in histaminergic content found in narcoleptic dogs may be due to the lack of excitatory input of hypocretin on TMN histaminergic neurons. Although loss of hypocretin input through Hcrtr 2 may also lead to decreases in catecholamine content, DA and NE contents were instead high in several brain structures tested. Thus, the imbalance of the neurotransmitter system may also be evident among different monoamine systems (Fig. 6). Considering the fact that catecholamine turn over was not reduced in these animals,[78] and together with the fact that compounds that enhance DA and NE transmission significantly improve symptoms of narcolepsy in these animals,[1] the increase in DA and NE contents may be compensatory, either mediated by Hcrtr 1 or by other neurotransmitter systems. Uncompensated low histamine levels in narcolepsy may therefore suggest that the hypocretin system is the major excitatory input to histaminergic neurons.

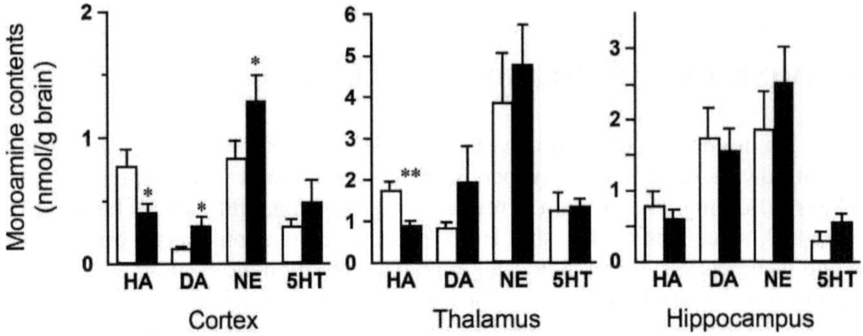

Figure. 5. Histamine, DA, NE and 5HT contents in the cortex, thalamus, and hippocampus in *Hcrtr 2*-mutated narcoleptic and control Dobermans. Histamine content in the cortex and thalamus was significantly lower in narcoleptic Dobermans (dark bars) compared to controls (open bars), while DA and NE were higher in these structures. Increases were statistically significant in cortex (by student's t-test).

This interpretation is also supported by hypocretin receptor distribution among different monoamine nuclei: TMN histaminergic neurons exclusively express Hcrtr 2, while LC NE neurons express Hcrtr 1 exclusively, and VTA DA neurons express both Hcrtr 1 and Hcrtr 2 (Fig. 6).[74]

We also measured histamine in the brain of 3 sporadic narcoleptic dogs and found that the histamine content in these animals were also as low as in *Hcrtr 2*-mutated narcoleptic Dobermans,[78] thus suggesting a decrease in histamine neurotransmission may also exist in ligand-deficit narcolepsy. Furthermore, two independent human studies showed a decease in histaminergic contents in the CSF in hypocretin-deficient narcolepsy,[83,84] suggesting altered histaminergic neurotransmission may not be limited to the Hcrtr 2-mutated narcolepsy. The histaminergic system may thus be a new target site for the treatment of EDS and other symptoms associated with narcolepsy, and the centrally active histaminommic compounds such as H3 autoreceptor antagonists, may have therapeutic effects on narcolepsy.[85]

9. HYPOCRETIN REPLACEMENT THERAPY

Finally, a canine narcolepsy model (both familial and sporadic narcoleptic dogs) was used for the evaluation of replacement/supplement therapies of hypocretin ligands.[26] Using a ligand deficit sporadic narcoleptic dog, we have assessed the effects on cataplexy of intravenous (IV) administration of hypocretin-1 (6 – 384 μg/kg). In a separate experiment, we found that a small portion of hypocretin-1 penetrates to the central nervous system (by assessing the increase in CSF hypocretin levels) after high doses (96-386 μg/kg) of IV administration of hypocretin-1.[26] However, hypocretin-1 at these high dose range only produced a short-lasting anticataplectic effects.[26] Thus a development of centrally penetrable hypocretin agonists (i.e., small molecular synthetic agonists), are likely to be necessary for the human application.

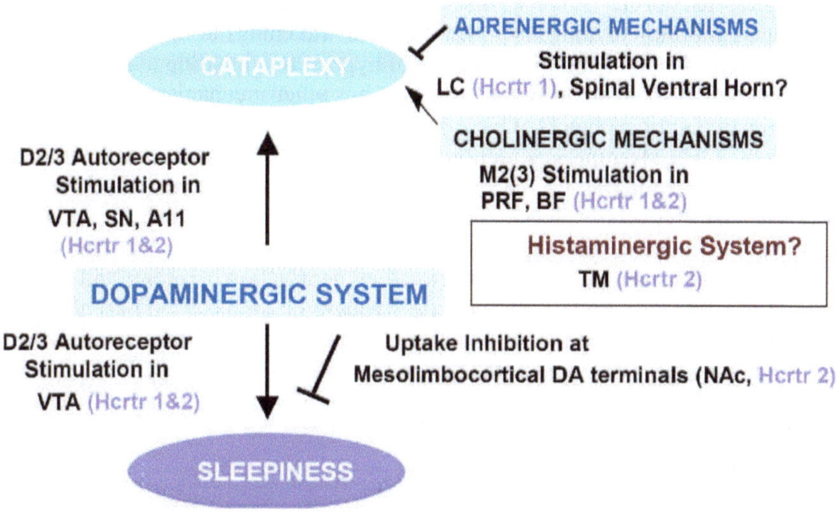

Figure 6. Monoaminergic and cholinergic control of sleepiness and cataplexy in relation to hypocretin input: a schematic perspective. The stimulation of adrenergic transmission by adrenergic uptake inhibitors potently reduces cataplexy; this pharmacological property is likely involved in the mode of action of currently used anticataplectic agents (e.g., tricyclic antidepressants). The fact that both presynaptic alpha-2 autoreceptor stimulation and postsynaptic alpha-1 blockade aggravate cataplexy is consistent with an inhibitory role of adrenergic transmission in the control of REM sleep atonia. Dopaminergic uptake inhibitors have no effect on cataplexy, although these compounds strongly induce EEG arousal. In contrast, D2/3 autoreceptor stimulation worsens both cataplexy and sleepiness. Since DA uptake inhibitors are reported to be mostly active at the level of mesocortical and mesolimbic DA terminals, DA projections to these regions may be more involved in mediating EEG arousal. Muscarinic M2 stimulation induces behavioral wakefulness and cortical desynchrony in control dogs, while it induces cataplexy in narcoleptic dogs. Although muscarinic antagonists significnatly reduce cataplexy in the canine model, attempts to use this class of compounds in humans have not been successful mainly due to the side effects. It was recently revealed that hypocretin-containing neurons project to these previously identified monoaminergic and cholinergic and cholinoceptive regions where hypocretin receptors are enriched. Impairments of hypocretin inputs may thus result in cholinergic and monoaminergic imbalance and in generation of narcoleptic symptoms. Hypocretin neurons also project densely to TMN histaminergic neurons and Hcrtr 2 are enriched in the TMN, and it is likely that some of narcolepsy symptoms may also be mediated by the histaminergic system. [78]

10. HYPOCRETIN DEFICIENCY AND NARCOLEPTIC PHENOTYPE

Human studies have demonstrated that the occurrence of cataplexy is tightly associated with hypocretin deficiency.[29] Furthermore, the hypocretin deficiency was already observed at very early stages of the disease (just after the onset of EDS), even before occurrences of clear cataplexy.[86] Occurrences of cataplexy are rare in acute symptomatic cases of EDS associated with a significant hypocretin deficiency.[87] Thus, it appears that chronic and selective deficit of hypocretin neurotransmission may be required for the occurrence of cataplexy. A possibility of an involvement of a secondary neurochemical change for the occurrence of cataplexy cannot be ruled out. If some of these changes are irreversible, hypocretin supplement therapy may only have limited effects on cataplexy.

Sleepiness in narcolepsy is most likely due to the difficulty in maintaining wakefulness as normal subjects do. The sleep pattern of narcoleptic subjects is also fragmented and they exhibit insomnia (frequent wakening) at night. This fragmentation occurs across 24 hours, and thus, the loss of hypocretin signaling are likely to play a role of this vigilance stage stability (see ref. 88), but other mechanism may also involved in EDS in narcoleptic subjects. One of the most important characteristics of EDS in narcolepsy is that sleepiness is reduced, and patients feel refreshed after a short nap, but this does not last long as they become sleepy within a short period of time. We have observed that hypocretin levels in the extracellular space and in the CSF of rats significantly fluctuate across 24 hours, and build-up toward the end of the active periods.[89] Several manipulations (such as sleep deprivation, exercise and long-term food deprivation) are also known to increase hypocretin tonus.[89-92] Thus, the lack of this hypocretin build-up caused by circadian time and various alerting stimulations may also play a role for EDS associated with hypocretin-deficient narcolepsy.[92]

11. CONCLUSION

Although the canine narcoleptic model significantly contributed to the understanding of the etiological and pathophysiological aspects of the disease, one of major missions (identifying the narcolepsy gene) using this animal model was over. The indispensable value of the sporadic hypocretin ligand deficit narcoleptic dogs is high, but it is difficult to obtain enough of these animals for systematic studies. *Hcrtr-2* mutated narcoleptic dogs are still useful for various pharmacological and physiological experiments, but some of the studies will be taken over by experiments using rodent models (especially for the experiments focusing on cell transplantation or gene therapy).

Before closing this chapter, I would like to emphasize again that the establishment of the animal model of narcolepsy was a long and steady effort (especially maintaining the colony and cloning of the gene), which finally lead to the first major breakthrough in over 100 years' history of narcolepsy research.

12. ACKNOWELDGEMENTS

The author acknowledges Dr. Nobuhiro Fujiki for the artwork and Daniel Wu for editing and formatting the manuscript.

13. REFERENCES

1. S. Nishino and E. Mignot, Pharmacological aspects of human and canine narcolepsy, *Prog Neurobiol.* **52**, 27-78 (1997).
2. M. M. Mitler, B. G. Boysen, L. Campbell and W. C. Dement, Narcolepsy-cataplexy in a female dog, *Exp Neurol.* **45**, 332-40 (1974).
3. C. D. Knecht, J. E. Oliver, R. Redding, R. Selcer and G. Johnson, Narcolepsy in a dog and a cat, *J Am Vet Med Assoc.* **162**, 1052-3 (1973).
4. A. Foutz, M. Mitler, L. Cavalli-Sforza and W. C. Dement, Genetic factors in canine narcolepsy, *Sleep.* **1**, 413-421 (1979).

5. L. Lin, J. Faraco, R. Li, H. Kadotani, W. Rogers, X. Lin, X. Qiu, P. J. de Jong, S. Nishino and E. Mignot, The sleep disorder canine narcolepsy is caused by a mutation in the hypocretin (orexin) receptor 2 gene, *Cell.* **98**, 365-76 (1999).
6. R. M. Chemelli, J. T. Willie, C. M. Sinton, J. K. Elmquist, T. Scammell, C. Lee, J. A. Richardson, S. C. Williams, Y. Xiong, Y. Kisanuki, T. E. Fitch, M. Nakazato, R. E. Hammer, C. B. Saper and M. Yanagisawa, Narcolepsy in orexin knockout mice: molecular genetics of sleep regulation, *Cell.* **98**, 437-451 (1999).
7. C. Peyron, J. Faraco, W. Rogers, B. Ripley, S. Overeem, Y. Charnay, S. Nevsimalova, M. Aldrich, D. Reynolds, R. Albin, R. Li, M. Hungs, M. Pedrazzoli, M. Padigaru, M. Kucherlapati, J. Fan, R. Maki, G. J. Lammers, C. Bouras, R. Kucherlapati, S. Nishino and E. Mignot, A mutation in a case of early onset narcolepsy and a generalized absence of hypocretin peptides in human narcoleptic brains, *Nat Med.* **6**, 991-7 (2000).
8. S. Nishino, B. Ripley, S. Overeem, G. J. Lammers and E. Mignot, Hypocretin (orexin) deficiency in human narcolepsy., *Lancet.* **355**, 39-40 (2000).
9. T. C. Thannickal, R. Y. Moore, R. Nienhuis, L. Ramanathan, S. Gulyani, M. Aldrich, M. Cornford and J. M. Siegel, Reduced number of hypocretin neurons in human narcolepsy, *Neuron.* **27**, 469-74. (2000).
10. S. Nishino, J. Riehl, J. Hong, M. Kwan, M. Reid and E. Mignot, Is narcolepsy REM sleep disorder? Analysis of sleep abnormalities in narcoleptic Dobermans., *Neuroscience Research.* **38**, 437-446 (2000).
11. K. I. Kaitin, T. S. Kilduff and W. C. Dement, Sleep fragmentation in genetically narcoleptic dogs., *Sleep.* **9**, 116-119 (1986).
12. K. I. Kaitin, T. S. Kilduff and W. C. Dement, Evidence for excessive sleepiness in canine narcoleptics, *Electroencephalogr Clin Neurophysiol.* **64**, 447-54 (1986).
13. T. L. Baker, A. S. Foutz, V. McNerney, M. M. Mitler and W. C. Dement, Canine model of narcolepsy: genetic and developmental determinants, *Exp Neurol.* **75**, 729-42 (1982).
14. E. Mignot, Genetic and familial aspects of narcolepsy, *Neurology.* **50**, S16-S22 (1998).
15. E. Mignot, C. Wang, C. Rattazzi, C. Gaiser, M. Lovett, C. Guilleminault, W. C. Dement and F. C. Grumet, Genetic linkage of autosomal recessive canine narcolepsy with an immunoglobulin heavy-chain switch-like segment, *Proc Natl Acad Sci U S A.* **88**, 3475-3478 (1991).
16. Y. Honda. (1988). Clinical features of Narcolepsy: Japanese Experience. In *HLA in Narcolepsy* (Honda, Y. & Juji, T., eds.), pp. 24-57.
17. J. Riehl, S. Nishino, R. Cederberg, W. C. Dement and E. Mignot, Development of cataplexy in genetically narcoleptic Dobermans, *Exp Neurol.* **152**, 292-302 (1998).
18. J. Riehl, S. Choi, E. Mignot and S. Nishino, Changes with age in severity of cataplexy and in sleep/wake fragmentation in narcoleptic Doberman pinschers, *Sleep.* **22S**, S3 (APSS abstract) (1999).
19. R. R. Dean, T. S. Kilduff, W. C. Dement and F. C. Grumet, Narcolepsy without unique MHC class II antigen association: Studies in the canine model, *Hum Immunol.* **25**, 27-35 (1989).
20. E. Mignot, S. Nishino, L. H. Hunt Sharp, J. Arrigoni, J. M. Siegel, M. S. Reid, D. M. Edgar, R. D. Ciaranello and W. C. Dement, Heterozygosity at the canarc-1 locus can confer susceptibility for narcolepsy: induction of cataplexy in heterozygous asymptomatic dogs after administration of a combination of drugs acting on monoaminergic and cholinergic systems., *J Neurosci.* **13**, 1057-1064 (1993).
21. A. S. Dutra, E. Mignot and J. M. Puck, Establishing FISH methodology to study synthenic regions between canine and human chromosomes, *Am J Hu Genet.* **57**, A112 (1995).
22. J. Faraco, X. Lin, R. Li, L. Hinton, L. Lin and E. Mignot, Genetic studies in narcolepsy, a disorder affecting REM sleep, *J Hered.* **90**, 129-32 (1999).
23. M. Hungs, J. Fan, L. Lin, X. Lin, R. A. Maki and E. Mignot, Identification and functional analysis of mutations in the hypocretin (orexin) genes of narcoleptic canines, *Genome Res.* **11**, 531-9. (2001).
24. B. Ripley, M. Okura, N. Fujiki, E. Mignot and S. Nishino, Measurement of CSF hypocretin-1 levels in familial and sporadic cases of canine narcolepsy, *Abstract, Society for Neuroscience.* **26**, in press (2000).
25. M. Tonokura, K. Fujita, M. Morozumi, Y. Yoshida, T. Kanbayashi and S. Nishino, Narcolepsy in a hypocretin/orexin-deficient Chihuahua, *Vet Rec.* **152**, 776-9 (2003).
26. N. Fujiki, B. Ripley, Y. Yoshida, E. Mignot and S. Nishino, Effects of IV and ICV hypocretin-1 (orexin A) in hypocretin receptor-2 gene mutated narcoleptic dogs and IV hypocretin-1 replacement therapy in a hypocretin ligand deficient narcoleptic dog, *Sleep.* **6**, 953-959 (2003).
27. S. J. Schatzberg, J. Barrett, K. l. Cutter, L. Ling and E. Mignot, Case Study: Effect of hypocretin replacement therapy in a 3-year-old Weimaraner with narcolepsy., *J Vet Internal Med.* in press. (2004).
28. G. Vogel, Studies in psychophysiology of dreams III. The dream of narcolepsy, *Arch Gen Psychiatry.* **3**, 421-428 (1960).

29. E. Mignot, G. J. Lammers, B. Ripley, M. Okun, S. Nevsimalova, S. Overeem, J. Vankova, J. Black, J. Harsh, C. Bassetti, H. Schrader and S. Nishino, The role of cerebrospinal fluid hypocretin measurement in the diagnosis of narcolepsy and other hypersomnias, *Arch Neurol.* **59**, 1553-62. (2002).
30. C. Guilleminault, Cataplexy, *Narcolepsy (Advances in Sleep Research Vol. 3).* 125-143 (1976).
31. M. Okura, J. Riehl, E. Mignot and S. Nishino, Sulpiride, a D2/D3 Blocker, Reduces Cataplexy but not REM Sleep in Canine Narcolepsy, *Neuropsychopharmacology.* **23**, 528-538 (2000).
32. J. M. Siegel. (1994). Brainstem mechanisms generating REM sleep. In *Principles and Practice of Sleep Medicine.* (Kryger, M. H., Roth, T. & Dement, W. C., eds.), pp. 125-144. W. B. Saunders Company, Philadelphia.
33. T. L. Baker and W. C. Dement. (1985). Canine narcolepsy-cataplexy syndrome: evidence for an inherited monoaminergic-cholinergic imbalance. In *Brain Mechanisms of Sleep* (McGinty, D. J., Drucker-Colin, R., Morrison, A. & Parmeggiani, P. L., eds.), pp. 199-233. Raven Press, New York.
34. M. E. Trulson and B. L. Jacobs, Raphe unit activity in freely moving cats: Correlation with level of behavioral arousal., *Brain Res.* **163**, 135-150 (1979).
35. G. Aston-Jones and F. E. Bloom, Activity of norepinephrine-containing locus coeruleus neurons in behaving rats anticipates fluctuations in the sleep-waking cycle., *J. Neurosci.* **1**, 876-886 (1981).
36. M. Wu, S. Gukyani, E. Mignot and J. Siegel, Activity of REM off cells during cataplexy in the narcoleptic dog, *Sleep Res.* **25**, 40 (1996).
37. J. D. Miller, J. Farber, P. Gatz, H. Roffwarg and D. C. German, Activity of mesencephalic dopamine and non-dopamine neurons across stages of sleep and waking in the rat., *Brain Res.* **273**, 133-141 (1983).
38. M. E. Trulson, D. W. Preussler and G. A. Howell, Activity of substantia nigra units across the sleep-waking cycle in freely moving cats, *Neurosci Lett.* **26**, 183-188 (1981).
39. E. Mignot, A. Renaud, S. Nishino, J. Arrigoni, C. Guilleminault and W. C. Dement, Canine cataplexy is preferentially controlled by adrenergic mechanisms: evidence using monoamine selective uptake inhibitors and release enhancers, *Psychopharmacology.* **113**, 76-82 (1993).
40. M. Peet and A. Coppen. (1979). The pharmacokinetics of antidepressant drugs: relevance to their therapeutic effect. In *Psychopharmacology of Affective Disorders* (Paykel, E. S. & Coppen, A., eds.), pp. 91-107. Oxford University Press, Oxford.
41. S. Nishino, J. Arrigoni, J. Shelton, W. C. Dement and E. Mignot, Desmethyl metabolites of serotonergic uptake inhibitors are more potent for suppressing canine cataplexy than their parent compounds, *Sleep.* **16**, 706-12 (1993).
42. H. Schmidt, R. Clark and R. Hyman, Protriptyline: an effective agent in the treatment of Narcolepsy-cataplexy syndrome and hypersomnia, *Am J Psychiatry.* **134**, 183-185 (1977).
43. M. J. Thorpy and M. Goswami. (1990). Treatment of narcolepsy. In *Handbook of Sleep Disorders* (Thorpy, M. J., ed.), pp. 235-258. Marcel Dekker, Inc., New York.
44. Y. Hishikawa, H. Ida, K. Nakai and Z. Kaneko, Treatment of narcolepsy with imipramine (tofranil) and desmethylimipramine (pertofran), *J Neuro Sci.* **3**, 453-461 (1965).
45. M. M. Mitler, R. Hajdukovic, M. Erman and J. A. Koziol, Narcolepsy, *J Clin Neurophysiol.* **7**, 93-118 (1990).
46. S. Nishino, B. Fruhstorfer, J. Arrigoni, C. Guilleminault, W. C. Dement and E. Mignot, Further characterization of the alpha-1 receptor subtype involved in the control of cataplexy in canine narcolepsy., *J Pharmacol Exp Ther.* **264**, 1079-1084 (1993).
47. S. Nishino, L. Haak, H. Shepherd, C. Guilleminault, T. Sakai, W. C. Dement and E. Mignot, Effects of central alpha-2 adrenergic compounds on canine narcolepsy, a disorder of rapid eye movement sleep., *J Pharmacol Exp Ther.* **253**, 1145-1152 (1990).
48. S. Nishino, J. Arrigoni, D. Valtier, J. D. Miller, C. Guilleminault, W. C. Dement and E. Mignot, Dopamine D2 mechanisms in canine narcolepsy, *J Neurosci.* **11**, 2666-2671 (1991).
49. M. S. Reid, M. Tafti, S. Nishino, R. Sampathkumaran, J. M. Siegel, W. C. Dement and E. Mignot, Local administration of dopaminergic drugs into the ventral tegmental area modulate cataplexy in the narcoleptic canine, *Brain Res.* **733**, 83-100 (1996).
50. B. Fruhstorfer, E. Mignot, S. Bowersox, S. Nishino, W. C. Dement and C. Guilleminault, Canine narcolepsy is associated with an elevated number of a_2 receptors in the locus coeruleus., *Brain Res.* **500**, 209-214 (1989).
51. S. Bowersox, T. Kilduff, K. Faul, W. C. Dement and R. D. Ciaranello, Brain dopamine receptor levels elevated in canine narcolepsy, *Brain Res.* **402**, 44-48 (1987).
52. R. Boehme, T. Baker, I. Mefford, J. Barchas, W. C. Dement and R. Ciaranello, Narcolepsy: cholinergic receptor changes in an animal model., *Life Sci.* **34**, 1825-1828 (1984).
53. T. Kanbayashi, S. Nishino, M. Tafti, Y. Hishikawa, W. C. Dement and E. Mignot, Thalidomide, a hypnotic with immune modulating properties, increases cataplexy in canine narcolepsy, *NeuroReport.* **12**, 1881-1886 (1996).

54. K. Honda, J. Riehl, E. Mignot and S. Nishino, Dopamine D3 agonists into the substantia nigra aggravate cataplexy but do not modify sleep, *NeuroReport.* **10**, 3717-24. (1999).
55. M. Okura, N. Fujiki, I. Kita, K. Honda, Y. Yoshida, E. Mignot and S. Nishino, The roles of midbrain and diencephalic dopamine cell groups in the regulation of cataplexy in narcoleptic Dobermans, *Neurobiol Dis.* **16**, 274-82 (2004).
56. S. Happe and C. Trenkwalder, Role of dopamine receptor agonists in the treatment of restless legs syndrome, *CNS Drugs.* **18**, 27-36 (2004).
57. R. Wittig, F. Zorick, P. Piccione, J. Sicklesteel and T. Roth, Narcolepsy and disturbed nocturnal sleep, *Clin Electroencephalogr.* **14**, 130-4 (1983).
58. M. Okura, N. Fujiki, B. Ripley, S. Takahashi, N. Amitai, E. Mignot and S. Nishino, Narcoleptic canines display periodic leg movements during sleep, *Psychiatry Clin Neurosci.* **55**, 243-4. (2001).
59. M. S. Reid, M. Tafti, J. Geary, S. Nishino, J. M. Siegel, W. C. Dement and E. Mignot, Cholinergic mechanisms in canine narcolepsy: I. Modulation of cataplexy via local drug administration into pontine reticular formation., *Neuroscience.* **59**, 511-522 (1994).
60. S. Nishino, M. Tafti, M. S. Reid, J. Shelton, J. M. Siegel, W. C. Dement and E. Mignot, Muscle atonia is triggered by cholinergic stimulation of the basal forebrain: implication for the pathophysiology of canine narcolepsy, *J Neurosci.* **15**, 4806-4814 (1995).
61. E. T. Rolls, M. K. Sanghera and A. Roper-Hall, The latency of activation of neurons in the lateral hypothalamus and substantia innominata during feeding in the monkey., *Brain Res.* **164**, 121-135 (1979).
62. S. Nishino, J. Mao, R. Sampathkumaran, K. Honda, W. C. Dement and E. Mignot, Differential effects of dopaminergic and noradrenergic uptake inhibitors on EEG arousal and cataplexy of narcoleptic canines, *Sleep Res.* **25**, 317 (1996).
63. E. Mignot, S. Nishino, C. Guilleminault and W. C. Dement, Modafinil binds to the dopamine uptake carrier site with low affinity, *Sleep.* **17**, 436-437 (1994).
64. J. P. Wisor, S. Nishino, I. Sora, G. H. Uhl, E. Mignot and D. M. Edgar, Dopaminergic role in stimulant-induced wakefulness, *J Neurosci.* **21**, 1787-94. (2001).
65. C. B. Saper and T. E. Scammell, Modafinil: A drug in search of a mechanism, *Sleep.* **27**, 11-12 (2004).
66. T. Gallopin, P. H. Luppi, F. A. Rambert, A. Frydman and P. Fort, Effect of the wake-promoting agent modafinil on sleep-promoting neurons from the ventrolateral preoptic nuclueus: an In vitro phramacologic study., *Sleep.* **27**, 19-25 (2004).
67. S. Nishino, J. Mao, R. Sampathkumaran, J. Shelton and E. Mignot, Increased dopaminergic transmission mediates the wake-promoting effects of CNS stimulants, *Sleep Research Online.* **1**, 49-61. http://www.sro.org/1998/Nishino/49/ (1998).
68. D. S. Segal and R. Kuczenski. (1994). Behavioral pharmacology of amphetamine. In *Amphetamine and its Analogs: Psychopharmacology, Toxicology and Abuse* (Cho, A. K. & Segal, D. S., eds.), pp. 115-150. Academic Press, San Diego.
69. T. Kanbayashi, K. Honda, T. Kodama, E. Mignot and S. Nishino, Implication of dopaminergic mechanisms in the wake-promoting effects of amphetamine: a study of D- and L-derivatives in canine narcolepsy, *Neuroscience.* **99**, 651-659 (2000).
70. M. N. Shouse, R. J. Staba, S. F. Saquib and P. R. Farber, Monoamines and sleep: microdialysis findings in pons and amygdala, *Brain Res.* **860**, 181-9 (2000).
71. D. Rye, D. L. Bliwize, B. Dihenia and P. Grecki, Daytime sleepiness in Parkinson's disease., *J Sleep Res.* **9**, 63-69 (2000).
72. S. Frucht, J. D. Rogers, P. E. Greene, M. F. Gordon and S. Fahn, Falling asleep at the wheel: motor vehicle mishaps in persons taking pramipexole and ropinirole, *Neurology.* **52**, 1908-10 (1999).
73. B. Ripley, N. Fujiki, M. Okura, E. Mignot and S. Nishino, Hypocretin levels in sporadic and familial cases of canine narcolepsy, *Neurobiol of Dis.* **8**, 525-534 (2001).
74. J. N. Marcus, C. J. Aschkenasi, C. E. Lee, R. M. Chemelli, C. B. Saper, M. Yanagisawa and J. K. Elmquist, Differential expression of orexin receptors 1 and 2 in the rat brain, *J. Comp. Neurol.* **435**, 6-25 (2001).
75. J. S. Lin, Brain structures and mechanisms involved in the control of cortical activation and wakefulness, with emphasis on the posterior hypothalamus and histaminergic neurons, *Sleep Med. Rev.* **4**, 471-503 (2000).
76. R. E. Brown, D. R. Stevens and H. L. Haas, The physiology of brain histamine, *Prog Neurobiol.* **63**, 637-72. (2001).
77. H. Haas and P. Panula, The role of histamine and the tuberomamillary nucleus in the nervous system, *Nat Rev Neurosci.* **4**, 121-30 (2003).
78. S. Nishino, N. Fujiki, B. Ripley, E. Sakurai, M. Kato, T. Watanabe, E. Mignot and K. Yanai, Decreased brain histamine contents in hypocretin/orexin receptor-2 mutated narcoleptic dogs, *Neurosci Lett.* **313**, 125-8. (2001).

79. L. Bayer, E. Eggermann, M. Serafin, B. Saint-Mleux, D. Machard, B. Jones and M. Muhlethaler, Orexins (hypocretins) directly excite tuberomammillary neurons, *Eur J Neurosci.* **14**, 1571-5. (2001).
80. K. S. Eriksson, O. Sergeeva, R. E. Brown and H. L. Haas, Orexin/hypocretin excites the histaminergic neurons of the tuberomammillary nucleus, *J Neurosci.* **21**, 9273-9. (2001).
81. A. Yamanaka, N. Tsujino, H. Funahashi, K. Honda, J. L. Guan, Q. P. Wang, M. Tominaga, K. Goto, S. Shioda and T. Sakurai, Orexins activate histaminergic neurons via the orexin 2 receptor, *Biochem Biophys Res Commun.* **290**, 1237-45. (2002).
82. J. T. Willie, R. M. Chemelli, C. M. Sinton, S. Tokita, S. C. Williams, Y. Y. Kisanuki, J. N. Marcus, C. Lee, J. K. Elmquist, K. A. Kohlmeier, C. S. Leonard, J. A. Richardson, R. E. Hammer and M. Yanagisawa, Distinct narcolepsy syndromes in Orexin receptor-2 and Orexin null mice: molecular genetic dissection of Non-REM and REM sleep regulatory processes, *Neuron.* **38**, 715-30 (2003).
83. T. Kanbayashi, T. Kodama, H. Hondo, S. Sato, N. Miyazaki, K. Kuroda, N. Abe, S. Nishino, Y. Inoue and T. Shimizu, CSF histamine and noradrenaline contents in narcolepsy and other sleep disorders, *Sleep.* **27**, A236 (2004).
84. S. Nishino, E. Sakurai, S. Nevisimalova, J. Vankova, Y. Yoshida, T. Watanabe, K. Yanai and E. Mignot, CSF histamine content is decreased in hypocretin-deficient human narcolepsy, *Sleep.* **25 (suppl)**, A476 (2002).
85. T. Shiba, N. Fujiki, J. Wisor, D. Edgar, T. Sakurai and S. Nishino, Wake promoting effects of thioperamide, a histamine H3 antagonist in orexin/ataxin-3 narcoleptic mice, *Sleep.* **(suppl)**, A241-242 (2004).
86. H. Tsukamoto, T. Ishikawa, Y. Fujii, M. Fukumizu, K. Sugai and T. Kanbayashi, Undetectable levels of CSF hypocretin-1 (orexin-A) in two prepubertal boys with narcolepsy, *Neuropediatrics.* **33**, 51-2. (2002).
87. H. Kubota, T. Kanbayashi, Y. Tanabe, J. Takanashi and Y. Kohno, A case of acute disseminated encephalomyelitis presenting hypersomnia with decreased hypocretin level in cerebrospinal fluid, *J Child Neurol.* **17**, 537-9. (2002).
88. C. B. Saper, T. C. Chou and T. E. Scammell, The sleep switch: hypothalamic control of sleep and wakefulness, *Trends Neurosci.* **24**, 726-31. (2001).
89. Y. Yoshida, N. Fujiki, T. Nakajima, B. Ripley, H. Matsumura, H. Yoneda, E. Mignot and S. Nishino, Fluctuation of extracellular hypocretin-1 (orexin A) levels in the rat in relation to the light-dark cycle and sleep-wake activities, *Eur J Neurosci.* **14**, 1075-81. (2001).
90. N. Fujiki, Y. Yoshida, B. Ripley, K. Honda, E. Mignot and S. Nishino, Changes in CSF hypocretin-1 (orexin A) levels in rats across 24 hours and in response to food deprivation, *NeuroReport.* **12**, 993-7. (2001).
91. P. J. Martins, V. D'Almeida, M. Pedrazzoli, L. Lin, E. Mignot and S. Tufik, Increased hypocretin-1 (orexin-a) levels in cerebrospinal fluid of rats after short-term forced activity, *Regul Pept.* **117**, 155-8 (2004).
92. S. Nishino, The hypocretin/orexin system in health and disease, *Biol Psychiatry.* **54**, 87-95 (2003).

DETAILED ANATOMY OF THE HYPOCRETINERGIC SYSTEM AND RELATED HYPOTHALAMIC CIRCUITS

ANATOMY OF THE HYPOCRETIN SYSTEM

Teresa L. Steininger and Thomas S. Kilduff *

1. INTRODUCTION

This chapter will discuss the anatomy of hypocretin (Hcrt) system, including the morphology and localization of the Hcrt cell bodies, axon terminal fields, and localization of Hcrt receptors. The hypocretin gene encodes two distinct peptides, hypocretin 1 and 2 (Hcrt1 and Hcrt2), that are derived from a single precursor molecule by proteolytic processing.[1,2] The peptides were named hypocretins due to the restricted hypothalamic expression and because of homology to the gut peptide secretin. This gene was reported independently[2] through screening of orphan G protein-coupled receptors (i.e., receptor genes having no identified endogenous ligand). Sakurai et al. used the term orexin to describe the same neuropeptides, which were later confirmed to be congruent with hypocretin.[2] Thus, the *prepro-hypocretin* gene is identical to *prepro-orexin,* Hcrt1 is equivalent to orexin-A, and Hcrt2 is identical to orexin-B. In this chapter, we will use the terms "Hcrt1" and "Hcrt2" to denote the two peptides and "Hcrt" to refer to the Hcrt/orexin-containing cells or to the *hcrt/orexin* gene. To refer to the receptors for these peptides, "HcrtR1" and "HcrtR2" will be used.

2. DISTRIBUTION AND MORPHOLOGY OF HCRT NEURONS

One of the most striking features of Hcrt is that its expression is restricted to neurons in the tuberal region of the hypothalamus, specifically, the perifornical region of the hypothalamus (PFH) and the lateral hypothalamic area (LHA). This hypothalamic region has been known for its role in feeding, self-stimulation reward and arousal.[3,4] We and others have described the distribution of Hcrt cells by *in situ* hybridization[1,2] and by immunohistochemistry using antibodies against prepro-hcrt,[5] Hcrt1[6-9] and Hcrt2.[7,10]

The number of Hcrt-containing neurons has been estimated to range between 1000 and 4000 in the rat[5,9,11] and 50,000-80,000 in humans.[12] Hcrt neurons are medium in size

* T. L Steininger and T.S. Kilduff. Biosciences Division, SRI International, Menlo Park, California 94025

(20-30 μm in diameter) and are multipolar or fusiform in shape, with 2-3 primary dendrites. Dendrites of Hcrt neurons are smooth or have a few spines. At the electron microscopic (EM) level, the cytoplasm of Hcrt neurons is rich in organelles and Hcrt immunoreactivity is associated with the rough endoplasmic reticulum, Golgi apparatus and cytoplasmic dense-core granules.[1,5]

Hcrt neurons are scattered throughout the tuberal lateral hypothalamus (LH) and are bilaterally symmetric.[5,6,8,13] The main concentration is immediately dorsal and lateral to the fornix (perifornical). The neurons extend medially toward the dorsomedial hypothalamic nucleus (DMH), with a few neurons entering this cell group and others located close to the midline third ventricle. The neurons extend laterally through the lateral hypothalamic area (LHA) to reach the medial edge of the internal capsule. The Hcrt neuronal population extends in the rostrocaudal direction from the level of the paraventricular nucleus to the level of the posterior hypothalamic area, although fewer neurons are seen at the extreme poles of this distribution.

Staining with antisera against Hcrt1 appears similar to immunohistochemistry against prepro-hcrt.[6] Neurons labeled by antisera against Hcrt1 or Hcrt2 appear to be co-localized in the hypothalamus,[7] however, Hcrt2- (but not Hcrt1) immunostaining has been described in neurons outside of the LH in the amygdala, bed nucleus of the stria terminalis, and olfactory system.[14,15] Since Hcrt gene expression has not been identified in these limbic regions and since these cells have not been recognized in other immunohistochemical studies, further studies need to be performed to determine whether these represent true Hcrt neurons (see Section 6 below).

3. COLOCALIZATION OF NEUROCHEMICALS IN HCRT NEURONS

Hcrt neurons contain several other neurotransmitters. Although the functional consequence of these additional neurotransmitters is not completely understood, phenotypic differences exist between the Hcrt ligand knockout mouse, in which only the Hcrt gene is deleted,[16] and the transgenic Hcrt/ataxin-3 mouse, in which Hcrt-expressing neurons degenerate.[17] Although a narcolepsy-like syndrome is seen in both animals, hypophagia and obesity occur only in the Hcrt/ataxin-3 mouse,[17] suggesting that other factors in the Hcrt neurons contribute to regulation of energy metabolism.

Following the identification of Hcrt neurons, colocalization studies were performed to determine whether other functionally important neurotransmitters are expressed in these neurons. Substances reported to colocalize with Hcrt can be classified into three major categories: neuropeptides and other neurotransmitters, receptors for neuromodulators, and other cellular factors. The neuropeptide melanin-concentrating hormone (MCH), originally isolated from the salmon pituitary, *does not* colocalize with Hcrt,[5,18,19] although it is expressed in a large population of perifornical and LHA neurons. Broberger *et al.* reported that the number of Hcrt neurons was equivalent to the number of MCH neurons at the level of the LH containing the Hcrt population, but that MCH neurons were more extensive in the rostral-caudal direction. The MCH neuron population extends rostrally to the level of the caudal suprachiasmatic nucleus (SCN) and caudally to the level of the mammillary recess.[18] In contrast, the Hcrt neuron population extends rostrally only to the level of the paraventricular nucleus and caudally only to the level of the posterior hypothalamic area.[5] Cocaine- and amphetamine-related transcript (CART), a neuropeptide found in a subpopulation of MCH neurons, also does not colocalize with Hcrt.[20] The gaseous neuromodulator nitric oxide is likely not a significant colocalizing factor, as the enzyme nitric oxide synthase (NOS) has been detected in roughly 4% of Hcrt neurons in the Long-Evans rat,[13] and no colocalization was seen in the Wistar rat.[21] In fact, nNOS seems not to be colocalized with MCH either, as only NOS-immunoreactive, but not MCH or Hcrt neurons, express the neuropeptide Y (NPY) Y1 receptor.[22]

The neuropeptide galanin was reported to be colocalized with some Hcrt neurons,[23] but the extent of colocalization among the Hcrt neuronal population was not reported. Hcrt neurons were found to be colocalized with *prodynorphin (dyn)A* mRNA. Double *in situ* hybridization and immunohistochemical experiments demonstrated a high degree (>94%) of correspondence between Hcrt- and DynA-expressing neurons,[24] although *dynA* mRNA was also found in adjacent hypothalamic nuclei that did not express Hcrt such as the supraoptic nucleus, the paraventricular nucleus, ventromedial hypothalamic nucleus, and the compact region of the DMH. Hcrt neurons were found to correspond to a neuronal population that was previously recognized by an ovine prolactin antiserum (prolactin-like immunoreactivity).[25] These prolactin-like immunoreactive neurons had been previously characterized as also containing bradykinin and dynorphin B.[26] After Hcrt was discovered, it was demonstrated that dynorphin B colocalized with Hcrt.[27] Hcrt neurons are likely to be glutamatergic as well, since glutamate[28,29] and also the excitatory amino acid transporter EAAT3[30] and vesicular glutamate transporters VGLUT1[31] and VGLUT2[31,32] have been found in Hcrt neurons and/or terminals.

Neurotransmitter receptors that have been reported to colocalize in Hcrt neurons include the $GABA_A$ receptor epsilon subunit,[33] the pancreatic polypeptide Y4 receptor,[34] and the adenosine A1 receptor.[35] In addition, the receptor for the adipose hormone leptin is found in Hcrt neurons.[23] The significance of these colocalizations will be discussed in Chapter 7 on afferent innervation of Hcrt neurons. Other cellular factors reported to be expressed in Hcrt neurons includes precursor-protein convertase[36], secretogranin II,[27,37] the transcription factor Stat-3,[23,38] and the neuronal pentraxin Narp, implicated in clustering of ionotropic glutamate receptors.[39]

4. DEVELOPMENT OF HCRT NEURONS

Although Northern analysis showed that *hcrt* mRNA was first detected at embryonic day (E)18,[1] a later study reported that *hcrt* mRNA was not detectable until postnatal day (P)1.[40] Our subsequent *in situ* hybridization studies confirmed *hcrt* mRNA expression as early as E18.[41] Hcrt-immunoreactivity was also detected at E18, although staining was faint. At E20, Hcrt neurons were small (4-7 μm diameter) and triangular or spindle-shaped with 2-3 short dendrites. The size of Hcrt cell bodies was found to reach adult levels by P16.[41,42] A few axons were seen in the hypothalamus, cortex and brainstem at this time. We also compared the development of Hcrt neurons with the neighboring MCH neuron population and found that MCH expression begins much earlier, prior to E16, and that axonal arborization is more advanced in MCH neurons. Peak terminal field density occurs at the same developmental age (P21) for both MCH and Hcrt axons.

5. COMPARATIVE STUDIES

The restricted hypothalamic distribution of Hcrt neurons appears to be similar among different species. Studies have confirmed the presence of Hcrt neurons in the human and monkey, various strains of mice, rats, and hamsters; in other mammals such as the cat, sheep, and pig; in the chicken, three species of amphibians, and the zebrafish.

In the human brain, Hcrt-immunoreactive neurons are localized to the perifornical area and DMH, with few neurons seen in the LHA[12,43] and Hcrt-immunoreactive axons are as widespread as those seen in rodents (see Section 6). In the monkey hypothalamus, the location of Hcrt neurons and the distribution of Hcrt axons is roughly similar to that seen in the rat, with Hcrt-immunoreactive neurons in the perifornical area and LHA and, to a lesser extent, in the DMH.[44] In the cat brain, Hcrt-1-immunoreactive neurons are present in the tuberal hypothalamus and have a roughly similar distribution to that seen in rodents, with the greatest concentrations of neurons seen dorsal and lateral to the fornix, although the neurons are slightly more widely scattered than in the rat with several neurons extending in the medial direction to the periventricular region and a few neurons as far ventral as the arcuate nucleus.[45,46] In the sheep hypothalamus, the distribution of Hcrt1-immunoreactive neurons differed from the rodent brain in that many Hcrt neurons were found in the DMH and perifornical area with fewer neurons scattered in the LHA.[47] Hcrt neurons in the Siberian and Syrian hamster, visualized by Hcrt1 immunohistochemistry, were present in the perifornical area and LHA. The morphology of Hcrt neurons in the hamster were piriform with 2-3 primary dendrites.[48,49] In the pig, Hcrt1- and Hcrt2-immunoreactive neurons have a similar distribution in the perifornical area and LHA.[50]

The Hcrt system has also been studied in non-mammalian vertebrates. In the frog *Rana ridibunda*, Hcrt2-immunoreactive neurons were detected in the diencephalon, with a high density of neurons found in the SCN and fewer neurons seen in the preoptic area, and the magnocellular and ventral hypothalamic nuclei.[51] The frog SCN is likely not directly analogous of the rodent SCN, since NPY neurons are found in the frog SCN whereas only NPY-positive fibers are seen in the rat SCN.[52] In the bullfrog *Rana catesbeiana*, additional Hcrt1-immunoreactive neurons were seen in the pituitary gland where they colocalized with frog prolactin-immunoreactivity.[53] In the toad *Xenopus laevis*, Hcrt neurons were found in the ventral hypothalamic nucleus. The distribution of immunoreactive fibers was widespread in the brain, similar to that seen in mammals.[54] In the chicken, *hcrt* mRNA and Hcrt immunoreactivity was seen in neurons in the periventricular hypothalamus and LHA.[55] The zebrafish (*Danio rerio*) was found to have two clusters of Hcrt-immunoreactive neurons: an anterior hypothalamic cluster and a preoptic cluster. However, only the hypothalamic cluster was considered to contain true Hcrt neurons, as mRNA was undetectable by *in situ* hybridization in the preoptic cluster.[56] Hcrt-immunoreactive axons were found to terminate among cholinergic and monoaminergic cells groups and Hcrt neurons were innervated by serotonergic, catecholaminergic, histaminergic and cholinergic fibers.[56]

6. DISTRIBUTION OF HCRT AND RECEPTORS IN NON-NEURAL TISSUES

The Hcrt peptides were originally thought to be synthesized exclusively in the brain, as peripheral tissues such as liver, spleen, thymus, testes and ovaries did not show *hcrt* mRNA expression.[1] A number of studies have since reported Hcrt immunoreactivity in non-neural brain tissues including the pineal gland[50] and olfactory epithelium,[15] although *hcrt* mRNA has not been localized in these structures to date. *Hcrt* mRNA was also detected in the ventricular ependymal cell lining.[57] Hcrt peptides and receptors have also been reported in several peripheral tissues including plasma,[58,59] sympathetic ganglia, myenteric plexus, endocrine cells of the gastrointestinal tract,[60] adrenal gland,[61-63] islet cells of the pancreas,[60,61] and placenta.[61] In the gut, the gastrin-containing endocrine cells of the stomach were Hcrt1 immunoreactive[64,65] and several Hcrt1-immunoreactive neurons were seen in the intestinal mucosa, a subpopulation of which also expressed serotonin.[64] *Hcrt* mRNA was shown by RT-PCR to be expressed in the testes,[63,66,67] but not in the ovaries.[63] The testes also apparently contain *hcrtr1* mRNA.[63,67] Merkel cells in the pig snout have also been reported to be Hcrt-immunoreactive.[68] Expression of *hcrt1* mRNA was also seen in the pituitary, kidney, adrenal, thyroid, jejunum,[63] olfactory epithelium,[15] and Merkel cells.[68] High levels of *hcrtr2* mRNA was seen in the adrenal gland (higher in male), and low levels were seen in the lung and pituitary.[63] HcrtR2 immunoreactivity was detected in the olfactory epithelium[15] and Merkel cells.[68]

7. DISTRIBUTION OF HCRT EFFERENT AXONS

The morphology of Hcrt-immunoreactive axons is heterogeneous and varies from thick, densely-varicose to fine, lightly-varicose.[10] Hcrt-immunoreactive axons are found throughout the brain, with the highest density of terminal fields seen in the hypothalamus (Fig. 1).[5,7,8]

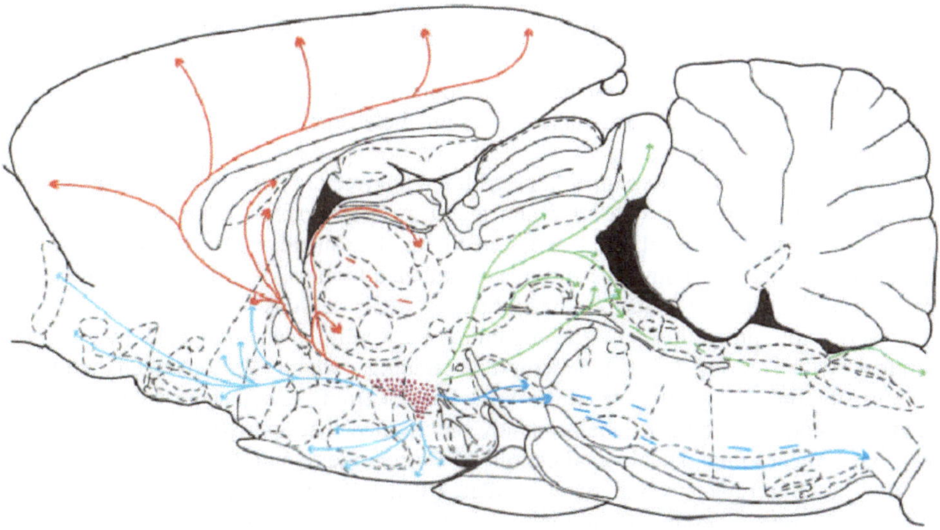

Figure 1. Schematic diagram of the pathways taken by Hcrt/orexin axons, which widely innervate the rat brain through both ascending and descending projections.[6] Copyright 1998 by the Society for Neuroscience.

Hypothalamic regions receiving Hcrt innervation include the LHA and posterior hypothalamic area (regions of Hcrt and MCH neuronal populations), the DMH, the paraventricular hypothalamic nucleus (parvocellular division), arcuate nucleus, and supramammillary nucleus. High Hcrt terminal densities were seen in the locus coeruleus (LC) and moderate-to-high densities in the septum, bed nucleus of the stria terminalis, thalamic paraventricular and reuniens nuclei, periaqueductal gray, substantia nigra, raphe, peribrachial pontine region, medullary reticular formation, and nucleus of the solitary tract, with lesser projections to the cortex, amygdala, hippocampus, and olfactory bulb.[5-7] Similar patterns were seen with antisera to Hcrt1 or Hcrt2.[7] Juxtaventricular (lateral and third ventricular) axon varicosities were noted in Hcrt1-immunostained material[6] and varicose terminals were seen in the circumventricular organs (subfornical organ and area postrema).[8] At the EM level, Hcrt-immunoreactive boutons in the periaqueductal gray contain densely-stained granular vesicles and make asymmetric contacts.[1] Similarly, Hcrt-immunoreactive axons in the LC make asymmetric, presumably excitatory, synapses with tyrosine hydroxylase-positive noradrenergic dendrites.[38]

Hcrt innervation is strongly targeted to ascending arousal systems including the noradrenergic LC, the cholinergic pedunculopontine and laterodorsal tegmental nuclei, the serotonergic dorsal and median raphe nuclei, and the histaminergic tuberomammillary nucleus.[5] Similarly, Hcrt1- and Hcrt2-immunoreactive axons in the cat brainstem were found at high density in the LC, raphe nuclei, and laterodorsal tegmental nucleus, as well as other brainstem regions.[69]

Hcrt innervation was found to strongly overlap with noradrenergic, dopamine beta-hydroxylase-immunoreactive fibers in several brain regions including the paraventricular thalamic nucleus, hypothalamus (mainly the region of Hcrt cell bodies), medial and central amygdala, the dorsal raphe nucleus (DR), and pedunculopontine tegmental nucleus.[70] Hcrt-immunoreactive axons were seen forming putative synaptic contacts with parvalbumin-immunoreactive neurons in the medial septum.[71] Hcrt varicose axons in the DR form putative synaptic contact with both serotonergic and GABAergic neurons as shown using double immunohistochemistry.[72] EM analysis of Hcrt-immunoreactive axons in the DR revealed axon terminals containing immunolabeled large dense core vesicles that made asymmetric contact with unlabeled raphe dendrites and rarely made synaptic contact with perikarya.[73]

Hcrt neurons also have connections with feeding-related hypothalamic regions. Hcrt is reciprocally connected with NPY and leptin receptor-positive neurons in the arcuate nucleus.[44] At the EM level, Hcrt axons in the arcuate nucleus also display axo-axonic synaptic contacts with both Hcrt and other unlabeled terminals.[44,74] Hcrt axons were also found to make synaptic contact with neighboring MCH neurons.[75]

Hcrt axon projections to the dopaminergic ventral tegmental area (VTA) were studied with retrograde and anterograde tracing, which demonstrated numerous retrogradely labeled neurons from the VTA but not from the adjacent medial substantia nigra.[76] Another retrograde labeling study determined that approximately 10% of Hcrt neurons project to the dorsal vagal complex.[9] Retrograde tracing from the nucleus of the solitary tract and nucleus ambiguous revealed that the percentage of Hcrt neurons that project to those sites comprise approximately 8% and 16%, respectively.[77]

In addition, Hcrt innervation of the spinal cord was seen in several studies using immunohistochemistry against Hcrt1,[78] Hcrt2,[10] or both.[79] Several immunoreactive fibers were seen in the dorsal horn of the spinal cord[78] and, at the EM level, these fibers were found to make both symmetric and asymmetric synapses on dendritic elements. Large dense-core vesicles seen in Hcrt axons were intensely immunoreactive.[78] Another study using antisera against Hcrt2 showed dense Hcrt fibers in the dorsal horn (marginal zone, lamina 1), intermediolateral column, and lamina 10, and strong innervation of the sacral spinal cord.[10] These projections were found to arise from the perifornical Hcrt neurons, as determined by retrograde tracing from the spinal cord.[10] Hcrt may also have effects on non-neural tissue in the brain, as Hcrt-immunoreactive fibers are seen in the region of the median eminence, ventricular ependymal lining,[5] and the pineal stalk of the rat[80] and pig.[50] Hcrt has also been found to project indirectly to brown adipose tissue, as shown using transsynaptic retrograde tracing with pseudorabies virus.[81]

8. LOCALIZATION OF HCRT RECEPTORS

The distribution of Hcrt receptors is largely consistent with Hcrt axon innervation patterns (Fig. 2). However, a differential distribution of the two Hcrt receptors, HcrtR1 and HcrtR2, has been demonstrated in several brain regions in both *in situ* hybridization[82-84] and immunohistochemical[85,86] studies. In the cerebral cortex, *hcrtr1* mRNA is expressed primarily in layers II, III and V, whereas *hcrtr2* mRNA is found at higher density in layers II and VI and more diffusely in other layers (Fig. 2). In the hippocampus, *hcrtr1* is expressed mainly in the CA2 region and medial dentate gyrus, while *hcrtr2* was most abundant in CA3. However, HcrtR1-immunoreactivity was found only in the dentate gyrus and not in CA2.[85] Other forebrain regions expressing moderate to high levels of *hcrtr1* mRNA are the tenia tecta, bed nucleus of the stria terminalis, horizontal limb of the diagonal band of Broca, and medial amygdala. In contrast, *hcrtr2* mRNA was prominent is the medial septum, vertical and horizontal limbs of the diagonal band of Broca, substantia innominata, and cortical amygdala.

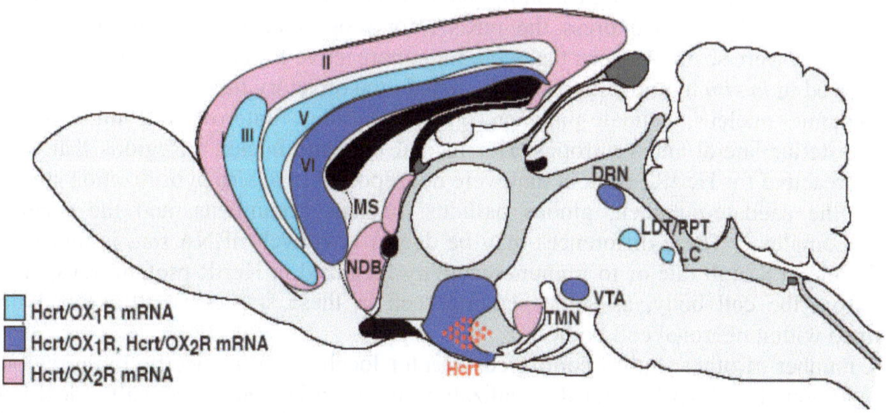

Figure 2. Distribution of Hcrt receptor 1 (HcrtR1) and receptor 2 (HcrtR2) mRNAs as described in Marcus *et al.*[84] for brain regions involved in arousal state regulation. *Abbreviations*: DRN, dorsal raphe nucleus; LC, locus coeruleus; LDT, laterodorsal tegmental nucleus; MS, medial septal nucleus; NDB, nuclei of the diagonal band of Broca; PPT, pedunculopontine nucleus; TMN, tuberomammillary nucleus; VTA, ventral tegmental area. I, III, V and VI represent cortical laminae. Modified from Ref 92.

In the hypothalamus, little or no *hcrtr1* mRNA expression was seen in the arcuate nucleus, paraventricular hypothalamic nucleus, LHA, or tuberomammillary nucleus, regions that prominently express *hcrtr2* mRNA. Hypothalamic regions with high expression of *hcrtr1* include the ventromedial hypothalamic nucleus and a specific region of anterior hypothalamic nucleus dorsolateral to the SCN.[83] In the thalamus, moderately dense levels of *hcrtr1 and hcrtr2* mRNAs were detected in the paraventricular thalamic nucleus and the intergeniculate leaflet, with lesser density in the rhomboid, reuniens, and other midline nuclei.

The subthalamic nucleus expressed both receptors, but preferentially *hcrtr1*. In the brainstem, the highest density of *hcrtr1* mRNA was found in the LC, which did not express *hcrtr2*. Other brainstem noradrenergic groups were noted as expressing *hcrtr1*. The cholinergic brainstem nuclei, the laterodorsal and pedunculopontine tegmental nuclei, expressed both receptors but expressed *hcrtr1* at a higher level.

The VTA, containing dopaminergic neurons, and the DR, containing serotonin neurons, expressed both receptors, although there was a medial-lateral preference in the DR, with *hcrtr1* medial and *hcrtr2* lateral. Moderate *hcrtr1* mRNA expression was seen in the pontine raphe, raphe magnus, raphe obscurus, and dorsal motor vagal complex. Moderate density of *hcrtr2* mRNA expression was described in ventral periaqueductal gray, midbrain reticular formation, dorsal interpeducular nucleus, Barrington's nucleus, the sensory trigeminal nucleus, ventrolateral medulla, and the dorsal vagal nucleus. Low levels of *hcrtr2* were seen in the facial motor nucleus, hypoglossal nucleus, and the external cuneate and gracile nuclei.[83,85,86] In addition, *hcrtr1* mRNA and HcrtR1 protein were identified within the spinal cord in all subdivisions of the gray matter.[85]

Immunohistochemical studies show HcrtR1 and HcrtR2 protein distribution in similar regions shown to express the mRNA for those receptors,[85,86] although some discrepancies were seen. Regions that were immunoreactive for HcrtR1 protein that were not reported in *in situ* hybridization studies include the olfactory tubercle, paraventricular hypothalamic nucleus, caudate-putamen, globus pallidus, anteroventral nucleus and ventroposterior lateral and ventroposterior medial thalamic nuclei.[85] Regions that were immunoreactive for HcrtR2 protein that were not reported in *in situ* hybridization studies include the caudate-putamen, globus pallidus, nucleus accumbens, and the superior olivary complex.[86] These differences may be due to low level mRNA that is translated into protein at a high rate or to immunoreactivity localized in HcrtR protein transported away from the cell body, as some staining seen in these studies is diffuse and not contained within neuronal cell bodies.

A number of other studies confirmed receptor localization in specific brain regions and, additionally, describe HcrtR localization in specific neurochemically-identified populations. Double-immunohistochemistry showed colocalization of HcrtR1 in NPY- and alpha-melanocyte-stimulating hormone (alpha-MSH)-immunoreactive neurons in the arcuate nucleus.[74,87,88] In the latter study, triple immunofluorescence determining the extent of colocalization of leptin and HcrtR1 receptors in NPY and alpha-MSH neurons revealed that alpha-MSH cells mainly contained both receptors, whereas NPY cells were more heterogeneous. HcrtR1 localization in the LC and adjacent mesencephalic trigeminal nucleus was seen in several studies.[89,90] HcrtR1-immunoreactivity was also confirmed in the tenia tecta, indusium griseum, septum, paraventricular thalamic nucleus, dorsal and median raphe,[88] bed nucleus of the stria terminalis,[88,89] medial septum and diagonal band, lateral septum, and substantia innominata,[89] pontine reticular formation, motor trigeminal nucleus, superior olive, ventral cochlear nucleus, Barrington's nucleus, and cholinergic neurons of the laterodorsal tegmental nucleus.[90] HcrtR2 receptor localization was confirmed in parvalbumin neurons in the medial septum,[71] the motor trigeminal nucleus, pontine reticular formation, Barrington's nucleus, and the ventral cochlear nucleus.[90] Both *hcrtr1* and *hcrtr2* mRNA[80] and immunoreactivity[50] were found in the pineal gland.

The composite distribution of the two Hcrt receptors strongly resembles the distribution of the MCH receptor.[91] MCH receptor mRNA distribution is similar to that of the HcrtR1 in the LC, amygdala, and other brainstem noradrenergic groups, whereas in other brain regions such as the septum, hypothalamus, and much of the brainstem, the distribution of MCH receptor mRNA resembles that of the HcrtR2.[91]

9. AFFERENT CONNECTIONS OF HCRT NEURONS

The afferent innervation of Hcrt neurons is less well understood. Colocalization of Hcrt with receptors for the neurotransmitters GABA[92] and NPY[34] suggests that Hcrt neurons are innervated by GABAergic and NPYergic neurons. GABAergic innervation may arise locally from GABAergic neurons that are abundant in the hypothalamus.[93] NPY, agouti-related peptide (AgRP), and alpha-MSH are neurotransmitters found in the arcuate nucleus of the hypothalamus, an important regulator of food intake and body weight.[94] Hcrt and MCH neurons receive innervation from NPY-, AgRP-, and alpha-MSH-immunoreactive fibers,[18,19] suggesting that Hcrt participates in some level of feeding regulation. Leptin, a hormone synthesized by adipose tissue, is thought to regulate body weight through leptin receptors in the arcuate nucleus.[95] (N.E. for a detailed description of intrahypothalamic afferents to hcrt neurons see Chapters 6 and 7). The presence of leptin receptors in Hcrt neurons suggests a role for Hcrt in feeding and body weight regulation.[23,96] In addition, regulation of feeding-related motivation from the nucleus accumbens is suggested by studies showing that pharmacological blockade of GABAergic accumbens neurons promotes Hcrt neuron activation, as indicated by c-Fos expression.[97]

Circadian modulation of the Hcrt neurons is suggested by the presence of a small number of direct projections from the SCN seen in the rat with anterograde tracing, as well as apposition with arginine-vasopressin (AVP)- and vasoactive intestinal peptide (VIP)-immunoreactive fibers, whose cell bodies are found in the SCN in the human brain.[29] Furthermore, Hcrt neurons receive innervation from the DMH, a region that receives dense SCN input that is GABAergic and glutamatergic.[98]

Indirect evidence for cholinergic and monoaminergic innervation of Hcrt neurons is suggested by *in vitro* slice electrophysiological experiments that demonstrate excitatory effects of acetylcholine and inhibitory effects of noradrenaline and serotonin on Hcrt neurons.[32,99] Evidence for noradrenergic innervation is supported by the presence of dopamine beta-hydroxylase-immunoreactive axons in putative contact with Hcrt neurons.[70] Evidence for serotonergic innervation is supported by 5-HT1A receptor-like immunoreactivity localization in Hcrt neurons and 5-HT transporter immunoreactive nerve endings found in close apposition to Hcrt neurons.[100]

Preliminary results of retrograde tracing experiments show putative afferents from limbic forebrain regions including the infralimbic cortex, nucleus accumbens, lateral septum, bed nucleus of the stria terminalis, preoptic area, amygdala, ventromedial hypothalamic nucleus, DMH, LHA, tuberomammillary nucleus, DR, substantia nigra, periaqueductal gray, parabrachial nucleus, and nucleus raphe magnus.[101] Further studies are necessary to confirm afferents that specifically innervate the Hcrt neurons.

10. CONCLUSIONS AND PERSPECTIVE

Although our understanding of the hypocretin/orexin system has been greatly advanced by examination of the morphology of these neurons, their development, efferent projections and comparative studies, it is clear that many aspects of the anatomy of this system remain unstudied. For example, although the Hcrt cells are a distinct neurochemical population of cells within the perifornical area and LHA, it is unclear whether these neurons represent a single, homogeneous population or whether distinct subpopulations exist. Further studies of the topography of the efferent projections of these cells as well as additional colocalization studies should be able to address this issue. Given the central role of the Hcrt system in the maintenance of wakefulness, identification of the inputs to these cells is a critical step to advancing our understanding of this system. Although the afferents to these cells have been little studied to date, it is likely that future neuroanatomical studies will involve use of both conventional and viral-based anatomical methodologies that utilize retrograde transsynaptic transport of marker molecules.[102-105] Lastly, a major question regards the apparent vulnerability of the Hcrt neurons that results in degeneration and, ultimately, narcolepsy in both humans and in some breeds of dogs (sporadic form). Although speculation and considerable experimentation has focused on potential immunological insults to the Hcrt cells leading to degeneration, one recent paper suggests that the Hcrt cells may be more susceptible to excitotoxins than the admixed MCH population in the LHA.[106] Anatomical tools will undoubtedly also be critical to the success of cell transplantation studies that will surely be undertaken in the coming years.

11. ACKNOWLEDGEMENTS

This work was supported by NIH R01 MH61755 and R01AG02584.

12. REFERENCES

1. L. de Lecea, T. S. Kilduff, C. Peyron, X.-B. Gao, P. E. Foye, P. E. Danielson, C. Fukuhara, E. L. F. Battenberg, V. T. Gautvik, F. S. Bartlett II, W. N. Frankel, A. N. van den Pol, F. E. Bloom, K. M. Gautvik and J. G. Sutcliffe, The hypocretins: hypothalamus-specific peptides with neuroexcitatory activity, *Proc Natl Acad Sci U S A.* **95**, 322-7. (1998).
2. T. Sakurai, A. Amemiya, M. Ishii, I. Matsuzaki, R. M. Chemelli, H. Tanaka, S. C. Williams, J. A. Richardson, G. P. Kozlowski, S. Wilson, J. R. Arch, R. E. Buckingham, A. C. Haynes, S. A. Carr, R. S. Annan, D. E. McNulty, W. S. Liu, J. A. Terrett, N. A. Elshourbagy, D. J. Bergsma and M. Yanagisawa, Orexins and orexin receptors: a family of hypothalamic neuropeptides and G protein-coupled receptors that regulate feeding behavior., *Cell.* **92**, 573-85 (1998).
3. L. L. Bernardis and L. L. Bellinger, The lateral hypothalamic area revisited: ingestive behavior, *Neurosci Biobehav Rev.* **20**, 189-287. (1996).
4. B. G. Hoebel, Hypothalamic self-stimulation and stimulation escape in relation to feeding and mating, *Fed Proc.* **38**, 2454-61 (1979).

5. C. Peyron, D. K. Tighe, A. N. van den Pol, L. de Lecea, H. C. Heller, J. G. Sutcliffe and T. S. Kilduff, Neurons containing hypocretin (orexin) project to multiple neuronal systems, *J Neurosci.* **18**, 9996-10015 (1998).
6. C. T. Chen, S. L. Dun, E. H. Kwok, N. J. Dun and J. K. Chang, Orexin A-like immunoreactivity in the rat brain, *Neurosci Lett.* **260**, 161-4 (1999).
7. Y. Date, Y. Ueta, H. Yamashita, H. Yamaguchi, S. Matsukura, K. Kangawa, T. Sakurai, M. Yanagisawa and M. Nakazato, Orexins, orexigenic hypothalamic peptides, interact with autonomic, neuroendocrine and neuroregulatory systems, *Proc Natl Acad Sci U S A.* **96**, 748-53 (1999).
8. T. Nambu, T. Sakurai, K. Mizukami, Y. Hosoya, M. Yanagisawa and K. Goto, Distribution of orexin neurons in the adult rat brain, *Brain Res.* **827**, 243-60 (1999).
9. T. A. Harrison, C. T. Chen, N. J. Dun and J. K. Chang, Hypothalamic orexin A-immunoreactive neurons project to the rat dorsal medulla, *Neurosci Lett.* **273**, 17-20 (1999).
10. A. N. van den Pol, Hypothalamic hypocretin (orexin): robust innervation of the spinal cord, *J Neurosci.* **19**, 3171-82 (1999).
11. T. S. Kilduff and C. Peyron, The hypocretin/orexin ligand-receptor system: Implications for sleep and sleep disorders, *Trends Neurosci.* **23**, 359-65 (2000).
12. T. C. Thannickal, R. Y. Moore, R. Nienhuis, L. Ramanathan, S. Gulyani, M. Aldrich, M. Cornford and J. M. Siegel, Reduced number of hypocretin neurons in human narcolepsy, *Neuron.* **27**, 469-74. (2000).
13. S. B. Cheng, S. Kuchiiwa, H. Z. Gao, T. Kuchiiwa and S. Nakagawa, Morphological study of orexin neurons in the hypothalamus of the Long-Evans rat, with special reference to co-expression of orexin and NADPH-diaphorase or nitric oxide synthase activities, *Neurosci Res.* **46**, 53-62 (2003).
14. J. Ciriello, M. P. Rosas-Arellano, L. P. Solano-Flores and C. V. de Oliveira, Identification of neurons containing orexin-B (hypocretin-2) immunoreactivity in limbic structures, *Brain Res.* **967**, 123-31 (2003).
15. M. Caillol, J. Aioun, C. Baly, M. A. Persuy and R. Salesse, Localization of orexins and their receptors in the rat olfactory system: possible modulation of olfactory perception by a neuropeptide synthetized centrally or locally, *Brain Res.* **960**, 48-61 (2003).
16. R. M. Chemelli, J. T. Willie, C. M. Sinton, J. K. Elmquist, T. Scammell, C. Lee, J. A. Richardson, S. C. Williams, Y. Xiong, Y. Kisanuki, T. E. Fitch, M. Nakazato, R. E. Hammer, C. B. Saper and M. Yanagisawa, Narcolepsy in orexin knockout mice: molecular genetics of sleep regulation, *Cell.* **98**, 437-51 (1999).
17. J. Hara, C. T. Beuckmann, T. Nambu, J. T. Willie, R. M. Chemelli, C. M. Sinton, F. Sugiyama, K. Yagami, K. Goto, M. Yanagisawa and T. Sakurai, Genetic ablation of orexin neurons in mice results in narcolepsy, hypophagia, and obesity, *Neuron.* **30**, 345-54. (2001).
18. C. Broberger, L. De Lecea, J. G. Sutcliffe and T. Hokfelt, Hypocretin/orexin- and melanin-concentrating hormone-expressing cells form distinct populations in the rodent lateral hypothalamus: relationship to the neuropeptide Y and agouti gene-related protein systems, *J Comp Neurol.* **402**, 460-74 (1998).
19. C. F. Elias, C. B. Saper, E. Maratos-Flier, N. A. Tritos, C. Lee, J. Kelly, J. B. Tatro, G. E. Hoffman, M. M. Ollmann, G. S. Barsh, T. Sakurai, M. Yanagisawa and J. K. Elmquist, Chemically defined projections linking the mediobasal hypothalamus and the lateral hypothalamic area, *J Comp Neurol.* **402**, 442-59 (1998).
20. C. F. Elias, C. E. Lee, J. F. Kelly, R. S. Ahima, M. Kuhar, C. B. Saper and J. K. Elmquist, Characterization of CART neurons in the rat and human hypothalamus, *J Comp Neurol.* **432**, 1-19. (2001).
21. D. J. Cutler, R. Morris, M. L. Evans, R. A. Leslie, J. R. Arch and G. Williams, Orexin-A immunoreactive neurons in the rat hypothalamus do not contain neuronal nitric oxide synthase (nNOS), *Peptides.* **22**, 123-8 (2001).
22. S. O. Fetissov, Z. Q. Xu, L. C. Byrne, H. Hassani, P. Ernfors and T. Hokfelt, Neuropeptide Y targets in the hypothalamus: nitric oxide synthesizing neurones express Y1 receptor, *J Neuroendocrinol.* **15**, 754-60 (2003).
23. M. Hakansson, L. de Lecea, J. G. Sutcliffe, M. Yanagisawa and B. Meister, Leptin receptor- and STAT3-immunoreactivities in hypocretin/orexin neurones of the lateral hypothalamus, *J Neuroendocrinol.* **11**, 653-63 (1999).
24. T. C. Chou, C. E. Lee, J. Lu, J. K. Elmquist, J. Hara, J. T. Willie, C. T. Beuckmann, R. M. Chemelli, T. Sakurai, M. Yanagisawa, C. B. Saper and T. E. Scammell, Orexin (hypocretin) neurons contain dynorphin, *J Neurosci.* **21**, RC168. (2001).
25. P. Y. Risold, B. Griffond, T. S. Kilduff, J. G. Sutcliffe and D. Fellmann, Preprohypocretin (orexin) and prolactin-like immunoreactivity are coexpressed by neurons of the rat lateral hypothalamic area, *Neurosci Lett.* **259**, 153-6 (1999).
26. B. Griffond, A. Deray, C. Jacquemard, D. Fellmann and C. Bugnon, Prolactin immunoreactive neurons of the rat lateral hypothalamus: immunocytochemical and ultrastructural studies, *Brain Res.* **635**, 179-86 (1994).

27. L. Bayer, G. Mairet-Coello, P. Y. Risold and B. Griffond, Orexin/hypocretin neurons: chemical phenotype and possible interactions with melanin-concentrating hormone neurons, *Regul Pept.* **104**, 33-9 (2002).
28. F. Torrealba, M. Yanagisawa and C. B. Saper, Colocalization of orexin a and glutamate immunoreactivity in axon terminals in the tuberomammillary nucleus in rats, *Neuroscience.* **119**, 1033-44 (2003).
29. E. E. Abrahamson, R. K. Leak and R. Y. Moore, The suprachiasmatic nucleus projects to posterior hypothalamic arousal systems, *Neuroreport.* **12**, 435-40 (2001).
30. M. Collin, M. Backberg, M. L. Ovesjo, G. Fisone, R. H. Edwards, F. Fujiyama and B. Meister, Plasma membrane and vesicular glutamate transporter mRNAs/proteins in hypothalamic neurons that regulate body weight, *Eur J Neurosci.* **18**, 1265-78 (2003).
31. D. L. Rosin, M. C. Weston, C. P. Sevigny, R. L. Stornetta and P. G. Guyenet, Hypothalamic orexin (hypocretin) neurons express vesicular glutamate transporters VGLUT1 or VGLUT2, *J Comp Neurol.* **465**, 593-603 (2003).
32. Y. Li, X. B. Gao, T. Sakurai and A. N. van den Pol, Hypocretin/Orexin excites hypocretin neurons via a local glutamate neuron-A potential mechanism for orchestrating the hypothalamic arousal system, *Neuron.* **36**, 1169-81 (2002).
33. N. Moragues, P. Ciofi, P. Lafon, G. Tramu and M. Garret, GABAA receptor epsilon subunit expression in identified peptidergic neurons of the rat hypothalamus, *Brain Res.* **967**, 285-9 (2003).
34. R. E. Campbell, M. S. Smith, S. E. Allen, B. E. Grayson, J. M. Ffrench-Mullen and K. L. Grove, Orexin neurons express a functional pancreatic polypeptide Y4 receptor, *J Neurosci.* **23**, 1487-97 (2003).
35. M. M. Thakkar, S. Winston and R. W. McCarley, Orexin neurons of the hypothalamus express adenosine A1 receptors, *Brain Res.* **944**, 190-4 (2002).
36. K. N. Nilaweera, P. Barrett, J. G. Mercer and P. J. Morgan, Precursor-protein convertase 1 gene expression in the mouse hypothalamus: differential regulation by ob gene mutation, energy deficit and administration of leptin, and coexpression with prepro-orexin, *Neuroscience.* **119**, 713-20 (2003).
37. B. Griffond, S. Grillon, J. Duval, C. Colard, C. Jacquemard, A. Deray and D. Fellmann, Occurrence of secretogranin II in the prolactin-immunoreactive neurons of the rat lateral hypothalamus: an in situ hybridization and immunocytochemical study, *J Chem Neuroanat.* **9**, 113-9 (1995).
38. T. L. Horvath, C. Peyron, S. Diano, A. Ivanov, G. Aston-Jones, T. S. Kilduff and A. N. van Den Pol, Hypocretin (orexin) activation and synaptic innervation of the locus coeruleus noradrenergic system, *J Comp Neurol.* **415**, 145-59 (1999).
39. I. M. Reti, R. Reddy, P. F. Worley and J. M. Baraban, Selective expression of Narp, a secreted neuronal pentraxin, in orexin neurons, *J Neurochem.* **82**, 1561-5 (2002).
40. A. N. Van Den Pol, P. R. Patrylo, P. K. Ghosh and X. B. Gao, Lateral hypothalamus: early developmental expression and response to hypocretin (orexin), *J Comp Neurol.* **433**, 349-63. (2001).
41. T. L. Steininger, T. S. Kilduff, M. Behan, R. M. Benca and C. F. Landry, Comparison of hypocretin/orexin and melanin-concentrating hormone neurons and axonal projections in the embryonic and postnatal rat brain, *J Chem Neuroanat.* **27**, 165-81 (2004).
42. Y. Yamamoto, Y. Ueta, Y. Hara, R. Serino, M. Nomura, I. Shibuya, A. Shirahata and H. Yamashita, Postnatal development of orexin/hypocretin in rats, *Brain Res Mol Brain Res.* **78**, 108-19 (2000).
43. R. Y. Moore, E. A. Abrahamson and A. Van Den Pol, The hypocretin neuron system: an arousal system in the human brain, *Arch Ital Biol.* **139**, 195-205 (2001).
44. T. L. Horvath, S. Diano and A. N. van den Pol, Synaptic interaction between hypocretin (orexin) and neuropeptide Y cells in the rodent and primate hypothalamus: a novel circuit implicated in metabolic and endocrine regulations, *J Neurosci.* **19**, 1072-87 (1999).
45. D. Wagner, R. Salin-Pascual, M. A. Greco and P. J. Shiromani, Distribution of hypocretin-containing neurons in the lateral hypothalamus and c-fos-immunoreactive neurons in the VLPO., *Sleep Res Online.* **3**, 35-42 (2000).
46. J. H. Zhang, S. Sampogna, F. R. Morales and M. H. Chase, Orexin (hypocretin)-like immunoreactivity in the cat hypothalamus: a light and electron microscopic study, *Sleep.* **24**, 67-76 (2001).
47. J. Iqbal, S. Pompolo, T. Sakurai and I. J. Clarke, Evidence that orexin-containing neurones provide direct input to gonadotropin-releasing hormone neurones in the ovine hypothalamus, *J Neuroendocrinol.* **13**, 1033-41 (2001).
48. P. A. McGranaghan and H. D. Piggins, Orexin A-like immunoreactivity in the hypothalamus and thalamus of the Syrian hamster (Mesocricetus auratus) and Siberian hamster (Phodopus sungorus), with special reference to circadian structures, *Brain Res.* **904**, 234-44 (2001).
49. E. M. Mintz, A. N. van den Pol, A. A. Casano and H. E. Albers, Distribution of hypocretin-(orexin) immunoreactivity in the central nervous system of Syrian hamsters (Mesocricetus auratus), *J Chem Neuroanat.* **21**, 225-38 (2001).
50. C. Fabris, B. Cozzi, A. Hay-Schmidt, B. Naver and M. Moller, Demonstration of an orexinergic central innervation of the pineal gland of the pig, *J Comp Neurol.* **471**, 113-27 (2004).

51. L. Galas, H. Vaudry, B. Braun, A. N. Van Den Pol, L. De Lecea, J. G. Sutcliffe and N. Chartrel, Immunohistochemical localization and biochemical characterization of hypocretin/orexin-related peptides in the central nervous system of the frog rana ridibunda [In Process Citation], *J Comp Neurol.* **429**, 242-52 (2001).
52. J. M. Danger, J. Guy, M. Benyamina, S. Jegou, F. Leboulenger, J. Cote, M. C. Tonon, G. Pelletier and H. Vaudry, Localization and identification of neuropeptide Y (NPY)-like immunoreactivity in the frog brain, *Peptides.* **6**, 1225-36 (1985).
53. T. Yamamoto, H. Suzuki, H. Uemura, K. Yamamoto and S. Kikuyama, Localization of orexin-A-like immunoreactivity in prolactin cells in the bullfrog (Rana catesbeiana) pituitary, *Gen Comp Endocrinol.* **135**, 186-92 (2004).
54. M. Shibahara, T. Sakurai, T. Nambu, T. Takenouchi, H. Iwaasa, S. I. Egashira, M. Ihara and K. Goto, Structure, tissue distribution, and pharmacological characterization of Xenopus orexins, *Peptides.* **20**, 1169-76 (1999).
55. T. Ohkubo, T. Boswell and S. Lumineau, Molecular cloning of chicken prepro-orexin cDNA and preferential expression in the chicken hypothalamus, *Biochim Biophys Acta.* **1577**, 476-80 (2002).
56. J. Kaslin, J. M. Nystedt, M. Ostergard, N. Peitsaro and P. Panula, The orexin/hypocretin system in zebrafish is connected to the aminergic and cholinergic systems, *J Neurosci.* **24**, 2678-89 (2004).
57. M. Kummer, S. J. Neidert, O. Johren and P. Dominiak, Orexin (hypocretin) gene expression in rat ependymal cells, *Neuroreport.* **12**, 2117-20 (2001).
58. Z. Arihara, K. Takahashi, O. Murakami, K. Totsune, M. Sone, F. Satoh, S. Ito and T. Mouri, Immunoreactive orexin-A in human plasma, *Peptides.* **22**, 139-42. (2001).
59. M. A. Dalal, A. Schuld, M. Haack, M. Uhr, P. Geisler, I. Eisensehr, S. Noachtar and T. Pollmacher, Normal plasma levels of orexin A (hypocretin-1) in narcoleptic patients, *Neurology.* **56**, 1749-51 (2001).
60. A. L. Kirchgessner and M. Liu, Orexin synthesis and response in the gut, *Neuron.* **24**, 941-51 (1999).
61. M. Nakabayashi, T. Suzuki, K. Takahashi, K. Totsune, Y. Muramatsu, C. Kaneko, F. Date, J. Takeyama, A. D. Darnel, T. Moriya and H. Sasano, Orexin-A expression in human peripheral tissues, *Mol Cell Endocrinol.* **205**, 43-50 (2003).
62. L. K. Malendowicz, C. Tortorella and G. G. Nussdorfer, Orexins stimulate corticosterone secretion of rat adrenocortical cells, through the activation of the adenylate cyclase-dependent signaling cascade, *J Steroid Biochem Mol Biol.* **70**, 185-8 (1999).
63. O. Johren, S. J. Neidert, M. Kummer, A. Dendorfer and P. Dominiak, Prepro-orexin and orexin receptor mRNAs are differentially expressed in peripheral tissues of male and female rats, *Endocrinology.* **142**, 3324-31 (2001).
64. A. L. Kirchgessner, Orexins in the brain-gut axis, *Endocr Rev.* **23**, 1-15 (2002).
65. M. J. de Miguel and M. A. Burrell, Immunocytochemical detection of orexin A in endocrine cells of the developing mouse gut, *J Histochem Cytochem.* **50**, 63-9 (2002).
66. T. Ohkubo, A. Tsukada and K. Shamoto, cDNA cloning of chicken orexin receptor and tissue distribution: sexually dimorphic expression in chicken gonads, *J Mol Endocrinol.* **31**, 499-508 (2003).
67. M. L. Barreiro, R. Pineda, V. M. Navarro, M. Lopez, J. S. Suominen, L. Pinilla, R. Senaris, J. Toppari, E. Aguilar, C. Dieguez and M. Tena-Sempere, Orexin 1 receptor messenger ribonucleic acid expression and stimulation of testosterone secretion by orexin-A in rat testis., *Endocrinology.* **145**, 2297-306 (2004).
68. A. Beiras-Fernandez, R. Gallego, M. Blanco, T. Garcia-Caballero, C. Dieguez and A. Beiras, Merkel cells, a new localization of prepro-orexin and orexin receptors, *J Anat.* **204**, 117-22 (2004).
69. J. H. Zhang, S. Sampogna, F. R. Morales and M. H. Chase, Distribution of hypocretin (orexin) immunoreactivity in the feline pons and medulla, *Brain Res.* **995**, 205-17 (2004).
70. B. A. Baldo, R. A. Daniel, C. W. Berridge and A. E. Kelley, Overlapping distributions of orexin/hypocretin- and dopamine-beta-hydroxylase immunoreactive fibers in rat brain regions mediating arousal, motivation, and stress, *J Comp Neurol.* **464**, 220-37 (2003).
71. M. Wu, Z. Zhang, C. Leranth, C. Xu, A. N. van den Pol and M. Alreja, Hypocretin increases impulse flow in the septohippocampal GABAergic pathway: implications for arousal via a mechanism of hippocampal disinhibition, *J Neurosci.* **22**, 7754-65 (2002).
72. R. J. Liu, A. N. van den Pol and G. K. Aghajanian, Hypocretins (orexins) regulate serotonin neurons in the dorsal raphe nucleus by excitatory direct and inhibitory indirect actions, *J Neurosci.* **22**, 9453-64 (2002).
73. Q. P. Wang, J. L. Guan, T. Matsuoka, Y. Hirayana and S. Shioda, Electron microscopic examination of the orexin immunoreactivity in the dorsal raphe nucleus, *Peptides.* **24**, 925-30 (2003).
74. J. L. Guan, R. Suzuki, H. Funahashi, Q. P. Wang, H. Kageyama, K. Uehara, S. Yamada, S. Tsurugano and S. Shioda, Ultrastructural localization of orexin-1 receptor in pre- and post-synaptic neurons in the rat arcuate nucleus, *Neurosci Lett.* **329**, 209-12 (2002).
75. J. L. Guan, K. Uehara, S. Lu, Q. P. Wang, H. Funahashi, T. Sakurai, M. Yanagizawa and S. Shioda, Reciprocal synaptic relationships between orexin- and melanin- concentrating hormone-containing

neurons in the rat lateral hypothalamus: a novel circuit implicated in feeding regulation, *Int J Obes Relat Metab Disord.* **26**, 1523-32. (2002).
76. J. Fadel and A. Y. Deutch, Anatomical substrates of orexin-dopamine interactions: lateral hypothalamic projections to the ventral tegmental area, *Neuroscience.* **111**, 379-87 (2002).
77. J. Ciriello, J. C. McMurray, T. Babic and C. V. de Oliveira, Collateral axonal projections from hypothalamic hypocretin neurons to cardiovascular sites in nucleus ambiguus and nucleus tractus solitarius, *Brain Res.* **991**, 133-41 (2003).
78. J. L. Guan, Q. P. Wang and S. Shioda, Immunoelectron microscopic examination of orexin-like immunoreactive fibers in the dorsal horn of the rat spinal cord, *Brain Res.* **987**, 86-92 (2003).
79. D. J. Cutler, R. Morris, V. Sheridhar, T. A. Wattam, S. Holmes, S. Patel, J. R. Arch, S. Wilson, R. E. Buckingham, M. L. Evans, R. A. Leslie and G. Williams, Differential distribution of orexin-A and orexin-B immunoreactivity in the rat brain and spinal cord, *Peptides.* **20**, 1455-70 (1999).
80. J. D. Mikkelsen, F. Hauser, L. deLecea, J. G. Sutcliffe, T. S. Kilduff, C. Calgari, P. Pevet and V. Simonneaux, Hypocretin (orexin) in the rat pineal gland: a central transmitter with effects on noradrenaline-induced release of melatonin, *Eur J Neurosci.* **14**, 419-25. (2001).
81. B. J. Oldfield, M. E. Giles, A. Watson, C. Anderson, L. M. Colvill and M. J. McKinley, The neurochemical characterisation of hypothalamic pathways projecting polysynaptically to brown adipose tissue in the rat, *Neuroscience.* **110**, 515-26 (2002).
82. P. Trivedi, H. Yu, D. J. MacNeil, L. H. Van der Ploeg and X. M. Guan, Distribution of orexin receptor mRNA in the rat brain, *FEBS Lett.* **438**, 71-5 (1998).
83. J. N. Marcus, C. J. Aschkenasi, C. E. Lee, R. M. Chemelli, C. B. Saper, M. Yanagisawa and J. K. Elmquist, Differential expression of orexin receptors 1 and 2 in the rat brain, *J Comp Neurol.* **435**, 6-25 (2001).
84. X. Y. Lu, D. Bagnol, S. Burke, H. Akil and S. J. Watson, Differential distribution and regulation of OX1 and OX2 orexin/hypocretin receptor messenger RNA in the brain upon fasting, *Horm Behav.* **37**, 335-44 (2000).
85. G. J. Hervieu, J. E. Cluderay, D. C. Harrison, J. C. Roberts and R. A. Leslie, Gene expression and protein distribution of the orexin-1 receptor in the rat brain and spinal cord, *Neuroscience.* **103**, 777-97 (2001).
86. J. E. Cluderay, D. C. Harrison and G. J. Hervieu, Protein distribution of the orexin-2 receptor in the rat central nervous system, *Regul Pept.* **104**, 131-44 (2002).
87. H. Funahashi, S. Yamada, H. Kageyama, F. Takenoya, J. L. Guan and S. Shioda, Co-existence of leptin- and orexin-receptors in feeding-regulating neurons in the hypothalamic arcuate nucleus-a triple labeling study, *Peptides.* **24**, 687-94 (2003).
88. R. Suzuki, H. Shimojima, H. Funahashi, S. Nakajo, S. Yamada, J. L. Guan, S. Tsurugano, K. Uehara, Y. Takeyama, S. Kikuyama and S. Shioda, Orexin-1 receptor immunoreactivity in chemically identified target neurons in the rat hypothalamus, *Neurosci Lett.* **324**, 5-8 (2002).
89. R. A. Espana, R. J. Valentino and C. W. Berridge, Fos immunoreactivity in hypocretin-synthesizing and hypocretin-1 receptor-expressing neurons: effects of diurnal and nocturnal spontaneous waking, stress and hypocretin-1 administration, *Neuroscience.* **121**, 201-17 (2003).
90. M. A. Greco and P. J. Shiromani, Hypocretin receptor protein and mRNA expression in the dorsolateral pons of rats, *Brain Res Mol Brain Res.* **88**, 176-82. (2001).
91. T. S. Kilduff and L. de Lecea, Mapping of the mRNAs for the hypocretin/orexin and melanin-concentrating hormone receptors: networks of overlapping peptide systems, *J Comp Neurol.* **435**, 1-5 (2001).
92. N. Moragues, P. Ciofi, P. Lafon, M. F. Odessa, G. Tramu and M. Garret, cDNA cloning and expression of a gamma-aminobutyric acid A receptor epsilon-subunit in rat brain, *Eur J Neurosci.* **12**, 4318-30 (2000).
93. S. R. Vincent, T. Hokfelt and J. Y. Wu, GABA neuron systems in hypothalamus and the pituitary gland. Immunohistochemical demonstration using antibodies against glutamate decarboxylase, *Neuroendocrinology.* **34**, 117-25 (1982).
94. J. K. Elmquist and J. S. Flier, Neuroscience. The fat-brain axis enters a new dimension, *Science.* **304**, 63-4 (2004).
95. J. K. Elmquist, Hypothalamic pathways underlying the endocrine, autonomic, and behavioral effects of leptin, *Int J Obes Relat Metab Disord.* **25 Suppl 5**, S78-82 (2001).
96. M. L. Hakansson, H. Brown, N. Ghilardi, R. C. Skoda and B. Meister, Leptin receptor immunoreactivity in chemically defined target neurons of the hypothalamus, *J Neurosci.* **18**, 559-72 (1998).
97. B. A. Baldo, L. Gual-Bonilla, K. Sijapati, R. A. Daniel, C. F. Landry and A. E. Kelley, Activation of a subpopulation of orexin/hypocretin-containing hypothalamic neurons by GABAA receptor-mediated inhibition of the nucleus accumbens shell, but not by exposure to a novel environment, *Eur J Neurosci.* **19**, 376-86 (2004).
98. T. C. Chou, T. E. Scammell, J. J. Gooley, S. E. Gaus, C. B. Saper and J. Lu, Critical role of dorsomedial hypothalamic nucleus in a wide range of behavioral circadian rhythms, *J Neurosci.* **23**, 10691-702 (2003).

99. A. Yamanaka, Y. Muraki, N. Tsujino, K. Goto and T. Sakurai, Regulation of orexin neurons by the monoaminergic and cholinergic systems, *Biochem Biophys Res Commun.* **303**, 120-9 (2003).
100. Y. Muraki, A. Yamanaka, N. Tsujino, T. S. Kilduff, K. Goto and T. Sakurai, Serotonergic regulation of the orexin/hypocretin neurons through the 5-HT1A receptor., *J Neurosci.* (in press).
101. K. Yoshida, R. A. Espana, S. L. McCormack, A. J. Crocker and T. E. Scammell, Afferents to the orexin neurons, *Sleep (Suppl.).* **27**, A13 (2004).
102. J. DeFalco, M. Tomishima, H. Liu, C. Zhao, X. Cai, J. D. Marth, L. Enquist and J. M. Friedman, Virus-assisted mapping of neural inputs to a feeding center in the hypothalamus, *Science.* **291**, 2608-13 (2001).
103. G. E. Pickard, C. A. Smeraski, C. C. Tomlinson, B. W. Banfield, J. Kaufman, C. L. Wilcox, L. W. Enquist and P. J. Sollars, Intravitreal injection of the attenuated pseudorabies virus PRV Bartha results in infection of the hamster suprachiasmatic nucleus only by retrograde transsynaptic transport via autonomic circuits, *J Neurosci.* **22**, 2701-10 (2002).
104. U. Maskos, K. Kissa, C. St Cloment and P. Brulet, Retrograde trans-synaptic transfer of green fluorescent protein allows the genetic mapping of neuronal circuits in transgenic mice, *Proc Natl Acad Sci U S A.* **99**, 10120-5 (2002).
105. C. A. Smeraski, P. J. Sollars, M. D. Ogilvie, L. W. Enquist and G. E. Pickard, Suprachiasmatic nucleus input to autonomic circuits identified by retrograde transsynaptic transport of pseudorabies virus from the eye, *J Comp Neurol.* **471**, 298-313 (2004).
106. H. Katsuki and A. Akaike, Excitotoxic degeneration of hypothalamic orexin neurons in slice culture, *Neurobiol Dis.* **15**, 61-9 (2004).

THE ANATOMY OF HYPOCRETIN NEURONS

Tamas L. Horvath*

1. INTRODUCTION

The discovery of hypocretins (Hcrt; also called orexins)[1,2] led to their detection in a unique subset of hypothalamic neurons with widespread distribution of their efferents in the central nervous system. This hypothalamic peptidergic system has emerged as critical regulator of arousal, alertness and associated autonomic, endocrine and metabolic functions.[2-8] Beyond the inherent beauty of this system from the perspective of a neuroanatomist in terms of its distribution, the specificity of the hypocretin system in supporting distinct autonomic, endocrine and behavioral mechanisms lies, to a great extent, within its afferent and efferent connectivity. Other chapters of this book detail the neurochemical and regional specificity of various afferents and efferent targets of the Hcrt system, therefore, the paragraphs below will focus predominantly on the basic neuroanatomical characteristics of these hypothalamic neurons from light and electron microscopic perspectives.

2. HCRT PERIKARYA

Neurons producing Hcrt peptides are distributed in a distinct region of the hypothalamus.[1,2] With the exception of one report on Hcrt in the peripheral nervous system,[9] there is no other place in the brain or periphery where Hcrt-producing neuronal perikarya can be found. Estimates on the number of Hcrt perikarya suggest that approximately 1000 perikarya of the bilateral hypothalamus produce Hcrt, a parameter that shows no apparent species or sex specificity.[10]

2.1. Light Microscopy

* Tamas L. Horvath, Department of Obstetrics, Gynecology & Reproductive Sciences and Department of Neurobiology, Yale University School of Medicine, New Haven CT 06520, USA

The region in which Hcrt neurons are located is frequently referred to as the lateral hypothalamus (LH). However, Hcrt cells are not limited to the LH proper; but rather, they occupy a space that includes parts of the LH and perifornical area with some cells present in the vicinity of the dorsomedial, paraventricular and ventromedial hypothalamic nuclei (Fig. 1).[10-13]

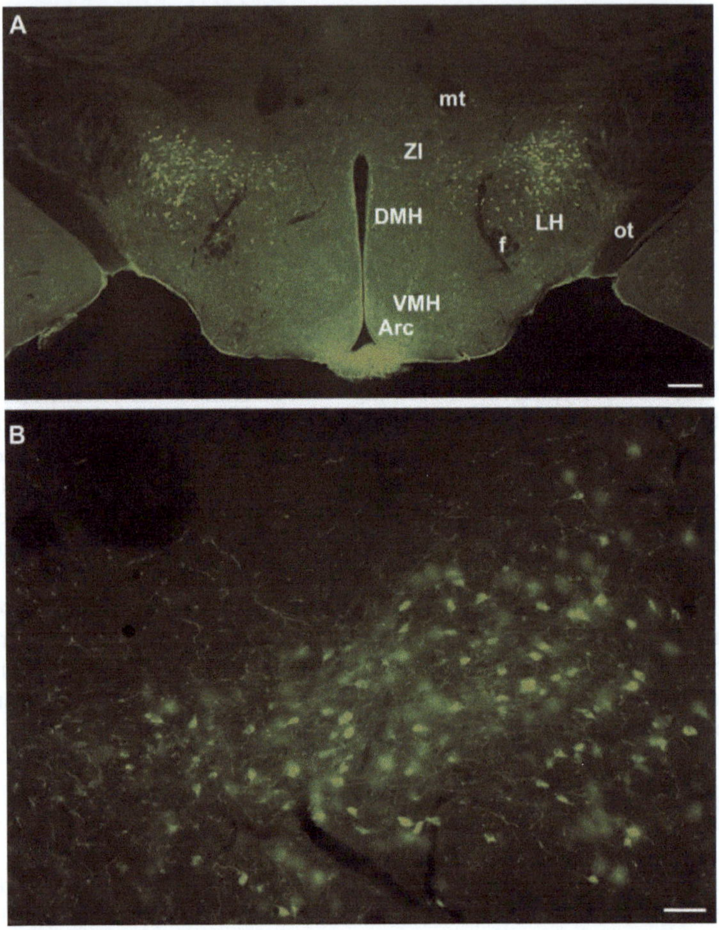

Figure 1. Low (A) and higher (B) power fluorescent images of Hcrt immunoreactive cells in the lateral hypothalamuc (LH)-perifornical region. Note that Hcrt cells are not limited to the LH in proper, but extend medially below the zona incerta (ZI). mt: mammillothalamic tract; ot: opti ctract; f: fornix; DMH: dorsomedial nucleus of the hypothalamus; VMH: ventromedial nucleus of the hypothalamus; Arc; arcuate nucleus. Bar scale on panels A and B represent 100 μm.

The Hcrt neuronal population is not homogeneous from the perspective of cell shape and size: it includes unipolar and bipolar, smaller round cells (Fig. 2a), bipolar fusiform neurons (Fig. 2b) as well as rectangular large neurons (Fig. 2c). The size of Hcrt perikarya is species dependent, but more in association with brain size of animals rather than evolutionary distance. Thus, for example, the average diameter of an Hcrt perikaryon is approximately 10-15 μm in the mouse, while this value is 20-25 μm in the rat and 40-50 μm in non-human primates (Rhesus or Vervet monkeys). From a cellular biological perspective, it is logical that in the case of a long projective system, such as the hypocretin circuitry (see below), species differences based on brain size will be more readily visible than the case might be for a local inter neuronal system.

When visualized by immunocytochemistry, Hcrt is most abundant in membranous structures of the perikaryon with numerous small, vesicle-like structures distributed evenly in the cytosol (Fig. 2a-c). Proximal dendtrites are frequently visible in immunocytochemical preparations (Fig. 2a-c), but, distal dendrites or dendritic spines cannot be detected by immunolabeling for Hcrt, even with the use of colchicine. Colchicine is a substance frequently used to enhance perikaryal labelling of peptides in fast firing neurons; it is interesting to note that its administration does not significantly increase the number of detectable Hcrt neurons.[13]

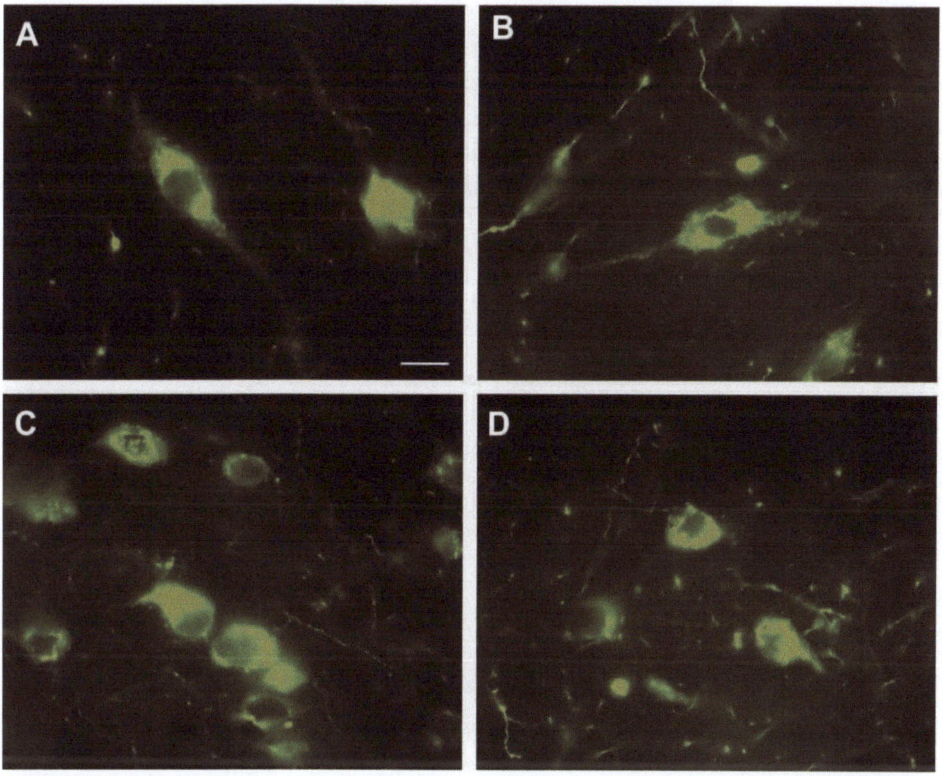

Figure 2. Fluorescent images of various shape Hcrt immunoreactive cells in the lateral hypothalamus perifornical region, including bipolar fusifom (A), rectangular (B) and bipolar and unipolar small round Hcrt cell (C and D). bar scale on A represent 10 μm for panels A-D.

2.2. Electron microscopy

Electron microscopic analysis of Hcrt-immunolabeled perikarya confirms that immunoprecipitation (representing Hcrt) is predominantly associated with a membranous structure, in particular, the Golgi apparatus, while diffuse labelling in the cytosol can also be detected (Fig. 3).[10,13] It is a characteristic feature that Hcrt neurons contain abundant endoplasmic reticuli and medium-sized mitochondria (Fig. 3). With regard to other cytoplasmic organelles, it is noteworthy to mention that a large number of Hcrt-producing perikarya exhibit an aggregation of apparent transport vesicles into a cluster that is commonly referred to as either a nematosome[14,15] or butrysome.[16] These organelles have been associated with synaptic plasticity of both the developing[16] and adult[14,15] brain, and this process may be a characteristic trait of the adult Hcrt system as well (see section 4 below).

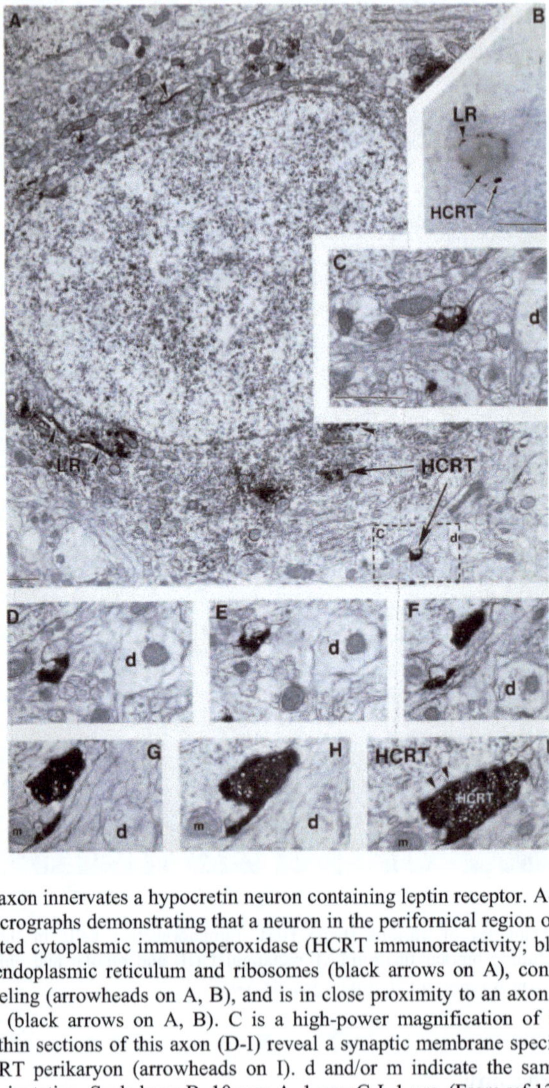

Figure 3. A hypocretin axon innervates a hypocretin neuron containing leptin receptor. A-I, Correlated light (B) and electron (A, C-I) micrographs demonstrating that a neuron in the perifornical region of a female rat contains homogeneously distributed cytoplasmic immunoperoxidase (HCRT immunoreactivity; black arrows on B) that is associated with the endoplasmic reticulum and ribosomes (black arrows on A), contains Golgi-associated leptin receptor (LR) labeling (arrowheads on A, B), and is in close proximity to an axon terminal that contains HCRT immunolabeling (black arrows on A, B). C is a high-power magnification of the boxed area on A. Consecutive serial ultrathin sections of this axon (D-I) reveal a synaptic membrane specialization between this HCRT bouton and HCRT perikaryon (arrowheads on I). d and/or m indicate the same dendrite (d) and/or mitochondrion (m) for orientation. Scale bars: B, 10 μm; A, 1 μm; C-I, 1 μm. (From ref 13)

An intriguing and potentially functionally important aspect of Hcrt neurons is the morphology of their nuclei: while most long projective excitatory neuronal systems, such as the pyramidal neurons of the hippocampal formation, possess round cell nuclei, Hcrt neurons frequently exhibit infoldings in the nuclear envelope (Fig. 3), which is much more characteristic of interneurons. That the Hcrt system is excitatory in nature has been described in the first publication on Hcrt[1] followed by other studies using both electrophysiology[17] and colocalization studies with vesicular glutamate transporters.[18,19]

3. AFFERENT INPUT TO THE HCRT NEURONS

To gain a deeper insight to the neurobiological regulation of Hcrt neurons, the organization of input to the Hcrt cells needs to be delineated. Some advances have been made on this front, although, the fine synaptological organization of Hcrt perikarya and dendrites still eludes the scientific community. However, new, emerging information on this topic reveals, yet again, how unique the Hcrt system is in relation to other known neuronal circuits of the central nervous system.

3.1. Light microscopy

3.1.1. Neuropeptides

The first attempts to determine the afferent inputs of Hcrt neurons were triggered by the initial indication that the Hcrt/orexin system is predominantly associated with energy metabolism regulation:[2] light microscopic evidence was provided to show that the two components of the arcuate nucleus melanocortin system a primum movens in appetite and food intake regulation), the neuropeptide Y (NPY)/Agouti-related peptide (AgRP)- and propopiomelanocortin- (POMC)- producing cells, provide putative inputs to the Hcrt neurons.[11-13] In addition to these feeding-associated inputs from the arcuate nucleus, another peptidergic system in the vicinity of the Hcrt neurons, the melanin-concentrating-hormone (MCH)-producing cells, were also shown to send efferents to within a close proximity to Hcrt perikarya.[20,21] Boutons containing corticotrophin releasing hormone (CRH), presumably arising from the paraventricular hypothalamic nucleus, were also found in close apposition to Hcrt perikarya. Hcrt afferent inputs containing NPY (which may arise from either the arcuate nucleus, brain stem or thalamus), MCH[20, 43] and CRH[22] were confirmed by electron microscopy, while to date no anatomical evidence exists that POMC or AgRP boutons establish synaptic connections with Hcrt neurons, although these are very likely possibilities.

3.1.2. Amino Acid Neurotransmitters

As a role emerged for Hcrt neurons as critical regulators of arousal, further clarification of their input was progressing. The fascinating revelation that Hcrt neurons are predominantly controlled by local excitatory interneurons[23] was made. Indeed, light microscopic observations revealed a robust interaction between vesicular glutamate transporter 2 (vGlut2)-containing, putative glutamatergic axon terminals and Hcrt perikarya.[23] Some of this excitatory input appears to be regulated by Hcrt itself, as Hcrt is present in boutons on Hcrt perikarya,[13] and, Hcrt administration has been shown to moderate presynaptic release of glutamate onto these neurons.[23] Strikingly, it was also observed that perikaryal GABA input onto the Hcrt neurons is very minimal, approximately 1/10th that of the glutamatergic input.[24] This makes these hypothalamic neurons unique in that there is no other example up till now, of long-projective, excitatory neurons being dominated by stimulatory input at the level of the perikaryon. Electron microscopic (see below) and electrophysiological analyses[24] confirmed the above light microscopic observations. Electrophysiological results indicate, however, that other afferents, including brain stem noradrenergic and serotoninergic inputs as well as afferent cholinergic pathways also play direct roles in the regulation of Hcrt neuronal activity.[23,25] Light microscopic evidence exists in support of serotoninergic innervation[26] of Hcrt perikarya, while synaptic confirmation of these and other monoaminergic interactions is yet to be provided.

Of course, there are other peptide and amino acid transmitters and modulators that may play critical roles in affecting the output of Hcrt neurons. Ongoing anatomical research will continuously provide clues and help develop a better blue print of the organization of the input to the Hcrt system. However, this work needs to be complemented by ultrastructural evidence (see below).

3.1.3. Humoral Signals

Humoral signals, such as leptin, glucose, glucocorticoids, insulin, and gonadal steroids may readily affect activity levels of Hcrt neurons both acutely and chronically. Leptin receptors have been described in Hcrt neurons,[13,27] and fasting, which is characterized by rapidly changing levels of circulating leptin, ghrelin, glucocorticoid, glucose, insulin and thyroid hormone, was shown to activate Hcrt neurons and their postsynaptic targets.[28]

3.2. Electron Microscopy

Confirmation of putative connectivity at the level of synapses is a critical step for developing a decent neurobiological comprehension of the regulatory relationship between pre- and postsynaptic elements. Thus, if the input organization and input stability and neurobiological characteristics of Hcrt neurons are to be achieved, classic immuno electron microscopy must be utilized in conjunction with light microscopy and physiological measurements.

As alluded to above, the input organization of Hcrt neurons provides previously unseen synaptological architecture of long projective, excitatory neurons. The overall frequency of synapses on Hcrt neurons is not that dissimilar (30-40 synapse /100 μm perikaryal membrane) from other analyzed hypothalamic peptidergic neurons.[29] However, at the heart of the synaptic organization of Hcrt afferents is the dominance of putative excitatory synaptic inputs[24] on the Hcrt perikarya based on the classification of synaptic inputs.[29] This method distinguishes between excitatory and inhibitory connections based on the symmetry of the synaptic membrane specializations. Synaptic connections in which the postsynaptic density is dominant are called asymmetric or Gray type I synapses, and are classified as stimulatory synaptic connections. When there is less of a difference between the thicknesses of the pre- and postsynaptic membrane specializations or there is a slight thickening of just the presynaptic membrane, the connection is referred to as symmetric or Gray type II and is characteristic of inhibitory synapses. When the synaptic input organization of Hcrt neurons was assessed based on symmetry, a startling finding was made that the vast majority of synapses established on Hcrt perikarya are asymmetric, Gray type I connections (Fig. 4) suggesting that Hcrt perikarya are dominated by excitatory inputs. The ratio of putative stimulatory to inhibitory contacts on Hcrt perikarya was approximately 10:1, which is unprecedented in the nervous system.[24] Electrophysiological analyses of mini excitatory and inhibitory postsynaptic currents confirmed this dominance of excitatory inputs onto Hcrt neurons.[24]

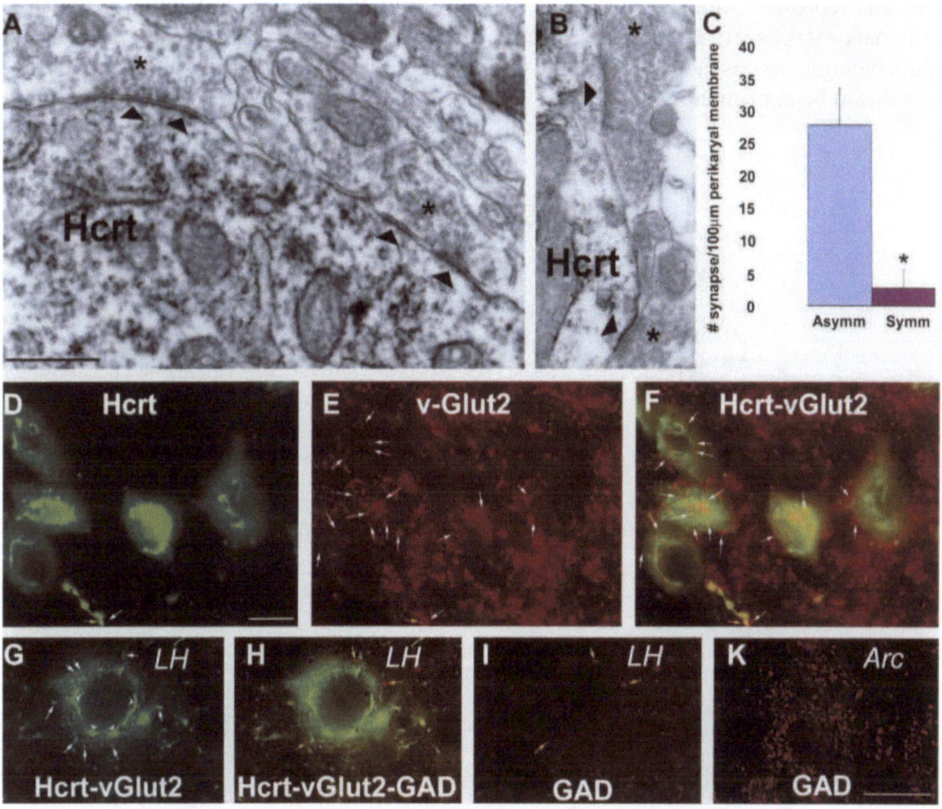

Figure. 4. A, B: Electron micrographs showing asymmetrical synaptic contacts between unidentified axon terminals and an Hcrt perikaryon (A) and a dendrite (B). Statistical analysis revealed that asymmetrical, putative excitatory axon terminals dominate inhibitory contacts on Hcrt perikarya (C). D-F: Double immunostaining for Hcrt (D) and vesicular glutamate transporter 2 (vGlut2; E) reveals robust interaction between red, vGlut2-immunopositive, putative excitatory axon terminals (white arrows) and green, Hcrt-containing cell bodies (F). Bar scale on D represents 10 μm for D-F. G-I: Multiple labeling for vGlut2, Hcrt (both labeled with green fluorescence; G and H) and glutamic acid decarboxylase (GAD, red fluorescence; H and I), which is the rate limiting enzyme in the biosynthesis of the inhibitory neurotransmitter GABA. In accordance with the electron microscopic analysis (A-C), vGlut2-immunolabeled terminals (white arrows on G) outnumber GAD-containing putative boutons (H and I) on this Hcrt-producing neuron (white arrowheads on G point to cytoplasmic membranous structures, a characteristic feature of Hcrt immunoreactivity). In contrast to Hcrt perikarya, most neuronal cell bodies and proximal dendrites in the hypothalamus (and elsewhere in the brain) are dominated by inhibitory GABA input as seen on an unlabeled arcuate nucleus neuron on panel J after GAD immunolabeling. Bar scale on J represents 10 μm for panels H-J. (From ref 24).

The light microscopic assessment of putative excitatory and inhibitory contacts on Hcrt cells by immunocytochemical visualization of vGlut2 (excitatory) and GABA (inhibitory) also indicated a predominance of stimulatory inputs, which was then confirmed by multiple labelling, correlated light and electron microscopic analyses (Fig. 5).[24] This analysis further strengthened the classification (by Gray) that asymmetric contacts represent stimulatory inputs, because all of the vGlut2-containing presynaptic terminals established asymmetric synaptic membrane specializations (Fig. 5). Whether glutamatergic axons account for the majority of excitatory inputs on Hcrt neurons remains to be confirmed.

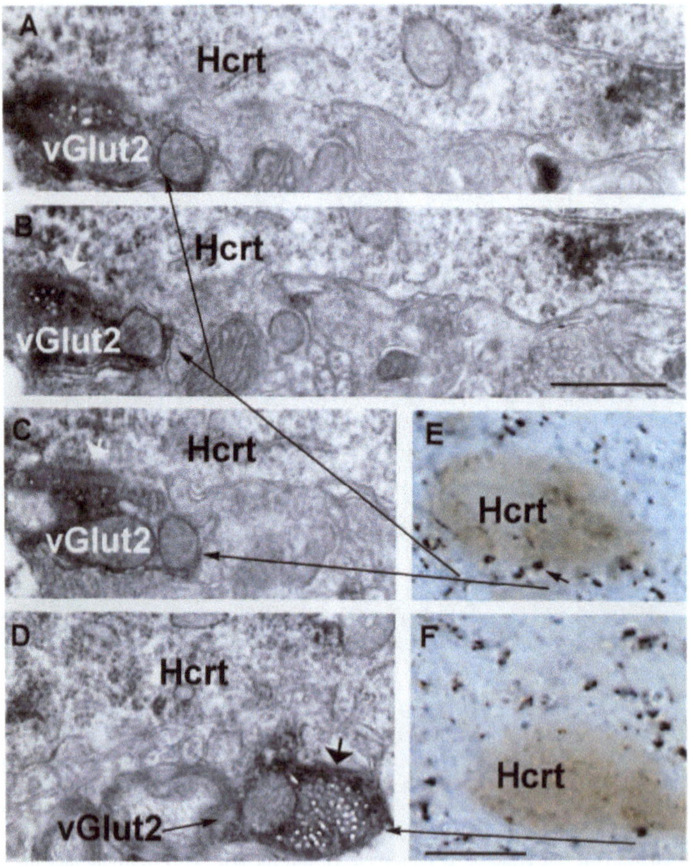

Figure 5. Correlated light and electron microscopy shows that vGlut2-immunoreactive axon terminals establish asymmetrical synaptic contacts with the perikarya of Hcrt neurons. A-C: Consecutive electron micrographs showing the establishment of an asymmetrical synapse by a vGlut2-immunoreactive axon terminal indicated on the light micrograph (black arrow on E). White arrows on B and C point to the typical postsynaptic density of asymmetrical synapses. D: Electron micrograph of the asymmetrical contact established by another vGlut2-immunoreactive bouton (indicated by the starting point of the arrow on panel F) on another Hcrt perikarya seen on panel F (light brown immunoprecipitation). Bar scale on B represents 1 μm for panels A-D. Bar scale on F represents 10 μm for E and F. (From ref 24)

Other afferents may also establish asymmetric contacts, including CRH-containing[22] and acetylcholine-expressing axon terminals. Additionally, since MCH-containing axon terminals (in which a fast acting synaptic transmitter is yet to be identified) also establish asymmetric contacts,[30] it is likely that the MCH input contributes to the asymmetric input ratio on Hcrt neurons. While the axo-somatic inputs on Hcrt neurons are unique in that excitatory inputs dwarf inhibitory contacts, it is also of interest to note that the presynaptic interaction of these excitatory inputs is very extensive (Fig. 6 and 7).

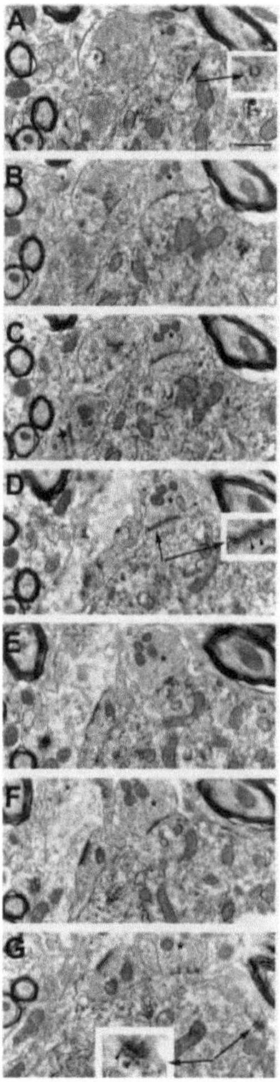

Figure 6. Electron micrographs (A-G) showing consecutive serial sections of a bouton (indicated by stars in A-G) that establishes an asymmetrical-like synaptic membrane specialization on an Hcrt-immunolabeled neuronal perikarya (diffuse immunoperoxidase precipitation in the cytosol). Arrowhead in enlarged insert of panel D point to typical postsynaptic density of asymmetrical contacts, which is established by the serially sectioned bouton. The synaptic membrane specialization appears to be ambiguous because the typical membrane specialization of an asymmetrical connection (also seen between another presynaptic bouton and the Hcrt perikarya on panel G) is only visible through a limited extent of the synapse. This together with the presence of a clathrine-coated pit in direct association with the postsynaptic side of the emerging synapse (A) suggest ongoing dynamic changes in postsynaptic membrane composition, and event that is supportive of the observed synaptic plasticity on Hcrt perikarya. Bar scale on A represents 1 µm for panels A-G. (From 24)

For example, it is not unusual to find that pre-synaptic terminals are adjacent to each other without glial interface sometimes manifesting synaptic membrane specializations (Fig. 6 and 7). Data from electrophysiological experiments led to the proposal that the presynaptic interaction affecting Hcrt neurons is orchestrated, at least in part, by Hcrt itself.[23] This proposition is supported by the fact that Hcrt is expressed in presynaptic boutons on Hcrt perikarya (Fig. 3).[13]

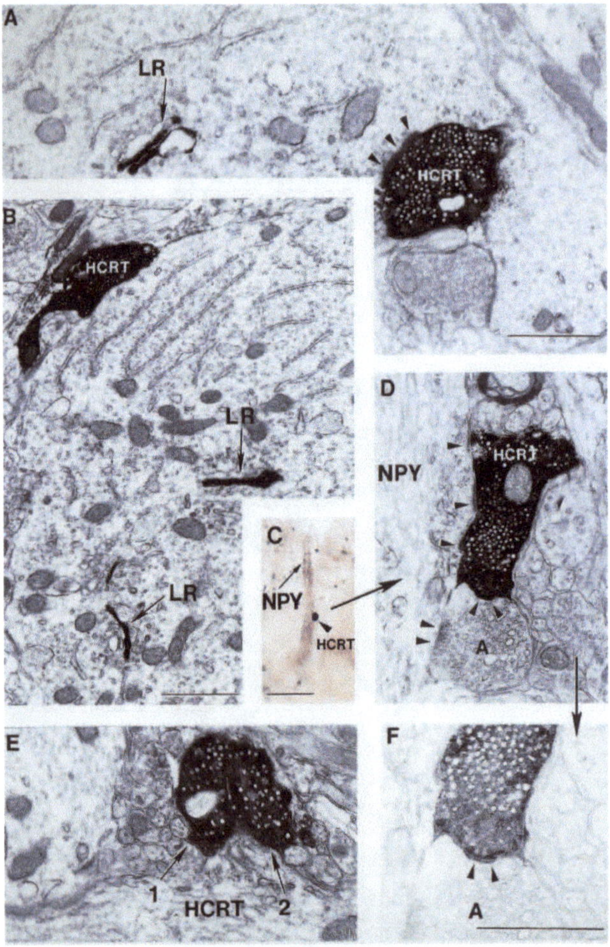

Figure 7. Synaptic interaction between hypocretin boutons and leptin receptor- or neuropeptide Y-producing neurons in the arcuate nucleus. A, B, Hypocretin (HCRT)-containing axon terminals establishing asymmetric (arrowheads on A) synaptic contacts with neuronal perikarya expressing leptin receptor (LR) immunolabeling associated with Golgi cisternae (black arrows) in the rat (A) and monkey (B) arcuate nucleus. C, D, F, Correlated light (C) and electron (D, F) micrographs demonstrating that the black HCRT axon terminal (arrowhead on C) in contacting an NPY-containing dendrite establishes an asymmetric synapse (arrowheads on D) on this NPY dendrite. On D, an unlabeled axon (A) next to the HCRT-labeled bouton also makes an asymmetric synapse on the same postsynaptic target. F is an underexposed, higher magnification of D revealing the intimate relationship between the HCRT-labeled and unlabeled (A) presynaptic terminals. Note the apparent membrane specialization between these boutons (arrowheads). E, Close apposition between two HCRT-immunolabeled axon terminals in the rat arcuate nucleus. Scale bars: C, 5 μm; A, D, E, 1 μm; B, F, 1 μm. (From 13)

The existence of nematosomes in Hcrt perikarya and the findings on other peptidergic neuronal populations of the hypothalamus[29] gave impetus to the intriguing possibility that the input organization of Hcrt neurons is not hard wired.[31] In further support of this idea are the observations from serial electron microscopic sections of the perikarya of Hcrt neurons. From this analysis, it was obvious that some of the synapses show immature characteristics such as incomplete postsynaptic densities that were frequently associated with clathrin-coated pits, a sign of active membrane reorganization (Fig. 7).[24] Furthermore, when animals were fasted, Hcrt perikarya, which were already dominated by excitatory inputs, recruited additional stimulatory synapses as revealed by both ultrastructural and electrophysiological analyses.[24]

Taken together, light and electron microscopic analyses of the synaptology of hypothalamic Hcrt neurons reveals a unique organization to their input in which excitatory contacts dominate inhibitory ones by several magnitudes. This synaptic organization makes it likely that, for the Hcrt system, noise is signal and this may be one of the reasons why the Hcrt system is easily activated. This input organization exhibits remarkable plasticity and, thus, may be an important foundation for easy arousal (from sleep), an ability that, for most species, is paramount to survival. Because the inhibitory input onto the Hcrt neuronal perikaryon is minimal, other mechanisms must be in place to prevent overexcitation. For this, an elegant electrophysiological explanation was provided which suggests that metabotropic glutamate receptors located in the terminals presynaptic to Hcrt neurons appear to moderate incoming excitation.[21] Additionally, an interesting possibility also exists that dynoprhin, which is produced in almost all Hcrt cells,[32] could be released from dendrites to down-regulate the activity of axcitatory inputs of Hcrt neurons.[33] Therefore, it is apparent that further delineation of the architecture and interaction of the presynaptic terminals of Hcrt neurons will be critical to better understand regulatory elements of Hcrt neuronal activity.

4. HCRT EFFERENTS

For an understanding of Hcrt's fundamental actions in specific brain functions, the projection field of Hcrt efferents provides critical clues. While much has been deciphered about the outputs of these hypothalamic peptidergic neurons, a lot must still be discerned with particular emphasis on the quantitative and qualitative relationship between Hcrt axon terminals and pre- and postsynaptic systems that are downstream of Hcrt action.

4.1. Light Microscopy

Hcrt fibers are present in various parts of the entire brain in agreement with the distribution pattern of the two hypocretin receptors.[2,34-37] In the vicinity of Hcrt perikarya, there is a homogeneous density of fibers in the tuberal region of the hypothalamus. A high density of fibers is also seen in other regions of the hypothalamus, although fewer fibers are seen in the medial preoptic nucleus, the anterior part of the ventromedial hypothalamic nucleus, and the paraventricular nucleus. While Hcrt fibers are present around the suprachiasmatic and supraoptic nuclei, few Hcrt axons (and most are of axons of passage) can be seen within these nuclei. Hcrt fibers and boutons innervate the arcuate nucleus. Hcrt fibers are not found in the mammillary bodies, but travel through the supramammillary nucleus and the posterior hypothalamic area.

Peyron et al.[10] divided the tracts that carry Hcrt efferents out of the hypothalamus into four different pathways: dorsal and ventral ascending pathways and dorsal and ventral descending pathways. The following is a summary of their description.

4.1.1. Dorsal Ascending Pathway

Hcrt neurons send axons through the zona incerta to the paraventricular nucleus of the thalamus, the central medial nucleus of the thalamus, and the lateral habenula. Hcrt fibers are present in the substantia innominata, the bed nucleus of the stria terminalis, the septal nuclei (medial and lateral), and the dorsal anterior nucleus of the olfactory bulb. Axons in these regions are long, thick and arborize into numerous boutons. Hcrt axons are not found in the caudate putamen and the globus pallidus but innervate the cortex. In the latter, Hcrt fibers are more extensive in the vicinity of the corpus callosum than in other layers. It is also noteworthy to mention that the other hypothalamic peptidergic system, consisting of the MCH neurons, sends more robust projections to the cortex where they are more diffusely distributed than the Hcrt efferents. Hcrt axons also project laterally through the zona incerta to enter the subincertal nucleus of the thalamus, and the subthalamic nucleus following the optic tract to the central, anterior, and medial amygdaloid nuclei.

4.1.2. Ventral Ascending Pathway

Hcrt fibers are present in the ventral pallidum, the vertical and horizontal limb of the diagonal band of Broca, the medial part of the accumbens nucleus, and the olfactory bulb. In the olfactory system, fibers are mainly seen in the anterior olfactory nuclei with no Hcrt axons seen in the mitral cell layer.

4.1.3. Dorsal Descending Pathway

Hcrt efferents travel through the mesencephalic central gray to innervate the superior and inferior colliculi and the pontine central gray. Dense Hcrt innervations of the locus coeruleus, the dorsal raphe nucleus, and the laterodorsal tegmental nucleus can be found. In this case, labelled fibers are long and thick with numerous boutons. Another bundle of Hcrt fibers passes through the dorsal tegmental area to the pedunculopontine nucleus, the parabrachial nucleus, and the subcoeruleus area. Avoiding the vestibular nuclei, Hcrt projections descend to the dorsolateral part of the nucleus of the solitary tract and the parvocellular reticular area and to the dorsal medullary region as well as the gelatinous layer of the caudal spinal trigeminal nucleus. In a separate description,[38] it was observed that the dorsal descending tract contains Hcrt fibers that travel caudally to innervate all levels of the spinal cord from cervical to sacral segments. Hcrt axons are found in the marginal zone (lamina 1), the intermediolateral column and lamina 10 as well as the caudal region of the sacral cord.

4.1.4. Ventral Descending Pathway

In this tract, Hcrt axons go through the interpeduncular nucleus, the ventral tegmental area, and the substantia nigra pars compacta. In the pons and medulla, Hcrt fibers are distributed through the raphe nuclei and the reticular formation in the pontis oralis, caudal and ventral, the ventral and a gigantocellular reticular nuclei, and the ventral medullary area. The Hcrt innervation of the raphe magnus, the lateral paragigantocellular nucleus, and the ventral sub-coeruleus area where the A5 noradrenergic cell group is located is particularly strong. On the other hand, Hcrt efferents avoid the red nucleus, the pontine nucleus, and the facial motor nucleus, as well as auditory structures such as the trapezoid body and the superior and inferior olive nuclei.

4.2. Electron Microscopy

The hypocretin system influences various brain functions by modulation of synaptic transmission. Electrophysiological data show that Hcrt can act both post-synaptically, by directly altering the membrane potential of neuronal perikarya, and pre-synaptically, by regulating the release of transmitters, such as GABA and glutamate, from adjacent axon terminals. This pre- and post-synaptic action of Hcrt can occur simultaneously on a particular target,[17] but examples have also been provided when Hcrt acts predominantly by altering pre-synaptic dynamics of the target neuron, for example, Hcrt neurons.[23] Regardless of the type of action, to appreciate the potential of Hcrt to regulate a putative target system, it is essential that the synaptic organization of Hcrt efferents on a postsynaptic target be revealed. It is important to demonstrate the type, location and extent of synaptic interaction between Hcrt efferents and their postsynaptic targets. Equally critical is the establishment of the hierarchical relationship between Hcrt boutons and other pre-synaptic inputs of a given target neuronal population. With the exception of some elementary observations (see below), the majority of these issues have not yet been addressed and, thus, they represent one of the essential areas that must be resolved in order to provide deeper insight to our understanding of Hcrt's regulatory role in the CNS.

As mentioned earlier, the Hcrt circuitry represents an excitatory system. In accordance with this, Hcrt axon terminals, regardless of the brain region, establish almost exclusively asymmetric synaptic contacts (Fig. 7 and 8). Hcrt-containing axon terminals are relatively large (0.6-1 mm diameter) and contain both large-core and small vesicles (Fig. 7). Hcrt-immunoreactivity in axon terminals is associated with the pool of large-core vesicles. Hcrt axon terminals frequently contain mitochondria, sometimes in unusually large numbers (Fig. 8).

An intriguing aspect of the synaptology of Hcrt efferents is that these peptidergic boutons are more often than not situated proximally on target cells. Thus, for example, Hcrt axon terminals on both the arcuate nucleus NPY cells, which are critical in mediating Hcrt's effect on food intake,[39] and on the locus coeruleus noradrenergic neurons, which are likely to be vital in mediating Hcrt's effect on arousal,[40,41] establish synapses predominantly on the perikaryal membrane or the membranes of dendrites located very proximally to the the perikaryon (Figs. 7, 8).[13,40] This appears to be the case also regarding Hcrt interaction with the medio-septal cholinergic neurons.[42] This proximal location of Hcrt boutons suggests a superior hierarchical positioning of Hcrt efferents to influence postsynaptic targets. This synaptic organization, together with the fact that in both the arcuate nucleus and locus coeruleus, almost all of the NPY and noradrenergic cells receive massive Hcrt innervation, helps explain the robust influence of Hcrt on both food intake and arousal. Besides this superior positioning of Hcrt boutons on post-synaptic targets, what further amplifies the influence of Hcrt axon terminals is

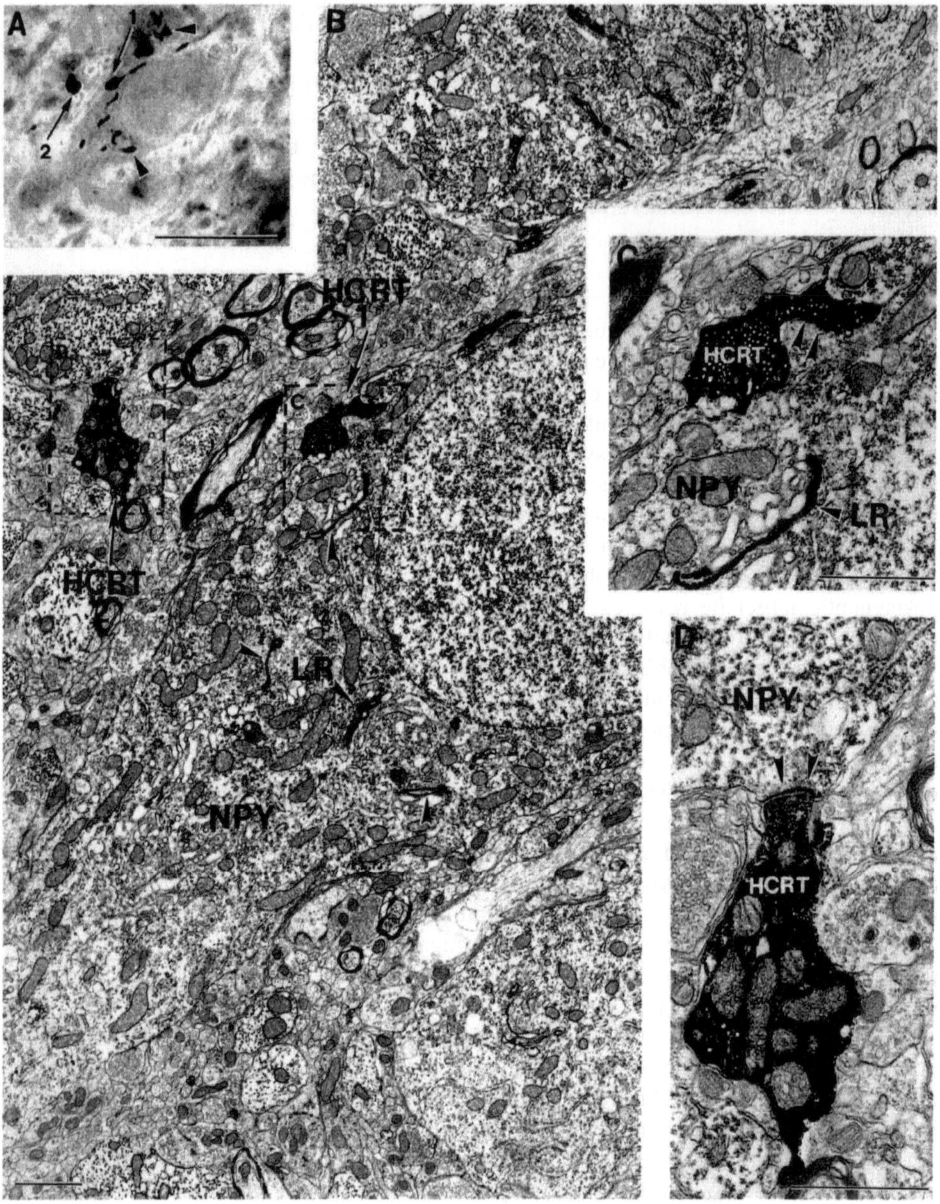

Figure 8. Hypocretin axons in synaptic contact with neuropeptide Y neurons containing leptin receptor in the monkey arcuate nucleus. A-D, Correlated light (A) and electron (B-D) micrographs demonstrating that in the monkey arcuate nucleus, two (arrows labeled 1 and 2 on A, B) black-colored hypocretin (HCRT) axon terminals are in close apposition to light stained neuropeptide Y (NPY)-containing neurons (curved arrows on A) HCRT axon 1 contacts a light NPY perikaryon that also expresses Golgi-associated black leptin receptor (LR) labeling (arrowheads on A-C). C and D are high-power magnifications of the boxed regions on B demonstrating that both HCRT boutons 1 and 2 establish asymmetric synaptic contacts (arrowheads on C, D) on the NPY- and leptin receptor-containing cell body (C) and on the NPY-containing dendrite (D). Note the abundance of mitochondria in the HCRT-containing axon terminal on D. Scale bars: A, 10 μm; B-D, 1 μm. (From 13)

their extensive interaction with other pre-synaptic boutons. These include interactions with putative excitatory as well as putative inhibitory pre-synaptic terminals,[13,24] and, they frequently exhibit synaptic-like membrane specializations between them (Fig. 6).

An important, albeit labor-intensive, task will be a detailed characterization of the qualitative and quantitative synaptology of Hcrt efferents with respect to both their post-synaptic and pre-synaptic elements. Given all of the limitations of anatomy, only through a thorough immuno-electron microscopic analysis of entire populations of Hcrt-targeted neurons will it be possible to place into perspective all of the data gathered by electrophysiological, molecular biological and physiological techniques.

5. SUMMARY

Much has been accomplished regarding the neuronal interaction between the hypocretin network and other circuits, but eve more lies ahead to be unveiled. The hypocretin circuitry has emerged as a key player in the regulation of arousal and associated autonomic, endocrine and metabolic regulation. Its unique afferent and efferent connectivity predicts that this system will further emerge as critical component of a broadening array of brain functions most of which provide necessary support for higher brain functions.

6. REFERENCES

1. L. de Lecea, T. S. Kilduff, C. Peyron, X. Gao, P. E. Foye, P. E. Danielson, C. Fukuhara, E. L. Battenberg, V. T. Gautvik, F. S. Bartlett 2nd, W. N. Frankel, A. N. van den Pol, F. E. Bloom, K. M. Gautvik and J. G. Sutcliffe, The hypocretins: hypothalamus-specific peptides with neuroexcitatory activity, Proc Natl Acad Sci U S A 95, 322-327 (1998).
2. T. Sakurai, A. Amemiya, M. Ishii, I. Matsuzaki, R. M. Chemelli, H. Tanaka, S. C. Williams, J. A. Richarson, G. P. Kozlowski, S. Wilson, J. R. Arch, R. E. Buckingham, A. C. Haynes, S. A. Carr, R. S. Annan, D. E. McNulty, W. S. Liu, J. A. Terrett, N. A. Elshourbagy, D. J. Bergsma and M. Yanagisawa, Orexins and orexin receptors: a family of hypothalamic neuropeptides and G protein-coupled receptors that regulate feeding behavior, Cell 92 573-585 (1998).
3. R. M. Chemelli, J. T. Willie, C. M. Sinton, J. K. Elmquist, T. Scammell, C. Lee, J. A. Richardson, S. C. Williams, Y. Xiong, Y. Kisanuki, T. E. Fitch, M. Nakazato, R. E. Hammer, C. B. Saper and M. Yanagisawa, Narcolepsy in orexin knockout mice: molecular genetics of sleep regulation, Cell 98, 437-451 (1999).
4. L. Lin, J. Faraco, R. Li, H. Kadotani, W. Rogers, X. Lin, X. Qiu, P. J. de Jong, S. Nishino and E. Mignot, The sleep disorder canine narcolepsy is caused by a mutation in the hypocretin (orexin) receptor 2 gene, Cell 98, 365-376 (1999).
5. C. Peyron, J. Faraco, W. Rogers, B. Ripley, S. Overeem, Y. Charnay, S. Nevsimalova, M. Aldrich, D. Reynolds, R. Albin, R. Li, M. Hungs, M. Pedrazzoli, M. Padigaru, M. Kucherlapati, J. Fan, R. Maki, G. J. Lammers, C. Bouras, R. Kucherlapati, S. Nishino and E. Mignot, A mutation in a case of early onset narcolepsy and a generalized absence of hypocretin peptides in human narcoleptic brains, Nat Med. 6, 991-997 (2000).
6. J. Hara, C. T. Beuckmann, T. Nambu, J. T. Willie, R. M. Chemelli, C. M. Sinton, F. Sugiyama, K. Yagami, K. Goto, M. Yanagisawa and T. Sakurai, Genetic ablation of orexin neurons in mice results in narcolepsy, hypophagia, and obesity, Neuron 30, 345-354 (2001).
7. J. G. Sutcliffe and L. de Lecea, The hypocretins: setting the arousal threshold, Nat Rev Neurosci. 3, 339-349 (2002).

8. A. Yamanaka, C. T. Beuckmann, J. T. Willie, J. Hara, N. Tsujino, M. Mieda, M. Tominaga, K. Yagami, F. Sugiyama, K. Goto, M. Yanagisawa and T. Sakurai, Hypothalamic orexin neurons regulate arousal according to energy balance in mice, Neuron 38, 701-713 (2003).
9. A. L. Kirchgessner and M. Liu, Orexin synthesis and response in the gut, Neuron 24, 941-951 (1999).
10. C. Peyron, D. K. Tighe, A. N. van den Pol, L. de Lecea, H. C. Heller, J. G. Sutcliffe and T. S. Kilduff, Neurons containing hypocretin (orexin) project to multiple neuronal systems, J Neurosci. 18, 9996-10015 (1998).
11. C. F. Elias, C. B. Saper, E. Maratos-Flier, N. A. Tritos, C. Lee, J. Kelly, J. B. Tatro, G. E. Hoffman, M. M. Ollmann, G. S. Barsh, T. Sakurai, M. Yanagisawa and J. K. Elmquist, Chemically defined projections linking the mediobasal hypothalamus and the lateral hypothalamic area, J Comp Neurol. 402, 442-459 (1998).
12. C. Broberger, L. de Lecea, J. G. Sutcliffe and T. Hokfelt, Hypocretin/orexin- and melanin-concentrating hormone-expressing cells form distinct populations in the rodent lateral hypothalamus: relationship to the neuropeptide Y and agouti gene-related protein systems, J Comp Neurol. 402, 460-474 (1998).
13. T. L. Horvath, S. Diano and A. N. van den Pol, Synaptic interaction between hypocretin (orexin) and neuropeptide Y cells in the rodent and primate hypothalamus: a novel circuit implicated in metabolic and endocrine regulations, J Neurosci. 19, 1072-1087 (1999).
14. C. Leranth, H. Sakamoto, N. J. MacLusky, M. Shanabrough and F. Naftolin, Estrogen responsive cells in the arcuate nucleus of the rat contain glutamic acid decarboxylase (GAD): an electron microscopic immunocytochemical study, Brain Res. 331, 376-381 (1985).
15. F. Naftolin, N. J. MacLusky, C. Z. Leranth, H. S. Sakamoto and L. M. Garcia-Segura, The cellular effects of estrogens on neuroendocrine tissues, J Steroid Biochem. 30, 195-207 (1988).
16. P. C. Kind, G. M. Kelly, H. J. Fryer, C. Blakemore and S. Hockfield S, Phospholipase C-beta1 is present in the botrysome, an intermediate compartment-like organelle, and is regulated by visual experience in cat visual cortex, J Neurosci. 17, 1471-1480 (1997).
17. A. N. van den Pol, X. B. Gao, K. Obrietan, T. S. Kilduff and A. B. Belousov, Presynaptic and postsynaptic actions and modulation of neuroendocrine neurons by a new hypothalamic peptide, hypocretin/orexins, J Neurosci. 18, 7962-7971 (1998).
18. M. Collin, M. Backberg, M. L. Ovesjo, G. Fisone, R. H. Edwards, F. Fujiyama and B. Meister, Plasma membrane and vesicular glutamate transporter mRNAs/proteins in hypothalamic neurons that regulate body weight, Eur J Neurosci. 18, 1265-1278 (2003).
19. D. L. Rosin, M. C. Weston, C. P. Sevigny, R. L. Stornetta and P. G. Guyenet PG, Hypothalamic orexin (hypocretin) neurons express vesicular glutamate transporters VGLUT1 or VGLUT2, J Comp Neurol. 465, 593-603 (2003).
20. J. L. Guan, K. Uehara, S. Lu, Q. P. Wang, H. Funahashi, T. Sakurai, M. Yanagizawa and S. Shioda, Reciprocal synaptic relationships between orexin- and melanin-concentrating hormone-containing neurons in the rat lateral hypothalamus: a novel circuit implicated in feeding regulation, Int J Obes Relat Metab Disord. 26, 1523-1532 (2002).
21. A. N. van den Pol, C. Acuna-Goycolea, K. R. Clark and P. K. Ghosh, Physiological properties of hypothalamic MCH neurons identified with selective expression of reporter gene after recombinant virus infection, Neuron 42, 635-652 (2004).
22. R. Winsky-Sommerer, A. Yamanaka, S. Diano, A. J. Roberts, T. Sakurai, E. Borok, T. S. Kilduff, T. L. Horvath and L. de Lecea, Interaction between the corticotropin-releasing factor system and hypocretins (orexins): a novel circuit mediating stress response (under review).
23. Y. Li, X. B. Gao, T. Sakurai and A. N. van den Pol, Hypocretin/Orexin excites hypocretin neurons via a local glutamate neuron-A potential mechanism for orchestrating the hypothalamic arousal system, Neuron 36, 1169-1181 (2002).
24. T. L. Horvath and X. B. Gao, Unorthodox input organization and plasticity of the hypocretin/orexin neurons: a synaptological basis for easy arousal (under review).
25. A. Yamanaka, Y. Muraki, N. Tsujino, K. Goto and T. Sakurai, Regulation of orexin neurons by the monoaminergic and cholinergic systems, Biochem Biophys Res Commun. 303, 120-129 (2003).
26. Y. Muraki, A. Yamanaka, N. Tsujino, T. S. Kilduff, K. Goto and T. Sakurai, Serotonergic regulation of the orexin/hypocretin neurons through the 5-HT1A receptor, J Neurosci. 24, 7159-7166 (2004).
27. M. Håkansson, L. de Lecea, J. G. Sutcliffe, M. Yanagisawa and B. Meister, Leptin receptor- and STAT3-immunoreactivities in hypocretin/orexin neurones of the lateral hypothalamus, J Neuroendocrinol. 11, 653-663 (1999).
28. S. Diano, B. Horvath, H. F. Urbanski, P. Sotonyi and T. L. Horvath, Fasting activates the non-human primate hypocretin (orexin) system and its postsynaptic targets, Endocrinology 144, 3774-3778 (2003).
29. S. Pinto, H. Liu, A. G. Roseberry, S. Diano, M. Shanabrough, X. Cai, J. M. Friedman and T. L. Horvath, Rapid re-wiring of arcuate nucleus feeding circuits by leptin, Science 304, 110-115 (2004).

30. J. C. Bittencourt, F. Presse, C. Arias, C. Peto, J. Vaughan, J. L. Nahon, W. Vale and P. E. Sawchenko, The melanin-concentrating hormone system of the rat brain: an immuno- and hybridization histochemical characterization, J Comp Neurol. 319, 218-245 (1992).
31. T. L. Horvath and S. Diano, The floating blueprint of hypothalamic feeding circuits. Nature Reviews Neuroscience 5, 662-667 (2004).
32. T. C. Chou, C. E. Lee, J. Lu, J. K. Elmquist, J. Hara, J. T. Willie, C. T. Beuckmann, R. M. Chemelli, T. Sakurai, M. Yanagisawa, C. B. Saper and T. E. Scammell, Orexin (hypocretin) neurons contain dynorphin, J Neurosci. 21, RC168 (2001).
33. M. L. Simmons, G. W. Terman, S. M. Gibbs and C. Chavkin, L-type calcium channels mediate dynorphin neuropeptide release from dendrites but not axons of hippocampal granule cells, Neuron 14, 1265-1272 (1995).
34. P. Trivedi, H. Yu, D. J. MacNeil, L. H. van der Ploeg and X. M. Guan, Distribution of orexin receptor mRNA in the rat brain, FEBS Lett. 438, 71-75 (1998).
35. G. J. Hervieu, J. E. Cluderay, D. C. Harrison, J. C. Roberts and R. A. Leslie, Gene expression and protein distribution of the orexin-1 receptor in the rat brain and spinal cord, Neuroscience 103, 777-797 (2001).
36. J. N. Marcus, C. J. Aschkenasi, C. E. Lee, R. M. Chemelli, C. B. Saper, M. Yanagisawa and J. K. Elmquist, Differential expression of orexin receptors 1 and 2 in the rat brain, J Comp Neurol. 435, 6-25 (2001).
37. M. Bäckberg, G. Hervieu, S. Wilson and B. Meister, Orexin receptor-1 (OX-R1) immunoreactivity in chemically identified neurons of the hypothalamus: focus on orexin targets involved in control of food and water intake, Eur J Neurosci. 15, 315-328 (2002).
38. A. N. van den Pol, Hypothalamic hypocretin (orexin): robust innervation of the spinal cord, J Neurosci. 19, 3171-3182 (1999).
39. M. R. Jain, T. L. Horvath, P. S. Kalra and S. P. Kalra, Evidence that NPY Y1 receptors are involved in stimulation of feeding by orexins (hypocretins) in sated rats, Regul Pept. 87, 19-24 (2000).
40. T. L. Horvath, C. Peyron, S. Diano, A. Ivanov, G. Aston-Jones, T. S. Kilduff and A. N. van Den Pol, Hypocretin (orexin) activation and synaptic innervation of the locus coeruleus noradrenergic system, J Comp Neurol. 415, 145-159 (1999).
41. J. J. Hagan, R. A. Leslie, S. Patel, M. L. Evans, T. A. Wattam, S. Holmes, C. D. Benham, S. G. Taylor, C. Routledge, P. Hemmati, R. P. Munton, T. E. Ashmeade, A. S. Shah, J. P. Hatcher, P. D. Hatcher, D. N. Jones, M. I. Smith, D. C. Piper, A. J. Hunter, R. A. Porter and N. Upton, Orexin A activates locus coeruleus cell firing and increases arousal in the rat, Proc Natl Acad Sci U S A. 96, 10911-6 (1999).
42. M. Wu, L. Zaborszky, T. Hajszan, A. N. van den Pol and M. Alreja, Hypocretin/orexin innervation and excitation of identified septohippocampal cholinergic neurons, J Neurosci. 24, 3527-3536 (2004).
43. L. Bayer, G. Mairet-Coello, P. Y. Risold and B. Griffond, Orexin/hypocretin neurons: chemical phenotype and possible interactions with melanin-concentrating hormone neurons, Regul Pept. 104, 33-39 (2002).

TRANSMITTER-IDENTIFIED NEURONS AND AFFERENT INNERVATION OF THE LATERAL HYPOTHALAMIC AREA

Focus on hypocretin and melanin-concentrating hormone

Christian Broberger and Tomas Hökfelt *

1. INTRODUCTION

The first insights into the functions of the lateral hypothalamic area (LHA) were the results of now classical investigations of localized lesions in the central nervous system. In a visionary paper, von Economo,[1] studying patients afflicted with encephalitis lethargica, drew the conclusion that the structures of the posterior hypothalamus (including the LHA) were essential for maintenance of the awake state. In support of this conclusion, Ranson[2] described how coma could result from lateral hypothalamic ablations in monkeys. In collaboration with Hetherington,[3] he then turned to the study of the neuroanatomical loci involved in regulating feeding behaviour, demonstrating that lesions placed in the medial hypothalamus result in pronounced over-eating and adiposity. In the same report, however, the authors noted in passing that destruction of more lateral portions of the hypothalamus caused the opposite effect. The classical papers by Anand and Brobeck[4, 5] (see also ref 6) followed up on this finding and described an often-fatal hypophagia and hypodipsia in response to bilateral electrolytic lesions of the LHA. These ablation studies revealed a prominent role for the LHA in both arousal and energy balance, positioning the region firmly within the neuronal circuitry controlling motivated behaviour.[7] However, the interconnectivity of the LHA to other brain regions within the circuitry and, importantly, the neurochemical coding underlying communication within the circuits remained unknown for a long time. While differences in cytoarchitectonics and connectivity (see refs 8,9) had allowed for some partitioning of the LHA into subdivisions, defining features that could be correlated to functional data remained elusive. The discovery of the melanin-concentrating hormone (MCH) and hypocretin (Hcrt) neuropeptide families, and the subsequent demonstrations that these

* C. Broberger and T. Hökfelt, Department of Neuroscience, Karolinska Institutet 171 77 Stockholm, Sweden

signalling molecules define two separate broadly projecting ascending systems, provided crucial tools for a new phase of pharmacological and histochemical investigation of the LHA. The present review summarizes what we know today of coexistence patterns of LHA neurons. Aspects of afferent input to the LHA, based on early tracing studies and with particular emphasis on subsequent histochemical identification of these pathways, will then be discussed. Functional characteristics of the Hcrt and MCH peptides are mentioned only briefly. For detailed discourses on this subject the reader is referred to in-depth reviews elsewhere in this volume.

2. MELANIN-CONCENTRATING HORMONE AND HYPOCRETIN DEFINE TWO SEPARATE NEURONAL POPULATIONS IN THE LHA

From comprehensive tracing studies by Saper[10] and others (for review see ref 8) it was evident that the LHA houses neurons that project globally over the brain similar to the classical cholinergic and monoaminergic ascending systems of the basal forebrain and brainstem. The discovery of MCH[11,12] (see ref 13) and the demonstration that expression of this peptide defines a large neuronal population in the LHA/zona incerta complex[14-15] first provided neurochemical identity to such a system. Indeed, MCH-immunoreactive (-ir) terminals were localized to a wealth of structures along the entirety of the neuraxis including the cerebral cortex as well as numerous key subcortical nuclei, although rather sparsely within the hypothalamus itself.[14,16] In contrast, cell bodies containing MCH-like immunoreactivity (-LI) and mRNA are exclusively found in the LHA, with the exception of two smaller accumulations of perikarya in the olfactory tubercle and pontine tegmentum.[16,1] The impressive dimensions of the MCH system strongly suggested its involvement in processes beyond the role in pigmentation inherent in the peptide name.[16]

[1] It soon became evident that the MCH-expressing neurons had actually been immunohistochemically identified earlier. In mapping studies using antisera generated against the pro-opiomelanocortin (POMC)-derived peptide α-melanocyte-stimulating hormone (α-MSH), immunoreactive perikarya had been observed in the LHA and zona incerta in addition to the cell bodies in the hypothalamic arcuate nucleus and the brainstem previously recognized to express POMC (e.g. 17. C. Köhler, L. Haglund and L. W. Swanson, A diffuse alpha MSH-immunoreactive projection to the hippocampus and spinal cord from individual neurons in the lateral hypothalamic area and zona incerta, *J Comp Neurol.* **223**, 501-14 (1984).· 18. N. Naito, I. Kawazoe, Y. Nakai, H. Kawauchi and T. Hirano, Coexistence of immunoreactivity for melanin-concentrating hormone and alpha-melanocyte-stimulating hormone in the hypothalamus of the rat, *Neurosci Lett.* **70**, 81-5 (1986).; C. Broberger, unpublished observations). Similar populations had been described with antisera against corticotropin-releasing hormone (CRH; e.g. 19. M. Kawata, K. Hashimoto, J. Takahara and Y. Sano, Immunohistochemical demonstration of the localization of corticotropin releasing factor-containing neurons in the hypothalamus of mammals including primates, *Anat Embryol (Berl).* **165**, 303-13 (1982).) and human growth hormone-releasing hormone (GRH; e.g. 20. I. Merchenthaler, S. Vigh, A. V. Schally and P. Petrusz, Immunocytochemical localization of growth hormone-releasing factor in the rat hypothalamus, *Endocrinology.* **114**, 1082-5 (1984).· 21. P. Y. Risold, D. Fellmann, D. Lenys and C. Bugnon, Coexistence of acetylcholinesterase-, human growth hormone-releasing factor(1-37)-, alpha-melanotropin- and melanin-concentrating hormone-like immunoreactivities in neurons of the rat hypothalamus: a light and electron microscope study, *Neurosci Lett.* **100**, 23-8 (1989).). This issue was later resolved with the demonstration that a portion of the prepro-MCH precursor, encoding the neuropeptide NEI, is recognized by several, if not most, α-MSH antisera as well as some CRH and GRH antisera22. J. L. Nahon, F. Presse, J. C. Bittencourt, P. E. Sawchenko and W. Vale, The rat melanin-concentrating hormone messenger ribonucleic acid encodes multiple putative neuropeptides coexpressed in the dorsolateral hypothalamus, *Endocrinology.* **125**, 2056-65 (1989). (see also 23. C. B. Saper, H. Akil and S. J. Watson, Lateral hypothalamic innervation of the cerebral cortex: immunoreactive staining for a peptide resembling but immunochemically distinct from pituitary/arcuate alpha-melanocyte stimulating hormone, *Brain Res Bull.* **16**, 107-20 (1986).).

In line with the lesion studies summarized above, behavioural studies also indicated a role in the regulation of energy balance. When administered into the brain, MCH peptides increased food intake,[24,25] providing an early candidate mediator of the proposed LHA feeding drive. However, limited pharmacological tools and the absence of any identified cotransmitter for MCH hampered further research into its functions.

In 1996, Sutcliffe and collaborators[26] took advantage of subtractive hybridisation to identify mRNA species enriched in the rat hypothalamus. Among these findings was a transcript tentatively named "clone 35" which showed a striking hybridisation pattern in the LHA, but which could not be detected outside of the hypothalamus. Continued efforts to characterize this clone led to the identification of the hypocretin family (Hcrt1 and 2) of peptides in 1998.[27] Independently, Yanagisawa and colleagues described two neuropeptide products deriving from one gene, which they named orexin A and B;[28] two orexin/hypocretin receptors were also described in the same paper. It soon became clear that Hcrt 1 and 2 are virtually identical to orexin A and B, respectively. Sakurai et al.'s study[28] further demonstrated that intracerebroventricular (icv) administration of either Hcrt1 or 2 increased light phase food consumption within an hour, and that preproHcrt mRNA levels increase following food deprivation. (Subsequently, a major role has been established for Hcrt peptides as essential promoters of wakefulness; see e.g.[29-30] The reader is referred to other chapters in this volume for detailed discussions of these results.)

The similarity in distribution of both cell bodies and projections[16,27] as well as in feeding effects raised the question of whether Hcrt and MCH are actually produced by the same group of neurons. However, in three independent reports,[31-32] (Figs. 1a-c, 2) which followed soon after the original characterizations, this was demonstrated not to be the case. The mRNAs for preproMCH and preproHcrt were detected (by double-label in situ hybridisation) in separate groups of cells, although Elias *et al.*[33] reported a few instances of overlap, representing less than 1% of either population (Fig. 1a-c). Rather than forming a completely intermingled mosaic, the Hcrt- and MCH neurons also, upon closer inspection, displayed individual topographies. The zona incerta e.g. is the exclusive domain of MCH cells. Both populations occupy the perifornical region, but are differentially located; Hcrt neurons cluster rostrally in the dorsal portion, whereas MHC neurons occupy a ventral, caudal position.[31, 33] From the perifornical region, the Hcrt population extends laterally towards the ventromedial tip of the cerebral peduncle, where it forms a distinct grouping centred in the magnocellular lateral hypothalamic nucleus. More caudally, the outer edges of the cerebral peduncle within the LHA are outlined by MCH neurons.[31,33] The dorsomedial nucleus and periventricular zone house scattered, smaller MCH neurons as well as Hcrt-ir perikarya.[16,31] Morphologically, Hcrt- and MCH-expressing neurons display no obvious differences in features.[32] Hcrt-ir cell bodies are comparatively large (20-30 μm) and multipolar or fusiform in shape, extending 2-4 primary dendrites.[27,32,34](Fig. 3d) Similar distribution patterns as well as the separation of MCH and Hcrt neurons into distinct populations is seen in the human brain.[33]

3. COEXISTENCE WITH OTHER SIGNALLING MOLECULES

Thus, with the discovery of MCH and Hcrt, two major populations of LHA projection neurons could be distinguished. There is now accumulating information on

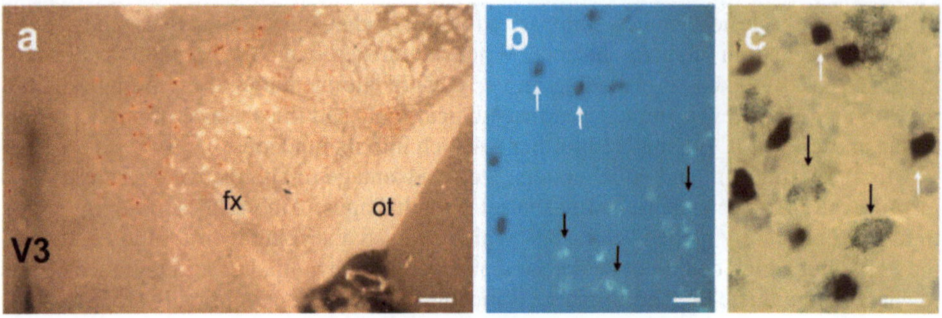

Figure 1. MCH- and Hcrt-expressing cells form separate populations in the LHA. Colour micrographs taken from sections of the rat lateral hypothalamus after double-label in situ hybridisation for MCH and Hcrt mRNAs. In a), an overview viewed in bright-field illumination, showing major anatomical landmarks. MCH mRNA-containing neurons are detected by the presence of a dark precipitate and are seen along the zona incerta extending into the LHA, with a smaller condensation of cells observed between the optic tract and the cerebral peduncle. Hcrt mRNA is detected as silver grains, which are seen clustered over cells stretching dorsally from the fornix to the medial tip of the cerebral peduncle. The absence of colocalisation of the two mRNAs (MCH indicated by white arrows; Hcrt indicated by black arrows) can be seen in dark-field illumination with epi-illumination (b), and in bright-field illumination at high magnification (c). fx, fornix; ot, optic tract; V3, third ventricle. Scale bar indicates 200 μm in (a), 50 μm in (b) and 25 μm in (c).

how other transmitters and neuromodulators within the LHA distribute in relationship to these two cell groups (see also ref 35). In this context, it may be noted that the expression of MCH and Hcrt is highly robust. In our own investigations, we have been struck by the fact that these peptides, unlike most others, can be visualized with ease in cell bodies without pre-treatment of the animal with the axonal transport inhibitor, colchicine. The mRNA levels are similarly high as evidenced e.g. by strikingly short exposure times in in situ hybridisation experiments.[31]

3.1. Cocaine- and Amphetamine-Regulated Transcript

Cocaine- and amphetamine-regulated transcript (CART) encodes several putative peptide fragments, some of which decrease feeding upon central administration.[36,37] (see ref 38). CART is expressed in numerous brain nuclei, and the actual site of its anorexigenic actions has not yet been fully determined. In the rat LHA, CART is detected in the lion's share of MCH neurons of the zona incerta and medial LHA, whereas most peripeduncular MCH cells do not contain CART.[39-40] In contrast, no Hcrt neurons stain for CART.[39-40] Notably, according to a preliminary report, some degree of Hcrt/CART coexistence can be seen in the human brain.[41] This would not be the first example of differential distribution of CART in primate and rodent brain; the transcript displays prominent discrepancies in expression e.g. between the human and rat thalamus.[42] Coexpression of orexigenic and anorexigenic neuromodulators within the same cell presents an intriguing paradox, which remains to be explained. Further elucidation of the pre- and postsynaptic actions of CART peptides may shed light on this issue; see ref 38.

3.2. Dynorphin

A near-total overlap is seen between Hcrt and the opioid neuropeptide, dynorphin, in LHA cell bodies.[43, 44] It will be interesting to learn more of the functional implications of such coexistence, since dynorphin, like Hcrt, stimulates feeding.[45] Notably, dynorphin and Hcrt act via similar pathways to presynaptically modulate GABA release.[46] Hcrt cotransmitters such as dynorphin may also play an important role in the regulation of arousal, since the narcoleptic phenotype of transgenic mice lacking Hcrt neurons recapitulates more closely the disease in humans than in animals with selective disruption of the Hcrt gene alone.[30]

3.3. Substance P

Substance P (SP)-LI has been detected in a small proportion (2.7%) of MCH cells.[47] However, this phenomenon has been observed only following administration of colchicine. In addition to its action as an inhibitor of axonal transport, colchicine is also known to affect the expression levels of some neuropeptide transcripts,[48] so this result may merit some caution. The possibility of SP/Hcrt coexistence has not been explored to date. However, a series of studies have been published on projections containing SP or the related peptide neurokinin (NK) B to the LHA; see ref 13. Such projections originate in a number of hypothalamic (including the preoptic and posterior nuclei as well as putative LHA interneurons) and extrahypothalamic (e.g. substantia innominata and periacqueductal gray area) regions.[47] It is not known which of these projections terminate specifically on Hcrt and MCH neurons. However, a majority of MCH neurons express the NK3 receptor[49] and these cells project heavily to the cerebral cortex (but only to a small extent to the spinal cord).[50] Intriguingly, there is evidence that this population is identical to the one coexpressing MCH/CART, suggesting an attractive convergence of histochemical and neuroanatomical specificity.[13] Activation of NK3 receptors results in increased transcription of MCH,[47] indicating that the SP/NKB projections may control cortical MCH tone. The NK1 receptor is also found in the LHA but does not coexist with MCH[47], opening up the possibility that it may be expressed together with Hcrt.

3.4. Galanin

The LHA harbours a population of neurons expressing the neuropeptide, galanin.[51, 52] These cells are largely void of Hcrt-LI, except for a few larger-sized cell bodies in the ventral perifornical region and the dorsolateral LHA.[53]

3.5. Neurotensin

A population of cells in LHA also expresses the neuropeptide, neurotensin.[54] It is not known if these cells express either Hcrt or MCH. It may be noted, however, that there is no coexpression of neurotensin and CART in the LHA,[40] which, by inference from the results summarized above, would exclude also any larger degree of neurotensin-MCH coexistence.

3.6. Other neuropeptides.

Virtually all Hcrt neurons stain for the neuropeptide precursor secretogranin II.[55, 56] The functional implications of this relationship are not known. - Thyrotropin-releasing hormone (TRH) is expressed in several hypothalamic nuclei, including cells in the LHA.[57] The possibility of TRH/Hcrt or TRH/MCH coexistence has not been explored. However, it is known that TRH mRNA and CART distribute in separate cell groups in this region, with TRH cells occupying a more ventral position.[39] Somatostatin is another neuropeptide which features prominently in several hypothalamic locations, including the LHA, where apparently separate populations are found in the zona incerta and surrounding the medial forebrain bundle.[58] Colocalisation studies involving somatostatin and Hcrt or MCH are not yet available. - Prior to the discovery of hypocretin peptides, a similar perifornical cell group had been identified as staining positive with an antiserum directed against prolactin.[59] It is now known that this staining likely represents crossreactivity with the Hcrt precursor protein,[60] and that prolactin is not expressed by Hcrt cells.

3.7. Nitric Oxide

A population of LHA neurons signal via the gaseous neurotransmitter nitric oxide (NO) as suggested by their expression of the enzyme NO synthase.[61] (NOS) These neurons do not contain Hcrt.[62] Similarly, other investigators found that a population of NOS-ir cells in the perifornical region, extending ventrally towards the ventromedial hypothalamic nucleus, and which display neuropeptide Y Y1 receptor-LI, is distinct from both Hcrt- and MCH neurons.[63]

3.8. Amino Acid Transmitters

A central question regards the identity of any classical, low molecular weight transmitters that may be costored with Hcrt and MCH. In general, the expression of neuropeptides in a neuron is accompanied by at least one classical transmitter.[64] It would appear likely that the LHA neurons, similar to what has been shown for most of the central nervous system, signal either via the excitatory amino acid glutamate (alternatively, aspartate) or via the inhibitory amino acid γ-amino-butyric acid (GABA; alternatively, glycine). Notably, in the medial hypothalamus, when signalling via these transmitters is blocked, fast, miniature, and evoked synaptic activity is almost completely abolished, in accordance with histochemical data.[65,66] Accumulating evidence suggests that Hcrt neurons contain an excitatory amino acid transmitter. Hcrt colocalizes with glutamate-LI[67] as well as immunoreactivity for excitatory amino acid transporter 3 and, to a smaller extent, vesicular glutamate transporters 1 and 2, membrane proteins responsible for glutamate reuptake.[68,74] In contrast, staining for glutamic acid decarboxylase-67 (GAD-67), a key enzyme in GABA biosynthesis, is completely absent from Hcrt neurons.[69] Lending further support to an excitatory phenotype, Hcrt-ir terminals form asymmetrical synaptic contacts both in rat[27] and cat.[82] Thus, it appears likely that Hcrt (in line with its excitatory electrophysiological actions[27]) coexists with glutamate. The identity of a cotransmitter for MCH is more controversial. Similar to the Hcrt neurons, both glutamate and glutamate transporters have been immunohistochemically detected in MCH cells.[67,68] However, the presence of GAD-67 mRNA in a large proportion of CART neurons in the LHA and zona incerta has also been reported.[40] Since CART coexists with MCH in these regions (see above), it would appear that at least a portion of MCH neurons have an inhibitory phenotype. Such a population of neurons could correspond to a previously described GABAergic projection from the zona incerta to the cerebral cortex.[70] Further research is needed to determine to what extent, if any, MCH neurons constitute a mixed population of inhibitory and excitatory cells.

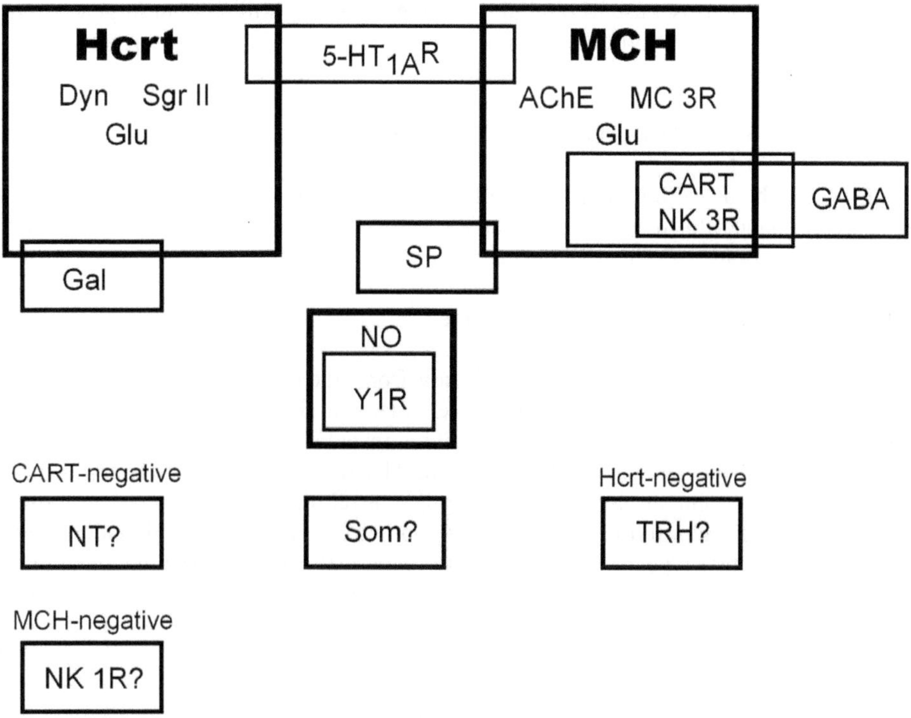

Figure 2. Schematic depiction of colocalization patterns of signalling molecules in the LHA. Hypocretin (Hcrt) and melanin-concentrating hormone (MCH)-expressing neurons form two mutually exclusive cell groups. A near total colocalisation has been reported between Hcrt and dynorphin (Dyn) and secretogranin II (Sgr II). These cells are also believed to be glutamatergic. A few Hcrt neurons contain galanin (Gal). MCH neurons, on the other hand, contain cholinergic receptors (MC 3R) as well as acetylcholine esterase (AChE). These cells also label for markers of a glutamatergic phenotype. A large subpopulation of these cells express cocaine- and amphetamine-regulated transcript (CART) and the neurokinin 3 receptors (NK 3R). Of these, a considerable proportion also contains GABAergic markers. A minor degree of MCH/Substance P (SP) colocalisation has been reported. Both Hcrt and MCH overlap with the serotonin 5-HT$_{1A}$ receptor. There may also be additional overlap between Hcrt and MCH and other neuromodulatory molecules. The possibilities of Hcrt coexisting with neurotensin (NT) or the neurokinin 1 receptor (NK 1R), of MCH coexisting with thyrotropin-releasing hormone (TRH), or of somatostatin (Som) coexisting with either Hcrt and/or MCH, remain to be investigated (indicated by "?"). Neurons that signal via nitric oxide (NO) and express the neuropeptide Y Y1 receptor form a population distinct from both Hcrt and MCH cells. See text for details and references. Box sizes are not scaled to sizes of the cell populations.

Thus, using the various transmitter combinations, several subpopulations of LHA neurons can now be distinguished (see Fig. 2 for schematic summary). Future investigations will likely add to the complexity of neurochemical phenotypes in this region, and it will be of interest to correlate the transmitter content to projection patterns. It should, however, be cautioned that the total number of cells that make up the LHA has been estimated to ca. 250,000.[71] This figure may be compared with available estimates of the Hcrt population (slightly above 1000)[32] and the MCH neurons (circa 90,000).[72] Needless to say, calculations of absolute neuron numbers have their inherent limitations. Yet, these figures serve as a useful reminder that the LHA may still harbour large clusters of cells whose transmitter and circuitry phenotype remains unaccounted for, and which may shed further light on the phenotype resulting from lateral hypothalamic ablations.

4. LOCAL CONNECTIONS WITHIN THE LHA

It may now be asked how neurons within the LHA interact with each other. Do the Hcrt neurons connect to themselves to provide autoregulatory feedback? Do they form reciprocal connections with the MCH neurons?

4.1. Hcrt-MCH: Anatomical Interactions

Intra-LHA connections were evident already from tracing studies; notably, a small degree of innervation contralateral to the tracer injection site could even be noted.[73,74] Accordingly, mappings of MCH- and Hcrt-ir fibres revealed dense local plexuses within the LHA.[16,32,75] Cell bodies containing MCH- and Hcrt-LI are frequently observed right next to each other, and at the light microscopic level the two cell types are seen contributing boutons apposed to the soma of the other.[56,76] Ultrastructural studies have confirmed that the two types of neurons supply reciprocal synaptic specializations onto each other.[34] However, the Hcrt-ir terminals, which form asymmetrical (i.e. excitatory) contacts onto MCH-ir somata and dendrites, appear comparatively more extensive than MCH-ir terminals onto Hcrt neurons.[34] (see also ref 77)

4.2. Distribution of Hcrt and MCH Receptors in the LHA

Initial reports on the expression patterns of Hcrt receptors reported no HcrtR1 mRNA in LHA, and only little HcrtR2 mRNA.[78] A subsequent detailed study[79] found a similar relationship, albeit these investigators, using highly sensitive technique, detected higher mRNA levels overall, suggesting some degree of HcrtR1 expression also in the LHA. When receptor-specific antisera became available, Hcrt1R-ir neurons were described in the perifornical region, and intense perikaryal staining was also found in the zona incerta.[80] Double-staining confirmed that HcrtR1-LI coexists with both Hcrt- and MCH-LI in the LHA.[81] The transmitter identity of HcrtR2-expressing neurons has not yet been reported. The two cloned MCH receptors are also expressed in the LHA. In the primate, MCHR2 appears to predominate,[82] but some expression is seen also of MCHR1 (also known as SLC-1), at least in the rat.[83,84] Again, the neurochemical phenotype of LHA MCHR-expressing cells remains to be described.

4.3. Hcrt-MCH: Functional Interactions

Application of Hcrt results in an excitation of MCH neurons in slice preparations (where these neurons normally exhibit little spontaneous activity), depolarising membrane potential and increasing discharge frequency.[76] Injection of Hcrt locally into the LHA also results in neuronal activation, determined as induction of expression of the immediate early gene c-Fos.[85] Furthermore, adding Hcrt to hypothalamic cultures of pre-adult rats increases MCH expression in acute isolates but has the opposite effect in ten day-old cultures.[56] These results suggest that in addition to more immediate effects on the electrical properties of the cell, local interactions within the LHA also involve long-term transcriptional changes. All in all, the predominant role of Hcrt in the regulation of MCH activity appears to be stimulatory. In contrast, MCH exerts inhibitory effects on LHA neurons.[86,87] While this appears to be a means for MCH neurons to regulate their own activity via inhibitory feedback,[88] it has not yet been functionally demonstrated that MCH also can inhibit the activity specifically of Hcrt-expressing cells. As we learn more of the physiological role of the Hcrt and, in particular, MCH systems in sleep and arousal, the implications of these local interactions between the two LHA populations will likely become clearer.

4.4. Hcrt Autoregulation

Local administration of Hcrt into the LHA itself results in increased feeding.[89] In addition to what has been described above concerning Hcrt-MCH interactions, this result may partly be explained by a positive feedback loop driving Hcrt neurons. This possibility is supported by the observation that at the light microscope level, a dense network of Hcrt-ir terminals can be seen surrounding, and often contacting, Hcrt-ir cell bodies in the LHA.[90,91,92] have recently described a local circuit providing feedback excitation of the Hcrt population, which may, however, involve an intermediary glutamatergic interneurons rather than direct autoinnervation. Specifically, it was shown that the depolarisation and increased firing rate observed in Hcrt neurons following application of Hcrt1 or 2 requires intact synaptic activity as well as intact ionotropic glutamatergic transmission. Furthermore, glutamatergic terminals were also shown to form close appositions suggestive of synaptic specializations onto Hcrt-expressing perikarya and dendrites. Since the evidence suggests that the Hcrt neurons are glutamatergic, a likely explanation for these findings is that Hcrt presynaptically facilitates release of glutamate onto the Hcrt neurons themselves. Indeed, as mentioned, Hcrt1R-LI has been observed in Hcrt neurons.[81] Yet, it cannot be excluded that these effects are mediated via local glutamatergic interneurons. The concept of interneurons modulating neurotransmission is well established in e.g. neocortical circuits but has been only scantily investigated in hypothalamic nuclei. Such putative interneurons also remain to be identified histochemically. Regardless of the mechanism, feedback excitation provides a powerful mechanism for amplifying the Hcrt tone. This mechanism may be of importance in stabilizing the behavioural state towards sleep or arousal by serving as a "flip-flop" device, which prevents the animal from drifting into intermediate states.[93]

5. AFFERENT INNERVATION OF THE LHA

The history of the elucidation of innervation of the LHA may be divided into two phases: before and after the discoveries of MCH and Hcrt. Comprehensive reviews of the characteristics of the LHA in the pre-Hcrt era are available and recommended.[8,9] The following overview focuses on hypothalamic afferent innervation of the LHA, in particular from the arcuate nucleus. Select extra-hypothalamic regions projecting to the

LHA are then described, with emphasis on projections likely involved in arousal. Innervation specifically of Hcrt and also of MCH neurons will be in focus. However, this section is not intended as an exhaustive catalogue documenting the full range of neuroanatomical inputs to the LHA. Most notably, several extra-hypothalamic projections to the LHA from e.g. the septum and amygdala[73,74] will not be covered.

When discussing pathways entering or leaving the LHA it should be kept in mind that through this region courses the medial forebrain bundle (MFB). The MFB comprises the ascending and descending fibres of more than 50 different cell groups, in addition to the LHA.[94,95] The intimate relationship between the LHA and MFB is exemplified by the fact that the extensively branched dendrites extending from LHA neurons infiltrate the bundle.[96,71] This relationship poses a technical challenge. Studying the afferent innervation of the LHA by local injection of retrograde tracers merits cautious interpretation since such substances to varying extent may also be taken up by the axons forming the bundle; see refs 97-98. Indeed, the "lateral hypothalamic syndrome" following LHA lesions came into question, when it was shown that localized destruction of the nigrostriatal dopamine pathway, which travels via the MFB, could recapitulate much of the symptoms.[99] It was not until the demonstration that excitotoxic lesions of the LHA, which eliminate cell bodies but spare fibres of passage, also result in hypophagia[6] that the concept of a feeding center in the LHA again came into favour; see ref 100. Such data benefit from confirmation via alternative approaches, e.g. injection of anterograde tracer substances in regions suspected to provide LHA innervation, or the demonstration of evoked electrical activity along a pathway.

6. INPUTS FROM THE ARCUATE NUCLEUS

6.1. Role in Feeding Behaviour

The first anatomically identified inputs to the Hcrt neurons were two projections from the arcuate nucleus, a bilaterally symmetric structure bordering the base of the third ventricle just dorsal to the median eminence; see refs 101-102. The arcuate nucleus is of particular interest with regard to energy metabolism, since it provides an entrance point for numerous anabolic and catabolic hormones into the brain, including leptin and insulin; see refs 103-104. Indeed, in the pioneering studies of hypothalamic lesions, it was noted that the arcuate nucleus was included in the ventromedial complex whose removal resulted in obesity[3]. Neurotoxic lesions, which more specifically target the arcuate nucleus result in the same phenotype.[105] The humoral metabolic information is transmitted further into the brain from the arcuate nucleus via two separate, but parallel, populations of ascending projection neurons; see ref 103. One group expresses the potently orexigenic neuropeptide Y[106-107] (NPY), the other POMC, which is cleaved into several melanocortin peptides, including α-MSH, which inhibit food intake.[108,109] These neurons are differentially regulated by the anorexigenic fat tissue-derived hormone leptin[110-111] and exert opposite postsynaptic effects, since NPY coexists with GABA[112, 113] (i.e. inhibitory neurons), whereas POMC neurons are likely glutamatergic[113,68] (i.e. excitatory).

6.2. Parallel Pathways from the Arcuate Nucleus to the LHA

By the late 1990's, it was thus evident that the arcuate nucleus plays a central role in the initiation of the body's response to starvation. Still, it was not known which brain regions lie downstream in the circuitry. Only the dorsomedial and paraventricular hypothalamic nucleus had been anatomically identified as direct targets of the arcuate nucleus NPY neurons.[114] Injection of NPY into these structures had also been reported to elicit feeding responses. However, the most potent stimulation of food intake behaviour was seen with highly localized injections into the perifornical area.[107] In addition, one often overlooked aspect of the hypothalamic lesion studies is the observation that the hyperphagic phenotype observed following ablations involving the arcuate nucleus disappears with subsequent destruction of the LHA,[4,5] indicating a hierarchical organization with the LHA operating downstream of the ventromedial complex. These results, in combination with the identification of two orexigenic peptides in separate LHA populations, suggested the existence of a direct arcuate-lateral hypothalamic pathway.

Histochemical evidence for such a projection was presented concurrently in two independent studies which demonstrated that NPY- as well as α-MSH-ir terminals cluster in dense plexuses in the rat LHA where they form close appositions onto both Hcrt and MCH neurons,[31, 33] see also refs 100, 91 (Fig. 3 a-d).

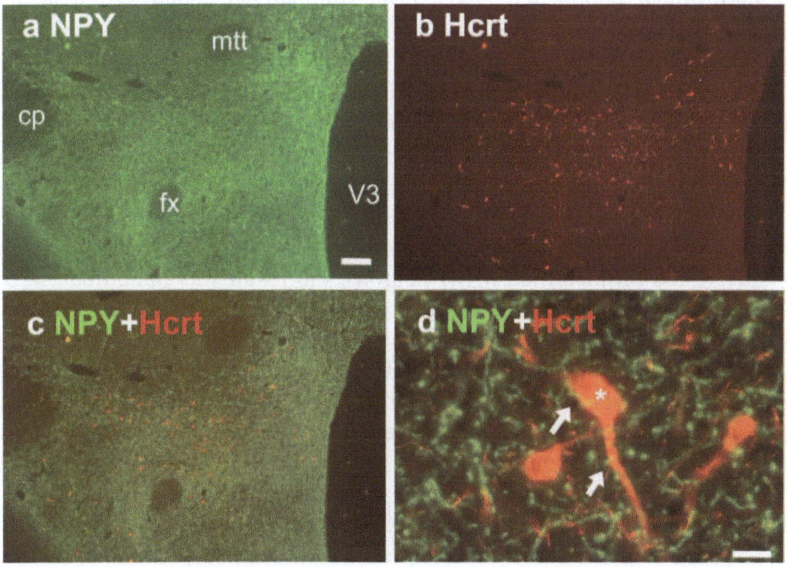

Figure 3. Hcrt neurons are contacted by NPY terminals. Fluorescence micrographs of sections of the rat lateral hypothalamus after staining with antisera against NPY (in green; a) and Hcrt (in red; b). (a) Note dense plexus of NPY-ir terminals extending over much of the hypothalamus, including the posterior, dorsomedial and lateral regions. (b) Hcrt-ir cell bodies are seen in a wide distribution over the LHA, in particular surrounding and dorsal to the fornix. Micrograph in (c) shows merged version of (a) and (b), and illustrates partially overlapping distribution patterns for NPY terminals and Hcrt cell bodies. In higher magnification (d), Hcrt cell bodies (asterisk) can be seen decorated with close appositions of NPY-ir terminals (arrows) along cell soma and primary dendrites. cp, cerebral peduncle; fx, fornix; mtt, mammillothalamic tract, V3, third ventricle. Scale bar in (a) indicates 200 μm for (a-c), and 20 μm in (d).

Elias *et al.*[33] could also show that similar patterns of innervation exist in human LHA. That the NPYergic terminals indeed derived from the arcuate nucleus and not from e.g. the brainstem, which also supplies the hypothalamus with ample NPY innervation,[115] was demonstrated by the presence of agouti-gene related peptide (AGRP) in the same terminals[31].

AGRP is exclusively expressed by NPY neurons in the arcuate nucleus and thus provides a selective marker for these cells.[31,116,139] It was also noted that the NPY and α-MSH innervation is not uniformly distributed.[35,33] The densest accumulation of terminals e.g. surrounds the fornix (similar to where behavioural studies locate the strongest feeding response to NPY).[107] The number of terminal appositions also appeared denser on Hcrt than on MCH cells (correlating to the density of local NPY-ir plexuses), while the MCH neurons in the periventricular zone were spared of any close relationships with NPY terminals.

Adding to the complexity of this hypothalamic connection, a reciprocal innervation from the Hcrt neurons to the arcuate nucleus has been found. Hcrt-ir terminals form synaptic specializations on both NPY[91] and POMC neurons.[117] Functional data suggest that intact signalling in either direction, LHA-arcuate nucleus-LHA, is important for proper food intake regulation. The prominent orexigenic effects of administration of NPY into the perifornical region, e.g., have been summarized above. Niimi et al.[118] also demonstrated that when animals are administered antiserum directed against Hcrt1, the feeding response to icv NPY is attenuated. Contrariwise, pharmacological blockade of the NPY Y1 receptor reduces the feeding stimulatory effect seen after icv administration of Hcrt 1.[119,120] There is also some question as to how the arcuate nucleus input shapes neural activity in the LHA. While the idea of orexigenic NPY stimulating LHA neurons to release the likewise orexigenic peptides Hcrt and MCH is appealing, this concept, as has been noted,[31] is not completely consistent with available pharmacological data. To date, all known NPY receptors elicit inhibitory responses (e.g. [121-122]). Indeed, when NPY was recently tested on identified MCH neurons, the result was inhibition via both pre- and postsynaptic mechanisms.[76] (The electrophysiological response of Hcrt neurons to NPY has not yet been reported.) Morphological data (reviewed above) suggesting that GABA is a cotransmitter for NPY and AGRP further support the inhibitory phenotype of these neurons. With regard to the POMCergic innervation of the LHA, no changes in electrophysiological properties of MCH neurons could be detected in response to a melanocortin agonis.t[76] It should, however, be kept in mind that neuropeptide receptors can mediate slower, transcriptional effects, in addition to more immediate changes in electrical properties. Thus, both NPY and AGRP increase the expression of c-Fos in Hcrt neurons,[118,123] although there are interesting examples where expression of NPY and Hcrt mRNAs are differentially regulated.[124] Furthermore, MCH mRNA levels increase following blockade of melanocortin receptors in vivo.[125] The physiological outcome of release of NPY/AGRP and melanocortin peptides as well as their costored classical transmitters onto LHA neurons is likely a composite of fast and slow events.

6.3. Functional Role Within the Feeding Circuitry

The identification of the arcuate-lateral hypothalamic pathway provides the metabolic signals from the periphery with a cortical channel and a neuroanatomical means of engaging cognitive processes in food intake behaviour. This information complements the previously dominant view that the paraventricular nucleus formed the major target for the arcuate nucleus, which focused on the neuroendocrine and autonomic response to starvation. However, the discovery of AGRP as a selective marker for arcuate nucleus NPY neurons also made it possible to map out the full extent of the projections emanating from these cells. By this approach it was demonstrated not only that the NPY and POMC neurons form almost completely parallel ascending pathways, but also that these extend well beyond the hypothalamus and include a wealth of structures, from the olfactory nuclei to the nucleus tractus solitarii.[126] In this manner, the metabolic signals are in a position to directly engage numerous specialized assemblies of neurons to participate in the decision to feed or not to feed. Such a concept of distributed systems has been invoked in systems-oriented models of the regulation of feeding behaviour.[127, 103,128,104] The circuitry can be compared to the basal ganglia, where incoming information has been proposed to disseminate to ensembles of "local experts"[129] before converging back to a common output signal.[130] In the feeding circuitry, the LHA could constitute one such "local expert", weighing in information of the arousal state or reward/hedonic qualities[131] and serving as a conduit for recruiting relevant parts of the cortex; see ref 103. Other such sites include the parabrachial complex, which could add e.g. gustatory information,[132] and the amygdala, which could contribute sensory cortical information pertaining to the food itself and the surrounding environment.[133] Interestingly, neural information from the gastrointestinal tract via the nucleus tractus solitarii feeds into largely the same nuclei as those targeted by the arcuate nucleus, including the LHA.[134-135] This relationship indicates that integration of humoral/metabolic and neural/visceral signals occurs at multiple sites, rather than collecting at a single bottleneck; see ref 103. However, the extensive reciprocal connections within this circuitry (between e.g. the LHA and the arcuate nucleus or the cerebral cortex) also suggest the existence of multiple neuronal loops, rather than a neat flow chart where information follows a straight path from sensory input into the CNS to motor output. (It should, however, be remembered that a true loop would require a signal to return to its point of origin, i.e. preferentially the very same neurons. The emerging pattern of subdivisions in the LHA may imply that this is not always the case.) Possibly, the 'feeding trace' reverberates within these loops before accumulating to a sufficient motor impetus. To summarize, while the neuroanatomy involved in integration of energy metabolism is becoming clearer, the true nature of integration within these circuits remains obscure.

7. OTHER INPUTS FROM THE HYPOTHALAMUS

7.1. The Suprachiasmatic Nucleus

The suprachiasmatic nucleus (SCh) is the major biological clock of the brain and provides an entrance for visual afferent signals to the circadian circuitry of the brain.[136, 137]; see ref 138 A direct projection from the SCh to both the Hcrt and MCH cells has been described in both rat and human brain;[67] see also ref 139. These projections derive from the shell (predominantly containing the neuropeptide, arginine-vasopressin) as well as the core (predominantly containing the neuropeptide, vasoactive intestinal peptide) of the SCh.[67] Notably, the diurnal changes in Hcrt expression and peptide content[140, 141] disappear following destruction of the SCh.[142] The major recipient of SCh innervation in the brain is the subparaventricular zone,[139, 143] which plays a role opposite to the LHA in arousal, since lesions targeting this region (in its ventral aspects) result in decreased sleep.[144] Available data suggest that, at least in part, topographically similar neurons in the SCh project to both the LHA and the subparaventricular zone.[139,67] It will therefore be of interest to elucidate the synaptology and signalling cascade through which SCh can orchestrate activity in these antagonistic brain regions. Curiously, the retinal ganglion cells, which play a major role in regulating the timing of SCh rhythmicity, also project directly to LHA neurons in a pathway whose functional role is not yet clear.[145]

7.2. Ventrolateral Preoptic Area

The ventrolateral preoptic area (VLPO) provides inhibitory innervation of major monoaminergic arousal systems, via neurons containing GABA and the neuropeptide galanin.[146,147] Several lines of evidence implicate the VLPO GABA/galanin cells as sleep-promoting neurons;[1] see ref 93. This region receives a small projection from both Hcrt and MCH neurons.[148] While a reciprocal pathway has not been described, there is functional data suggesting that Hcrt neurons may be under control of the VLPO. When neurons in the preoptic area are silenced by pharmacological activation of $GABA_A$ receptors, animals dramatically decrease the time spent asleep. Concomitant with this effect, c-Fos is induced in Hcrt neurons ipsilateral to drug infusion, indicative of a relatively direct neural pathway.[149] Possibly, the extensive galanin-ir fibre plexus which is found in the perifornical region[52] may in part originate in the VLPO; results from tracing experiments support this hypothesis.[73] However, definitive evidence of an inhibitory GABA/galanin connection from the VLPO to Hcrt neurons is still lacking.

7.3. Dorsomedial Hypothalamic Nucleus

Tracing studies have revealed LHA interconnections with the dorsomedial hypothalamic nucleus (DMH). These connections display subnuclear topography and are partially concentrated around a reciprocal innervation of medial parts of the LHA and the DMH.[74,94,150-151] The DMH has been implicated in several homeostatic functions, in particular the stress response; see ref 152. Intriguingly, destruction of the DMH results in the disruption of circadian rhythms.[153] In this context, a prominent GABAergic projection from the SCh to the DMH[143] has generated much interest. Signals from the SCh to the DMH may be further transmitted to the LHA Hcrt neurons by way of glutamatergic neurons which also contain TRH, as demonstrated recently by Chou *et al.*[153] Thus, in addition to the direct SCh-LHA pathway described above, signals from the circadian pacemaker may also regulate Hcrt neurons by inhibiting excitation deriving from the DMH. Adding to the complexity is the existence of a GABAergic projection from the DMH to the VLPO, which could serve to promote arousal drive by disinhibition of Hcrt neurons.[148,153]

7.4. Ventromedial Hypothalamic Nucleus

In contrast to the DMH, direct interconnections between the LHA and the ventromedial hypothalamic nucleus (VMH) appear relatively modest.[150,154] Like the LHA, however, the VMH connects bidirectionally to the DMH (see refs 8 and 152), and it is possible that the interconnection between these three nuclei is involved partly in the phenotypes observed after ventromedial and lateral hypothalamic lesions.

8. INPUTS FROM OTHER AROUSAL SYSTEMS

The LHA projects extensively to the major monoaminergic nuclei.[18, 37, 94] Excitatory effects have been described for Hcrt in each of these regions, as well in the cholinergic dorsal tegmentum[46,155-156] There is now ample evidence that these arousal systems in turn provide reciprocal innervation which may contribute to coordinate transitions between sleep and waking; see refs 93, 157 and 158.

8.1. Noradrenergic and Adrenergic Innervation

One of the earliest discoveries to stem from the development of the Falck-Hillarp technique to visualize bioaminergic neurons[159] was the existence of extensive plexuses of monoaminergic and serotonergic nerve fibres in the LHA.[160,161] Direct projections from the noradrenergic nexus in the locus coeruleus to the LHA have since been confirmed with several techniques; e.g.[74,73,162,163]. Recently, it was shown that both somatic and dendritic compartments of Hcrt cells are targeted by noradrenergic afferents (identified as immunoreactive for the biosynthetic enzyme, dopamine-β-hydroxylase).[164] Both α and β-adrenoceptors are expressed in the LHA[165] and noradrenalin hyperpolarizes both Hcrt and MCH cells via postsynaptic mechanisms.[88,166] These data suggest that Hcrt neurons are regulated via inhibitory feedback from the locus coeruleus. Lastly, there is a parallel, but less extensive, input to the LHA from adrenergic neurons in the lower brain stem.[167]

8.2. Serotonergic Innervation

Serotonin (5-hydroxytryptamine; 5-HT)-ir terminals provide dense innervation of the LHA,[168,35] originating in the dorsal raphe nucleus.[74] Immunoreactivity for the 5-HT_{1A} receptor has been reported in both Hcrt and MCH neurons, closely associated with serotonergic terminals.[169] 5-HT hyperpolarizes Hcrt as well as MCH cells through postsynaptic mechanisms.[92,166,76] Similar to the noradrenergic situation, the serotonergic projection to the Hcrt cells thus appears to serve as an inhibitory feedback signal. It is uncertain if 5-HT also plays a more long-term role in the regulation of LHA peptides; no transcriptional regulation of Hcrt was observed in response to administration of the 5-HT uptake inhibitor fluoxetine in obese Zucker rats.[170]

8.3. Cholinergic innervation

Unlike the monoaminergic arousal systems, the cholinergic neurons based in the pedunculopontine and laterodorsal tegmental nuclei are at their most active during rapid eye movement (REM) sleep;[171] see ref 158. This phenomenon has led to the suggestion that cholinergic mechanisms underlie REM atonia, a physiological phenomenon that, in narcolepsy, occurs anomalously in the awake state as cataplexy; see ref 158. The discovery that loss of Hcrt-producing neurons is a key feature of this disease[172,173] has therefore fuelled interest in Hcrt-cholinergic interactions. Cholinoceptive neurons are found in the LHA;[174] this population partially overlaps with the MCH neurons,[21] which also are contacted by ascending cholinergic projections.[175] The synaptic relationship between Hcrt neurons and cholinergic terminals remains to be elucidated. Intriguingly, and again in contrast to the monoaminergic effects, Hcrt neurons are excited and depolarise beyond firing threshold when the cholinergic agonist carbachol is applied[166] (but see also ref 92). This effect involves postsynaptic mechanisms and is blocked by atropine, an antagonist to the metabotropic muscarine subtype of cholinergic receptors. However, in MCH neurons, which have been shown to express the M3 muscarinic receptor,[176] muscarine itself exerts inhibitory postsynaptic effects, while simultaneously increasing neurotransmitter release from terminals innervating the MCH neuron via presynaptic mechanisms.[76] In slice cultures, carbachol stimulation rapidly increases MCH transcription.[175] Thus, the role of cholinergic mechanisms in the regulation of MCH is at present ambiguous. For Hcrt, the evidence suggests the existence of an excitatory loop between the LHA and cholinergic ascending systems. It remains, however, to reconcile such a pathway with the symptomatology of narcolepsy.

8.4. Histaminergic innervation

Histamine has relatively recently received considerable interest in its role as a central neurotransmitter and is produced by neurons in the hypothalamic tuberomammillary nucleus (TMN; see ref 177). Histaminergic fibres fan out similarly to the other arousal systems with dense innervation of the LHA.[178] In contrast to the growing literature on interactions between the LHA and monoaminergic and cholinergic systems, though, little is known about the relationship of histamine and this region. Histamine commonly elicits slow, depolarising responses (e.g. [179]; see ref 177; but see also ref 180). However, in Hcrt neurons in slice preparations, no change in membrane properties could be detected following application of histamine[92] (no published reports describe the response of MCH neurons to histamine stimulation). It is possible that the TMN exerts more prominent regulation of LHA via GABA, which acts as a cotransmitter for histamine. Such effects may not be fully expressed in a slice preparation, which contains only TMN terminals in isolation, and where spontaneous GABA/histamine release is likely low.

9. INPUTS FROM THE BASAL GANGLIA

Kelley and colleagues have studied a pathway from the nucleus accumbens to the LHA. Such a connection was early demonstrated with tracing techniques.[73,74] Subsequent studies revealed that GABAergic spiny projection neurons in the shell region provide a diffuse innervation of the LHA in its entire rostrocaudal length.[181] (In contrast, the accumbal core region supplies terminals only in the LHA region bordering on the entopeduncular nucleus).[182] When glutamatergic excitation is blocked in the shell region, animals respond with increased food intake.[183] This behaviour was not expressed when the LHA was simultaneously silenced with GABAergic stimulation. Moreover, under these circumstances, c-Fos is also induced in Hcrt neurons, but not in MCH neurons.[184, 185] These results indicate that the accumbens-LHA projection negatively regulates activity in the Hcrt system. Such a pathway has been proposed to provide an indirect route for the cortex (which projects heavily to the nucleus accumbens) in inhibiting LHA-mediated feeding when competing stimuli appear.[186] However, the data do not exclude the involvement of indirect connections. The numerous parallel direct and indirect pathways connecting the LHA with e.g. the SCh, VLPO and DMH (reviewed above) serve as a reminder that extensive feedback links often contribute in the fine-tuning of Hcrt-related behaviour.

10. CORTICAL INPUTS

Prominent efferent projections extend from the LHA to the cerebral cortex.[187,10] Quantitatively, the density of hypothalamic afferents to the cortical mantle is second only to the thalamus among extracortical inputs.[10] The LHA-cortical pathway is reviewed elsewhere in this volume. Not surprisingly, there is reciprocal innervation of the LHA from cortical regions.[73,74,187,188] While the efferent LHA innervation of the cortex covers the cortex relatively broadly and diffusely,[10,16,32] the cortico-LHA pathway has mainly been described originating in frontal regions, especially the prefrontal cortex.[74,187,188] In both directions, however, the tracts follow chiefly ipsilateral routes. Curiously, both pyramidal layer V neurons and non-pyramidal neurons in layer IV appear to contribute to this pathway.[187] Through these connections, the LHA may interact directly with the cortex to switch it between sleep and awake states, in addition to indirect regulation via tuning of activity in other arousal systems. It is a matter for future investigation to determine what role, if any, such a direct pathway plays in modulating state-specific activity in cortical circuits.[189]

11. CONCLUDING REMARKS

The discovery of Hcrt and MCH has directed a great deal of attention to the LHA. Subsequent histochemical efforts have confirmed and refined our understanding of previous tracing data, and aided in the elucidation of the physiological role of the LHA systems. The connections reviewed above tie the LHA into previously recognized feeding and arousal circuits, and indicate a position at the interface of these two behavioural phenomena. It is not surprising then that the LHA is targeted by so many regulatory influences. Adding to the described neural inputs, Hcrt neurons may also be regulated by direct hormonal input and, importantly, by local glucose concentrations.[190] This new knowledge makes it possible to pose new questions, which will help to define the exact role of Hcrt and MCH in the control of metabolic and sleep-wake state, and to what extent these systems diverge functionally.

12. ACKNOWLEDGEMENTS

Support provided by the Marianne and Marcus Wallenberg Foundation, Wenner-Gren Foundation, Jeanson's Foundation, Åke Wiberg's Foundation, Magnus Bergvall's Foundation, the Teodor Nerander Fund, Lars Hiertas Minne, the Swedish Medical Society to C.B. and the Swedish Research Council to C.B. and T.H.

13. REFERENCES

1. C. von Economo, Sleep as a problem of localization, *J. Nerv. Ment. Dis.* **71**, 249-259 (1930).
2. S. W. Ranson, Somnolence caused by hypothalamic lesions in the monkey, *Arch. Neurol. Psychiatry.* **41**, 1-23 (1939).

3. A. W. Hetherington and S. W. Ranson, Hypothalamic lesions and adiposity in the rat, *Anat. Rec.* **78**, 149-172 (1940).
4. B. K. Anand and J. R. Brobeck, Hypothalamic control of food intake in rats and cats, *Yale J Biol Med.* **24**, 123-40 (1951a).
5. B. K. Anand and J. R. Brobeck, Localization of a "feeding center" in the hypothalamus of the rat, *Proc Soc Exp Biol Med.* **77**, 323-4 (1951b).
6. E. M. Stricker, A. F. Swerdloff and M. J. Zigmond, Intrahypothalamic injections of kainic acid produce feeding and drinking deficits in rats, *Brain Res.* **158**, 470-3 (1978).
7. E. Stellar, The physiology of motivation, *Psychol Rev.* **61**, 5-22 (1954).
8. L. L. Bernardis and L. L. Bellinger, The lateral hypothalamic area revisited: neuroanatomy, body weight regulation, neuroendocrinology and metabolism, *Neurosci Biobehav Rev.* **17**, 141-93 (1993).
9. L. L. Bernardis and L. L. Bellinger, The lateral hypothalamic area revisited: ingestive behavior, *Neurosci Biobehav Rev.* **20**, 189-287 (1996).
10. C. B. Saper, Organization of cerebral cortical afferent systems in the rat. II. Hypothalamocortical projections, *J Comp Neurol.* **237**, 21-46 (1985).
11. H. Kawauchi, I. Kawazoe, M. Tsubokawa, M. Kishida and B. I. Baker, Characterization of melanin-concentrating hormone in chum salmon pituitaries, *Nature.* **305**, 321-3 (1983).
12. J. M. Vaughan, W. H. Fischer, C. Hoeger, J. Rivier and W. Vale, Characterization of melanin-concentrating hormone from rat hypothalamus, *Endocrinology.* **125**, 1660-5 (1989).
13. B. Griffond and B. I. Baker, Cell and molecular cell biology of melanin-concentrating hormone, *Int Rev Cytol.* **213**, 233-77 (2002).
14. G. Skofitsch, D. M. Jacobowitz and N. Zamir, Immunohistochemical localization of a melanin concentrating hormone-like peptide in the rat brain, *Brain Res Bull.* **15**, 635-49 (1985).
15. F. Presse, G. Hervieu, T. Imaki, P. E. Sawchenko, W. Vale and J. L. Nahon, Rat melanin-concentrating hormone messenger ribonucleic acid expression: marked changes during development and after stress and glucocorticoid stimuli, *Endocrinology.* **131**, 1241-50 (1992).
16. J. C. Bittencourt, F. Presse, C. Arias, C. Peto, J. Vaughan, J. L. Nahon, W. Vale and P. E. Sawchenko, The melanin-concentrating hormone system of the rat brain: an immuno- and hybridization histochemical characterization, *J Comp Neurol.* **319**, 218-45 (1992).
17. C. Köhler, L. Haglund and L. W. Swanson, A diffuse alpha MSH-immunoreactive projection to the hippocampus and spinal cord from individual neurons in the lateral hypothalamic area and zona incerta, *J Comp Neurol.* **223**, 501-14 (1984).
18. N. Naito, I. Kawazoe, Y. Nakai, H. Kawauchi and T. Hirano, Coexistence of immunoreactivity for melanin-concentrating hormone and alpha-melanocyte-stimulating hormone in the hypothalamus of the rat, *Neurosci Lett.* **70**, 81-5 (1986).
19. M. Kawata, K. Hashimoto, J. Takahara and Y. Sano, Immunohistochemical demonstration of the localization of corticotropin releasing factor-containing neurons in the hypothalamus of mammals including primates, *Anat Embryol (Berl).* **165**, 303-13 (1982).
20. I. Merchenthaler, S. Vigh, A. V. Schally and P. Petrusz, Immunocytochemical localization of growth hormone-releasing factor in the rat hypothalamus, *Endocrinology.* **114**, 1082-5 (1984).
21. P. Y. Risold, D. Fellmann, D. Lenys and C. Bugnon, Coexistence of acetylcholinesterase-, human growth hormone-releasing factor(1-37)-, alpha-melanotropin- and melanin-concentrating hormone-like immunoreactivities in neurons of the rat hypothalamus: a light and electron microscope study, *Neurosci Lett.* **100**, 23-8 (1989).
22. J. L. Nahon, F. Presse, J. C. Bittencourt, P. E. Sawchenko and W. Vale, The rat melanin-concentrating hormone messenger ribonucleic acid encodes multiple putative neuropeptides coexpressed in the dorsolateral hypothalamus, *Endocrinology.* **125**, 2056-65 (1989).
23. C. B. Saper, H. Akil and S. J. Watson, Lateral hypothalamic innervation of the cerebral cortex: immunoreactive staining for a peptide resembling but immunochemically distinct from pituitary/arcuate alpha-melanocyte stimulating hormone, *Brain Res Bull.* **16**, 107-20 (1986).
24. D. Qu, D. S. Ludwig, S. Gammeltoft, M. Piper, M. A. Pelleymounter, M. J. Cullen, W. F. Mathes, R. Przypek, R. Kanarek and E. Maratos-Flier, A role for melanin-concentrating hormone in the central regulation of feeding behaviour, *Nature.* **380**, 243-7 (1996).
25. M. Rossi, S. J. Choi, D. O'Shea, T. Miyoshi, M. A. Ghatei and S. R. Bloom, Melanin-concentrating hormone acutely stimulates feeding, but chronic administration has no effect on body weight, *Endocrinology.* **138**, 351-5 (1997).
26. K. M. Gautvik, L. de Lecea, V. T. Gautvik, P. E. Danielson, P. Tranque, A. Dopazo, F. E. Bloom and J. G. Sutcliffe, Overview of the most prevalent hypothalamus-specific mRNAs, as identified by directional tag PCR subtraction, *Proc Natl Acad Sci U S A.* **93**, 8733-8 (1996).

27. L. de Lecea, T. S. Kilduff, C. Peyron, X. Gao, P. E. Foye, P. E. Danielson, C. Fukuhara, E. L. Battenberg, V. T. Gautvik, F. S. Bartlett, 2nd, W. N. Frankel, A. N. van den Pol, F. E. Bloom, K. M. Gautvik and J. G. Sutcliffe, The hypocretins: hypothalamus-specific peptides with neuroexcitatory activity, *Proc Natl Acad Sci U S A.* **95**, 322-7 (1998).
28. T. Sakurai, A. Amemiya, M. Ishii, I. Matsuzaki, R. M. Chemelli, H. Tanaka, S. C. Williams, J. A. Richarson, G. P. Kozlowski, S. Wilson, J. R. Arch, R. E. Buckingham, A. C. Haynes, S. A. Carr, R. S. Annan, D. E. McNulty, W. S. Liu, J. A. Terrett, N. A. Elshourbagy, D. J. Bergsma and M. Yanagisawa, Orexins and orexin receptors: a family of hypothalamic neuropeptides and G protein-coupled receptors that regulate feeding behavior, *Cell.* **92**, 1 page following 696 (1998).
29. R. M. Chemelli, J. T. Willie, C. M. Sinton, J. K. Elmquist, T. Scammell, C. Lee, J. A. Richardson, S. C. Williams, Y. Xiong, Y. Kisanuki, T. E. Fitch, M. Nakazato, R. E. Hammer, C. B. Saper and M. Yanagisawa, Narcolepsy in orexin knockout mice: molecular genetics of sleep regulation, *Cell.* **98**, 437-51 (1999).
30. J. Hara, C. T. Beuckmann, T. Nambu, J. T. Willie, R. M. Chemelli, C. M. Sinton, F. Sugiyama, K. Yagami, K. Goto, M. Yanagisawa and T. Sakurai, Genetic ablation of orexin neurons in mice results in narcolepsy, hypophagia, and obesity, *Neuron.* **30**, 345-54 (2001).
31. C. Broberger, L. De Lecea, J. G. Sutcliffe and T. Hokfelt, Hypocretin/orexin- and melanin-concentrating hormone-expressing cells form distinct populations in the rodent lateral hypothalamus: relationship to the neuropeptide Y and agouti gene-related protein systems, *J Comp Neurol.* **402**, 460-74 (1998a).
32. C. Peyron, D. K. Tighe, A. N. van den Pol, L. de Lecea, H. C. Heller, J. G. Sutcliffe and T. S. Kilduff, Neurons containing hypocretin (orexin) project to multiple neuronal systems, *J Neurosci.* **18**, 9996-10015 (1998).
33. C. F. Elias, C. B. Saper, E. Maratos-Flier, N. A. Tritos, C. Lee, J. Kelly, J. B. Tatro, G. E. Hoffman, M. M. Ollmann, G. S. Barsh, T. Sakurai, M. Yanagisawa and J. K. Elmquist, Chemically defined projections linking the mediobasal hypothalamus and the lateral hypothalamic area, *J Comp Neurol.* **402**, 442-59 (1998).
34. J. L. Guan, K. Uehara, S. Lu, Q. P. Wang, H. Funahashi, T. Sakurai, M. Yanagizawa and S. Shioda, Reciprocal synaptic relationships between orexin- and melanin-concentrating hormone-containing neurons in the rat lateral hypothalamus: a novel circuit implicated in feeding regulation, *Int J Obes Relat Metab Disord.* **26**, 1523-32 (2002).
35. E. E. Abrahamson and R. Y. Moore, The posterior hypothalamic area: chemoarchitecture and afferent connections, *Brain Res.* **889**, 1-22 (2001).
36. P. Kristensen, M. E. Judge, L. Thim, U. Ribel, K. N. Christjansen, B. S. Wulff, J. T. Clausen, P. B. Jensen, O. D. Madsen, N. Vrang, P. J. Larsen and S. Hastrup, Hypothalamic CART is a new anorectic peptide regulated by leptin, *Nature.* **393**, 72-6 (1998).
37. P. D. Lambert, P. R. Couceyro, K. M. McGirr, S. E. Dall Vechia, Y. Smith and M. J. Kuhar, CART peptides in the central control of feeding and interactions with neuropeptide Y, *Synapse.* **29**, 293-8 (1998).
38. C. Broberger, Cocaine- and amphetamine-regulated transcript (CART) and food intake: Behavior in search of anatomy, *Drug. Dev. Res.* **51**, 124-142 (2000).
39. C. Broberger, Hypothalamic cocaine- and amphetamine-regulated transcript (CART) neurons: histochemical relationship to thyrotropin-releasing hormone, melanin-concentrating hormone, orexin/hypocretin and neuropeptide Y, *Brain Res.* **848**, 101-13 (1999).
40. C. F. Elias, C. E. Lee, J. F. Kelly, R. S. Ahima, M. Kuhar, C. B. Saper and J. K. Elmquist, Characterization of CART neurons in the rat and human hypothalamus, *J Comp Neurol.* **432**, 1-19 (2001).
41. C. Peyron and Y. Charnay, Hypocretin (orexin) and CART peptides are co-expressed in human but not in rodents' hypothalamus, *FENS Meet Abstr.* (2002).
42. Y. L. Hurd and P. Fagergren, Human cocaine- and amphetamine-regulated transcript (CART) mRNA is highly expressed in limbic- and sensory-related brain regions, *J Comp Neurol.* **425**, 583-98 (2000).
43. B. Griffond, A. Deray, D. Fellmann, P. Ciofi, D. Croix and C. Bugnon, Colocalization of prolactin- and dynorphin-like substances in a neuronal population of the rat lateral hypothalamus, *Neurosci Lett.* **156**, 91-5 (1993).
44. T. C. Chou, C. E. Lee, J. Lu, J. K. Elmquist, J. Hara, J. T. Willie, C. T. Beuckmann, R. M. Chemelli, T. Sakurai, M. Yanagisawa, C. B. Saper and T. E. Scammell, Orexin (hypocretin) neurons contain dynorphin, *J Neurosci.* **21**, RC168 (2001).
45. B. A. Gosnell, J. E. Morley and A. S. Levine, Opioid-induced feeding: localization of sensitive brain sites, *Brain Res.* **369**, 177-84 (1986).
46. K. S. Eriksson, O. A. Sergeeva, O. Selbach and H. L. Haas, Orexin (hypocretin)/dynorphin neurons control GABAergic inputs to tuberomammillary neurons, *Eur J Neurosci.* **19**, 1278-84 (2004).

47. V. Cvetkovic, F. Poncet, D. Fellmann, B. Griffond and P. Y. Risold, Diencephalic neurons producing melanin-concentrating hormone are influenced by local and multiple extra-hypothalamic tachykininergic projections through the neurokinin 3 receptor, *Neuroscience.* **119**, 1113-45 (2003).
48. R. Cortes, S. Ceccatelli, M. Schalling and T. Hokfelt, Differential effects of intracerebroventricular colchicine administration on the expression of mRNAs for neuropeptides and neurotransmitter enzymes, with special emphasis on galanin: an in situ hybridization study, *Synapse.* **6**, 369-91 (1990).
49. B. Griffond, P. Ciofi, L. Bayer, C. Jacquemard and D. Fellmann, Immunocytochemical detection of the neurokinin B receptor (NK3) on melanin-concentrating hormone (MCH) neurons in rat brain, *J Chem Neuroanat.* **12**, 183-9 (1997).
50. F. Brischoux, V. Cvetkovic, B. Griffond, D. Fellmann and P. Y. Risold, Time of genesis determines projection and neurokinin-3 expression patterns of diencephalic neurons containing melanin-concentrating hormone, *Eur J Neurosci.* **16**, 1672-80 (2002).
51. G. Skofitsch and D. M. Jacobowitz, Immunohistochemical mapping of galanin-like neurons in the rat central nervous system, *Peptides.* **6**, 509-46 (1985).
52. T. Melander, T. Hokfelt and A. Rokaeus, Distribution of galaninlike immunoreactivity in the rat central nervous system, *J Comp Neurol.* **248**, 475-517 (1986).
53. M. Håkansson, L. de Lecea, J. G. Sutcliffe, M. Yanagisawa and B. Meister, Leptin receptor- and STAT3-immunoreactivities in hypocretin/orexin neurons of the lateral hypothalamus, *J Neuroendocrinol.* **11**, 653-63 (1999).
54. M. J. Alexander, M. A. Miller, D. M. Dorsa, B. P. Bullock, R. H. Melloni, Jr., P. R. Dobner and S. E. Leeman, Distribution of neurotensin/neuromedin N mRNA in rat forebrain: unexpected abundance in hippocampus and subiculum, *Proc Natl Acad Sci U S A.* **86**, 5202-6 (1989).
55. B. Griffond, S. Grillon, J. Duval, C. Colard, C. Jacquemard, A. Deray and D. Fellmann, Occurrence of secretogranin II in the prolactin-immunoreactive neurons of the rat lateral hypothalamus: an in situ hybridization and immunocytochemical study, *J Chem Neuroanat.* **9**, 113-9 (1995).
56. L. Bayer, G. Mairet-Coello, P. Y. Risold and B. Griffond, Orexin/hypocretin neurons: chemical phenotype and possible interactions with melanin-concentrating hormone neurons, *Regul Pept.* **104**, 33-9 (2002).
57. R. M. Lechan and I. M. Jackson, Immunohistochemical localization of thyrotropin-releasing hormone in the rat hypothalamus and pituitary, *Endocrinology.* **111**, 55-65 (1982).
58. O. Johansson, T. Hokfelt and R. P. Elde, Immunohistochemical distribution of somatostatin-like immunoreactivity in the central nervous system of the adult rat, *Neuroscience.* **13**, 265-339 (1984).
59. L. Paut-Pagano, J. L. Valatx, K. Kitahama and M. Jouvet, [Prolactin-secreting neurons in the dorsolateral hypothalamus in Sprague-Dawley rats], *C R Acad Sci III.* **309**, 369-76 (1989).
60. P. Y. Risold, B. Griffond, T. S. Kilduff, J. G. Sutcliffe and D. Fellmann, Preprohypocretin (orexin) and prolactin-like immunoreactivity are coexpressed by neurons of the rat lateral hypothalamic area, *Neurosci Lett.* **259**, 153-6 (1999).
61. S. R. Vincent and H. Kimura, Histochemical mapping of nitric oxide synthase in the rat brain, *Neuroscience.* **46**, 755-84 (1992).
62. D. J. Cutler, R. Morris, M. L. Evans, R. A. Leslie, J. R. Arch and G. Williams, Orexin-A immunoreactive neurons in the rat hypothalamus do not contain neuronal nitric oxide synthase (nNOS), *Peptides.* **22**, 123-8 (2001).
63. S. O. Fetissov, Z. Q. Xu, L. C. Byrne, H. Hassani, P. Ernfors and T. Hökfelt, Neuropeptide y targets in the hypothalamus: nitric oxide synthesizing neurons express Y1 receptor, *J Neuroendocrinol.* **15**, 754-60 (2003).
64. T. Hökfelt, O. Johansson, A. Ljungdahl, J. M. Lundberg and M. Schultzberg, Peptidergic neurons, *Nature.* **284**, 515-21 (1980).
65. A. N. van den Pol, J. P. Wuarin and F. E. Dudek, Glutamate, the dominant excitatory transmitter in neuroendocrine regulation, *Science.* **250**, 1276-8 (1990).
66. C. Decavel and A. N. Van den Pol, GABA: a dominant neurotransmitter in the hypothalamus, *J Comp Neurol.* **302**, 1019-37 (1990).
67. E. E. Abrahamson, R. K. Leak and R. Y. Moore, The suprachiasmatic nucleus projects to posterior hypothalamic arousal systems, *Neuroreport.* **12**, 435-40 (2001).
68. M. Collin, M. Backberg, M. L. Ovesjo, G. Fisone, R. H. Edwards, F. Fujiyama and B. Meister, Plasma membrane and vesicular glutamate transporter mRNAs/proteins in hypothalamic neurons that regulate body weight, *Eur J Neurosci.* **18**, 1265-78 (2003).
69. D. L. Rosin, M. C. Weston, C. P. Sevigny, R. L. Stornetta and P. G. Guyenet, Hypothalamic orexin (hypocretin) neurons express vesicular glutamate transporters VGLUT1 or VGLUT2, *J Comp Neurol.* **465**, 593-603 (2003).
70. C. S. Lin, M. A. Nicolelis, J. S. Schneider and J. K. Chapin, A major direct GABAergic pathway from zona incerta to neocortex, *Science.* **248**, 1553-6 (1990).

71. M. Palkovits and H. Van Cuc, Quantitative light and electron microscopic studies on the lateral hypothalamus in rat. Cell and synaptic densities, *Brain Res Bull.* **5**, 643-7 (1980).
72. K. M. Knigge, D. Baxter-Grillo, J. Speciale and J. Wagner, Melanotropic peptides in the mammalian brain: the melanin-concentrating hormone, *Peptides.* **17**, 1063-73 (1996).
73. F. C. Barone, M. J. Wayner, S. L. Scharoun, R. Guevara-Aguilar and H. U. Aguilar-Baturoni, Afferent connections to the lateral hypothalamus: a horseradish peroxidase study in the rat, *Brain Res Bull.* **7**, 75-88 (1981).
74. H. Kita and Y. Oomura, An HRP study of the afferent connections to rat lateral hypothalamic region, *Brain Res Bull.* **8**, 63-71 (1982a).
75. Y. Date, Y. Ueta, H. Yamashita, H. Yamaguchi, S. Matsukura, K. Kangawa, T. Sakurai, M. Yanagisawa and M. Nakazato, Orexins, orexigenic hypothalamic peptides, interact with autonomic, neuroendocrine and neuroregulatory systems, *Proc Natl Acad Sci U S A.* **96**, 748-53 (1999).
76. A. N. van den Pol, C. Acuna-Goycolea, K. R. Clark and P. K. Ghosh, Physiological properties of hypothalamic MCH neurons identified with selective expression of reporter gene after recombinant virus infection, *Neuron.* **42**, 635-52 (2004).
77. J. H. Zhang, S. Sampogna, F. R. Morales and M. H. Chase, Orexin (hypocretin)-like immunoreactivity in the cat hypothalamus: a light and electron microscopic study, *Sleep.* **24**, 67-76 (2001).
78. P. Trivedi, H. Yu, D. J. MacNeil, L. H. Van der Ploeg and X. M. Guan, Distribution of orexin receptor mRNA in the rat brain, *FEBS Lett.* **438**, 71-5 (1998).
79. J. N. Marcus, C. J. Aschkenasi, C. E. Lee, R. M. Chemelli, C. B. Saper, M. Yanagisawa and J. K. Elmquist, Differential expression of orexin receptors 1 and 2 in the rat brain, *J Comp Neurol.* **435**, 6-25 (2001).
80. G. J. Hervieu, J. E. Cluderay, D. C. Harrison, J. C. Roberts and R. A. Leslie, Gene expression and protein distribution of the orexin-1 receptor in the rat brain and spinal cord, *Neuroscience.* **103**, 777-97 (2001).
81. M. Bäckberg, G. Hervieu, S. Wilson and B. Meister, Orexin receptor-1 (OX-R1) immunoreactivity in chemically identified neurons of the hypothalamus: focus on orexin targets involved in control of food and water intake, *Eur J Neurosci.* **15**, 315-28 (2002).
82. A. W. Sailer, H. Sano, Z. Zeng, T. P. McDonald, J. Pan, S. S. Pong, S. D. Feighner, C. P. Tan, T. Fukami, H. Iwaasa, D. L. Hreniuk, N. R. Morin, S. J. Sadowski, M. Ito, A. Bansal, B. Ky, D. J. Figueroa, Q. Jiang, C. P. Austin, D. J. MacNeil, A. Ishihara, M. Ihara, A. Kanatani, L. H. Van der Ploeg, A. D. Howard and Q. Liu, Identification and characterization of a second melanin-concentrating hormone receptor, MCH-2R, *Proc Natl Acad Sci U S A.* **98**, 7564-9 (2001).
83. G. J. Hervieu, J. E. Cluderay, D. Harrison, J. Meakin, P. Maycox, S. Nasir and R. A. Leslie, The distribution of the mRNA and protein products of the melanin-concentrating hormone (MCH) receptor gene, slc-1, in the central nervous system of the rat, *Eur J Neurosci.* **12**, 1194-216 (2000).
84. Y. Saito, M. Cheng, F. M. Leslie and O. Civelli, Expression of the melanin-concentrating hormone (MCH) receptor mRNA in the rat brain, *J Comp Neurol.* **435**, 26-40 (2001).
85. M. A. Mullett, C. J. Billington, A. S. Levine and C. M. Kotz, Hypocretin I in the lateral hypothalamus activates key feeding-regulatory brain sites, *Neuroreport.* **11**, 103-8 (2000).
86. X. B. Gao and A. N. van den Pol, Melanin concentrating hormone depresses synaptic activity of glutamate and GABA neurons from rat lateral hypothalamus, *J Physiol.* **533**, 237-52 (2001).
87. X. B. Gao and A. N. van den Pol, Melanin-concentrating hormone depresses L-, N-, and P/Q-type voltage-dependent calcium channels in rat lateral hypothalamic neurons, *J Physiol.* **542**, 273-86 (2002).
88. X. B. Gao, P. K. Ghosh and A. N. van den Pol, Neurons synthesizing melanin-concentrating hormone identified by selective reporter gene expression after transfection in vitro: transmitter responses, *J Neurophysiol.* **90**, 3978-85 (2003).
89. M. G. Dube, S. P. Kalra and P. S. Kalra, Food intake elicited by central administration of orexins/hypocretins: identification of hypothalamic sites of action, *Brain Res.* **842**, 473-7 (1999).
90. A. N. van den Pol, X. B. Gao, K. Obrietan, T. S. Kilduff and A. B. Belousov, Presynaptic and postsynaptic actions and modulation of neuroendocrine neurons by a new hypothalamic peptide, hypocretin/orexin, *J Neurosci.* **18**, 7962-71 (1998).
91. T. L. Horvath, S. Diano and A. N. van den Pol, Synaptic interaction between hypocretin (orexin) and neuropeptide Y cells in the rodent and primate hypothalamus: a novel circuit implicated in metabolic and endocrine regulations, *J Neurosci.* **19**, 1072-87 (1999a).
92. Y. Li, X. B. Gao, T. Sakurai and A. N. van den Pol, Hypocretin/Orexin excites hypocretin neurons via a local glutamate neuron-A potential mechanism for orchestrating the hypothalamic arousal system, *Neuron.* **36**, 1169-81 (2002).
93. C. B. Saper, T. C. Chou and T. E. Scammell, The sleep switch: hypothalamic control of sleep and wakefulness, *Trends Neurosci.* **24**, 726-31 (2001).
94. C. B. Saper, L. W. Swanson and W. M. Cowan, An autoradiographic study of the efferent connections of the lateral hypothalamic area in the rat, *J Comp Neurol.* **183**, 689-706 (1979).

95. J. G. Veening, L. W. Swanson, W. M. Cowan, R. Nieuwenhuys and L. M. Geeraedts, The medial forebrain bundle of the rat. II. An autoradiographic study of the topography of the major descending and ascending components, *J Comp Neurol.* **206**, 82-108 (1982).
96. W. J. S. Krieg, The hypothalamus of the albino rat, *J. Comp. Neurol.* **55**, 12-89 (1932).
97. L. W. Swanson. (1987). The hypothalamus. In *The handbook of chemical neuroanatomy* (Björklund, A., Hökfelt, T. & Swanson, L. W., eds.), Vol. 5, "Integrated systems of the CNS", pp. 1-124. Elsevier, Amsterdam.
98. R. B. Simerly. (1995). Anatomical substrates of hypothalamic integration. In *The rat nervous system* 2 edit. (Paxinos, G., ed.), pp. 353-376. Academic Press, San Diego.
99. U. Ungerstedt, Is interruption of the nigro-striatal dopamine system producing the "lateral hypothalamus syndrome"?, *Acta Physiol Scand.* **80**, 35A-36A (1970).
100. P. E. Sawchenko, Toward a new neurobiology of energy balance, appetite, and obesity: the anatomists weigh in, *J Comp Neurol.* **402**, 435-41 (1998).
101. B. M. Chronwall, Anatomy and physiology of the neuroendocrine arcuate nucleus, *Peptides.* **6 Suppl 2**, 1-11 (1985).
102. B. J. Everitt, B. Meister, T. Hökfelt, T. Melander, L. Terenius, A. Rökaeus, E. Theodorsson-Norheim, G. Dockray, J. Edwardson, C. Cuello, R. Elde, M. Goldstein, H. Hemmings, C. Ouimet, I. Walaas, P. Greengard, W. Vale, E. Weber, J.-Y. Wu and K.-J. Chang, The hypothalamic arcuate nucleus-median eminence complex: immunohistochemistry of transmitters, peptides and DARPP-32 with special reference to coexistence in dopamine neurons, *Brain Res.* **396**, 97-155 (1986).
103. C. Broberger and T. Hökfelt, Hypothalamic and vagal neuropeptide circuitries regulating food intake, *Physiol Behav.* **74**, 669-82 (2001).
104. H. J. Grill and J. M. Kaplan, The neuroanatomical axis for control of energy balance, *Front Neuroendocrinol.* **23**, 2-40 (2002).
105. J. W. Olney, Brain lesions, obesity, and other disturbances in mice treated with monosodium glutamate, *Science.* **164**, 719-21 (1969).
106. J. T. Clark, P. S. Kalra, W. R. Crowley and S. P. Kalra, Neuropeptide Y and human pancreatic polypeptide stimulate feeding behavior in rats, *Endocrinology.* **115**, 427-9 (1984).
107. B. G. Stanley, W. Magdalin, A. Seirafi, W. J. Thomas and S. F. Leibowitz, The perifornical area: the major focus of (a) patchily distributed hypothalamic neuropeptide Y-sensitive feeding system(s), *Brain Res.* **604**, 304-17 (1993).
108. R. Poggioli, A. V. Vergoni and A. Bertolini, ACTH-(1-24) and alpha-MSH antagonize feeding behavior stimulated by kappa opiate agonists, *Peptides.* **7**, 843-8 (1986).
109. W. Fan, B. A. Boston, R. A. Kesterson, V. J. Hruby and R. D. Cone, Role of melanocortinergic neurons in feeding and the agouti obesity syndrome, *Nature.* **385**, 165-8 (1997).
110. T. W. Stephens, M. Basinski, P. K. Bristow, J. M. Bue-Valleskey, S. G. Burgett, L. Craft, J. Hale, J. Hoffmann, H. M. Hsiung, A. Kriauciunas and et al., The role of neuropeptide Y in the antiobesity action of the obese gene product, *Nature.* **377**, 530-2 (1995).
111. C. F. Elias, C. Aschkenasi, C. Lee, J. Kelly, R. S. Ahima, C. Bjorbaek, J. S. Flier, C. B. Saper and J. K. Elmquist, Leptin differentially regulates NPY and POMC neurons projecting to the lateral hypothalamic area, *Neuron.* **23**, 775-86 (1999).
112. T. L. Horvath, I. Bechmann, F. Naftolin, S. P. Kalra and C. Leranth, Heterogeneity in the neuropeptide Y-containing neurons of the rat arcuate nucleus: GABAergic and non-GABAergic subpopulations, *Brain Res.* **756**, 283-6 (1997).
113. M. L. Ovesjö, M. Gamstedt, M. Collin and B. Meister, GABAergic nature of hypothalamic leptin target neurons in the ventromedial arcuate nucleus, *J Neuroendocrinol.* **13**, 505-16 (2001).
114. F. L. Bai, M. Yamano, Y. Shiotani, P. C. Emson, A. D. Smith, J. F. Powell and M. Tohyama, An arcuato-paraventricular and -dorsomedial hypothalamic neuropeptide Y-containing system which lacks noradrenaline in the rat, *Brain Res.* **331**, 172-5 (1985).
115. P. E. Sawchenko, L. W. Swanson, R. Grzanna, P. R. Howe, S. R. Bloom and J. M. Polak, Colocalization of neuropeptide Y immunoreactivity in brainstem catecholaminergic neurons that project to the paraventricular nucleus of the hypothalamus, *J Comp Neurol.* **241**, 138-53 (1985).
116. T. M. Hahn, J. F. Breininger, D. G. Baskin and M. W. Schwartz, Coexpression of Agrp and NPY in fasting-activated hypothalamic neurons, *Nat Neurosci.* **1**, 271-2 (1998).
117. J. L. Guan, T. Saotome, Q. P. Wang, H. Funahashi, T. Hori, S. Tanaka and S. Shioda, Orexinergic innervation of POMC-containing neurons in the rat arcuate nucleus, *Neuroreport.* **12**, 547-51 (2001).
118. M. Niimi, M. Sato and T. Taminato, Neuropeptide Y in central control of feeding and interactions with orexin and leptin, *Endocrine.* **14**, 269-73 (2001).
119. M. R. Jain, T. L. Horvath, P. S. Kalra and S. P. Kalra, Evidence that NPY Y1 receptors are involved in stimulation of feeding by orexins (hypocretins) in sated rats, *Regul Pept.* **87**, 19-24 (2000).

120. A. Yamanaka, K. Kunii, T. Nambu, N. Tsujino, A. Sakai, I. Matsuzaki, Y. Miwa, K. Goto and T. Sakurai, Orexin-induced food intake involves neuropeptide Y pathway, *Brain Res.* **859**, 404-9 (2000).
121. D. A. Ewald, P. C. Sternweis and R. J. Miller, Guanine nucleotide-binding protein Go-induced coupling of neuropeptide Y receptors to Ca2+ channels in sensory neurons, *Proc Natl Acad Sci U S A.* **85**, 3633-7 (1988).
122. Q. Q. Sun, J. R. Huguenard and D. A. Prince, Neuropeptide Y receptors differentially modulate G-protein-activated inwardly rectifying K+ channels and high-voltage-activated Ca2+ channels in rat thalamic neurons, *J Physiol.* **531**, 67-79 (2001).
123. H. Zheng, M. M. Corkern, S. M. Crousillac, L. M. Patterson, C. B. Phifer and H. R. Berthoud, Neurochemical phenotype of hypothalamic neurons showing Fos expression 23 h after intracranial AgRP, *Am J Physiol Regul Integr Comp Physiol.* **282**, R1773-81 (2002).
124. B. Beck, S. Richy, T. Dimitrov and A. Stricker-Krongrad, Opposite regulation of hypothalamic orexin and neuropeptide Y receptors and peptide expressions in obese Zucker rats, *Biochem Biophys Res Commun.* **286**, 518-23 (2001).
125. R. Hanada, M. Nakazato, S. Matsukura, N. Murakami, H. Yoshimatsu and T. Sakata, Differential regulation of melanin-concentrating hormone and orexin genes in the agouti-related protein/melanocortin-4 receptor system, *Biochem Biophys Res Commun.* **268**, 88-91 (2000).
126. C. Broberger, J. Johansen, C. Johansson, M. Schalling and T. Hökfelt, The neuropeptide Y/agouti gene-related protein (AGRP) brain circuitry in normal, anorectic, and monosodium glutamate-treated mice, *Proc Natl Acad Sci U S A.* **95**, 15043-8 (1998b).
127. A. G. Watts, Understanding the neural control of ingestive behaviors: helping to separate cause from effect with dehydration-associated anorexia, *Horm Behav.* **37**, 261-83 (2000).
128. H. R. Berthoud, Multiple neural systems controlling food intake and body weight, *Neurosci Biobehav Rev.* **26**, 393-428 (2002).
129. R. A. Jacobs, M. I. Jordan, S. J. Nowlan and G. E. Hinton, Adapative mixtures of local experts, *Neuronal Comput.* **3**, 79-87 (1991).
130. A. M. Graybiel, T. Aosaki, A. W. Flaherty and M. Kimura, The basal ganglia and adaptive motor control, *Science.* **265**, 1826-31 (1994).
131. D. L. Margules and J. Olds, Identical "feeding" and "rewarding" systems in the lateral hypothalamus of rats, *Science.* **135**, 374-5 (1962).
132. G. E. Hermann and R. C. Rogers, Convergence of vagal and gustatory afferent input within the parabrachial nucleus of the rat, *J Auton Nerv Syst.* **13**, 1-17 (1985).
133. L. W. Swanson and G. D. Petrovich, What is the amygdala?, *Trends Neurosci.* **21**, 323-31 (1998).
134. H. Herbert, M. M. Moga and C. B. Saper, Connections of the parabrachial nucleus with the nucleus of the solitary tract and the medullary reticular formation in the rat, *J Comp Neurol.* **293**, 540-80 (1990).
135. G. J. Ter Horst, P. de Boer, P. G. Luiten and J. D. van Willigen, Ascending projections from the solitary tract nucleus to the hypothalamus. A Phaseolus vulgaris lectin tracing study in the rat, *Neuroscience.* **31**, 785-97 (1989).
136. R. Y. Moore and V. B. Eichler, Loss of a circadian adrenal corticosterone rhythm following suprachiasmatic lesions in the rat, *Brain Res.* **42**, 201-6 (1972).
137. F. K. Stephan and I. Zucker, Circadian rhythms in drinking behavior and locomotor activity of rats are eliminated by hypothalamic lesions, *Proc Natl Acad Sci U S A.* **69**, 1583-6 (1972).
138. R. M. Buijs and A. Kalsbeek, Hypothalamic integration of central and peripheral clocks, *Nat Rev Neurosci.* **2**, 521-6 (2001).
139. A. G. Watts, L. W. Swanson and G. Sanchez-Watts, Efferent projections of the suprachiasmatic nucleus: I. Studies using anterograde transport of Phaseolus vulgaris leucoagglutinin in the rat, *J Comp Neurol.* **258**, 204-29 (1987).
140. S. Taheri, D. Sunter, C. Dakin, S. Moyes, L. Seal, J. Gardiner, M. Rossi, M. Ghatei and S. Bloom, Diurnal variation in orexin A immunoreactivity and prepro-orexin mRNA in the rat central nervous system, *Neurosci Lett.* **279**, 109-12 (2000).
141. Y. Yoshida, N. Fujiki, T. Nakajima, B. Ripley, H. Matsumura, H. Yoneda, E. Mignot and S. Nishino, Fluctuation of extracellular hypocretin-1 (orexin A) levels in the rat in relation to the light-dark cycle and sleep-wake activities, *Eur J Neurosci.* **14**, 1075-81 (2001).
142. S. Zhang, J. M. Zeitzer, Y. Yoshida, J. P. Wisor, S. Nishino, D. M. Edgar and E. Mignot, Lesions of the Suprachiasmatic Nucleus Eliminate the Daily Rhythm of Hypocretin-1 Release, *Sleep.* **27**, 619-627 (2004).
143. R. M. Buijs, Y. X. Hou, S. Shinn and L. P. Renaud, Ultrastructural evidence for intra- and extranuclear projections of GABAergic neurons of the suprachiasmatic nucleus, *J Comp Neurol.* **340**, 381-91 (1994).

144. J. Lu, Y. H. Zhang, T. C. Chou, S. E. Gaus, J. K. Elmquist, P. Shiromani and C. B. Saper, Contrasting effects of ibotenate lesions of the paraventricular nucleus and subparaventricular zone on sleep-wake cycle and temperature regulation, *J Neurosci.* **21**, 4864-74 (2001).
145. H. Kita and Y. Oomura, An anterograde HRP study of retinal projections to the hypothalamus in the rat, *Brain Res Bull.* **8**, 249-53 (1982b).
146. J. E. Sherin, P. J. Shiromani, R. W. McCarley and C. B. Saper, Activation of ventrolateral preoptic neurons during sleep, *Science.* **271**, 216-9 (1996).
147. J. E. Sherin, J. K. Elmquist, F. Torrealba and C. B. Saper, Innervation of histaminergic tuberomammillary neurons by GABAergic and galaninergic neurons in the ventrolateral preoptic nucleus of the rat, *J Neurosci.* **18**, 4705-21 (1998).
148. T. C. Chou, A. A. Bjorkum, S. E. Gaus, J. Lu, T. E. Scammell and C. B. Saper, Afferents to the ventrolateral preoptic nucleus, *J Neurosci.* **22**, 977-90 (2002).
149. S. Satoh, H. Matsumura, A. Fujioka, T. Nakajima, T. Kanbayashi, S. Nishino, Y. Shigeyoshi and H. Yoneda, FOS expression in orexin neurons following muscimol perfusion of preoptic area, *Neuroreport.* **15**, 1127-31 (2004).
150. P. G. Luiten and P. Room, Interrelations between lateral, dorsomedial and ventromedial hypothalamic nuclei in the rat. An HRP study, *Brain Res.* **190**, 321-32 (1980).
151. R. H. Thompson and L. W. Swanson, Organization of inputs to the dorsomedial nucleus of the hypothalamus: a reexamination with Fluorogold and PHAL in the rat, *Brain Res Brain Res Rev.* **27**, 89-118 (1998).
152. L. L. Bernardis and L. L. Bellinger, The dorsomedial hypothalamic nucleus revisited: 1998 update, *Proc Soc Exp Biol Med.* **218**, 284-306 (1998).
153. T. C. Chou, T. E. Scammell, J. J. Gooley, S. E. Gaus, C. B. Saper and J. Lu, Critical role of dorsomedial hypothalamic nucleus in a wide range of behavioral circadian rhythms, *J Neurosci.* **23**, 10691-702 (2003).
154. G. J. Ter Horst and P. G. Luiten, Phaseolus vulgaris leuco-agglutinin tracing of intrahypothalamic connections of the lateral, ventromedial, dorsomedial and paraventricular hypothalamic nuclei in the rat, *Brain Res Bull.* **18**, 191-203 (1987).
155. T. L. Horvath, C. Peyron, S. Diano, A. Ivanov, G. Aston-Jones, T. S. Kilduff and A. N. van Den Pol, Hypocretin (orexin) activation and synaptic innervation of the locus coeruleus noradrenergic system, *J Comp Neurol.* **415**, 145-59 (1999b).
156. R. J. Liu, A. N. van den Pol and G. K. Aghajanian, Hypocretins (orexins) regulate serotonin neurons in the dorsal raphe nucleus by excitatory direct and inhibitory indirect actions, *J Neurosci.* **22**, 9453-64 (2002).
157. E. Mignot, S. Taheri and S. Nishino, Sleeping with the hypothalamus: emerging therapeutic targets for sleep disorders, *Nat Neurosci.* **5 Suppl**, 1071-5 (2002).
158. J. G. Sutcliffe and L. de Lecea, The hypocretins: setting the arousal threshold, *Nat Rev Neurosci.* **3**, 339-49 (2002).
159. B. Falck, N. Å. Hillarp, G. Thieme and A. Torp, Fluorescence of catechol amines and related compounds condensed with formaldehyde, *J. Histochem. Cytochem.* **10**, 348-354 (1962).
160. N. E. Anden, A. Dahlstrom, K. Fuxe and K. Larsson, Mapping out of catecholamine and 5-hydroxytryptamine neurons innervating the telencephalon and diencephalon, *Life Sci.* **4**, 1275-9 (1965).
161. K. Fuxe, Evidence for the Existence of Monoamine Neurons in the Central Nervous System. 3. The Monoamine Nerve Terminal, *Z Zellforsch Mikrosk Anat.* **65**, 573-96 (1965).
162. L. W. Swanson and B. K. Hartman, The central adrenergic system. An immunofluorescence study of the location of cell bodies and their efferent connections in the rat utilizing dopamine-beta-hydroxylase as a marker, *J Comp Neurol.* **163**, 467-505 (1975).
163. S. F. Leibowitz and L. L. Brown, Histochemical and pharmacological analysis of catecholaminergic projections to the perifornical hypothalamus in relation to feeding inhibition, *Brain Res.* **201**, 315-45 (1980).
164. B. A. Baldo, R. A. Daniel, C. W. Berridge and A. E. Kelley, Overlapping distributions of orexin/hypocretin- and dopamine-beta-hydroxylase immunoreactive fibers in rat brain regions mediating arousal, motivation, and stress, *J Comp Neurol.* **464**, 220-37 (2003).
165. S. F. Leibowitz, M. Jhanwar-Uniyal, B. Dvorkin and M. H. Makman, Distribution of alpha-adrenergic, beta-adrenergic and dopaminergic receptors in discrete hypothalamic areas of rat, *Brain Res.* **233**, 97-114 (1982).
166. A. Yamanaka, Y. Muraki, N. Tsujino, K. Goto and T. Sakurai, Regulation of orexin neurons by the monoaminergic and cholinergic systems, *Biochem Biophys Res Commun.* **303**, 120-9 (2003b).
167. T. Hökfelt, K. Fuxe, M. Goldstein and O. Johansson, Immunohistochemical evidence for the existence of adrenaline neurons in the rat brain, **66**, 235-251 (1974).

168. H. W. Steinbusch and R. Nieuwenhuys, Localization of serotonin-like immunoreactivity in the central nervous system and pituitary of the rat, with special references to the innervation of the hypothalamus, *Adv Exp Med Biol.* **133**, 7-35 (1981).
169. M. Collin, M. Backberg, K. Onnestam and B. Meister, 5-HT1A receptor immunoreactivity in hypothalamic neurons involved in body weight control, *Neuroreport.* **13**, 945-51 (2002).
170. A. Gutierrez, G. Saracibar, L. Casis, E. Echevarria, V. M. Rodriguez, M. T. Macarulla, L. C. Abecia and M. P. Portillo, Effects of fluoxetine administration on neuropeptide y and orexins in obese zucker rat hypothalamus, *Obes Res.* **10**, 532-40 (2002).
171. M. el Mansari, K. Sakai and M. Jouvet, Unitary characteristics of presumptive cholinergic tegmental neurons during the sleep-waking cycle in freely moving cats, *Exp Brain Res.* **76**, 519-29 (1989).
172. C. Peyron, J. Faraco, W. Rogers, B. Ripley, S. Overeem, Y. Charnay, S. Nevsimalova, M. Aldrich, D. Reynolds, R. Albin, R. Li, M. Hungs, M. Pedrazzoli, M. Padigaru, M. Kucherlapati, J. Fan, R. Maki, G. J. Lammers, C. Bouras, R. Kucherlapati, S. Nishino and E. Mignot, A mutation in a case of early onset narcolepsy and a generalized absence of hypocretin peptides in human narcoleptic brains, *Nat Med.* **6**, 991-7 (2000).
173. T. C. Thannickal, R. Y. Moore, R. Nienhuis, L. Ramanathan, S. Gulyani, M. Aldrich, M. Cornford and J. M. Siegel, Reduced number of hypocretin neurons in human narcolepsy, *Neuron.* **27**, 469-74 (2000).
174. J. H. Haring and J. N. Davis, Acetylcholinesterase neurons in the lateral hypothalamus project to the spinal cord, *Brain Res.* **268**, 275-83 (1983).
175. L. Bayer, P. Y. Risold, B. Griffond and D. Fellmann, Rat diencephalic neurons producing melanin-concentrating hormone are influenced by ascending cholinergic projections, *Neuroscience.* **91**, 1087-101 (1999).
176. M. Yamada, T. Miyakawa, A. Duttaroy, A. Yamanaka, T. Moriguchi, R. Makita, M. Ogawa, C. J. Chou, B. Xia, J. N. Crawley, C. C. Felder, C. X. Deng and J. Wess, Mice lacking the M3 muscarinic acetylcholine receptor are hypophagic and lean, *Nature.* **410**, 207-12 (2001).
177. H. Haas and P. Panula, The role of histamine and the tuberomamillary nucleus in the nervous system, *Nat Rev Neurosci.* **4**, 121-30 (2003).
178. T. Watanabe, Y. Taguchi, S. Shiosaka, J. Tanaka, H. Kubota, Y. Terano, M. Tohyama and H. Wada, Distribution of the histaminergic neuron system in the central nervous system of rats; a fluorescent immunohistochemical analysis with histidine decarboxylase as a marker, *Brain Res.* **295**, 13-25 (1984).
179. D. A. McCormick and A. Williamson, Modulation of neuronal firing mode in cat and guinea pig LGNd by histamine: possible cellular mechanisms of histaminergic control of arousal, *J Neurosci.* **11**, 3188-99 (1991).
180. K. H. Lee, C. Broberger, U. Kim and D. A. McCormick, Histamine modulates thalamocortical activity by activating a chloride conductance in ferret perigeniculate neurons, *Proc Natl Acad Sci U S A.* **101**, 6716-21 (2004).
181. G. J. Mogenson, L. W. Swanson and M. Wu, Neural projections from nucleus accumbens to globus pallidus, substantia innominata, and lateral preoptic-lateral hypothalamic area: an anatomical and electrophysiological investigation in the rat, *J Neurosci.* **3**, 189-202 (1983).
182. L. Heimer, D. S. Zahm, L. Churchill, P. W. Kalivas and C. Wohltmann, Specificity in the projection patterns of accumbal core and shell in the rat, *Neuroscience.* **41**, 89-125 (1991).
183. C. S. Maldonado-Irizarry, C. J. Swanson and A. E. Kelley, Glutamate receptors in the nucleus accumbens shell control feeding behavior via the lateral hypothalamus, *J Neurosci.* **15**, 6779-88 (1995).
184. B. A. Baldo, L. Gual-Bonilla, K. Sijapati, R. A. Daniel, C. F. Landry and A. E. Kelley, Activation of a subpopulation of orexin/hypocretin-containing hypothalamic neurons by GABAA receptor-mediated inhibition of the nucleus accumbens shell, but not by exposure to a novel environment, *Eur J Neurosci.* **19**, 376-86 (2004).
185. H. Zheng, M. Corkern, I. Stoyanova, L. M. Patterson, R. Tian and H. R. Berthoud, Peptides that regulate food intake: appetite-inducing accumbens manipulation activates hypothalamic orexin neurons and inhibits POMC neurons, *Am J Physiol Regul Integr Comp Physiol.* **284**, R1436-44 (2003).
186. A. E. Kelley, Ventral striatal control of appetitive motivation: role in ingestive behavior and reward-related learning, *Neurosci Biobehav Rev.* **27**, 765-76 (2004).
187. H. Kita and Y. Oomura, Reciprocal connections between the lateral hypothalamus and the frontal complex in the rat: electrophysiological and anatomical observations, *Brain Res.* **213**, 1-16 (1981).
188. R. M. Beckstead, An autoradiographic examination of corticocortical and subcortical projections of the mediodorsal-projection (prefrontal) cortex in the rat, *J Comp Neurol.* **184**, 43-62 (1979).
189. M. Steriade, D. A. McCormick and T. J. Sejnowski, Thalamocortical oscillations in the sleeping and aroused brain, *Science.* **262**, 679-85 (1993).

190. A. Yamanaka, C. T. Beuckmann, J. T. Willie, J. Hara, N. Tsujino, M. Mieda, M. Tominaga, K. Yagami, F. Sugiyama, K. Goto, M. Yanagisawa and T. Sakurai, Hypothalamic orexin neurons regulate arousal according to energy balance in mice, *Neuron.* **38**, 701-13 (2003a).

PHYSIOLOGICAL CONSEQUENCES OF HYPOCRETIN ACTIVATION

PHYSIOLOGICAL CHARACTERISTICS OF HYPOCRETIN/OREXIN NEURONS

Anthony N. van den Pol*

1. INTRODUCTION

The discovery of a novel hypothalamic neuropeptide, hypocretin/orexin, in 1998 by two groups working independently[1,2] has generated growing levels of interest due to the role of these cells in arousal and narcolepsy, as well as energy homeostasis. The preprohypocretin protein is cleaved into two neuroactive peptides, hypocretin 1 and hypocretin 2. In this chapter, the hypocretin nomenclature will be used, and it is viewed as synonymous with orexin nomenclature; hypocretin 1 is orexin A, and hypocretin 2 is orexin B. A large number of studies have analyzed the effects of the hypocretins on neuronal physiology, and on behavior.

Within the CNS, the hypocretin neurons are located exclusively in the lateral hypothalamus/perifornical area where they are randomly distributed among neurons of several other identified and unidentified phenotypes. Morphologically, the cells are typical of the lateral hypothalamus, with three or four proximal dendrites that branch distally, and long axons that innervate both local (Figure 1) and distant regions of the brain. Hypocretin axons innervate many areas of the brain and spinal cord[3,4] and similarly, the peptide receptors are distributed widely within the brain.[5,6] The lack of a clear morphological signature has made it difficult to identify live hypocretin neurons for the purpose of electrophysiological characterization. Sakurai et al. identified and studied the gene encoding for hypocretin and its upstream promoter region.[69] By integrating the hypocretin promoter upstream from a sequence coding for a codon-corrected jellyfish green fluorescent protein (GFP) and generating transgenic mice expressing this sequence, the hypocretin neurons could be identified by their green fluorescence when stimulated by blue light. Importantly, all cells that expressed GFP in hypocretin GFP transgenic mice also showed immunoreactivity for hypocretin, indicating the selective expression of the reporter gene only in hypocretin neurons.[21,27]

* Department of Neurosurgery, Yale University School of Medicine, New Haven, CT 06520

Prior to the use of GFP-expressing transgenic mice, an initial view of the responses of hypocretin cells was achieved using histological approaches, based either on c-fos expression, or on changes in gene expression detected with in situ hybridization in fixed sections containing hypocretin neurons. These types of experiments often focused on the relation of hypocretin neurons to the metabolic state of the animal, as described below.

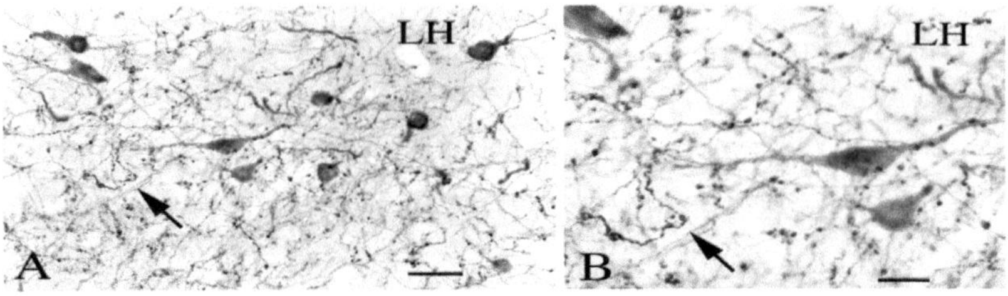

Figure 1. Hypocretin neurons and axons. A. In this photomicrograph, hypocretin neurons and axons are seen in the lateral hypothalamus (LH) after immunostaining with an antiserum against hypocretin-2. B. Higher magnification of the same region (arrow) showing terminal hypocretin axons and boutons on and near hypocretin cell bodies. Scale bar, A-25 μm, B- 12 μm. (From 42).

The primary focus of the present paper is on the physiological characteristics of hypocretin cells. Hypocretin cells contain glutamate[7,8] and may express glutamate vesicular transporters[9] that are typical of hypothalamic glutamatergic neurons,[10] suggesting glutamate may function as a fast amino acid transmitter in hypocretin neurons. Hypocretin cells may also contain the opioid peptide dynorphin.[11] Hypocretin cells express mu opioid receptors, and show a change in gene expression during morphine withdrawal.[12] As the hypocretin cells may contain other neuroactive substances, it is important to note that the hypocretin peptide itself plays a crucial role in the signaling of these cells. The functional disturbance associated with the absence or substantive reduction of hypocretin neurons in narcoleptic patients[13,14] could be explained either by the loss of hypocretin or by the loss of another cotransmitter in the same cells. However, the selective knockout of the hypocretin peptide in mice that does not eliminate the neurons but only the peptide,[15] lack of transport of hypocretin to neurosecretory vesicles in a human case[13], and loss of the hypocretin receptor in dogs[16] all lead to narcolepsy, and can best be explained solely by the absence of hypocretin or its receptors.[17] The complete loss of the hypocretin neuron may lead to a greater deficit in arousal,[18] possibly due to the loss of the other cotransmitters. The importance of hypocretin is further underlined by the absence of the peptide in the majority of cases of human narcolepsy.[19]

2. HYPOCRETIN NEURONS SHOW SPONTANEOUS REGULAR ACTION POTENTIALS

One of the striking features of hypocretin neurons is their regular firing pattern (Figure 2), with a spontaneous spike frequency of 2-5 Hz and a fairly constant interspike interval. This regular spike pattern continues even when synaptic actions are blocked, although the spike frequency is decreased in the absence of excitatory synaptic input.

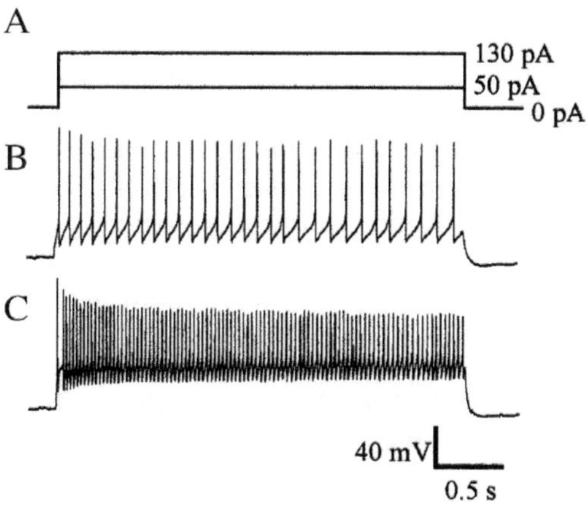

Figure 2. Hypocretin cells show regular spikes with little spike frequency adaptation when depolarized. When 50 or 130 pA current (A) was injected into hypocretin cells, they showed a sustained depolarization and regular firing pattern with little spike frequency adaptation (B,C). (From Li et al [20]).

When depolarized, hypocretin neurons can fire at very rapid rates, with all cells tested firing over 100 Hz, and some exceeding 200 Hz.[20] Spikes were eliminated with the sodium channel blocker tetrodotoxin (TTX). The regular firing rate of hypocretin cells is similar to the regular spontaneous firing rate of noradrenergic cells of the locus coeruleus.[21,22] In unidentified cells recorded from the lateral hypothalamus in live rats, some cells showed similar 2-5 Hz spike frequencies,[23] suggesting that the spike frequency in brain slices may be representative of the normal spiking behavior of these cells in the brain. The regular spontaneous pattern of hypocretin neurons is not typical of all cells in the same area of the lateral hypothalamus. Neurons that synthesize melanin concentrating hormone (MCH) are intermixed with hypocretin cells; single cells do not express both peptides.[3,24,25] MCH neurons were identified by GFP expression after recombinant adeno-associated virus infection and in transgenic mice, in each case with the reporter gene controlled by the MCH promoter. MCH cells show a very different pattern of electrical activity than hypocretin cells. Most MCH cells are silent, or show only low levels of action potentials with a mean spike frequency of 0.15 Hz.[26] Spike frequency adaptation, a phenomenon where continued depolarization leads to a slowing of spike frequency, is not a common attribute of hypocretin cells (Figure 2[20], but is a characteristic of the nearby MCH neurons. The resting membrane potential of hypocretin neurons is positive to that of MCH neurons. We recorded a mean resting potential of –57 mV and an input resistance of 356 MΩ in hypocretin cells[20] and a mean membrane potential of –61 mV in MCH cells. The spike threshold for mouse hypocretin neurons, based on the start of the rapid depolarization, was –47 mV. Another group recorded from mouse hypocretin cells and found a membrane potential a few mV negative to that which we found.[27] One lab recorded from neurons of the rat lateral hypothalamus and after labeling the recorded cells with a tracer, identified them post hoc by immunocytochemistry; they reported a more positive membrane potential (-46 mV) of hypocretin cells in rats than reported in mice; this difference in membrane potential could be due to species differences, or in part could be explained by the uncompensated junction potential of about 10 mV in the rat work or to differences in the ionic milieu of the respective experiments.[28]

Hypocretin cells show a membrane rectification characterized by a sag when receiving hyperpolarizing current pulses, suggestive of an I_h current. Hypocretin cells show low threshold spikes when depolarized from a hyperpolarized state, and this is followed by a slow afterdepolarization; MCH cells do not show these characteristics. Substituting NMDG or choline chloride for sodium chloride, or intracellular perfusion with BAPTA, or treatment with flufenamate eliminated the slow afterdepolarizing potential, suggesting involvement of a calcium-activated non-selective cation current.[28]

3. SYNAPTIC INPUT TO HYPOCRETIN NEURONS IS MEDIATED BY GLUTAMATE AND GABA.

Neuroactive peptides receive substantial attention in the hypothalamus, and play important roles in the modulation of neuronal activity. But the primary synaptic transmitters in the hypothalamus, similar to elsewhere in the brain, are glutamate and GABA.[29,30] All hypocretin cells tested showed excitatory responses to glutamate and the glutamate agonists NMDA and AMPA; in current clamp glutamate agonists depolarized the cells and increased spike frequency, and in voltage clamp generated an inward current. Immunoreactive hypocretin cells were contacted by axons containing vesicular glutamate transporter 2, a marker for glutamatergic axons, providing further substantiation for a glutamate input.[20] All hypocretin cells showed inhibitory responses to GABA or the GABA-A receptor agonist, muscimol, characterized by hyperpolarization of the membrane potential, reduction in spikes, and an outward current.[20,27] More importantly, the glutamate receptor antagonists AP5 and CNQX together blocked excitatory synaptic activity, and the GABA-A receptor antagonist bicuculline blocked inhibitory synaptic activity.[20] Together, blocking both glutamate and GABA ionotropic receptors eliminated synaptic activity, suggesting that glutamate and GABA released by axonal terminals in synaptic contact with hypocretin cells account for the great majority of synaptic actions.

4. INHIBITORY RESPONSE TO NOREPINEPHRINE AND SEROTONIN

Hypocretin neurons project to the dorsal raphe and locus coeruleus[3], two systems involved in aspects of arousal.[21] In the locus coeruleus, hypocretin axons make direct synaptic contact with asymmetrical-type synapses.[22] Hypocretin excites locus coeruleus neurons by several mechanisms, in large part by activation of the hypocretin receptor 1 (Ox1R).[31,22,32,33,34] Hypocretin increased action potentials and spike synchrony in the norepinephrine containing cells of the mouse locus coeruleus (Figure 3G).[32]

Similarly, hypocretin increases the activity of serotonin cells in the raphe, primarily by activation of non-selective cation currents.[35,36] Axons containing norepinephrine or serotonin are found in the lateral hypothalamus in the same region in which hypocretin cells are located, and boutons immunoreactive for dopamine beta hydroxylase, a marker for norepinephrine (and epinephrine), are reported to contact hypocretin neurons.[37]

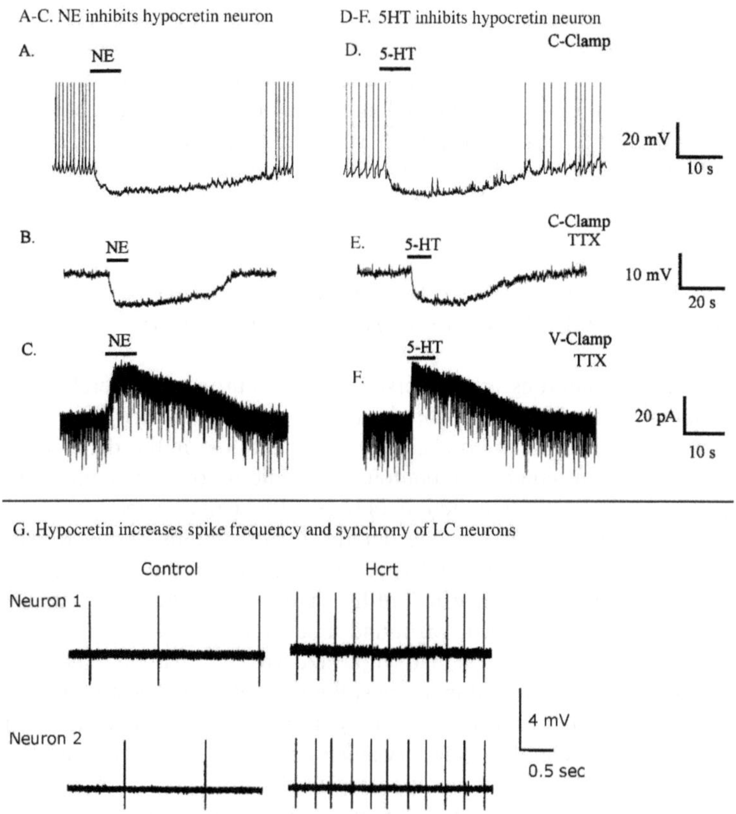

Figure 3. Norepinephrine and serotonin inhibit hypocretin neurons. A. Norepinephrine (NE; 50 μM) eliminated spikes in this hypocretin cell. B. In TTX, NE hyperpolarized the membrane potential. C. In TTX, NE evoked at outward current. D. Serotonin (5HT; 100 μM) blocked spikes. E. In TTX, 5HT evoked hyperpolarization. F. In TTX, 5HT evoked an outward current. G. In locus coeruleus (LC) neurons that synthesize NE, hypocretin evoked an increase in spike frequency. In these two LC cells recorded simultaneously, in the presence of hypocretin the spikes of the two neurons show a higher degree of synchronization. (From Li et al.[20])

Depending on the receptor subtypes expressed by target cells, serotonin and norepinephrine can exert either excitatory or inhibitory actions. As both serotonin and norepinephrine have been postulated to enhance arousal, one might expect that both transmitters would excite hypocretin neurons. But instead, both transmitters exerted inhibitory actions on hypocretin neurons. In current clamp, both serotonin and norepinephrine reduced spike frequency or completely blocked all spikes. In voltage clamp, both evoked an outward current, in the presence or absence of TTX, suggesting a direct effect on serotonin and norepinephrine receptors expressed by hypocretin neurons (Figure 3A-F).[20]

Hypocretin cells excite neurons in regions of the brain where cholinergic cells are involved in arousal such as the dorsal tegmentum[38] and the basal forebrain.[39,40] When a moderate level of acetylcholine (10 μM) was applied to hypocretin cells, only a minority showed an excitatory response detected as repeatable inward currents under voltage clamp.[20] However, when higher concentrations of acetylcholine or an agonist, carbachol (100 μM) were used, most cells showed an excitatory response that could be blocked by the metabotropic antagonist, atropine,[27] suggesting activation of a metabotropic-type actetylcholine receptor. Although hypocretin increases the activity of histamine neurons,[41] histamine had little detectable effect on hypocretin cells.[20,27]

5. HYPOCRETIN ACTIVATES HYPOCRETIN NEURONS BY AN INDIRECT EXCITATION OF GLUTAMATERGIC INTERNEURONS

The hypocretin neurons are dispersed among neurons of several phenotypes in the lateral hypothalamus. If the hypocretin cells are involved in some type of arousal function, then some means of orchestrating the actions of the dispersed cells may be needed. As hypocretin- containing boutons terminate on or near hypocretin cells (Figure 1,[42,22] there may be some interaction between the cells. Although hypocretin exerts substantial excitatory effects in the lateral hypothalamus,[43] we found little direct postsynaptic effect of hypocretin on hypocretin neurons; in the presence of TTX to block action potentials, we found little effect of hypocretin 1 or 2 on either membrane potential or input resistance;[20] in contrast, we found a substantial and direct excitatory effect of hypocretin on the nearby neurons that synthesize MCH.[26] Although no substantial direct effect was found, hypocretin did increase spike frequency, suggesting that hypocretin did lead indirectly to excitation of hypocretin neurons. The excitatory effect appeared to be mediated by an increase in the excitatory synaptic tone to hypocretin neurons. Hypocretin increased release of glutamate onto hypocretin cells from presynaptic excitatory boutons (Figure 4), and in the presence of glutamate receptor antagonists, little excitatory action of hypocretin on hypocretin cells could be detected. This suggested a potential mechanism for orchestrating the diffuse hypocretin cell population by hypocretin-mediated enhancement of glutamate release from local excitatory neurons. Part of the effect could be due to hypocretin receptors on glutamatergic axons that innervate hypocretin cells; hypocretin increased the frequency of miniature excitatory postsynaptic currents, with little change in the amplitude distribution, suggestive of a presynaptic action of hypocretin.[20]

6. GROUP 3 METABOTROPIC GLUTAMATE RECEPTORS INHIBIT HYPOCRETIN NEURONS

Hypocretin neurons, like other neurons throughout the brain, show excitatory responses to ionotropic glutamate receptor activation. In contrast, metabotropic glutamate receptors (mGluR) can elicit either excitatory or inhibitory actions, determined in large part by their coupling to secondary messenger systems. The mGluRs are divided into three groups; group 3 mGluRs are activated by the agonist L-AP4. L-AP4 had no direct effect on hypocretin neurons in the absence of synaptic activity. In contrast, L-AP4 reduced spontaneous excitatory postsynaptic currents; these EPSCs generally increase the activity of hypocretin cells.

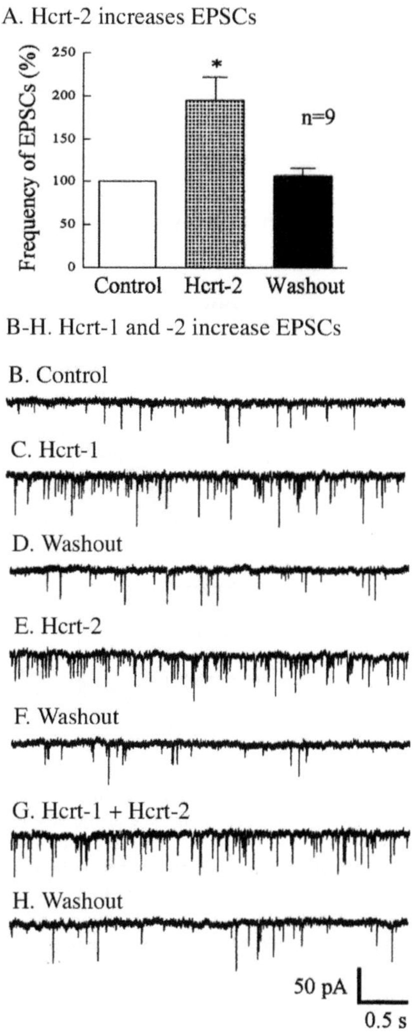

FIigure 4. Hypocretin increases excitatory synaptic activity in hypocretin neurons. (A) In the presence of bicuculline to block GABA receptors, hypocretin-2 increased EPSCs that recovered after washout. (B-H) Traces from a single neuron, showing both Hcrt-1 and -2 increase spontaneous EPSCs in the same cell. (B) In the presence of bicuculline (30 μM) EPSCs were recorded from hypocretin neurons clamped at –60 mV. C. Hcrt-1 (1 μM) increased frequency of EPSCs. D. Washout. E. Hcrt-2 (2 μM) increased frequency of EPSCs. F. Washout. G. Hcrt-1 plus Hcrt-2 increased frequency of EPSCs again. H. Washout. (From Li et al.[20])

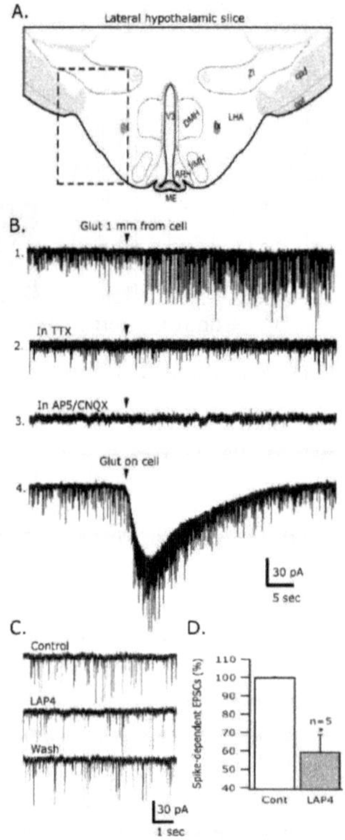

Figure 5. Glutamate neurons in lateral hypothalamus excite hypocretin neurons. A. In a microslice, indicated by square, consisting of only one side of the lateral hypothalamus, a microdrop of glutamate was applied to the slice 1 mm away from the recorded hypocretin cell. B1. Microdrop application increased excitatory synaptic activity. B2. TTX blocked the synaptic increase. B3. Glutamate receptor antagonists blocked the increase in microdrop-evoked EPSCs. B4. When the microdrop was applied directly to the recorded cell, a direct depolarization resulted. C. L-AP4 (100 μM) reduced the frequency of EPCSs in a typical hypocretin cell. D. When only large action potential-mediated EPSCs were examined, L-AP4 caused a substantial decrease in frequency. (From Acuna-Goycolea *et al.*[48]).

To further study these inhibitory mGluRs, a lateral hypothalamic microslice was used consisting solely of the lateral hypothalamus on one side of the brain (Figure 5). Excitatory synaptic actions could be evoked by electrical stimulation of axons within the slice, but these actions could be due to either local projections or to the distal ends of long axons cut during slice preparation. To stimulate local cells, a microdrop of glutamate was used which excites the cell body leading to action potentials, but does not excite axonal shafts directly. The microdrop-evoked synaptic excitation could be eliminated by ionotropic glutamate receptor antagonists suggesting that locally projecting axons from cells within the lateral hypothalamus synapsed on hypocretin neurons.[44] TTX also blocked the synaptic actions of the microdrop, indicating that the actions were synaptic and not due to a direct effect of the microdrop on hypocretin cells (Figure 5). In lateral hypothalamus microslices, L-AP4 reduced the frequency of spike-dependent EPSCs.

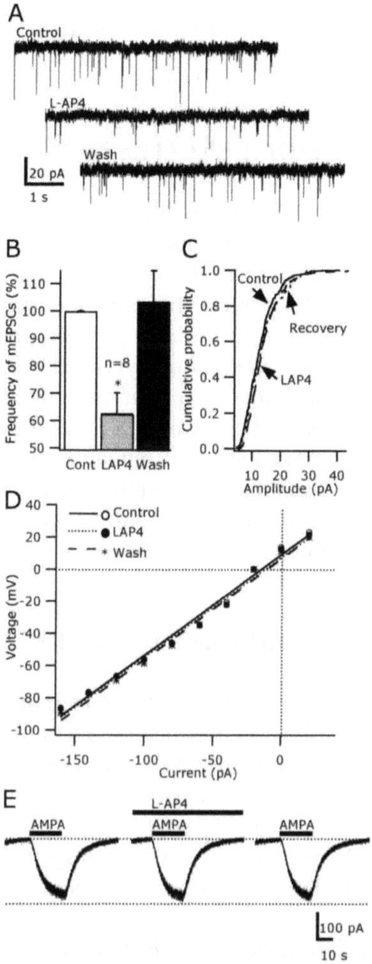

Figure. 6. Presynaptic mechanism- group 3 mGluRs attenuate miniature EPSC frequency. A. mEPSCs were recorded from hypocretin cells held at –60 mV in the presence of 0.5 μM TTX. L-AP4 (100 μM) induced a reversible decrease in the frequency of mEPSCs. B. Summary bar graph of data from five experiments illustrating a significant effect of L-AP4 on the frequency (mean±SEM) of mEPSCs. C. Cumulative probability histogram showing a lack of effect of L-AP4 on mEPSC amplitude. D. L-AP4 did not alter the input resistance of hypocretin cells. I-V relations before, during and after application show no direct effect of the group 3 mGluR agonist. E. L-AP4 did not alter the response of postsynaptic AMPA receptors, as shown in the representative traces of AMPA-evoked currents in hypocretin neurons before (left), during (middle), and after (right) bath application of 100 μM L-AP4. These results support the view that L-AP4 presynaptically attenuates excitatory transmission to hypocretin cells. (From Acuna-Goycolea *et al*)[48].

L-AP4 had no effect on application of AMPA onto hypocretin cells, suggesting that the actions of L-AP4 were probably not based on postsynaptic modulation of AMPA-type glutamate receptors (Figure 6E). TTX effectively isolates the axon terminal from signals from its cell body by blocking action potentials that originate in the cell body. In the presence of TTX, L-AP4 still depressed the frequency of miniature EPSCs, without changing the amplitude (Figure 6A-D). This suggests that glutamate release onto hypocretin neurons is inhibited by activation of mGluRs on excitatory axon terminals that synapse with hypocretin neurons.

7. GLUCAGON-LIKE PEPTIDE 1 EXCITES HYPOCRETIN NEURONS

Glucagon-like peptide 1 (GLP-1) is a peptide synthesized primarily in the caudal part of the brain, particularly in the area of the nucleus of the solitary tract. Axons from GLP-1 neurons are found throughout the brain, with substantive innervation of the lateral hypothalamus; similarly, GLP-1 receptors that are widespread in the CNS are also found in the lateral hypothalamus.[45-47] As the cellular physiological actions of GLP-1 had not been studied, we examined the responses of hypocretin cells.[48] GLP-1 and its agonists exerted excitatory actions on hypocretin neurons, increasing spike frequency and depolarizing the membrane potential even in TTX, suggesting a direct effect. GLP-1 also increased the frequency of spontaneous and miniature EPSCs, with little effect on the amplitude of the mEPSCs, suggesting presynaptic enhancement of transmitter release at glutamate synapses onto hypocretin cells. In contrast, GLP-1 agonists had no detectable effect on the nearby MCH neurons.

8. METABOLIC SIGNALS MODULATE HYPOCRETIN NEURON ACTIVITY

In early studies of hypocretin[2] reported that injections of the peptide into the CNS increased feeding. Food deprivation enhanced hypocretin mRNA, [2] leptin blocked the fasting-mediated rise in hypocretin mRNA,[49] and 2-deoxyglucose reduced hypocretin mRNA.[50] Hypocretin mRNA is also decreased in the genetically obese mice *ob/ob* and *db/db*.[51] Hypocretin may alter insulin secretion.[52] Insulin induced hypoglycemia enhances hypocretin mRNA and c-Fos.[53,54] The finding in hypocretin neurons that gene expression was shifted by changes in energy balance suggested the cells might respond to signals of metabolic state.

Using acutely isolated hypocretin cells lacking axons and dendrites, Yamanaka *et al.*,[55] reported that high glucose levels (30 mM) reduced spike frequency, and low glucose increased activity. Ghrelin, a signaling peptide from the gut, and also found in the hypothalamus, increased activity; in contrast, leptin, a signal of fat stores from the periphery, hyperpolarized the cells and reduced spike frequency. Whereas normal mice showed an increased wakefulness and exploration when fasted, ataxin-hypocretin mice that lack hypocretin neurons failed to show this increased arousal; these results therefore tie together the wake-promoting actions of the hypocretin cells with decreases in food.[55]

9. HYPOCRETIN CELLS ARE EXCITATORY

Hypocretin axons make asymmetrical type synaptic specializations, typical of excitatory synapses.[56] The release of hypocretin from axon terminals of hypocretin cells results in excitatory actions. Hypocretin mediated excitation has been found in a large number of brain regions, many of them involved in arousal. Hypocretin increases activity in the hypothalamus,[42,43] locus coeruleus,[31,22,33] dorsal raphe,[35,36] dorsal horn of the spinal cord,[57] spinal motoneurons,[58] nucleus of the solitary tract,[59] dorsal motor nucleus,[60] dorsolateral tegmentum,[38] tuberomammillary nucleus,[41] basal forebrain,[39,40] midline thalamus,[61] and trigeminal nucleus.[62] A number of mechanisms of excitation have been elucidated, and include enhancement of presynaptic neurotransmitter release, [42] activation of non-selective cation channels,[35,36,63] activation of sodium-calcium exchanger,[41,64,65] attenuation of potassium channels[66,61] and raising intracellular calcium.[42,67] Many of the targets of hypocretin excitation such as the locus coeruleus or dorsal raphe have long and widespread axonal arbors, and hypocretin release would thereby have further secondary influence on additional targets. Although hypocretin generally appears to be excitatory, consistent with the colocalization of glutamate and vesicular glutamate transporters, in a number of brain regions hypocretin has been reported to activate GABA or glycine circuits which would ultimately result in depression in targets downstream from activated inhibitory neurons.[42,57,35,64,67,68]

10. ACKNOWLEDGEMENTS

This work was supported by NIH NS 41454, NS37788, NS34887. I thank Drs Ying Li, Guido Wollmann, and Claudio Acuna-Goycolea for helpful suggestions on the manuscript.

11. REFERENCES

1. L. de Lecea, T. S. Kilduff, C. Peyron, X. Gao, P. E. Foye, P. E. Danielson, C. Fukuhara, E. L. Battenberg, V. T. Gautvik, F. S. Bartlett, 2nd, W. N. Frankel, A. N. van den Pol, F. E. Bloom, K. M. Gautvik and J. G. Sutcliffe, The hypocretins: hypothalamus-specific peptides with neuroexcitatory activity, *Proc Natl Acad Sci U S A.* **95**, 322-7 (1998).
2. T. Sakurai, A. Amemiya, M. Ishii, I. Matsuzaki, R. M. Chemelli, H. Tanaka, S. C. Williams, J. A. Richardson, G. P. Kozlowski, S. Wilson, J. R. Arch, R. E. Buckingham, A. C. Haynes, S. A. Carr, R. S. Annan, D. E. McNulty, W. S. Liu, J. A. Terrett, N. A. Elshourbagy, D. J. Bergsma and M. Yanagisawa, Orexins and orexin receptors: a family of hypothalamic neuropeptides and G protein-coupled receptors that regulate feeding behavior, *Cell.* **92**, 573-85 (1998).
3. C. Peyron, D. K. Tighe, A. N. van den Pol, L. de Lecea, H. C. Heller, J. G. Sutcliffe and T. S. Kilduff, Neurons Containing Hypocretin (Orexin) Project to Multiple Neuronal Systems, *J Neurosci.* **18**, 9996-10015 (1998).
4. A. N. van den Pol, Hypothalamic hypocretin (orexin): robust innervation of the spinal cord, *J. Neurosci.* **19**, 3171-82 (1999).
5. J. N. Marcus, C. J. Aschkenasi, C. E. Lee, R. M. Chemelli, C. B. Saper, M. Yanagisawa and J. K. Elmquist, Differential expression of orexin receptors 1 and 2 in the rat brain, *J Comp Neurol.* **435**, 6-25. (2001).
6. P. Trivedi, H. Yu, D. J. MacNeil, L. H. Van der Ploeg and X. M. Guan, Distribution of orexin receptor mRNA in the rat brain, *FEBS Lett.* **438**, 71-5 (1998).
7. E. E. Abrahamson, R. K. Leak and R. Y. Moore, The suprachiasmatic nucleus projects to posterior hypothalamic arousal systems, *Neuroreport.* **12**, 435-40. (2001).
8. F. Torrealba, M. Yanagisawa and C. B. Saper, Colocalization of orexin a and glutamate immunoreactivity in axon terminals in the tuberomammillary nucleus in rats, *Neuroscience.* **119**, 1033-44 (2003).
9 D. L. Rosin, M. C. Weston, C. P. Sevigny, R. L. Stornetta and P. G. Guyenet, Hypothalamic orexin (hypocretin) neurons express vesicular glutamate transporters VGLUT1 or VGLUT2, *J Comp Neurol.* **465**, 593-603 (2003).
10. R. T. Fremeau, Jr., M. D. Troyer, I. Pahner, G. O. Nygaard, C. H. Tran, R. J. Reimer, E. E. Bellocchio, D. Fortin, J. Storm-Mathisen and R. H. Edwards, The expression of vesicular glutamate transporters defines two classes of excitatory synapse, *Neuron.* **31**, 247-60 (2001).
11. T. C. Chou, A. A. Bjorkum, S. E. Gaus, J. Lu, T. E. Scammell and C. B. Saper, Afferents to the ventrolateral preoptic nucleus, *J Neurosci.* **22**, 977-90 (2002).

12. D. Georgescu, V. Zachariou, M. Barrot, M. Mieda, J. T. Willie, A. J. Eisch, M. Yanagisawa, E. J. Nestler and R. J. DiLeone, Involvement of the lateral hypothalamic peptide orexin in morphine dependence and withdrawal, *J Neurosci.* **23**, 3106-11 (2003).
13. C. Peyron, J. Faraco, W. Rogers, B. Ripley, S. Overeem, Y. Charnay, S. Nevsimalova, M. Aldrich, D. Reynolds, R. Albin, R. Li, M. Hungs, M. Pedrazzoli, M. Padigaru, M. Kucherlapati, J. Fan, R. Maki, G. J. Lammers, C. Bouras, R. Kucherlapati, S. Nishino and E. Mignot, A mutation in a case of early onset narcolepsy and a generalized absence of hypocretin peptides in human narcoleptic brains, *Nat Med.* **6**, 991-997 (2000).
14. T. C. Thannickal, R. Y. Moore, R. Nienhuis, L. Ramanathan, S. Gulyani, M. Aldrich, M. Cornford and J. M. Siegel, Reduced number of hypocretin neurons in human narcolepsy, *Neuron.* **27**, 469-74. (2000).
15. R. M. Chemelli, J. T. Willie, C. M. Sinton, J. K. Elmquist, T. Scammell, C. Lee, J. A. Richardson, S. C. Williams, Y. Xiong, Y. Kisanuki, T. E. Fitch, M. Nakazato, R. E. Hammer, C. B. Saper and M. Yanagisawa, Narcolepsy in orexin knockout mice: molecular genetics of sleep regulation, *Cell.* **98**, 437-51 (1999).
16. L. Lin, J. Faraco, R. Li, H. Kadotani, W. Rogers, X. Lin, X. Qiu, P. J. de Jong, S. Nishino and E. Mignot, The sleep disorder canine narcolepsy is caused by a mutation in the hypocretin (orexin) receptor 2 gene, *Cell.* **98**, 365-76 (1999).
17. A. N. van den Pol, Narcolepsy: a neurodegenerative disease of the hypocretin system?, *Neuron.* **27**, 415-8. (2000).
18. J. Hara, C. T. Beuckmann, T. Nambu, J. T. Willie, R. M. Chemelli, C. M. Sinton, F. Sugiyama, K. Yagami, K. Goto, M. Yanagisawa and T. Sakurai, Genetic ablation of orexin neurons in mice results in narcolepsy, hypophagia, and obesity, *Neuron.* **30**, 345-54. (2001).
19. S. Nishino, B. Ripley, S. Overeem, G. J. Lammers and E. Mignot, Hypocretin (orexin) deficiency in human narcolepsy [letter] [In Process Citation], *Lancet.* **355**, 39-40 (2000).
20. Y. Li, X. B. Gao, T. Sakurai and A. N. van den Pol, Hypocretin/Orexin excites hypocretin neurons via a local glutamate neuron-A potential mechanism for orchestrating the hypothalamic arousal system, *Neuron.* **36**, 1169-81 (2002).
21. M. Usher, J. D. Cohen, D. Servan-Schreiber, J. Rajkowski and G. Aston-Jones, The role of locus coeruleus in the regulation of cognitive performance, *Science.* **283**, 549-54 (1999).
22. T. L. Horvath, C. H. Warden, M. Hajos, A. Lombardi, F. Goglia and S. Diano, Brain uncoupling protein 2: uncoupled neuronal mitochondria predict thermal synapses in homeostatic centers, *J Neurosci.* **19**, 10417-27 (1999).
23. M. N. Alam, H. Gong, T. Alam, R. Jaganath, D. McGinty and R. Szymusiak, Sleep-waking discharge patterns of neurons recorded in the rat perifornical lateral hypothalamic area, *J Physiol.* **538**, 619-31 (2002)..
24. C. Broberger, L. de Lecea, J. G. Sutcliffe and T. Hökfelt, Hypocretin/orexin- and melanin-concentrating hormone expressing cells form distinct populations in the rodent lateral hypothalamus: relationship to neuropeptide Y innervation, *J. Comp. Neurol.* **402**, 460-474 (1998).
25. C. F. Elias, C. B. Saper, E. Maratos-Flier, N. A. Tritos, C. Lee, J. Kelly, J. B. Tatro, G. E. Hoffman, M. M. Ollmann, G. S. Barsh, T. Sakurai, M. Yanagisawa and J. M. Elmquist, Chemically defined projections linking the mediobasal hypothalamus and the lateral hypothalamic area, *J Comp Neurol.* **402**, 442-459 (1998).
26. A. N. Van Den Pol, C. Acuna-Goycolea, K. R. Clark and P. K. Ghosh, Physiological Properties of Hypothalamic MCH Neurons Identified with Selective Expression of Reporter Gene after Recombinant Virus Infection, *Neuron.* **42**, 635-52 (2004).
27. A. Yamanaka, C. T. Beuckmann, J. T. Willie, J. Hara, N. Tsujino, M. Mieda, M. Tominaga, K. Yagami, F. Sugiyama, K. Goto, M. Yanagisawa and T. Sakurai, Hypothalamic orexin neurons regulate arousal according to energy balance in mice, *Neuron.* **38**, 701-13 (2003).
28. E. Eggermann, L. Bayer, M. Serafin, B. Saint-Mleux, L. Bernheim, D. Machard, B. E. Jones and M. Muhlethaler, The wake-promoting hypocretin-orexin neurons are in an intrinsic state of membrane depolarization, *J Neurosci.* **23**, 1557-62 (2003).
29. A. N. van den Pol, J. P. Wuarin and F. E. Dudek, Glutamate, the dominant excitatory transmitter in neuroendocrine regulation, *Science.* **250**, 1276-8 (1990).
30. C. Decavel and A. N. Van den Pol, GABA: a dominant neurotransmitter in the hypothalamus, *J Comp Neurol.* **302**, 1019-37 (1990).
31. J. J. Hagan, R. A. Leslie, S. Patel, M. L. Evans, T. A. Wattam, S. Holmes, C. D. Benham, S. G. Taylor, C. Routledge, P. Hemmati, R. P. Munton, T. E. Ashmeade, A. S. Shah, J. P. Hatcher, P. D. Hatcher, D. N. Jones, M. I. Smith, D. C. Piper, A. J. Hunter, R. A. Porter and N. Upton, Orexin A activates locus coeruleus cell firing and increases arousal in the rat, *Proc Natl Acad Sci U S A.* **96**, 10911-6 (1999).

32. A. N. van den Pol, P. K. Ghosh, R. J. Liu, Y. Li, G. K. Aghajanian and X. B. Gao, Hypocretin (orexin) enhances neuron activity and cell synchrony in developing mouse GFP-expressing locus coeruleus, *J Physiol.* **541**, 169-85 (2002).
33. P. Bourgin, S. Huitrón-Reséndiz, A. Spier, V. Fabre, B. Morte, J. Criado, J. Sutcliffe, S. Henriksen and L. de Lecea, Hypocretin-1 modulates REM sleep through activation of locus coeruleus neurons, *J. Neurosci.* **20**, 7760-5 (2000).
34. A. Ivanov and G. Aston-Jones, Hypocretin/orexin depolarizes and decreases potassium conductance in locus coeruleus neurons, *Neuroreport.* **11**, 1755-8. (2000).
35. R. J. Liu, A. N. van den Pol and G. K. Aghajanian, Hypocretins (orexins) regulate serotonin neurons in the dorsal raphe nucleus by excitatory direct and inhibitory indirect actions, *J Neurosci.* **22**, 9453-64 (2002).
36. R. E. Brown, O. A. Sergeeva, K. S. Eriksson and H. L. Haas, Convergent excitation of dorsal raphe serotonin neurons by multiple arousal systems (orexin/hypocretin, histamine and noradrenaline), *J Neurosci.* **22**, 8850-9 (2002).
37. B. A. Baldo, R. A. Daniel, C. W. Berridge and A. E. Kelley, Overlapping distributions of orexin/hypocretin- and dopamine-beta-hydroxylase immunoreactive fibers in rat brain regions mediating arousal, motivation, and stress, *J Comp Neurol.* **464**, 220-37 (2003).
38. S. Burlet, C. J. Tyler and C. S. Leonard, Direct and indirect excitation of laterodorsal tegmental neurons by Hypocretin/Orexin peptides: implications for wakefulness and narcolepsy, *J Neurosci.* **22**, 2862-72 (2002).
39. E. Eggermann, M. Serafin, L. Bayer, D. Machard, B. Saint-Mleux, B. E. Jones and M. Muhlethaler, Orexins/hypocretins excite basal forebrain cholinergic neurones, *Neuroscience.* **108**, 177-81 (2001).
40. M. Wu, L. Zaborszky, T. Hajszan, A. N. van den Pol and M. Alreja, Hypocretin/orexin innervation and excitation of identified septohippocampal cholinergic neurons, *J Neurosci.* **24**, 3527-36 (2004).
41. K. S. Eriksson, O. Sergeeva, R. E. Brown and H. L. Haas, Orexin/hypocretin excites the histaminergic neurons of the tuberomammillary nucleus, *J Neurosci.* **21**, 9273-9 (2001).
42. A. N. van den Pol, X. B. Gao, K. Obrietan, T. S. Kilduff and A. B. Belousov, Presynaptic and postsynaptic actions and modulation of neuroendocrine neurons by a new hypothalamic peptide, hypocretin/orexin, *J Neurosci.* **18**, 7962-71 (1998).
43. A. N. van den Pol, P. R. Patrylo, P. K. Ghosh and X. B. Gao, Lateral hypothalamus: Early developmental expression and response to hypocretin (orexin), *J Comp Neurol.* **433**, 349-363. (2001)..
44. C. Acuña-Goycolea, Y. Li and A. N. Van Den Pol, Group III metabotropic glutamate receptors maintain tonic inhibition of excitatory synaptic input to hypocretin/orexin neurons, *J Neurosci.* **24**, 3013-22 (2004).
45. S. L. Jin, V. K. Han, J. G. Simmons, A. C. Towle, J. M. Lauder and P. K. Lund, Distribution of glucagonlike peptide I (GLP-I), glucagon, and glicentin in the rat brain: an immunocytochemical study, *J Comp Neurol.* **271**, 519-32 (1988).
46. H. Yamamoto, T. Kishi, C. E. Lee, B. J. Choi, H. Fang, A. N. Hollenberg, D. J. Drucker and J. K. Elmquist, Glucagon-like peptide-1-responsive catecholamine neurons in the area postrema link peripheral glucagon-like peptide-1 with central autonomic control sites, *J Neurosci.* **23**, 2939-46 (2003).
47. I. Merchenthaler, M. Lane and P. Shughrue, Distribution of pre-pro-glucagon and glucagon-like peptide-1 receptor messenger RNAs in the rat central nervous system, *J Comp Neurol.* **403**, 261-80 (1999).
48. C. Acuna-Goycolea, Y. Li and A. N. Van Den Pol, Group III metabotropic glutamate receptors maintain tonic inhibition of excitatory synaptic input to hypocretin/orexin neurons, *J Neurosci.* **24**, 3013-22 (2004).
49. M. López, L. Seoane, M. C. García, F. Lago, F. F. Casanueva, R. Senaris and C. Diéguez, Leptin regulation of prepro-orexin and orexin receptor mRNA levels in the hypothalamus, *Biochem Biophys Res Commun.* **269**, 41-5 (2000).
50. L. Bayer, C. Colard, N. U. Nguyen, P. Y. Risold, D. Fellmann and B. Griffond, Alteration of the expression of the hypocretin (orexin) gene by 2- deoxyglucose in the rat lateral hypothalamic area, *Neuroreport.* **11**, 531-3 (2000).
51. Y. Yamamoto, Y. Ueta, Y. Date, M. Nakazato, Y. Hara, R. Serino, M. Nomura, I. Shibuya, S. Matsukura and H. Yamashita, Down regulation of the prepro-orexin gene expression in genetically obese mice, *Brain Res Mol Brain Res.* **65**, 14-22 (1999).
52. K. W. Nowak, P. Mackowiak, M. M. Switonska, M. Fabis and L. K. Malendowicz, Acute orexin effects on insulin secretion in the rat: in vivo and in vitro studies, *Life Sci.* **66**, 449-54 (2000).
53. B. Griffond, P. Y. Risold, C. Jacquemard, C. Colard and D. Fellmann, Insulin-induced hypoglycemia increases preprohypocretin (orexin) mRNA in the rat lateral hypothalamic area, *Neurosci Lett.* **262**, 77-80 (1999).
54. T. Moriguchi, T. Sakurai, T. Nambu, M. Yanagisawa and K. Goto, Neurons containing orexin in the lateral hypothalamic area of the adult rat brain are activated by insulin-induced acute hypoglycemia, *Neurosci Lett.* **264**, 101-4 (1999).

55. A. Yamanaka, C. T. Beuckmann, J. T. Willie, J. Hara, N. Tsujino, M. Mieda, M. Tominaga, K. Yagami, F. Sugiyama, K. Goto, M. Yanagisawa and T. Sakurai, Hypothalamic orexin neurons regulate arousal according to energy balance in mice, *Neuron.* **38**, 701-13 (2003).
56. T. L. Horvath, C. H. Warden, M. Hajos, A. Lombardi, F. Goglia and S. Diano, Brain uncoupling protein 2: uncoupled neuronal mitochondria predict thermal synapses in homeostatic centers, *J Neurosci.* **19**, 10417-27 (1999).
57. T. J. Grudt, A. N. van Den Pol and E. R. Perl, Hypocretin-2 (orexin-B) modulation of superficial dorsal horn activity in rat, *J Physiol.* **538**, 517-525 (2002).
58 . J. Yamuy, S. J. Fung, M. Xi and M. H. Chase, Hypocretinergic control of spinal cord motoneurons, *J Neurosci.* **24**, 5336-45 (2004).
59. B. N. Smith, S. F. Davis, A. N. Van Den Pol and W. Xu, Selective enhancement of excitatory synaptic activity in the rat nucleus tractus solitarius by hypocretin 2, *Neuroscience.* **115**, 707-14 (2002).
60. L. L. Hwang, C. T. Chen and N. J. Dun, Mechanisms of orexin-induced depolarizations in rat dorsal motor nucleus of vagus neurones in vitro, *J Physiol.* **537**, 511-20 (2001).
61. L. Bayer, C. Colard, N. U. Nguyen, P. Y. Risold, D. Fellmann and B. Griffond, Alteration of the expression of the hypocretin (orexin) gene by 2- deoxyglucose in the rat lateral hypothalamic area, *Neuroreport.* **11**, 531-3 (2000).
62. J. H. Peever, Y. Y. Lai and J. M. Siegel, Excitatory effects of hypocretin-1 (orexin-A) in the trigeminal motor nucleus are reversed by NMDA antagonism, *J Neurophysiol.* **89**, 2591-600 (2003).
63. B. Yang and A. V. Ferguson, Orexin-A depolarizes nucleus tractus solitarius neurons through effects on nonselective cationic and K+ conductances, *J Neurophysiol.* **89**, 2167-75 (2003).
64. M. F. Wu, J. John, N. Maidment, H. A. Lam and J. M. Siegel, Hypocretin release in normal and narcoleptic dogs after food and sleep deprivation, eating, and movement, *Am J Physiol Regul Integr Comp Physiol.* **283**, R1079-86 (2002).
65. D. Burdakov, B. Liss and F. M. Ashcroft, Orexin excites GABAergic neurons of the arcuate nucleus by activating the sodium--calcium exchanger, *J Neurosci.* **23**, 4951-7 (2003).
66. K. A. Kohlmeier, T. Inoue and C. S. Leonard, Hypocretin/Orexin peptide signalling in the ascending arousal system: Elevation of intracellular calcium in the mouse dorsal raphe and laterodorsal tegmentum, *J Neurophysiol.* (2004).
67. T. M. Korotkova, K. S. Eriksson, H. L. Haas and R. E. Brown, Selective excitation of GABAergic neurons in the substantia nigra of the rat by orexin/hypocretin in vitro, *Regul Pept.* **104**, 83-89 (2002).
68. S. F. Davis, K. W. Williams, W. Xu, N. R. Glatzer and B. N. Smith, Selective enhancement of synaptic inhibition by hypocretin (orexin) in rat vagal motor neurons: implications for autonomic regulation, *J Neurosci.* **23**, 3844-54 (2003).
69. T. Sakurai, T. Moriguchi, K, Furuya, N. Kajiwara, T. Nakamura, M. Yanagisawa and K. Goto, Structure and function of human prepro-orexin gene, *J. Biol. Chem.,* **274:** 17771-17776 (1999).

THE NE SYSTEM AS A TARGET FOR HYPOCRETIN NEURONS:

Implications for regulation of arousal

Gary Aston-Jones, J. Patrick Card, Yan Zhu, Mónica González and Elizabeth Haggerty*

1. INTRODUCTION

As described in detail elsewhere in this volume, hypocretin (Hcrt, also called orexin) is a neuropeptide discovered recently in brain by subtractive hybridization technology [1] and orphan receptor analysis.[2] Anatomical studies revealed that this peptide is made only in a select group of neurons in the hypothalamus, which project widely throughout the neuraxis.[3] Functional analyses initially revealed a potential role for Hcrt in food intake, but later studies showed an important role of this neuropeptide system in regulation of arousal and sleep-waking states. It now seems clear that interference with Hcrt neurotransmission at any of several steps in either animals or humans leads to a narcoleptic-like syndrome.[4,5]

Although Hcrt projections ramify widely in the CNS, they are particularly target brain systems associated with arousal and regulation of sleep and waking. Among these areas, the nucleus locus coeruleus (LC) is especially heavily innervated by Hcrt fibers [3] (Fig. 1). Studies have shown that noradrenergic neurons of the LC receive direct Hcrt inputs via asymmetrical synapses, and that Hcrt in slices directly depolarizes and activates LC neurons.[6-9] Activation by Hcrt is also seen in 5HT[10,11] histamine,[12,13] and cholinergic brain neurons implicated in arousal mechanisms.[14] Together with results showing that loss of Hcrt is associated with narcolepsy, these findings have led to the general view that Hcrt function is important in regulation or maintenance of arousal or waking.[4,5,15,16]

Here we review our recent work on the effects of this peptide on physiological properties of LC neurons, and the innervation of locus coeruleus in rat and monkey by

* Gary Aston-Jones, J. Patrick Card, Yan Zhu, Mónica González and Elizabeth Haggerty, Department of Psychiatry, University of Pennsylvania, Philadelphia, PA 19104.

Hcrt fibers. We also review recent work showing Hcrt projections from the dorsomedial hypothalamus (DMH) area to the LC that may convey circadian information to this arousal system. These findings confirm that the LC is a prominent target for Hcrt actions, but also indicate that the functional connection between Hcrt and LC is more complex than originally conceived

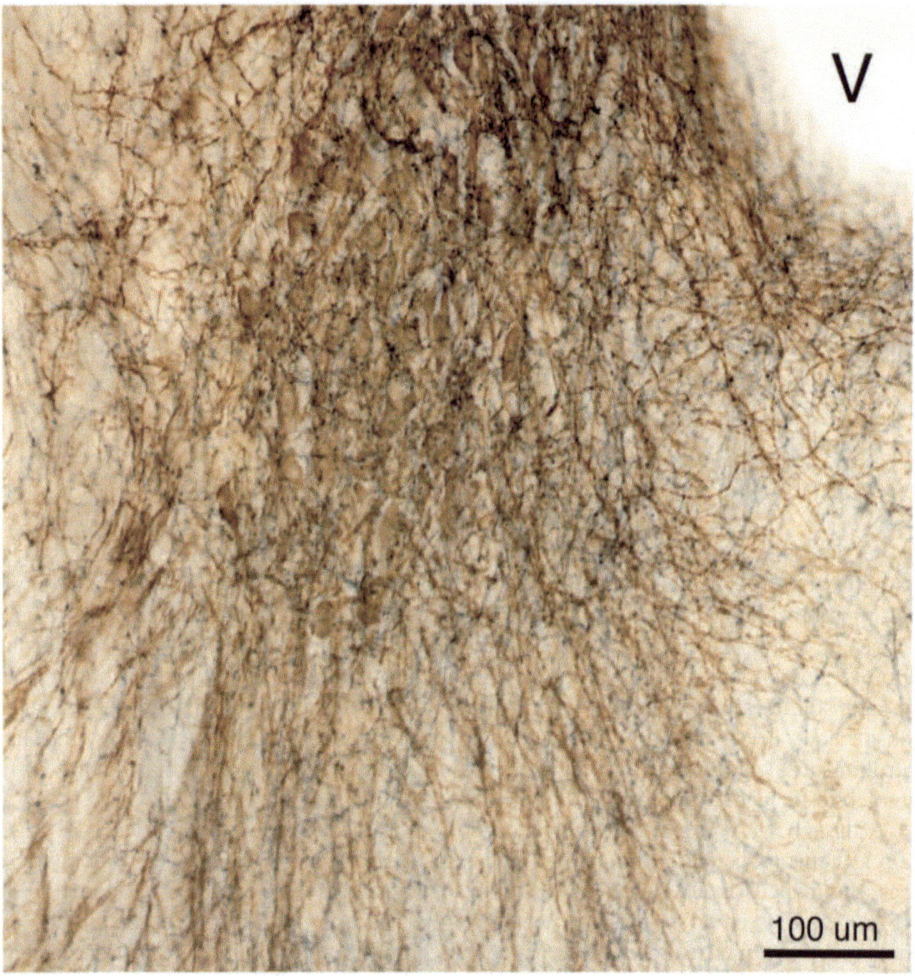

Figure 1. Hcrt-1 immunoreactive fibers in rat LC and peri-LC. Hcrt+ fibers and boutons (black) are densely located not only in the LC nucleus proper (upper) but also among TH+ LC dendrites (TH+ elements stained brown, lower). Frontal section. Medial is to the right, dorsal is at top. V – Fourth ventricle.

2. EXCITATORY EFFECT OF HCRT ON LC NEURONS.

The prominent innervation of LC neurons by Hcrt fibers[6] indicates that Hcrt has an important role in modulating the activity of the LC noradrenergic system. It was found that orexin-A administered intracerebroventricularly activated LC neurons.[8] To better understand the cellular-level effects of Hcrt on LC activity, we studied the effect of Hcrt administration in single LC neurons recorded intracellularly in brain slices.

As we described recently,[6] LC neurons were consistently activated by Hcrt in the presence of TTX (1 uM; Fig. 2), indicating that the increase in LC discharge was directly produced by activation of Hcrt receptors on LC neurons. The increased spike frequency produced by Hcrt was associated with a reduced amplitude of the slow component of the after hyperpolarization (AHP) following single action potentials ($-20.3 \pm 4.9\%$, $p<0.05$; Fig. 2). This enhanced repolarization appeared to contribute to the reduction in interspike interval. Hcrt also suppressed the AHP of LC neurons in the presence of TTX+Ba^{2+} to a similar extent ($-23.2 \pm 11.7\%$). At the same time, Hcrt did not change the amplitude or shape of Ca^{2+}-spikes recorded in the presence of TTX + Ba^{2+}.

In addition to these changes in spike frequency and AHP, other studies showed that bath-applied Hcrt consistently evoked a small depolarization of LC neurons (Fig. 3; 3.5 ± 0.3 mV, $p< .05$).[7] Puff application of Hcrt (50 - 100 μM in the puff pipette) depolarized LC neurons more clearly (6.9 ± 0.8 mV, n = 5 cells; $p < .02$). Application of Co^{2+} (1 mM) to the ACSF containing TTX (1 μM) effectively suppressed Ca^{2+} -spikes but not the depolarization evoked by Hcrt (7.2 ± 0.7 mV, n= 12). Cd^{2+} (1 mM) also was tested with similar results. Importantly, the apparent input resistance of these neurons was significantly higher during Hcrt -induced depolarization (174 ± 13 mΩ, n=5) then before Hcrt (150 ± 9.0 mΩ; $p<0.05$, n=5).

These findings indicate that HCRT–evoked depolarization was produced by channels other than TTX-sensitive Na^+ channels, or Co^{2+}- or Cd^{2+}-sensitive Ca^{2+} channels, and that Hcrt decreased conductance in LC neurons. Additional studies revealed that HCRT generated smaller depolarizations in LC neurons when they were hyperpolarized 15 – 20 mV from the resting membrane potential (5.2 ± 1.0 mV) compared to responses observed without such hyperpolarization (9.2 ± 1.3 mV; n=4, $p<0.05$, Fig. 3).[7] Moreover, the HCRT-evoked depolarization was not observed in three LC neurons where MP was shifted to –90 mV (near the K equilibrium potential), but a depolarizing response appeared when the MP was returned to rest (Fig 2).

Together, the above results indicate that stimulation of Hcrt receptors on LC neurons may decrease a resting K^+ conductance in LC neurons, leading to depolarization and increased spike rate.[7] Our results cannot exclude the participation the other ions in the Hcrt response because in some cells the depolarization developed without any change in apparent input resistance and could not be completely suppressed by strong hyperpolarization. However, the latter might result from Hcrt actions on distal dendrites of LC neurons where hyperpolarized current applied to the soma is much less effective in altering membrane potential. Additional experiments are needed to determine if a decrease in K+ conductance is only the mechanism of HCRT action on LC neurons, or is a part of more complex ionic mechanisms.

3. HCRT INNERVATION OF THE LC.

As previously reported, the rat and monkey LC receives a dense innervation by Hcrt fibers. As shown in Fig. 1, fibers and terminals are dense not only throughout the LC nucleus proper at all levels, but also in the peri-LC area that contains a large number of distal LC dendrites.[17] It is notable that the LC contains by far more Hcrt fiber and

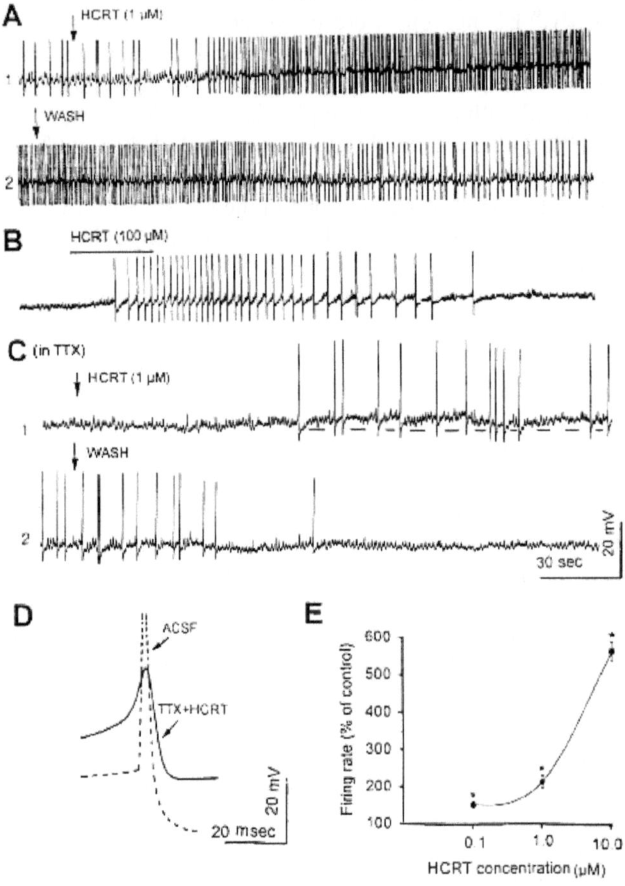

Figure 2. Hcrt activates LC neurons. **A:** Bath application of Hcrt-2 activates a typical LC neuron recorded in a slice preparation (1 μM Hcrt-2 applied beginning at arrow in trace 1; wash begins at arrow in trace 2). Both traces from same LC neuron. **B:** ''Puff'' application of Hcrt-2 to an LC neuron from a micropipette (30-second duration of application indicated by the line above the record). **C**1: Hcrt-2 (1 μM) in the presence of TTX (1 μM) generates a small depolarization and initiates spontaneous discharge (apparently Ca2+ spikes; see panel D) 2: Washout of Hcrt-2 in the continued presence of TTX. Both traces from the same LC neuron. **D:** A spike obtained during Hcrt-2 application in the presence of TTX at a faster sweep speed. Note broad, slow spike waveform typical of Ca2+ spikes compared with a typical Na+ spike found before TTX application. **E:** Average effects of different concentrations of Hcrt-2 on the spontaneous discharge rate of LC neurons recorded intracellularly in slices. Action potential frequency was significantly increased by all concentrations of Hcrt-2. An increase in spike frequency was seen even in the presence of TTX (1 μM) when Hcrt-2 was applied. Taken from ref 6.

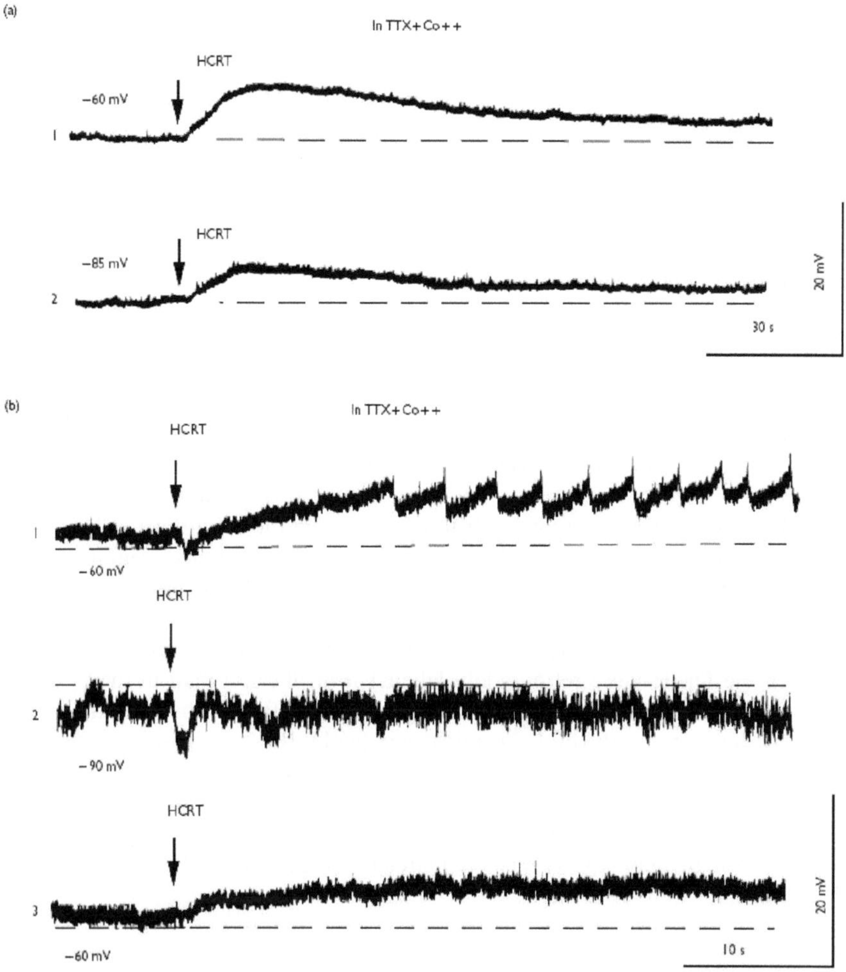

Figure 3. Hcrt-evoked depolarization demonstrates voltage dependence. (a1,2) Two depolarizing responses to a puff application of Hcrt-2 (as in Fig. 1). Both responses were recorded in the same cell, either at resting membrane potential (a1), or 10 min later when the cell was hyperpolarized to -85mV (a2). Note a smaller depolarization in the latter. (b) Three puff applications of Hcrt-2 (30 s application) to the same cell, at 10 min intervals and different membrane potentials as indicated. The first application evoked a clear depolarization at resting membrane potential (b1). The second application, after the membrane potential was shifted to approx. -90 mV, produced no depolarization (b2). The third application again depolarized the cell when the membrane potential was returned to the previous resting level (b3). TTX (1 uM) and Co2+ (1 mM) were added to the bath 3 min before trace b1 was recorded, and were continuously present during the following records. Note that Ca2+ spikes, and associated membrane oscillations, but not the Hcrt-induced depolarization, were blocked by Co2+. The dotted lines with the numbers below indicate the level of membrane potential in all records. a and b represent LC neurons from different slices. Taken from ref 7.

terminal innervation than other nearby regions, indicating that it is a highly favored target.

Although Hcrt neurons are located within a circumscribed area of the hypothalamus, it was not known if the dense LC innervation originated in a specific region or cluster of

Hcrt neurons. Thus, we sought to determine if the Hcrt neurons were composed of projection specific subpopulations. As a first step in that effort we analyzed the distribution and density of Hcrt neurons in the rat and monkey hypothalamus to determine if there was evidence of topographical subdivisions. As shown in Fig. 4, in rat we delineated 3 subregions of the hypothalamus area that contains Hcrt neurons: the dorsomedial, perifornical and lateral hypothalamus (mediorostral to laterocaudal, respectively). In monkey, four clusters of Hcrt neurons were evident along the rostrocaudal axis of the hypothalamus. These rough delineations were based upon apparent groupings of Hcrt neurons in the rostrocaudal and mediolateral axes, and proximity of Hcrt neurons to hypothalamic landmarks.

To determine if the topographical delineations that we observed represented segregation of projection specific populations of neurons, we conducted tract-tracing studies with the beta subunit of cholera toxin (CTb), a retrograde tracer. Injections of CTb that encompassed the core LC nucleus and peri-LC dendritic zone yielded numerous retrogradely labeled neurons in the hypothalamus of both rats and monkeys. Many of these retrogradely labeled LC afferent neurons also stained for Hcrt in both species, confirming the hypothalamic input that is presumed to underlie the Hcrt innervation of the LC (Fig. 4). Notably, there were differences among the subgroups of Hcrt neurons in the numbers and percentages of cells that were retrogradely labeled from the LC. In rat, a significantly higher percentage of LC afferents in the DMH and perifornical region stained for Hcrt than in the lateral hypothalamus (16%, 15% and 10%, respectively; $p<0.05$). Analyzed differently, we found that a significantly higher percentage of Hcrt neurons in the DMH were retrogradely labeled from the LC than in the perifornical or lateral hypothalamus (18%, 6% and 5%, respectively; $p<0.05$). These results indicate that the bulk of Hcrt input to the rat LC originates in the DMH and perifornical hypothalamus, and that Hcrt neurons in the DMH are more likely to project to the LC than Hcrt neurons in other hypothalamic subfields. The functional significance of this topography of Hcrt projections to the LC warrants further investigation, but may correspond to suprachiasmatic nucleus (SCN)-mediated circadian regulation of LC activity and arousal, as described later in this chapter.

4. GABA INTERNEURONS IN THE PERI-LC: TARGET FOR HCRT INPUTS

We recently examined inputs to the LC nucleus from the neurons in the dendritic field of the LC using microinjection of retrograde tract-tracers into the LC nucleus proper [18]. Focal microinjections of the sensitive tracer wheat germ agglutinin conjugated to apo (inactivated) horseradish peroxidase and coupled to colloidal gold (WGA-Au) labeled a discrete population of small neurons in the peri-LC surround. Double labeling studies revealed that these neurons often stained for the GABA synthetic enzyme glutamate decarboxylase (GAD; Fig. 5). Counts of neurons in 12 sections from 3 rats demonstrated that approximately 45% of WGA-Au-labeled afferents to the LC in the peri-LC zone also stained for GAD. Thus, these local neurons contribute to the dense GABA fiber input known to innervate the LC nucleus.[18]

In this same study, we also used the retrograde transynaptic tracer Pseudorabies virus (PRV) to study the possible inputs to distal LC dendrites in the peri-LC area.[17,18] This tracer is particularly well suited for such a study because following focal injection into a cell body area such as the LC it is transported out distal LC dendrites where it

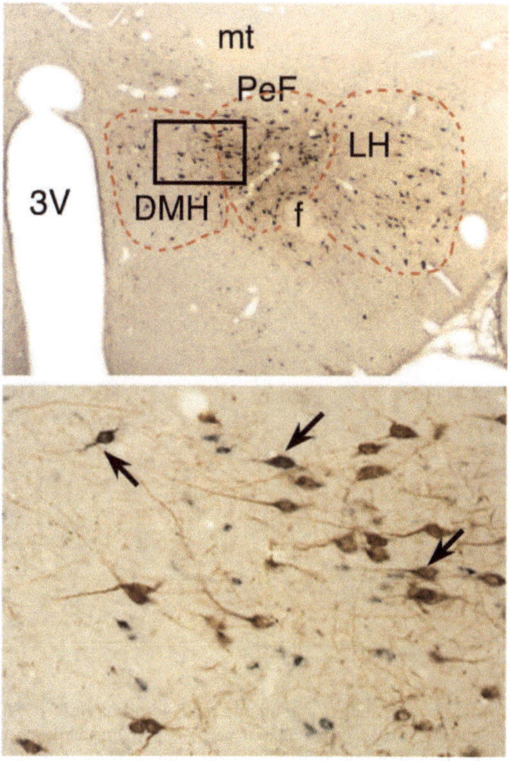

Figure 4. Upper: Frontal sections at low-power through the rat hypothalamus double stained for Hcrt (brown) and CTb (retrogradely transported from the LC, black). These images depict the different areas that were defined for counting Hcrt+ neurons: DMH - dorsomedial hypothalamus; PeF - perifornical hypothalamus; LH - lateral hypothalamus. Other abbreviations: 3V - third ventricle; mt - mammilothalmic tract; f - fornix. Lower: High-power photo taken from box in upper panel, showing doubly labeled neurons in the DMH and PeF regions.

transsynaptically labels afferents to those dendrites.[19] Following focal PRV injections into the LC, we found that neurons in areas known to be direct afferents to the LC (e.g., the nucleus paragigantocellularis)[20] became labeled with PRV at about 23 hr of survival. Many neurons in the peri-LC also exhibited PRV labeling at this time, consistent with the observations above made with WGA-Au indicating direct inputs to the LC proper. However, a time-course analysis revealed that PRV+ neurons became much more numerous in the peri-LC at 35 hr of survival (188.2 $\pm$ 20.5 vs. 74.8 $\pm$ 9.7 cells per section, respectively). These results indicate that peri-LC labeling is mediated by PRV transport and replication rather than spread of injection outside of the LC. We interpret the differing magnitude of infection of peri-LC neurons at 23 hours and 35 hours as evidence for two populations of neurons with differing synaptic relations with the LC. Specifically, we postulated that the differing temporal course of infection of the two populations of neurons reflected either first-order infection through projections into the LC nuclear core (23 hour group) or second-order infection resulting from transneuronal

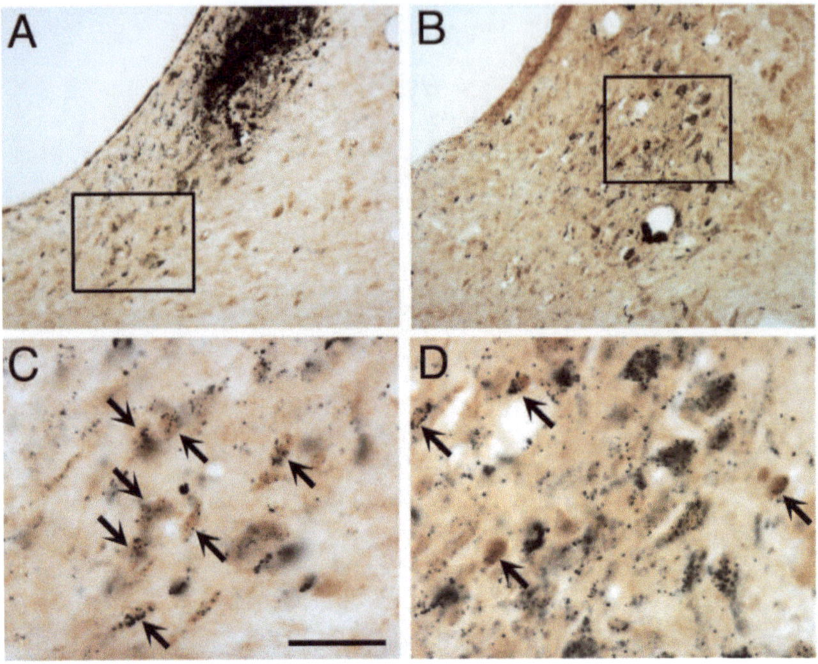

Figure 5. LC afferents in peri-LC zone stain for glutamic acid decarboxylase (GAD). Photomicrographs of frontal sections through the ventral LC and ventromedial peri-LC showing neurons labeled for GAD and the retrograde tracer WGA-Au after injection into the LC nuclear core. Arrows indicate neurons stained for both WGA-Au and GAD. *A, B*, Bright-field photomicrographs of mid- (*A*) and rostral (*B*) peri-LC areas stained for both GAD (brown) and the retrograde tracer WGA-Au (black particulate) in a case with a focal WGA-Au injection in the LC nuclear core. *C, D*, High-power photos taken from areas in *A* and *B*. Note that WGA-Au-labeled LC afferents (filled with black particles) often also stained for GAD (diffuse brown stain). Arrows indicate neurons stained for both WGA-Au and GAD. For all panels, ventral is down, medial is left. Scale bars: (in *A*) *A, B*, 250 um; *C, D*, 50 um. Taken from.[18]

passage of virus from dendrites of infected LC neurons (35 hours group) [18]. This time course is consistent with the replication cycle of PRV.[21]

We extended these studies using electron microscopy (EM) to ultrastructurally examine the innervation of LC neurons by these peri-LC GABA neurons. In counts of EM material, we found that 101/131 (77%) PRV-labeled neurons in the peri-LC stained for GABA. Moreover, GABA terminals frequently made symmetrical synaptic contacts onto PRV+ or TH+ dendrites in the peri-LC region. Some of these GABA terminals contained PRV particles, as did some of the TH+ dendrites that received input from GABA terminals (Fig. 6). These results are consistent with the interpretation that PRV virions were transported into peri-LC GABA neurons from LC dendrites, indicating a synaptic connection from local peri-LC GABA neurons to LC distal dendrites. We also found that GABA+ somata and dendrites in the peri-LC received numerous symmetrical and asymmetrical synaptic inputs, indicating that these GABA neurons integrate numerous inputs to this region.

As noted above, the peri-LC dendritic zone that contains these GABA interneurons also contains dense Hcrt innervation. In recent studies, we carried out ultrastructural

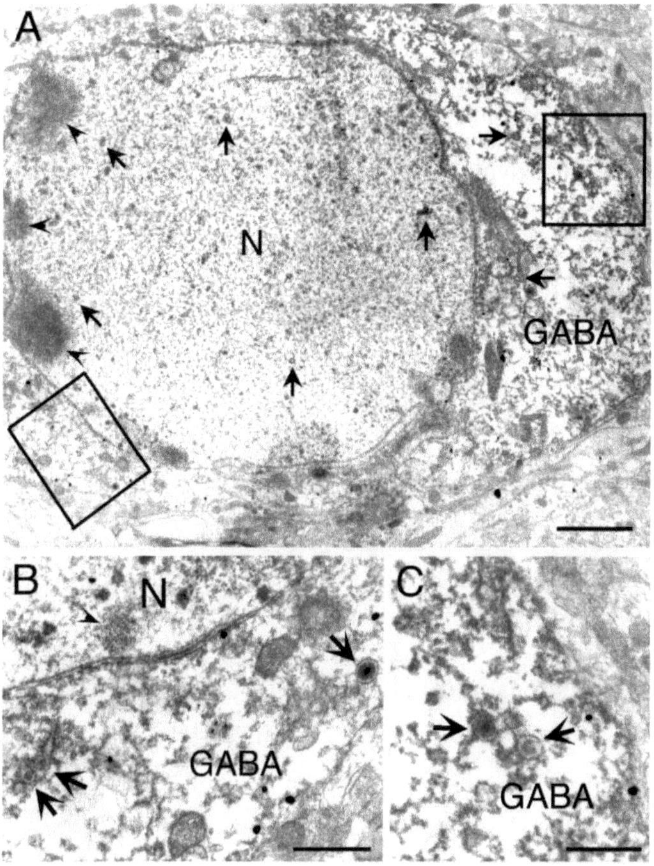

Figure 6. Ultrastructure of GABAergic LC afferents. EM photomicrograph of a neuron in the rostroventral peri-LC immunolabeled for PRV (stained with DAB) and GABA (stained with silver-intensified gold). High-power images in *B* and *C* are taken from areas indicated in the low-power image in *A*. Note the virions in the nucleus (N) at arrows in *A* and the nuclear inclusions in *A* and *B*, indicative of PRV infection (at arrowheads). Note also the cytoplasmic virions (indicated by arrows in *B* and *C*), and abundant GABA staining in the cytoplasm. Scale bars: *A*, 2 um; *B, C*, 0.5 um. Taken from ref 18.

studies using immunocytochemistry to examine whether these GABA interneurons receive direct Hcrt innervation.[22] For this, we examined Hcrt and GABA staining in the peri-LC of rats that had received focal PRV injections into the LC. Results revealed that a high percentage of Hcrt terminals contacted GABA+ profiles, and virtually all of the Hcrt-GABA synaptic contacts were asymmetrical. This high prevalence of Hcrt-GABA contacts indicates that GABA elements are a prominent Hcrt target in the peri-LC region. The majority of such contacts were onto dendrites. In cases with PRV injections into the LC nuclear core, PRV profiles were frequently seen in GABA+ dendrites or somata that were post-synaptic to Hcrt+ terminals (Fig. 7). Again, the prevalent synaptic arrangement was asymmetrical. These results indicate that Hcrt fibers not only make contact onto noradrenergic LC neurons as previously reported, but also make frequent contact onto neighboring GABAergic neurons that in turn contact LC neurons. The finding that the bulk of these Hcrt-GABA contacts are asymmetrical indicates that Hcrt may be excitatory on these GABA interneurons.

As GABA is strongly inhibitory on LC neurons,[23,24] these results in turn indicate that Hcrt has both direct excitatory effects on LC neurons (as reviewed above) and inhibitory effects on LC neurons via a feedforward circuit involving local GABA interneurons. Thus, the effect of activity in Hcrt neurons on LC function is more complex than originally envisioned. As these GABA interneurons that synapse onto LC cells integrate a number of different inputs, the effect of Hcrt inputs on LC activity may be gated or modulated by other afferents onto these GABA neurons. Additional studies are called for to understand the functional consequences of this additional level of integration for Hcrt influences on LC activity.

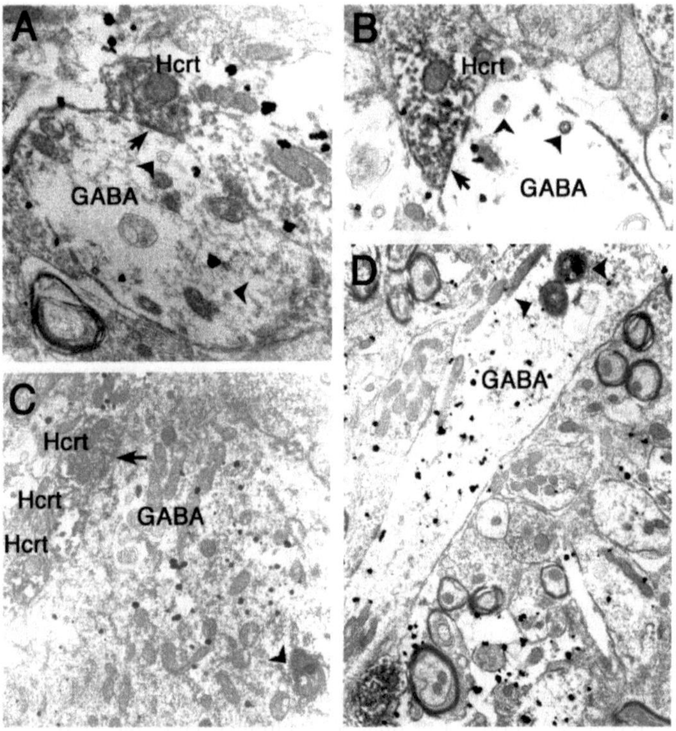

Figure 7. Electron micrographs from the peri-LC area in a case in which PRV was injected into the LC nuclear core. Tissue was immunolabeled for Hcrt (DAB reaction product) and GABA (silver-intensified gold). PRV profiles (indicated at arrowheads) were present within GABA+ profiles (cells and dendrites) either as individual virions (panels A and B) or sequestered within multivesicular bodies (panels C and D). The morphology of virions and their presence within multivesicular bodies is consistent with the life cycle previously established for PRV in the rodent CNS (Card et al., 1993). Arrows indicate synaptic contacts between Hcrt+ and GABA+ profiles infected with PRV.

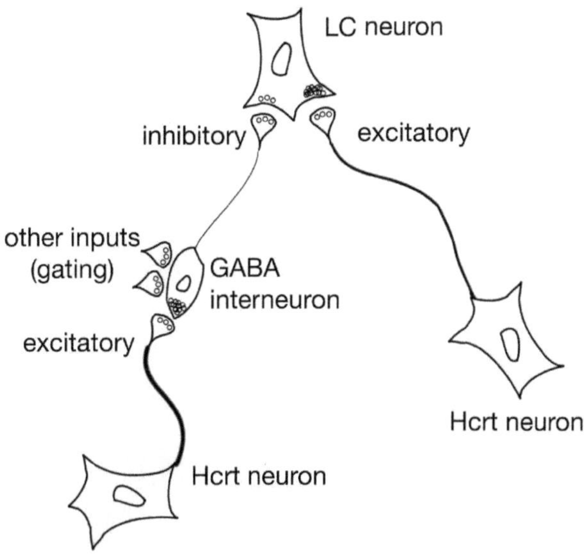

Figure 8. Illustration of direct excitatory Hcrt innervation of LC neurons, and of feedforward inhibitory Hcrt influence on LC via GABA interneurons that project to LC.

5. A CIRCUIT FROM THE SCN TO THE LC AND CIRCADIAN REGULATION OF AROUSAL: A POSSIBLE ROLE FOR HCRT?

Neurons in the hypothalamic SCN exhibit an endogenous circadian rhythm of activity that plays a prominent role in the temporal organization of behaviors. However, the projections of the SCN are largely confined to the hypothalamus and there is no direct projection to the LC. Given that both the SCN and the LC play prominent roles in the regulation of behavioral state we sought to determine if the SCN influences LC activity through polysynaptic connections. Toward that end we demonstrated that focal microinjection of PRV into the rat LC infected a large number of neurons in the SCN.[25] This labeling occurred at a survival time of greater than 44 hr, indicating an indirect circuit connection, consistent with prior work indicated that there is no direct projection from the SCN to the LC. To identify the relay through which the SCN became infected we conducted a series of double labeling anatomical and lesion studies. Our results showed that the dorsomedial nucleus of the hypothalamus (DMH) was a major relay in this circuit. For example, as shown in Fig. 9 cell body-specific lesions of the DMH substantially decreased the number of PRV-labeled neurons in the SCN ipsilateral to the injected LC.[25]

These results, in view of the role of the SCN as the brain's primary regulator of circadian rhythmicity, indicated that the LC may have a circadian rhythm. This

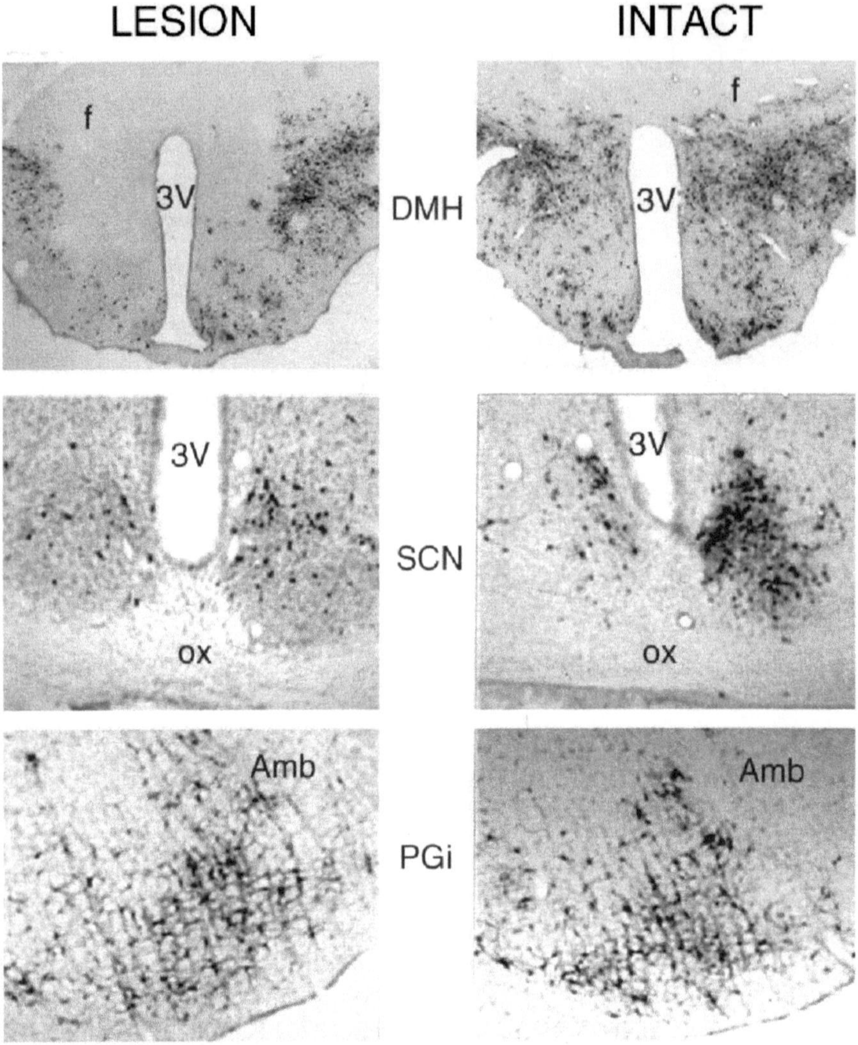

Figure 9. PRV labeling following lesions of the dorsomedial hypothalamic nucleus (DMH). Labeled neurons following PRV injection into the LC, in DMH-lesioned (left) and intact animals (right). The DMH was lesioned bilaterally with ibotenic acid, and two weeks later, the animal received an injection of PRV into the LC. Both intact and lesioned animals were perfused 56 h after the PRV injection. Top, DMH stained for PRV immunoreactivity. Note the lack of PRV staining bilaterally in the DMH of the lesioned animal. Middle, note decreased number of labeled neurons in the SCN of the lesioned animal compared to the intact animal. Bottom, there was no difference in PRV labeling between these two animals in the nucleus paragigantocellularis lateralis, a direct LC afferent in the rostral ventral medulla. Taken from ref 25.

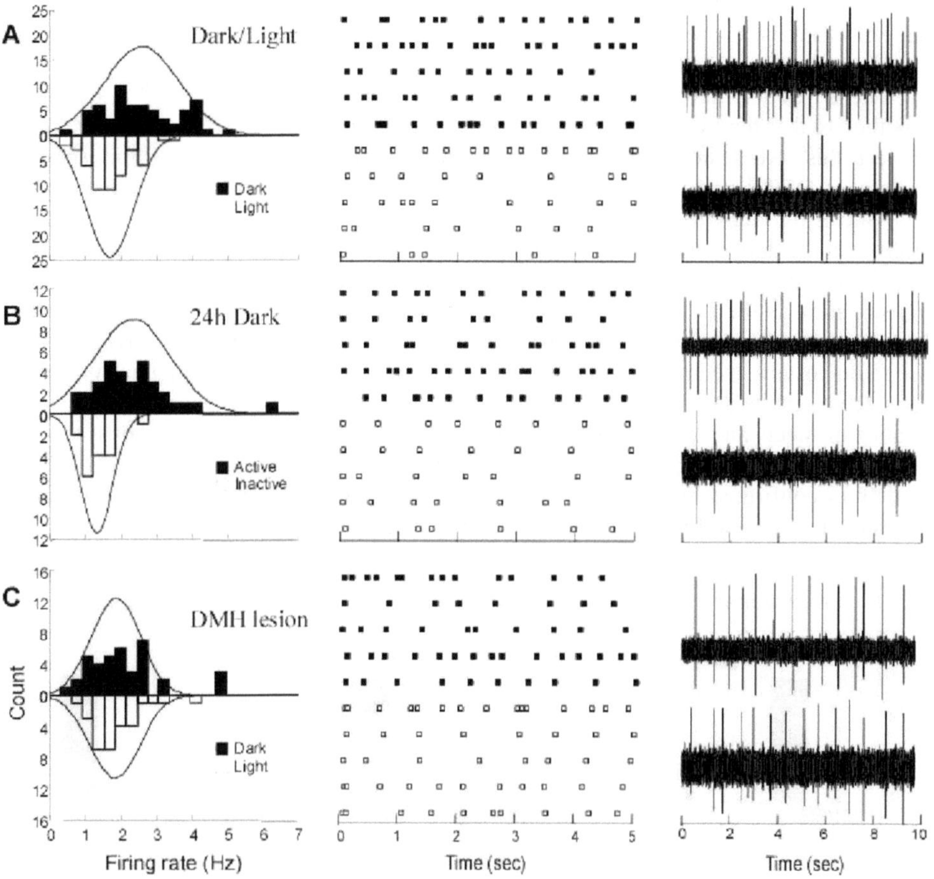

Figure 10. LC impulse activity during the circadian cycle. Histograms, rasters of impulse activity and electrophysiological traces showing the distributions of LC firing rates during different epochs of the circadian cycle. Three paired groups of rats were maintained in either 12-h/12-h dark/light cycles (**a**, **c**) or 24 h of darkness (**b**). The raster and oscilloscope traces were taken from the group of rats labeled in the corresponding histograms to the left. Rasters are composed of 5 sec epochs from five randomly chosen LC neurons from each corresponding group. One paired group received bilateral ibotenic acid lesions of the DMH (**c**). Impulse activity was recorded in LC neurons during either the dark or light photoperiod (**a**, **c**) or the active or inactive epoch of the rat's circadian cycle in 24-h darkness (**b**). There was a significant difference between LC firing rates during dark and light periods (**a**), and between LC firing rates during active versus inactive periods in continuous darkness (**b**; $p < 0.0001$ for each). Lesions of the DMH eliminated the difference in LC firing rates during the dark versus light photoperiods (**c**). Scale on the *y*-axis of the histograms corresponds only to the histogram bars. Solid lines on each of the graphs, best-fit normal distribution curves for the histogram data. Taken from ref 25.

possibility had not been previously examined without the confound of sleep and waking changes, which also are known to strongly influence LC activity. We examined LC activity in Spague Dawley rats taken from the light (rest) or dark (active) cycle and placed under halothane anesthesia. In this way, LC activity was recorded during a similar state (anesthesia) during different periods of the circadian cycle. Results showed that LC neurons exhibit a circadian rhythm with the highest activity in the active period, and that this rhythm depends upon an intact DMH (Fig. 10). This set of findings confirmed that the SCN-DMH-LC circuit revealed in our anatomical studies was functional and provided circadian information to the LC.

These results, and the role of the LC in regulation of arousal,[26,27] led us to hypothesize that the LC system plays a role in circadian regulation of sleep and waking.[28] To test this, we lesioned LC projections with systemic injections of the selective neurotoxin DSP-4 and recorded sleep and waking in behaving rats using telemetry for EEG and EMG activities. Results showed that lesioned rats exhibited less waking than controls in the dark (active) period, but more waking than controls in the rest period (Fig. 11).[29] The former result is expected with lesion of a system whose activity produces arousal, such as the LC. However, why would these lesions also produce more waking than normal in the rest period? We propose that this results from decreased drive from the sleep homeostatic mechanism, in response to less waking in the preceding active period. Thus, these results demonstrate that although LC lesions produce no overall change in the amount of sleep and waking over a 24r hr period (consistent with previous reports), damage to this system decreases the circadian rhythm of the sleep-waking cycle. This result was confirmed by a cosinor analysis, showing that the circadian rhythms of all stages of sleep and waking were decreased by 27% to 28%.[29]

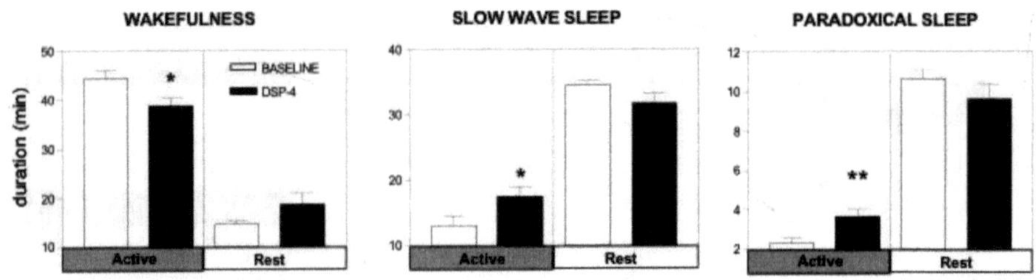

Figure 11. Hourly mean duration (in min) of each vigilance state during the active period (night) and rest period (day) under 12:12 LD, n=6. Open bars - before DSP4 lesion; filled bars - after DSP4 lesion. * $P < 0.05$, ** $P < 0.01$. Note decreased waking but increased sleep in the active period following DSP4 lesions, indicating a decreased circadian rhythm of sleep-waking cycles.

These results are important not only for showing that the LC is involved in circadian regulation of sleep and waking, but by demonstrating a specific neural circuit for the circadian regulation of arousal. Notably for the present article, these results demonstrate that the DMH is an important component of this circuit, and are the first to indicate a role for the DMH in either sleep-wake or circadian function. Recent results by Chou et al. [30] confirm that lesions of the DMH substantially interfere with circadian rhythms of sleep and waking.

Our recent results show that numerous Hcrt+ neurons project to the LC from the DMH and adjacent hypothalamic regions (reviewed above). This, combined with the role of the DMH as a relay between the SCN and the LC, suggests the possibility that Hcrt projections to the LC may convey circadian information to the LC from the SCN. While admittedly speculative, this view is consistent with the role of Hcrt as a wake-promoting neuropeptide. Thus, activity of GABA neurons in the SCN (the most prominent cell type there) during the light cycle could, in nocturnal animals, inhibit Hcrt neurons that normally activate arousal-inducing LC cells. Decreased activity in the SCN during the dark cycle would disinhibit LC neurons by this same circuit. In this way, Hcrt projections to the LC from the DMH could participate in circadian regulation of arousal. Such a function is also compatible with results showing that Hcrt deficits are associated with narcolepsy. Lacking a sustained stimulation of LC neurons during the active period, subjects with Hcrt deficits may have unstable waking states that would lead to inappropriate sleep epochs such as seen in narcoleptics. More work is needed to examine this possible function for Hcrt innervation of the LC.

6. ACKNOWLEDGEMENTS

This work was supported by PHS grant NS24698.

7. REFERENCES

1. L. de Lecea, T. S. Kilduff, C. Peyron, X. Gao, P. E. Foye, P. E. Danielson, C. Fukuhara, E. L. Battenberg, V. T. Gautvik, F. S. Bartlett, 2nd, W. N. Frankel, A. N. van den Pol, F. E. Bloom, K. M. Gautvik and J. G. Sutcliffe, The hypocretins: hypothalamus-specific peptides with neuroexcitatory activity, *Proc Natl Acad Sci U S A.* **95**, 322-7 (1998).
2. T. Sakurai, A. Amemiya, M. Ishii, I. Matsuzaki, R. M. Chemelli, H. Tanaka, S. C. Williams, J. A. Richardson, G. P. Kozlowski, S. Wilson, J. R. Arch, R. E. Buckingham, A. C. Haynes, S. A. Carr, R. S. Annan, D. E. McNulty, W. S. Liu, J. A. Terrett, N. A. Elshourbagy, D. J. Bergsma and M. Yanagisawa, Orexins and orexin receptors: a family of hypothalamic neuropeptides and G protein-coupled receptors that regulate feeding behavior [see comments], *Cell.* **92**, 573-85 (1998).
3. C. Peyron, D. K. Tighe, A. N. van den Pol, L. de Lecea, H. C. Heller, J. G. Sutcliffe and T. S. Kilduff, Neurons containing hypocretin (orexin) project to multiple neuronal systems, *J Neurosci.* **18**, 9996-10015 (1998).
4. S. Taheri, J. M. Zeitzer and E. Mignot, The role of hypocretins (orexins) in sleep regulation and narcolepsy, *Annu Rev Neurosci.* **25**, 283-313 (2002).
5. J. G. Sutcliffe and L. de Lecea, The hypocretins: setting the arousal threshold, *Nat Rev Neurosci.* **3**, 339-49 (2002).
6. T. L. Horvath, C. Peyron, D. Sabrina, A. Ivanov, G. Aston-Jones, T. Kilduff and A. N. van den Pol, Strong hypocretin (orexin) innervation of the locus coeruleus activates noradrenergic cells, *J. Comp. Neurol.* **415**, 145-159 (1999).
7. A. Ivanov and G. Aston-Jones, Hypocretin/orexin depolarizes and decreases potassium conductance in locus coeruleus neurons, *Neuroreport.* **11**, 1755-1758 (2000).
8. J. J. Hagan, R. A. Leslie, S. Patel, M. L. Evans, T. A. Wattam, S. Holmes, C. D. Benham, S. G. Taylor, C. Routledge, P. Hemmati, R. P. Munton, T. E. Ashmeade, A. S. Shah, J. P. Hatcher, P. D. Hatcher, D. N. Jones, M. I. Smith, D. C. Piper, A. J. Hunter, R. A. Porter and N. Upton, Orexin A activates locus coeruleus cell firing and increases arousal in the rat, *Proc Natl Acad Sci U S A.* **96**, 10911-6 (1999).
9. P. Bourgin, S. Huitron-Resendiz, A. D. Spier, V. Fabre, B. Morte, J. R. Criado, J. G. Sutcliffe, S. J. Henriksen and L. de Lecea, Hypocretin-1 modulates rapid eye movement sleep through activation of locus coeruleus neurons, *J Neurosci.* **20**, 7760-5 (2000).
10. R. E. Brown, O. A. Sergeeva, K. S. Eriksson and H. L. Haas, Convergent excitation of dorsal raphe serotonin neurons by multiple arousal systems (orexin/hypocretin, histamine and noradrenaline), *J Neurosci.* **22**, 8850-9 (2002).
11. R. J. Liu, A. N. van den Pol and G. K. Aghajanian, Hypocretins (orexins) regulate serotonin neurons in the dorsal raphe nucleus by excitatory direct and inhibitory indirect actions, *J Neurosci.* **22**, 9453-64 (2002).
12. K. S. Eriksson, O. Sergeeva, R. E. Brown and H. L. Haas, Orexin/hypocretin excites the histaminergic neurons of the tuberomammillary nucleus, *J Neurosci.* **21**, 9273-9 (2001).
13. L. Bayer, E. Eggermann, M. Serafin, B. Saint-Mleux, D. Machard, B. Jones and M. Muhlethaler, Orexins (hypocretins) directly excite tuberomammillary neurons, *Eur J Neurosci.* **14**, 1571-5 (2001).

14. E. Eggermann, M. Serafin, L. Bayer, D. Machard, B. Saint-Mleux, B. E. Jones and M. Muhlethaler, Orexins/hypocretins excite basal forebrain cholinergic neurones, *Neuroscience.* **108**, 177-81 (2001).
15. E. Mignot, S. Taheri and S. Nishino, Sleeping with the hypothalamus: emerging therapeutic targets for sleep disorders, *Nat Neurosci.* **5 Suppl**, 1071-5 (2002).
16. M. Mieda and M. Yanagisawa, Sleep, feeding, and neuropeptides: roles of orexins and orexin receptors, *Curr Opin Neurobiol.* **12**, 339-45 (2002).
17. M. T. Shipley, L. Fu, M. Ennis, W. Liu and G. Aston-Jones, Dendrites of locus coeruleus neurons extend preferentially into two pericoerulear zones, *J. Comp. Neurol.* **365**, 56-68 (1996).
18. G. Aston-Jones, Y. Zhu and P. Card, Numerous GABAergic afferents to locus coeruleus in the pericoerulear dendritic zone: possible interneuronal pool, *J. Neurosci.* **24**, 2313 – 2321 (2004).
19. G. Aston-Jones and J. P. Card, Use of Pseudorabies virus to delineate multisynaptic circuits in brain: opportunities and limitations, *J. Neurosci. Meth.* **103**, 51-61 (2000).
20. G. Aston-Jones, M. Ennis, V. A. Pieribone, W. T. Nickell and M. T. Shipley, The brain nucleus locus coeruleus: restricted afferent control of a broad efferent network, *Science.* **234**, 734-737 (1986).
21. L. W. Enquist, P. J. Husak, B. W. Banfield and G. A. Smith, Infection and spread of alphaherpesviruses in the nervous system, *Advances in Viral Research.* **51**, 237-347 (1999).
22. Y. Zhu, J. P. Card and G. Aston-Jones, Frequent hypocretin innervation of GABA neurons in the peri-locus coeruleus region: ultrastructural studies, *Soc. Neurosci. Abstr.* **29**, 931.15 (2003).
23. S. S. Osmanovic and S. A. Shefner, gamma-Aminobutyric acid responses in rat locus coeruleus neurones in vitro: a current-clamp and voltage-clamp study, *J. Physiol.* **421**, 151-70 (1990).
24. M. Ennis and G. Aston-Jones, GABA-mediated inhibiton of locus coeruleus from the dorsomedial rostral medulla, *J. Neurosci.* **9**, 2973-2981 (1989).
25. G. Aston-Jones, S. Chen, Y. Zhu and M. Oshinsky, A neural circuit for circadian regulation of arousal, *Nature Neurosci.* **4**, 732-738 (2001).
26. G. Aston-Jones, S. L. Foote and F. E. Bloom. (1984). Anatomy and physiology of locus coeruleus neurons: functional implications. In *Norepinephrine (Frontiers of Clinical Neuroscience, Vol. 2)* (Ziegler, M. & Lake, C. R., eds.), pp. 92-116. Williams and Wilkins, Baltimore.
27. C. W. Berridge and B. D. Waterhouse, The locus coeruleus-noradrenergic system: modulation of behavioral state and state-dependent cognitive processes, *Brain Res Brain Res Rev.* **42**, 33-84 (2003).
28. G. Aston-Jones, M. M. Gonzalez and S. M. Doran. (in press). Role of the Locus Coeruleus-Norepinephrine System in Arousal and Circadian Regulation of the Sleep-Waking Cycle. In *Norepinephrine: Neurobiology and Therapeutics for the 21st Century* (Ordway, G. A., Schwartz, M. & Frazer, A., eds.).
29. M. C. Gonzalez, W. Lu and G. Aston-Jones, Role of the noradrenergic locus coeruleus system in sleep and waking: circadian factors, *Soc. Neurosci. Abstr.* **28**, 871.11 (2002).
30. T. C. Chou, T. E. Scammell, J. J. Gooley, S. E. Gaus, C. B. Saper and J. Lu, Critical role of dorsomedial hypothalamic nucleus in a wide range of behavioral circadian rhythms, *J Neurosci.* **23**, 10691-702 (2003).

HYPOCRETIN/OREXIN ACTIONS ON MESOPONTINE CHOLINERGIC SYSTEMS CONTROLING BEHAVIORAL STATE

Christopher S. Leonard, Christopher J. Tyler, Sophie Burlet, Shigeo Watanabe and Kristi A. Kohlmeier*

1. INTRODUCTION

In the short time since the hypocretin/orexin peptide system was discovered,[1,2] a remarkable body of evidence has accumulated indicating this system's important role in regulating feeding, energy metabolism and arousal.[3-6] Considerable progress has also been made identifying the CNS targets and actions of the two hypocretin/orexin peptides (Hcrt/Orx-A and -B) [cf 7]. We have been interested in understanding the actions of Hcrt/Orx on brainstem cholinergic and monoaminergic systems since these systems are involved with controlling arousal and sleep. Considering the extensive innervation by Hcrt/Orx afferents[8] and the robust expression of orexin receptor message in these structures,[9] it is likely that important components of the wake-promoting and REM suppressing abilities of Hcrt/Orx[10,11] are mediated through actions on these systems.[12-14] Here we discuss some of our recent findings concerning Hcrt/Orx actions on both the cholinergic and non-cholinergic neurons of the laterodorsal tegmental (LDT) nucleus which form part of the ascending cholinergic reticular system.[15,16]

The laterodorsal (LDT) nucleus and neighboring pedunculopontine (PPT) tegmental nucleus contain mesopontine cholinergic (MPCh) neurons which are intermixed with numerous non-cholinergic neurons – some of which are GABAergic.[17] These MPCh neurons express high levels of the enzyme nitric oxide synthase NOS,[18,19,20] which serves as a convenient marker and functions to produce nitric oxide (NO) in response to neural activity both locally[21] and at their terminals.[22] These MPCh neurons have wide-spread projections which include key targets in the thalamus [for review 23], pontine reticular formation (mPRF[24,25-27] and midbrain dopamine regions).[28,29]

* C. S. Leonard, C. J. Tyler, S. Burlet, S. Watanabe and K. A. Kohlmeier Dept. of Physiology, New York Medical College, Valhalla, NY 10595

Functionally, stimulation of MPCh neurons promotes long-lasting EEG desynchronization via thalamic afferents[30] and the release of dopamine from the VTA and SN neurons.[31,32] In addition, MPCh neurons play a significant role in generating REM sleep and the associated muscle atonia [for review 33] via projections to the mPRF[34] where cholinergic stimulation produces a REM-like state.[35-38] These sites are of particular interest since muscarinic stimulation of this region in narcoleptic dogs evokes cataplexy at doses that are ineffective in normal dogs.[39] Moreover, microdialysis of atropine into the mPRF blocks the effects of systemic physostigmine to increase attack frequency[39] and mPRF acetylcholine levels are elevated during these attacks,[40] suggesting that exaggerated cholinergic transmission to the mPRF is a key trigger of cataplexy.

Based on these observations and the strong evidence that disruption of Hcrt/Orx signaling produces narcolepsy/cataplexy,[41,42] we initially hypothesized that Hcrt/Orx would inhibit MPCh neurons. Contrary to our original hypothesis, we found that Hcrt/Orx excites mouse MPCh neurons through both direct and indirect mechanisms and potently stimulates the elevation of intracellular calcium in these neurons.

2. EXPERIMENTAL

Our methods for preparation of brainslices, whole-cell recordings, immunocytochemistry[43] and calcium imaging[44,45] have been previously described in detail and are only briefly summarized here.

Brain slices were prepared from 14-32 day old C57/Bl6 mice in ice-cold artificial cerebrospinal fluid (ACSF) which contained (in mM): 121 NaCl, 5 KCl, 1.2 $NaH2PO_4$, 2.7 $CaCl_2$, 1.2 $MgSO_4$, 26 $NaHCO_3$, 20 dextrose, 4.2 lactic acid and was oxygenated by bubbling with carbogen (95% O_2 and 5% CO_2). Slices containing the LDT or DR were incubated at 35°C for 15mins in oxygenated ACSF, and were then stored at room temperature, until they were used.

Neurons were visualized using a video camera and DIC optics of a fixed-stage microscope (Olympus BX50WI). Neurons were recorded in voltage clamp or current clamp mode using an Axoclamp 2A or Axopatch series amplifier (Axon Instruments). Giga-seal whole cell recordings were made with pipettes containing (in mM) 144 K-Gluconate, 0.2 EGTA, 3 $MgCl_2$, 10 HEPES, 0.3 NaGTP, 4 Na_2ATP. To study IPSCs, an internal solution with KCl substituted for K-Gluconate was used so that IPSCs were inward at resting membrane potential. Biocytin (0.1%) and Na-GTP were added to the pipette solution just before use. Pipettes used for extracellular recordings were filled with extracellular solution. PSCs were compared using Kolmogorov-Smirnov statistics with Mini Analysis software. Additional analysis and figure preparation was done using Igor Pro (Wavemetrics) software. Differences between means were compared using a two-sample, two-tailed, t-test corrected for multiple comparisons (when necessary) and repeated measure ANOVA using DataDesk 6 software (Data Description, Inc). Numerical results are reported as mean±SEM.

Fluorescence related to intracellular calcium concentration was measured from neurons in slices that had been either bulk-loaded with fura2-AM or individually filled with bis-fura2 from patch pipettes. Ca^{2+} transients were generally monitored by measuring the emission at 515 nm resulting from excitation of fura-2 with 380 nm (F_{380}). Optical recordings were made using a frame-transfer cooled CCD camera system (EEV 57 chip, Micromax System, Roper Scientific) that was controlled with custom software

(TI Workbench) running on a Mac OS computer. Images were either acquired discontinuously every 1-4 seconds with the shutter closed between images (600ms exposure) or continuously (~52ms/frame), with the shutter open for the entire epoch. Changes in intracellular calcium concentration were estimated as changes in delta F/F (dF/F) where F is the fluorescence at rest within a ROI following subtraction of background fluorescence. delta F is the change in fluorescence following subtraction of the F prior to stimulation. dF/F was usually corrected for photobleaching. Since rises in Ca^{2+} produce a decrease in F_{380} with fura2, all dF/F measures are inverted so positive-going traces indicate elevation of $[Ca^{2+}]_i$. Differences between means were determined by utilization of a paired Student's t test or a one-way ANOVA. Non-parametric comparisons were conducted utilizing Chi-Square analysis.

Cholinergic neurons in the LDT selectively colocalize the enzyme brain nitric oxide synthase bNOS, [20] and were identified in this study by immunohistochemical labeling for bNOS using conventional immunocytochemistry following slice fixation in 4% paraformaldehyde. Cryoprotected brain slices were re-sectioned at 40 m for histology. Filled neurons were visualized with avidin-Texas Red and bNOS was visualized using a monoclonal bNOS antibody (Sigma; 1:400) tagged with FITC-labeled secondary antibody. Mounted sections were imaged using appropriate filter sets with a color CCD camera (Coolsnap, Roper Scientific) mounted on an epi-fluorescence microscope (BX60; Olympus).

3. HYPOCRETIN/OREXIN STIMULATES THE FIRING OF LDT NEURONS

As a first step to investigating the actions of Hcrt/Orx on mesopontine cholinergic neurons, Burlet et al. bath applied either Hcrt/Orx-A or –B while recording extracellularly from visualized neurons within the confines of the LDT (Fig 1). This insured that extracellularly recorded action potentials arose from the selected soma rather than from fibers or dendrites in the area. LDT neurons typically had no or a low spontaneous firing rate. However following Hcrt/Orx application, firing was initiated and/or the rate was greatly increased. Moreover, while firing eventually recovered to control levels, recovery was slow and required prolonged superfusion with control ACSF. In this example (Fig 1), following recovery, the patch pipette was exchanged for one with our standard intracellular solution (K-Gluconate) and the same neuron was patched again and recorded in whole-cell mode. After filling the neuron with biocytin, the tissue was processed for bNOS immunocytochemistry. This revealed that the recorded neuron was immunopositive for bNOS and hence was cholinergic. This pattern of action was observed for all cells tested in this manner in the LDT (n=15). Thus, Hcrt/Orx evokes prolonged firing of LDT neurons and is capable of bringing even quiescent neurons to fire repetitively.

4. HYPOCRETIN/OREXIN STIMULATES CHOLINERGIC AND NON-CHOLINERGIC LDT NEURONS BY DIRECT AND INDIRECT MEANS

To investigate the mechanisms underlying Hcrt/Orx-mediated firing of LDT neurons, whole-cell patch clamp experiments were conducted under voltage-clamp conditions with the neurons held near their normal resting membrane potential (~ –

60mV). This approach enabled us to observe Hcrt/Orx actions on subthreshold membrane currents.

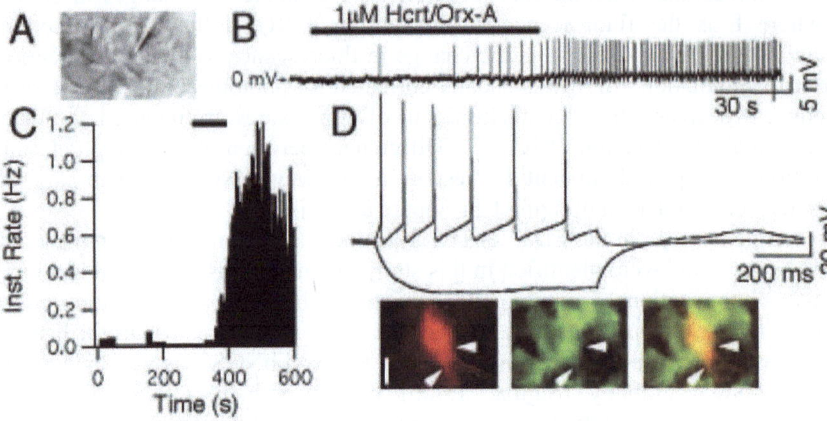

Figure 1. Hcrt/Orx excites LDT neurons. A. IR-DIC image of the neuron from which data in B, C and D were obtained. B. Loose-patch recording from the neuron showing the induction of firing by Hcrt/Orx. C. Histogram detailing the increase in firing rate induced by Hcrt/Orx. The bar over the histogram indicates the application of Hcrt/Orx. D. The same neuron recorded in whole-cell mode with a pipette containing biocytin. Injection of positive DC current fires action potentials. Panels below show the biocytin filled neuron from which recordings were obtained (Texas red), the same tissue immunoprocessed for bNOS showing a field of cholinergic neurons (FITC, green) and the two images merged revealing that the recorded neuron was bNOS+ (yellow).

.In addition, since these experiments were conducted in the presence of $GABA_A$ and Glycine receptor blockers, we were able to evaluate the actions of Hcrt/Orx on spontaneous excitatory synaptic currents (sEPSCs).

Consistent with the pattern of firing observed by extracellular recording, we found that Hcrt/Orx application evoked a slow inward current in LDT neurons. In addition, we found that this depolarizing inward current was often accompanied by an increase in the frequency of sEPSCs (Fig 2). While this current was relatively small (mean ~25pA), it nevertheless appears sufficient to drive the spiking observed in extracellular recordings. Indeed, application of Hcrt/Orx in the presence of ionotropic glutamate receptor blockers produced spiking in extracellularly recorded LDT neurons.

Since the LDT contains both cholinergic and non-cholinergic neurons, we filled recorded neurons with biocytin and processed the slices for bNOS immunocytochemistry following the recordings. We found that the actions of Hcrt/Orx were not limited to either cell type and that Hcrt/Orx had comparable actions on both bNOS –immunopositive and –immunonegative LDT neurons. Thus, as might have been predicted from the extracellular recording experiments in which all studied neurons responded to Hcrt/Orx, both bNOS positive and bNOS negative neurons of the LDT are excited by Hcrt/Orx.

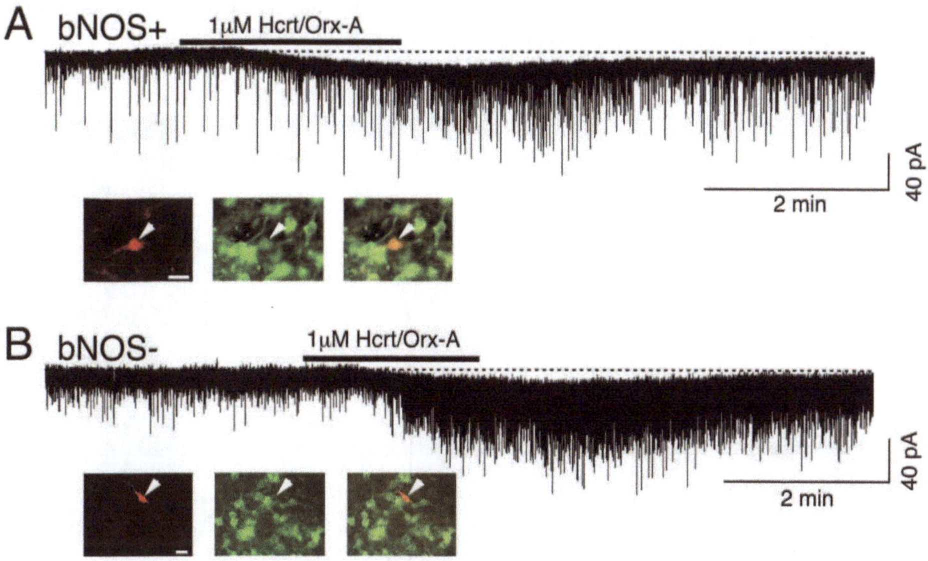

Figure 2. Hcrt/Orx induces a slow inward current in cholinergic and non cholinergic LDT neurons A. Top. Membrane current from an LDT neuron clamped near its resting potential before and after bath superfusion of Hcrt/Orx. Bottom left panel in A and B shows the neuron from which membrane current was obtained (Biocytin, red). Middle panel in A and B shows bnos immunoreactivity in the same field. Left panel in A and B shows the two images merged. The merged image in A reveals that this neuron was bNOS+ B. Membrane current induced by Hcrt/Orx in a neuron that was determined by immunocytochemistry to be in the LDT but bNOS-. Scale bar in A and B bottom left is 20 μm.

5. HYPOCRETIN/OREXIN STIMULATES EXCITATORY AFFERENTS TO LDT

Experiments to further examine the effect of Hcrt/Orx on excitatory synaptic input to LDT neurons were conducted by Burlet et al (2002). These experiments investigated spontaneous-, miniature- and electrically-evoked EPSCs and reached three main conclusions. The first conclusion was that the ability of Hcrt/Orx to increase both the frequency and amplitude of spontaneous EPSCs (sEPSCs) in LDT neurons required TTX-sensitive action potentials. This indicates that Hcrt/Orx acts on presynaptic glutamatergic neurons to fire action potentials, which then release transmitter upon the recorded LDT neurons. The site of spike generation in these presynaptic neurons is not clear. One possibility is that the somata of presynaptic glutamate releasing neurons are located in the slice and that Hcrt/Orx causes firing by acting on their soma-dendritic regions. Alternatively, it is possible that Hcrt/Orx causes ectopic action potentials in the terminal regions of these glutamatergic afferents as has been recently been described for thalamocortical afferents.[46]

The second main conclusion was that Hcrt/Orx does not enhance the sensitivity of post-synaptic glutamate receptors since it did not increase the mean amplitude of mEPSCs. These are the EPSC that occur in the absence of spike-evoked release and probably correspond to the release of single vesicles. Since the amplitude distribution of these events were unchanged by Hcrt/Orx, it is unlikely that Hcrt/Orx alters ionotropic glutamate receptor sensitivity in LDT, as has been suggested in other neurons.[47]

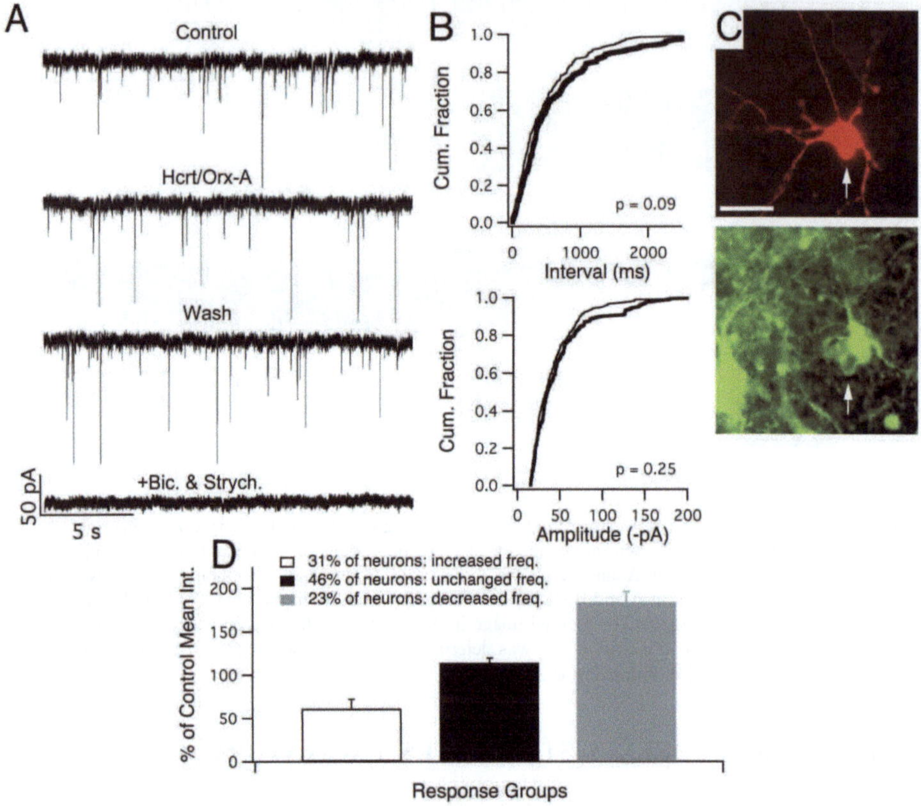

Figure 3. Hcrt/Orx induced an increase in synaptic activity. A Whole cell recordings of membrane current showing the increase in EPSCs following the application of Hcrt/Orx and the blockade of these currents when Hcrt/Orx is applied in the presence of bicuculline and strychnine. B Cummulative distributions of sEPSC intervals and amplitudes from before(thin line) and after (thick line) application of Hcrt/Orx show that both amplitude and frequency of sEPSCs were significantly increased by Hcrt/Orx. C Top panel shows the cell from which recordings were obtained and Bottom panel shows the same field with bNOS+ immunocytochemistry. These data show that the recorded neuron was bNOS+ and hence cholinergic. D Population means of sEPSC inter-event intervals from 13 LDT neurons expressed as a percentage of control values.

The third conclusion was that EPSCS evoked by local stimulation were increased in amplitude following Hcrt/Orx application. These results indicated again that Hcrt/Orx was acting presynaptically and suggested that it either increased excitability of glutamatergic afferents or increased release probability from glutamatergic afferents.

6. HYPOCRETIN/OREXIN HAD INCONSISTENT ACTIONS ON INHIBITORY AFFERENTS TO LDT

Previous studies of Hcrt/Orx actions in hypothalamic cultures[48] indicate that Hcrt/Orx stimulates synaptic release of both glutamate and GABA. Burlet et al. therefore also examined the effect of Hcrt/Orx on sIPSCs in LDT neurons. Functionally this could be important since ultimately the excitability of these neurons will be profoundly influenced by the balance of excitatory and inhibitory input. Moreover, considerable evidence indicates sleep related activity of monoaminergic systems are under important inhibitory control.[49]

Under recording conditions where sIPSCs were pharmacologically isolated, inhibitory afferents to LDT neurons showed mixed responses to application of Hcrt/Orx. Figure 3 illustrates an example of spontaneous IPSCs (inward deflections in A) recorded from a bNOS immunoreactive LDT neuron. As can be seen in the last trace in A, the IPSCs were entirely abolished following bath application of bicuculline and strychnine. As can also be seen in this example, Hcrt/Orx had no effect on the frequency or amplitude distributions of these IPSCs. This profile was representative of most LDT neurons (46%). A smaller fraction of LDT neurons showed an increased (decreased mean inter-event interval; 23%) or decreased frequency (31%) of IPSCs.

Thus, while Hcrt/Orx consistently stimulated excitatory synaptic input to LDT neurons (>85% of neurons) it had a less widespread and consistent action on inhibitory inputs to LDT neurons.

7. HYPOCRETIN/OREXIN EVOKED A NOISY CATION CURRENT IN LDT NEURONS

Experiments conducted by Burlet et al. also provided some early clues about the slow inward current evoked by Hcrt/Orx in LDT. That current was not blocked by TTX or antagonists to ionotropic glutamate, GABA or glycine receptors and was resistant to the lowering extracellular calcium. Since extracellular calcium was reduced sufficiently to block electrically-evoked synaptic transmission, these data strongly argued that the slow inward current was mediated by activation of Hcrt/Orx receptors located on the recorded LDT neurons.

The precise nature of the current(s) that excite LDT neurons has been difficult to elucidate, in part, because of its small size under whole-cell recording conditions. Evidence from the Burlet et al. study indicated that the inward current is accompanied by an increase in membrane current noise and an increase in membrane conductance, in most cases. Since, the current noise increased with membrane hyperpolarization, it was suggested that the inward current was partly mediated by activation of a cation current.

Data from more recent voltage clamp experiments in our lab have confirmed this idea.[50] We found that the inward current is insensitive to extracellular Cs (2-3mM), which effectively blocks a hyperpolarization-activated cation current (h-current) present in many LDT neurons. Using ACSF that contains TTX, synaptic blockers and Cs, and ramp voltage commands, we have observed that Hcrt/Orx activates a current that is quite linear in the range of −100 to −40mV. Interestingly, this current has greater noise at more negative potentials and has a predicted reversal potential between 0 and +30mV (Fig 4).

These properties strongly suggest that Hcrt/Orx receptor activation on LDT neurons activates a cation current.

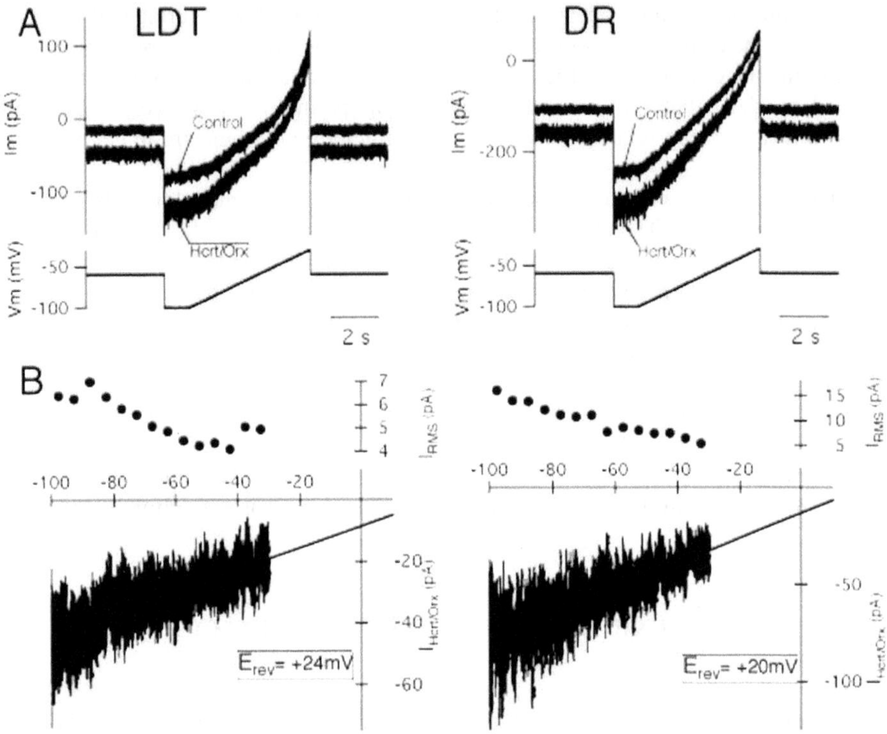

Figure 4. The inward current activated by Hcrt/Orx is a cation current. A. Voltage ramps (-100 mV to –40 mV) performed prior to Hcrt/Orx application and after Hcrt/Orx application from a cell in the LDT (right) and a neuron in the dorsal raphe (left). B. Top Hcrt/Orx induced membrane noise as a function of membrane potential indicates that noise decreases as the membrane potential of the neuron nears the estimated reversal potential of the Hcrt/Orx induced current. Bottom. Current-voltage relationship of the Hcrt/Orx-induced currents generated by subtracting the control ramp from the Hcrt/Orx ramp for each cell type. Reversal potentials of the Hcrt/Orx current are extrapolated from the best fit line between –100mV and –40mV.

Additional recordings using a patch solution with Cs-Gluconate substituted for K-Gluconate indicated that the Hcrt/Orx-evoked current was not blocked and, in fact, had a similar estimated reversal potential to that observed with K-Gluconate. Since the estimated reversal potential was too negative for a Na-selective current ($E_{Na} \sim +70mV$) and the reversal potential didn't appear to shift in a positive direction after substituting intracellular K^+ with Cs^+, the activated channels are likely to have similar permeabilities to these cations. This suggests that this inward current is mediated by noisy and non-specific cation channels.

Since LDT cholinergic neurons are strongly inhibited by serotonin[51,52] and serotonergic input to MPCh neurons arises from the dorsal raphe DR,[53] we also examined the depolarizing current evoked by Hcrt/Orx in DR neurons (Fig 4). Under identical conditions to those used in recording LDT neurons, Hcrt/Orx evoked a cation current with similar properties[50,54,55] the current was not reduced by low extracellular calcium and was not blocked by either internal or external Cs^+ ions. Moreover, it's current–voltage relationship was linear between –100 and –30mV having a predicted reversal potential near zero mV. This current was also accompanied by a large increase in membrane current noise, which was larger at more negative membrane potentials when the driving force was largest, as seen in LDT neurons. Thus, direct Hcrt/Orx-excitation of DR neurons is mediated by a noisy non-specific cation current that is similar to the Hcrt/Orx evoked cation current in LDT neurons. Based on these parallel studies, the current evoked in DR appears to be ~2-3 time larger than in LDT neurons.

These findings suggest that the physiological release of Hcrt/Orx in the LDT and DR, which likely happens during active waking,[56] would excite neurons in both nuclei. Paradoxically, since 5-HT inhibits cholinergic LDT neurons and suppresses firing of REM-on LDT neurons,[57] the net effect of Hcrt/Orx release at these sites may be inhibition of LDT cholinergic neurons and suppression of REM-on firing.

Finally, it should be noted that while Hcrt/Orx activates an inward cation current in both DR and LDT neurons, persuasive evidence indicates that Hcrt/Orx activates an inward current carried by the Na^+/Ca^{2+} exchanger in cholinergic and GABAergic septo-hippocampal neurons[58,59] and tuberomammillary neurons.[60] Thus, it is likely that multiple effectors can mediate the actions of Hcrt/Orx.

8. HYPOCRETIN/OREXIN ELEVATES INTRACELLULAR CALCIUM IN LDT NEURONS

Since many cation channels are permeable to calcium and orexin receptor activation releases calcium from intracellular stores in transfected cells[61,62] we have also examined the possibility that Hcrt/Orx mobilizes intracellular calcium in LDT and DR neurons[45,63] This issue is of particular interest since intracellular calcium is a ubiquitous messenger that can couple membrane events to both cytoplasmic and nuclear processes.

In our initial studies of calcium regulation by Hcrt/Orx, we bulk-loaded slices containing the LDT and DR with fura2-AM. The strength of this approach is that it enabled us to image multiple cells in a given slice without the associated problem of dialysis from a patch pipette, which removes important cytoplasmic factors. Of course, the disadvantage of this approach is that the membrane potential can't be monitored or controlled. Nevertheless, under these conditions, we found that large fractions of dye-loaded cells in the LDT and DR (65.7% LDT; 68.4% DR) displayed calcium transients in response to bath application of Hcrt/Orx. An example of the dF/F changes observed in the LDT is shown in Figure 5. Surprisingly, we found no evidence for Hcrt/Orx mediated release from intracellular stores as was originally described in cells transfected with recombinant orexin receptors.[2,62] Rather, we found that the calcium transients required extracellular calcium and were not blocked at selective concentrations of a drug that interferes with Na^+/Ca^{2+} exchange. These data suggested that the calcium influx was mediated by activation of a calcium-permeable ion channel.

But which channels? Experiments with the L-type calcium channel blocker-nifedipine reduced or abolished the calcium influx, and experiments with Bay K 8644 (5-10 μM), potentiated the Hcrt/Orx-evoked influx, suggesting involvement of L-type calcium channels.

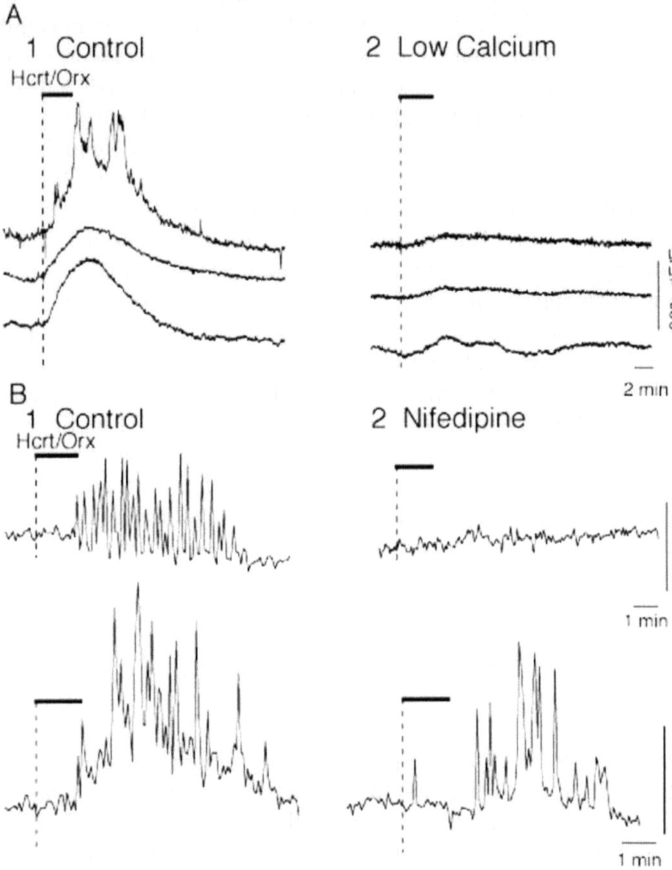

Figure 5. Hcrt/Orx-induced rises in [Ca2+]i were attenuated by lowering extracellular calcium or by blocking L-type calcium channels. A1 Ca-transients induced by Hcrt/Orx in fura 2-AM loaded LDT neurons are indicated by changes in dF/F. A2 These transients were significantly reduced by lowering Ca2+ in the extracellular space. B Nifedipine, a L-type calcium channel inhibitor, in the bath attenuated or abolished Ca2+ responses to Hcrt/Orx.

Moreover, indirect experiments suggested that the calcium entry did not result from depolarization alone, raising the possibility that the calcium permeable channels were up-modulated. Nevertheless, in most cases, nifedipine did not abolish the calcium entry suggesting that one or more additional calcium entry pathways were involved. Of course, one possibility is that the cation channel(s), which drives depolarization in these cells, is itself calcium permeableTo address this possibility and to determine if L-channels are modulated by Hcrt/Orx, we have recently conducted experiments to image intracellular calcium levels during simultaneous whole-cell recordings from LDT and DR neurons. This has turned out to be remarkably difficult, suggesting that the Hcrt/Orx signaling system is sensitive to one or more of the experimental procedures.

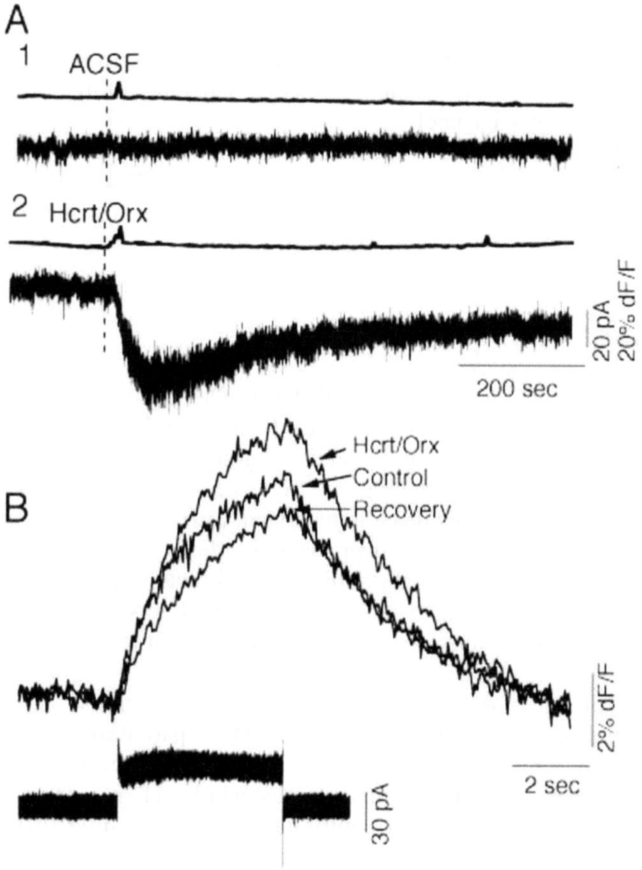

Figure 6. The inward current activated by Hcrt/Orx at resting membrane potential does not provide a major entry-route for Ca^{2+} however Hcrt/Orx enhanced calcium transients resulting from voltage-operated calcium channels. A1, 2 Top traces are changes in dF/F from one LDT neuron and bottom traces are membrane current from the same cell clamped near its resting membrane potential. In A1, control ACSF is being delivered to the cell to monitor artifact in the dF/F trace due to the solution change. In A2, Hcrt/Orx is applied and induces an inward current that is not accompanied by a change in dF/F other than the brief artifact associated with solution change. B. Rises in calcium induced by voltage jumps from –60mV to –30mV were enhanced by Hcrt/Orx application suggesting that Hcrt/Orx enhances calcium entry through voltage-operated Ca^{2+} channels.

Nevertheless, we have now found that during the Hcrt/Orx–evoked inward current with membrane potential clamped near rest (-60mV holding), there are minor, if any, rises of intracellular calcium (Fig 6A). This suggests that the Hcrt/Orx-evoked cation current, does not provide a significant route of calcium influx. Nevertheless during current-clamp recordings, either an authentic Hcrt/Orx–evoked depolarization or an Hcrt/Orx–like depolarization, produced by injecting a previously recorded Hcrt/Orx current, evoked calcium transients reminiscent of those previously observed in AM-loaded neurons.

Finally, by monitoring somatic calcium transients, during voltage jumps from –60mV to –30m, we found that Hcrt/Orx application could reversibly increase these transients (Fig 6B). This is consistent with the indirect evidence from fura2-AM loaded cells and suggests that Hcrt/Orx is up-modulating calcium influx through voltage-dependent calcium permeable channels.

Results from these new whole-cell calcium imaging experiments indicate that the Hcrt/Orx-evoked cation current is not a significant source of calcium influx but provides the depolarization necessary to active voltage-dependent calcium influx pathways – which themselves appear up-regulated by Hcrt/Orx actions.

9. CONCLUSIONS

In summary, our recent studies indicate that Hcrt/Orx peptides excite mesopontine cholinergic and non-cholinergic neurons of the LDT via multiple actions that are likely to have a powerful impact on their activity and integrative properties. Application of Hcrt/Orx causes sufficient depolarization of LDT neurons to drive even quiescent neurons into repetitive firing. This drive is mediated, at least partly, by activation of a noisy cation current and by excitation of presynaptic glutamate-releasing afferents. In contrast, Hcrt/Orx effects on presynaptic inhibitory afferents were more diverse. This may reflect a less important influence of Hcrt/Orx on inhibitory inputs or, more likely, may indicate that subpopulations of LDT neurons are differentially regulated by Hcrt/Orx.

Hcrt/Orx also produces a long-lasting increase in intracellular calcium in LDT neurons that arises from a calcium influx, mediated in part through L-type calcium channels rather than by release from intracellular stores or through reverse operation of the Na^+/Ca^{2+} exchanger. Based on simultaneous whole-cell recording and calcium imaging, the cation current appears to contribute surprisingly little to the Hcrt/Orx-evoked calcium influx. Rather, it appears the cation current provides the depolarizing drive necessary to activate voltage-operated calcium-permeable channels, including L-type channels. These calcium channels, themselves appear to be up-modulated by Hcrt/Orx, perhaps through activation of PKC.[45]

10. FUNCTIONAL IMPLICATIONS

Collectively, the picture that is emerging is one in which hypothalamic Hcrt/Orx neurons are capable of activating all elements of the brain arousal system. While the precise nature and timing of this Hcrt/Orx drive is not yet clear for each structure, hypothalamic Hcrt/Orx neurons are in an excellent position to coordinate the activity and interplay between arousal system elements. Consequently, it is expected that the loss of Hcrt/Orx peptides in narcolepsy would disfacilitate the entire arousal system and promote the excessive daytime sleepiness of this disorder.

As described above, considerable evidence also indicates MPCh neurons play specific roles REM sleep and REM atonia. Based on this evidence, we initially hypothesized that Hcrt/Orx might inhibit MPCh neurons and that the loss of Hcrt/Orx in narcolepsy would disinhibit MPCh neurons and result in exaggerated Ach release in the mPRF, which would promote cataplexy. Our findings appear to rule out this simple scenario and suggest that MPCh neurons are disfacilitated in the absence of Hcrt/Orx. The consequences of sustained disfacilitation are not known but might include an upregulation of post-synaptic mAchRs in the mPRF, as suggested to account for muscarinic receptor changes in narcoleptic dogs[64] or changes MPCh neuron excitability or neurotransmitter-related enzymes.[65]

Our findings also suggest that the REM suppression by ICV injection of Hcrt/Orx occurs despite Hcrt/Orx mediated excitation of MPCh neurons. Several obvious possibilities to consider are: 1) that the REM on cholinergic neurons are the small fraction of MPCh neurons not excited by Hcrt/Orx; 2) that the excitation of MPCh neurons is overwhelmed by inhibition from elsewhere like the DR, which is more strongly excited by Hcrt/Orx that are MPCh neurons; and 3) that Hcrt/Orx works downstream of MPCh neurons to inhibit REM. Future physiological studies of mouse Hcrt/Orx-mutants should be of particular value in addressing these and other possibilities.

11. ACKNOWLEDGEMENTS

We would like to thank Stephen Grupke, Mike Kalogiannis and Dr. Morten Kristensen for their many useful contributions in support of this work. Research was supported by NIH grants NS10877, NS27881 and HL64150

12. REFERENCES

1. L. de Lecea, T. S. Kilduff, C. Peyron, X. Gao, P. E. Foye, P. E. Danielson, C. Fukuhara, E. L. Battenberg, V. T. Gautvik, F. S. Bartlett, 2nd, W. N. Frankel, A. N. van den Pol, F. E. Bloom, K. M. Gautvik and J. G. Sutcliffe, The hypocretins: hypothalamus-specific peptides with neuroexcitatory activity, *Proc Natl Acad Sci U S A.* **95**, 322-7 (1998).
2. T. Sakurai, A. Amemiya, M. Ishii, I. Matsuzaki, R. M. Chemelli, H. Tanaka, S. C. Williams, J. A. Richarson, G. P. Kozlowski, S. Wilson, J. R. Arch, R. E. Buckingham, A. C. Haynes, S. A. Carr, R. S. Annan, D. E. McNulty, W. S. Liu, J. A. Terrett, N. A. Elshourbagy, D. J. Bergsma and M. Yanagisawa, Orexins and orexin receptors: a family of hypothalamic neuropeptides and G protein-coupled receptors that regulate feeding behavior, *Cell.* **92**, 1 page following 696 (1998).
3. J. G. Sutcliffe and L. de Lecea, The hypocretins: setting the arousal threshold, *Nat Rev Neurosci.* **3**, 339-49 (2002).
4. T. Sakurai, Orexin: a link between energy homeostasis and adaptive behaviour, *Curr Opin Clin Nutr Metab Care.* **6**, 353-60 (2003).
5. J. T. Willie, R. M. Chemelli, C. M. Sinton and M. Yanagisawa, To eat or to sleep? Orexin in the regulation of feeding and wakefulness, *Annu Rev Neurosci.* **24**, 429-58 (2001).
6. M. Mieda and M. Yanagisawa, Sleep, feeding, and neuropeptides: roles of orexins and orexin receptors, *Curr Opin Neurobiol.* **12**, 339-45 (2002).
7. J. P. Kukkonen, T. Holmqvist, S. Ammoun and K. E. Akerman, Functions of the orexinergic/hypocretinergic system, *Am J Physiol Cell Physiol.* **283**, C1567-91 (2002).
8. C. Peyron, D. K. Tighe, A. N. van den Pol, L. de Lecea, H. C. Heller, J. G. Sutcliffe and T. S. Kilduff, Neurons containing hypocretin (orexin) project to multiple neuronal systems, *J Neurosci.* **18**, 9996-10015 (1998).
9. J. N. Marcus, C. J. Aschkenasi, C. E. Lee, R. M. Chemelli, C. B. Saper, M. Yanagisawa and J. K. Elmquist, Differential expression of orexin receptors 1 and 2 in the rat brain, *J Comp Neurol.* **435**, 6-25 (2001).
10. D. C. Piper, N. Upton, M. I. Smith and A. J. Hunter, The novel brain neuropeptide, orexin-A, modulates the sleep-wake cycle of rats, *Eur Journal of Neuroscience.* **12**, 726-30 (2000).
11. M. I. Smith, D. C. Piper, M. S. Duxon and N. Upton, Evidence implicating a role for orexin-1 receptor modulation of paradoxical sleep in the rat, *Neurosci Lett.* **341**, 256-8 (2003).
12. J. J. Hagan, R. A. Leslie, S. Patel, M. L. Evans, T. A. Wattam, S. Holmes, C. D. Benham, S. G. Taylor, C. Routledge, P. Hemmati, R. P. Munton, T. E. Ashmeade, A. S. Shah, J. P. Hatcher, P. D. Hatcher, D. N.

Jones, M. I. Smith, D. C. Piper, A. J. Hunter, R. A. Porter and N. Upton, Orexin A activates locus coeruleus cell firing and increases arousal in the rat, *Proceedings of the National Academy of Science (USA)*. **96**, 10911-6 (1999).
13. P. Bourgin, S. Huitron-Resendiz, A. D. Spier, V. Fabre, B. Morte, J. R. Criado, J. G. Sutcliffe, S. J. Henriksen and L. de Lecea, Hypocretin-1 modulates rapid eye movement sleep through activation of locus coeruleus neurons, *J Neurosci.* **20**, 7760-5 (2000).
14. M. Xi, F. R. Morales and M. H. Chase, Effects on sleep and wakefulness of the injection of hypocretin-1 (orexin-A) into the laterodorsal tegmental nucleus of the cat, *Brain Res.* **901**, 259-64. (2001).
15. C. C. Shute and P. R. Lewis, The ascending cholinergic reticular system: neocortical, olfactory and subcortical projections, *Brain.* **90**, 497-520 (1967).
16. M. M. Mesulam, E. J. Mufson, B. H. Wainer and A. I. Levey, Central cholinergic pathways in the rat: an overview based on an alternative nomenclature (Ch1-Ch6), *Neuroscience.* **10**, 1185-201 (1983).
17. K. J. Maloney, L. Mainville and B. E. Jones, c-Fos expression in GABAergic, serotonergic, and other neurons of the pontomedullary reticular formation and raphe after paradoxical sleep deprivation and recovery, *J Neurosci.* **20**, 4669-79 (2000).
18. D. S. Bredt, P. M. Hwang and S. H. Snyder, Localization of nitric oxide synthase indicating a neural role for nitric oxide, *Nature.* **347**, 768-70 (1990).
19. S. R. Vincent and H. Kimura, Histochemical mapping of nitric oxide synthase in the rat brain, *Neuroscience.* **46**, 755-84 (1992).
20. S. R. Vincent, K. Satoh, D. M. Armstrong and H. C. Fibiger, NADPH-Diaphorase: A selective histochemical marker for the cholinergic neurons of the pontine reticular formation, *Neuroscience Letters.* **43**, 31-36 (1983).
21. C. S. Leonard, E. K. Michaelis and K. M. Mitchell, Activity–dependent nitric oxide concentration dynamics in the laterodorsal tegmental nucleus *in vitro*, *Journal of Neurophysiology.* **86**, 2159-2172. (2001).
22. J. A. Williams, S. R. Vincent and P. B. Reiner, Nitric oxide production in rat thalamus changes with behavioral state, local depolarization, and brainstem stimulation, *J Neurosci.* **17**, 420-7 (1997).
23. B. H. Wainer and M.-M. Mesulam. (1990). Ascending cholinergic pathways in the rat brain. In *Brain Cholinergic Systems* (Steriade, M. & Biesold, D., eds.), pp. 65-119. Oxford University Press, New York.
24. A. Mitani, K. Ito, A. E. Hallanger, B. H. Wainer, K. Kataoka and R. W. McCarley, Cholinergic projections from the laterodorsal and pedunculopontine tegmental nuclei to the pontine gigantocellular tegmental field in the cat, *Brain Res.* **451**, 397-402 (1988).
25. B. E. Jones, Immunohistochemical study of choline acetyltransferase-immunoreactive processes and cells innervating the pontomedullary reticular formation in the rat, *J Comp Neurol.* **295**, 485-514 (1990).
26. K. Semba, Aminergic and cholinergic afferents to REM sleep induction regions of the pontine reticular formation in the rat, *J Comp Neurol.* **330**, 543-56 (1993).
27. K. Semba, P. B. Reiner and H. C. Fibiger, Single cholinergic mesopontine tegmental neurons project to both the pontine reticular formation and the thalamus in the rat, *Neuroscience.* **38**, 643-54 (1990).
28. S. A. Oakman, P. L. Faris, P. E. Kerr, C. Cozzari and B. K. Hartman, Distribution of pontomesencephalic cholinergic neurons projecting to substantia nigra differs significantly from those projecting to ventral tegmental area, *Journal of Neuroscience.* **15**, 5859-69. (1995).
29. S. A. Oakman, P. L. Faris, C. Cozzari and B. K. Hartman, Characterization of the extent of pontomesencephalic cholinergic neurons' projections to the thalamus: comparison with projections to midbrain dopaminergic groups, *Neuroscience.* **94**, 529-47 (1999).
30. R. Curro Dossi, D. Pare and M. Steriade, Short-lasting nicotinic and long-lasting muscarinic depolarizing responses of thalamocortical neurons to stimulation of mesopontine cholinergic nuclei, *J Neurophysiol.* **65**, 393-406 (1991).
31. G. L. Forster and C. D. Blaha, Laterodorsal tegmental stimulation elicits dopamine efflux in the rat nucleus accumbens by activation of acetylcholine and glutamate receptors in the ventral tegmental area, *Eur Journal of Neuroscience.* **12**, 3596-604. (2000).
32. C. D. Blaha and P. Winn, Modulation of dopamine efflux in the striatum following cholinergic stimulation of the substantia nigra in intact and pedunculopontine tegmental nucleus-lesioned rats, *Journal of Neuroscience.* **13**, 1035-44 (1993).
33. M. Steriade and R. W. McCarley. (1990). *Brainstem Control of Wakefulness and Sleep*, Plenum, New York.
34. J. J. Quattrochi, A. N. Mamelak, R. D. Madison, J. D. Macklis and J. A. Hobson, Mapping neuronal inputs to REM sleep induction sites with carbachol-fluorescent microspheres, *Science.* **245**, 984-6 (1989).
35. H. A. Baghdoyan, M. L. Rodrigo-Angulo, R. W. McCarley and J. A. Hobson, Site-specific enhancement and suppression of desynchronized sleep signs following cholinergic stimulation of three brainstem regions, *Brain Res.* **306**, 39-52 (1984).
36. J. W. Gnadt and G. V. Pegram, Cholinergic brainstem mechanisms of REM sleep in the rat, *Brain Res.* **384**, 29-41 (1986).

37. P. Bourgin, P. Escourrou, C. Gaultier and J. Adrien, Induction of rapid eye movement sleep by carbachol infusion into the pontine reticular formation in the rat, *Neuroreport.* **6**, 532-6 (1995).
38. J. Velazquez-Moctezuma, P. J. Shiromani and J. C. Gillin, Acetylcholine and acetylcholine receptor subtypes in REM sleep generation, *Prog Brain Res.* **84**, 407-13 (1990).
39. M. S. Reid, M. Tafti, J. N. Geary, S. Nishino, J. M. Siegel, W. C. Dement and E. Mignot, Cholinergic mechanisms in canine narcolepsy--I. Modulation of cataplexy via local drug administration into the pontine reticular formation, *Neuroscience.* **59**, 511-22 (1994).
40. M. S. Reid, J. M. Siegel, W. C. Dement and E. Mignot, Cholinergic mechanisms in canine narcolepsy--II. Acetylcholine release in the pontine reticular formation is enhanced during cataplexy, *Neuroscience.* **59**, 523-30 (1994).
41. R. M. Chemelli, J. T. Willie, C. M. Sinton, J. K. Elmquist, T. Scammell, C. Lee, J. A. Richardson, S. C. Williams, Y. Xiong, Y. Kisanuki, T. E. Fitch, M. Nakazato, R. E. Hammer, C. B. Saper and M. Yanagisawa, Narcolepsy in orexin knockout mice: molecular genetics of sleep regulation, *Cell.* **98**, 437-51 (1999).
42. L. Lin, J. Faraco, R. Li, H. Kadotani, W. Rogers, X. Lin, X. Qiu, P. J. de Jong, S. Nishino and E. Mignot, The sleep disorder canine narcolepsy is caused by a mutation in the hypocretin (orexin) receptor 2 gene, *Cell.* **98**, 365-76 (1999).
43. S. Burlet, C. J. Tyler and C. S. Leonard, Direct and indirect excitation of laterodorsal tegmental neurons by Hypocretin/Orexin peptides: implications for wakefulness and narcolepsy, *J Neurosci.* **22**, 2862-72 (2002).
44. C. S. Leonard, S. R. Rao and T. Inoue, Serotonergic inhibition of action potential evoked calcium transients in NOS-containing mesopontine cholinergic neurons, *J Neurophysiol.* **84**, 1558-72 (2000).
45. K. A. Kohlmeier, T. Inoue and C. S. Leonard, Hypocretin/orexin peptide signaling in the ascending arousal system: elevation of intracellular calcium in the mouse dorsal raphe and laterodorsal tegmentum, *J Neurophysiol.* **92**, 221-35 (2004).
46. E. K. Lambe and G. K. Aghajanian, Hypocretin (orexin) induces calcium transients in single spines postsynaptic to identified thalamocortical boutons in prefrontal slice, *Neuron.* **40**, 139-50 (2003).
47. B. Yang, W. K. Samson and A. V. Ferguson, Excitatory effects of orexin-A on nucleus tractus solitarius neurons are mediated by phospholipase C and protein kinase C, *J Neurosci.* **23**, 6215-22 (2003).
48. A. N. van den Pol, X. B. Gao, K. Obrietan, T. S. Kilduff and A. B. Belousov, Presynaptic and postsynaptic actions and modulation of neuroendocrine neurons by a new hypothalamic peptide, hypocretin/orexin, *J Neurosci.* **18**, 7962-71 (1998).
49. D. Nitz and J. Siegel, GABA release in the dorsal raphe nucleus: role in the control of REM sleep, *Am J Physiol.* **273**, R451-5 (1997).
50. C. J. Tyler, S. Burlet and C. S. Leonard, Hypocretin/Orexin-A (Hcrt/Orx) excites cholinergic laterodorsal tegmental (LDT) neurons by activating a cation current, *Soc. Neuroscience Absts.* **27**, 411.2 (2001).
51. C. S. Leonard and R. Llinas, Serotonergic and cholinergic inhibition of mesopontine cholinergic neurons controlling REM sleep: an in vitro electrophysiological study, *Neuroscience.* **59**, 309-30 (1994).
52. J. I. Luebke, R. W. Greene, K. Semba, A. Kamondi, R. W. McCarley and P. B. Reiner, Serotonin hyperpolarizes cholinergic low-threshold burst neurons in the rat laterodorsal tegmental nucleus in vitro, *Proc Natl Acad Sci U S A.* **89**, 743-7 (1992).
53. T. L. Steininger, B. H. Wainer, R. D. Blakely and D. B. Rye, Serotonergic dorsal raphe nucleus projections to the cholinergic and noncholinergic neurons of the pedunculopontine tegmental region: a light and electron microscopic anterograde tracing and immunohistochemical study, *J Comp Neurol.* **382**, 302-22 (1997).
54. R. E. Brown, O. A. Sergeeva, K. S. Eriksson and H. L. Haas, Convergent excitation of dorsal raphe serotonin neurons by multiple arousal systems (orexin/hypocretin, histamine and noradrenaline), *J Neurosci.* **22**, 8850-9 (2002).
55. R. J. Liu, A. N. van den Pol and G. K. Aghajanian, Hypocretins (orexins) regulate serotonin neurons in the dorsal raphe nucleus by excitatory direct and inhibitory indirect actions, *J Neurosci.* **22**, 9453-64 (2002).
56. L. I. Kiyashchenko, B. Y. Mileykovskiy, N. Maidment, H. A. Lam, M. F. Wu, J. John, J. Peever and J. M. Siegel, Release of hypocretin (orexin) during waking and sleep states, *J Neurosci.* **22**, 5282-6 (2002).
57. M. M. Thakkar, R. E. Strecker and R. W. McCarley, Behavioral state control through differential serotonergic inhibition in the mesopontine cholinergic nuclei: a simultaneous unit recording and microdialysis study, *J Neurosci.* **18**, 5490-7. (1998).
58. M. Wu, Z. Zhang, C. Leranth, C. Xu, A. N. van den Pol and M. Alreja, Hypocretin increases impulse flow in the septohippocampal GABAergic pathway: implications for arousal via a mechanism of hippocampal disinhibition, *J Neurosci.* **22**, 7754-65 (2002).
59. M. Wu, L. Zaborszky, T. Hajszan, A. N. van den Pol and M. Alreja, Hypocretin/orexin innervation and excitation of identified septohippocampal cholinergic neurons, *J Neurosci.* **24**, 3527-36 (2004).

60. K. S. Eriksson, O. Sergeeva, R. E. Brown and H. L. Haas, Orexin/hypocretin excites the histaminergic neurons of the tuberomammillary nucleus, *J Neurosci.* **21**, 9273-9. (2001).
61. P. E. Lund, R. Shariatmadari, A. Uustare, M. Detheux, M. Parmentier, J. P. Kukkonen and K. E. Akerman, The orexin OX1 receptor activates a novel Ca2+ influx pathway necessary for coupling to phospholipase C, *J Biol Chem.* **275**, 30806-12 (2000).
62. D. Smart, J. C. Jerman, S. J. Brough, S. L. Rushton, P. R. Murdock, F. Jewitt, N. A. Elshourbagy, C. E. Ellis, D. N. Middlemiss and F. Brown, Characterization of recombinant human orexin receptor pharmacology in a Chinese hamster ovary cell-line using FLIPR, *Br J Pharmacol.* **128**, 1-3 (1999).
63. K. A. Kohlmeier, S. Watanabe and C. S. Leonard, Hypocretin/orexin (H/O)-evoked calcium transients in laterodorsal tegmentum (LDT) and dorsal raphe (DR) neurons studied by simultaneous whole cell recording and calcium imaging in mouse brain slices, *Soc. Neurosci. Abstracts* **30** (2004).
64. T. S. Kilduff, S. S. Bowersox, K. I. Kaitin, T. L. Baker, R. D. Ciaranello and W. C. Dement, Muscarinic cholinergic receptors and the canine model of narcolepsy, *Sleep* **9**, 102-106 (1986).
65. S. L. Grupke, M. Kalogiannis and C. S. Leonard, Immunocytochemical evidence for differential regulation of choline acetyl transferase in brainstem of double orexin receptor knockout (DKO) mice, *Soc. Neurosci. Abstracts* **30** (2004).

THE AMINERGIC SYSTEMS AND THE HYPOCRETINS

Oliver Selbach and Helmut L. Haas*

1. INTRODUCTION

Hypocretin neurons[1-13] in the feeding and defense center of the posterior hypothalamus exhibit an intricate relationship with the noradrenergic, serotonergic, dopaminergic, histaminergic, and cholinergic, arousal control systems in the brainstem, hypothalamus and basal forebrain.[14-24] They densely innervate and potently activate aminergic cells via direct and indirect post- and presynaptic mechanisms[25]. Aminergic, glutamatergic and GABAergic signals, in turn, vary according to behavioral state and imprint inhibitory feedback and excitatory feed forward control[26,27] on the intrinsic activity[28] of hypocretinergic neurons. In the orchestra of biogenic amines and inhibitory and excitatory amino acids hypocretins are conductors linking food, mood, sleep, motor function and memory. This is relevant for basic body and brain functions both in health and in disease. Since the discovery of the hypocretin system[1,2] numerous reviews on various aspects have been published.[2-10,25,29] Here, we focus on interactions of hypocretins with histaminergic neurons in the tuberomamillary nucleus (TMN),[14-19,21] serotonergic neurons in the dorsal raphe (DR),[20] and dopaminergic neurons in the ventral tegmental area and substantia nigra (VTA/SN),[22-24] as well as on the orchestration of noradrenaline, acetylcholine, glutamate and GABA signaling in cortical targets, particularly the hippocampus and its putative role in sustained cortical activation, plasticity, and memory functions.[29-31]

2. THE HYPOCRETIN SYSTEM

Hypocretins (Hcrt-1 and Hcrt-2) are derived from a common pre-pro-hypocretin gene expressed in a small group of hypothalamic neurons located in the perifornical area. Hypocretin neurons have widespread projections,[32,33] densely accumulating in infralimbic central representations of the sympathetic nervous system,[9,12,34] and intero- and nociception.[35-37] In their targets within[38] and even outside the nervous system Hcrt-1 and Hcrt-2 both act on two G-Protein coupled receptors exhibiting differential agonist affinity and expression throughout and even outside the nervous system.[38] They couple to

*O. Selbach and H.L. Haas. Heinrich-Heine-University. Institute of Physiology D-40225 Düsseldorf

pertussistoxin-sensitive and -insensitive, PLC-dependent, calcium-sensitive and calcium-insensitive signal transduction mechanisms, some of which exhibit rather unusual effector pathways[25] (reviewed by Kukkonen and Akerman in this book). These include Hcrt2-receptor mediated activation of electrogenic sodium-calcium exchangers (NCX)[14,15,18,39,40] or Hcrt1-receptor mediated activation of non-specific cationic conductances.[9,20,25] Other signal transduction pathways include activation of L-type voltage-dependent calcium channels,[25] closure of G-protein activated inward rectifier potassium channels (GIRK),[41] and activation of plasticity-related kinases,[30] including PKC, PKA,[22,42] and MAPK,[25] as well as release of mitochondrial pro-apoptotic factors.[43]

Single cell RT-PCR and whole cell patch clamp recordings of acutely isolated aminergic neurons revealed a striking coexpression of serotonin 5-HT2C receptors and NCX in TMN neurons,[18] and a high but subunit- and cell-type specific coexpression of non-specific cationic conductances of the transient receptor potential channel family (TRPC) in all aminergic but not locus coeruleus neurons (Figure 1).[19] TRPC, evolutionary conserved cellular sensors for a variety of stimuli,[44] and NCX, set points for cellular calcium homeostasis and survival,[45] thus link hypocretin and aminergic signaling.

2.1. Glutamate and GABA

Glutamate and gamma-aminobutyric acid (GABA) are the most abundant excitatory and inhibitory neurotransmitters in the nervous system.[6,26,35] Although hypothalamic hypocretin neurons *in vitro* may be in an intrinsic state of depolarization,[28] their activity is also driven by a glutamatergic excitatory feed forward mechanism.[26] In fact, hypocretin neurons exhibit glutamate immunoreactivity and express specific vesicular glutamate transporters, suggesting that they release glutamate by themselves.[46] In turn, the activity of Hcrt neurons *in vitro* is limited by inhibitory constraints including tonic activation of presynaptic type III metabotropic glutamate autoreceptors (mGluR)[47] and negative feedback through GABAergic,[48] catecholaminergic, serotonergic, and purinergic signals activating postsynaptic metabotropic $GABA_B$, catecholamine (α_2), serotonin ($5\text{-}HT_{1A}$), and adenosine (A_1) receptors. Coupled to $G_{i/o}$-proteins these receptors all converge on inward rectifier potassium (GIRK) channels, which inhibit hypocretin neurons by hyperpolarization.[26,49] A major inhibitory GABA-galaninergic pathway to the posterior hypothalamus emanates from sleep-active neurons in the ventrolateral preoptic area (VLPO).[50-52] Mutual inhibition between these neurons and the ascending aminergic systems has been proposed to form a bistable flip-flop sleep-switch gating rapid transitions between behavioral states[6,13,53] and other homeostatic body functions.[35]

A common feature of hypocretins apart from their predominant neuroexcitatory actions[1] is that they increase impulse flow in GABAergic (inter)neurons[16,39,40,54] and facilitate the release of glutamate in many if not all targets.[26,30,55-57] Whereas the excitatory direct effects of hypocretins on aminergic and GABAergic cells depend on activation of postsynaptic Hcrt-receptors,[9,20,25] modulation of glutamate, GABA,[16,39,40] and amine[30,56,58] release relies on presynaptic mechanisms, involving modulation of presynaptic calcium homeostasis and/or sensitization of parts of the release machinery such as Rab3, implicated in both behavioral state control and synaptic plasticity.[59] Dissociated pre- and postsynaptic inhibitory and excitatory hypocretin effects promote network bistability[6,60] and negative feedback protecting from excitotoxicity.[43,45,61]

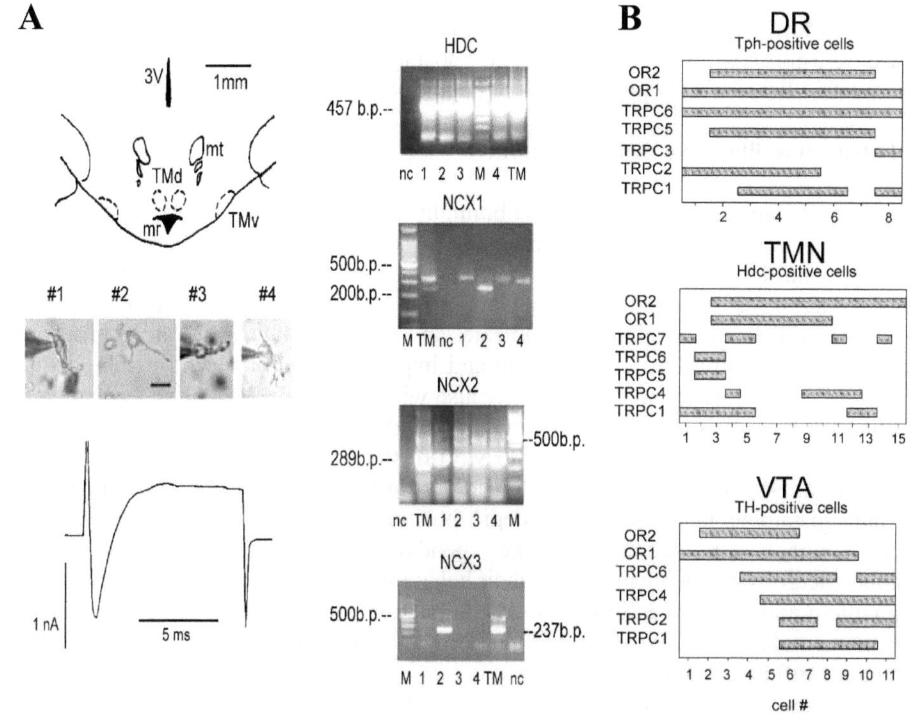

Figure 1. Molecular convergence of hypocretin signaling in aminergic cells. Single cell RT-RCR and whole cell recordings in aminergic neurons. (**A**) Expression patterns of mRNA encoding for histidine decarboxylase (HDC) and Na^+/Ca^{2+} exchanger subtypes (NCX) in the TMN. (**B**) Overlapping co-expression of hypocretin receptors (OR1 and OR2) and TRPC channels in TMN, DR, and VTA. Modified from Sergeeva *et al.*[17,19]

3. THE AMINERGIC SYSTEMS

The relatively small number of aminergic neurons located in the locus coeruleus (noradrenaline), dorsal raphe (serotonin), ventral tegmental area/substantia nigra (dopamine), lateral dorsal tegmentum/basal forebrain (acetylcholine), and tuberomamillary nucleus (histamine) project with multifold arborisations to most regions of the central nervous system, partially overlapping with the projections of hypocretin neurons.[32] Apart from some distinctions they display comparable morphological features, intrinsic electrophysiological properties, and according to the reciprocal-interaction model of REM and non-REM sleep alterations, behavioral state-dependent activity patterns.[21,62,63] All form mutual connections, acting in concert to control synchrony of selected cell populations throughout the entire nervous system.[30,31,42,64-66] Released rarely from synaptically specialized structures (dopamine) but mostly from varicosities at some distance from their target receptors, they act by volume transmission on post- and presynaptic receptors, utilizing common or convergent signal transduction pathways[19,30] to control the metabolic[67] and electrical state[62,63,68] of the brain.

Biogenic amines switch higher brain functions and hormonal states during hunger and satiety, during waking and sleep, during pleasure and aversion, during offense and defense, during reward and pain, during dosing and attention, during stress and contemplative life, during reproduction and cognition, during learning and memory.

3.1. Tuberomamillary Nucleus (Histamine)

Among the aminergic systems in the brain, histamine has received the least attention although it is an equally important regulator of many homeostatic body functions,[21] including control of behavioral state (maintenance of wakefulness),[69-71] appetite and energy metabolism, neuroendocrine regulation, nociception, and learning or memory.[21] Indeed, several lines of evidence indicate an exceptionally close anatomical and functional relationship between histamine and hypocretin neurons, who exhibit mutual connectivity, cooperativity and associativity also with other neuroendocrine systems.[72]

The only source of neuronal histamine in the adult brain is the wake-active neurons located in the tuberomamillary nucleus (TMN) of the posterior hypothalamus, close or adjacent to the wall of the third ventricle and perifornically close to hypocretin neurons. This puts histamine neurons in a strategic position suited to relay remote humoral-paracrine signals e.g. secreted by the ependym[52,73] or circumventricular organs. Histamine neurons can be identified by their behavioral state-dependent (pacemaker-like) firing pattern and their exceptionally strong immunoreactivity to adenosine deaminase,[21] a putative molecular firewall shielding wake-active histamine neurons from the somnogenic influences of adenosine.

During prolonged waking adenosine accumulates in the concourse of heightened activity in basal forebrain cholinergic arousal centers.[74] Hypocretin neurons bear low affinity adenosine A1R and are sensitive (increasing c-fos immunoreactivity) to stimulant doses of the adenosine receptor antagonist caffeine, the most commonly used wake-promoting drug worldwide.[7,75-77] Adenosine is also released by activation of leptomeningeal prostaglandin PGD_2 receptors, localized exclusively at the ventrorostral surface of the basal forebrain, in the vicinity of sleep-active neurons in the ventrolateral preoptic area (VLPO).[52] VLPO neurons release GABA and galanin, a peptide co-expressed in some histamine neurons and exerting inhibitory and neuroprotective effects.[21] In contrast, PGE_2 infusion into the posterior hypothalamus and TMN promotes wakefulness, increases body temperature and heart rate.[52] Thus, adenosine-mediated homeostatic sleep-drive may work by A1R-mediated inhibition of wake promoting cholinergic and hypocretin but not histamine neurons, and high affinity A2R-mediated excitation of VLPO neurons. In contrast, histamine neurons *in vitro*[16,17] and *in vivo*,[69,70,78] strongly interact with cholinergic arousal systems[70] and are controlled by constitutive histamine H3R-autoreceptors, galanin, and GABA.[21]

Histamine neurons express $GABA_A$ receptors with distinct subunit composition, pharmacology, and electrophysiology.[17] Whole-cell recordings and single-cell RT-PCR from acutely isolated rat tuberomamillary neurons revealed differences in the expression levels of GABA receptor gamma-subunits, sensitivity to GABA and zinc, as well as in the modulation of IPSC-decay times by GABAergic drugs, such as zolpidem, a partial benzodiazepine agonist with hypnotic but little or no amnesiac and addictive properties. Functional heterogeneity of $GABA_A$ receptors in the arousal systems may segregate the sedative and addictive effects of GABAergic drugs, including alcohol. In fact, the GABA-galaninergic pathway from sleep-active neurons in the ventrolateral preoptic area[51] to the TMN[16] has been identified as a (the) major and specific target for the sedative effects of GABAergic anesthetics.[78]

Pharmacological intervention studies in living animals, including EEG recordings in mice lacking the histamine synthesizing enzyme histidine-decarboxylase (HDC), or histamine receptors[69,70] provide additional and compelling evidence, that activity of histaminergic neurons is a prerequisite for maintained wakefulness, particularly in the context of environmental challenges.[69,70] HDC knockout mice exhibit normal overall sleep and wake amounts under undisturbed conditions, but wake fragmentation, increased REM sleep, slower EEG activity while awake, and an inability to maintain awake in novel environments. Similar results have been obtained from transgenic mice lacking H1 receptors.[52,79] The central histamine system may thus be a key component of sustained waking and adaptive coping according to nutritional-metabolic (foraging) or noxious-immunological challenges, flight-fight stress responses, and social (resident-intruder) conflicts, which have been implicated with functions of the hypocretin system as well.[11]

Accumulating evidence supports a mutual positive feedback between the hypocretin and histamine system, mediated by Hcrt2- and H1-receptors, respectively. Both, the effect of hypocretins on wakefulness[52] as that of leptin[80] or amylin[81] on body weight, are blocked by H1-receptor antagonists and blunted in H1-receptor knockout mice.[52] Likewise, drugs interfering with H1R functions,[21] such as commonly used antihistamines or certain psychotropic drugs[58] have well known sedative and/or metabolic side-effects. H1-receptor deficient mice have lower total brain hypocretin levels,[82] whereas Hcrt2-receptor-deficient narcoleptic dogs have decreased histamine, but elevated dopamine/noradrenaline levels in the thalamus and cortex.[82,83] Conversely, decreased dopaminergic tone in Parkinson's disease is associated with elevated histamine levels.[84] This suggests that dopaminergic signaling is inversely intervened in the feedback loop between the histamine and hypocretin system.

Electrophysiological recordings from histamine and hypocretin neurons in hypothalamic slices also support the view of a positive feedback loop between the two systems. Hypocretin neurons exhibit close contact to histamine-immunoreactive fibers in hypothalamic slices.[14] However, recordings from identified hypocretin neurons expressing green fluorescent protein did not reveal direct excitatory actions when exposed to histamine,[26] although this does not preclude H1R-dependent long loop feed forward excitation or disinhibition through infralimbic pathways.[39,85] TMN neurons (Figure 2) preferentially express Hcrt2 receptors and are densely innervated by Hcrt-containing axons forming somatic and dendritic contacts. Hypocretins, acting on postsynaptic Hcrt2 receptors depolarize histaminergic TMN neurons by activation of an electrogenic sodium-calcium exchanger (NCX) and a Ca^{2+} current associated with a small decrease in input resistance and increases in spontaneous firing.[14] NCX in TMN neurons is also a convergent target for excitatory actions of serotonin on 5-HT2C receptors[15,18] playing a role in hypocretin-induced stress-related behavioral responses (Figure 1).[86] In contrast, Hcrt1 receptors in the TMN, convergent with α_2-adrenoreceptor[78,87] activation, increase GABA outflow from the endogenous sleep pathway of the VLPO by presynaptic mechanisms (Figure 3).[16,51] Dynorphin in turn, which is highly co-expressed in hypocretin neurons and suggested to play a role in accumbens-dependent feeding responses and increased body mass index in narcolepsy,[72,88] strongly suppressed. GABAergic inputs to the TMN, overwriting the facilitatory effect of hcrt-1.[16]

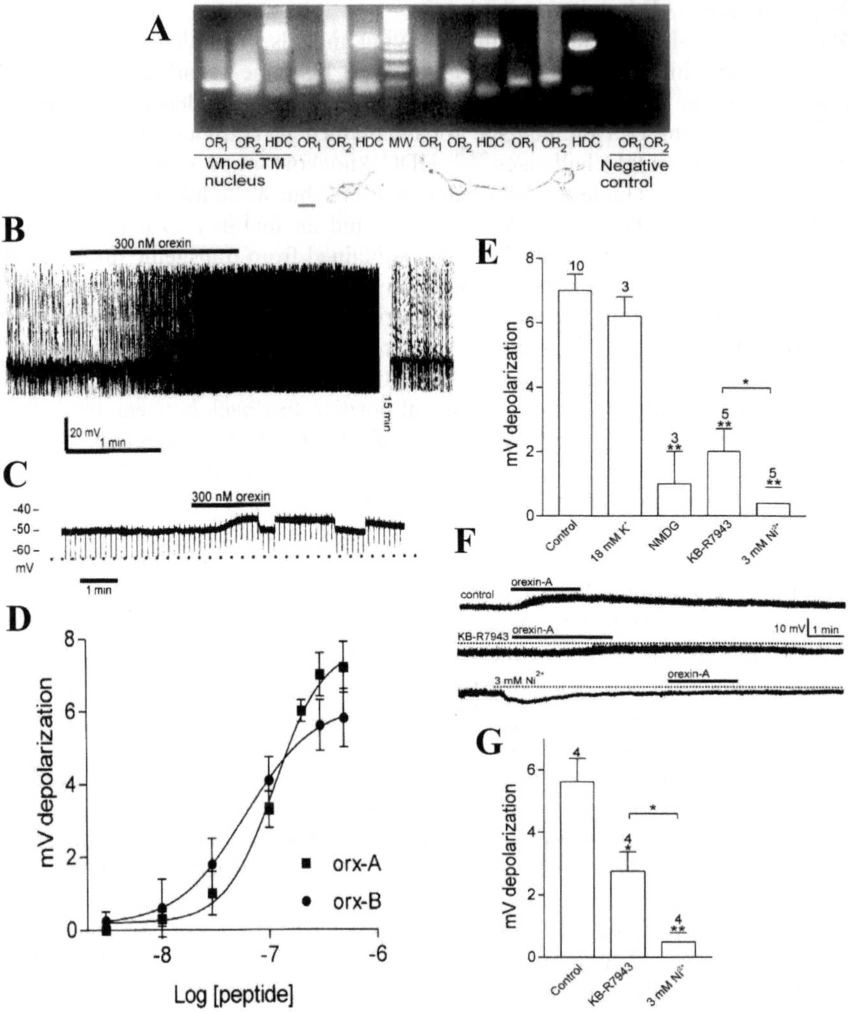

Figure 2. Molecular and electrophysiological interactions of hypocretins with histamine neurons. (**A**) Co-expression of histidine-decarboxylase (HDC) and hypocretin receptors (OR1 and OR2; single cell RT-PCR). (**B**) Hypocretin-1/Orexin-A (300 nM) induced increases in spontaneous firing and (**C**) tetrodotoxin-insensitive depolarization in TMN neurons (intracellular sharp electrode recordings). (**D**) Dose dependent postsynaptic depolarization by Orexin-A/B. (**E-G**) Block of hypocretin-induced depolarization by inhibition of sodium-calcium-exchangers (NCX). Modified from Eriksson *et al.*[14]

Thus, hypocretin/dynorphin containing neurons excite TMN neurons directly via Hcrt2 receptors and through presynaptic disinhibition gated by dynorphin. Unfortunately, the mechanisms controlling co-expression and -release of dynorphins in hypocretin neurons are unknown. However, it may be transcriptional mechanisms, intrinsic to hypocretin neurons and imprinted by metabolic[3,89,90] and circadian state cues[8,91,92] and aminergic influences that gate sleep-wake transitions[6,13] and basic body functions.[35,36,93]

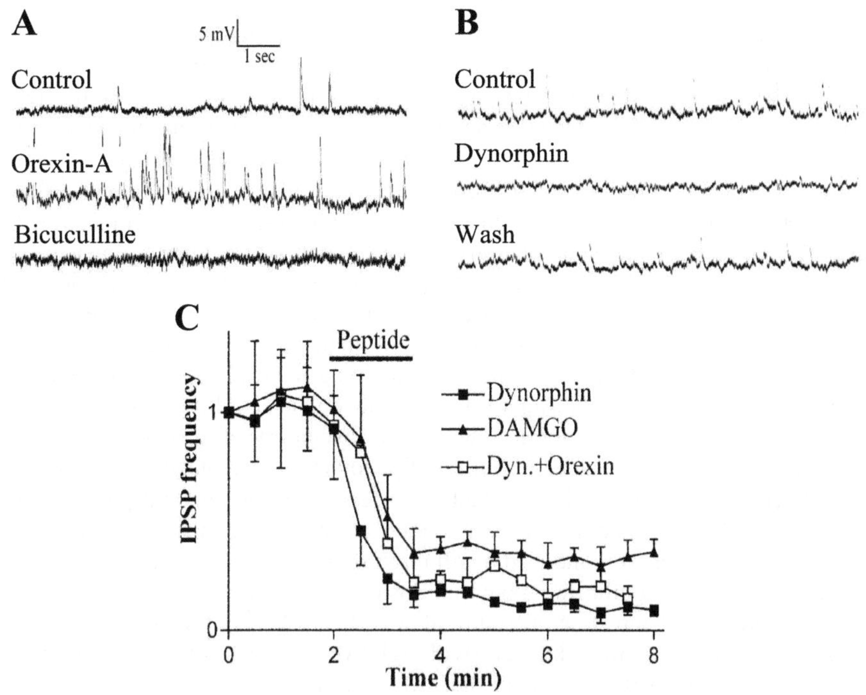

Figure 3. Hypocretin and dynorphin modulate GABA input to histamine neurons. (**A**) Orexin-A increases spontaneous IPSPs (blocked by bicuculline, 20μM). Note the bicuculline-insensitive increase in membrane noise induced by Orexin-A. (**B**) Activation of opioid receptors by the κ-receptor agonist dynorphin and the μ-receptor agonist (D-Ala, N-Me-Phe, glycinol)-enkephalin (DAMGO; 1 μM) suppress spontaneous IPSPs. (**C**) Co-application of Orexin-A and dynorphin overwrites Orexin-A effects. Modified from Eriksson et al.[16]

The symptoms of human narcolepsy[10] support a segregation of histamine and hypocretin/dynorphin neuronal function in the control of behavioral state stability, [53] wakefulness,[69] energy metabolism, [89,90,94] and motor activity.[10] Studies in Hcrt2 receptor-deficient narcoleptic dogs on cataplexy, [71] the sudden loss of postural muscle tone elicited by emotionally arousing primary rewards, such as food presentation, showed that histamine neurons, in contrast to other monoaminergic "REM-off" cell groups, are active during cataplexy. Activity of histamine neurons is thus linked to the maintenance of waking, whereas that of noradrenergic and serotonergic neurons is tightly coupled to the maintenance of muscle tone in waking and its loss in REM sleep and cataplexy.

3.2. Dorsal Raphe (Serotonin)

Serotonin (5-HT) neurons of the dorsal raphe in the brainstem are part of modulatory ascending and descending pathways that gate sleep-wake states[62,63] and nociception.[37,95] Opposing dopaminergic functions, serotonin mediates behavioral suppression, promoting satiety and inhibition of a variety of motivated behaviors. It plays a role in anxiety, migraine, depression, obesity and narcolepsy. Selective serotonin reuptake inhibitors are widely prescribed antidepressants but also used in sleep disorders.[7,96] Specific 5-HT1B/D receptor agonists are gold standard in the therapy of migraine attacks.[37,95]

Serotonergic neurons express high amounts of Hcrt1 and Hcrt2 receptors[20] and a dense network of hypocretinergic fibers surrounds neurons in the dorsal raphe (DR).[32] Hypocretins but not MCH strongly depolarize serotonin neurons in brain slices convergent with histamine H1, and adrenergic $\alpha 1$ receptors (Figure 3).[20] At higher concentrations (EC50, approximately 450-600 nm) they also increase impulse flow in local GABAergic interneurons in the dorsal raphe.[54] Single-cell PCR demonstrated the presence of both receptors in tryptophan hydroxylase-positive neurons. Agonists of three arousal-related systems impinging on dorsal raphe neurons (hypocretin, histamine and the noradrenaline systems) caused an inward current and increase in current noise in whole-cell patch-clamp recordings. The inward current induced by all three agonists was significantly reduced in extracellular solution containing reduced sodium and insensitive to protein kinase inhibitors or thapsigargin. In extracellular recordings, all three agonists increased the firing rate of serotonin neurons; the excitatory effects of histamine and Hcrt-1 were occluded by previous application of phenylephrine. Antagonists of histamine H1 and $\alpha 1$ receptors, respectively, blocked the effects of histamine and phenylephrine, suggesting that all three systems act via common effector mechanisms. Three types of current-voltage responses were induced by all three agonists but in no case did the current reverse at the potassium equilibrium potential. Instead, in many cases the Hcrt-1 induced current reversed in calcium-free medium at a value (-23 mV) consistent with the activation of a mixed cation channel (with relative permeabilities for sodium and potassium of 0.43 and 1, respectively) of the TRPC family.[19]

In contrast, hypocretin neurons in hypothalamic slices are strongly inhibited (hyperpolarized) by serotonin[49] and catecholamines[26] through activation of postsynaptic 5-HT1A and/or α_2-adrenoreceptors. Relieving hypocretin neurons from their negative serotonergic feedback increases locomotor activity in rodents.[49] They have also been implicated in leptin-independent hyperphagia, diabetes,[97] and suicidal behavior.[98] However, hypocretin-induced hyperlocomotion, grooming, face washing, wet dog shaking, and chewing was only partly inhibited by the serotonin antagonists, but abolished by haloperidol, suggesting a segregation in the control of hypocretin-induced stress-related behaviors and locomotion by dopamine and serotonin pathways.[99]

Serotonin 5-HT transporter immunoreactivity in nerve endings in the vicinity of hypocretin neurons, together with the well established role of 5-HT1A receptors in anxiety suggests a role of the hypothalamic hypocretin in mood disorders, similar to what has been proposed for the MCH system.[100] Clinical symptoms and dysfunction of hypocretinergic signaling in some patients suffering from depression or attention-deficit hyperactivity disorder (ADHD) as well as animal models of these diseases support a role for hypocretins in disorders of mood.[101,102] Moreover, narcolepsy exhibits a high degree of comorbidity with migraine.[37,95,103] Antidepressants, used to treat mood, sleep and pain disorders, uniformly exhibit delayed mood stabilizing effects associated with a tonic activation and downregulation of 5-HT1A autoreceptors, which may have impact on 5-HT1A-mediated inhibition of hypocretinergic activity. Chronic fluoxetine administration in obese Zucker rats reduced body weight gain, food intake, adipocyte size, fat mass, and body protein via hypocretin-independent pathways,[104] suggesting that side-effects of selective serotonin-reuptake inhibitors on body weight are hypocretin-independent. In contrast, sibutramine, a mixed norepinephrine, serotonin and dopamine reuptake inhibitor is used for treatment of obesity but also relieves cataplexy in human narcolepsy.[105] The pharmacological profile of psychostimulants may thus determine both, interactions with the hypocretin system [58] and drug-specificity.

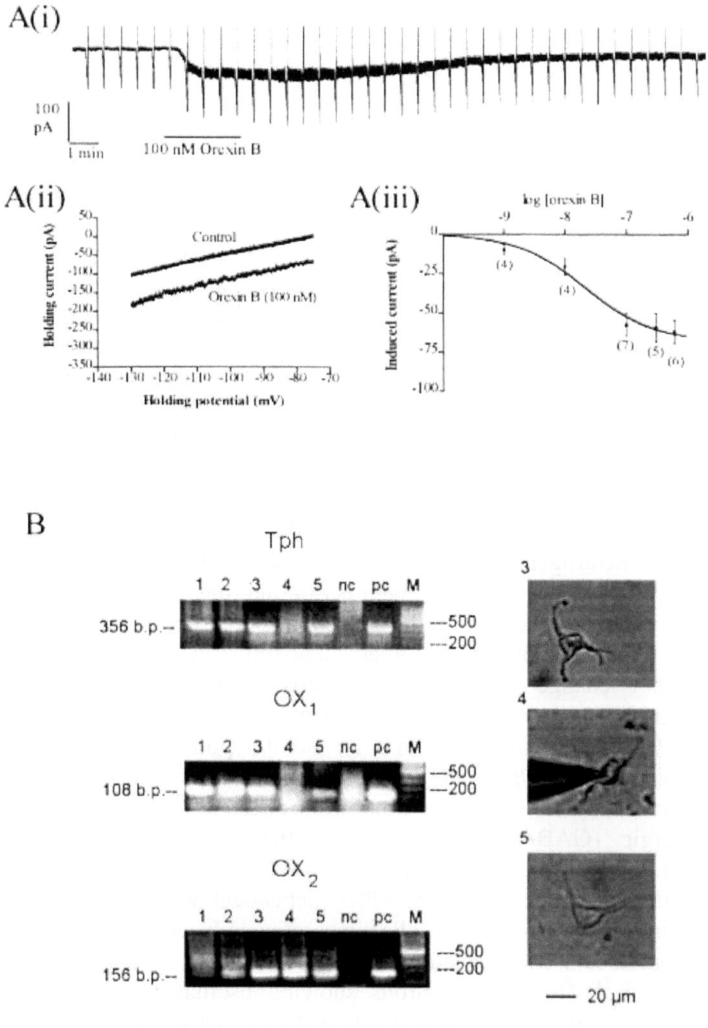

Figure 4. pocretin effects on serotoninergic dorsal raphe neurons. (**A**) Chart recordings under voltage clamp. (**Ai**) Downward deflections represent the responses to voltage ramps from the holding potential of -75 to -130 mV. Hypocretin-2 / Orexin-B induced inward current and increased current noise. (**Aii**), Current-voltage plot control versus Orexin B. (**Aiii**) Dose-response curve for Orexin B. (**B**), Co-expression of tryptophan hydroxylase (*Tph*) and hypocretin receptors (OX_1 and OX_2) (single-cell RT-PCR). Modified from Brown *et al.*[20]

3.3. Ventral tegmental area / Substantia Nigra (Dopamine)

Dopamine signaling serves motor functions and primary natural rewards,[106] such as food and sex, and in its usurpation movement disorders, compulsion, addiction, and psychosis.[58,107-109] Evidence from narcolepsy-cataplexy,[7,8,96] attention deficit hyperactivity disorder,[101] and sleep disturbances in Parkinson's disease,[110] suggest that dopamine-hypocretins interactions may link arousal, hedonic set points[107,111], and motor activity.

Psychostimulants such as amphetamine, cocaine, and methylphenidate interfere with dopamine re-uptake and/or release and are the most potent wake promoting and reinforcing drugs.[109,112] In contrast, midbrain dopaminergic lesions impair wakefulness in response to behavioral stimulation. Likewise, dopamine transporter knockout mice have sleep abnormalities, are hyperactive, and are insensitive to the normally robust wake-promoting actions of amphetamines or modafinil,[112] a stimulant approved for the treatment of excessive daytime sleepiness in narcolepsy. Modafinil, whose pharmacological profile has still not been resolved, selectively activates histaminergic and hypocretinergic cells of the hypothalamic arousal system.[113,114] Similarly amphetamines and atypical antipsychotics[58] also activate these systems but to a different degree and with preference for prefrontal cortical targets. There hypocretins in turn facilitate presynaptic release of catecholamines,[56,58] suggesting hyporcetins may be endogenous "amphetamines".

Dopaminergic drugs have differential effects depending on their affinity for dopaminergic receptor subtypes and for presynaptic versus postsynaptic receptors. Sleep itself can be controlled via D2/3 presynaptic receptor modulation of the ventral tegmental area (VTA), but not substantia nigra (SN),[112] suggesting a functional and pharmacological heterogeneity of dopaminergic cell groups. Recent work indicates the existence of sleep-state dependent dopaminergic neurons in the ventral periaqueductal gray (PAG),[115] a rostral extension of the VTA that is contiguous with caudal diencephalic dopaminergic groups. Contrary to other monoaminergic cell groups, dopaminergic neurons do not change their activity greatly throughout the sleep-wake cycle. Mesocorticolimbic dopaminergic and non-dopaminergic neurons in the VTA are densely innervated by Hcrt-fibers emanating from the lateral hypothalamus.[116]

Single-unit extracellular and whole-cell patch-clamp recordings revealed that hypocretins but not MCH excited neurons from different groups of A10 dopamine and non-dopaminergic (GABAergic) neurons in the VTA by tetrodotoxin-insensitive depolarization and tetrodotoxin-sensitive increased cell firing (Figure 5).[22,24] The signal transduction implicates Ca^{2+}-, PLC-, and PKC-dependent pathways.[117]

Notably, in some dopaminergic cells hypocretins induced burst firing, which is rarely seen *in vitro* but *in vivo* encodes unexpected appearance of rewards or stimuli-predicting reward e.g. food.[106] Dopamine neurons showing oscillatory activity in response to hypocretins had smaller afterhyperpolarizations than other groups of neurons. Single-cell PCR experiments showed that Hcrt receptors were expressed in both dopaminergic and non-dopaminergic neurons exhibiting a selective co-expression with the neuroprotective calcium binding protein calbindin. In contrast, dopaminergic cells in the SN were not excited by hypocretins but a much greater portion of GABAergic cells in the SN is innervated[116] and activated by hypocretins in a thapsigargin and PKA-sensitive manner, convergent with histamine acting on H1 receptors.[23]

Excitation of dopaminergic and GABAergic neurons in the VTA may promote arousal, locomotor activity, and search for food during states of hunger or metabolic challenges that increase activity of glucose-responsive and leptin receptor-positive hypocretin neurons.[3,89,90,94,118] This idea is supported by the known effects of hypocretins on wakefulness, food intake and locomotor activity as well as by the effect of food reward in Hcrt2 receptor-deficient narcoleptic dogs, which is a reliable trigger of cataplexy.[96] These dogs have elevated dopamine/noradrenaline but decreased histamine levels in the cortex and thalamus and display periodic leg movements during sleep.[83,96,119] Disinhibition of dopaminergic influences descending from the telencephalon to the spinal cord may thus contribute to abnormal movements during sleep and provide a link to periodic leg movement or restless leg syndrome in humans.[7,8] Hcrt1 induced hyper-locomotion in animals is blocked and other stress-related stereotypic behaviors such as grooming, face washing and chewing[86] are attenuated by drugs that antagonize dopamine receptors, such as haldol.[99]

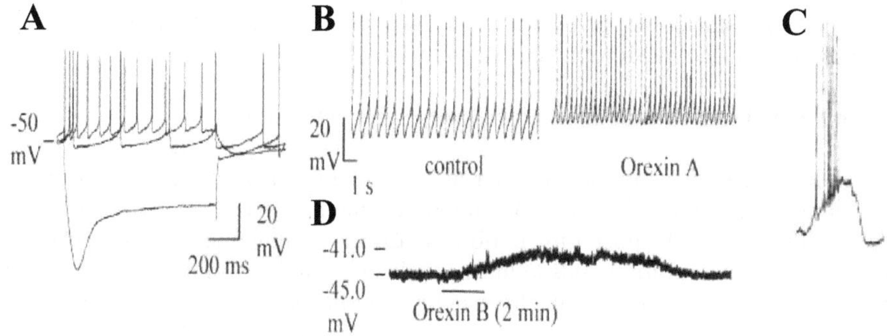

Figure 5. Electrophysiological properties of dopaminergic neurons and their response to hypocretins/orexins. (**A**) Voltage responses to current pulses (-0.4, 0, +0.1 pA). (**B**) Spontaneous action potentials before and after application of Orexin A (100 nM), chart recording of membrane potential (current clamp). (**C**) Burst firing induced by Orexin B. (**D**) Tetrodotoxin-insensitive depolarization by Orexin B (100 nM). Modified from Korotkova et al.[22]

Haldol also blocked some of the memory enhancing effects of Hcrt1 on passive avoidance learning in rodents.[120] Dopamine receptors have been implicated in protein synthesis-dependent synaptic plasticity and memory consolidation[107]. Hypocretinergic modulation of dopamine signaling may thus be important for normal memory as well as compulsion, addiction and psychosis.[58,107-109]

Finally, hypocretins also interfere with dopaminergic control of neuroendocrine functions, like prolactin release. Central administration of Hcrt1 suppressed basal and D2R-antagonist (domperidone) stimulated plasma prolactin release,[121] while fasting, a prominent activator of the hypocretin system[2,3,5,89,90,94,118] increased tuberoinfundibular dopamine outflow and inhibited prolactin secretion in estrogen-primed ovariectomized rats through up-regulation of the central hypocretin system.[122]

3.4. Locus Coeruleus (Noradrenaline)

The strongest projections of hypocretin-fibers and the highest density of Hcrt1 receptors among all aminergic cell groups is found in noradrenergic neurons of the locus coeruleus.[32,123,124] Action potential firing and temporal synchrony of noradrenergic LC neurons is increased by Hcrt1 receptor-dependent direct postsynaptic and indirect glutamatergic mechanisms.[66,123,125] In addition, hypocretins facilitate noradrenaline and glutamate release in cortical and subcortical targets.[56,57] Catecholamines in turn, convergent with serotonin and GABA, provide negative inhibitory feedback to the hypothalamic hypocretin neurons *in vitro*.[26] Interestingly, preliminary evidence suggests a dynamic control of noradrenergic feedback on hypocretin neurons according to aminergic tone and behavioral state.[126]

The interplay between hypocretins and catecholamines has implications not only for control of behavioral state,[124,127] nurture, plasticity[30,128] and memory retrieval[129] but probably also for a number of side-effects of sympathomimetics and sympatholytics.[29] Van den Pol and Aston-Jones give a detailed description of interactions between the noradrenergic system and hypocretin neurons in this book.

3.5. Laterodorsal Tegmentum / Basal forebrain (Acetylcholine)

The cholinergic systems in the laterodorsal tegmentum (LdT)[130] and basal forebrain are major targets of hypocretinergic innervation and modulation[39,85,130,131](reviewed by Leonard in this book). Hypocretins uniformly excite cholinergic and GABAergic neurons. Acetylcholine in turn, opposite to catecholamines and serotonin but convergent with glutamate, excites hypothalamic hypocretin neurons *in vitro*[26,27] providing feed forward mechanisms implicated in muscle tone control, gating of REM sleep, cortical activation, arousal, attentional set shifting, binding, and reversal learning.[132,133]

Notably, acetylcholine becomes the major excitatory transmitter in the absence of glutamate in the hypothalamus (but not in the cortex) *in vitro*, providing a plasticity mechanism invoking glutamate/GABA and excitation/inhibition balance.[134] Likewise, cholinergic tone *in vivo* in the thalamus and cortex, concurrent with large scale oscillatory brain activities, varies according to behavioral state and opposite or complementary to the activity of other aminergic systems.[62,63]

4. HIPPOCAMPUS AND CORTEX

Hypocretins and biogenic amines and their cognate receptors densely accumulate in infralimbic structures including the amygdala, hippocampus, and insular, cingulated, and prefrontal cortex. Apart from convergent, direct excitatory effects of hypocretins on non-specific thalamic relay neurons[135] and intracortically projecting layer 6b neurons,[136] the orchestration of aminergic, glutamatergic, and GABAergic signals provides a powerful mechanism promoting cortical activation, plasticity, and probably memory functions.[129,133,137]

The prefrontal cortex is an organizer of attentional set shifting, cognitive appraisal, executive fuctions, and autonomic responses to hypoglycemia, hunger, and drowsiness.[35,36,93] Hypocretins facilitate the release of catecholamines[56,58] and through glutamatergic mechanisms they induce calcium transients in dendritic spines of distinct prefrontal thalamocortical synapses.[138] Action potential backfiring in presynaptic afferents of non-specific thalamic neurons[135,138] is considered to promote synchronous behavioral state-dependent ensemble activity.[68]

Likewise, in the hippocampus, an integrator of multimodal sensorimotor functions and neuroendocrine-autonomic control,[139] hypocretins, again in concert with biogenic amines, promote synchronous network activity and synaptic plasticity.[30,31,64,65] Hcrt receptors are expressed in all subregions of the hippocampus but most densely in the dentate gyrus,[30] the main entrance pathway conveying information from the entorhinal cortex to the hippocampus.

Injection of Hcrt1 into the dentate gyrus[140] or locus coeruleus,[128] a preferred target of hypocretinergic modulation,[32,123] promotes the induction of glutamate-dependent LTP *in vivo*. Co-application of adrenergic receptor antagonists in the locus coeruleus blocked Hcrt-induced synaptic plasticity in the dentate gyrus.[128] Immunoreactivity for the native Hcrt-1 peptide is largely restricted to Schaffer-collateral-CA1[30] and septohippocampal pathways. The latter "gates" hippocampal disinhibition and theta rhythm through convergent facilitatory actions of hypocretins and biogenic amines.[39,85,141] Theta rhythm is a behavioral state-dependent large scale oscillatory brain activity associated with exploratory behavior, novelty, stress, REM sleep, and bidirectional, spike-timing-dependent plasticity.[68] In contrast, so-called Schaffer-collateral axons provide excitatory drive from the CA3 region, an autoassociative sharp wave generator,[68] to the CA1 region, the major output of the hippocampus. Sharp waves, in contrast to theta rhythm, occur during consummatory behavior, rest, immobility, and slow wave sleep, thus representing an off-line state of the brain promoting hippocampal-cortical dialogue and memory consolidation. Histamine and hypocretins are powerful modulators of these oscillations in vitro[21,30] and in vivo.[31,64,65]

Brief bath application of Hcrt1 to hippocampal slices triggers theta-rhythm patterned sharp wave-concurrent field burst activity in the CA3 region, capable to induce a slow onset long-term potentiation of synaptic transmission (LTP_{OX}) in CA1 region.[30,142] Consistent with the absence of direct hypocretin effects on hippocampal neurons in culture[1], as well as with the mechanisms of activation of thalamocortical synapses[138] this form of plasticity is expressed presynaptically. More important, it requires coordinated signaling of glutamate, GABA, noradrenaline, and acetylcholine (Figure 6).

Analysis of the molecular signature of LTP_{OX} revealed a requirement for co-activation of multiple plasticity-related kinases and synthesis of new proteins.[142] Blockade of a few decisive signal transduction pathways, including metabotropic glutamate $mGluR_1$ receptors, CaMKII, and PKA, or mTOR-dependent protein synthesis converted LTP_{OX} into a long-term depression (LTD), indicating that hypocretins, in line with theta-patterned stimulus conditions, promote bidirectional modifications of synaptic strength. Interestingly, the direction of Hcrt-induced plasticity is dependent on the age of the animals, suggesting a developmental switch[60] and explaining seemingly contradictory findings in rats.[143] The molecular signature of hypocretin-induced synaptic plasticity matches that of conditioned taste aversion.[132]

Collectively, hypocretins, in concert with biogenic amines, promote protein synthesis-dependent LTP, as well as protein synthesis-independent LTD.[30] This binary mode of action may promote network bistability[6] and homeostatic functions[60], outbalancing electrical activities, while in parallel providing set points for structural rewiring and rescaling of neural circuits.

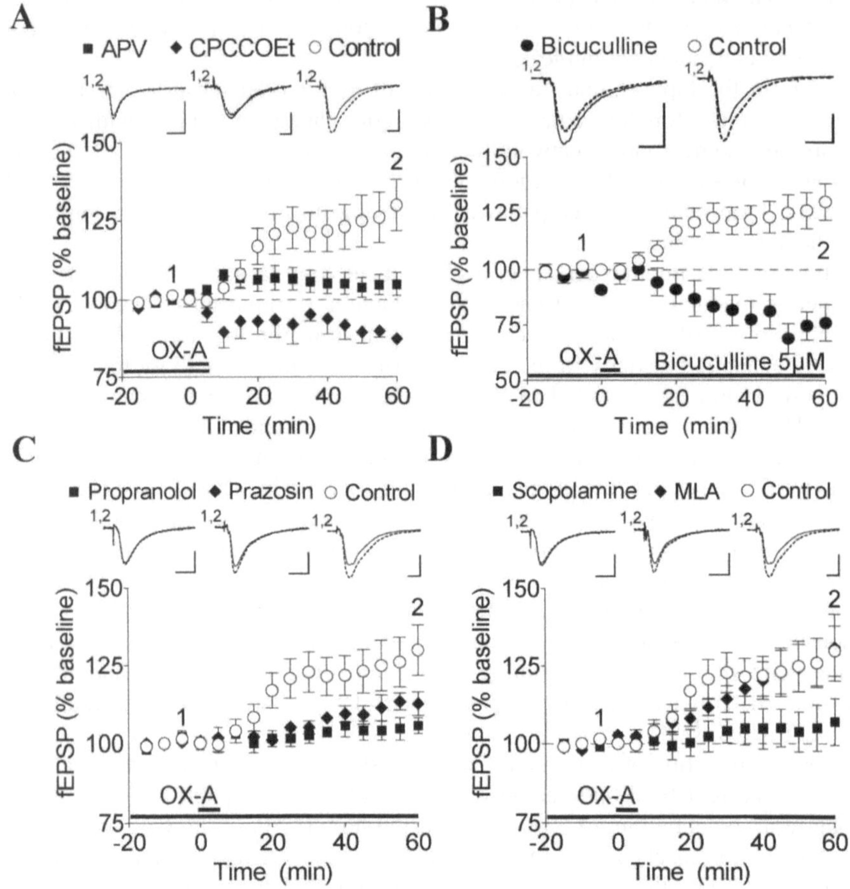

Figure 6. Hypocretin-induced synaptic plasticity (LTP_{OX}) in mouse hippocampal slices requires co-activation of glutamate, GABA, noradrenaline, and acetylcholine receptors. Samples (insets; calibration: 0.5 mV, 5 ms) and time course (normalized averaged data) of extracellular field responses recorded from the CA1 region of mouse hippocampal slices. (A) LTP_{OX} (○) is blocked by antagonists of glutamate NMDA (APV 50μM; ■) and group I $mGluR_1$ (CPCCOEt 50μM; ♦), (B) $GABA_A$ (bicuculline, 5μM; •), (C) beta- (propranolol, 1μM; ■) and 1-adrenergic (prazosin, 1μM; ♦), (D) muscarinic (scopolamine, 10μM; ■) but not -7 bungaro-toxin-sensitive nicotinic AChRs (MLA, 10 nM; ♦) receptors. Note the long-term depression induced by OX-A in the presence of $mGluR_1$- and $GABA_A$-antagonists. Modified from Selbach *et al.*[30]

Hypocretins through convergent signal transduction pathways imprint internal (metabolic)[3,89,90] and external (circadian)[8,92] state cues on neural circuits and likely behavioral state-dependent memory functions.[129,133,137] Conditions that affect sleep alter the expression of molecules associated with synaptic plasticity.[137] Likewise, learning-dependent alterations in large scale network oscillations, such as increases in sleep spindle density have been reported both in animals and humans.[62,63] Hypocretin-induced plasticity may thus act as a floating blueprint[3] for both, the control of memory[133] and behavioral state.[62,63,92] Short-lasting, protein synthesis-independent synaptic plasticity may link episodic "working" memory and REM-nonREM-alternations, whereas protein synthesis-dependent long-term plasticity consolidates sleep-wake states[92] and memory.[133] Narcolepsy is characterized by defects in behavioral state stability and sleep-wake consolidation.[8,92] Encoding deficits observed in narcolepsy[8] may thus not only be secondary to disordered states of vigilance but also result from impaired hypocretinergic modulation of synaptic transmission in the hippocampus. Moreover, modulation of hippocampal plasticity may serve as a set point in food and drug seeking behavior[109] and neuroendocrine-autonomic feedback[139] and thus integrate in behavioral state-dependent control of basic body functions.

In contrast, evidence providing direct support for a role of hypocretins in memory is still limited and pharmacological experiments in animals revealed ambiguous results. Intracerebroventricular injection of Hcrt1 in rats impaired spatial memory tested in the Morris water-maze[143] while it facilitated contextual forms of memory[120,144] in rodents. Post-training administration of Hcrt-1 improved retention in active T-maze foot shock avoidance and one trial step-down passive avoidance in control and SAMP8 mice,[144] an animal model of Alzheimer's disease, exhibiting age-related learning deficits that correlate with increased levels of brain beta amyloid and impaired sensitivity to memory-enhancing compounds. At least, this suggests that Hcrt-1 may relieve functional impairments in neurodegenerative conditions, such as Alzheimer's and Huntington's disease.[61] Expression of hypocretin receptors in human CD34+ stem cells[145] and various cancer cell lines[43] emphasizes a general role of hypocretins in cellular plasticity.

5. CONCLUSION

Aminergic systems, orchestrated by hypocretin/dynorphin neurons, mediate complementary aspects of behavioral state and body homeostasis control, such as maintenance of waking in the context of environmental challenges and stress (histamine), satiety, behavioral suppression and sensory gating (serotonin), primary reward and motor activity (dopamine), arousal, postural muscle tone, and sympathetic outflow (noradrenaline, acetylcholine). Synchronization of biogenic amine and amino acid signaling according to environmental, nutritional and circadian influences by the hypocretin system provides a global mechanism controlling metabolic and electrical state of the whole brain.

6. REFERENCES

1. L. de Lecea, T.S. Kilduff, C. Peyron, X. Gao, P.E. Foye, P.E. Danielson, C. Fukuhara, E.L. Battenberg, V.T. Gautvik, F.S. Bartlett, W.N. Frankel, A.N. van den Pol, F.E. Bloom, K.M. Gautvik and J.G. Sutcliffe, The hypocretins: hypothalamus-specific peptides with neuroexcitatory activity, *Proc.Natl.Acad.Sci.U.S.A* **95**, 322-327 (1998).
2. T. Sakurai, Orexin: a link between energy homeostasis and adaptive behaviour, *Curr.Opin.Clin.Nutr.Metab Care* **6**, 353-360 (2003).
3. T.L. Horvath and S. Diano, Opinion: The floating blueprint of hypothalamic feeding circuits, *Nat.Rev.Neurosci.* **5**, 662-667 (2004).
4. J.G. Sutcliffe and L. De Lecea, The hypocretins: setting the arousal threshold, *Nat.Rev.Neurosci.* **3**, 339-349 (2002).
5. J.T. Willie, R.M. Chemelli, C.M. Sinton and M. Yanagisawa, To eat or to sleep? Orexin in the regulation of feeding and wakefulness, *Annu. Rev. Neurosci.* **24**, 429-458 (2001).

6. C.B. Saper, T.C. Chou and T.E. Scammell, The sleep switch: hypothalamic control of sleep and wakefulness, *Trends Neurosci* **24**, 726-731 (2001).
7. E. Mignot, S. Taheri and S. Nishino, Sleeping with the hypothalamus: emerging therapeutic targets for sleep disorders, *Nature Neuroscience* **5 Suppl**, 1071-1075 (2002).
8. S. Taheri, J.M. Zeitzer and E. Mignot, The role of hypocretins (orexins) in sleep regulation and narcolepsy, *Annu. Rev. Neurosci.* **25**, 283-313 (2002).
9. A.V. Ferguson and W.K. Samson, The orexin/hypocretin system: a critical regulator of neuroendocrine and autonomic function, *Front Neuroendocrinol.* **24**, 141-150 (2003).
10. J.M. Siegel, Hypocretin (orexin): role in normal behavior and neuropathology, *Annual Review of Psychology* **55**, 125-148 (2004).
11. Y. Kayaba, A. Nakamura, Y. Kasuya, T. Ohuchi, M. Yanagisawa, I. Komuro, Y. Fukuda and T. Kuwaki, Attenuated defense response and low basal blood pressure in orexin knockout mice, *Am.J Physiol Regul.Integr.Comp Physiol* **285**, R581-R593 (2003).
12. J.C. Geerling, T.C. Mettenleiter and A.D. Loewy, Orexin neurons project to diverse sympathetic outflow systems, *Neuroscience* **122**, 541-550 (2003).
13. D. McGinty and R. Szymusiak, Hypothalamic regulation of sleep and arousal, *Front Biosci.* **8**, s1074-s1083 (2003).
14. K.S. Eriksson, O. Sergeeva, R.E. Brown and H.L. Haas, Orexin/hypocretin excites the histaminergic neurons of the tuberomammillary nucleus, *J. Neurosci.* **21**, 9273-9279 (2001).
15. K.S. Eriksson, D.R. Stevens and H.L. Haas, Serotonin excites tuberomammillary neurons by activation of Na(+)/Ca(2+)-exchange, *Neuropharmacology* **40**, 345-351 (2001).
16. K.S. Eriksson, O.A. Sergeeva, O. Selbach and H.L. Haas, Orexin (hypocretin)/dynorphin neurons control GABAergic inputs to tuberomammillary neurons, *Eur.J Neurosci.* **19**, 1278-1284 (2004).
17. O.A. Sergeeva, K.S. Eriksson, I.N. Sharonova, V.S. Vorobjev and H.L. Haas, GABA(A) receptor heterogeneity in histaminergic neurons, *Eur.J Neurosci.* **16**, 1472-1482 (2002).
18. O.A. Sergeeva, B.T. Amberger, K.S. Eriksson, A. Scherer and H.L. Haas, Co-ordinated expression of 5-HT2C receptors with the NCX1 Na+/Ca2+ exchanger in histaminergic neurones, *J Neurochem.* **87**, 657-664 (2003).
19. O.A. Sergeeva, T.M. Korotkova, A. Scherer, R.E. Brown and H.L. Haas, Co-expression of non-selective cation channels of the transient receptor potential canonical family in central aminergic neurones, *J. Neurochem.* **85**, 1547-1552 (2003).
20. R.E. Brown, O.A. Sergeeva, K.S. Eriksson and H.L. Haas, Convergent excitation of dorsal raphe serotonin neurons by multiple arousal systems (orexin/hypocretin, histamine and noradrenaline), *J. Neurosci.* **22**, 8850-8859 (2002).
21. H. Haas and P. Panula, The role of histamine and the tuberomamillary nucleus in the nervous system, *Nat.Rev.Neurosci.* **4**, 121-130 (2003).
22. T.M. Korotkova, O.A. Sergeeva, K.S. Eriksson, H.L. Haas and R.E. Brown, Excitation of ventral tegmental area dopaminergic and nondopaminergic neurons by orexins/hypocretins, *J. Neurosci.* **23**, 7-11 (2003).
23. T.M. Korotkova, H.L. Haas and R.E. Brown, Histamine excites GABAergic cells in the rat substantia nigra and ventral tegmental area in vitro, *Neuroscience Letters* **320**, 133-136 (2002).
24. T.M. Korotkova, K.S. Eriksson, H.L. Haas and R.E. Brown, Selective excitation of GABAergic neurons in the substantia nigra of the rat by orexin/hypocretin in vitro, *Regulatory Peptides* **104**, 83-89 (2002).
25. J.P. Kukkonen, T. Holmqvist, S. Ammoun and K.E. Akerman, Functions of the orexinergic/hypocretinergic system, *Am J Physiol Cell Physiol* **283**, C1567-C1591 (2002).
26. Y. Li, X.B. Gao, T. Sakurai and A.N. van den Pol, Hypocretin/Orexin excites hypocretin neurons via a local glutamate neuron-A potential mechanism for orchestrating the hypothalamic arousal system, *Neuron* **36**, 1169-1181 (2002).
27. A. Yamanaka, Y. Muraki, N. Tsujino, K. Goto and T. Sakurai, Regulation of orexin neurons by the monoaminergic and cholinergic systems, *Biochem. Biophys. Res. Commun.* **303**, 120-129 (2003).
28. E. Eggermann, L. Bayer, M. Serafin, B. Saint-Mleux, L. Bernheim, D. Machard, B.E. Jones and M. Muhlethaler, The wake-promoting hypocretin-orexin neurons are in an intrinsic state of membrane depolarization, *J. Neurosci.* **23**, 1557-1562 (2003).
29. O. Selbach, K.S. Eriksson and H.L. Haas, Drugs to interfere with orexins (hypocretins), *Drug News Perspect.* **16**, 669-681 (2003).
30. O. Selbach, N. Doreulee, C. Bohla, K.S. Eriksson, O.A. Sergeeva, W. Poelchen, R.E. Brown and H.L. Haas, Orexins/hypocretins cause sharp wave- and theta-related synaptic plasticity in the hippocampus via glutamatergic, gabaergic, noradrenergic, and cholinergic signaling, *Neuroscience* **127**, 519-528 (2004).
31. A.A. Ponomarenko, J.S. Lin, O. Selbach and H.L. Haas, Temporal pattern of hippocampal high-frequency oscillations during sleep after stimulant-evoked waking, *Neuroscience* **121**, 759-769 (2003).

32. C. Peyron, D.K. Tighe, A.N. van den Pol, L. De Lecea, H.C. Heller, J.G. Sutcliffe and T.S. Kilduff, Neurons containing hypocretin (orexin) project to multiple neuronal systems, *J. Neurosci.* **18**, 9996-10015 (1998).
33. B.A. Baldo, R.A. Daniel, C.W. Berridge and A.E. Kelley, Overlapping distributions of orexin/hypocretin- and dopamine-beta-hydroxylase immunoreactive fibers in rat brain regions mediating arousal, motivation, and stress, *J Comp Neurol.* **464**, 220-237 (2003).
34. Y. Date, Y. Ueta, H. Yamashita, H. Yamaguchi, S. Matsukura, K. Kangawa, T. Sakurai, M. Yanagisawa and M. Nakazato, Orexins, orexigenic hypothalamic peptides, interact with autonomic, neuroendocrine and neuroregulatory systems, *Proc.Natl.Acad.Sci.U.S.A* **96**, 748-753 (1999).
35. C.B. Saper, THE CENTRAL AUTONOMIC NERVOUS SYSTEM: Conscious Visceral Perception and Autonomic Pattern Generation, *Annu. Rev. Neurosci.* **25**, 433-469 (2002).
36. A.D. Craig, How do you feel? Interoception: the sense of the physiological condition of the body, *Nat.Rev.Neurosci.* **3**, 655-666 (2002).
37. T. Bartsch, M.J. Levy, Y.E. Knight and P.J. Goadsby, Differential modulation of nociceptive dural input to [hypocretin] orexin A and B receptor activation in the posterior hypothalamic area, *Pain* **109**, 367-378 (2004).
38. J.N. Marcus, C.J. Aschkenasi, C.E. Lee, R.M. Chemelli, C.B. Saper, M. Yanagisawa and J.K. Elmquist, Differential expression of orexin receptors 1 and 2 in the rat brain, *J. Comp Neurol.* **435**, 6-25 (2001).
39. M. Wu, Z. Zhang, C. Leranth, C. Xu, A.N. van den Pol and M. Alreja, Hypocretin increases impulse flow in the septohippocampal GABAergic pathway: implications for arousal via a mechanism of hippocampal disinhibition, *J Neurosci.* **22**, 7754-7765 (2002).
40. D. Burdakov, B. Liss and F.M. Ashcroft, Orexin excites GABAergic neurons of the arcuate nucleus by activating the sodium--calcium exchanger, *J. Neurosci.* **23**, 4951-4957 (2003).
41. Q.V. Hoang, D. Bajic, M. Yanagisawa, S. Nakajima and Y. Nakajima, Effects of orexin (hypocretin) on GIRK channels, *J. Neurophysiol.* **90**, 693-702 (2003).
42. T.M. van den, M.F. Nolan, K. Lee, P.J. Richardson, R.M. Buijs, C.H. Davies and D. Spanswick, Orexins induce increased excitability and synchronisation of rat sympathetic preganglionic neurones, *J.Physiol* **549**, 809-821 (2003).
43. P. Rouet-Benzineb, C. Rouyer-Fessard, A. Jarry, V. Avondo, C. Pouzet, M. Yanagisawa, C. Laboisse, M. Laburthe and T. Voisin, Orexins acting at native OX1 receptor in colon cancer and neuroblastoma cells or at recombinant OX1 receptor suppress cell growth by inducing apoptosis, *J Biol.Chem.* (2004).
44. D.E. Clapham, TRP channels as cellular sensors, *Nature* **426**, 517-524 (2003).
45. M.P. Mattson, S.L. Chan and W. Duan, Modification of brain aging and neurodegenerative disorders by genes, diet, and behavior, *Physiol Rev.* **82**, 637-672 (2002).
46. F. Torrealba, M. Yanagisawa and C.B. Saper, Colocalization of orexin a and glutamate immunoreactivity in axon terminals in the tuberomammillary nucleus in rats, *Neuroscience* **119**, 1033-1044 (2003).
47. C. Acuna-Goycolea, Y. Li and A.N. van den Pol, Group III metabotropic glutamate receptors maintain tonic inhibition of excitatory synaptic input to hypocretin/orexin neurons, *J Neurosci.* **24**, 3013-3022 (2004).
48. S. Satoh, H. Matsumura, A. Fujioka, T. Nakajima, T. Kanbayashi, S. Nishino, Y. Shigeyoshi and H. Yoneda, FOS expression in orexin neurons following muscimol perfusion of preoptic area, *Neuroreport* **15**, 1127-1131 (2004).
49. Y. Muraki, A. Yamanaka, N. Tsujino, T.S. Kilduff, K. Goto and T. Sakurai, Serotonergic regulation of the orexin/hypocretin neurons through the 5-HT1A receptor, *J Neurosci.* **24**, 7159-7166 (2004).
50. T. Gallopin, P. Fort, E. Eggermann, B. Cauli, P.H. Luppi, J. Rossier, E. Audinat, M. Muhlethaler and M. Serafin, Identification of sleep-promoting neurons in vitro, *Nature* **404**, 992-995 (2000).
51. T.C. Chou, A.A. Bjorkum, S.E. Gaus, J. Lu, T.E. Scammell and C.B. Saper, Afferents to the ventrolateral preoptic nucleus, *Journal of Neuroscience* **22**, 977-990 (2002).
52. O. Hayaishi and Z.L. Huang, Role of orexin and prostaglandin E(2) in activating histaminergic neurotransmission, *Drug News Perspect.* **17**, 105-109 (2004).
53. T. Mochizuki, A. Crocker, S. McCormack, M. Yanagisawa, T. Sakurai and T.E. Scammell, Behavioral state instability in orexin knock-out mice, *J Neurosci.* **24**, 6291-6300 (2004).
54. R.J. Liu, A.N. van den Pol and G.K. Aghajanian, Hypocretins (orexins) regulate serotonin neurons in the dorsal raphe nucleus by excitatory direct and inhibitory indirect actions, *J. Neurosci.* **22**, 9453-9464 (2002).
55. A.N. van den Pol, X.B. Gao, K. Obrietan, T.S. Kilduff and A.B. Belousov, Presynaptic and postsynaptic actions and modulation of neuroendocrine neurons by a new hypothalamic peptide, hypocretin/orexin, *J Neurosci.* **18**, 7962-7971 (1998).
56. K. Hirota, T. Kushikata, M. Kudo, T. Kudo, D.G. Lambert and A. Matsuki, Orexin A and B evoke noradrenaline release from rat cerebrocortical slices, *Br. J. Pharmacol.* **134**, 1461-1466 (2001).
57. J. John, M.F. Wu, T. Kodama and J.M. Siegel, Intravenously administered hypocretin-1 alters brain amino acid release: an in vivo microdialysis study in rats, *J.Physiol* **548**, 557-562 (2003).

58. J. Fadel, M. Bubser and A.Y. Deutch, Differential activation of orexin neurons by antipsychotic drugs associated with weight gain, *J. Neurosci.* **22**, 6742-6746 (2002).
59. D. Kapfhamer, O. Valladares, Y. Sun, P.M. Nolan, J.J. Rux, S.E. Arnold, S.C. Veasey and M. Bucan, Mutations in Rab3a alter circadian period and homeostatic response to sleep loss in the mouse, *Nature Genetics* **32**, 290-295 (2002).
60. G.G. Turrigiano and S.B. Nelson, Homeostatic plasticity in the developing nervous system, *Nat. Rev. Neurosci.* **5**, 97-107 (2004).
61. A. Petersen, J. Gil, M.L.C. Maat-Schieman, M. Bjorkqvist, P. Mohapel, R. Araujo, N. Smith, N. Popovic, N. Wierup, P. Norlen, J. Li, R.A.C. Ross, F. Sundler, H. Mulder and P. Brundin, Orexin loss - More evidence for hypothalamic dysfunction in Huntington´s disease, *Soc Neurosci Abstr.* **564.1**, (2004).
62. J.A. Hobson and E.F. Pace-Schott, The cognitive neuroscience of sleep: neuronal systems, consciousness and learning, *Nat.Rev.Neurosci.* **3**, 679-693 (2002).
63. E.F. Pace-Schott and J.A. Hobson, The neurobiology of sleep: genetics, cellular physiology and subcortical networks, *Nat.Rev.Neurosci.* **3**, 591-605 (2002).
64. A. Knoche, H. Yokoyama, A. Ponomarenko, C. Frisch, J. Huston and H.L. Haas, High-frequency oscillation in the hippocampus of the behaving rat and its modulation by the histaminergic system, *Hippocampus* **13**, 273-280 (2003).
65. A.A. Ponomarenko, A. Knoche, T.M. Korotkova and H.L. Haas, Aminergic control of high-frequency (approximately 200 Hz) network oscillations in the hippocampus of the behaving rat, *Neuroscience Letters* **348**, 101-104 (2003).
66. A.N. van den Pol, P.K. Ghosh, R.J. Liu, Y. Li, G.K. Aghajanian and X.B. Gao, Hypocretin (orexin) enhances neuron activity and cell synchrony in developing mouse GFP-expressing locus coeruleus, *J Physiol* **541**, 169-185 (2002).
67. A. Peters, U. Schweiger, L. Pellerin, C. Hubold, K.M. Oltmanns, M. Conrad, B. Schultes, J. Born and H.L. Fehm, The selfish brain: competition for energy resources, *Neuroscience and Biobehavioral Reviews* **28**, 143-180 (2004).
68. G. Buzsaki and A. Draguhn, Neuronal oscillations in cortical networks, *Science* **304**, 1926-1929 (2004).
69. R. Parmentier, H. Ohtsu, Z. Djebbara-Hannas, J.L. Valatx, T. Watanabe and J.S. Lin, Anatomical, physiological, and pharmacological characteristics of histidine decarboxylase knock-out mice: evidence for the role of brain histamine in behavioral and sleep-wake control, *J Neurosci* **22**, 7695-7711 (2002).
70. J.S. Lin, Y. Hou, K. Sakai and M. Jouvet, Histaminergic descending inputs to the mesopontine tegmentum and their role in the control of cortical activation and wakefulness in the cat, *J Neurosci.* **16**, 1523-1537 (1996).
71. J. John, M.F. Wu, L.N. Boehmer and J.M. Siegel, Cataplexy-active neurons in the hypothalamus: implications for the role of histamine in sleep and waking behavior, *Neuron* **42**, 619-634 (2004).
72. S.P. Kalra, M.G. Dube, S. Pu, B. Xu, T.L. Horvath and P.S. Kalra, Interacting appetite-regulating pathways in the hypothalamic regulation of body weight, *Endocrine Reviews* **20**, 68-100 (1999).
73. M. Kummer, S.J. Neidert, O. Johren and P. Dominiak, Orexin (hypocretin) gene expression in rat ependymal cells, *Neuroreport* **12**, 2117-2120 (2001).
74. T. Porkka-Heiskanen, R.E. Strecker, M. Thakkar, A.A. Bjorkum, R.W. Greene and R.W. McCarley, Adenosine: a mediator of the sleep-inducing effects of prolonged wakefulness, *Science* **276**, 1265-1268 (1997).
75. J.A. Murphy, S. Deurveilher and K. Semba, Stimulant doses of caffeine induce c-FOS activation in orexin/hypocretin-containing neurons in rat, *Neuroscience* **121**, 269-275 (2003).
76. M.M. Thakkar, S. Winston and R.W. McCarley, Orexin neurons of the hypothalamus express adenosine A1 receptors, *Brain Research* **944**, 190-194 (2002).
77. H.L. Haas and O. Selbach, Functions of neuronal adenosine receptors, *Naunyn Schmiedebergs Arch. Pharmacol.* **362**, 375-381 (2000).
78. L.E. Nelson, T.Z. Guo, J. Lu, C.B. Saper, N.P. Franks and M. Maze, The sedative component of anesthesia is mediated by GABA(A) receptors in an endogenous sleep pathway, *Nature Neuroscience* **5**, 979-984 (2002).
79. I. Inoue, K. Yanai, D. Kitamura, I. Taniuchi, T. Kobayashi, K. Niimura, T. Watanabe and T. Watanabe, Impaired locomotor activity and exploratory behavior in mice lacking histamine H1 receptors, *Proc. Natl. Acad Sci U. S A* **93**, 13316-13320 (1996).
80. H. Yoshimatsu, E. Itateyama, S. Kondou, D. Tajima, K. Himeno, S. Hidaka, M. Kurokawa and T. Sakata, Hypothalamic neuronal histamine as a target of leptin in feeding behavior, *Diabetes* **48**, 2286-2291 (1999).
81. A. Mollet, T.A. Lutz, S. Meier, T. Riediger, P.A. Rushing and E. Scharrer, Histamine H1 receptors mediate the anorectic action of the pancreatic hormone amylin, *Am. J Physiol Regul. Integr. Comp Physiol* **281**, R1442-R1448 (2001).

82. L. Lin, J. Wisor, T. Shiba, S. Taheri, K. Yanai, S. Wurts, X. Lin, M. Vitaterna, J. Takahashi, T.W. Lovenberg, M. Koehl, G. Uhl, S. Nishino and E. Mignot, Measurement of hypocretin/orexin content in the mouse brain using an enzyme immunoassay: the effect of circadian time, age and genetic background, *Peptides* **23**, 2203-2211 (2002).
83. L. Lin, J. Faraco, R. Li, H. Kadotani, W. Rogers, X. Lin, X. Qiu, P.J. de Jong, S. Nishino and E. Mignot, The sleep disorder canine narcolepsy is caused by a mutation in the hypocretin (orexin) receptor 2 gene, *Cell* **98**, 365-376 (1999).
84. J.O. Rinne, O.V. Anichtchik, K.S. Eriksson, J. Kaslin, L. Tuomisto, H. Kalimo, M. Roytta and P. Panula, Increased brain histamine levels in Parkinson's disease but not in multiple system atrophy, *J Neurochem.* **81**, 954-960 (2002).
85. M. Wu, L. Zaborszky, T. Hajszan, A.N. van den Pol and M. Alreja, Hypocretin/orexin innervation and excitation of identified septohippocampal cholinergic neurons, *J Neurosci.* **24**, 3527-3536 (2004).
86. M.S. Duxon, J. Stretton, K. Starr, D.N. Jones, V. Holland, G. Riley, J. Jerman, S. Brough, D. Smart, A. Johns, W. Chan, R.A. Porter and N. Upton, Evidence that orexin-A-evoked grooming in the rat is mediated by orexin-1 (OX1) receptors, with downstream 5-HT2C receptor involvement, *Psychopharmacology (Berl)* **153**, 203-209 (2001).
87. D.R. Stevens, A. Kuramasu, K.S. Eriksson, O. Selbach and H.L. Haas, alpha(2)-Adrenergic receptor-mediated presynaptic inhibition of GABAergic IPSPs in rat histaminergic neurons, *Neuropharmacology* **46**, 1018-1022 (2004).
88. T.C. Chou, C.E. Lee, J. Lu, J.K. Elmquist, J. Hara, J.T. Willie, C.T. Beuckmann, R.M. Chemelli, T. Sakurai, M. Yanagisawa, C.B. Saper and T.E. Scammell, Orexin (hypocretin) neurons contain dynorphin, *J. Neurosci.* **21**, RC168 (2001).
89. X.H. Liu, R. Morris, D. Spiller, M. White and G. Williams, Orexin a preferentially excites glucose-sensitive neurons in the lateral hypothalamus of the rat in vitro, *Diabetes* **50**, 2431-2437 (2001).
90. A. Yamanaka, C.T. Beuckmann, J.T. Willie, J. Hara, N. Tsujino, M. Mieda, M. Tominaga, K. Yagami, F. Sugiyama, K. Goto, M. Yanagisawa and T. Sakurai, Hypothalamic orexin neurons regulate arousal according to energy balance in mice, *Neuron* **38**, 701-713 (2003).
91. I.V. Estabrooke, M.T. McCarthy, E. Ko, T.C. Chou, R.M. Chemelli, M. Yanagisawa, C.B. Saper and T.E. Scammell, Fos expression in orexin neurons varies with behavioral state, *J. Neurosci.* **21**, 1656-1662 (2001).
92. J.M. Zeitzer, C.L. Buckmaster, K.J. Parker, C.M. Hauck, D.M. Lyons and E. Mignot, Circadian and homeostatic regulation of hypocretin in a primate model: implications for the consolidation of wakefulness, *J. Neurosci.* **23**, 3555-3560 (2003).
93. D. Teves, T.O. Videen, P.E. Cryer and W.J. Powers, Activation of human medial prefrontal cortex during autonomic responses to hypoglycemia, *Proc.Natl.Acad.Sci.U.S.A* **101**, 6217-6221 (2004).
94. J. Hara, C.T. Beuckmann, T. Nambu, J.T. Willie, R.M. Chemelli, C.M. Sinton, F. Sugiyama, K. Yagami, K. Goto, M. Yanagisawa and T. Sakurai, Genetic ablation of orexin neurons in mice results in narcolepsy, hypophagia, and obesity, *Neuron* **30**, 345-354 (2001).
95. S. Overeem, J.A. van Vliet, G.J. Lammers, F.G. Zitman, D.F. Swaab and M.D. Ferrari, The hypothalamus in episodic brain disorders, *Lancet Neurol.* **1**, 437-444 (2002).
96. S. Nishino and E. Mignot, Pharmacological aspects of human and canine narcolepsy, *Progress in Neurobiology* **52**, 27-78 (1997).
97. K. Nonogaki, A.M. Strack, M.F. Dallman and L.H. Tecott, Leptin-independent hyperphagia and type 2 diabetes in mice with a mutated serotonin 5-HT2C receptor gene, *Nat. Med.* **4**, 1152-1156 (1998).
98. C. Schmauss, Serotonin 2C receptors: suicide, serotonin, and runaway RNA editing, *Neuroscientist.* **9**, 237-242 (2003).
99. I. Matsuzaki, T. Sakurai, K. Kunii, T. Nakamura, M. Yanagisawa and K. Goto, Involvement of the serotonergic system in orexin-induced behavioral alterations in rats, *Regulatory Peptides* **104**, 119-123 (2002).
100. B. Borowsky, M.M. Durkin, K. Ogozalek, M.R. Marzabadi, J. DeLeon, B. Lagu, R. Heurich, H. Lichtblau, Z. Shaposhnik, I. Daniewska, T.P. Blackburn, T.A. Branchek, C. Gerald, P.J. Vaysse and C. Forray, Antidepressant, anxiolytic and anorectic effects of a melanin-concentrating hormone-1 receptor antagonist, *Nat. Med.* **8**, 825-830 (2002).
101. J.M. Swanson, Role of executive function in ADHD, *J. Clin. Psychiatry* **64 Suppl 14**, 35-39 (2003).
102. R.M. Salomon, B. Ripley, J.S. Kennedy, B. Johnson, D. Schmidt, J.M. Zeitzer, S. Nishino and E. Mignot, Diurnal variation of cerebrospinal fluid hypocretin-1 (Orexin-A) levels in control and depressed subjects, *Biol. Psychiatry* **54**, 96-104 (2003).
103. M. Monda, A.N. Viggiano, A.L. Viggiano, F. Fuccio and L. De, V, Cortical spreading depression blocks the hyperthermic reaction induced by orexin A, *Neuroscience* **123**, 567-574 (2004).

104. A. Gutierrez, G. Saracibar, L. Casis, E. Echevarria, V.M. Rodriguez, M.T. Macarulla, L.C. Abecia and M.P. Portillo, Effects of fluoxetine administration on neuropeptide y and orexins in obese zucker rat hypothalamus, *Obesity Research* **10**, 532-540 (2002).
105. L.E. Krahn, W.R. Moore and S.I. Altchuler, Narcolepsy and obesity: remission of severe cataplexy with sibutramine, *Sleep Med.* **2**, 63-65 (2001).
106. A.E. Kelley and K.C. Berridge, The neuroscience of natural rewards: relevance to addictive drugs, *J Neurosci.* **22**, 3306-3311 (2002).
107. R.A. Wise, Dopamine, learning and motivation, *Nat.Rev.Neurosci.* **5**, 483-494 (2004).
108. R.J. DiLeone, D. Georgescu and E.J. Nestler, Lateral hypothalamic neuropeptides in reward and drug addiction, *Life Sciences* **73**, 759-768 (2003).
109. B. Boutrel, P.J. Kenny, S.E. Specio, A. Markou, G.F. Koob and L. De Lecea, Hypocretin regulates brain reward function and reinstatement of cocaine seeking behavior in rats, *Soc Neurosci Abstr.* **573.2**, (2004).
110. X. Drouot, S. Moutereau, J.P. Nguyen, J.P. Lefaucheur, A. Creange, P. Remy, F. Goldenberg and M.P. d'Ortho, Low levels of ventricular CSF orexin/hypocretin in advanced PD, *Neurology* **61**, 540-543 (2003).
111. B. Boutre, P.J. Kenny, S.E. Specio, A. Markou, G.F. Koob and L. De Lecea, Hypocretin regulates brain reward function and reinstatement of cocaine seeking behavior in rats, *Soc Neurosci Abstr.* **573.2**, (2004).
112. J.P. Wisor, S. Nishino, I. Sora, G.H. Uhl, E. Mignot and D.M. Edgar, Dopaminergic role in stimulant-induced wakefulness, *J. Neurosci.* **21**, 1787-1794 (2001).
113. R.M. Chemelli, J.T. Willie, C.M. Sinton, J.K. Elmquist, T. Scammell, C. Lee, J.A. Richardson, S.C. Williams, Y. Xiong, Y. Kisanuki, T.E. Fitch, M. Nakazato, R.E. Hammer, C.B. Saper and M. Yanagisawa, Narcolepsy in orexin knockout mice: molecular genetics of sleep regulation, *Cell* **98**, 437-451 (1999).
114. T.E. Scammell, I.V. Estabrooke, M.T. McCarthy, R.M. Chemelli, M. Yanagisawa, M.S. Miller and C.B. Saper, Hypothalamic arousal regions are activated during modafinil-induced wakefulness, *J. Neurosci.* **20**, 8620-8628 (2000).
115. M.M. Thakkar, R.E. Strecker and R.W. McCarley, Phasic but not tonic REM-selective discharge of periaqueductal gray neurons in freely behaving animals: relevance to postulates of GABAergic inhibition of monoaminergic neurons, *Brain Research* **945**, 276-280 (2002).
116. J. Fadel and A.Y. Deutch, Anatomical substrates of orexin-dopamine interactions: lateral hypothalamic projections to the ventral tegmental area, *Neuroscience* **111**, 379-387 (2002).
117. K. Uramura, H. Funahashi, S. Muroya, S. Shioda, M. Takigawa and T. Yada, Orexin-a activates phospholipase C- and protein kinase C-mediated Ca2+ signaling in dopamine neurons of the ventral tegmental area, *Neuroreport* **12**, 1885-1889 (2001).
118. S. Diano, B. Horvath, H.F. Urbanski, P. Sotonyi and T.L. Horvath, Fasting activates the nonhuman primate hypocretin (orexin) system and its postsynaptic targets, *Endocrinology* **144**, 3774-3778 (2003).
119. S. Nishino, N. Fujiki, B. Ripley, E. Sakurai, M. Kato, T. Watanabe, E. Mignot and K. Yanai, Decreased brain histamine content in hypocretin/orexin receptor-2 mutated narcoleptic dogs, *Neuroscience Letters* **313**, 125-128 (2001).
120. G. Telegdy and A. Adamik, The action of orexin A on passive avoidance learning. Involvement of transmitters, *Regulatory Peptides* **104**, 105-110 (2002).
121. S.H. Russell, M.S. Kim, C.J. Small, C.R. Abbott, D.G. Morgan, S. Taheri, K.G. Murphy, J.F. Todd, M.A. Ghatei and S.R. Bloom, Central administration of orexin A suppresses basal and domperidone stimulated plasma prolactin, *J. Neuroendocrinol.* **12**, 1213-1218 (2000).
122. Y.C. Hsueh, S.M. Cheng and J.T. Pan, Fasting stimulates tuberoinfundibular dopaminergic neuronal activity and inhibits prolactin secretion in oestrogen-primed ovariectomized rats: involvement of orexin A and neuropeptide Y, *J. Neuroendocrinol.* **14**, 745-752 (2002).
123. J.J. Hagan, R.A. Leslie, S. Patel, M.L. Evans, T.A. Wattam, S. Holmes, C.D. Benham, S.G. Taylor, C. Routledge, P. Hemmati, R.P. Munton, T.E. Ashmeade, A.S. Shah, J.P. Hatcher, P.D. Hatcher, D.N. Jones, M.I. Smith, D.C. Piper, A.J. Hunter, R.A. Porter and N. Upton, Orexin A activates locus coeruleus cell firing and increases arousal in the rat, *Proc.Natl.Acad.Sci.U.S.A* **96**, 10911-10916 (1999).
124. P. Bourgin, S. Huitron-Resendiz, A.D. Spier, V. Fabre, B. Morte, J.R. Criado, J.G. Sutcliffe, S.J. Henriksen and L. De Lecea, Hypocretin-1 modulates rapid eye movement sleep through activation of locus coeruleus neurons, *J. Neurosci.* **20**, 7760-7765 (2000).
125. T.L. Horvath, C. Peyron, S. Diano, A. Ivanov, G. Aston-Jones, T.S. Kilduff and A.N. van den Pol, Hypocretin (orexin) activation and synaptic innervation of the locus coeruleus noradrenergic system, *J Comp Neurol.* **415**, 145-159 (1999).
126. J. Grivel, I. Tobler, M. Muhlethaler and M. Serafin, Following sleep deprivation the excitation of rat hypocretin/orexin neurons by noradrenaline reverses to inhibition, *Soc Neurosci Abstr.* **318.8.**, (2004).
127. G. Aston-Jones, S. Chen, Y. Zhu and M.L. Oshinsky, A neural circuit for circadian regulation of arousal, *Nat. Neurosci.* **4**, 732-738 (2001).

128. S.G. Walling, D.J. Nutt, M.D. Lalies and C.W. Harley, Orexin-A infusion in the locus ceruleus triggers norepinephrine (NE) release and NE-induced long-term potentiation in the dentate gyrus, *J Neurosci.* **24**, 7421-7426 (2004).
129. C.F. Murchison, X.Y. Zhang, W.P. Zhang, M. Ouyang, A. Lee and S.A. Thomas, A distinct role for norepinephrine in memory retrieval, *Cell* **117**, 131-143 (2004).
130. S. Burlet, C.J. Tyler and C.S. Leonard, Direct and indirect excitation of laterodorsal tegmental neurons by Hypocretin/Orexin peptides: implications for wakefulness and narcolepsy, *J. Neurosci.* **22**, 2862-2872 (2002).
131. E. Eggermann, M. Serafin, L. Bayer, D. Machard, B. Saint-Mleux, B.E. Jones and M. Muhlethaler, Orexins/hypocretins excite basal forebrain cholinergic neurones, *Neuroscience* **108**, 177-181 (2001).
132. D.E. Berman and Y. Dudai, Memory extinction, learning anew, and learning the new: dissociations in the molecular machinery of learning in cortex, *Science* **291**, 2417-2419 (2001).
133. L.A. Graves, K. Hellman, S. Veasey, J.A. Blendy, A.I. Pack and T. Abel, Genetic evidence for a role of CREB in sustained cortical arousal, *J Neurophysiol.* **90**, 1152-1159 (2003).
134. A.B. Belousov, B.F. O'Hara and J.V. Denisova, Acetylcholine becomes the major excitatory neurotransmitter in the hypothalamus in vitro in the absence of glutamate excitation, *J Neurosci.* **21**, 2015-2027 (2001).
135. L. Bayer, E. Eggermann, B. Saint-Mleux, D. Machard, B.E. Jones, M. Muhlethaler and M. Serafin, Selective action of orexin (hypocretin) on nonspecific thalamocortical projection neurons, *J. Neurosci.* **22**, 7835-7839 (2002).
136. L. Bayer, M. Serafin, E. Eggermann, B. Saint-Mleux, D. Machard, B.E. Jones and M. Muhlethaler, Exclusive postsynaptic action of hypocretin-orexin on sublayer 6b cortical neurons, *J Neurosci.* **24**, 6760-6764 (2004).
137. C. Cirelli, C.M. Gutierrez and G. Tononi, Extensive and divergent effects of sleep and wakefulness on brain gene expression, *Neuron* **41**, 35-43 (2004).
138. E.K. Lambe and G.K. Aghajanian, Hypocretin (orexin) induces calcium transients in single spines postsynaptic to identified thalamocortical boutons in prefrontal slice, *Neuron* **40**, 139-150 (2003).
139. R. Lathe, Hormones and the hippocampus, *J Endocrinol.* **169**, 205-231 (2001).
140. M.J. Wayner, D.L. Armstrong, C.F. Phelix and Y. Oomura, Orexin-A (Hypocretin-1) and leptin enhance LTP in the dentate gyrus of rats in vivo, *Peptides* **25**, 991-996 (2004).
141. D. Gerashchenko, R. Salin-Pascual and P.J. Shiromani, Effects of hypocretin-saporin injections into the medial septum on sleep and hippocampal theta, *Brain Research* **913**, 106-115 (2001).
142. O. Selbach, C. Bohla, N. Doreulee, K.S. Eriksson, O.A. Sergeeva and H.L. Haas, Rapid protein synthesis-dependent synaptic plasticity in the hippocampus by orexins/hypocretins., *FENS Abstr.* **A084.22.**, (2004).
143. S. Aou, X.L. Li, A.J. Li, Y. Oomura, T. Shiraishi, K. Sasaki, T. Imamura and M.J. Wayner, Orexin-A (hypocretin-1) impairs Morris water maze performance and CA1-Schaffer collateral long-term potentiation in rats, *Neuroscience* **119**, 1221-1228 (2003).
144. L.B. Jaeger, S.A. Farr, W.A. Banks and J.E. Morley, Effects of orexin-A on memory processing, *Peptides* **23**, 1683-1688 (2002).
145. U. Steidl, S. Bork, S. Schaub, O. Selbach, J. Seres, M. Aivado, T. Schroeder, U.P. Rohr, R. Fenk, S. Kliszewski, C. Maercker, P. Neubert, S.R. Bornstein, H.L. Haas, G. Kobbe, D.G. Tenen, R. Haas and R. Kronenwett, Primary human CD34+ hematopoietic stem and progenitor cells express functionally active receptors of neuromediators, *Blood* **104**, 81-88 (2004).

EFFECTS OF HYPOCRETIN/OREXIN ON THE THALAMOCORTICAL ACTIVATING SYSTEM

Evelyn K. Lambe and George K. Aghajanian*

1. INTRODUCTION

The midline and intralaminar thalamic neurons are unique in that they coordinate activity levels broadly across cerebral cortex and support attention and awareness, rather than simply relaying one type of sensory information to one particular cortical region. The peptides, hypocretin 1 and 2 (orexin A and B), selectively excite thalamic midline-intralaminar neurons and also their projections in prefrontal cortex. Distinct hypocretin projections to the thalamus and prefrontal cortex suggest that the ability of hypocretin to excite both the cell body and nerve terminal may not be redundant, but instead may be involved in co-ordinating networks of bursting neurons. Midline-intralaminar neurons are unusual in that they burst during wakefulness, when hypocretin levels are elevated, instead of during slow wave sleep like specific sensory thalamic neurons. The ability of hypocretin to excite the final synapse of the ascending arousal system suggests a role for prefrontal release of hypocretin in executive tasks such as flexible or divided attention.

2. THALAMOCORTICAL ACTIVATING SYSTEM

The midline-intralaminar nuclei of the thalamus are part of the ascending arousal system. Functional imaging studies have shown that activation of these nuclei leads to higher levels of arousal and attention.[1,2] They have long been considered "nonspecific" since they project broadly within frontal cortex and can recruit other areas of cortex to become active.[3] However, the different groups of nuclei appear to have fairly specific domains within which they promote attention and arousal: viscero-limbic functions (dorsal group), cognitive functions (lateral group), multimodal sensory processing

* Evelyn K. Lambe, Department of Psychiatry, Yale University School of Medicine, New Haven CT 06508. George K. Aghajanian, Departments of Psychiatry and Pharmacology, Yale University School of Medicine, New Haven CT 06508.

(ventral group), and limbic motor functions (posterior group).[4] These nuclei are shown in Figure 1.

Lesions of midline-intralaminar thalamic nuclei give rise to disturbances of attention, initiative, and executive functions similar to those seen with prefrontal cortex lesions. These deficits fit with the heavy projection from these thalamus nuclei to prefrontal cortex. Midline-intralaminar neurons have monosynaptic projections which terminate on the apical dendrites of cortical pyramidal neurons.[5] Anterograde labeling studies have shown that these axons terminate predominantly in layers I and V of

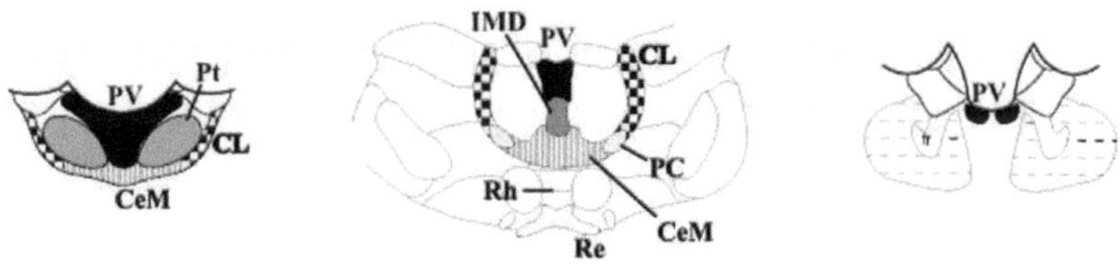

Figure 1. Midline and intralaminar nuclei of the thalamaus shown in coronal section from most anterior to most posterior from left to right (adapted from Van der Werf et al., 2002).[4] These nuclei include paraventricular (PV), paratenial (Pt), central lateral (CL), central medial (CeM), interomediodorsal (IMD), paracentral (PC), rhomboid (Rh), reuniens (Re) and the caudal intralaminar nucleus (the most ventral region shown with dotted lines on the far right image).

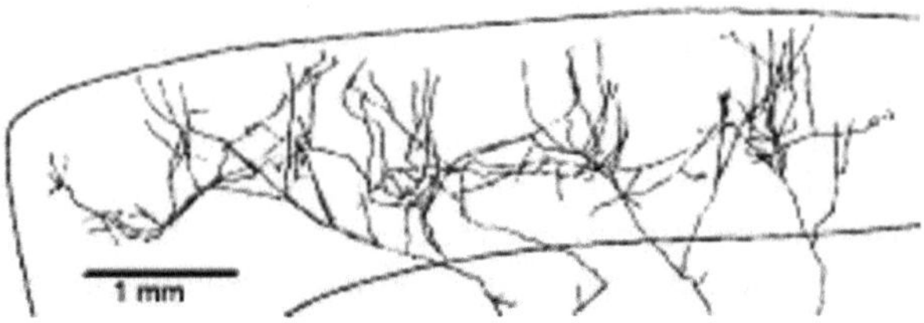

Figure 2. The cortical axonal arbor of a single central lateral thalamic neuron, shown in sagittal section of dorsal prefrontal cortex (adapted from Deschenes et al., 1996).[6] The most anterior part of cortex is on the left. Such extensive arborization means that midline-intralaminar neurons innervate overlapping regions of cortex.

prefrontal cortex.[3] Single axons have been shown to arborize extensively,[6] as illustrated by the arbor from a single central lateral neuron in a sagittal section of prefrontal cortex shown in Figure 2.

Thalamocortical axons are unusual in that they are susceptible to the generation of terminal spikes.[7,8] It has been suggested that the induction of ectopic spikes in the overlapping terminal fields would co-ordinate bursting of groups of thalamic neurons.[9,10] Midline-intralaminar thalamic neurons are unusual in that they displaying bursting behavior during waking,[11] the opposite pattern to that observed in sensory neurons which burst during slow wave sleep. It is not understood what regulates the bursting of midline-intralaminar thalamic neurons during waking.

3. SELECTIVE HYPOCRETIN PROJECTIONS

Hypocretin neurons in the lateral hypothalamus project to many brain regions involved in wakefulness, as depicted in Figure 3; however, their projections terminating in thalamus and in cortex show unusual selectivity. In the thalamus, these axons predominantly target the midline-intralaminar nuclei.[12] It is perhaps not surprising that these nuclei are targeted by a wakefulness-promoting peptide given their role in arousal and attention. Hypocretin projections to cortex are generally sparse, except for medial prefrontal cortex which receives a denser projection[13] that appears to originate from a unique population of hypocretin neurons with minimal collaterals.[14]

4. HYPOCRETIN EXCITES MIDLINE -INTRALAMINAR THALAMIC NEURONS

Hypocretin has been shown selectively to depolarize midline-intralaminar thalamic neurons and not sensory thalamic neurons.[16] Such specificity is congruent with previous observations that hypocretin fibers avoid specific sensory thalamic nuclei, terminating predominantly in midline-intralaminar nuclei such as rhomboid and central medial.[12] Midline-intralaminar nuclei show prominent expression of the hypocretin receptor 2,[17,18] which is very sensitive to both hypocretin peptides.

Bayer and colleagues examined the electrophysiolgical effects of hypocretin-1 and 2 peptides on thalamic neurons in brain slice with whole cell recording.[16] Both peptides potently depolarized neurons in midline-intralaminar nuclei with EC_{50} less than 20 nanomolar but did not depolarize neurons in sensory thalamic nuclei (Figure 4).

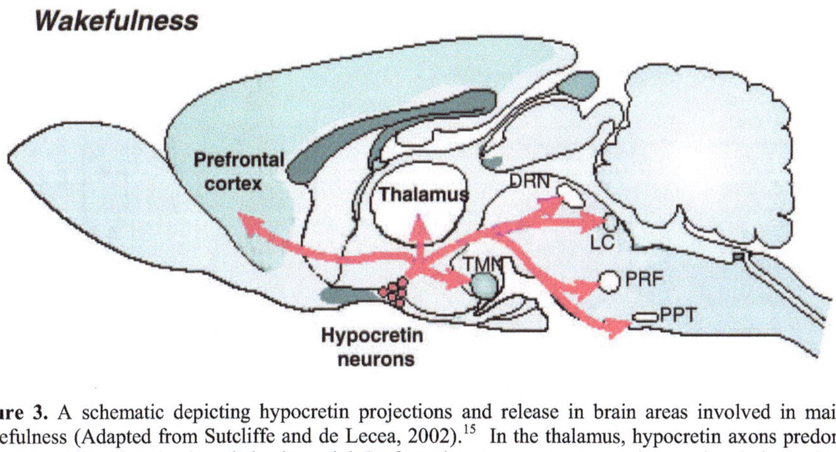

Figure 3. A schematic depicting hypocretin projections and release in brain areas involved in maintaining wakefulness (Adapted from Sutcliffe and de Lecea, 2002).[15] In the thalamus, hypocretin axons predominantly project to midline-intralaminar thalamic nuclei. Prefrontal cortex appears to receive a uniquely heavy hypocretin projection compared to other cortical areas. Hypocretin neurons also project to many lower brain areas involved in wakefulness, including tuberomammilary nucleus (TMN), dorsal raphe nucleus (DRN), locus coeruleus (LC), pontine reticular formation (PRF), and prepontine tegmental area (PPT).

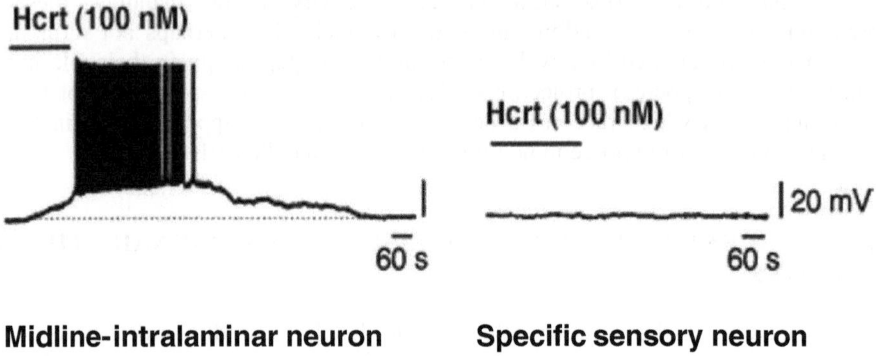

Figure 4. Left, hypocretin-2 peptide potently depolarizes an intralaminar thalamic neuron, but it does not change the resting potential of a sensory relay neuron, right (adapted from Bayer et al., 2002).[16]

The G_q-coupled hypocretin receptor 2 have been shown to depolarize neurons in a number of other brain regions through inhibiting a potassium current open at rest,[16,19] by activating a calcium current[20] or a nonspecific cation current,[21] or stimulating the electrogenic sodium-calcium exchanger.[22,23] Bayer and colleagues suggest the former may be responsible for the depolarization seen in midline-intralaminar thalamic neurons.[16]

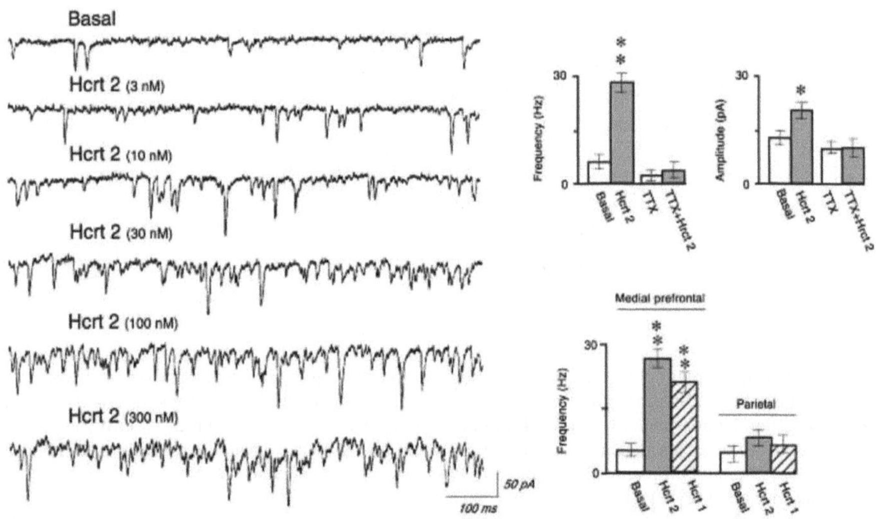

Figure 5. Bath application of hypocretin induces a large increase in spontaneous excitatory postsynaptic currents (sEPSCs) in prefrontal layer V pyramidal neurons. These are mediated by glutamate and are action-potential dependent, as they can be suppressed by TTX. Both hypocretin peptides elicit an increase in sEPSCs, and this increase is much greater in medial prefrontal cortex than in other cortical areas (adapted from Lambe and Aghajanian, 2003).[24]

5. HYPOCRETIN EXCITES THALAMOCORTICAL TERMINALS IN PREFRONTAL CORTEX

5.1 Pharmacology and lesion studies

The susceptibility of thalamocortical terminals to the generation of terminal spikes would promote bursting in the projecting thalamic neurons through antidromic transmission. It has been suggested that neurotransmitters released in the terminal field could thus influence the bursting of groups of neurons in a distant brain region. Our recent work suggests that hypocretin released in prefrontal cortex could act in this manner.[24]

In examining the effects of hypocretin in prefrontal cortex, we found that there was very little direct effect of hypocretin on prefrontal neurons. In prefrontal slice, we recorded from layer V pyramidal neurons by whole cell patch clamp and applied hypocretin in the bath. Hypocretin had little effect on the resting membrane potential or the holding current, but potently increased synaptic release of glutamate onto these cells. This effect was observed in voltage clamp as a large increase in spontaneous excitatory postsynaptic currents (sEPSCs) illustrated in Figure 5. Blocking action potentials with tetrodotoxin (TTX) eliminated the hypocretin-induced increase in sEPSCs. This indirect effect of hypocretin occurred preferentially in medial prefrontal cortex, a region which has been shown to receive a greater density of hypocretin projections than other cortical regions.[14]

We started to suspect that hypocretin was able to induce terminal spikes in thalamocortical terminals because prior thalamic lesions greatly suppressed this effect, as shown in Figure 6. Furthermore, the hypocretin-elicited increase in sEPSCs could be suppressed by agonists of mu-opioid receptors, such as DAMGO. Mu-opioid agonists are known to be able to suppress thalamocortical[25] but not cortico-cortical transmission.[26]

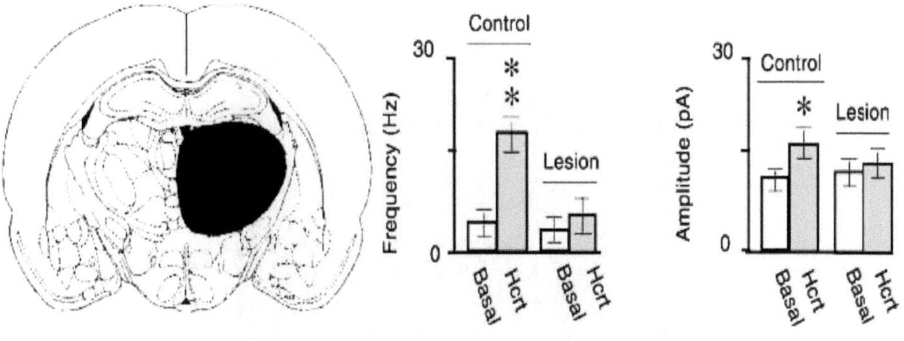

Figure 6. A prior unilateral thalamic lesion almost eliminates hypocretin-induced sEPSCs in ipsilateral cortex. By contrast, hypocretin-elicited sEPSCs remain robust in the medial prefrontal cortex in the opposite hemisphere (** P < 0.01; * P < 0.05). Adapted from Lambe and Aghajanian, 2003.[24]

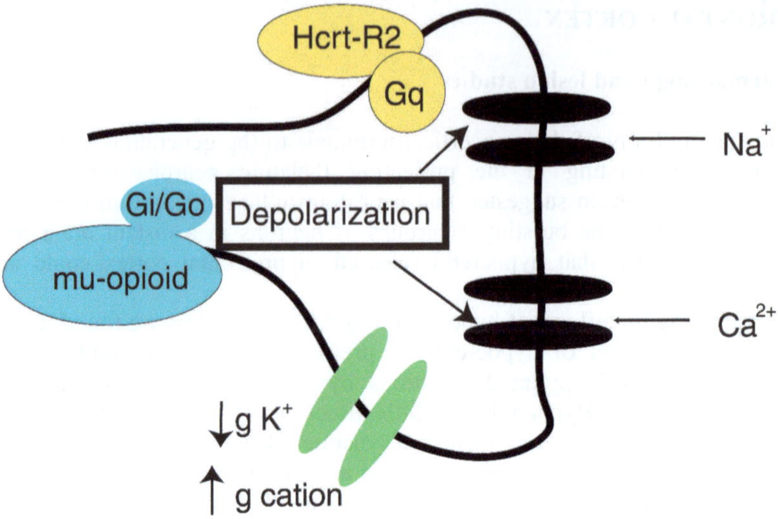

Figure 7. A schematic showing how G_q-coupled hypocretin receptors can induce depolarization to threshold in thalamocortical terminals, possibly through inhibiting a potassium conductance (g K^+) or enhancing a cation conductance (g cation). In vivo studies have shown that these terminals are particularly susceptible to the induction of terminal spikes that propagate back down the axon to the distant cell body. G_i/G_o-coupled mu-opioid receptors would physiologically oppose the depolarization induced by hypocretin.

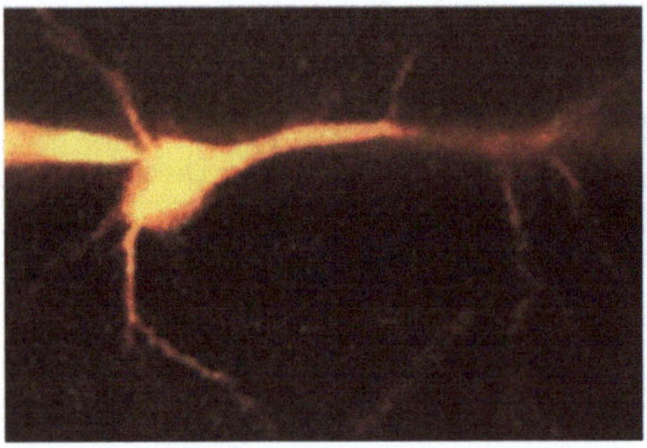

Figure 8. A prefrontal layer V pyramidal neuron being filled with Oregon green BAPTA-1 and Alexa 594 through the patch pipette on the left. This combination of dyes makes the neuron appear orange when the red and green detection channels are merged.

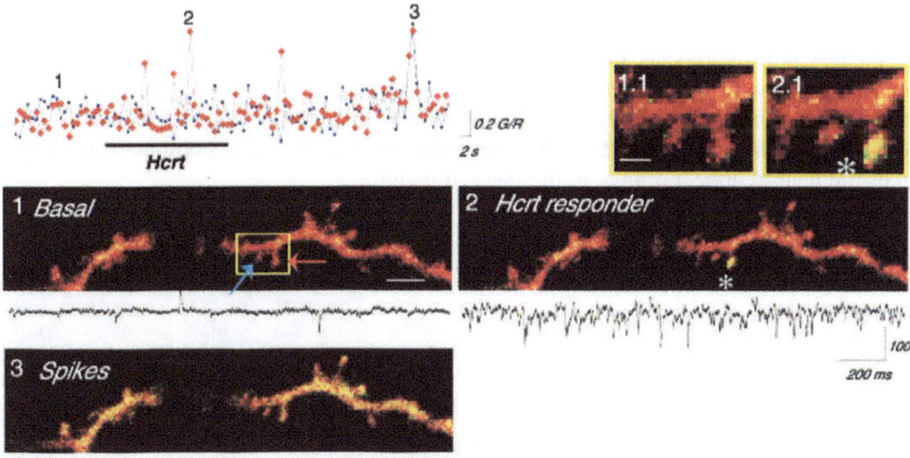

Figure 9. The graph in red at the top shows how bath application of hypocretin changes the ratio of green to red intensity at a single spine. The graph in blue shows these measurements from a neighboring spine that did not show calcium transients during hypocretin. The images shown correspond to the points on the graph marked 1,2, and 3. The last shows the overall calcium increase in the dendrite and all spines with a depolarization-induced burst of spikes (adapted from Lambe and Aghajanian, 2003).[24]

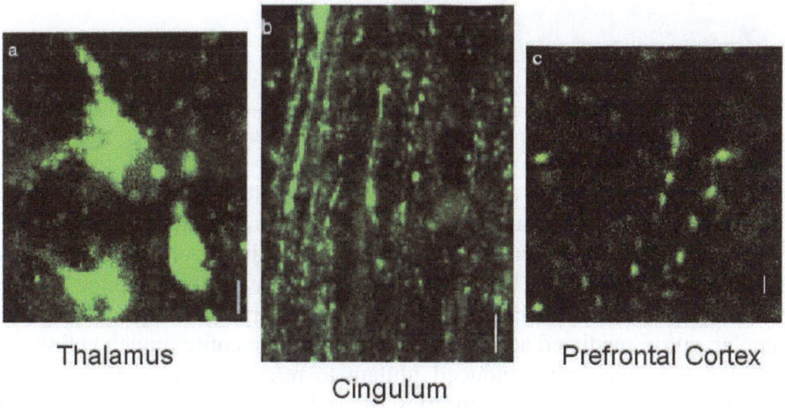

Figure 10. The above images show (a) thalamic neurons and (b, c) axons filled with the anterograde tracer *Phaseolus vulgaris* conjegated to the green fluorescent marker Alexa 488. Thalamic axons pass through the cingulum (b), as shown in horizontal slice, on their way to prefrontal cortex (c). Adapted from Lambe and Aghajanian, 2003.[24]

Our model of this process is shown in Figure 7. Hypocretin receptor 2 is abundant in the midline-intralaminar thalamic neurons which project to prefrontal cortex.[17,18] Hypocretin stimulation appears to depolarize thalamocortical terminals to threshold for spiking. This effect can be physiologically opposed by stimulation of the G_i/G_o-coupled mu-opioid receptors.

5.2 Two-photon calcium imaging studies

To test this hypothesis more rigorously, we used two-photon imaging to identify synapses at which hypocretin induced glutamate release. First, we filled layer V pyramidal neurons with two dyes - a green calcium indicator, Oregon green BAPTA-1 and a red control dye, Alexa 595. As illustrated in Figure 8, this combination of dyes makes the neuron appear orange in the merged image.

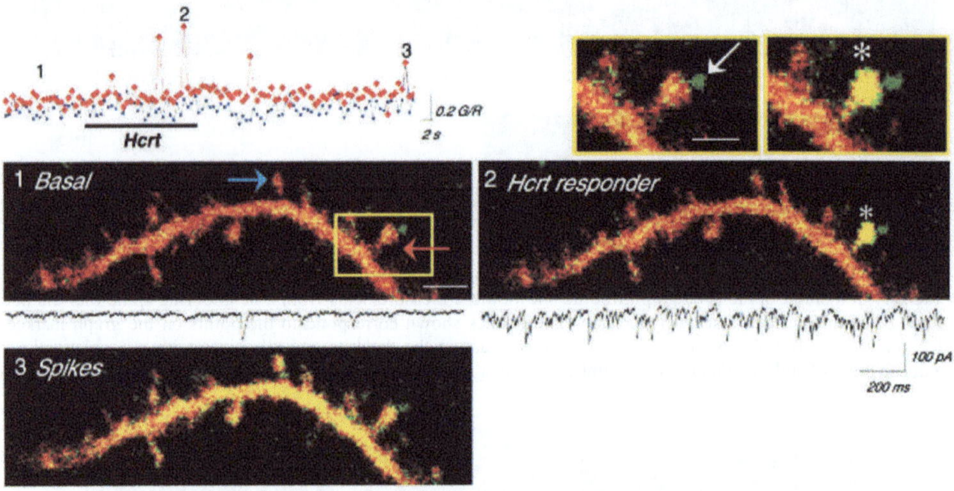

Figure 11. Spines in apposition to labeled thalamic boutons were significantly more likely to show calcium transients in response to hypocretin (adapted from Lambe and Aghajanian, 2003).[24]

This combination of dyes allows us to image postsynaptic calcium transients at single dendritic spines. Under baseline conditions, the dendrite and spines appear orange, but they can turn yellow (indicative of an increase in the green to red ratio) when an influx of calcium increases the brightness of the green calcium indicator. For example, Figure 9 shows an NMDA-mediated calcium transient turning a single spine yellow, and then a depolarization-mediated action potential turning the entire branch yellow.

We found that hypocretin induced glutamate release onto a small percentage of spines on branches of the apical dendrites of prefrontal layer V pyramidal neurons.[24] These spines tended to be on higher-order branches located in midlayer V or in layer I of medial prefrontal cortex. Although the apical dendritic location of the responding spines would be congruent with excitation of direct thalamic input, these findings do not identify which glutamatergic projections are excited by hypocretin.[24]

To further test our hypothesis that hypocretin selectively excites thalamocortical synapses within prefrontal cortex, we labeled thalamic projections to prefrontal cortex by electroporating the thalamus in vivo with the anterograde tracer *Phaseolus vulgaris* conjugated to a green fluorescent marker, Alexa 488. Several days later, slices were made to examine thalamocortical labeling, as illustrated in Figure 10.

After filling layer V pyramidal neurons with the combination of dyes described before, we found scattered appositions in the *x-y* plane between green-labeled thalamic boutons and orange-labeled dendritic spines.[24] Just as only a fraction of spines responded to hypocretin in our previous experiments, apparent appositions between labeled boutons and spines were infrequent and occurred primarily on higher order branches of the apical dendrites in layers V and I. We imaged these spines to observe the effects of hypocretin, as shown in Figure 11.

Hypocretin induced calcium transients at 7 out of 8 spines appearing to be in apposition to labeled thalamic boutons but only at 2 out of 26 control spines that were in focus but not in apparent apposition to a labeled bouton.[24] This dramatic difference showed that spines with thalamic appositions were a significantly different population (Chi-squared test, $P < 0.0001$). These experiments provide strong convergent evidence that hypocretin excites thalamocortical synapses in prefrontal cortex.

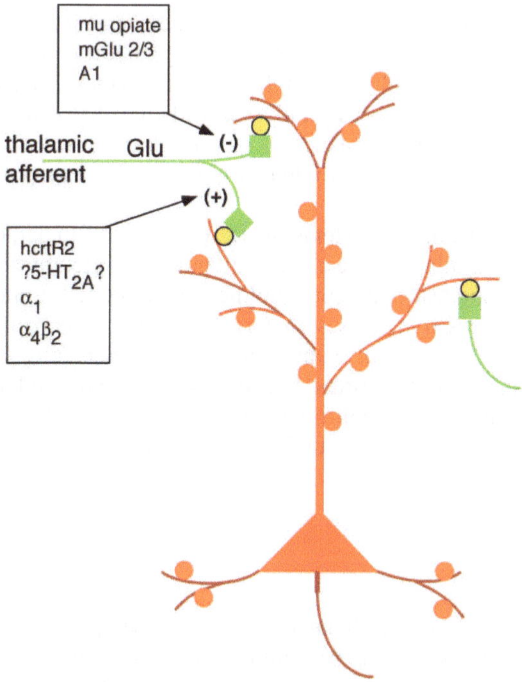

Figure 12. A schematic showing thalamic inputs to a prefrontal layer V pyramidal neuron can be excited different neurotransmitters: hypocretin excites through hypocretin receptor 2, serotonin through 5-HT_{2A} receptors (although unclear whether these receptors are pre- or post-synaptic), norepinephrine through α_1 receptors, and acetylcholine through the high-affinity $\alpha_4\beta_2$ nicotinic receptors. This synapse can also be inhibited by enkephalin through mu-opioid receptors, glutamate through group 2/3 metabotropic glutamate (mGlu) receptors, and adenosine through A1 adenosine receptors.

6. AROUSAL AND ATTENTION

Hypocretin levels has been shown to correlate with states of attention and arousal,[15] with high levels observed during alertness and periods of physical activity.[27-29] By inducing ectopic spikes in thalamocortical terminals, prefrontal hypocretin has the potential to affect alertness and attention in vivo. Our results raise questions about the regulation of hypocretin within the prefrontal cortex and in the midline-intralaminar thalamus in healthy subjects. For example, does hypocretin increase in prefrontal cortex during tasks that require attention?

Recently it has been shown that patients with narcolepsy – who lack hypocretin – have selective deficits in the prefrontal or executive attention network.[30] While older literature suggests that cognitive deficits in narcolepsy may simply be a result of tiredness, Rieger and colleagues show that the selective deficit in divided attention persists affect controlling for alertness.[30] Furthermore, another recent study has showed abnormal frontal cortex physiology in narcoleptics during pre-attentive and attentive tasks.[31]

Similar to hypocretin, nicotine excites high-affinity nicotinic acetylcholine receptors on thalamocortical terminals and results in a large increase in glutamate release onto prefrontal layer V pyramidal neurons.[32] Interestingly, nicotine has repeatedly been shown to enhance attention. While this effect has mainly been demonstrated with systemic administration of nicotine, a recent study showed that direct infusion of nicotine into prefrontal cortex improves performance of rats on a task of divided attention.[33] It would be fascinating to study the effect of hypocretin infused into prefrontal cortex in this manner.

Multiple neurotransmitter systems work together at thalamocortical synapses in prefrontal cortex to regulate alertenss and attentional state. In addition to hypocretin, there is evidence to suggest that acetylcholine,[32] serotonin,[34-36] and norepinephrine[35] all can excite thalamocortical synapses. Neurotransmitters that can inhibit this synapse include enkephalin,[34] adenosine,[37] and glutamate.[38] Interestingly, hypocretin neurons project not only to prefrontal cortex but also to the raphe and locus coeruleus,[12] the source of serotonin and norepinephrine respectively. The interrelationships among these different neurotransmitter systems under varying physiological states remains to be determined.

7. CONCLUSIONS

Hypocretin can excite the thalamocortical arousal pathway at two levels. It can selectively depolarize neurons in the midline-intralaminar nuclei of the thalamus, but it can also directly excite their axon terminals within prefrontal cortex. Since these two regions appear to receive projections from different populations of hypocretin neurons, further understanding of what governs hypocretin release in these two areas will yield greater insight into thalamocortical involvement in arousal and attention.

8. REFERENCES

1. S. Kinomura, J. Larsson, B. Gulyas and P. E. Roland, Activation by attention of the human reticular formation and thalamic intralaminar nuclei, *Science.* **271**, 512-5 (1996).
2. C. M. Portas, G. Rees, A. M. Howseman, O. Josephs, R. Turner and C. D. Frith, A specific role for the thalamus in mediating the interaction of attention and arousal in humans, *J Neurosci.* **18**, 8979-89 (1998).
3. H. W. Berendse and H. J. Groenewegen, Restricted cortical termination fields of the midline and intralaminar thalamic nuclei in the rat, *Neuroscience.* **42**, 73-102 (1991).
4. Y. D. Van der Werf, M. P. Witter and H. J. Groenewegen, The intralaminar and midline nuclei of the thalamus. Anatomical and functional evidence for participation in processes of arousal and awareness, *Brain Res Brain Res Rev.* **39**, 107-40 (2002).
5. G. Marini, L. Pianca and G. Tredici, Thalamocortical projection from the parafascicular nucleus to layer V pyramidal cells in frontal and cingulate areas of the rat, *Neurosci Lett.* **203**, 81-4 (1996).

6. M. Deschenes, J. Bourassa and A. Parent, Striatal and cortical projections of single neurons from the central lateral thalamic nucleus in the rat, *Neuroscience.* **72**, 679-87 (1996).
7. M. J. Gutnick and D. A. Prince, Thalamocortical relay neurons: antidromic invasion of spikes from a cortical epileptogenic focus, *Science.* **176**, 424-6 (1972).
8. D. Pinault and R. Pumain, Antidromic firing occurs spontaneously on thalamic relay neurons: triggering of somatic intrinsic burst discharges by ectopic action potentials, *Neuroscience.* **31**, 625-37 (1989).
9. D. Pinault, Backpropagation of action potentials generated at ectopic axonal loci: hypothesis that axon terminals integrate local environmental signals, *Brain Res Brain Res Rev.* **21**, 42-92 (1995).
10. D. A. McCormick and D. Contreras, On the cellular and network bases of epileptic seizures, *Annu Rev Physiol.* **63**, 815-46 (2001).
11. M. Steriade, R. Curro Dossi and D. Contreras, Electrophysiological properties of intralaminar thalamocortical cells discharging rhythmic (approximately 40 HZ) spike-bursts at approximately 1000 HZ during waking and rapid eye movement sleep, *Neuroscience.* **56**, 1-9 (1993).
12. C. Peyron, D. K. Tighe, A. N. van den Pol, L. de Lecea, H. C. Heller, J. G. Sutcliffe and T. S. Kilduff, Neurons containing hypocretin (orexin) project to multiple neuronal systems, *J Neurosci.* **18**, 9996-10015 (1998).
13. J. Fadel and A. Y. Deutch, Anatomical substrates of orexin-dopamine interactions: lateral hypothalamic projections to the ventral tegmental area, *Neuroscience.* **111**, 379-87 (2002).
14. J. Fadel, M. Bubser and A. Y. Deutch, Differential activation of orexin neurons by antipsychotic drugs associated with weight gain, *J Neurosci.* **22**, 6742-6 (2002).
15. J. G. Sutcliffe and L. de Lecea, The hypocretins: setting the arousal threshold, *Nat Rev Neurosci.* **3**, 339-49 (2002).
16. L. Bayer, E. Eggermann, B. Saint-Mleux, D. Machard, B. E. Jones, M. Muhlethaler and M. Serafin, Selective action of orexin (hypocretin) on nonspecific thalamocortical projection neurons, *J Neurosci.* **22**, 7835-9 (2002).
17. J. N. Marcus, C. J. Aschkenasi, C. E. Lee, R. M. Chemelli, C. B. Saper, M. Yanagisawa and J. K. Elmquist, Differential expression of orexin receptors 1 and 2 in the rat brain, *J Comp Neurol.* **435**, 6-25 (2001).
18. P. Trivedi, H. Yu, D. J. MacNeil, L. H. Van der Ploeg and X. M. Guan, Distribution of orexin receptor mRNA in the rat brain, *FEBS Lett.* **438**, 71-5 (1998).
19. R. Brown, O. Sergeeva, K. Eriksson and H. Haas, Orexin A excites serotonergic neurons in the dorsal raphe nucleus of the rat., *Neuropharmacolgy.* **40**, 457-459 (2001).
20. K. S. Eriksson, O. Sergeeva, R. E. Brown and H. L. Haas, Orexin/hypocretin excites the histaminergic neurons of the tuberomammillary nucleus, *J Neurosci.* **21**, 9273-9 (2001).
21. R. Liu, A. van den Pol and G. Aghajanian, Hypocretins (orexins) regulate serotonin neurons in the dorsal raphe nucleus by excitatory direct and inhibitory indirect actions., *J Neurosci.* **22**, 9453-9464 (2002).
22. M. Wu, Z. Zhang, C. Leranth, C. Xu, A. van den Pol and M. Alreja, Hypocretin increases impulse flow in the septohippocampal GABAergic pathway: implications for arousal via a mechanism of hippocampal disinhibition., *J Neurosci.* **22**, 7754-7765 (2002).
23. D. Burdakov, B. Liss and F. Ashcroft, Orexin excites GABAergic neurons of the arcuate nucleus by activating the sodium--calcium exchanger., *J Neurosci.* **23**, 4951-4947 (2003).
24. E. Lambe and G. Aghajanian, Hypocretin (orexin) induces calcium transients in single spines postsynaptic to identified thalamocortical boutons in prefrontal slice., *Neuron.* **40**, 139-150 (2003).
25. G. J. Marek, R. A. Wright, J. C. Gewirtz and D. D. Schoepp, A major role for thalamocortical afferents in serotonergic hallucinogen receptor function in the rat neocortex, *Neuroscience.* **105**, 379-92 (2001).
26. E. Tanaka and R. A. North, Opioid actions on rat anterior cingulate cortex neurons in vitro, *J Neurosci.* **14**, 1106-13 (1994).
27. L. I. Kiyashchenko, B. Y. Mileykovskiy, N. Maidment, H. A. Lam, M. F. Wu, J. John, J. Peever and J. M. Siegel, Release of hypocretin (orexin) during waking and sleep states, *J Neurosci.* **22**, 5282-6 (2002).
28. M. F. Wu, J. John, N. Maidment, H. A. Lam and J. M. Siegel, Hypocretin release in normal and narcoleptic dogs after food and sleep deprivation, eating, and movement, *Am J Physiol Regul Integr Comp Physiol.* **283**, R1079-86 (2002).
29. Y. Yoshida, N. Fujiki, T. Nakajima, B. Ripley, H. Matsumura, H. Yoneda, E. Mignot and S. Nishino, Fluctuation of extracellular hypocretin-1 (orexin A) levels in the rat in relation to the light-dark cycle and sleep-wake activities, *Eur J Neurosci.* **14**, 1075-81 (2001).
30. M. Rieger, G. Mayer and S. Gauggel, Attention deficits in patients with narcolepsy, *Sleep.* **26**, 36-43 (2003).
31. A. Naumann, J. Bierbrauer, H. Przuntek and I. Daum, Attentive and preattentive processing in narcolepsy as revealed by event-related potentials (ERPs), *Neuroreport.* **12**, 2807-11 (2001).
32. E. K. Lambe, M. R. Picciotto and G. K. Aghajanian, Nicotine induces glutamate release from thalamocortical terminals in prefrontal cortex, *Neuropsychopharmacology.* **28**, 216-25 (2003).

33. B. Hahn, M. Shoaib and I. P. Stolerman, Involvement of the prefrontal cortex but not the dorsal hippocampus in the attention-enhancing effects of nicotine in rats, *Psychopharmacology (Berl).* (2003).
34. G. J. Marek and G. K. Aghajanian, 5-Hydroxytryptamine-induced excitatory postsynaptic currents in neocortical layer V pyramidal cells: suppression by mu-opiate receptor activation., *Neuroscience.* **86**, 485-497 (1998).
35. G. J. Marek and G. K. Aghajanian, 5-HT2A receptor or alpha1-adrenoceptor activation induces excitatory postsynaptic currents in layer V pyramidal cells of the medial prefrontal cortex., *Eur J Pharmacol.* **367**, 197-206 (1999).
36. E. K. Lambe and G. K. Aghajanian, The role of Kv1.2-containing potassium channels in serotonin-induced glutamate release from thalamocortical terminals in rat frontal cortex, *J Neurosci.* **21**, 9955-63 (2001).
37. G. E. Stutzmann, G. J. Marek and G. K. Aghajanian, Adenosine preferentially suppresses serotonin2A receptor-enhanced excitatory postsynaptic currents in layer V neurons of the rat medial prefrontal cortex., *Neuroscience.* **105**, 55-69 (2001).
38. G. J. Marek, R. A. Wright, D. D. Schoepp, J. A. Monn and G. Aghajanian, Physiological antagonism between 5-hydroxytryptamine(2A) and group II metabotropic glutamate receptors in prefrontal cortex., *J Pharmacol Exp Ther.* **292**, 76-87 (2000).

PHARMACOLOGY OF THE HYPOCRETINS AND DRUG DESIGN

IN VIVO PHARMACOLOGY OF OREXIN (HYPOCRETIN) RECEPTORS

Neil Upton*

1. INTRODUCTION

The discovery of the orexin peptides (also known as hypocretins) and receptors and the subsequent rapid progress in identifying the key role of this system in normal behaviour and in neuropathology represents one of the most remarkable advances of modern neuroscience. The orexin peptides, termed orexin-A (hypocretin1) (OxA) and orexin-B (hypocretin2) (OxB), and their cognate receptors, orexin-1 (OX_1R or hcrtr1) and orexin-2 (OX_2R, or hcrtr2), were first reported in 1998.[1,2] Because the orexin peptides were originally isolated from hypothalamic tissue and aware of the well known role of this brain region in appetite regulation, Sakurai and colleagues[2] went on to evaluate the effects of centrally administered OxA and OxB on food intake in rats. They found that both peptides were able to stimulate feeding. Within a few years numerous other pre-clinical studies utilising direct injection of OxA (and to a lesser extent OxB) into the brain had led to suggestions that the orexins may have wide-ranging actions including regulation of hormone secretion, autonomic homeostasis, gastrointestinal function and perhaps most importantly, arousal state.[3] Regarding effects on arousal, the most dramatic findings were that defects in the orexin system accounted for the sleep disorder narcolepsy in genetically-susceptible canines (mutation of the *hcrtr2* gene[4] and also many humans (loss of orexin peptides in hypothalamus and reduced OxA levels in cerebrospinal fluid (CSF).[5-7]

Progress to delineate the orexin receptor subtype(s) underpinning the many observed actions of the orexins has been somewhat slower, hampered in particular by a paucity of selective orexin receptor antagonists. This situation is gradually changing and a number of orexin receptor antagonists with sufficient bioavailability to support *in vivo* administration have now been synthesised. The studies with such tool agents are crucial if

* Neil Upton, GlaxoSmithKline, NFSP North, Third Avenue, Harlow, Essex, England

we are to fully unlock the potential therapeutic benefit that could be gained from modulating the orexin system in human disease.

2. PHARMACOLOGICAL TOOLS FOR CHARACTERIZING THE OREXIN PEPTIDE-RECEPTOR SYSTEM *IN VIVO*

2.1. Orexin Receptor Agonists

OxA and OxB are 33 and 28 amino acid peptides respectively, derived from a common 130 amino acid precursor peptide, prepro-orexin, by proteolytic cleavage.[2] OxA is fully conserved across mammalian species, whilst mouse/rat OxB differs by 2 amino acids compared to the human form.[8] These peptides bind to two G-protein-coupled receptors (GPCRs), termed OX_1R and OX_2R. The receptors are 64% homologous and highly conserved across species (e.g. the rat and human receptors display 94% and 95% homology for OX_1R and OX_2R, respectively.[2]

Table 1. Agonist and antagonist pharmacology at Hcrtr1/OX_1R and Hcrtr2/OX_2R

	OX_1R		OX_2R	
	pK_i	pK_b/*pEC_{50}	pK_i	pK_b/* pEC_{50}
Hcrt1/OxA	7.7[a]	*8.0[b]	7.4[a]	*8.2[b]
Hcrt2/OxB	6.4[a]	*7.3[b]	7.4[a]	*8.4[b]
[A^{11}] OxB	-	*5.6[c]	-	*7.6[c]
SB-334867	7.2[d]	7.5[d]	-	5.7[d]
SB-408124	-	7.8[c]	-	5.8[c]

[a] Sakurai et al. 1998
[b] Smart et al. 1999
[c] Gartlon et al. 2001
[d] Smart et al. 2001

Radioligand-binding studies using ^{125}I-orexin-A have shown OxA to have equal affinity at OX_1R and OX_2R, whereas OxB has ~10-fold greater affinity at OX_2R than OX_1R (Table 1). The agonist pharmacology of the orexin receptors determined in recombinant systems by measurement of $[Ca^{2+}]_i$ is comparable with that determined by radioligand binding with OxA and OxB being equipotent at OX_2R and OxB displaying moderate selectivity for OX_2R over OX_1R (Table 1). A number of groups have now undertaken structural modifications of the orexin peptides and determined the impact on activity at the orexin receptor subtypes. Asahi and colleagues[9] have synthesised a modified form of human OxB (alanine for leucine at position 11, [A^{11}]OxB,) which retains agonist activity but is markedly more selective (100-fold, Table 1) than native OxB for OX_2R over OX_1R. This synthetic peptide has now been used in a limited number of *in vivo* studies (see Section 3.2.2). Recently, Lang and coworkers[10] described several OxB analogues with >1000-fold selectivity for OX_2R and a truncated form of hcrt1/OxA ($OxA_{2\text{-}23}$) with modest OX_1R preference which may prove useful as pharmacological probes for the future.

The vast majority of studies delineating *in vivo* actions of the orexin peptides have utilized direct administration of OxA into the brain. Following intracerebroventricular

(ICV) injection, OxA activates neurons (as indicated by c-Fos immunoreactivity[11] in a pattern consistent with the distribution of orexin peptide and receptors (see Section 3). While an important research strategy, this approach suffers from the real concern that doses found to produce significant behavioural changes may be supra-physiological. OxB has proven to be less popular as an *in vivo* tool, probably because it is considered to be more metabolically labile than OxA.[12] Very few studies have employed systemic (intravenous; IV) injection of OxA and there is conflicting evidence as to whether the peptide can enter into the brain following administration by this route.[12,13]

Many authors have ascribed orexin-induced actions *in vivo* to specific orexin receptor subtypes based solely on the relative efficacies/potencies of OxA and OxB. Unfortunately, this approach is somewhat misguided given that OxB is at best only moderately selective for OX2R and is also metabolically unstable. Studies employing selective antagonists should help clarify this situation.

2.2. Orexin Receptor Antagonists

SB-334867

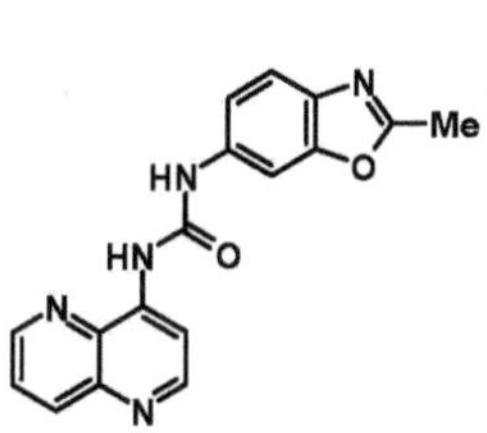

SB-408124

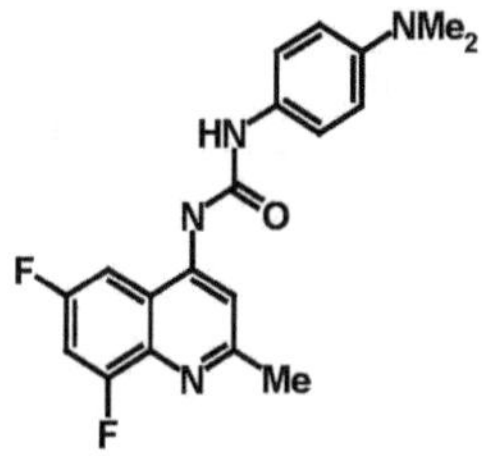

Figure 1. Structures of SB-334867 and SB-408124

The first reported antagonists of orexin receptors were derived from a series of 1,3-Biarylureas, exemplified by SB-334867 (1-(2-methylbenzoxazol-6-yl)-3-[1,5]napthyridin-4-yl urea; Figure 1).[14,15,16] This compound has an affinity of 40nM at OX1R and is >50-fold selective over OX2R (Table 1) and a wide range of other GPCRs and ion channels.[16] Importantly for *in vivo* studies, SB-334867 is also systemically bioavailable and brain penetrant. Thus, following intraperitoneal (IP) administration at 10mg/kg, μM brain levels of parent compound are attained within 30 minutes and maintained for at least 2 hours (Table 2). Due to this overall profile, SB-334867 has proven to be the most useful pharmacological tool for defining the physiological role of OX1R to date. Dose selection for *in vivo* studies with this agent has been further aided by the early observation that SB-334867 (1-10mg/kg IP) was able to inhibit an hcrt1/OxA-evoked behavioural (grooming) response in rats.[14]

Subsequent lead optimisation work led to the discovery of SB-408124 (1-(6,8-difluoro-2-methyl-quinolin-4-yl)-3-(4-dimethylamino-phenyl)-urea; Figure 1), an agent with slightly enhanced antagonist potency and greater OX1R selectivity than SB-334867 (Table 1). Again, SB-408124 is brain penetrant (Table 2) and produces dose-related (1-10mg/kg) and complete reversal of OxA-induced grooming, in this case, following oral

(PO) administration.[17] SB-408124 has most often been used to confirm initial findings with SB-334867.

An interesting recent development is the identification and *in vitro* characterisation of *N*-Acyl 6,7-dimethoxy-1,2,3,4-tetrahydroisoquinoline as the first OX2R selective antagonist.[70] Studies are now eagerly awaited to determine whether this compound or related analogues can be utilised for assessing the role of OX2R *in vivo*.

3. IN VIVO PHARMACOLOGY OF OREXIN RECEPTORS

Within the central nervous system (CNS), neurons expressing orexin are mainly concentrated in the lateral and posterior hypothalamic areas.[18,2,19] However, these orexin neurones send projections to regions throughout the CNS including the locus coeruleus (LC) and raphe nucleus of the brainstem, the cholinergic neurons of the brainstem and forebrain, histaminergic cells of the posterior hypothalamus, cerebral cortex, limbic system[18,2,11,20] and spinal cord.[21,22,23,13] Orexin receptors are located in a pattern consistent with the innervated regions but the distribution of OX1R and OX2R are on the whole strikingly different. OX1R are densely expressed in the prefrontal and intralimbic cortex, hippocampus, paraventricular thalamic nucleus, ventromedial hypothalamic nucleus, dorsal raphe nucleus, LC, spinal cord and dorsal root ganglia. OX2R are prominent in the cerebral cortex, septal nuclei, hippocampus, medial thalamic groups, raphe nuclei, various nuclei of the hypothalamus and spinal cord, but are absent in the LC.[24,25,13,26,27,28,29] Outside of the brain, both orexin peptide and receptors have been identified in the enteric nervous system and pancreas.[30,31]

Table 2. Brain levels of SB-334867 and SB-408124 in rats

Compound	Dose (mg/kg, route)	Time Post-Dose (hours)	Brain Level (μM)
SB-334867	10 IP	0.5	13.7[a]
		1	8.8[a]
		2	3.8[b]
		4	<0.03[b]
	30 IP	0.5	25.1[c]
		2	20.0[c]
		4	11.8[c]
		8	0.4[c]
		12	0.05[c]
		≥ 24	NQ[c]
SB-408124	3 PO	0.5	0.05[a]
		1	<0.01[a]

NQ: Non quantifiable
[a] Unpublished observation
[b] Porter et al. 2001
[c] Ishii et al. 2003

The apparent diffuse nature of orexin and its receptors provided one of the first clues of the multiple potential physiological roles that were soon to be discovered for this unique system. In addition, the differential distribution of OX1R and OX2R suggested that distinct orexin receptor subtypes may well subserve specific actions of orexin.

Pharmacological and genetic evidence that sheds light on this possibility is described below.

3.1. Feeding and Appetite

The first described behavioural action of orexin resulted from studies by Sakurai and colleagues[2] who found that ICV administration of either OxA or OxB produced a dose-dependent stimulation of food intake in rats. This seminal observation led the authors to specifically name the peptides as 'orexins' after the Greek word 'orexis' meaning appetite. A flurry of research activity followed and soon numerous groups had replicated the orexigenic action of OxA, although the initial finding with OxB subsequently proved to be less reproducible.[32,33] Conversely, knockout of the *orexin* gene[34] or ablation of orexin neurons[35] created mice that underate. Considered alongside the neuranatomical localisation of orexin peptide and receptors in known CNS feeding centres and evidence that manipulations of nutritional state induce pronounced changes in

Table 3. Effects of OxA and SB-334867 on the BSS in rats

Treatment 1 (dose, µg ICV)	Treatment 2 (dose, mg/kg IP)	Transition Time[a] (minutes)
Vehicle	Vehicle	23[b]
OxA (10)	Vehicle	37
OxA (10)	SB-334867 (3)	35
OxA (10)	SB-334867 (10)	31[b]
OxA (10)	SB-334867 (30)	19[b]

Adapted from Rodgers et al, 2001

[a] Time taken for the transition from mainly eating to mainly resting in male rats. The delay in this transition point following hcrt1/OxA is inhibited by SB-334867

[b] Significantly different from OxA alone

orexin neurons, it has become increasingly clear that the orexin neuropeptide sytem plays a significant role in the regulation of feeding and energy homeostasis.[34,32,33]

Rodgers and coworkers[32,71] have conducted an elegant behavioural analysis of precisely how OxA increases feeding in rats, utilising the behavioural satiety sequence (BSS), a concept describing the normal transition from eating → grooming → resting.[36] They found that OxA (3-10µg ICV) preserves the microstructure of the BSS but, crucially, delays its onset resulting in increased feeding. This group then went on to show that the delayed transition from mainly eating to predominantly resting produced by OxA (10µg ICV) could be dose-dependently (3-30mg/kg IP) and completely inhibited by the OX_1R selective antagonist SB-334867[34] (Table 3). These findings are in keeping with those from less-detailed earlier studies where SB-334867 (30mg/kg IP) blocked the hyperphagic response evoked by a higher dose of OxA (30µg ICV).[37]

Perhaps of even greater significance, when administered alone, acute SB-334867 (10-30mg/kg IP) suppresses food intake in rats following either prior fasting [37] or under conditions when free-feeding levels are high.[37,71] The anorectic effect of SB-334867 is specifically attributable to an acceleration of behavioural satiety (Figure 2) rather than non-specific actions such as sedation or malaise.[38,71] Fascinatingly, in addition to decreasing food intake, a single dose of SB-334867 (10-30mg/kg IP) has also been found to reduce bodyweight gain in rats for periods lasting at least 3 to 5 days post-treatment,[38,71] despite the fact that parent compound is virtually undetectable in brain (or blood) beyond 12 hours of dosing (Table 1). This suggests that acute antagonism of the OX1R may initiate a sequence of downstream events that extend beyond the period of receptor blockade.[32] Consistent with the acute anorectic action of SB-334867, repeated administration of the OX1R antagonist (30mg/kg IP, once and then twice daily) reduced the cumulative food intake and body weight gain over 14 days in genetically obese (*ob/ob*) mice.[39] These changes were accompanied by decreased total fat mass, increased metabolic rate and decreased fasting blood glucose and plasma insulin.

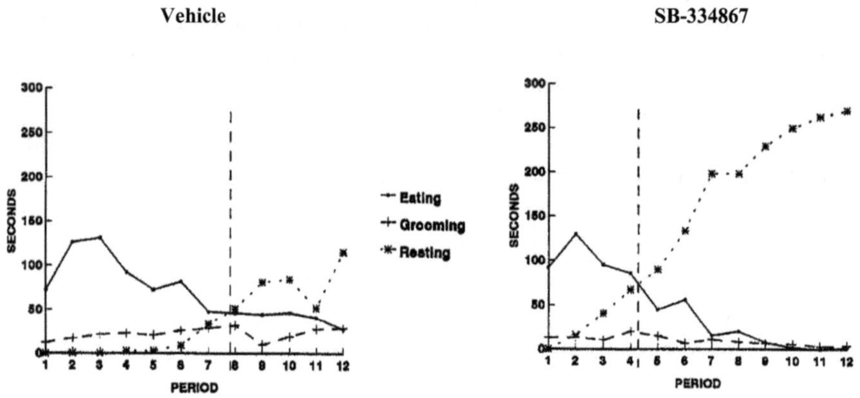

Figure 2. Effect of SB-334867 (30mg/kg IP) on BSS in rats. Data are from Rodgers et al. (2001) (reproduced with permission of Blackwell Science). When given alone, the OX1R antagonist advances the transition from mainly eating to mainly resting (vertical dashed line) resulting in reduced feeding.

Taken collectively, there is now compelling pharmacological evidence for a physiological role of endogenous orexin and OX1R in the modulation of appetite and potentially also energy expenditure. The neural sites subserving OxA-induced hyperphagia appear to be restricted to several discrete nuclei within the hypothalamus.[32] OX1R are widely distributed throughout the hypothalamus and are found on both orexin-expressing neurons and also neurons expressing other neuropeptides known to regulate appetite, including neuropeptide Y and melanin-concentrating hormone.[40] Thus, OX1R are ideally situated to influence multiple neuropeptide signalling pathways implicated in feeding. Whether OX2R can play a similar role is presently undetermined. However, observations that dogs with non-functional OX2R (see Section 3.2) exhibit no overt feeding or metabolic abnormalities[34] and that manipulations of nutritional status can change expression of OX1R, but not OX2R,[41] provides an initial indication that the OX1R subtype may play the dominant role in controlling appetite.

3.2. Arousal and Sleep

3.2.1. Orexins and Narcolepsy

Although *orexin* peptide gene knockout mice proved to be hypophagic, the most striking, but at the time unexpected, finding with these animals was that they exhibited a phenotype remarkably similar to the human sleep disorder narcolepsy.[42] Continuous behavioural monitoring revealed that homozygous knockouts underwent sudden periods of inactivity, especially during the dark period when mice are normally most awake and active. Electroencephalographic (EEG) recordings confirmed that these episodes were not related to epilepsy, and that the mice suffered from cataplectic attacks – a hallmark of narcolepsy. In addition, the mutant mice spent increased time in rapid eye movement (REM) sleep during the dark phase and showed episodes of direct transition from wakefulness to REM sleep, another event that is characteristic of narcolepsy. At the same time, Lin and colleagues[4] found, using positional cloning, that mutations in the *hcrtr2* gene were responsible for inherited canine narcolepsy. As a consequence of the mutations, hcrtr2s are non-functional being either unable to localize properly to the cell membrane or bind the hcrt/orexin peptides.[4,43] Confirmation of disruptions of the hypocretin system in human narcolepsy was very soon to follow. Two independent research groups reported few or absent hypocretin-producing neurons in the brains of narcoleptic patients,[6,7] a finding consistent with observed losses of hcrt1 in the CSF of a significant proportion of patients with narcolepsy.[5,44] Furthermore, a single patient with unusually severe early onset narcolepsy was identified as carrying a mutation of the *hypocretin* gene itself.[6]

Workers from the laboratory of Yanagisawa have subsequently created transgenic mice[35] and rats[45] in which hypocretin/orexin neurons are genetically ablated. In keeping with the *orexin* peptide knockout mice, both species exhibit the key EEG and behavioural features of human narcolepsy. This group has also gone on to study in detail the phenotype of mice in which the genes for either or both orexin receptors have been deleted. As might be expected, *OX2R* knockout mice have a narcolepsy phenotype[46,47] but it is less severe than that of the *orexin* peptide knockouts. *OX1R* knockouts appear behaviourally normal and show only mild sleep fragmentation.[48] Most interestingly, the double *OX1R* and *OX2R* mutants fully recapitulate the *orexin* peptide knockout narcolepsy phenotype,[48] suggesting that activation of both orexin receptor subtypes contributes to the physiological control of arousal state.

3.2.2. Orexins are Key Regulators of Arousal

Inspired by evidence that the LC, a key site in the determination of attentional state, received dense orexinergic innervation and increased its firing rate in response to OxA *in vitro*, Hagan and colleagues[49,50] became the first to show that exogenously administered OxA could influence arousal in conscious animals. Using EEG and electromyographic recordings they found that ICV injections of OxA (1-30μg) at light onset (the major sleep period) produced a dose-dependent increase in the time rats spent awake. The enhancement of arousal was accompanied by a marked reduction in paradoxical sleep (PS ≡REM) and deep slow wave sleep at the highest dose (Figure 3). Furthermore, OxA prolonged the latency to the first occurrence of PS, an effect that could be completely inhibited by the OX1R selective antagonist SB-334867 (10 & 30mg/kg IP).[51] Local administration of Hcrt1 (but not Hcrt2) into the LC of rats similarly increased wakefulness and reduced both PS and deep slow wave sleep, concomitant with increased neuronal firing in this region.[52] These effects were also blocked by an immunoneutralizing antibody to Hcrt1.[52] Subsequent *in vitro* electrophysiological studies have confirmed the importance of OX1R in mediating orexin-induced excitation of the LC.[53]

At the same doses as those modifying the sleep-wake cycle, ICV OxA (1-30μg) evokes several other changes indicative of an enhanced state of arousal in rats, including intense whole body grooming and sustained hyperlocomotion.[49,14,54] OxB (30μg ICV) induces a more subtle head grooming response but again markedly increases locomotor activity with minimum effective doses of only 3-10μg ICV.[54] Similarly, the synthetically modified form of human OxB, [A[11]]OxB, which is highly OX2R selective (Table 1), also potently (3-30μg ICV) stimulates locomotion.[55] In terms of the underlying pharmacology of these behavioural effects, the grooming response to ICV OxA is dose-dependently and

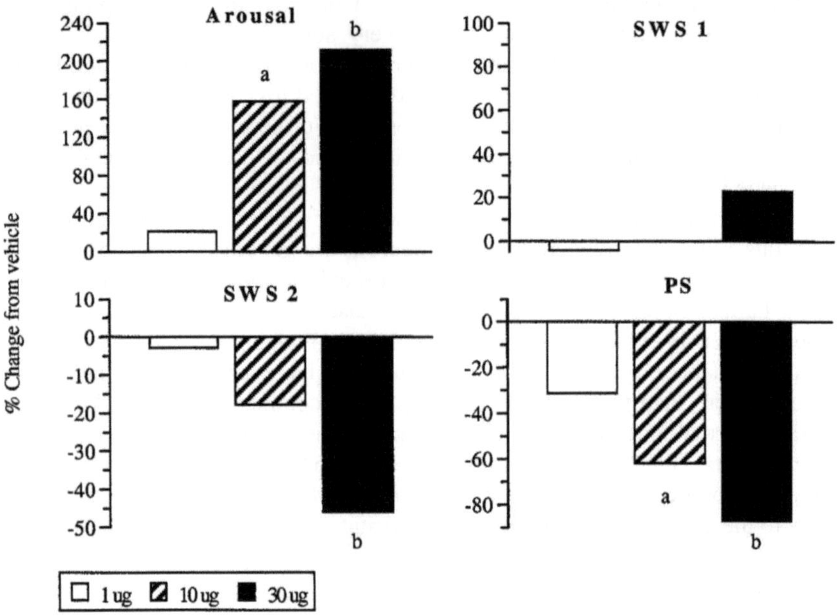

Figure 3. Effect of hcrt1/OxA (1-30μg ICV), administered to rats at the beginning of the sleep phase, on the proportion of time spent in arousal, slow wave sleep (SWS) 1 and 2 and PS. [a]P<0.05 and [b]P<0.01 compared to vehicle-treated controls. Data are adapted from ref 50.

completely inhibited by both SB-334867 (1-10mg/kg IP);[14] and SB-408124 (1-10mg/kg PO)[17] indicating sole mediation by OX1R. In contrast, SB-334867 (10-30mg/kg IP) and SB-408124 (10mg/kg PO) have either no effect or only partially inhibit the hyperactivity produced by OxA (3μg ICV), OxB (10μg ICV) and [A[11]]OxB (10μg ICV) (Table 4) implicating a role for OX1R and, by inference, OX2R in modulating behavioural arousal. Studies have now been undertaken to elucidate the neurotransmitter pathways involved in OxA-induced grooming and hyperlocomotion, downstream from orexin receptors. Monoaminergic systems appear to be particularly important with both dopamine (D1 & D2;)[56,57] and serotonin receptor (5HT2;)[14,57] antagonists found to attenuate the grooming and/or locomotor response to ICV OxA.

Thus, there is now an impressive body of data supporting a key role of OX1R and OX2R in controlling arousal under normal and/or pathophysiological conditions. As described above, there is already direct evidence that OX1R within the LC are involved in this respect.[52,53] It is highly likely that excitation of orexin receptors localized in regions such as tuberomammillary nucleus, lateral preoptic area and basal forebrain cholinergic area are equally important given that all of these brain centres are exquisitely sensitive to the wake-promoting action of hcrt1/OxA in conscious animals.[58,59,60]

Table 4. Effect of OX1R antagonists on the hyperlocomotor responses to hcrt1/OxA, OxB and $[A^{11}]$OxB in rats

	% Increase in Locomotor Activity [a]		
Pretreatment [b] (Dose, mg/kg)	**hcrt1/OxA (3µg ICV)**	**OxB (10µg ICV)**	**$[A^{11}]$ OxB (10µg ICV)**
Vehicle	148	211	
SB-334867 (10 IP)	167[c]	89[d]	
SB-334867 (30 IP)	39[d]		
Vehicle	136	225	63
SB-408124 (10 PO)	164[c]	100[d]	80[c]
SB-408124 (30 PO)	173[c]		

[a] Increase in spontaneous motor activity (total infrared beam breaks) in the 60 minute period immediately following peptide administration, relative to vehicle-treated control rats
[b] Pretreatments were given 30 minutes prior to peptide administration
[c] Non significantly different and [d] $P<0.05$ compared to orexin peptide alone
Data from Jones et al. (2001), Gartlon et al. (2001) and unpublished observations

3.3. Pain Modulation

The distribution of the orexins and their receptors in areas of the brain (e.g. periaqueductal gray, hypothalamus and thalamus) and spinal cord associated with nociceptive processing (see Section 3 for references), triggered speculation that the orexinergic system may be involved in modulating pain. Initial support for this idea came from a study in which ICV hcrt1/OxA (3-30µg) produced dose-related analgesia in a rat hotplate test.[13] Systemic (IV) administration of hcrt1/OxA also proved to be effective in this rat model and in the equivalent mouse paradigm at doses between 3-30mg/kg. In other mouse models, IV OxA inhibited visceral nociception (abdominal constriction; 10 & 30mg/kg) and thermal hyperalgesia (intraplantar carrageenan; 3-30mg/kg), with an efficacy equivalent to the opioid analgesic morphine.[13] The analgesic activity of hcrt1/OxA (30mg/kg IV) in the mouse hotplate test was substantially attenuated by SB-334867 (10 & 30mg/kg IP) but not by the opioid receptor antagonist naloxone.[13] These findings indicate that the action of IV hcrt1/OxA is mediated, at least in part, by OX1R without involvement of the opioid system. Although when given by itself, SB-334867 had no effect on acute nociceptive thresholds in the mouse hotplate model (10 & 30mg/kg IP), the selective OX1R antagonist was pro-hyperalgesic in the mouse carrageenan assay (3 & 10mg/kg IP). Considered together with the demonstration of orexin containing fibres projecting from the hypothalamus to the spinal cord,[22] these data were taken as evidence of an endogenous descending orexin inhibitory drive that is present under some conditions of inflammation.[13]

The site(s) of analgesic action of OxA following IV administration remains to be established. Detailed pharmacokinetic studies have failed to detect the presence of hcrt1/OxA in brain tissue following IV dosing of the peptide in rats or mice.[13] This suggests that systemically administered OxA is likely to exert its analgesic effects at a peripheral site, although the degree of penetration into the spinal cord is not known.

Indeed, a subsequent study has shown that intrathecal (spinal) injection of OxA (but not OxB) produces marked analgesic activity in rat models of inflammatory and thermal pain.[61] These effects were inhibited by intrathecal SB-334867 clearly illustrating that activation of spinal OX1R can induce analgesia.[61]

3.4. Other actions

A number of studies have demonstrated sympathomimetic effects of the hypocretins/orexins. ICV,[62] intracisternal,[63] intrathecal[64] or rostro venterolateral medullary[63] injections of OxA have all been shown to modify cardiovascular function in both conscious and anaeasthetized rats. Hirota and colleagues[65] recently reported that OxA(~20 μg ICV)-evoked increases in heart rate and blood pressure in anaesthetized rats correlate with elevations in plasma noradrenaline levels. These effects of OxA are markedly inhibited by SB-334867 (50nM ICV) suggesting that activation of central OX1R regulates sympathetic activity. This notion is in keeping with the ability of hcrt1/OxA to facilitate noradrenaline release from cerebrocortical slices *in vitro via* an action at OX1R.[65] The dense localisation of orexin nerve fibres and OX1R in brain regions (e.g. hypothalamic and brainstem nuclei) related to autonomic function (see Section 3 for references) provide an anatomical framework for these pharmacological findings.

Workers from the same laboratory have also provided evidence that orexinergic modulation of noradrenergic transmission may be an important mediator of barbiturate anaesthesia.[66] Thus, in rat cerebrocortical slices several barbiturate anaesthetics inhibit hcrt1/OxA-induced noradrenaline release at clinically relevant concentrations. In addition, OxA (~4 & 20 μg ICV) significantly decreases barbiturate anaesthesia time in rats, an action reversed by SB-334867 (50nM ICV).[66] Of particular interest, SB-334867 (0.5-50nM ICV) concentration-dependently potentiates barbiturate sleep time in its own right,[66] indicating tonic activation of OX1R in this test situation.

Relatively few studies have examined the role of orexin in the enteric nervous system. Implantation of electrodes in the serosa of the small intestine has allowed investigation of the effects of OxA on small bowel motor control in fasted conscious rats.[67] IV infusions of OxA (100 & 500pmol/kg/min) significantly prolong the myoelectric motor complex (MMC) cycle length, resulting in a more fed-like motor pattern. Conversely, peripheral administration of SB-334867 alone slows the MMC cycle length at the same dose (10mg/kg IV bolus) that abolishes the effect of hcrt1/OxA on fasting motility. These findings highlight a physiological role for OX1R in regulating rat small bowel function. The precise site(s) at which SB-334867 exerts its effects on gut motility remains to be established but a local action is possible as OX1R (and OX2R) are widespread throughout the gastrointestinal tract.[31,67]

Table 5. *In vivo* pre-clinical evidence implicating a role for OX1R and/or OX2R in orexin-mediated behaviours

Behavioural Response	OX1R	OX2R
Feeding[a]	• hcrt1/OxA-induced hyperphagia is completely inhibited by SB-334867 • SB-334867 attenuates feeding and reduces body weight gain in its own right	
Sleep-wakefulness[b]	• OX1R knockouts show mildly fragmented sleep • Increased wakefulness and reduced PS evoked by hcrt1/OxA is blocked by SB-334867 and an immunoneutralising antibody to OX1R	• Non-functional OX2R are responsible for inherited canine narcolepsy • OX2R knockout mice exhibit a narcolepsy phenotype
Grooming[c]	• hcrt1/OxA-induced grooming is prevented by SB-334867 and SB-408124	
Hyperactivity[c]	• The hyperlocomotor response to OxB is partially inhibited by SB-334867 and SB-408124	• The increase in locomotor activity produced by the OX2R selective agonist $[A^{11}]$ OxB is not antagonised by SB-408124[d]
Pain[e]	• The analgesic effect of hcrt1/OxA is substantially attenuated by SB-334867 • SB-334867 alone is pro-hyperalgesic in some pain states	
Cardiovascular function[f]	• hcrt1/OxA-induced elevations of heart rate and blood pressure are inhibited by SB-334867	
Gut motility[f]	• The prolongation of MMC[g] cycle length evoked by hcrt1/OxA is abolished by SB-334867 • SB-334867 shortens MMC[3] cycle length in its own right	

[a] Section 3.1
[b] Section 3.2.1
[c] Section 3.2.2
[d] Further studies with OX2R antagonists, when available, are required to confirm the role of OX2R in this behavioural response
[e] Section 3.3
[f] Section 3.4
[g] Myoelectric Motor Complex

4. CONCLUSIONS AND THERAPEUTIC OPPORTUNITIES FOR THE FUTURE

There is a growing consensus among researchers in the field that the orexins play a pivotal role in orchestrating the complex behavioural and physiologic responses underpinning the complementary homeostatic processes of feeding and sleep-wakefulness. Emerging genetic and pharmacological data are now beginning to elucidate the orexin receptor subtype(s) mediating these varied effects of the orexins (Table 5) and highlight a number of exciting therapeutic possibilities for agents that selectively modulate this novel system. The dramatic loss of orexin peptides in many narcolepsy sufferers (see Section 3.2.1) suggests that orexin receptor agonists may be of benefit as replacement therapy in this disabilitating neurological disorder. Indeed, ICV administration of OxA rescues the narcolepsy-cataplexy phenotype of orexin neuron-ablated mice.[68] Furthermore, modafinil, which has been approved and marketed for narcolepsy, promotes wakefulness in rats in association with activation of orexin neurones.[69] Importantly, hcrt1/OxA-induced arousal in animals is not followed by periods of rebound sleep,[49,50] often a confounding factor in therapies for excessive daytime sleepiness based on classical psychostimulants. Pre-clinical studies also suggest that OX_1R agonists may be of value in circumstances where acute injury and pain may be present, such as post-operatively, and in conditions such as arthritis and neuropathic pain where hyperalgesia is a significant factor.[13] However, a major challenge to delivering this therapeutic promise remains the development of small molecule orexin receptor agonists that are bioavailable and CNS penetrant.

In order for orexin receptor antagonists to be of therapeutic benefit it is implicit that there must be a degree of endogenous orexinergic tone at the orexin receptors. It is already evident that orexin receptor antagonists do not always produce the opposite effect as exogenously administered orexin peptides. For example, OxA elevates blood pressure and heart rate in rats but the OX_1R selective antagonist SB-334867 has no haemodynamic effects in its own right.[65] The most striking evidence that orexin receptor antagonists can be pharmacologically active when given by themselves comes from experiments showing that SB-334867 reduces food intake and body weight gain while increasing energy expenditure in rodents (see Section 3.1). This suggests that OX_1R antagonists may have potential in the treatment of obesity and metabolic disorders. Counter to the envisaged role for orexin receptor agonist to promote arousal in narcoleptics, the most exciting therapeutic prospect for orexin receptor antagonists is predicted to be in the treatment of insomnia. Studies to determine whether OX_1R and/or OX_2R antagonists can induce physiologically balanced sleep have not yet been reported but are eagerly awaited. However, of some interest in this regard is the observation that in contrast to clinically used hypnotic benzodiazepines, SB-334867 is not overtly sedating in circumstances when rats are normally active.[14,54,35,38] This at least suggests that orexin receptor antagonists should be devoid of the unwanted hangover effects classically associated with benzodiazepines.

Hopefully, the rapidity with which we have come to understand the unique physiological and pathophysiological roles fulfilled by the orexin peptides and receptors will now be matched by the emergence of novel orexin system therapeutics over the next decade.

5. REFERENCES

1. L. de Lecea, T. S. Kilduff, C. Peyron, X. Gao, P. E. Foye, P. E. Danielson, C. Fukuhara, E. L. Battenberg, V. T. Gautvik, F. S. Bartlett, 2nd, W. N. Frankel, A. N. van den Pol, F. E. Bloom, K. M. Gautvik and J. G. Sutcliffe, The hypocretins: hypothalamus-specific peptides with neuroexcitatory activity, *Proc Natl Acad Sci U S A*. **95**, 322-7 (1998).
2. T. Sakurai, A. Amemiya, M. Ishii, I. Matsuzaki, R. M. Chemelli, H. Tanaka, S. C. Williams, J. A. Richardson, G. P. Kozlowski, S. Wilson, J. R. Arch, R. E. Buckingham, A. C. Haynes, S. A. Carr, R. S. Annan, D. E. McNulty, W. S. Liu, J. A. Terrett, N. A. Elshourbagy, D. J. Bergsma and M. Yanagisawa, Orexins and orexin receptors: a family of hypothalamic neuropeptides and G protein-coupled receptors that regulate feeding behavior, *Cell*. 92, 573-85 (1998).
3. W. K. Samson and Z. T. Resch, The Hypocretin/Orexin Story, *Trends Endocrinol Metab*. **11**, 257-262 (2000).
4. L. Lin, J. Faraco, R. Li, H. Kadotani, W. Rogers, X. Lin, X. Qiu, P. J. de Jong, S. Nishino and E. Mignot, The sleep disorder canine narcolepsy is caused by a mutation in the hypocretin (orexin) receptor 2 gene, *Cell*. **98**, 365-76 (1999).
5. S. Nishino, B. Ripley, S. Overeem, G. J. Lammers and E. Mignot, Hypocretin (orexin) deficiency in human narcolepsy, *Lancet*. **355**, 39-40 (2000).
6. C. Peyron, J. Faraco, W. Rogers, B. Ripley, S. Overeem, Y. Charnay, S. Nevsimalova, M. Aldrich, D. Reynolds, R. Albin, R. Li, M. Hungs, M. Pedrazzoli, M. Padigaru, M. Kucherlapati, J. Fan, R. Maki, G. J. Lammers, C. Bouras, R. Kucherlapati, S. Nishino and E. Mignot, A mutation in a case of early onset narcolepsy and a generalized absence of hypocretin peptides in human narcoleptic brains, *Nat Med*. **6**, 991-997 (2000).
7. T. C. Thannickal, R. Y. Moore, R. Nienhuis, L. Ramanathan, S. Gulyani, M. Aldrich, M. Cornford and J. M. Siegel, Reduced number of hypocretin neurons in human narcolepsy, *Neuron*. **27**, 469-74. (2000).
8. D. Smart, J. C. Jerman, S. J. Brough, S. L. Rushton, P. R. Murdock, F. Jewitt, N. A. Elshourbagy, C. E. Ellis, D. N. Middlemiss and F. Brown, Characterisation of recombinant human orexin receptor pharmacology in a Chinese hamster ovary cell-line using FLIPR, *Br J Pharmacol*. **128**, 1-3 (1999).
9. S. Asahi, S. Egashira, M. Matsuda, H. Iwaasa, A. Kanatani, M. Ohkubo, M. Ihara, T. Sakurai and H. Morishima, Structure-activity relationship studies on the novel neuropeptide orexin, in: *Pept Sci 1999*, N. Fujii, ed., pp37-40 (2000).
10. M. Lang, R. Söll, F. Dürrenberger, F. M. Dautzenberg and A. G. Beck-Sickinger, Structure-activity studies of orexin A and orexin B at the human orexin 1 and orexin 2 receptors led to orexin 2 receptor selective and orexin 1 receptor preferring ligands, *J Med Chem*. **47**, 1153-1160 (2004).
11. Y. Date, Y. Ueta, H. Yamashita, H. Yamaguchi, S. Matsukura, K. Kangawa, T. Sakurai, M. Yanagisawa and M. Nakazato, Orexins, orexigenic hypothalamic peptides, interact with autonomic, neuroendocrine and neuroregulatory systems, *Proc Natl Acad Sci U S A* **96**, 748-753 (1999).
12. A. J. Kastin and V. Akerstrom, Orexin A but not orexin B rapidly enters brain from blood by simple diffusion, *J Pharmacol Exp Ther*. **289**, 219-223 (1999).
13. S. Bingham, P. T. Davey, A. J. Babbs, E. A. Irving, M. J. Sammons, M. Wyles, P. Jeffrey, L. Cutler, I. Riba, A. Johns, R. A. Porter, N. Upton, A. J. Hunter and A. A. Parsons, Orexin-A, an hypothalamic peptide with analgesic properties, *Pain* **92**, 81-90 (2000).
14. M. S. Duxon, J. Stretton, K. Starr, D. N. C. Jones, V. Holland, G. Riley, J. Jerman, S. Brough, D. Smart, A. Johns, W. Chan, R. A. Porter and N. Upton, Evidence that orexin-A-evoked grooming in the rat is mediated by orexin-1 (OX_1) receptors, with downstream 5-HT_{2C} receptor involvement, *Psychopharmacology*. **153**, 203-209 (2001).
15. R. A. Porter, W. N. Chan, S. Coulton, A. Johns, M. S. Hadley, K. Widdowson, J. C. Jerman, S. J. Brough, M. Coldwell, D. Smart, F. Jewitt, P. Jeffrey and N. Austin, 1,3-Biarylureas as selective non-peptide antagonists of the orexin-1 receptor, *Bioorg Med Chem Lett*. **11**, 1907-1910 (2001).
16. D. Smart, C. Sabido-David, S. J. Brough, F. Jewitt, A. Johns, R. A. Porter and J. C. Jerman, SB-334867-A: the first selective orexin-1 receptor antagonist, *Br J Pharmacol*. **132**, 1179-1182 (2001).
17. N. Upton, J. Stretton, M. S. Duxon, D. Howlett, J. P. Pilleux, J. D. Martin and R. Porter, Evidence that orexin-A evoked grooming, but not water intake, in the rat is mediated by the orexin-1 (OX1) receptor, *Soc Neurosci Abst*. **27**: Program no. 320-321 (2001).
18. C. Peyron, D. K. Tighe, A. N. van den Pol, L. de Lecea, H. C. Heller, J. G. Sutcliffe and T. S. Kilduff, Neurons containing hypocretin (orexin) project to multiple neuronal systems, *J Neurosci*. **18**, 9996-10015 (1998).
19. S. Taheri, M. Mahmoodi, J. Opacka-Juffry, M. A. Ghatei and S. R. Bloom, Distribution and quantification of immunoreactive orexin A in rat tissues, *FEBS Lett*. **457**, 157-161 (1999).

20. T. Nambu, T. Sakurai, K. Mizukami, Y. Hosoya, M. Yanagisawa, and K. Goto, Distribution of orexin neurons in the adult rat brain, *Brain Res.* **827,** 243-260 (1999).
21. D. J. Cutler, R. Morris, V. Sheridhar, T. A. K. Wattam, S. Holmes, S. Patel, J. R. S. Arch, S. Wilson, R. E. Buckingham, M. L. Evans, R. A. Leslie and G. Williams, Differential distribution of orexin-A and orexin-B immunoreactivity in the rat brain and spinal cord, *Peptides* **20**, 1455-1470 (1999).
22. A. N. van den Pol, Hypothalamic hypocretin (orexin): robust innervation of the spinal cord, *J Neurosci.* **19**, 3171-3182 (1999).
23. Y. Date, M. S. Mondal, S. Matsukura and M. Nakazato, Distribution of orexin-A and orexin-B (hypocretins) in the rat spinal cord, *Neurosci Lett.* **288**, 87-90 (2000).
24. P. Trivedi, H. Yu, D. J. MacNeil, L. H. van der Ploeg and X. M. Guan, Distribution of orexin receptor mRNA in the rat brain, *FEBS Lett.* **438**, 71-75 (1998).
25. X.-Y. Lu, D. Bagnol, S. Burke, H. Akil, and S. J. Watson, Differential distribution and regulations of OX1 and OX2 orexin/hypocretin receptor messenger RNA in the brain upon fasting, *Horm Behav.* **27**, 335-344 (2000).
26. G. J. Hervieu, J. E. Cluderay, D. C. Harrison, J. C. Roberts and R. A. Leslie, Gene expression and protein distribution of the orexin-1 receptor in the rat brain and spinal cord, *Neuroscience* **103**, 777-797 (2001).
27. M. A. Greco and P. J. Shiromani, Hypocretin receptor protein and mRNA expression in the dorsolateral pons of rats, *Brain Res. Mol. Brain Res.* **88**, 176-182 (2001).
28. J. N. Marcus, C. J. Aschkenasi, C. E. Lee, R. M. Chemelli, C. B. Saper, M. Yanagisawa and J. K. Elmquist, Differential expression of orexin receptors 1 and 2 in the rat brain, *J Comp Neurol.* **435**, 6-25 (2001).
29. J. E. Cluderay, D. C. Harrison, and G. J. Hervieu, Protein distribution of the orexin-2 receptor in the rat central nervous system, *Regul. Pept.* **104,** 131-144 (2002).
30. A. L. Kirchgessner and M. Liu, Orexin synthesis and response in the gut, *Neuron* **24**, 941-951 (1999).
31. E. Näslund, M. Ehrström, J. Ma, P. M. Hellström and A. L. Kirchgessner, Localization and effects of orexin on fasting motility in the rat duodenum, *Am J Physiol Gastrointest Liver Physiol.* **282**, 470-479 (2002).
32. R. J. Rodgers, Y. Ishii, J. C. G. Halford and J. E. Blundell, Orexins and appetite regulation, *Neuropeptides* **36**, 303-325 (2002).
33. D. Smart and J. C. Jerman, The physiology and pharmacology of the orexins, *Pharamcol Ther.* **94**, 51-61 (2002).
34. J. T. Willie, R. M. Chemelli, C. M. Sinton and M. Yanagisawa, To eat or to sleep? Orexin in the regulation of feeding and wakefulness, *Annu Rev Neurosci.* **24**, 429-58 (2001).
35. J. Hara, C. T. Beuckmann, T. Nambu, J. T. Willie, R. M. Chemelli, C. M. Sinton, F. Sugiyama, K. Yagami, K. Goto, M. Yanagisawa and T. Sakurai, Genetic ablation of orexin neurons in mice results in narcolepsy, hypophagia, and obesity, *Neuron* **30**, 345-354 (2001).
36. J. Antin, J. Gibbs, J. Holt, R. C. Young and G. P. Smith, Cholecystokinin elicits the complete behavioural sequence of satiety in rats, *J Comp Physiol Psychol.* **89**, 748-760 (1975).
37. A. C, Haynes, B. Jackson, H. Chapman, M. Tadayyon, A. Johns, R. A. Porter and J. R. S. Arch, A selective orexin-1 receptor antagonist reduces food consumption in male and female rats, *Regul Pept.* **96**, 45-51 (2002).
38. Y. Ishii, J. E. Blundell, J. C. G. Halford, N. Upton, R. Porter, A. Johns and R. J. Rodgers, Differential effects of the selective orexin-1 receptor antagonist SB-334867 and lithium chloride on the behavioural satiety sequence in rats, *Physiol Behav.* **81**, 129-140 (2004).
39. A. C. Haynes, H. Chapman, C. Taylor, G. B. T. Moore, M. A. Cawthorne, M. Tadayyon, J. C. Clapham and J. R. S. Arch, Anorectic, thermogenic and anti-obesity activity of a selective orexin-1 receptor antagonist in ob/ob mice, *Regul Pept.* **104**, 153-159 (2002).
40. M. Bäckberg, G. Hervieu, S. Wilson and B. Meister, Orexin receptor-1 (OX-R1) immunoreactivity in chemically identified neurons of the hypothalamus: focus on orexin targets involved in control of food and water intake, *Eur J Neurosci.* **15**, 315-328 (2002).
41. M. López, L. Seoane, M. C. García, F. Lago, F. F. Casanueva, R. Senaris and C. Diéguez, Leptin regulation of prepro-orexin and orexin receptor mRNA levels in the hypothalamus, *Biochem Biophys Res Commun.* **269**, 41-5 (2000).
42. R. M. Chemelli, J. T. Willie, C. M. Sinton, J. K. Elmquist, T. Scammell, C. Lee, J. A. Richardson, S. C. Williams, Y. Xiong, Y. Kisanuki, T. E. Fitch, M. Nakazato, R. E. Hammer, C. B. Saper and M. Yanagisawa, Narcolepsy in orexin knockout mice: molecular genetics of sleep regulation, *Cell* **98**, 437-451 (1999).
43. M. Hungs, J. Fan, L. Lin, X. Lin, R. Maki and E. Mignot, Identification and functional analysis of mutations in the hypocretin (orexin) genes of narcoleptic canines, *Genome Res.* **11**, 531-539 (2001).
44. B. Ripley, S. Overeem, N. Fujiki, S. Nevsimalova, M. Uchino, J. Yesavage, D. Di Monte, K. Dohi, A. Melberg, G. J. Lammers, Y. Nishida, F. W. Roelandse, M. Hungs, E. Mignot, and S. Nishino, CSF hypocretin/orexin levels in narcolepsy and other neurological conditions, *Neurology* **57**,2253-2258 (2001).

45. C. T. Beuckmann, C. M. Sinton, S. C. Williams, J. A. Richardson, R. E. Hammer, T. Sakurai and M. Yanagisawa, Expression of a poly-glutamine-ataxin-3 transgene in orexin neurons induced narcolepsy-cataplexy in the rat, *J. Neurosci.* **24**, 4469-4477 (2004).
46. R. M. Chemelli, C. M. Sinton and M. Yanagisawa, Polysomnographic characterization of Orexin-2 receptor knockout mice, *Sleep (suppl).* **23**, 296-297 (2000).
47. S. Tokita, R. M. Chemelli, J. T. Willie and M. Yanagisawa, Behavioral characterization of orexin-2 receptor (OX_2R) knockout mice, *Sleep* **24**, A20-21 (2001).
48. Y. Y. Kisanuki, R. M. Chemelli, C. M. Sinton, S. C. Williams, J. A. Richardson, R. E. Hammer and M. Yanagisawa, The role of orexin receptor type-1 (OXIR) in the regulation of sleep, *Sleep* **23,** *(suppl)* 91 (2000).
49. J. J. Hagan, R. A. Leslie, S. Patel, M. L. Evans, T. A. Wattam, S. Holmes, C. D. Benham, S. G. Taylor, C. Routledge, P. Hemmati, R. P. Munton, T. E. Ashmeade, A. S. Shah, J. P. Hatcher, P. D. Hatcher, D. N. Jones, M. I. Smith, D. C. Piper, A. J. Hunter, R. A. Porter and N. Upton, Orexin A activates locus coeruleus cell firing and increases arousal in the rat, *Proc Natl Acad Sci U S A.* **96**, 10911-6 (1999).
50. D. C. Piper, N. Upton, M. I. Smith, and A. J. Hunter, The novel brain neuropeptide, orexin-A, modulates the sleep-wake cycle of rats, *Eur. J. Neurosci.* **12,** 726-730 (2000).
51. M. I. Smith, C. D. Piper, M. S. Duxon and N. Upton, Evidence implicating a role for orexin-1 receptor modulation of paradoxical sleep in the rat, *Neuroscience Lett.* **341**, 256-258 (2003).
52. P. Bourgin, S. Huitrón-Reséndiz, A. Spier, V. Fabre, B. Morte, J. Criado, J. G. Sutcliffe, S. Henriksen and L. de Lecea, Hypocretin-1 modulates REM sleep through activation of locus coeruleus neurons, *J. Neurosci.* **20**, 7760-5 (2000).
53. E. M. Soffin, M. L. Evans, C. H. Gill, M. H. Harries, C. D. Benham and C. H. Davies, SB-334867-A antagonises orexin mediated excitation in the locus coeruleus, *Neuropharmacology* **42**, 127-133 (2002).
54. D. N. C. Jones, J. Gartlon, F. Parker, S. G. Taylor, C. Routledge, P. Hemmati, R. P. Munton, T. E. Ashmeade, J. P. Hatcher, A. Johns, R. A. Porter, J. J. Hagan, A. J. Hunter, and N. Upton, Effects of centrally administered orexin-B and orexin-A: a role for orexin-1 receptors in orexin-B-induced hyperactivity, *Psychopharmacology* **153**, 210-218 (2001).
55. J. E. Gartlon, M. Duxon, R. Porter, J. P. Pilleux, J. J. Hagan, A. J. Hunter, N. Upton, and D. N. C. Jones, Role of OX1 and OX2 receptors in the motor activity response to the orexins, *Soc Neurosci Abst.* **27**, Program no.320.2 (2001).
56. T. Nakamura, K. Uramura, T. Nambu, T. Yada, K. Goto, M. Yanagisawa and T. Sakurai, Orexin-induced hyperlocomotion and stereotypy are mediated by the dopaminergic system, *Brain Res.* **873**, 181-87 (2000).
57. I. Matsuzaki, T. Sakurai, K. Kunii, T. Nakamura, M. Yanagisawa, and K. Goto, Involvement of the serotonergic system in orexin-induced behavioral alterations in rats, *Regul Pep.* **104**, 119-123 (2002).
58. M. M. Methippara, M. N. Alam, R. Szymusiak and D. McGinty, Effects of lateral preoptic area application of orexin-A on sleep-wakefulness, *Neuroreport* **11**, 3423-3426 (2000).
59. R. A. España, B. A. Baldo, A. E. Kelley, and C. W. Berridge, Wake-promoting and sleep-suppressing actions of hypocretin (orexin): Basal forebrain sites of action. *Neuroscience* **106,** 699-715 (2001).
60. Z. L. Huang, W. M. Qu, W. D. Li, T. Mochizuki, N. Eguchi, T. Watanabe, Y. Urade, and O. Hayaishi, Arousal effect of orexin A depends on activation of the histaminergic system, *Proc.Natl.Acad.Sci.U.S.A.* **98**, 9965-9970 (2001).
61. T. Yamamoto, N. Nozaki-Taguchi and T. Chiba, Analgesic effect of intrathecally administered –orexin-A in the rat formalin test and in the rat hot plate test, *Br J Pharmacol.* **137**, 170-176 (2002).
62. T. Shirasaka, M. Nakazato, S. Matsukura, M. Takasaki, and H. Kannan, Sympathetic and cardiovascular actions of orexins in conscious rats, *Am. J. Physiol* **277,** R1780-R1785 (1999).
63. C. T. Chen, L. L. Hwang, J. K. Chang and N. J. Dun, Pressor effects of orexins injected intracisternally and to rostral ventrolateral medualla of anesthetized rats, *Am J Physiol.* **278**, 692-697 (2000).
64. V. R. Antunes, G. C. Brailoiu, E. H. Kwok, P. Scruggs and N. J. Dun, Orexins/hypocretins excite rat sympathetic preganglionic neurons in vivo and in vitro, *Am J Physiol Regulatory Integrative Comp Physiol.* **281**, 1801-1807 (2001).
65. K. Hirota, T. Kushikata, M. Kudo, T. Kudo, D. Smart, and A. Matsuki, Effects of central hypocretin-1 administration on hemodynamic responses in young-adult and middle-aged rats, *Brain Research.* **981**, 143-150 (2003).
66. T. Kushikata, K. Hirota, H. Yoshida, M. Kudo, D. G. Lambert, D. Smart, J. C. Jerman and A. Matsuki, Orexinergic neurons and barbiturate anesthesia, *Neuroscience* **121**, 855-863 (2003).
67. M. Ehrström, E. Näslund, J. Ma, A. L. Kirchgessner and P. M. Hellström, Physiological regulation and NO-dependent inhibition of migrating myoelectric complex in the rat small bowel by hcrt1/OxA, *Am J Physiol Gastrointest Liver Physiol.* **285**, 688-695 (2003).

68. M. Mieda, J. T. Willie, J. Hara, C. M. Sinton, T. Sakurai and M. Yanagisawa, Orexin peptides prevent cataplexy and improve wakefulness in an orexin neuron-ablated model of narcolepsy in mice, *Proc Natl Acad Sci USA.* **101**, 4649-4654 (2004).
69. T. E. Scammell, I. V. Estabrooke, M. T. McCarthy, R. M. Chemelli, M. Yanagisawa, M. S. Miller and C. B. Saper, Hypothalamic arousal regions are activated during modafinil-induced wakefulness, *J Neurosci.* **20**, 8620-8628 (2000).
70. M. Hirose, S. Egashira, Y. Goto, T. Hashihayata, N. Ohtake, H. Iwaasa, M. Hata, T. Fukami, A. Kanatani and K. Yamada, N-acyl 6,7-dimethoxy-1,2,3,4-tetrahydroisoquinoline: the first orexin-2 receptor selective non-peptidic antagonist, *Bioorg Medl Chem Lett.* **13**, 4497-4499 (2003).
71. R. J. Rodgers, J. C. G. Halford, R. L. Nunes de Souza, A. L. Canto de Souza, D. C. Piper, J. R. S. Arch, N. Upton, R. A. Porter, A. Johns and J. E. Blundell, SB-334867, a selective orexin-1 receptor antagonist, enhances behavioural satiety and blocks the hyperphagic effect of orexin-A in rats, *Eur J Neurosci.* **13**, 1444-1452 (2001).

INTRACELLULAR SIGNAL PATHWAYS UTILIZED BY THE HYPOCRETIN/OREXIN RECEPTORS

Jyrki P. Kukkonen and Karl E. O. Åkerman*

1. INTRODUCTION

Since their discovery, the unique roles of the hypocretin/orexin system in the regulation of sleep, metabolism and feeding have been established. Meanwhile, the mechanisms by which hypocretins act at the cellular level are far from clear. Hypocretins mediate their effects through interaction with two G-protein-coupled receptors commonly called hcrtr-1/OX_1R and hcrtr-2/OX_2R. Hypocretins cause neuronal excitation in apparently all the areas of CNS where the receptors are expressed. In most situations this excitation is associated with a small and slowly developing depolarization by about 10 mV. The signal pathways leading to this depolarization and putative other mechanism of excitation need further clarification. It was initially suggested that hypocretins depolarize cells through inhibition of K^+ channels. This suggestion was based on a frequently observed increase in the input resistance. Other studies suggest that the depolarization is at least in part mediated by activation of nonselective cation channels (NSCC). Hcrtr are also expressed outside the nervous system, in particular in endocrine cells. In these cells, hcrtr elicit stimulation or inhibition of hormone synthesis and/or secretion. The available knowledge of particular signal pathways mediated by hcrtr in these systems is summarized in this chapter.

2. CELLULAR SIGNALING PATHWAYS

2.1. G-proteins

Hypocretin receptors belong to the superfamily of G-protein-coupled receptors and are therefore supposed to signal mainly via heterotrimeric G-proteins. There are four

* Jyrki P. Kukkonen, Uppsala University, Uppsala, Sweden. Karl E. O. Åkerman, Uppsala University, Uppsala, Sweden and University of Kuopio, Kuopio, Finland

different G-protein families, the G_q, $G_{i/o}$, G_s, and $G_{12/13}$, whose mechanisms of signaling and role in the physiology of hcrtr will be discussed in the following.

2.1.1. G-protein Pathways

The G_q family of G-proteins – G_q, G_{11}, G_{14} and $G_{15/16}$ – are thought to mainly signal via the Gα subunit to the phosphatidylinositol-specific phospholipase Cβ (PI-PLCβ), which then hydrolyzes phosphatidylinositol species (PI, PIP and PIP_2) to diacylglycerol and inositol phosphates (IP_1, IP_2 and IP_3). Therewith PI-PLCβ activation connects receptors to Ca^{2+} release (via IP_3 [phosphatidylinositol-1,4,5-trisphosphate]) and to e.g. protein kinase C (via diacylglycerol).

The $G_{i/o}$ family incorporates G_{i1}, G_{i2}, G_{i3}, G_{o1}, G_{o2}, G_z, $G_{t\text{-rod}}$, $G_{t\text{-cone}}$ and G_{gust}. The first 6 of these are ubiquitously expressed, and therefore discussed further. $G\alpha_{i1\text{-}3}$ and G_z inhibit – at least certain – adenylyl cyclase isoforms. Even G_o may be able to inhibit adenylyl cyclase in some cases.[1] Gβγ signaling is also usually considered to originate from $G_{i/o}$ proteins. Effectors for Gβγ include for instance PI-PLC, voltage-gated K^+ and Ca^{2+} channels, some adenylyl cyclase isoforms and PI3-kinase γ. When investigating the coupling to $G_{i/o}$ proteins one or the other of two criteria is often employed: i) the ability to inhibit cAMP generation and ii) the ability of the receptor signal to be inhibited by pertussis toxin, which rather selectively, at least among heterotrimeric G-proteins, inactivates $G_{i/o}$ family G-proteins, except for G_z.

The G_s family of G-proteins consists of G_{olf} and the more ubiquitous $G_{s\text{-long}}$ and $G_{s\text{-short}}$. G_s-proteins are thought to signal almost exclusively to stimulation of membrane bound adenylyl cyclase isoforms, all 9 of which respond to at least $G\alpha_s$. However, cAMP elevation is not a certain proof for activation of G_s, since many other factors, such as Ca^{2+}/calmodulin, protein kinase C (PKC) and Gβγ can stimulate certain adenylyl cyclase isoforms.

The $G_{12/13}$ family contains G_{12} and G_{13}. Various effectors have been suggested for these novel G-proteins, e.g. Na^+/H^+ exchanger and Lsc/p115 RhoGEF, but the effectors and the role in signaling seems to vary between cell types and studies.

Since several G-protein effectors are regulated by multiple inputs, a downstream response to G-protein-coupled receptor activation cannot easily be attributed to a particular G-protein. For instance, both different $G\alpha_q$ and Gβγ can directly activate PI-PLCβ. Other mechanisms for increased PI-PLC activity would include cAMP- and Epac-dependent activation of PI-PLCε, Ca^{2+}-dependent activation of PI-PLCδ or -ζ or receptor tyrosine kinase-transactivation-dependent activation of PI-PLCγ. The same applies to cAMP levels, which are regulated both positively and negatively by many downstream G-protein signals. Some methods, such as GTPγS- and GTP-azidoanilide-labeling combined with immunoprecipitation can be used to directly address receptor activation of specific G-proteins. However, these methods do not alone enlighten the role of these G-proteins in the signaling. There are further aspects complicating these issues. Firstly, there are putatively several important effectors of G-proteins, such as ion channels and exchangers, the mechanism of regulation of which by G-protein cascades is unknown. Secondly, G-protein-coupled receptors have been shown to interact and signal via proteins other than G-proteins (reviewed in Brady and Limbird;[2] Heuss and Gerber [3]).

2.1.2. Hypocretin Receptors' Coupling to G-proteins

Hypocretin receptors were originally – and also still very often are – thought to couple to G_q, since they strongly elevate cytosolic Ca^{2+} in many systems and also activate PI-PLC. However, the only studies directly addressing the G-protein coupling of hcrtr have been performed on human adrenal gland. In these studies, hcrtr-2 receptors have been seen to be capable of coupling to G_q, G_i and G_s, but not G_o using the GTP-azidoanilide labeling method, while the coupling to $G_{12/13}$ was not investigated.[4,5] Hypocretin1-stimulated *in situ* GTPγS binding in the rat brain stem[6,7] suggest the ability of hcrtr to couple to $G_{i/o}$ proteins, which are the G-proteins primarily seen with this technique. Indirect evidence of the coupling of the hcrtr to $G_{i/o}$ proteins has been obtained in some studies using pertussis toxin (see below).

2.2. Hypocretin Receptor Signaling in Neurons

Increased electrical activity of neurons was one of the two first responses observed for hypocretins.[8] This has since then been verified in most neuronal systems and more mechanistical evidence has been gathered. Hcrtr may elicit both presynaptic – e.g. increased Ca^{2+} influx – and postsynaptic effects – e.g. inhibition of K^+ channels, activation of non-selective cation channels or Na^+/Ca^{2+} exchanger. The available information concerning the signaling mechanisms mediating these effects – based on the use of ion substitution or blockers of particular signal pathways – is rather discrepant. The large variation in results obtained can in part be explained by the fact that different studies have been performed in different brain centra and by different experimental conditions such as the concentration of hcrtr. An increase in membrane resistance likely resulting from an inhibition of K^+ channels has been seen in several studies with hcrtr (Table 1). In the case of other G-protein-coupled receptors, GIRK (G-protein-regulated inward rectifier) channels have been shown to be inhibited through the PLC pathway either via activated PKC or reduced PIP_2 level see e.g. [9] Since hcrtr have been shown to activate PI-PLCβ pathway in many cells, this pathway could explain the inhibition of K^+ channels. In rat locus ceruleus neurons, GIRK channels are inhibited by hypocretin1, but the mechanism of this is unknown.[10] Upon heterologous expression of hcrtr-1 or -2 together with GIRK1 and -2 subunits in HEK-293 cells hcrtr stimulation leads to a transient stimulation and more delayed and long-lasting inhibition of the GIRK current.[10] The former is mostly sensitive to pertussis toxin whereas the latter is not. Activation of GIRK is thus thought to be mediated by Gβγ from $G_{i/o}$ G-proteins and the inhibition might relate to the above-mentioned PLC pathway, likely via $G\alpha_q$. In rat nucleus tractus solitarius, whole-cell K^+ currents are inhibited by hypocretins in a manner dependent on PKC.[11] However, this current does not correspond to the GIRK family since it is activated by depolarization. PKC inhibitors have also been able to inhibit other responses to hypocretins. This has in particular been the case in studies on hypocretin-activated Ca^{2+} elevation.[12-14] Interestingly, in all of these studies the Ca^{2+} elevation has been shown to be sensitive to blockers of voltage-gated Ca^{2+} channels (VGCC). One mechanism of action of hypocretins could thus be a PKC-mediated depolarization via reduction in K^+ conductance, leading to opening of voltage-gated Na^+ and Ca^{2+} channels.

Table 1. Summarized data on hypocretin responses in CNS neurons. The table represents a subset of studies; mainly studies where some inhibitors and ionic substitutions have shed light on the mechanisms have been included.

Area	Inhibitor	$-Ca^{2+}$	$-Na^+$	PK etc. inhibition	Response observed	Conc. (nM)	Ref.
SN-PR	thapsigargin			PKA	Firing ↑	100	15
VTA	DHP	+		D609	$[Ca^{2+}]_i$ ↑	100	12
	ωCTx			PKC			
TMN	Ni^{2+}	+	+	KB-R	Depolarization	30	16
CCx		+			NE release ↑	100	17
ArcN		0			Firing ↑	100	18
ArcN		+		KB-R			19
AP					NSC current ↑	0.1-10	11
DMNV			+	GDPβS	NSC current ↑		20
DMNV					I_K ↓	30-300	21
LDTN		0			Inward current ↑	300-1000	22
LDTN	DHP	+		PKC	$[Ca^{2+}]_i$ ↑	30-1000	23
DR	DHP	+		PKC	$[Ca^{2+}]_i$ ↑	30-1000	23
DR		0	+		Inward current	1-3000	24,25
LC					I_K ↓?	1000	26,27
HTh	Cd^{2+}	+		PKC	$[Ca^{2+}]_i$ ↑	1000	28
NTS				D609	Firing ↑		11
				PKC	NSC current ↑		
					I_K ↓		
S-HC	Ni^{2+}	0	+	KB-R	Firing ↑	10-30000	29,30
					NSC current ↑		

CNS loci are referred as: AP, area postrema; ArcN, arcuate nucleus; CCx, cerebral cortex; DMNV, dorsal motor nucleus of vagus; DR, dorsal Raphe nucleus; HTh, hypothalamus; LDTN, laterodorsal tegmental nucleus; LC, locus cereleus; NTS, nucleus tractus solitarius; S-HC, septum-hippocampus; SN-PR, substantia nigra pars reticulata; TMN, tuberomamillary nucleus; VTA, ventral tegmental area. For ion substitutions, + indicates the ability to block the response and 0 lack of effect. NE stands for norepinephrine and NSC for nonselective cation. Thapsigargin is a depleter of the endoplasmic reticulum Ca^{2+} store. D609 is a putative PC-PLC (phosphatidylcholine-specific phospholipase C) inhibitor. PK refers to protein kinases. PKC inhibitors used are bisindolylmaleimides and chelerytrin. K-BR refers to K-BR7943. DHP refers to dihydropyridines (an L-type VGCC blockers) and ωCTx to ω-conotoxin (an N-type VGCC blocker).

The effect of hypocretin in substantia nigra pars reticulate was interestingly shown to be sensitive to an inhibitor of protein kinase A (PKA), which mediates effects of cAMP, but insensitive to blockers of PKC.[15] So far this is the only study implicating a role of the G_s system, or other ways of elevating cAMP, in the hcrtr signaling in the CNS.

Some studies have failed to observe evidence for inhibition of K^+ channels in response to hypocretin challenge. Instead, an increase in inward cation currents has been observed suggesting that hypocretins activate nonselective cation channels (NSCC).[11,24,31] In rat nucleus tractus solitarius, NSCC activation by hypocretins depends on PKC.[11] In other cases the mechanisms are not known. A family of cation channels called TRP with several subfamilies (TRPC, TRPM, TRPP and TRPV) has been cloned; when the products of these genes are expressed heterologously, they produce nonselective cation channels.[32] TRP channels show fairly specific expression patterns in hypocretin-responding cells.[33] This, together with some biophysical properties of the channels, suggests that TRP channels may be likely targets for some hcrtr signaling in neurons.

The depolarization by hypocretins seen in some systems is at least partially sensitive to the inhibitor of the Na^+/Ca^{2+} exchanger, KB-R7943.[19,34] However, there is some reason to be careful with the interpretation of this data. Firstly, there is little information available on the mechanisms, which regulate these exchangers, aside from the electrochemical gradients of these ions. Secondly, at low concentrations (1-10 μM), KB-R7943 seems to be relatively specific inhibitor of the reverse function (Ca^{2+} influx–Na^+ efflux) of all the three isoforms, NCX1-3.[35,36] The reverse function rather hyperpolarizes than depolarizes the cells. In contrast, higher concentrations (30 μM) of KB-R7943 show a clear inhibition of the normal function (Na^+ influx–Ca^+ efflux) of the NCX1 isoform, but the inhibitory potency of K-BR7943 on this function of NCX2 and -3 are not known. In addition, KB-R7943 may also inhibit ion channels.[36]

An inhibitor of the putative phosphatidylcholine-specific PLC (PC-PLC), D609, has been shown to block Ca^{2+} and electrophysiological responses to hypocretin stimulation in some neuronal preparations.[11,12] There is very little information about the specificity of this blocker and the regulation and the role of this enzyme in cellular signaling, so the significance of these findings is also uncertain at present.

2.3. Hypocretin Receptor Signaling in Endocrine Systems

Hypocretin peptide and receptor distribution has been extensively mapped in endocrine organs. The receptors have been demonstrated using immunohistochemistry and mRNA detection in the adrenal gland (probably both cortex and medulla), testis, pineal gland, endocrine pancreas, pituitary gland and gastro-intestinal endocrine cells reviewed in [37] In some cases functional effects on secretion have been observed, but the cellular mechanisms are usually unclear. Also, some contradictions exist between different studies on receptor subtype expression and responses. In the rat testis, orexin receptors stimulate testosterone production.[38] The mechanism was not investigated but it could be related to IP_3 synthesis shown in another study on rat testis.[39] In the rat and human adrenal gland, cAMP elevation is able to stimulate glucocorticoid synthesis in a manner similar to adrenocorticotropic hormone,[40,41] which would thus be attractive to explain by an effect of G_s.[5] It would be interesting to know, whether hcrt also can supply similar trophic support to adrenal cortex as adrenocorticotropic hormone. In addition, PI-PLC activation, also observed in adrenal gland,[5,41] could theoretically stimulate aldosterone synthesis in *zona glomerulosa*.[42] In human pheochromocytomas, orexins also activate PI-PLC,[41] something that might increase synthesis and secretion of catecholamines.[42] In ovine somatotropes hypocretins are suggested to increase growth hormone secretion via PKC-dependent activation of L-type Ca^{2+} channels.[43] In STC-1 intestinal endocrine cell line, instead, L-type Ca^{2+} channels are activated secondary to depolarization, leading to secretion of cholecystokinin.[44]

2.4. Hypocretin Receptor Signaling in Heterologous Expression Systems

2.4.1. Ca^{2+} and PLC Signaling

Hcrtr have been stably expressed in CHO-K1 (Chinese hamster ovary), PC12 (rat pheochromocytoma), neuro-2a (mouse neuroblastoma) and BIM (neuroblastoma?) cells.[45-49] In all the cell lines, and via both receptors, hcrt stimulation causes prominent Ca^{2+} elevations, suggestive of activation of PI-PLC cascade, probably via G_q proteins. Direct measurements of PI-PLC activity in CHO cells expressing hcrtr-1, and neuro-2a and PC12 cells expressing both receptor subtypes verify the involvement IP_3 in Ca^{2+} release.[47,48] Interestingly, pertussis toxin-pretreatment has not been performed to our knowledge in any cell line except the BIM cells, where the Ca^{2+} elevation is insensitive to this treatment, but the strong Ca^{2+} elevation suggests that $G\alpha_q$-type proteins, rather than $G\beta\gamma$, would mediate the response. PI-PLC-dependent Ca^{2+} signaling links heterologous expression systems to – at least – endocrine cells but also probably to neurons (see above).

Ca^{2+} signaling in CHO cells expressing hcrtr-1 is not only dependent on PI-PLC activation and subsequent IP_3-mediated Ca^{2+} release. In both cell lines the Ca^{2+} responses mediated by both hcrtr require 100-fold higher hypocretin concentrations in the absence of extracellular Ca^{2+}. Further investigations show that the primary Ca^{2+} response in these cells is Ca^{2+} influx via an unknown channel, probably a non-selective cation channel, which also somehow amplifies the PLC signaling.[47,50] Similar coupling is suggested to occur in the hcrtr-2-expressing CHO cells and in neuro-2a and PC12 cells expressing either receptor subtype.[48,51] This suggests that hcrtr possess an intrinsic ability to couple to some Ca^{2+} channels, linking these results in heterologous expression systems to especially the neuronal signaling of hypocretins (see above). The mechanisms of the hcrtr–Ca^{2+} channel coupling have not been investigated in the recombinant expression systems so far.

2.4.2. Signaling to Adenylyl Cyclase

Hcrtr-2 have been shown to inhibit cAMP generation in recombinant BIM cells in a pertussis toxin-sensitive manner.[49] High efficacy coupling to G_i seems to be a general propensity of both hypocretin receptors, as similar effects are seen even in other cell lines (Kukkonen et al., unpublished). Probably the most interesting aspect of this coupling is its efficacy, which appears some orders of magnitude higher than the efficacy of coupling to Ca^{2+} elevation.[49] This suggests that the $G_{i/o}$ coupling persists even at very low receptor expression levels or at very low agonist levels where other couplings would not be able to exist. Pertussis toxin-sensitive G-proteins also seem to be involved in hcrt signaling in neurons,[49,52,53] though this issue has been addressed rather seldom, mainly because long preincubations required for the effect of pertussis toxin are unsuitable for slice preparations. Remarkably, in one of the studies[53] the potency of hypocretin1 appears very high, comparable to the inhibition of cAMP production in recombinant systems. It would be interesting to know whether the responses in the other studies reporting high hypocretin potency e.g. [12] would be sensitive to pertussis toxin.

cAMP elevations in response to hcrtr activation have not been reported in recombinant cells, in contrast to cAMP elevations in endocrine cells (see above). However, adenylyl cyclase is positively and negatively regulated by many intracellular signal molecules, most importantly by $G\alpha_{i/o}$, $G\alpha_s$, $G\beta\gamma$, PKC and Ca^{2+}/calmodulin, and it is only a matter of time before more cAMP effects are observed in recombinant and endogenous preparations. cAMP mediates its effects in the cell mainly via protein kinase A, but cAMP can also directly activate ion channels (CNG and HCN channels) and the Rap GDP/GTP exchange factor (GEF) Epac. Hcrtr could thus be speculated to affect for instance smooth muscle contraction, cell growth and metabolism and hormone release via regulation of cAMP levels.

2.4.3. Cell Plasticity

G-protein-coupled receptors have for long been known to regulate the growth and maintenance of the endocrine organs adrenal cortex, thyroid and the gonads. During the last decade, the mechanisms of this have started to gradually unravel. It has been clearly shown that the regulation of plastic events is by no means restricted to these endocrine organs, but can be seen in essentially all cells expressing G-protein-coupled receptors. The pathways utilized by G-protein-coupled receptors are mostly similar to the pathways used by receptor tyrosine kinases, e.g. PKC, small G-proteins (e.g. Ras), PI3-kinase, Src etc.[54,55] In some cases G-protein-coupled receptors "hijack" receptor tyrosine kinases (transactivation) and in other cases they target these effectors independently of receptor tyrosine kinases. Mainly G-proteins but also other interaction partners of the receptors are engaged in this signaling, but how these signals then engage e.g. Src is not always clear.

For hypocretin receptors, these aspects of signaling have barely been addressed. However, there are reports of the ability of hcrtr-1 to strongly activate the p42/44 MAP-kinase (ERK) pathway in recombinant CHO cells.[37,56] It is difficult to speculate on the connection of these responses to the physiological role of hypocretins. What concerns the brain, the anatomy of the hcrt-ergic neurites suggests that hypocretins are in part released in a paracrine fashion, leading to long-term effects – e.g. modulation of plasticity – of hypocretins. In a recent study on hippocampal slices, hypocretin pretreatment causes a long-lasting potentiation of the electrical activity in the CA1 region,[57] suggesting that hypocretins induce plastic changes. It is also tempting to speculate that hypocretins could play a similar role in development and maintenance of the adrenal cortex as the adrenocorticotropic hormone (see above).

3. CONCENTRATION-RESPONSE RELATIONSHIPS

Very variable concentrations of hypocretin – ranging from 1 pM to 1 μM – have been applied on cell preparates in different studies. Only in a few studies the concentration-response relationships have been measured. It is not known, in which amounts hypocretins are released in the CNS, but it appears likely that micromolar concentrations might be attained in the synaptic cleft and immediately outside of it, whereas upon the much more long-distance diffusion of hypocretin from the putative non-synaptic release sites, the concentrations would dramatically fall (it should be remembered that the concentration falls 1000-fold upon 10-fold increase in the distance in a spherical space). In the light of the very preliminary data on the concentration-dependence of different hcrtr responses, it appears likely that different signal pathways would be utilized by the synaptic and paracrine hypocretin.

4. RECEPTOR SUBTYPE DIFFERENCES IN SIGNALING?

It is tempting to speculate that hcrtr-1 and -2 would display not only distinct pharmacology with respect to hypocretin-2 but also distinct signaling properties. Despite their high homology in the transmebrane segments, hcrtr-1 and -2 display clear differences in the intracellular parts, suggesting that there would be a prerequisite for distinct signaling. With the current tools, or rather the lack of some central ones, we consider that the only systems, where comparisons can be performed with high enough reliability are recombinant expression systems. Even in these systems one has to make sure that the clonal selection does not distort the results. The only report published so far on differences in signaling suggests that hcrtr-2 but not hcrtr-1 receptors are able to couple to G_i proteins in recombinant BIM cells.[49] However, this is not in agreement with our results (Kukkonen et al., unpublished) with other cell lines, suggesting that either the expression profile of different G_i subtypes in BIM cells allows distinction not possible in other systems, or that clonal selection has separated the genetic background of BIM-hcrtr-1 and -2 cell lines. At the moment it is thus unclear whether there are any differences in the signaling of the two hypocretin receptor subtypes. However, very few studies have addressed this issue, and we thus consider the question to be unresolved.

5. FUTURE PERSPECTIVES

The information available on signaling of hypocretins at the cellular level is at present very limited, and clearly much effort should be concentrated here. However, very interesting and prominent signals have already been identified in native cells at the same time as hcrtr have been shown to couple to members of most, if not all, G-protein families. The future work also consists of the investigation of the link between the activation of different G-proteins and possible other effectors and the downstream signals. A third important task is to bridge together the signaling at cellular, organ and organism level.

6. REFERENCES

1. J. Näsman, J. P. Kukkonen, T. Holmqvist and K. E. Åkerman, Different roles for Gi and Go proteins in modulation of adenylyl cyclase type-2 activity, *J Neurochem.* **83**, 1252-61 (2002).
2. A. E. Brady and L. E. Limbird, G protein-coupled receptor interacting proteins: emerging roles in localization and signal transduction, *Cell Signal.* **14**, 297-309 (2002).
3. C. Heuss and U. Gerber, G-protein-independent signaling by G-protein-coupled receptors, *Trends Neurosci.* **23**, 469-75 (2000).
4. E. Karteris, H. S. Randeva, D. K. Grammatopoulos, R. B. Jaffe and E. W. Hillhouse, Expression and coupling characteristics of the CRH and orexin type 2 receptors in human fetal adrenals, *J Clin Endocrinol Metab.* **86**, 4512-9 (2001).

5. H. S. Randeva, E. Karteris, D. Grammatopoulos and E. W. Hillhouse, Expression of orexin-A and functional orexin type 2 receptors in the human adult adrenals: implications for adrenal function and energy homeostasis, *J Clin Endocrinol Metab.* **86**, 4808-13 (2001).
6. R. Bernard, R. Lydic and H. A. Baghdoyan, Hypocretin-1 activates G proteins in arousal-related brainstem nuclei of rat, *Neuroreport.* **13**, 447-50 (2002).
7. R. Bernard, R. Lydic and H. A. Baghdoyan, Hypocretin-1 causes G protein activation and increases ACh release in rat pons, *Eur J Neurosci.* **18**, 1775-85 (2003).
8. L. de Lecea, T. S. Kilduff, C. Peyron, X. Gao, P. E. Foye, P. E. Danielson, C. Fukuhara, E. L. Battenberg, V. T. Gautvik, F. S. Bartlett, 2nd, W. N. Frankel, A. N. van den Pol, F. E. Bloom, K. M. Gautvik and J. G. Sutcliffe, The hypocretins: hypothalamus-specific peptides with neuroexcitatory activity, *Proc Natl Acad Sci U S A.* **95**, 322-7 (1998).
9. B. M. Ances, J. H. Greenberg and J. A. Detre, Effects of variations in interstimulus interval on activation-flow coupling response and somatosensory evoked potentials with forepaw stimulation in the rat, *J Cereb Blood Flow Metab.* **20**, 290-7 (2000).
10. Q. V. Hoang, D. Bajic, M. Yanagisawa, S. Nakajima and Y. Nakajima, Effects of orexin (hypocretin) on GIRK channels, *J Neurophysiol.* **90**, 693-702 (2003).
11. B. Yang and A. V. Ferguson, Orexin-A depolarizes nucleus tractus solitarius neurons through effects on nonselective cationic and K+ conductances, *J Neurophysiol.* **89**, 2167-75 (2003).
12. K. Uramura, H. Funahashi, S. Muroya, S. Shioda, M. Takigawa and T. Yada, Orexin-a activates phospholipase C- and protein kinase C-mediated Ca2+ signaling in dopamine neurons of the ventral tegmental area, *Neuroreport.* **12**, 1885-9. (2001).
13. A. N. Van Den Pol, P. R. Patrylo, P. K. Ghosh and X. B. Gao, Lateral hypothalamus: Early developmental expression and response to hypocretin (orexin), *J Comp Neurol.* **433**, 349-363. (2001).
14. K. A. Kohlmeier, T. Inoue and C. S. Leonard, Hypocretin/orexin peptide signaling in the ascending arousal system: elevation of intracellular calcium in the mouse dorsal raphe and laterodorsal tegmentum, *J Neurophysiol.* **92**, 221-35 (2004).
15. T. M. Korotkova, K. S. Eriksson, H. L. Haas and R. E. Brown, Selective excitation of GABAergic neurons in the substantia nigra of the rat by orexin/hypocretin in vitro, *Regul Pept.* **104**, 83-89 (2002).
16. R. E. Brown, O. Sergeeva, K. S. Eriksson and H. L. Haas, Orexin A excites serotonergic neurons in the dorsal raphe nucleus of the rat, *Neuropharmacology.* **40**, 457-9. (2001).
17. K. Hirota, T. Kushikata, M. Kudo, T. Kudo, D. G. Lambert and A. Matsuki, Orexin A and B evoke noradrenaline release from rat cerebrocortical slices, *Br J Pharmacol.* **134**, 1461-6 (2001).
18. M. Rauch, T. Riediger, H. A. Schmid and E. Simon, Orexin A activates leptin-responsive neurons in the arcuate nucleus, *Pflugers Arch.* **440**, 699-703. (2000).
19. D. Burdakov, B. Liss and F. M. Ashcroft, Orexin excites GABAergic neurons of the arcuate nucleus by activating the sodium--calcium exchanger, *J Neurosci.* **23**, 4951-7 (2003).
20. J. J. Hwang, V. J. Dzau and C. C. Liew, Genomics and the pathophysiology of heart failure, *Curr Cardiol Rep.* **3**, 198-207 (2001).
21. G. Grabauskas and H. C. Moises, Gastrointestinal-projecting neurones in the dorsal motor nucleus of the vagus exhibit direct and viscerotopically organized sensitivity to orexin, *J Physiol.* **549**, 37-56 (2003).
22. S. Burlet, C. J. Tyler and C. S. Leonard, Direct and indirect excitation of laterodorsal tegmental neurons by Hypocretin/Orexin peptides: implications for wakefulness and narcolepsy, *J Neurosci.* **22**, 2862-72 (2002).
23. K. A. Kohlmeier, T. Inoue and C. S. Leonard, Hypocretin/Orexin peptide signalling in the ascending arousal system: Elevation of intracellular calcium in the mouse dorsal raphe and laterodorsal tegmentum, *J Neurophysiol.* (2004).
24. R. E. Brown, O. A. Sergeeva, K. S. Eriksson and H. L. Haas, Convergent excitation of dorsal raphe serotonin neurons by multiple arousal systems (orexin/hypocretin, histamine and noradrenaline), *J Neurosci.* **22**, 8850-9 (2002).
25. R. J. Liu, A. N. van den Pol and G. K. Aghajanian, Hypocretins (orexins) regulate serotonin neurons in the dorsal raphe nucleus by excitatory direct and inhibitory indirect actions, *J Neurosci.* **22**, 9453-64 (2002).
26. T. L. Horvath, C. Peyron, S. Diano, A. Ivanov, G. Aston-Jones, T. S. Kilduff and A. N. van Den Pol, Hypocretin (orexin) activation and synaptic innervation of the locus coeruleus noradrenergic system, *J Comp Neurol.* **415**, 145-59 (1999).
27. A. Ivanov and G. Aston-Jones, Hypocretin/orexin depolarizes and decreases potassium conductance in locus coeruleus neurons, *Neuroreport.* **11**, 1755-8. (2000).
28. A. N. van den Pol, X. B. Gao, K. Obrietan, T. S. Kilduff and A. B. Belousov, Presynaptic and postsynaptic actions and modulation of neuroendocrine neurons by a new hypothalamic peptide, hypocretin/orexin, *J Neurosci.* **18**, 7962-71 (1998).

29. L. I. Kiyashchenko, B. Y. Mileykovskiy, N. Maidment, H. A. Lam, M. F. Wu, J. John, J. Peever and J. M. Siegel, Release of hypocretin (orexin) during waking and sleep states, *J Neurosci.* **22**, 5282-6 (2002).
30. M. Wu, L. Zaborszky, T. Hajszan, A. N. van den Pol and M. Alreja, Hypocretin/orexin innervation and excitation of identified septohippocampal cholinergic neurons, *J Neurosci.* **24**, 3527-36 (2004).
31. L. L. Hwang, C. T. Chen and N. J. Dun, Mechanisms of orexin-induced depolarizations in rat dorsal motor nucleus of vagus neurones in vitro, *J Physiol.* **537**, 511-20 (2001).
32. D. E. Clapham, TRP channels as cellular sensors, *Nature.* **426**, 517-24 (2003).
33. O. A. Sergeeva, T. M. Korotkova, A. Scherer, R. E. Brown and H. L. Haas, Co-expression of non-selective cation channels of the transient receptor potential canonical family in central aminergic neurones, *J Neurochem.* **85**, 1547-52 (2003).
34. K. S. Eriksson, O. Sergeeva, R. E. Brown and H. L. Haas, Orexin/hypocretin excites the histaminergic neurons of the tuberomammillary nucleus, *J Neurosci.* **21**, 9273-9 (2001).
35. B. Linck, Z. Qiu, Z. He, Q. Tong, D. W. Hilgemann and K. D. Philipson, Functional comparison of the three isoforms of the Na+/Ca2+ exchanger (NCX1, NCX2, NCX3), *Am J Physiol.* **274**, C415-23 (1998).
36. T. Iwamoto and S. Kita, Development and application of Na+/Ca2+ exchange inhibitors, *Mol Cell Biochem.* **259**, 157-61 (2004).
37. J. P. Kukkonen, T. Holmqvist, S. Ammoun and K. E. Akerman, Functions of the orexinergic/ hypocretinergic system, *Am J Physiol Cell Physiol.* **283**, C1567-91 (2002).
38. M. L. Barreiro, R. Pineda, V. M. Navarro, M. Lopez, J. S. Suominen, L. Pinilla, R. Senaris, J. Toppari, E. Aguilar, C. Dieguez and M. Tena-Sempere, Orexin 1 receptor messenger ribonucleic acid expression and stimulation of testosterone secretion by orexin-A in rat testis, *Endocrinology.* **145**, 2297-306 (2004).
39. E. Karteris, J. Chen and H. S. Randeva, Expression of human prepro-orexin and signaling characteristics of orexin receptors in the male reproductive system, *J Clin Endocrinol Metab.* **89**, 1957-62 (2004).
40. L. K. Malendowicz, C. Tortorella and G. G. Nussdorfer, Orexins stimulate corticosterone secretion of rat adrenocortical cells, through the activation of the adenylate cyclase-dependent signaling cascade, *J Steroid Biochem Mol Biol.* **70**, 185-8 (1999).
41. G. Mazzocchi, L. K. Malendowicz, L. Gottardo, F. Aragona and G. G. Nussdorfer, Orexin A stimulates cortisol secretion from human adrenocortical cells through activation of the adenylate cyclase-dependent signaling cascade, *J Clin Endocrinol Metab.* **86**, 778-82. (2001).
42. T. Nanmoku, K. Isobe, T. Sakurai, A. Yamanaka, K. Takekoshi, Y. Kawakami, K. Goto and T. Nakai, Effects of orexin on cultured porcine adrenal medullary and cortex cells, *Regul Pept.* **104**, 125-130 (2002).
43. R. Xu, Q. Wang, M. Yan, M. Hernandez, C. Gong, W. C. Boon, Y. Murata, Y. Ueta and C. Chen, Orexin-A augments voltage-gated Ca2+ currents and synergistically increases growth hormone (GH) secretion with GH-releasing hormone in primary cultured ovine somatotropes, *Endocrinology.* **143**, 4609-19 (2002).
44. K. P. Larsson, K. E. Akerman, J. Magga, S. Uotila, J. P. Kukkonen, J. Nasman and K. H. Herzig, The STC-1 cells express functional orexin-A receptors coupled to CCK release, *Biochem Biophys Res Commun.* **309**, 209-16 (2003).
45. D. Smart, J. C. Jerman, S. J. Brough, W. A. Neville, F. Jewitt and R. A. Porter, The hypocretins are weak agonists at recombinant human orexin-1 and orexin-2 receptors, *Br J Pharmacol.* **129**, 1289-91 (2000).
46. T. Sakurai, A. Amemiya, M. Ishii, I. Matsuzaki, R. M. Chemelli, H. Tanaka, S. C. Williams, J. A. Richardson, G. P. Kozlowski, S. Wilson, J. R. Arch, R. E. Buckingham, A. C. Haynes, S. A. Carr, R. S. Annan, D. E. McNulty, W. S. Liu, J. A. Terrett, N. A. Elshourbagy, D. J. Bergsma and M. Yanagisawa, Orexins and orexin receptors: a family of hypothalamic neuropeptides and G protein-coupled receptors that regulate feeding behavior, *Cell.* **92**, 573-85 (1998).
47. P. E. Lund, R. Shariatmadari, A. Uustare, M. Detheux, M. Parmentier, J. P. Kukkonen and K. E. Akerman, The Orexin OX1 Receptor Activates a Novel Ca2+ Influx Pathway Necessary for Coupling to Phospholipase C, *J Biol Chem.* **275**, 30806-30812 (2000).
48. T. Holmqvist, K. E. Akerman and J. P. Kukkonen, Orexin signaling in recombinant neuron-like cells, *FEBS Lett.* **526**, 11-4 (2002).
49. Y. Zhu, Y. Miwa, A. Yamanaka, T. Yada, M. Shibahara, Y. Abe, T. Sakurai and K. Goto, Orexin receptor type-1 couples exclusively to pertussis toxin-insensitive G-proteins, while orexin receptor type-2 couples to both pertussis toxin-sensitive and -insensitive G-proteins, *J Pharmacol Sci.* **92**, 259-66 (2003).
50. J. P. Kukkonen and K. E. Akerman, Orexin receptors couple to Ca2+ channels different from store-operated Ca2+ channels, *Neuroreport.* **12**, 2017-20 (2001).
51. S. Ammoun, T. Holmqvist, R. Shariatmadari, H. B. Oonk, M. Detheux, M. Parmentier, K. E. Akerman and J. P. Kukkonen, Distinct recognition of OX1 and OX2 receptors by orexin peptides, *J Pharmacol Exp Ther.* **305**, 507-14 (2003).
52. Q. V. Hoang, P. Zhao, S. Nakajima and Y. Nakajima, Orexin (Hypocretin) Effects on Constitutively Active Inward Rectifier K+ Channels in Cultured Nucleus Basalis Neurons, *J Neurophysiol.* (2004).

53. S. Muroya, H. Funahashi, A. Yamanaka, D. Kohno, K. Uramura, T. Nambu, M. Shibahara, M. Kuramochi, M. Takigawa, M. Yanagisawa, T. Sakurai, S. Shioda and T. Yada, Orexins (hypocretins) directly interact with neuropeptide Y, POMC and glucose-responsive neurons to regulate Ca 2+ signaling in a reciprocal manner to leptin: orexigenic neuronal pathways in the mediobasal hypothalamus, *Eur J Neurosci.* **19**, 1524-34 (2004).
54. J. S. Gutkind, Regulation of mitogen-activated protein kinase signaling networks by G protein-coupled receptors, *Sci STKE.* **2000**, RE1 (2000).
55. T. Gudermann, R. Grosse and G. Schultz, Contribution of receptor/G protein signaling to cell growth and transformation, *Naunyn Schmiedebergs Arch Pharmacol.* **361**, 345-62 (2000).
56. S. Hilairet, M. Bouaboula, D. Carriere, G. Le Fur and P. Casellas, Hypersensitization of the Orexin 1 receptor by the CB1 receptor: evidence for cross-talk blocked by the specific CB1 antagonist, SR141716, *J Biol Chem.* **278**, 23731-7 (2003).
57. O. Selbach, N. Doreulee, C. Bohla, K. S. Eriksson, O. A. Sergeeva, W. Poelchen, R. E. Brown and H. L. Haas, Orexins/hypocretins cause sharp wave- and theta-related synaptic plasticity in the hippocampus via glutamatergic, gabaergic, noradrenergic, and cholinergic signaling, *Neuroscience.* **127**, 519-28 (2004).

THE HYPOCRETINS IN NARCOLEPSY AND AROUSAL

THE HYPOCRETINS AND NARCOLEPSY

Pathophysiology and Diagnosis

Wynne Chen, Jamie M. Zeitzer, Emmanuel Mignot *

1. INTRODUCTION

C.C. Westphal (1877) was the first to report a convincing case of a patient with sleepiness and emotionally induced episodes of muscle weakness — a condition Gélineau (1880) later called "narcolepsy", a name derived from the Greek words for "stupor" or "numbness" (narke) and "to take hold of" (lambanein). In 1960, it was recognized that narcolepsy was associated with the premature onset of REM sleep.[1] Now recognized to be a key biological marker for the disorder, such sleep onset REM periods (soREMPs) underlie the utility of the multiple sleep latency test (MSLT), a polysomnographic test used to confirm the clinical diagnosis of narcolepsy.[2] Most commonly idiopathic, in rare instances, the symptoms of narcolepsy may result from pathologic conditions related to the central nervous system, almost always involving the hypothalamus.

The cause of narcolepsy has been a mystery for over one hundred years (Table 1). In 1998, the hypothalamic neuropeptide hypocretin (Hcrt) (also called orexin) was simultaneously discovered by two, independent laboratories.[3,4] Soon thereafter, it was discovered that mutations in one of the hypocretin receptors (HcrtR) was the underlying cause of narcolepsy in a canine model[5] and targeted disruption of the hcrt gene caused a narcolepsy-like phenotype in mice.[6] Human narcolepsy has since been shown to be commonly associated with a loss of both hypocretin peptide in cerebrospinal fluid (CSF) and a loss of the hypothalamic neurons that produce hypocretin.[7-9]

Two neurophysiologic abnormalities underlie the symptoms of narcolepsy, an inability to maintain wakefulness (excessive daytime sleepiness, EDS) and intrusions of predominantly REM sleep phenomena into wakefulness. The latter manifests clinically as cataplexy (muscle weakness triggered by emotions), hypnagogic hallucinations (dream-like experiences occurring at sleep onset), and sleep paralysis (the inability to move while

* W. Chen, J. M. Ziotzer and E.Mignot. Stanford University Center for Narcolepsy 701-B Welch Road, Room 146, MC: 5742 Palo Alto, CA 94304 U.S.A

falling asleep or upon awakening). Furthermore, severe sleep disruption usually develops over time during the course of the disorder, and is often considered a fifth symptom of the condition (narcolepsy pentad).[10] Narcolepsy is not typically associated with excessive amounts of sleep in a given 24-hour period but, rather, an inability to consolidate sleep and vigilance. It can be best considered as a dysregulation of vigilance state boundary control due to an Hcrt deficiency.[11]

Table 1. Milestones in narcolepsy research and therapy

1877	First description by Westphal in the medical literature
1880	Gélineau called the disorder "narcolepsy"
1902	Loëwenfeld coined the term "cataplexy"
1917	Von Economo recognizes the importance of the posterior hypothalamus and mesencephalon in the mediation of daytime sleepiness in encephalitis lethargica
1935	First use of amphetamines in the treatment of narcolepsy
1960	Description of sleep onset REM periods in narcoleptic subjects
1970	Description of the Multiple Sleep Latency Test (MSLT)
1973	First report of a narcoleptic dog [12]
1983	Association of narcolepsy with HLA-DR2 in Japanese patients [13]
1985	Monoaminergic and cholinergic imbalance in canine narcolepsy [14]
1992	Association of human narcolepsy with HLA-DQB1*0602 [15]
1993	Adrenergic reuptake inhibition mediates the effect of anticataplectic antidepressants [16]
1998	Dopaminergic mediation of stimulant effect identified [17]
1998	Identification of hypocretins/orexins and their receptors by two groups [3,4]
1999	Hypocretin receptor-2 (Hcrtr2) mutations found to cause narcolepsy in dogs [5]
1999	Hypocretin knockout mice have narcolepsy [6]
2000	A single case of a mutation in hypocretin gene causing human narcolepsy is reported [8]
2000	HLA- associated, sporadic human narcolepsy cases are also found to be associated with hypocretin deficiency [7-9]
2001	Complex HLA-DQ effects, not only HLA-DQB1*0602, modulate narcolepsy susceptibility[18]
2001	A transgenic hypocretin/ataxin-3 mouse model mimics hypocretin neuronal degeneration, which is thought to be the basis of human narcolepsy [19]

The discovery that narcolepsy is strongly associated with a specific human leukocyte antigen (HLA) subtype, HLA-DQB1*0602[20] has lead to the hypothesis that there may be an immunologic component to the development of this disorder. Indeed, the few narcoleptic postmortem brains that have been examined reveal absent or profoundly decreased numbers of hypocretin-producing cells,[21,22] though it is currently unknown if this destruction is due to an auto-immune-mediated process.

2. CLINICAL ASPECTS OF NARCOLEPSY

Narcolepsy remains an under- and misdiagnosed condition, in part due to the misperception that it is an extremely rare disorder. However, the prevalence of narcolepsy-cataplexy approximates that of Multiple Sclerosis (MS) and Parkinson's Disease (PD)[23] and may be as high as 20-60 per 100,000 people in the Western countries.[24] Furthermore, as symptoms often emerge during adolescence, the development of social skills and self-esteem, as well as academic achievement may be adversely affected. Moreover, individuals may be reluctant to disclose symptoms fully, given their often bizarre nature and, as such, life-long morbidity may prove significant. The population prevalence of narcolepsy without cataplexy is unknown.

Severe daytime sleepiness, often combined with nocturnal sleep fragmentation, leads to a feeling of constant tiredness in narcoleptics. This chronic sleep pressure coupled with poor sleep/wake state boundary control may underlie the commonly observed sudden, uncontrollable bouts of sleepiness or "sleep attacks" during the day. Cataplexy (Gk., "to strike down"), another cardinal symptom of narcolepsy, was first used by Loëwenfeld (1902) to describe the sudden, bilateral loss of muscle tone *without* loss of consciousness. Triggered by any situation that requires sudden action or strong emotion (laughing and mirth are the most commonly reported triggers), these attacks may be partial (isolated muscle groups) or complete (causing complete collapse), but rarely involve all muscles simultaneously. The knees (buckling) and head/neck muscles (head bobbing, jaw sagging and slurred speech) are most commonly affected, with such episodes lasting for seconds to (rarely) minutes. Muscle weakness escalates progressively over several seconds and significant bodily injury typically does not occur. Cataplexy is pathognomonic for narcolepsy. Another common symptom of narcolepsy is hypnagogic hallucinations, which are dream-like episodes that occur as one is falling asleep and may be particularly terrifying when associated with sleep paralysis. Both hypnagogic hallucinations and sleep paralysis, however, are not specific to narcolepsy, and are commonly reported in the general population, especially in association with severe, chronic sleep deprivation. Other symptoms of narcolepsy may include automatic behavior and sleep-talking, which also reflect abnormal vigilance state boundary control. In addition, other sleep disorders such as periodic limb movement disorder (PLMD), REM-Behavior Disorder (RBD), the sleep-related breathing disorders (SRBD) and rarely, sleep-walking, are seen in the context of narcolepsy.[25,26]

Until recently, the diagnosis of narcolepsy has been primarily based on clinical symptomatology, with cataplexy being the most specific symptom and diagnostic predictor for the disorder.[27] Cataplexy is, however, sometimes difficult to distinguish clinically from normal experiences; these may include feelings of weakness when laughing hysterically, in the context of exciting athletic activities, during sex or when extremely angry. Atonic seizures or other forms of hypotonia may also be difficult to differentiate from cataplexy.[26] The MSLT is therefore commonly used to objectively quantify sleepiness and the occurrence of soREMPs, which are both consistent with the diagnosis of narcolepsy. In the MSLT, narcoleptics typically exhibit a short mean sleep latency (≤ 8 minutes) and more than two transitions into REM sleep (or soREMPs). The MSLT is typically performed after a night in the sleep laboratory, where nocturnal sleep is studied to exclude other causes of EDS or nocturnal sleep disruption, including sleep-related breathing disorders (SRBDs). Subjects undergoing an MSLT also must be free from psychotropic drug use, such as antidepressants and stimulants, as these will confound interpretation of the MSLT.

In most international classifications, narcolepsy is defined by the presence of sleepiness and cataplexy or by the polysomnographic documentation of REM sleep abnormalities. The diagnosis of narcolepsy has been complicated by the realization that not all people described as having narcolepsy manifest all of the aforementioned classical symptoms. Additionally, the association with HLA and hypocretin deficiency is very tight only in idiopathic ("primary") narcolepsy with typical cataplexy. Other cases, such as those associated with atypical cataplexy or without any cataplexy, constitute a grey area, which has been called the "narcolepsy clinical borderland".[28]

Table 2 Proposed Icd2 (International Statistical Classification Of Diseases And Related Health Problems) For Hypersomnia Not Due To Sleep Related Breathing Disorders

Diagnosis	Diagnostic Criteria	Pathophysiology and Clinical/Pathological Subtypes
Narcolepsy with Typical Cataplexy	► EDS occurring almost daily for at least 3 months. ► Definite history of cataplexy ► No other medical or mental disorder which accounts for symptoms. ► Confirmation by: 1. Nocturnal PSG followed by MSLT (SL ≤8 min, ≥2 soREMPs) 2. CSF Hcrt-1 level (≤110 pg/mL)	*= 95% with Hcrt deficiency and DQB1*0602* 1. Hypocretin gene mutations: Only one case of a pre- pro-hypocretin mutation causing narcolepsy has been reported, in an individual who was HLA negative and in whom symptoms began unusually early (atypical case). 2. Narcolepsy-cataplexy with normal CSF hcrt levels: up to 10% of cases with typical cataplexy in reported cases 3. Multiplex/familial subtypes 4. HLA DQB1*0602 negative cases: These individuals generally have normal CSF Hcrt-1 levels and may occur more frequently in multiplex families 5. Late onset narcolepsy-cataplexy: It is rare but possible for symptoms of narcolepsy to appear after 40 years of age 6. Isolated cataplexy: Very rare cases of familial isolated cataplexy with very early onset have been reported and may represent a distinct clinical entity. Isolated cata plexy is also occasionally observed in children at time of disease onset; sleepiness usually develops within one year of the onset of cataplexy.
Narcolepsy without Typical Cataplexy	► EDS occurring almost daily for at least 3 months **► No or doubtful cataplexy** ► No other medical or mental disorder which accounts for symptoms ► MSLT: SL ≤ 8 min, ≥ 2 soREMPs	*= Unknown, heterogeneous 16% Hcrt-1 deficiency, 40% HLA-DQB1*0602*
Narcolepsy associated with a known Physiologic Condition	► EDS occurring almost daily for at least 3 months **► History of definite cataplexy, atypical cataplexy, or no cataplexy Concurrent medical/neurological disorder** ► MSLT : SL ≤ 8 min, ≥ 2 soREMPs	*= With or without Hcrt-1 deficiency; caused by various disorders*
Idiopathic Hypersomnia	► EDS occurring almost daily for at least 3 months **► No cataplexy** **► MSLT: No soREMPs**	*= Unknown, likely heterogeneous etiology*

The symptoms of narcolepsy may also bee seen in association with, or caused by other medical and neurologic conditions, which has been referred to as "secondary" or "symptomatic" narcolepsy. In his classic studies of the encephalitis lethargica pandemic (1917-1923), Von Economo (1931) described individuals with hypothalamic and upper brainstem lesions in whom sleepiness was a primary symptom.

He went on to hypothesize that the posterior hypothalamus was critical to the promotion of wakefulness, a premise consistent with various reports of hypersomnia secondary to structural lesions in that region.[25] These recent advances have led to a reclassification of narcolepsy into more distinct subcategories as described in Table 2.

3. GENETIC ASPECTS OF NARCOLEPSY

A genetic component to narcolepsy was suggested by Westphal's (1877) initial report, in which both the patient and his mother were afflicted with the condition. Recent studies have, though, indicated that narcolepsy-cataplexy is only rarely a familial disorder. Only 1 to 2% of first-degree narcolepsy relatives ever develop the disorder;[29] it is interesting to note however, that this 1 to 2% increase in risk represents a 20 to 40-fold increased risk over the prevalence in the general population (0.02-0.18%), suggesting genetic effects or shared environmental factors, as this increase in risk is too low for simple Mendelian inheritance. However, single families without HLA-DQB1*0602 and a high familial transmission have been also been reported, suggesting genetic heterogeneity.[29] Therefore, narcolepsy must be considered a complex, multigenic disorder that is in part, environmentally determined.

Only 25 to 31% of monozygotic twins are concordant for narcolepsy.[29] In one case, low and high CSF hypocretin-1 levels were reported in monozygotic twins with and without narcolepsy, respectively.[30] Symptomatic concordance in monozygotic twins has been reported in the face of HLA-DQB1*0602 positivity with normal hypocretin levels[31] or in the context of HLA negativity.[29] This suggests that some non-HLA, non-hypocretin-mediated cases of narcolepsy may have a yet undetermined genetic basis. As discordant twins can develop narcolepsy later in life in association with environmental triggers such as emotional stress or chronic sleep deprivation,[32] additional, non-genetic factors are likely to be associated with the onset of the disease.

A familial tendency for narcolepsy has also been suggested by animal studies. Narcolepsy was first reported in canines in 1973[12] and, like humans, most canine cases are sporadic. However, in 1976, a genetic form of canine narcolepsy was discovered in Dobermans and Labradors,[33] from which a colony of narcoleptic canines was established. Extensive neurophysiologic and neuro-pharmacological studies were conducted using this canine model, demonstrating clinical similarities with the human disorder;[14] affected dogs display emotionally-triggered cataplexy often elicited by food or by play, fragmented sleep, and a short sleep latency. In Doberman Pinschers, Labrador Retrievers and in a family of Dachshunds, narcolepsy is transmitted as a single autosomal recessive trait with high penetrance. In 1999, the gene causing this familial form of canine narcolepsy in the former two breeds was localized within an 800-kb region of canine chromosome 12 and positionally cloned;[34] this was followed two years later by the identification of a single novel point mutation in the Hcrtr2 gene in a narcoleptic Dachshund multiplex family (E54K).[35] Hypocretin gene mutations have not been detected in the sporadic form of canine narcolepsy, but these dogs do have an absence of the hypocretin peptide in the CSF and the brain[36] as in human narcoleptics, suggesting a

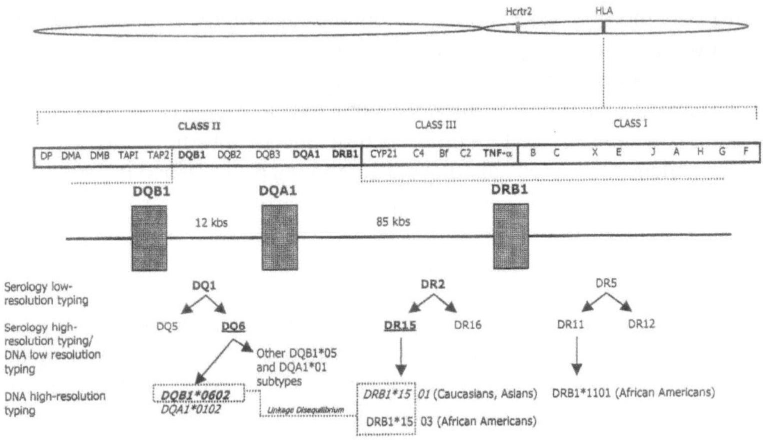

Figure 1 HLA DR and HLA DQ alleles most typically observed in narcoleptic patients. Modified from [25 37]. The genes located on the short arm of chromosome 6 encode heterodimeric HLA proteins. The HLA DQB1 and DQA1 genes located 85kb centromeric to the gene DRB1, on chromosome 6p21. Initial serological typing of the 80' revealed the DR2, DR5, and DQ1 subtypes. These HLA antigens have been further split into two allelic subtypes, DR15 and DR16 on the one hand, and DQ5 and DQ6 on the other hand, and with the use of high-resolution serological typing, all narcoleptic patients are positive for DR15 and DQ6. Subsequent high-resolution DNA sequencing or oligotyping have identified molecular subtypes of DR15 and DQ6: Caucasians and Asians are DRB1*1501, DQA1*0102-DQB1*0602, while Blacks are more often DRB1*1503, DQA1*0102-DQB1*0602. In addition, some African Americans are negative for DR2 but generally carry the DR11 subtype DRB1*1101.

similar pathogenesis. It is of interest to note however, that sporadic canines do not share a single dog leukocyte antigen DLA-DQ allele.[38,39]

After the discovery of the Hcrtr2 gene mutation in dogs, several studies in human narcoleptics looked for similar mutations in the genes for Hcrt and its receptors (Hcrtr1 and Hcrtr2). Peyron and colleagues[21] screened 74 patients with narcolepsy, using familial cases and otherwise "unusual" cases of narcolepsy-cataplexy with the expectation that mutations would be more readily detected, but no hypocretin receptor mutations specific to narcolepsy were found. Furthermore, while frequent Hcrtr1 and Hcrtr2 polymorphisms were identified, they were also not associated with the disorder;[40] other corroborating studies have since been published.[39,41] However, in one atypical case of narcolepsy (cataplexy presenting at the age of 6 months), a heterozygous mutation within the secretory signal sequence of the prepro-hypocretin (peptide precursor) gene was observed.[21]. Functional studies indicated that this mutation impaired trafficking and processing of the peptide, leading to intracellular accumulation, and presumably to cell death.[21]

Honda, Juji, and colleagues[42] in Japan were the first to report that 100% of narcoleptics with cataplexy were positive for HLA-DR2 (Figure 1), a tight association subsequently confirmed in Europe and North America.[25] About 25% of the normal Caucasian and Japanese population is positive for HLA-DR2 versus 90 to 100% of individuals with narcolepsy-cataplexy. Further studies have indicated that the association extends to the HLA-DQ region, most specifically in the context of the DR15, DQ6 (DR2, DQ1- Dw2) haplotype.[43] Interestingly, however, there is a lower DR2 association in African American patients.[44] Sequencing of HLA DRB1, DQA1 and DQB1 genes later indicated that all Caucasian and Japanese narcoleptic subjects have the same alleles, now called DRB1*1501, DQA1*0102 and DQB1*0602.[45] In African American subjects, however, a primary association with DQA1*0102 and DQB1*0602 was observed,[15,46,47] explaining the lower HLA-DR2 association. DQB1*0602, an allele found in only 12%, 25% and 38% of the control Japanese, Caucasian and African American populations respectively, but in 90% of those with narcolepsy-cataplexy, is now considered the primary HLA susceptibility gene. The allele is found in only 40% of cases with narcolepsy without cataplexy,[48] suggesting more heterogeneity.

Even in cases with cataplexy, the HLA association observed in narcolepsy is complex and not simply the result of a dominant effect of DQB1*0602. HLA-DQB1*0602 homozygosity doubles or quadruples the risk for developing narcolepsy.[49] Additionally, the relative risk for narcolepsy varies in heterozygous subjects according to the allele associated with DQB1*0602.[18,50] These alleles convey either an increased or decreased risk for developing narcolepsy; for example, a higher risk has been observed in heterozygotes co-expressing DQB1*0301, while those carrying either DQB1*0601 or DQB1*0501 have a lower risk.[18] Additional small effects are suspected for various DRB1 and DQA1 alleles but have not been yet fully characterized.

The finding that approximately 25% of familial narcoleptic cases are negative for DQB1*0602[25,51] points towards the existence of highly penetrant non-HLA narcolepsy genes. Systematic genome screening in extended narcoleptic families has revealed a significant link in 4q13-q21 (lod score of 3.09) in 8 small multigenerational families of narcoleptics. Furthermore, reported associations between narcolepsy and monoamine oxidase-A (MAO-A),[52] catechol-O-methyl transferase (COMT)[53] and tumor necrosis factor-alpha (TNF-alpha)[54] polymorphisms suggest that other susceptibility genes may be found, in addition to or independently of, the HLA and/or hypocretin system. COMT is a key enzyme in dopaminergic and noradrenergic transmission, and while no association has been found between genotype or allele frequency in narcoleptics, a sexual dimorphism and a strong effect of COMT genotype on disease phenotype has been reported. Female narcoleptics with high COMT activity fall asleep twice as fast on the MSLT as women with low COMT activity, while the opposite is true in narcoleptic men. The COMT genotype also strongly affects the presence of sleep paralysis and the number of REM sleep onsets during the MSLT. TNF-alpha is also interesting in that its intravenous or intracerebral administration produces sleepiness,[55] and it is an HLA class III gene which is physically located within the susceptibility region of the HLA class II on chromosome 6 in humans. While initial studies failed to find any mutations or polymorphisms in the TNF-alpha gene or its promoter, the analysis of single-nucleotide polymorphisms (SNPs) in DRB1*1501 positive Japanese narcoleptics has revealed an association between the disorder and the rare chromosomal recombinant TNF-alpha gene (TNF alpha(-857T)), independent of DRB1*1501.[54] Furthermore, a single-nucleotide polymorphism in the TNF receptor 2 gene (TNFR2-196R)[56] has also been associated with narcolepsy, suggesting that the genetic impairment in the TNF-alpha pathway may interfere directly with the phenotype of narcolepsy, and that an inflammatory mechanism may play a role in the development of the disorder.[39] This inflammatory mechanism may be in part, related to environmental factors.

A severe physiological stress, divorce, bereavement, change in the sleep-wakefulness cycle, accident (including head trauma), illness or pregnancy, often precedes the appearance of the symptoms of narcolepsy (sleepiness and/or cataplexy) by a few weeks or months.[25] Patients with narcolepsy are also more likely to be born in March and less frequently in September, suggesting perinatal environmental effects.[57] However, because of its very tight HLA allele association, it is likely that it is ultimately an autoimmune-based degeneration of hypocretin neurons that leads to the narcolepsy phenotype in humans. The association of autoimmune diseases with various MHC proteins, particularly HLA class II antigens, is well established;[39] examples include diseases such as Rheumatoid Arthritis (DR4), Diabetes Mellitus type I (IDDM) (DR3, DR4, DQB1*0302, DQB1*0201, DQB1*0602), and Multiple Sclerosis (DRB1*1501, DQB1*0602). In these autoimmune disorders, susceptibility HLA proteins from particular alleles are believed to bind peptide motifs derived from processed foreign antigen, leading to the HLA presentation of self-antigens and a sustained immune response causing tissue and cell destruction.[39] The interaction of particular HLA proteins with processed autoantigens then determines whether tolerance or autoimmunity occurs. Because HLA-DQB1*0602 confers disease susceptibility while the very similar DQB1*0601 is protective, very minor variations in the peptide binding pockets of these molecules determine the disease occurrence, but how these minor changes damage hypocretin neurons in the hypothalamus is unclear.[39] Indeed, studies of narcoleptic brains do suggest a degeneration of hypocretin neurons, as evidenced by the absence of prepro-orexin mRNA and orexin immunoreactivity from the lateral hypothalamic area of human brains.[21,22] However, hypocretin neurons may also express other neurotransmitters, and the pathophysiology of narcolepsy may be affected by the additional loss of these as yet unknown factors.

4. HYPOCRETIN DEFICIENCY IN NARCOLEPSY

Nishino *et al.*[58] found lumbar CSF Hcrt-1 levels below the detection limit (40 pg/mL) in narcoleptics, while control subjects had an average of 280 pg/mL. This study provided the first clear link between hypocretin dysfunction and human narcolepsy. These findings have been extended and replicated by other groups[59] (see below). It has thus far not been possible to measure Hcrt-2 in the CSF, as Hcrt-2 is likely degraded rapidly or is unstable in CSF.

Further pathologic evidence of selective hypocretin deficiency has come from immunohistochemical and *in situ* hybridization studies which have indicated a global loss of hypocretin in the brains of human narcoleptics (Figure 2).[21,22] This loss of hypocretin cells seems selective, as melanin-concentrating hormone (MCH) neurons, which are normally located in the same region in the hypothalamus, are not downregulated.[21] This same pathological phenotype has been found in the few cases of sporadic canine narcolepsy as well. An increased number of astrocytes in the lateral hypothalamus but not the thalamus of narcoleptic brains has also been detected,[22] suggesting lateral hypothalamic area-specific gliosis.

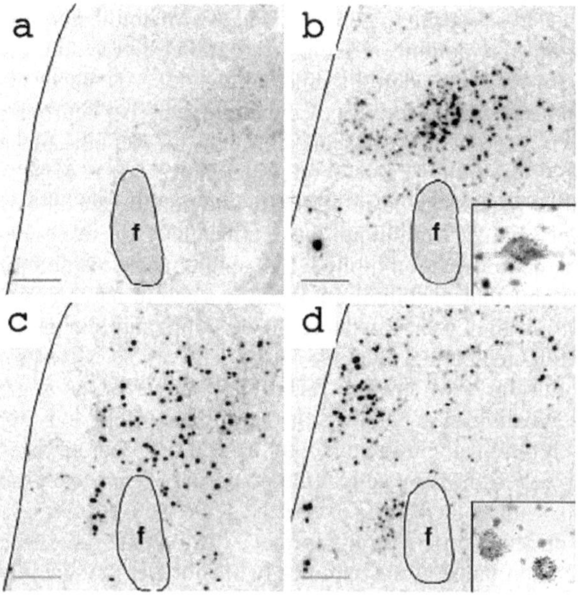

Figure 2 Absence of hypocretin transcripts in the lateral hypothalamus of narcoleptic patients (A) versus controls (B). MCH transcripts are detected in the same region in both narcoleptics (C) and controls (D), suggesting that the loss of hypocretin-containing neurons is selective. From Peyron *et al* [21]. F= fornix

It is clear, however, that hypocretin stabilizes, rather than generates vigilance states.[40] Animal studies have shown that hypocretin signaling is crucial in maintaining wakefulness and regulating REM sleep, and likely involves both aminergic and cholinergic neurons. While still incomplete, there are several hypotheses regarding the function(s) of hypocretin and how their loss leads to the symptoms of EDS and REM dissociation seen in narcolepsy. Hypocretinergic neurons send excitatory innervation to a variety of nuclei involved in the maintenance or generation of wake, including the noradrenergic neurons of the locus coeruleus (LC), the serotonergic neurons of the dorsal raphe (DR), and the histaminergic neurons of the tuberomammillary nucleus (TMN). The histaminergic neurons may play a particularly important role as histamine concentrations are low in narcoleptic dogs with the mutant Hcrtr2 receptor[60] as well as in human narcoleptics.[61] Additionally, mice lacking the H1 histamine receptor appear to have no waking response to intraventricular injections of hypocretin-1.[62] While dopaminergic signaling also promotes wakefulness (dopamine antagonists can induce sleep and amphetamines promote wakefulness by increasing extracellular concentrations of dopamine), interactions between the hypocretin and dopaminergic systems are poorly understood. Finally, hypocretin may also promote wakefulness by exciting cholinergic neurons of the laterodorsal and pedunculopontine tegmental (LDT/PPT) nuclei. Most LDT/PPT neurons are active during wakefulness and promote electroencephalographic desynchrony through projections to the thalamus.

How hypocretin deficiency mediates REM sleep abnormalities and cataplexy in narcolepsy is more controversial. Like REM sleep, cataplexy is modulated by monoaminergic (particularly D2/D3 dopaminergic and/or alpha-1(b) adrenergic) tone and cholinergic (M2/M3 muscarinic receptor) systems. Anatomically, cholinergic neurons in the basal forebrain and pontine reticular formation (PRF), and LDT/PPT, as well as dopaminergic neurons in the ventral tegmental area (VTA) appear to be involved. REM sleep atonia is induced by the activity of the cholinoceptive neurons in the PRF (nucleus reticularis pontis); these neurons descend through the medulla and inhibit motoneurons through activation of inhibitory glycinergic interneurons. It is presumed that connections between the limbic system, the basal forebrain and pontine nuclei explain how emotional events can trigger cataplexy, although these tracts have yet to be identified. The REM-active LDT/PPT neurons are inhibited by amines such as norepinephrine, and it is believed that the apparent disinhibition of REM sleep in narcolepsy is due to decreased aminergic activity from a loss of hypocretin signaling, and less aminergic activity would disinhibit the REM-generating neurons of the LDT/PPT.[63] Therefore, hypocretin cells would be silent during REM sleep in this model. However, an alternative hypothesis is that hypocretin signaling may actually promote REM sleep, and therefore, both models may be correct, in that hypocretin cells may exhibit different patterns of activity — with some promoting wakefulness by activating aminergic brain regions and wake-active cells of the LDT/PPT, while others may excite the REM-active cells of the LDT/PPT. The finding that the dopaminergic neurons in the VTA mediate the experience of pleasure is also of potential importance in the narcolepsy phenotype, though an exact link has yet to be established.

Finally, hypocretin is most likely involved in the integration of sleep regulation and metabolic status, as narcoleptics are slightly more overweight than the general population.[64,65] Hypocretin cells can respond electrophysiologically to common circulating signals of metabolic state, including glucose, leptin and ghrelin,[66] which may possibly increase wakefulness when needing to search for food in the fasting state. Therefore, it is likely that hypocretin plays an as yet to be defined role, in the complex interactions between the hypothalamic peptidergic systems in the regulation of appetite, feeding, and metabolism.

5. ROLE OF CSF HYPOCRETIN IN THE DIAGNOSIS OF NARCOLEPSY

These new neurobiologic (hypocretin deficiency) and genetic (HLA) findings have lead to a redefinition of narcolepsy. Whereas the MSLT has been classically used to distinguish, for example, narcolepsy without cataplexy from idiopathic hypersomnia, the expanded disease continuum of the "narcolepsy clinical borderland" has made the diagnosis of narcolepsy more difficult. As such, the use of CSF Hcrt-1 levels as a biologically-based diagnostic test for narcolepsy has been particularly useful in establishing a new pathophysiologically-based criterion for its classification (Table 2). Indeed, the MSLT is not entirely specific or sensitive, in that 15% of narcoleptic patients have a negative test,[67,68] whereas normal subjects or patients with other sleep disorders associated with EDS may also have abnormally short sleep latencies on the MSLT. Quantitative Receiver Operating Curves (QROC) analysis in a large number of lumbar CSF Hcrt-1 samples has indicated that the value of 110 pg/ml, one third of normal control values in our laboratory, is the best cut off value for diagnostic purposes. Using this cut off, CSF Hcrt-1 measurements (Figure 3) increase the sensitivity of the diagnostic process, as 16% of hypocretin-deficient narcoleptic patients do not have a positive

MSLT.[59] Furthermore, the MSLT may be modified by psychotropic drugs, whereas treatment does not dramatically influence the CSF Hcrt-1 concentration.[59]

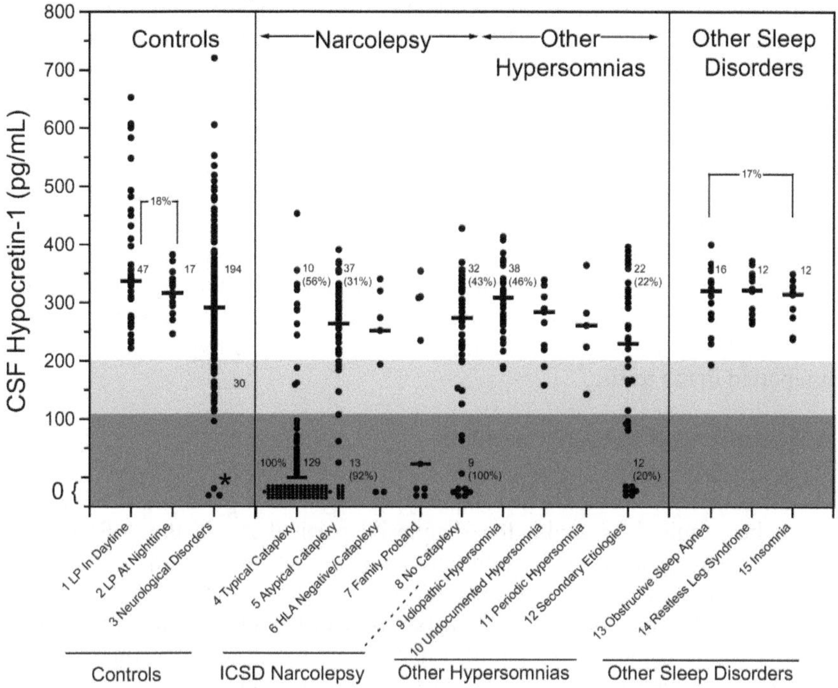

Figure 3 Cerebrospinal fluid hypocretin-1 concentrations are plotted for individuals across various conditions and disorders, including sleep disorders. Modified from Mignot *et al* [59].Each point represents the crude Hcrt-1 concentration in one individual. Horizontal bars indicate the median value for that clinical group. The horizontal dashed lines delineate the cut-off values for normal (> 200 pg/mL) and low < 110 pg/mL) Hcrt-1 levels in the CSF. The numbers by each cluster of points within each value range represent the total number of subjects in that respective groups, with the percentage of those individuals who were HLA DQB1*0602 positive indicated in parentheses. **Clinical Categories:** 1. LP Daytime = Healthy control subjects, lumbar puncture (LP) performed between 9am-7pm; 2. LP Nighttime = Healthy control subjects, LP performed between 11pm-6am.; 3. Neurological disorders = Patients with various neurological disorders, including Guillain-Barré (of whom two had undetectable Hcrt-1 concentrations, indicated by the *); 4. Narcolepsy with typical cataplexy = History of cataplexy triggered by joking or laughing; 5. Narcolepsy with atypical cataplexy = History of cataplexy triggered by events other than joking or laughing; 6. HLA negative/Cataplexy = History of typical cataplexy, but HLA DQB1*0602 negative; 7. Family proband = 4 probands from HLA DQB1*0602 negative families; 8. No cataplexy = No cataplexy, MSL <8 minutes plus 2 or more soREMPs; 9. Idiopathic Hypersomnia = According to ICSD criteria; 10. Undocumented Hypersomnia = Complaint of hypersomnia without confirmatory sleep testing; 11. Periodic Hypersomnia = According to ICSD criteria (the CSF of one subject found to have a Hcrt-1 concentration < 200 pg/mL was obtained during an episode of hypersomnia); 12. Secondary Etiologies = Narcolepsy and hypersomnia related to various disorders, including trauma, tumor, infection, degenerative diseases and genetic diseases; 13. Obstructive Sleep Apnea = Sleep-disordered breathing with RDI >10/hr; 14. Restless Legs Syndrome = According to ICSD criteria; 15. Insomnia= According to ICSD criteria.

Furthermore, CSF concentrations of Hcrt-1 are relatively stable throughout the 24-hour day,[69] in contrast to locally released, brain hypocretin levels.[70]

Unfortunately, while a plasma-based assay would be more convenient, available assays to date have been unreliable in measuring plasma levels of hypocretin peptides. Several studies have revealed disparate results, using commercially available radioimmunoassay (RIA) kits, which have not been accompanied by chromatographic analysis of the detected radioimmunoreactivity.[71-73] It is possible that interfering substances in the plasma or those produced through plasma extraction procedures have resulted in incorrect detection in the various RIAs.[74] Furthermore, these studies have not generally shown a decrease in plasma Hcrt-1-like immunoreactivity in human narcolepsy, which suggests that these assays may be measuring cross-reacting molecules rather than the Hcrt-1 ligand itself. The only notable exception is a report by Higuchi *et al.*[75] who demonstrated that decreased levels of plasma Hcrt-1 in narcoleptic patients using HPLC (reverse-phase SEP-PAK C18) separation to confirm that their signal included at least partially, genuine Hcrt-1. However, the source of the hypocretin measured in these blood assays is still unclear, as outside the hypothalamus, prepro-hypocretin mRNA has also been reported in the testis.[76]

5.1 Narcolepsy with Definite Cataplexy

In all patients with clear-cut cataplexy, it might be argued that CSF Hcrt-1 levels may not be required to make the diagnosis, particularly if the MSLT is positive. However, obtaining such information may be important in situations as discussed above (*e.g.,* when patients are taking medication which may lower the specificity of the MSLT). In a sample of 274 patients with various sleep disorders (171 with narcolepsy) and 296 controls, a cutoff value of 110 pg/mL (30% of the mean control values) was the most predictive of narcolepsy, and most samples had undetectable levels (< 40 pg/mL in most assays), while a few samples had detectable, but very reduced levels.[59] The cutoff of 110 pg/mL was particularly predictive in cases with cataplexy (99% specific, 87% sensitive), which is more specific than the MSLT. Therefore, measuring CSF Hcrt-1 levels may be especially useful in those with definite cataplexy but negative MSLT's.

5.2 Narcolepsy without Cataplexy or with Atypical Cataplexy

CSF Hcrt-1 levels have a more limited predictive power in cases of narcolepsy without cataplexy or with atypical cataplexy. While specificity remains high at 99%, the sensitivity decreases to 16% in these individuals, with most having normal hypocretin levels.[59] These cases may therefore represent distinct clinical entities that do not involve abnormal hypocretin neurotransmission. Furthermore, HLA typing may a more prudent first step as almost all cases of narcolepsy with low CSF Hcrt-1 levels are also HLA-DQB1*0602 positive.[59,77,78] Indeed, there have only been three exceptions reported to date, including an individual with very mild and atypical cataplexy.[59,79] It is estimated that the probability of finding a low hypocretin level (< 110 pg/mL) in an HLA negative individual with no cataplexy to be less than one percent.[80] Many of these cases are children in whom cataplexy may develop later in the course of the disease. Therefore, CSF Hcrt-1 measurements may be useful in young children with EDS but without clear cataplexy, as well as in those in whom cataplexy is suspected but not accurately reported by parents.

5.3. Narcolepsy Associated with a Known Physiological Condition

5.3.1. Intermediate CSF Hcrt-1 Levels

In contrast to typical narcoleptics, healthy individuals and most of those with other sleep disorders (SRBD, RLS, insomnia) are usually found to have CSF Hcrt-1 levels above 200 pg/mL (Figure 3).[59] However, in extremely rare cases of narcolepsy or other causes of hypersomnia, Hcrt-1 levels have been detected in between 200 and 110 pg/mL, suggesting the possibility of a 'partial hypocretin deficiency' state. As such, the significance of such an intermediate value should be considered with caution, as up to 13% of individuals with hypersomnia related to another neurologic condition, have been found to have CSF Hcrt-1 values within this intermediate range (Figure 3);[59] most cases represent severe brain pathology, particularly head trauma, encephalitis and subarachnoid hemorrhage.[81] In these situations, the decrease in Hcrt-1 may reflect attenuated Hcrt-1 transmission or changes in CSF flow. It has also been shown that CSF Hcrt-1 decreases slightly (<10%) with the use of a serotonin reuptake inhibitor, but never to near-detectable levels as in narcolepsy. [69]

5.3.2. Secondary Narcolepsy

CSF Hcrt-1 measurements may also prove useful in narcolepsy associated with trauma, tumor, infection, degenerative diseases, as well as genetic disorders. Indeed, symptoms of narcolepsy have been reported after traumatic brain injury,[82] acute disseminated encephalomyelitis (ADEM), [83] hypothalamic sarcoidosis[84] or Histiocytosis X,[85] multiple sclerosis (MS),[86] and Parkinson's Disease (PD).[80] In some cases, lesions of the hypothalamic hypocretin centers have been clearly identified using MRI, as in bilateral MS plaques in the hypothalamus, and tumors of the third ventricle.[87,88] Cataplexy may be present in these cases, and the CSF Hcrt-1 levels may be either in the narcolepsy range (<110 pg/mL) or in the intermediate range.[87,88] These neurological disorders may cause damage to nearby hypocretin projection sites, with adequate preservation of cell bodies to maintain detectable levels or, in some cases, their presence may simply be coincidental.[80]

Genetic disorders such as Neimann-Pick Type C,[89] Coffin-Lowry Syndrome[90] and Norrie's Disease[91] have been reported to be associated with daytime sleepiness and/or cataplexy. CSF Hcrt-1 has been measured in cases of Neimann Pick Type C and intermediate levels have been found in some cases with cataplexy.[59,89,92] Some diseases are associated with the development of both narcolepsy and sleep-related disordered breathing such as myotonic dystrophy and Prader-Willi syndrome.[80] In these patients, some but not all have very low CF Hcrt-1 levels (<110 pg/mL), suggesting hypocretin deficiency.[59,93] Similarly, in one case of late-onset Congenital Hypoventilation Syndrome (a disorder with reported hypothalamic abnormalities), very low CSF Hcrt-1 levels were found in an individual with otherwise unexplained sleepiness and cataplexy-like episodes,[59] with an excellent response to anticataplectic therapy.[80] Thus, CSF Hcrt-1 levels may be particularly helpful in complex clinical situations in which the history, polysomnographic and/or MSLT data are difficult to interpret.

6. HYPOCRETIN DEFICIENCY AND PHARMACOLOGIC CORRELATES

The current treatment of narcolepsy is symptomatically based. In the past, cataplexy was treated using tricyclic medications (such as protriptyline and cloimpramine) or higher dose serotonin-reuptake inhibitors (such as fluoxetine). Adrenergic reuptake inhibition has been shown to be most effective in reducing cataplexy in animal models[16] and clinical experience with selective noradrenergic reuptake inhibitors such as reboxetine suggest very positive anticataplectic effects. More recently dual serotonergic-adrenergic reuptake inhibitors such as venlafaxine have been more commonly used as first-line therapy for cataplexy. Experience with newer medications such atomoxetine, a selective adrenergic reuptake inhibitor similar to reboxetine, is more limited but promising effects have been observed in clinical practice. As serotonin reuptake inhibitors also reduce REM sleep, this may suggest a preferential adrenergic control of REM atonia.

Sleepiness is commonly treated using amphetamine-like stimulants (*L*- or *D*-amphetamine, methylphenidate or modafinil). These compounds produce wakefulness by presynaptic augmentation of dopaminergic neurotransmission, either via enhancement of dopamine release (amphetamine) or reuptake inhibition (methylphenidate).[17,94] However, the mode of action of modafinil remains controversial.

GHB (gamma-hydroxybutyrate) was approved by the FDA in 2002 for treatment of cataplexy (Schedule III controlled substance).[95] GHB is administered at night and consolidates nocturnal sleep by increasing slow wave sleep while also decreasing REM sleep. Chronic use also decreases daytime cataplexy and possibly sleepiness, usually within 1-3 months after initiation of therapy. However, this compound has a very short half-life and must be administered twice during the night. Current experience suggests a favorable side effect/efficacy profile, but the margin of safety for the compound is small and approval for GHB was complicated by its popularity as a drug of abuse. Pharmacological studies are being performed to evaluate efficacy on EDS.

Anecdotal information is available on the treatment of hypnagogic hallucinations, sleep paralysis, and disturbed nocturnal sleep. The first two symptoms partially respond to antidepressants, while disturbed nocturnal sleep can be treated using GHB or benzodiazepine hypnotics but the latter of these does not improve daytime sleepiness.

7. FUTURE PROSPECTS

While much has been learned about the cause of narcolepsy-cataplexy in humans and animal models we still do not known whether narcolepsy is an autoimmune disorder resulting from the degeneration of hypocretin neurons. Indeed, studies showing the absence of prepro-hypocretin mRNA and hypocretin immunoreactivity from the lateral hypothalamic area of human brains,[21,22] are consistent with this possibility. However, attempts to confirm this autoimmune hypothesis directly[96,97] have been, to date, unsuccessful.

Studies in peripheral blood have failed to reveal any evidence of humoral or cellular autoimmune processes in narcoleptics. Using ELISA, no IgG autoantibodies directed against Hcrt-1, Hcrt-2, or prepro-hypocretin overlapping peptides have been detected in the serum of narcolepsy patients. However, the absence of anti-hypocretin antibodies does not preclude the possibility of other auto-antibodies directed against other antigens expressed by hypocretin neurons. For example, in Myasthenia Gravis (MG, a peripheral nervous system autoimmune disorder directed against the neuromuscular junction), pathogenic anti-acetylcholine receptor autoantibodies can be detected in the peripheral blood of most patients and can transfer the disease when injected into mice. However, 15% of patients with MG have no detectable anti-acetylcholine receptor antibody in their blood, yet their serum can still transfer the disease, which raises the possibility that other pathogenic (neutralizing) auto-antibodies exist, specific for other, as of yet undetermined auto-antigens.

Mono-lymphocytic secretion of TNF-alpha and other proinflammatory cytokines such as IL-beta, IL-1 RA, IL-2 and TNF-beta have also been studied and found to be no different between HLA DR2 positive narcoleptic patients and controls.[98] Only IL-6 secretion was found to be higher in narcoleptics. While a major peripheral cellular inflammatory reaction in narcolepsy is also unlikely given the lack of difference between T-cell subsets and natural killer (NK) activity in these two populations, this does not exclude the presence of a local proinflammatory cellular process within the central nervous system. While studies of hypocretin neurons in narcoleptic brains have not detected any inflammatory infiltrates in the hypothalamus,[21,22] non-specific gliosis has been observed.[22] Furthermore, while CNS microglial class II expression is not significantly different between narcoleptics and control subjects, brains used for these studies may have contained inflammatory infiltrates which were present at the time of disease onset, but which have since abated. Based on measurements of CSF Hcrt-1, the disease process is likely rapidly progressive, given the already low or absent Hcrt-1 levels in patients who are diagnosed early in the disease process.

Indeed, further research is required to better characterize the circumstances surrounding the onset of narcolepsy symptoms, including environmental factors that may trigger the disease in those so genetically predisposed. Furthermore, atypical cases of narcolepsy should be carefully investigated in relation to CSF Hcrt-1 levels, in order to better describe the spectrum of the "narcolepsy borderland". Whether CSF Hcrt-1 levels correlate with progression of narcolepsy symptoms also requires further study, as this may provide the opportunity to either halt or possibly reverse the progression of disease with immunosuppressive medications. While there has been one report[99] of an unsuccessful attempt using high-dose prednisone, CSF Hcrt-1 levels were already undetectable at presentation in this individual, suggesting that the physiological process responsible for his narcolepsy had likely already reached an irreversible stage. More recently, immune modulation using intravenous immunoglobulin has been shown to be temporarily effective in one patient treated very close to disease onset.[100] This exciting result suggests that it may be possible to stop the development of narcolepsy in some cases using immunomodulation.

Finally, it would be hoped that hypocretin-based therapies could be devised, and replace the symptomatic treatments currently available. The effects of Hcrt-1 administration on sleep and narcolepsy symptoms have been studied.[101,102] Intraventricular administration of Hcrt-1 in wild type rodents and normal canines is strongly wake-promoting.[102] This effect is likely mediated in part by the Hcrtr2, the same dose of Hcrt-1 has no effect in Hcrtr2-mutated narcoleptic dogs.[103] Problematically however, hypocretin does not cross the blood brain barrier significantly and a centrally penetrable agonist will most likely need to be devised to be effective. It may also be that the field will benefit from current research in the area of stem cell research and transplantation. In type I Diabetes Mellitus, an autoimmune disorder where insulin-producing cells have been destroyed, transplantation of islet cells in the liver can lead to a permanent cure, providing donor cells can be identified. One day it may be possible to transplant hypocretin cells into the CNS, effectively reversing the deficiency of hypocretin. Transplantation studies using fetal hypothalamic cells and hypocretin-deficient rodent models are now being conducted to test the feasibility of this theory.

8. REFERENCES

1. G. Vogel, Studies in psychophysiology of dreams. III. The dream of narcolepsy, *Arch Gen Psychiatry.* **3**, 421-8 (1960).
2. M. Carskadon, W. Dement, M. Mitler, T. Roth, P. Westbrook and S. Keenan, Guidelines for the Multiple Sleep Latency Test (MSLT): a standardized measure of sleepiness, *Sleep.* **9**, 519-524 (1986).
3. L. de Lecea, T. S. Kilduff, C. Peyron, X. Gao, P. E. Foye, P. E. Danielson, C. Fukuhara, E. L. Battenberg, V. T. Gautvik, F. S. Bartlett, 2nd, W. N. Frankel, A. N. van den Pol, F. E. Bloom, K. M. Gautvik and J. G. Sutcliffe, The hypocretins: hypothalamus-specific peptides with neuroexcitatory activity, *Proc Natl Acad Sci U S A.* **95**, 322-7 (1998).
4. T. Sakurai, A. Amemiya, M. Ishii, I. Matsuzaki, R. M. Chemelli, H. Tanaka, S. C. Williams, J. A. Richardson, G. P. Kozlowski, S. Wilson, J. R. Arch, R. E. Buckingham, A. C. Haynes, S. A. Carr, R. S. Annan, D. E. McNulty, W. S. Liu, J. A. Terrett, N. A. Elshourbagy, D. J. Bergsma and M. Yanagisawa, Orexins and orexin receptors: a family of hypothalamic neuropeptides and G protein-coupled receptors that regulate feeding behavior, *Cell.* **92**, 573-85 (1998).
5. L. Lin, J. Faraco, R. Li, H. Kadotani, W. Rogers, X. Lin, X. Qiu, P. J. de Jong, S. Nishino and E. Mignot, The sleep disorder canine narcolepsy is caused by a mutation in the hypocretin (orexin) receptor 2 gene, *Cell.* **98**, 365-76 (1999).
6. R. M. Chemelli, J. T. Willie, C. M. Sinton, J. K. Elmquist, T. Scammell, C. Lee, J. A. Richardson, S. C. Williams, Y. Xiong, Y. Kisanuki, T. E. Fitch, M. Nakazato, R. E. Hammer, C. B. Saper and M. Yanagisawa, Narcolepsy in orexin knockout mice: molecular genetics of sleep regulation, *Cell.* **98**, 437-51 (1999).
7. S. Nishino, B. Ripley, S. Overeem, G. J. Lammers and E. Mignot, Hypocretin (orexin) deficiency in human narcolepsy, *Lancet.* **355**, 39-40 (2000).
8. C. Peyron, J. Faraco, W. Rogers, B. Ripley, S. Overeem, Y. Charnay, S. Nevsimalova, M. Aldrich, D. Reynolds, R. Albin, R. Li, M. Hungs, M. Pedrazzoli, M. Padigaru, M. Kucherlapati, J. Fan, R. Maki, G. J. Lammers, C. Bouras, R. Kucherlapati, S. Nishino and E. Mignot, A mutation in a case of early onset narcolepsy and a generalized absence of hypocretin peptides in human narcoleptic brains, *Nat Med.* **6**, 991-997 (2000).
9. T. C. Thannickal, R. Y. Moore, R. Nienhuis, L. Ramanathan, S. Gulyani, M. Aldrich, M. Cornford and J. M. Siegel, Reduced number of hypocretin neurons in human narcolepsy, *Neuron.* **27**, 469-74. (2000).
10. C. Guilleminault. (1976). Cataplexy. In *Proceedings of the First International Symposium on Narcolepsy.* (Guilleminault, C., Dement, W. & Passouant, P., eds.), pp. 125-144. Spektrum Publications, New York.
11. R. Broughton, V. Valley, M. Aguirre, J. Roberts, W. Suwalski and W. Dunham, Excessive daytims sleepiness and the pathophysiology of narcolepsy-cataplexy: a laboratory perspective, *Sleep.* **9**, 205-215 (1986).
12. C. Knecht, J. Oliver, R. Redding, R. Selcer and G. Johnson, Narcolepsy in a dog and a cat., *J Am Vet Med Assoc.* **162**, 1052-3 (1973).
13. Y. Honda, A. Asake, Y. Tanaka and T. Juji, Discrimination of narcolepsy byusing genetic markers and HLA, *Sleep Res.* **1**, 254 (1983).
14. S. Nishino and E. Mignot, Pharmacological aspects of human and canine narcolepsy, *Prog Neurobiol.* **52**, 27-78 (1997).
15. E. Mignot, X. Lin, J. Arrigoni, C. Macaubas, F. Olive and J. Hallmayer, DQB1*0602 and DQA1*0102 (DQ1) are better markers than DR2 for narcolepy in Causcasians and Black Americans, *Sleep.* **17**, S60-7 (1994).
16. E. Mignot, A. Renaud, S. Nishino, J. Arrigoni, C. Guilleminault and W. Dement, Canine cataplexy is preferentially controlled by adrenergic mechanisms: evidence using monoamine selectvie uptake inhibitors and release enhancers, *Psychopharmacology.* **113**, 76-82 (1993).
17. S. Nishino, J. Mao, R. Sampathkumaran, J. Shelton and E. Mignot, Increased dopaminergic transmission mediates the wake-promoting effects of CNS stimulants, *Sleep Res Online.* **1**, 49-61 (1998).
18. E. Mignot, L. Lin, W. Rigers, Y. Honda, X. Qiu, X. Lin, M. Okun, H. Hohjoh, T. Miki, S. Hsu, M. Leffell, F. Grumet, M. Fernandez-Vina, M. Honda and N. Risch, Complex HLA-DR and -DQ interactions confer risk of narcolepsy-cataplexy in three ethnic groups, *Am J Hum Genet.* **68**, 686-699 (2001).

19. J. Hara, C. Beuckmann, T. Nambu, J. Willie, R. Chemelli, C. Sinton, F. Sugiyama, K. Yagami, K. Goto, M. Yanagisawa and T. Sakurai, Genetic ablation of orexin neurons in mice results in narcolespy, hypophagia, and obesity, *Neuron.* **30**, 345-354 (2001).
20. K. Matsuki, F. Grumet, X. Lin, C. Guilleminault, W. Dement and E. Mignot, DQ rather than DR gene marks susceptibility to narcolepsy, *Lancet.* **339**, 1052 (1992).
21. C. Peyron, J. Faraco and W. Rogers, A mutation in a case of early onset narcolepsy and a generalized absence of hypocretin peptides in humans narcoleptic brains, *Nat Med.* **6**, 991-997 (2000).
22. T. Thannickal, R. Moore and R. Nienhuis, Reduced number of hypocretin neurons in human narcolepsy, *Neuron.* **27**, 469-474 (2000).
23. J. Emard, J. Thouez and D. Gauvreau, Neurodegenerative diseases and risk factors: a literature review, *Soc Sci Med.* **40**, 847-858 (1995).
24. S. Overeem, E. Mignot, J. vanDijk and G. Lammers, Narcolepsy: clinical features, new pathophysiologic insights, and future perspectives, *J Clin Neurophysiol.* **18**, 78-105 (2001).
25. Y. Dauvilliers, M. Billiard and J. Montplaisir, Clinical aspects and pathophysiology of narcolepsy, *Clin Neurophysiol.* **114**, 2000-2017 (2003).
26. S. Taheri, J. Zeitzer and E. Mignot, The role of hypocretins (orexins) in sleep regulation and narcolepsy, *Annu Rev Neurosci.* **25**, 283-313 (2002).
27. E. Mignot, A Commentary on the Neurobiology of the Hypocretin/Orexin System, *Neuropsychopharmacology.* **25**, S5-S13 (2001).
28. C. Bassetti, M. Gugger, M. Bischof, J. Mathis, C. Sturzenegger, E. Werth, B. Radanov, B. Ripley, S. Nishino and E. Mignot, The narcoleptic borderland: a multimodal diagnostic approach including cerebrospinal fluid levels of hypocretin-1 (orexin A), *Sleep Med.* **4**, 7-12 (2003).
29. E. Mignot, Genetic and familial aspects of narcolepsy, *Neurology.* **50**, S16-S22 (1998).
30. Y. Dauvilliers, S. Maret, C. Bassetti, B. Carlander, M. Billiard, J. Touchon and M. Tafti, A monozygotic twin pair discordant for narcolepsy and CSF hypocretin-1, *Neurology.* **62**, 2137-2138 (2004).
31. R. Khatami, S. Maret, E. Werth, J. Retey, D. Schmid, F. Maly, M. Tafti and C. Bassetti, Monozygotic twins concordant for narcolepsy-cataplexy without any detectable abnormality in the hypocrtein (orexin) pathways, *Lancet.* **363**, 1199-1200 (2004).
32. M. Honda, Y. Honda, S. Uchida, S. Miyazaki and K. Tokunaga, Monozygotic twins incompletely concordant for narcolepsy, *Biol Psychiatry.* **49**, 943-947 (2001).
33. M. Mitler, O. Soave and W. Dement, Narcolepsy in several dogs, *J Am Vet Med Assoc.* **168**, 1036-1038 (1976).
34. L. Lin, J. Faraco and R. Li, The sleep disorder canine narcolepsy is caused by a mutation in the hypocretin (orexin) receptor 2 gene, *Cell.* **98**, 365-376 (1999).
35. M. Hungs, J. Fan, L. Lin, X. Lin, R. Maki and E. Mignot, Identification and functional analysis of mutations in the hypocretin (orexin) genes of narcoleptic canines, *Genome Res.* **11**, 531-539 (2001).
36. B. Ripley, N. Fujiki and M. Okura, Hypocretin levels in sporadic and familial cases of canine narcolepsy, *Neurobiol Dis.* **8**, 525-534 (2001).
37. L. Lin, M. Hungs and E. Mignot, Narcolepsy and the HLA region, *J Neuroimmunology.* **117**, 9-20 (2001).
38. J. Wagner, R. Storb, B. Storer and E. Mignot, DLA-DQB1 alleles and bone marrow transplantation experiments in narcoleptic dogs, *Tissue Antigens.* **56**, 223-31 (2000).
39. D. Chabas, S. Taheri, C. Renier and E. Mignot, The Genetics of Narcolepsy, *Annu Rev Genomics Hum Genet.* **4**, 459-83 (2003).
40. S. Overeem, T. Scammell and G. Lammers, Hypocretin/orexin and sleep: implications for the pathophysiology and diagnosis of narcolepsy, *Curr Opin Neurology.* **15**, 739-745 (2002).
41. M. Gencik, N. Dahmen, S. Wieczorek and M. Kasten, A prepro-orexin gene polymorphism is associated wtih narcolepsy, *Neurology.* **56**, 115-117 (2001).
42. T. Juji, M. Satake, Y. Honda and Y. Doi, HLA antigens in Japanese patients with narcolepsy, *Tissue Antigens.* **24**, 316-19 (1984).
43. W. H. O. (WHO), WHO: Nomenclature Committee for factors of the HLA system, 1989, *Immunogenetics.* **31**, 131-40 (1990).
44. S. Neely, R. Rosenberg, J. Spire, J. Antel and B. Arnason, HLA antigens in narcolepsy, *Neurology.* **37**, 1858-60 (1987).
45. S. Kuwata, K. Tokunaga, F. Jin, T. Juji, T. Sasaki and Y. Honda, Letter to the editor after Dr. Aldrich's review on narcolepsy, *NEJM.* **324**, 271-2 (1991).
46. K. Matsuki, F. Grumet, X. Lin, C. Guilleminault, W. Dement and E. Mignot, HLA DQB1*0602 rather than HLA DRw15 (DR2) is the disease susceptibility gene in Black narcolepsy, *Lancet.* **339**, 1052 (1992).
47. E. Mignot, A. Kimura, A. Lattermann, X. Lin, S. Yasunuga and G. Mueller-Eckhardt, Extensive HLA class II studies in 58 non DRB1*15 (DR2) narcoleptic patients with cataplexy, *Tissue Antigens.* **49**, 329-41 (1997).

48. E. Mignot, R. Hayduk, J. Black and F. Grumet, HLA DQB1*0602 is associated with cataplexy in 509 narcoleptic patients, *Sleep.* **20**, 1012-1020 (1997).
49. Z. Pelin, C. Guilleminault, N. Risch, F. Grumet and E. Mignot, HLA DQB1*0602 homozygosity increases relative risk for narcolepsy but not disease severity in two ethnic groups. US Modafinil in Narcolepsy Multicenter Study Group, *Tissue Antigens.* **51**, 96-100 (1998).
50. Y. Dauvilliers, M. Bazin, B. Ondze, O. Bera, M. Bazin and A. Besset, Severity of narcolepsy among Frech of different ethnic origins (South of France and Martinique). *Sleep.* **25**, 50-5 (2002).
51. E. Mignot, J. Meeyan, F. Grumet, J. Hallmeyer, C. Guilleminault and P. Hesla, HLA class II and narcolepys in thirty-three multiplex families, *Sleep Res Online.* **25**, 303 (1996).
52. H. Koch, I. Craig, M. Dahlitz, R. Denney and D. Parkes, Analysis of the monoamine oxidase genes and the Norrie disease gene locus in narcolepsy, *Lancet.* **353**, 645-646 (1999).
53. Y. Dauvilliers, E. Neidhart, M. Lecendreux, M. Billiard and M. Tafti, MAO-A and COMT polymorphisms and gene effects in narcolepsy, *Mol Psychiatry.* **6**, 367-72 (2001).
54. H. Hohjoh, T. nakayama, J. Ohashi, T. Miyagawa, H. Tanaka, T. Akaza, Y. Honda, T. Juji and K. Tokunaga, Signficant association of a single nucleotide polymorphism in the tumor necrosis factor-alpha (TNF-alpha) gene promoter with human narcolepsy, *Tissue Antigens.* **54**, 138-145 (1999).
55. S. Shoham, D. Davenne, A. Cady, C. Dinarello and J. Krueger, Recombinant tumor necrosis factor and interleukin 1 enhance slow-wave sleep, *Am J Physiol.* **253**, 42-9 (1987).
56. H. Hohjoh, N. Terada, M. Kawashima, Y. Honda and K. Tokunaga, Significant association of the tmor necrosis factor receptor 2 (TNFR2) gene wiht human narcolepsy, *Tissue Antigens.* **56**, 446-48 (2000).
57. Y. Dauvilliers, B. Carlander, N. Molinari, A. Desautels, M. Okun, M. Tafti, J. Montplaisir, E. Mignot and M. Billiard, Month of birth as a risk factor for narcolepsy, *Sleep.* **26**, 663-5 (2003).
58. S. Nishino, B. Ripley and S. Overeem, Hypocretin (orexin) deficiency in human narcolepsy, *Lancet.* **355**, 39-40 (2000).
59. E. Mignot, G. Lammers, B. Ripley, M. Okun, S. Nevsimalova, S. Overeem, J. Vankova, J. Black, J. Harsh, C. Bassetti, H. Schrader and S. Nishino, The role of CSF hypocretin measurements in the diagnosis of narcolepsy and other hypersomnias, *Arch Neurol.* **59**, 1553-1562 (2002).
60. S. Nishino, N. Fujiki and B. Ripley, Decreased brain histamine content in hypocretin/orexin receptor-2 mutated narcoleptic dogs, *Neurosci Lett.* **313**, 125-128 (2001).
61. T. Kanbayashi, T. Kodama, H. Kondo, S. Satoh, N. Miyazaki, K. Kuroda, M. Abe, S. Nishino, Y. Inoue and T. Shimizu, CSF Histamine and Noradrenaline Contest in Narcolepsy and Other Sleep Disorders, *Sleep.* **27**, A236 (2004).
62. Z. Huang, W. Qu and W. Li, Arousal effect of orexin A depends on activation of the histaminergic system, *Proc Natl Acad Sci USA.* **98**, 9965-9970 (2001).
63. C. Saper, T. Chou and T. Scammell, The sleep switch: hypothalamic control of sleep and wakefulness, *Trends Neurosci.* **24**, 726-731 (2001).
64. A. Schuld, J. Hebebrand, F. Geller and T. Pollmacher, Increased body-mass index in patients with narcolepsy, *Lancet.* **355**, 1274-1275 (2000).
65. M. Okun, L. Lin, Z. Pelin, S. Hong and E. Mignot, Clinical aspects of narcolepsy-catalexy across ethnic groups, *Sleep.* **25**, 27-35 (2002).
66. A. Yamanaka, C. Beuckmann and J. Willie, Hypothalamic orexin neurons regulate arousal according to energy balance in mice, *Neuron.* **38**, 701-13 (2003).
67. M. Aldrich, R. Chervin and B. Malow, Value of the multiple sleep latency test (MSLT) for the diagnosis of narcolepsy, *Sleep.* **20**, 620-629 (1997).
68. A. Moscovitch, M. Pertinen and C. Guilleminault, The positive diagnosis of narcolepsy and narcolepsy's borderland, *Neurology.* **43**, 55-60 (1993).
69. R. Salomon, B. Ripley, J. Kennedy, B. Johnson, D. Schmidt, J. Zeitzer, S. Nishino and E. Mignot, Diurnal variation of cerebrospinal fluid hypocretin-1 (Orexin-A) levels in control and depressed subjects, *Biol Psychiatry.* **54**, 96-104 (2003).
70. Y. Yoshida, N. Fujiki, T. Nakajima, B. Ripley, H. Matsumura, H. Yoneda, E. Mignot and S. Nishino, Fluctuation of extracellular hypocretin-1 (orexin A) leves in the rat in relation to the light-dark cycle and sleep-wake activities, *Eur J Neurosci.* **14**, 1075-1081 (2001).
71. M. Dalal, A. Schuld, M. Haack, M. Uhr, P. Geisler, I. Eisensehr, S. Noachtar and T. Pollmacher, Normal plasma levels of orexin A (hypocretin-1) in narcoleptic patients, *Neurology.* **56**, 1749-1751. Erraturm in : Neurology 2002; 58:334. (2001).
72. N. Igarashi, K. Tatsumi, A. Nakamura, S. Sakao, Y. Takiguchi, T. Nishikawa and T. Kuriyama, Plasma orexin-A levels in obstructive sleep apnea-hypopnea syndrome, *Chest.* **124**, 1381-1385 (2003).
73. S. Sakurai, T. Nishijima, Z. Arihara and K. Takahashi, Plasma orexin-A levels in obstructive sleep apnea-hypopnea syndrome, *Chest.* **125**, 1963 (2004).

74. S. Taheri, J. M. Zeitzer and E. Mignot, The role of hypocretins (orexins) in sleep regulation and narcolepsy, *Annu Rev Neurosci.* **25**, 283-313 (2002).
75. S. Higuchi, A. SUsui, M. Murasaki, S. Matsushita, N. Nishioka, A. Yoshino, T. Matsui, H. Muraoka, Y. Ishizuka, S. Kanba and T. Sakurai, Plasma orexin-A lower is lower in patients with narcolepsy, *Neurosci Lett.* **318**, 61-64 (2002).
76. E. Karteris, J. Chen and H. Randeva, Expression of human prepro-orexin and signaling charcteristics or orexin receptors in the male reproductive system, *J Clin Endocrinol Metab.* **89**, 1381-1385 (2004).
77. L. Krahn, VS, L. Oliver, B. Boeve and M. SIlber, Hypocretin (orexin) levels in cerebrospinal fluid of patients with narcolepsy: relationship to cataplexy and HLA DQB1*0602 status, *Sleep.* **25**, 733-736 (2002).
78. T. Kanbayashi, Y. Inoue and S. Chiba, CSF hypocretin-1 (orexin A) concentrations in narcolepsy with and without cataplexy and idiopathic hypersomnia, *J Sleep Res.* **11**, 91-93 (2002).
79. M. Dalal, A. Schuld and T. Pollmacher, Undetectable CSF level of orexin A (hypocretin-1) in a HLA-DR2 negative patient with narcolepsy-cataplexy, *J Sleep Res.* **11**, 273 (2002).
80. E. Mignot, W. Chen and J. Black, On the value of measuring CSF hypocretin-1 in diagnosing narcolepsy, *Sleep.* **26**, 646-9 (2003).
81. B. Ripley, S. Overeem and N. Fujiki, CSF hypocretin/orexin levels in narcolepsy and other neurological conditions, *Neurology.* **57**, 2253-2258 (2001).
82. C. Guilleminault, K. FAull, L. Miles and J. van den Hoed, Posttraumatic excessive daytime sleepiness: a review of 20 patients, *Neurology.* **33**, 1584-1589 (1983).
83. R. Gledhill, P. Bartel, Y. Yoshida, S. Nishino and T. Scammell, Narcolepsy caused by acute disseminated encephalomyelitits, *Arch Neurol.* **61**, 758-60 (2004).
84. I. Rubinstein, T. Gray, H. Moldofsky and V. Hoffstein, Neurosarcoidosis associated with hypersomnolence treated with corticosteroids and brain irradiation, *Chest.* **94**, 205-6 (1988).
85. G. Rosen, A. Bendel, J. Neglia, C. Moertel and M. Mahowald, Sleep in children with neoplasms of the central nervous system: case review of 14 children, *Pediatrics.* **112**, e46-54 (2003).
86. A. Autret, B. Lucas, K. Mondon, C. Hommet, P. Corcia, D. Saudeau and B. de Toffol, Sleep and brain lesions: a critical review of the literature and additional new cases, *Neurophysiol Clin.* **31**, 356-75 (2001).
87. J. Arii, T. Kanbayashi and Y. Tanabe, A hypersomnolent girl with decreased CSF hypocretin level afte removal of a hypothalamic tumor, *Neurology.* **56**, 1775-1776 (2001).
88. T. Scammell, S. Nishino, E. Mignot and C. Saper, Narcolepsy and low CSF orexin (hypocretin) concentration after a diencephalic stroke, *Neurology.* **56**, 1751-1753 (2001).
89. J. Vankova, I. Stepanova, R. Jech, M. Elleder, L. Ling, E. Mignot, S. Nishino and S. Nevsimalova, Sleep disturbances and hypocretin deficiency in Niemann-Pick disease type C, *Sleep.* **26**, 427-430 (2003).
90. H. Fryssira, S. Kountoupi, J. Delaunoy and L. Thomaidis, A female with Coffin-Lowry syndrome and "cataplexy", *Genet Couns.* **13**, 405-9 (2002).
91. J. Parkes, Genetic factors in human sleep disorders with special reference to Norrie disease, Prader-Willi syndrome and Moebius syndrome, *J Sleep Res.* **8S**, 14-22 (1999).
92. T. Kanbayashi, M. Abe, S. Fujimoto, T. Miyachi, T. Takahashi, T. Yano, Y. Sawaishi, J. Arii, G. Szilagyi and T. Shimizu, Hypocretin deficiency in Niemann-Pick type C with cataplexy, *Neuropediatrics.* **34**, 52-53 (2003).
93. J. Martinez-Rodriguez, L. Lin, A. Iranzo, D. Genis, M. Marti, J. Santamaria and E. Mignot, Decreased hypocretin-1 (Orexin-A) levels in the cerebrospinal fluid of patients with myotonic dystrophy and excessive daytime sleepiness, *Sleep.* **26**, 287-290 (2003).
94. J. Wisor, S. Nishino, I. Sora, G. Uhl, E. Mignot and D. Edgar, Dopaminergic role in stimulant-induced wakefulness, *J Neurosci.* **21**, 1787-1794 (2001).
95. Xyrem approved for muscle problems in narcolepsy, *FDA Consum.* **36**, 7 (2002).
96. B. Carlander, J. Eliaou and M. Billiard, Autoimmune hypothesis in narcolepsy, *Neurophysiol Clin.* **23**, 15-22 (1993).
97. E. Mignot, M. Tafti, W. Dement and F. Grumet, Narcolepsy and immunity, *Adv Neuroimmunol.* **5**, 23-37 (1995).
98. D. Hinze-Selch, T. Wetter, Y. Zhang, H. Lu and E. Albert, In vivo and in vitro immune variables in patients with narcolespy and HLA-DR2 matched controls, *Neurology.* **50**, 1149-1152 (1998).
99. M. Hecht, L. Ling, C. Kushida, D. Umetsu, S. Taheri, M. Einen and E. Mignot, Report of a case of immunosuppression with prednisone in an 8-year old boy with an acute onset of hypocretin-deficiency narcolepsy, *Sleep.* **26**, 809-810 (2003).
100. M. Lecendreux, S. Maret, C. Bassetti, M. Mouren and M. Tafti, Clinical efficacy of high-dose intravenous immunoglobulins near the onset of narcolepsy in a 10-year-old boy, *J Sleep Res.* **12**, 347-8 (2003).
101. J. John, M. Wu and J. Siegel, Systemic administration of hypocretin-1 reduces catapexy and normalizes sleep and waking durations in narcoleptic dogs, *Sleep Res Online.* **3**, 23-28 (2000).

102. S. Nishino, N. Fujiki, Y. Yoshida and E. Mignot, The effects of hypocretin-1 in hypocretin receptor-2 mutated and hypocretin deficient narcoleptic dogs, *Sleep.* **26**, A287 (2003).
103. Y. Yoshida, N. Fujiki, R. Maki, D. Schwarz and S. Nishino, Differential kinetics of hypocretins in the cerebrospinal fluid after intracerebroventricular administration in rats, *Neurosci Lett.* **346**, 182-6 (2003).

AN APPROACH TO DETERMINING THE FUNCTIONS OF HYPOCRETIN (OREXIN)

Jerome M. Siegel *

1. INTRODUCTION

Many papers on hypocretins/orexins begin with a list of the functions they regulate. Included in these lists are some of the following: sleep, food intake, water intake, gastric acid secretion, blood pressure, heart rate, movement, muscle tone, arousal, release of lutenizing hormone, corticosterone, insulin, growth hormone and prolactin. It is also reported that hypocretin/orexin coordinates monoamine, acetylcholine and amino acid release. Finally it is well established that the loss of hypocretin neurons is linked to narcolepsy (reviewed in ref. 1). One can anticipate that with further research the "laundry list" of hypocretin functions will continue to grow.

Although the finding of links between hypocretin/orexin and a wide range of behaviors and physiological changes is useful, it is obvious that a simple catalog of hypocretin/orexin relations does not provide a fundamental insight into its function(s). It is analogous to saying that the biceps muscle is involved in eating, drinking, motor activity, sexual behavior; sleep etc., because its activity is strongly modulated during all of these behaviors. In terms of its putative arousal functions, it is not sufficient to say that hypocretin/orexin is arousal related. A number of brain systems are active during arousal.[2,3] Does hypocretin/orexin play a unique role? How does hypocretin's role in each of these behaviors or control mechanisms differ from that of other neurotransmitter systems? How does the loss of hypocretin/orexin explain the symptoms of narcolepsy? Is abnormal hypocretin/orexin function involved in all cases of human narcolepsy, including those with normal hypocretin/orexin levels in the CSF and no mutation of the hypocretin/orexin system?[4,5,6] The theme of this chapter is that some answers are beginning to emerge from the hypocretin/orexin literature, although much remains to be done.

* Jerome M. Siegel VA GLAHS Sepulveda and Dept. of Psychiatry, UCLA Medical Center, Los Angeles CA

2. ARE HYPOCRETIN CELLS HOMOGENEOUS?

One key question that must be addressed is whether all hypocretin/orexin neurons have the same function or whether there may be subcategories of hypocretin/orexin neurons dedicated to different physiological or behavioral functions. An analogy may be drawn to the dorsal raphe serotonergic, posterior hypothalamic histaminergic and locus coeruleus noradrenergic neurons. Each one of these cell groups has a fairly homogeneous population of cells in terms of size and neurochemical phenotype. Existing evidence suggests that all these cell groups show a similar "sleep-off" pattern of discharge, i.e. they discharge tonically during waking, greatly reduce activity in nonREM sleep and cease activity in REM sleep.[7] Although many cells in these groups send multiple axonal projections to more rostral and more caudal regions, there is some specificity. For example more caudally placed locus coeruleus cells are more likely to have caudal projections than more rostrally located cells.[8,9] It should also be noted that although cells in these cell groups may have similar projections, local presynaptic mechanisms may strongly modulate release,[10,11] so that for example serotonergic cells may release 5HT on one side of the brain but not on the other.[12]

We need to consider the possibility that the hypocretin/orexin cell population may also have subgroups with different projection and perhaps even different activity patters and behavioral/physiological relations. In our early human work we measured cell size and found that the different hypocretin/orexin subpopulations had differing mean sizes. For example the lateral hypocretin/orexin cells were approximately 80 % larger in cross-sectional area than the dorsomedial cell group (Nienhuis and Siegel, unpublished data). hypocretin/orexin release may be modulated at target zones independently of discharge rate, as is the case with dopamine, serotonin containing and other cell groups. Currently the only studies of hypocretin/orexin release have looked at CSF levels or microdialysates in specific brain regions.[13] These observations necessarily reflect the overall changes in hypocretin/orexin release. However, individual hypocretin/orexin neurons could have discharge or release patterns which differ from this overall pattern. Until we can identify individual hypocretin/orexin neurons *in vivo* we will not be able to directly address these questions. Nevertheless studies of hypocretin/orexin release provide important clues to the nature of hypocretin/orexin release.

3. REGULATION OF HYPOCRETIN/OREXIN RELEASE

In studies of normal and of narcoleptic dogs, we have found that hypocretin/orexin level is not simply a property of a given animal or a function of time of day. Rather, it is closely tied to behavior.

Because of the link between hypocretin/orexin and narcolepsy, we first investigated the effects of sleep deprivation and consequent sleepiness on hypocretin/orexin level. We found that sleep deprivation for 24 hours, executed by walking the dogs whenever they began to go to sleep produced a 70% increase in hypocretin/orexin levels relative to dogs whose exercise was "yoked" to that of the experimental animals.[14] We measured activity actigraphically and found, not surprisingly, that the experimental animals had fewer periods of extended inactivity. To control for the increased activity of the sleep deprived animals, we compared animals that were kept awake for a 2 hour period, an interval that does not require forced locomotion, to animals that were active in a yard for the same 2 hour period. We found that this manipulation produced the same elevation of hypocretin/orexin level produced by sleep deprivation (Figure 1). We found the same increase in hypocretin/orexin level with activity in the normal cat.[13] Thus, the most parsimonious conclusion is that the activity rather than the sleep loss was the cause of the elevation of hypocretin/orexin level after sleep deprivation.

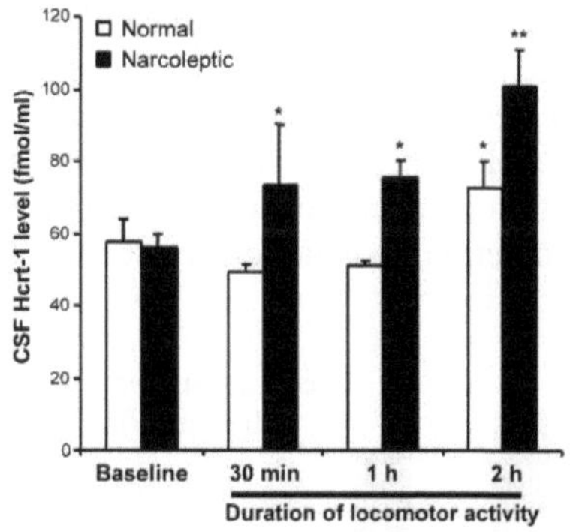

Figure 1 Exercise elevates hypocretin/orexin levels measured in the cerebrospinal fluid in both normal and narcoleptic dogs.[14]

We next subjected normal dogs to 48 h of food deprivation (we used dogs in these studies because of the ease of drawing adequate volumes of CSF from dogs in comparison to smaller animals). "Orexigenic" compounds (i.e. compounds stimulating eating), such as neuropeptide Y increase in concentration with food deprivation (reviewed in ref 1). We found no such increase in hypocretin/orexin level with food deprivation (Figure 2).

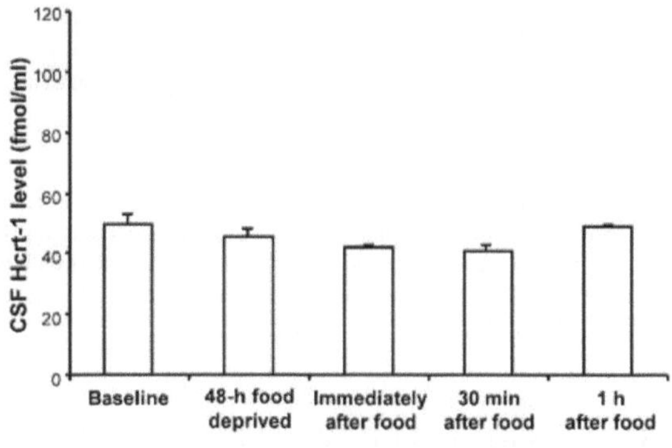

Figure 2. Food deprivation does not alter hypocretin level, and eating after food deprivation does not significantly alter hypocretin/orexin level in contrast to other "orexigenic" compounds whose levels are significantly elevated with food deprivation.[14] A similar elevation of hypocretin/orexin level with motor activity was observed in normal cats. [13]

We also saw no significant change in hypocretin/orexin level after feeding at the end of the deprivation period.[14] These results are inconsistent with the hypothesis that hypocretin/orexin release is tightly linked to food intake. However, in the context of our activity findings, if a food-deprived animal became more active under conditions of food deprivation, it would be predicted that hypocretin/orexin level would rise. Other data that is inconsistent with an orexigenic role of hypocretin/orexin is the lack of anorexia or reduced weight in the hypocretin/orexin ligand knockout mouse,[15] the obesity of the ataxin mutant mouse in which hypocretin/orexin cells degenerate postnatally[16] and the obesity tendency in unmedicated human narcoleptics.[17] Both the narcoleptic human and ataxin mutant animals gain weight despite *reduced* food intake.[1] This obesity can be explained by the reduced activity and consequent caloric expenditure, but is inconsistent with the hypothesis that hypocretin/orexin deficient mice would be anorexic.

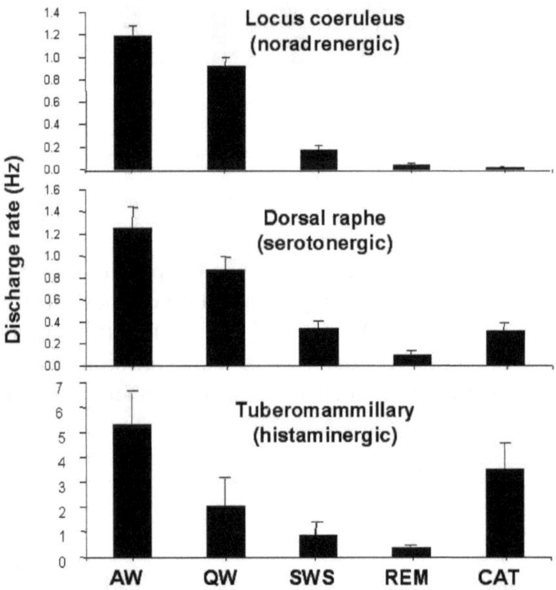

Figure 3. Sleep waking cycle discharge of monoaminergic cells. Monoaminergic cells behave similarly across normal sleep cycles, with all showing maximal discharge in active waking, decreased discharge in quiet waking, greatly reduced discharge in sow wave sleep (SWS) and minimal discharge in REM sleep. In cataplexy, noradrenergic locus coeruleus cells are "off," whereas histaminergic cells are "on" and serotonergic dorsal raphe cells have an intermediate pattern (From ref 7).

4. ARE HYPOCRETINS ASSOCIATED WITH LOCOMOTOR ACTIVITY?

A final point of reference for developing hypotheses as to the underlying function(s) of hypocretin/orexin cells is careful observation of animals and humans without hypocretin/orexin neurons or with mutations affecting hypocretin/orexin release or postsynaptic response to the peptide. These are narcoleptic animals. Genetically narcoleptic dogs do not weigh less than age and breed matched controls (unpublished observations (John, Wu and Siegel). It is unclear if they move less than controls, but clearly they do not exhibit the prolonged periods of activity that characterize normal animals. A typical symptom of human narcolepsy is periods of daytime immobility (i.e. naps) coupled with interrupted nighttime sleep.[18] A second symptom of narcolepsy is sudden losses of muscle tone, without loss of consciousness (cataplexy). These losses are linked to cessation of activity in locus coeruleus cells),[19] reduced activity in serotonergic cells[20] and maintained or increased activity in histaminergic cells (Figure 3). They are also linked to activation of medial medullary motor inhibitory cells (Figure 4).[21]

All of these motor links support an underlying connection of hypocretin/orexin release to motor activity. This link can explain many of the phenomena attributed to hypocretin/orexin cells. It is important to measure motor activity and muscle tone in any in vivo examination of hypocretin/orexin release correlates. Only such measurements can separate motor activity from other putative correlates of hypocretin/orexin activity.

Although the data discussed above link hypocretin/orexin release to motor activity, they still leave important questions unanswered. What aspect of motor activity is most closely linked to hypocretin/orexin release? Is it the types of movement, rhythmic vs.

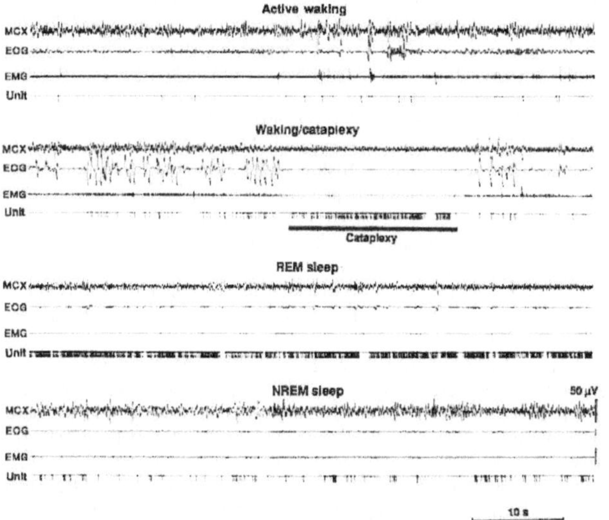

Figure 4. Medullary cataplexy-on cells. Most brainstem cells are active in waking and REM sleep. However the "cataplexy-on" cell type illustrated is inactive during waking with movements but is maximally active in REM sleep and immediately prior to and during cataplexy attacks in narcolepsy. These cells are likely to trigger cataplexy by active inhibition of motoneruons, acting in concert with disfacilitation produced by the cessation of activity in noradrenergic cells during cataplexy and REM sleep, as is shown in Figure 3 (From ref 21).

exploratory, rapid vs. slow? Is muscle tone itself the key variable? Is it simply the level of muscle tone? Is it the velocity of movement? Is it the continuity of movement? Is it the emotions that accompany activity? Is it the alertness that accompanies movement? The last two questions might be addressed by comparing the hypocretin/orexin release in animals moving at differing speeds on a treadmill. If movement per se is the key element there should be a lawful relationship. If it is the accompanying excitement, one might expect small or even absent increases in hypocretin/orexin release with prolonged rhythmic movement.

More data are needed to separate these and other potential correlates of hypocretin/orexin cell activity. Current in vivo studies of such phenomena have been performed by CSF extraction of by microdialysis. Such techniques do not permit a fine-grained analysis of hypocretin/orexin cell activity. Thus an important advance would be the in vivo recording of identification of hypocretin/orexin cells in relation to behavior. Techniques to identify such cells in vivo using the results of pioneering in vitro work are becoming available. We can look forward to a clarification of the underlying function(s) of hypocretin/orexin cells once these techniques are successful.

5. REFERENCES

1 J. M. Siegel, Hypocretin (orexin). role in normal behavior and neuropathology, *Annual Rev.of Psychol.* **55**:125-148 (2004).
2. B. E. Jones, Basic Mechanisms of Sleep-Wake States. In: *Principles and Practice of Sleep Medicine*, edited by M.H. Kryger, et al, pp. 134-154. W.B. Saunders, Philadelphia (2000).
3. J. M. Siegel, Brainstem mechanisms generating REM sleep, In: *Principles and Practice of Sleep Medicine*, edited by M.H. Kryger, et al, pp. 112-133. W.B. Saunders Company, Philadelphia. (2000).
4.S. Nishino, B. Ripley, S. Overeem, G. J. Lammers and E. Mignot, Hypocretin (orexin) deficiency in human narcolepsy, *Lancet.* **355**, 39-40 (2000).
5. C. Peyron, J. Faraco, W. Rogers, B. Ripley, S. Overeem, Y. Charnay, S. Nevsimalova, M. Aldrich, D. Reynolds, R. Albin, R. Li, M. Hungs, M. Pedrazzoli, M. Padigaru, M. Kucherlapati, J. Fan, R. Maki, G. J. Lammers, C. Bouras, R. Kucherlapati, S. Nishino and E. Mignot, A mutation in a case of early onset narcolepsy and a generalized absence of hypocretin peptides in human narcoleptic brains, *Nat Med.* **6**, 991-997 (2000).
6. T. C. Thannickal, R. Y. Moore, R. Nienhuis, L. Ramanathan, S. Gulyani, M. Aldrich, M. Cornford and J. M. Siegel, Reduced number of hypocretin neurons in human narcolepsy, *Neuron.* **27**, 469-74 (2000).
7. J. John, M.-F. Wu, L. N. Boehmer and J. M. Siegel, Cataplexy-active neurons in the posterior hypothalamus: implications for the role of histamine in sleep and waking behavior, *Neuron.* **42**:619 (2004).
8. G. Aston-Jones, M. Ennis, V. A. Pieribone, W. T. Nickell and M. T. Shipley, The brain nucleus locuscoeruleus: Restricted afferent control of a broad efferent network, *Science.* **234**:734-737 (1986).
9. G. Aston-Jones, S. L. Foote and F. E. Bloom, Anatomy and physiology of locus coeruleus neurons: functional implications. In: *Norepinephrine*, edited by M. Ziegler, et al, pp. 92-116. Williams and Wilkins, Baltimore (1984).
10. G. Di Chiara, G. Tanda, and E. Carboni, Estimation of in-vivo neurotransmitter release by brain microdialysis: the issue of validity, *Behav.Pharmacol.* **7**:640-657 (1996).
11. D. L. Marshall, P. H. Redfern and S. Wonnacott, Presynaptic nicotinic modulation of dopamine release in the three ascending pathways studied by in vivo microdialysis: comparison of naive and chronic nicotine-treated rats, *J Neurochem.* **68**:1511-9 (1997).
12. L. R. J. Baxter, E. C. Clark, R. F. Ackermann, G. Lacan and W. P. Melega, Brain Mediation of Anolis Social Dominance Displays. ii. differential forebrain serotonin turnover, and effects of specific 5-ht receptor agonists, *Brain Behav.Evol.* 57:184-201(2001).
13. L. I. Kiyashchenko, B. Y. Mileykovskiy, N. Maidment, H. A. Lam, M. F. Wu, J. John, J. Peever and J. M. Siegel, Release of hypocretin (orexin) during waking and sleep states, *J Neurosci.* **22**: 5282-6 (2002).
14. M. F. Wu, J. John, N. Maidment, H. Lam and J. Siegel, Hypocretin release in normal and narcoleptic dogs after food and sleep deprivation, eating, and movement, *Am J Physiol Regul Integr Comp Physiol.* **283**:R1079-86 (2002).
15. R. M. Chemelli, J. T. Willie, C. Sinton, J. Elmquist, T. Scammell, C. Lee, J. Richardson, S. Williams, Y. Xiong, Y. Kisanuki, T. Fitch, M. Nakazato, R. Hammer, C. Saper, and M. Yanagisawa, Narcolepsy in *orexin* knockout mice: Molecular genetics of sleep regulation, *Cell.* **98**:437-451 (1999).
16. J. Hara, C. T. Beuckmann, T. Nambu, J. T. Willie, R. M. Chemelli, C. M. Sinton, F. Sugiyama, K. Yagami, K. Goto, M. Yanagisawa and T. Sakurai, Genetic ablation of orexin neurons in mice results in narcolepsy, hypophagia, and obesity, *Neuron.* **30**:345-354 (2001).

17. A. Schuld, J. Hebebrand, F. Geller and T. Pollmacher, Increased body-mass index in patients with narcolepsy, *Lancet.* 355:1274-5 (2000).
18. M. S. Aldrich, Diagnostic aspects of narcolepsy, *Neurology.* **50**:S2-7 (1998)
19. M. F. Wu, S. A. Gulyani, E. Yau, E. Mignot, B. Phan and J. M. Siegel, Locus coeruleus neurons: cessation of activity during cataplexy, *Neuroscience.* **91**:89-99 (1999).
20. M. F. Wu, J. John, G. B. Nguyen, and J. M. Siegel, Serotonergic dorsal raphe REM-off cells reduce discharge but do not shut off during cataplexy, *Sleep.* **23**:A2-A3(2000).
21. J. M. Siegel, R. Nienhuis, H. M. Fahringer, R. Paul, P. Shiromani, W. C. Dement, E. Mignot and C. Chiu, Neuronal activity in narcolepsy: identification of cataplexy-related cells in the medial medulla, *Science.* **252**:1315-8 (1991).

HYPOCRETIN IN NEUROPSYCHIATRIC DISORDERS

Patrice Bourgin and Yves Dauvilliers*

1. INTRODUCTION

Narcolepsy is a disabling sleep disorder affecting 1 per 2000 individuals and characterized by two major symptoms, excessive daytime sleepiness and cataplexy. Narcolepsy was early shown to be tightly associated with HLA-DR2 and DQB1*0602 suggesting a possible autoimmune mechanism.[1] Recently, a major discovery in its physiopathology was made through the identification of hypocretin (hcrt) deficiency. It was first demonstrated in animal models that lack of hypocretin ligand or efficient hcrt receptor2 cause narcolepsy.[2,3] While preprohypocretin gene mutation in human narcolepsy remains an exception, a dramatic decrease of hcrt in the brain[4,5] and undetectable cerebrospinal fluid (CSF) levels of hcrt-1 are observed in most of the sporadic narcoleptic patients with typical cataplexy and HLA DQB1*0602 positive.[6-14]

Then, several studies extended the measurement of hcrt in the CSF to numerous neurological or psychiatric disorders. Several pathologies were found to be associated with a decrease in CSF hypocretin-1 levels when compared to normal controls, although the importance of these findings remains controversial. This review summarizes the actual knowledge on hcrt in neurological and psychiatric disorders, except those treated in a specific chapter of the present book i.e. the neurological disorder narcolepsy and, for the psychiatric part, drug addiction and stress.

2. HYPERSOMNIAS (EXCEPT TYPICAL NARCOLEPSY-CATAPLEXY)

2.1. Atypical Narcolepsy

Low CSF hypocretin-1 levels are reported in more than 90% of sporadic narcoleptic patients with cataplexy and HLA DQB1*0602 positive.[6-14] The results are more difficult

* Patrice Bourgin, Center for Narcolepsy Research, Stanford University, Palo Alto, California, 94304. Yves Dauvilliers, Hôpital Gui de Chauliac, Montpellier, INSERM E0361, Montpellier, France.

to interpret for the atypical forms of the disease. Narcolepsy without cataplexy and HLA DQB1*0602 negative is almost systematically associated with normal levels of hcrt-1 in the CSF and normal or low levels of hcrt-1 can be observed in narcoleptic patients with only one of the two following characteristics: cataplexy or HLA DQB1*0602 positive. Indeed, several narcoleptic patients with clear-cut cataplexy and normal CSF hypocretin-1 levels have been reported as well as narcoleptic patients without cataplexy but with low CSF hypocretin-1 levels.[7,10-14] Moreover, a dramatic decrease in the number of hypocretin-containing neurons of the hypothalamus has been reported in post mortem brains of both narcoleptic patients with and without cataplexy.[4,5] CSF hypocretin-1 level is also tightly associated with the presence of HLA DQB1*0602 allele since several DQB1*0602 negative narcoleptic-cataplectic patients had normal levels of hypocretin-1 in the CSF [7,10,11,14]. In addition, normal or intermediate hypocretin levels of CSF hcrt-1 were reported among narcoleptic patients with relatives affected either with narcolepsy-cataplexy or isolated daytime naps or lapses into sleep, suggesting that hcrt deficiency as measured by CSF hcrt levels is weakly associated to the familial form of the disease. These findings suggest that the DQB1*0602 allele may have a major role in conferring low or absent hypocretin neurotransmission.[7,10,11,14] However, the recent report of the first HLA-DQB1*0602-positive monozygotic twin pair discordant for narcolepsy with undetectable CSF hypocretin-1 in the affected twin and normal level in the unaffected co-twin suggests that environmental factors play a key role in the development of the disease. [15] Together these data suggest the presence of different mechanisms involved in the physiopathology of narcolepsy. However, an alteration of the hcrt system is still possible in the atypical form not associated with low level of hcrt-1 since a defect in hcrt-1 or 2 receptors as well as in any downstream element in the hcrt pathway or only partial hcrt lesions without low hcrt-1 consequences cannot be excluded.

2.2. Idiopathic Hypersomnia

Idiopathic hypersomnia is characterized by an excessive daytime somnolence with possibility of prolonged daytime sleep episodes of non REM sleep. There are no REM sleep abnormalities as cataplexy and no other sleep disorders that may alter the sleep at night causing an excessive daytime sleepiness. Monosymptomatic and polysymptomatic idiopathic hypersomnias are mainly associated with normal CSF hypocretin levels or high levels in a few cases.[11-14] This finding is of interest since hypocretin dysfunction appears not to be the final common pathway of the pathophysiology of most hypersomnia syndrome and allow the hypothesis of a continuum in the physiopathology of narcolepsy without cataplexy and idiopathic hypersomnia.

2.3. Post -Traumatic Hypersomnia

Head trauma is another rare cause of hypersomnia in which CSF hypocretin levels were investigated in a few cases.[11,13,14] The results including either normal or low CSF hypocretin-1 levels may depend on the severity of the trauma condition and especially the localization of the lesions. Although the involvement of hypocretin transmission abnormality remains uncertain in this condition, the long delay between brain injury, the occurrence of excessive daytime somnolence and the time of lumbar puncture may confound the results.

2.4. Obstructive Sleep Apnea Syndrome

Obstructive sleep apnea syndrome is characterized by repetitive episodes of upper airway obstruction that occur during sleep, usually associated with a reduction in blood oxygen saturation. Arousals are associated with the respiratory events leading to a sleep fragmentation and an excessive daytime somnolence. Because of its high prevalence the sleep apnea syndrome is the major cause of excessive daytime sleepiness disorders. The CSF hypocretin-1 levels in obstructive sleep apnea syndrome are within the control range.[11,16] In addition, patients affected with sleep apnea syndrome and more than 2 sleep onset REM periods on the multiple sleep latency tests, one of the diagnostic criteria for narcolepsy, were also associated with normal CSF hypocretin–1 levels.[14] These CSF hcrt-1 results do not reflect the excessive daytime sleepiness itself and do not support the involvement of hypocretin transmission in abnormal regulation of REM sleep in this condition.

2.5. Hypersomnias: Conclusion

Finally hypocretin ligand deficiency appears not to be the major cause for hypersomnia other than narcolepsy-cataplexy. A continuum in the pathophysiology of narcolepsy without cataplexy and idiopathic hypersomnia may be hypothezised.[11,13,14] However, a dysfunction in hypocretin-1 or 2 receptor, an ineffective signal neurotransmission or only partial hypocretin lesions without low CSF hypocretin-1 consequences cannot be definitely excluded in those disorders.

3. IMMUNE DISORDERS

Narcolepsy is due to the deficiency of the hypocretins and is associated with a loss of hypocretin-containing cells. The peripubertal onset of the disease, the tight HLA DQB1*0602 association, the low concordance rate in monozygotic twins and the complex genetic susceptibility in family studies suggest an autoimmune mechanism. The observation, in postmortem narcoleptic brains, of a gliosis in the posterior hypothalamus where the hypocretinergic cell bodies are located, as well as the possibility of efficacy of immunosuppressive treatment administered immediately after the onset of the disease, strongly support the hypothesis of a putative immune process targeting the hypocretinergic neurons with irreversible damage.[17] To test whether other neuroimmune disorders could target the hypocretin cells, the CSF hcrt-1 level has been evaluated in neuroimmune disorders.

3.1. Immune Polyneuropathies

The first evaluation has been done in acute immune polyneuropathies i.e. the Guillain-Barre syndrome, the Miller-Fisher syndrome that is a variant of the previous one, and the chronic inflammatory demyelinating polyneuropathy. In this latter disease, hcrt-1 levels in the CSF are normal, except in one case. In contrast, it has been confirmed, mostly in Japanese, that a subset of patients with Guillain-Barre syndrome have undetectable or low level of CSF hcrt-1 and that almost half of the Miller-Fisher syndrome have low hcrt-1 level.[7,18,19] This result has not been clearly confirmed in

Caucasian patients.[14,20] The patients with Guillain-Barre syndrome or Miller-Fisher syndrome who display low levels of hcrt-1 do not have cataplexy or sleep onset REM periods, two main symptoms of narcolepsy that are considered to be related to hcrt deficiency. Thus, further evaluations including polysomnography and multiple sleep latency tests are needed in order to research sleep abnormalities in these patients.

3.2. Encephalitis And Demyelinating Disorders

It has been recently discovered that a subset of patients with anti-Ma2 encephalitis, an uncommon autoimmune disorder, display narcoleptic symptoms with cataplexy depending on the localization of the encephalitis. If narcoleptic patients do not have anti-Ma2 antibodies, patients with anti-Ma2 encephalitis and symptomatic narcolepsy have undetectable CSF hcrt-1 when excessive daytime sleepiness is associated.[21,22] This latter observation suggests an extension of the autoimmune process to the lateral hypothalamus, in particular to the hcrt neurons. Indeed, anti-Ma2 encephalitis affects specific brain regions although Ma antigens are present in all neurons. The analogy to narcolepsy raises the question of the possibility of a ubiquitous antigen present on hcrt cells that would be preferentially targeted and damaged.

Multiple sclerosis is a chronic demyelinating disorder of the central nervous system and viral and autoimmune etiologies are postulated. In patients with multiple sclerosis CSF hcrt-1 levels are in the normal range.[7,14,23,24] This measurement was important to check since the physiopathology of multiple sclerosis has been proposed to have similarities with that of narcolepsy, sharing an autoimmune process and especially the same HLA haplotype DR2, DQB1*0602.[24] Only one patient with multiple sclerosis had low level of CSF hcrt-1.[25] This latter case was particular in that he had severe hypersomnia and bilateral hypothalamic lesions as documented by the MRI. The hypersomnia and the radiological hypothalamic abnormalities were corrected under treatment suggesting that, if the damage targets the hcrt-containing cells, it was still reversible.

Low CSF hcrt-1 levels with associated hypersomnia have been mentioned in other demyelinating disorder such as the acute disseminated encephalomyelitis but these case reports remain anecdotal.[26,27]

3.3. Immune Disorders: Conclusion

Hypocretin deficiency in narcolepsy is most probably due to hypocretin neuronal death as documented by the analysis of post-mortem brains from narcoleptic patients. It would be informative to evaluate the hypocretin system in post-mortem brains from patients with neuroimmune disorders. Indeed, in contrast to narcolepsy, full recovery of the hcrt levels associated with a clinical improvement occurs under treatment in some immune disorders such as many cases of Guillain-Barre syndrome,[18] suggesting a functional alteration of he hcrtergic system. Thus, CSF hcrt-1 levels in such disease might also predict sleep symptoms and the course of the disease. A subset of neuroimmune disorders might also be a good model for studying possible autoimmune alteration of the hcrt system.

4. OTHER NEUROLOGICAL DISORDERS

Low hypocretin levels are also observed in a large range of other neurological conditions. In these disorders the alteration of the hcrt level may reflect focal lesions in the hypothalamus or transient or chronic dysfunction of the hypothalamus.

4.1. Kleine-Levin Syndrome (KLS)

KLS is a rare cause of hypersomnia characterized by a recurrent alteration of sleep-wake regulation and frequent abnormal feeding and sexual behaviors. Recent association reported with HLA DQB1*0201 together with the young age of onset, the recurrence of symptoms, and the frequent infectious precipitating factors suggest an autoimmune etiology for KLS.[14] Since the role of hypocretin neuropeptides in both feeding behavior and sleep-wake regulation is now well established, an abnormality of the hypocretin neurotransmission has been proposed in KLS patients. CSF hypocretin-1 has been measured in few cases with normal results for most of them (5 out of 6).[11,14,28] However, the most striking finding is the two fold decrease in CSF hypocretin-1 level during the symptomatic episode in the only case with CSF available during both asymptomatic and symptomatic period, suggesting a possible intermittent alteration of the hypocretin system.[14] We may suggest a focal, transient, and immune-modulated process in the hypothalamic hypocretin system in KLS, resulting in altered hypocretin neurotransmission only during symptomatic episodes.

4.2. Prader Willi Syndrome

Prader Willi syndrome is a complex genetic disorder characterized by hypotonia, short stature, hypogonadism, mental retardation, behavioral troubles and hyperphagia, which result in excessive obesity. Intermediate CSF hypocretin levels have been reported in the literature in two cases of Prader-Willi syndrome.[11,14] This finding might represent a biological marker of this syndrome that could be also associated with Kleine-Levin syndrome.

4.3. Niemann-Pick Disease

Niemann-Pick disease type C is an inherited metabolic disorder in which harmful quantities of a fatty substance accumulate in several tissues including the brain. Subjects with Niemann-Pick disease type C have been reported to display narcolepsy like symptoms, including cataplexy. CSF hypocretin-1 levels were measured in four patients with short mean sleep latencies on the MSLT for three patients and sleep onset REM periods in one patient with cataplexy and HLA DQB1*0602 positive. CSF hypocretin-1 levels were reduced in 2 patients including one patient with cataplexy while, in the 2 other patients, the levels were at the lower range of the normal values. Those findings suggest that lysosomal storage abnormalities that are involved in the physiopathology of the disease may alter the functioning of the hypothalamus and, more specifically, hypocretin-containing cells. The hypocretin system dysfunction observed may explain the sleep abnormalities and cataplexy reported in Niemann-Pick disease type C.[29]

4.4. Myotonic Dystrophy

Myotonic dystrophy type 1 is a multisystem disorder with myotonia, muscle weakness, cataracts, endocrine dysfunction, and intellectual impairment caused by a CTG triplet expansion in the 3' untranslated region of the DMPK gene on 19q13. Myotonic dystrophy type 1 is frequently associated with an excessive daytime sleepiness in particular a short sleep latency and the presence of sleep onset REM periods during the MSLT, two characteristics observed in narcolepsy. CSF hypocretin-1 levels were measured in 7 patients in the literature with a mean average of CSF hcrt-1 significantly lower than that in normal controls.[14,30] However, the significance of this latter result needs to be confirmed on a larger sample because only one patient had a low level and three had intermediate levels. Hypocretin-1 levels did not correlate clinically with disease severity or duration or with subjective or objective sleepiness reports. We therefore may propose that a dysfunction of the hypothalamic hypocretin system may mediate sleepiness and abnormal regulation of REM in a subgroup of patients with myotonic dystrophy type 1.

4.5. Hypothalamic Lesions

A low level of CSF hypocretin-1 has been reported in several cases of alteration of the hypothalamus. For example, a decreased CSF hypocretin level was reported in a hypersomnia case after hypothalamic tumor removal[31] as well as in a young man who developed narcolepsy after a large hypothalamic stroke.[32] In contrast, a 60-year-old man with acromegaly, who developed narcolepsy with cataplexy two weeks after completing radiotherapy for a pituitary adenoma had normal CSF hypocretin-1 levels,[33] indicating that other factors may be important in the development of this condition.

We may therefore hypothesize that a loss of orexin neurons or their relevant targets may underlie the neuropathology of several secondary cases of excessive daytime sleepiness with or without cataplexy. In addition, a common hypothalamic, hypocretin-independent dysfunction may also be present in some of hypersomnia conditions.

5. MOVEMENT DISORDERS

5.1. Parkinson Disease

Parkinson's disease (PD) is a neurodegenerative disease, mainly of the dopaminergic cells, and is a movement disorder characterized by tremor, rigidity and bradykynesia. Sleep disturbances are frequently observed in PD characterized by difficulties to maintain sleep and excessive daytime sleepiness. Recent attention focused on this latter symptom to report more specifically 'sleep attacks' that can be induced by all dopaminergic drugs. The dopaminergic ventral tegmental area and substantia nigra is one the main target of the hypocretins. Dopamine and hypocretins have both a crucial role in control of arousal[34-36] and the projections of these two arousal-related systems, originating in distinct brain areas, jointly target several forebrain regions and brainstem monoaminergic nuclei involved in regulating motivational processes.[37] Because an alteration of the hypocretinergic system might major the excessive daytime sleepiness related to dopamine deficiency and because it is not known whether dopamine projections may influence in return the hypocretin tone, several studies examined CSF hcrt-1 levels in PD patients and controls.[11,14,38,39] Two of them focused on the relation ship between the disease characteristics and the hcrt levels. Overeem et al.,[38] reported normal lumbar CSF hypocretin-1 levels in a few mild PD patients with excessive daytime somnolence whereas, on a larger sample, Drouot et al.,[39] showed low levels of ventricular CSF hypocretin-1 in advanced form of the disease as compared to neurological controls. Despite the methodological differences, this discrepancy might be explained by the fact that hcrt levels may be altered only in a restrictive subgroup of PD patients; especially in advanced forms when a highly developed degenerative process may extend to hypocretin neurons. Further studies focusing on hypocretins in different forms of the disease are required in order to determine whether a subgroup of patients[40] display low level of hctr and whether these low levels would be related to an hcrt cell degeneration, which would need to be considered for the therapeutic management of these patients.

5.2. Restless Legs Syndrome

Restless legs syndrome is a common disorder (5 to 15% of the general population) characterized by dysesthesias in the legs occurring mainly at night and accompanied by a compelling urge to move the legs leading often the patient to walk. Most patients also have periodic leg movements during sleep that are associated with microarousals leading to sleep fragmentation and excessive daytime somnolence in many cases. The cause of the disease still remains unknown except secondary cases due to iron deficiency, side effects of certain drugs, peripheral neuropathies, uremia or others. The efficiency of L-dopa and D2/D3 agonists for the treatment of RLS as well as PLM suggests the involvement of the dopaminergic system in the pathophysiology of the disease. Hypocretins are wakefulness promoting peptides and stimulate the dopaminergic neurons, both systems providing joint innervation to arousal systems. There is no consensus on hcrt levels in RLS. Allen et al., first reported increase CSF levels of hcrt-1 in the early-onset form of the disease that have a high familial aggregation and slow evolution as compared to the late-onset form that is associated with progressive severity and low familial aggregation.[41] This result has not been confirmed in two other studies.[11,42] They did not find any correlation with the severity of the RLS or PLM or sleep parameters. One possible explanation for the lack of replication of this result might be due to an alteration of the hcrt neurotransmission only in a time-fashion during the symptomatic period of the day, which might explain the differences between the studies since the collection of samples were performed at different time of the day. Further explorations are needed in order to test this hypothesis.

6. DEMENTIA

Alzheimer disease is the most frequent cause of dementia with frequent sleep/wake cycle disturbances and alterations of circadian rhythms. Patients affected with Alzheimer disease had normal levels of CSF hypocretin-1.[7,14] In addition, no correlation between the CSF hypocretin levels and normal ageing has been reported. Dementia with Lewy bodies (DLB) is a rare cause of dementia associated with a progressive cognitive decline, fluctuations in cognition, recurrent visual hallucinations and parkinsonism. Sleep-wake abnormalities were frequent in DLB including excessive daytime sleepiness, insomnia, and REM sleep behavior disorder. The hcrt-1 levels in DLB patients are within the normal range.[20] Considering the similar observation of normal CSF hypocretin-1 levels in Parkinson disease, it is likely that the sleep-wake disturbances reported in these two neurodegenerative disorders are primarily unrelated to a dysfunction in the hypocretinergic neurotransmission.

Prion disorders are frequently associated with sleep abnormalities especially in fatal familial insomnia and Creutzfeldt-Jakob disease. Fatal familial insomnia (FFI) is a rare hereditary prion disease clinically characterized by inability to sleep, dysautonomia and motor disturbances, caused by a mutation at codon 178 of the prion protein (PrP) gene. Previous studies in FFI showed that the dramatic reduction of sleep pattern and total sleep time was related to selective lesion of mediodorsal and anterior ventral nuclei of the thalamus with a probable associated functional hypothalamus alteration. However, one published case report[43] describing a FFI homozygote at codon 129 and a personal observation (on FFI heterozygote at codon 129) failed to report any abnormality in CSF hypocretin-1 level. The same results were found in Creutzfeldt-Jakob disease.[14,43] The abnormal regulation of the sleep-wake cycle in prion disorders seems unrelated to an alteration in the hypocretinergic transmission pathway.

Finally, all these data together suggest that the hcrt neurotransmission is not involved in dementia.

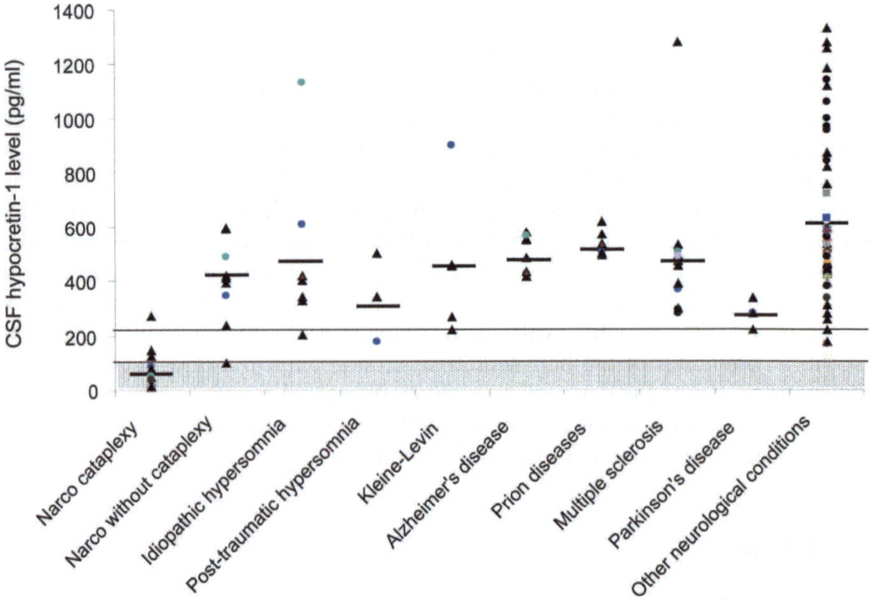

Figure 1. CSF hypocretin-1 levels in various etiologies of hypersomnia (n=50) and in major neurological disorders (n=88). Each dot represents a single patient. Thick bars represent mean values for each group. The bottom dotted line represents the cut-off point for low level (less than 110pg/ml) and the upper dotted line the cut-off for intermediate level (between 110 and 200 pg/ml)

Low CSF hypocretin-1 levels were observed in 23 out of 26 narcolepsy-cataplexy patients with one narcoleptic patient with cataplexy with normal hypocretin level (HLA-DQB1*0602 negative) and the two others (familial forms) were within the intermediate level. Only one out of 9 narcoleptic patients without cataplexy had a low level. Idiopathic hypersomniac patients had normal hypocretin-1 levels and one out of three patients affected with post-traumatic hypersomnia had also intermediate levels. The four Kleine-Levin patients had normal hypocretin levels during their asymptomatic periods (but one KLS patient investigated during both symptomatic and asymptomatic periods showed a two-fold decrease from 221 to 111 pg/ml in CSF hypocretin-1 during a symptomatic episode).[14] The 88 non-hypersomnia patients included Alzheimer's disease, prion disease, multiple sclerosis, Parkinson's disease and others various disease categories: none of them had low level and only three had intermediate levels including two patients with normal pressure hydrocephalus and one with unclear central vertigo.

7. PAIN DISORDERS

The regulation of pain perception is another function in which the hypothalamus plays a part. The destruction of the posterior hypothalamus caused transient hyperalgesia in rats and the electrical stimulation of this structure can relieve pain. Functional abnormalities of the hypothalamus have been proposed in the initiation of central pain disorders such as in cluster headache and migraine. In addition, hypocretin neurons have mu-opioid receptors and respond to chronic morphine administration and opiate antagonist-precipitated morphine withdrawal.[44] Induction of c-Fos and the hypocretin gene itself is observed in orexin cells during morphine withdrawal. Hypocretin knock-out mice develop attenuated morphine dependence, as indicated by a less severe antagonist-precipitated withdrawal syndrome.[44] Finally, hypocretin-1 induces analgesia in acute and inflammatory pain models.

We may therefore hypothezised that the hypocretin system interacts with acute or chronic pain disorders. CSF hypocretin-1 was measured in only few central pain case reports in the literature with result within the normal range.[11,14] However, no lumbar puncture was performed during asymptomatic and symptomatic episodes. The relationship between central pain disorders (including migraine and cluster headache) and the dysfunction of the hypocretin system remains to be determined.

8. PSYCHIATRIC DISORDERS

Many years of research have demonstrated the ability of the hypothalamus to control the homeostatic state and physiology in animals. Specific lesions experiments have characterized sub-nuclei of the hypothalamus that are important in sleep, feeding, autonomic control, sex drive and motivational behavior. The hypocretin neurons that are located in the dorso-lateral hypothalamus provide a dense innervation to the other nuclei within the hypothalamus as well as the monoaminergic systems, in particular key areas for behavior control. Hypocretins are involved in several hypothalamic functions as control of vigilance state, feeding and energy balance but also other as neuroendocrine system function.

Because of the recent interest in hypocretin dysfunction in neurological diseases and because of the functions of the hcrts, recent studies have evaluated whether altered hypocretin neurotransmission might be also involved in some psychiatric disorders. We will focus in the present review on depression and schizophrenia. The involvement of the hypocretins in stress and drug addiction is also today documented and we recommend to the readers to refer to the specific chapters in the book devoted to these two aspects of behavior. The hcrt system is also currently evaluated in other psychiatric disorders such as eating disorders.

8.1. Depressive Syndrome

Sleep disturbances are common and early symptoms in depression. The relationship between depression, sleep and circadian rhythms has been strongly documented. In some cases insomnia, stress and depression are even seen as continuum and a significant number of narcoleptic patients suffer from depression. Depression is associated with short REM sleep latencies in many cases. REM sleep is suppressed by almost all antidepressant medications and sleep deprivation. REM sleep deprivation has antidepressant effect. The neurochemistry of depression has focused on monoaminergic dysfunction, especially noradrenergic and serotoninergic pathways. The hypocretins, two neuropeptides that promote wakefulness and inhibit REM sleep are upregulated under REM sleep deprivation[45] and might be involved in his antidepressant effect. In addition, the projections distribution of the hypocretins is coherent with a direct involvement in depression since these projections are noted in aminergic cell groups i.e. the noradrenergic locus coeruleus, the serotoninergic dorsal raphe, the histaminergic tuberomamillary nucleus and the dopaminergic substantia nigra and ventral tegmental area, and also cholinergic groups with a neuroexcitatory effect.[46] Finally, hypocretins enhance the monoaminergic tone,[35,47] as the antidepressant agents, and also activate the hypothalamic-pituitary-adrenal-axis whose changes in diurnal activity are observed in depression.

In animals, the Wistar-Kyoto (WKY) rats have increased REM sleep from early in development and are hypoactive, hypophagic and less sensitive to antidepressant drugs leading to the use of this model as a model of depression. Even considering the intrinsic restrictions of an animal model of depression as compared to the complexity of a depressive syndrome, it is interesting to notice that this latter model has a brain hypocretin deficiency especially observed in control areas of sleep and emotion.[48]

In human, the group headed by E. Mignot examined hypocretin-1 levels in the CSF of 14 control and 15 depressed patients.[49] The mean baseline values in control subjects were not different from patients indicating that hypocretin deficiency is an unlikely cause for depression. They also explored whether antidepressant treatment modified hypocretin-1 levels in depressed subjects. They found that treatment with sertraline but not bupropion, was associated with decreased hypocretin levels, suggesting a small serotoninergic influence of the 5-HT system on the hypocretinergic tone which is supported by anatomical data showing tight contacts between serotoninergic fibers and hypocretin-containing cells. More interestingly, they observed a small slight but significant reduction in the amplitude of circadian rhythm of hypocretin-1 levels in depressed subjects as compared to controls. Surprisingly, the diurnal variation in hypocretin-1 levels was the opposite of that observed in rats and monkeys and the circadian variation was low which may be partialy explained by the fact that CSF was drawn continuously in supine subjects who were confined to bed rest.[49] Diminished circadian rhythms of behaviors, physiologic measures and peripheral neuroendocrine functions can be observed in depression[50] and circadian hypocretin variation might be one of the reduced oscillating parameter. Even if a slight improvement of the amplitude was observed in treated patients, it is impossible to conclude today because of the small sample size. It needs to be replicate and more interestingly in different subtypes of depression. Indeed, depression includes different depressive syndromes whose physiopathology is probably different even if sharing some common biological mechanisms. Under this light, it would be of great interest to study the hypocretin circadian variation in seasonal affective disorder, a subtype of depression with chronobiological abnormalities. Anatomical and physiological data support this hypothesis since hypocretins neurons are indirectly under the control of the central circadian pacemaker, the suprachiasmatic nucleus, via a relay in the dorsomedial hypothalamus and the lesion of the SCN eliminate the daily rhythm of hypocretin-1 release.[51] It would be also interesting to examine the level of expression of hypocretin in mania.

8.2. Schizophrenia

CSF hypocretin levels have also been studied in a psychotic disorder, schizophrenia. The hypocretins have been shown to have a neuroexcitatory effect on the midbrain dopaminergic neurons. Central administration of hypocretins increases the number of stereotypy as well as locomotor activity, an effect that is prevented by the administration of dopaminergic D2 receptor antagonists, a class of compounds used in the treatment of schizophrenia.[52] Hypocretin has also been shown to activate the hypothalamo- pituitary adrenal axis leading to an increase in release of corticotropin releasing hormone (CRH).[53,54] Because abnormalities of the HPA axis and especially of the dopaminergic transmission are reported in schizophrenia and because sleep abnormalities are commonly observed in schizophrenic patients, Nishino and colleagues examined the hypocretin-1 levels in the CSF of 13 patients and 12 controls.[55] Hypocretin dysfunction is unlikely to be involved or to mediate dopaminergic dysfunction observed in schizophrenia since CSF hypocretin levels were similar in the two groups. The only significant result reported by the authors was a positive correlation in schizophrenic patients, but not in control subjects, of the hypocretin levels with sleep latency, one of the most consistent sleep disturbances observed in schizophrenia. Dalal et al., evaluated the effect of antipsychotic drugs on the hcrt levels and they reported lower hcrt levels in patients treated with an haloperidol (antidopaminergic).[56] All these results suggest that complementary experiments are necessary on a larger population of patients and controls, even if an altered hypocretin neurotransmission do not seem consistently to be involved in the pathophysiology of schizophrenia.

9. LIMITATIONS AND PERSPECTIVES

The hcrt in human CSF is measured by radioimmunoassay (RIA) and only the hcrt type 1 is detected. Hcrt-1 levels are not different in respect to gender or age.[57] Normal levels are higher than 200 pg/ml and decreased levels corresponds to dosage < 110 pg/ml.

However, the significance of intermediate (between 110-200 pg/ml) and high hypocretin (>500 pg/ml) levels is still unclear and needs further investigations.

Based on studies in narcolepsy, it is clear that undetectable CSF hcrt-1 reflects the lack of hcrt in the brain,[4,5] at least in hcrt deficiency–related diseases. In rats, an average loss of 73% of hcrt neurons induced following the injection of hypocretin-2-saporin into the hypothalamus produces a 50% decrease in CSF hcrt levels.[58] However, do CSF measurements always reflect levels of hypocretin that are present at the active projection areas of the brain? The evaluation of the hcrt system remains today very restricted in human since the hcrt-1 or 2 receptors function as well as the downstream hcrt pathways or only partial hcrt lesions without low level consequences are not estimate.

Hcrt-1 levels in the CSF increase under specific conditions such as sleep deprivation and decrease under REM sleep rebound in rats suggesting that REM sleep deprivation and rebound activate and inhibit respectively the hcrtergic system.[45] Increased hcrtergic tone during REM sleep deprivation may be important to interpret changes, or lack of changes, in hcrt dosages in the CSF of patients since REM sleep deprivation might be a secondary indirect consequence of a sleep disorder. This might explain some of the discrepancies between studies. The detailed analysis of the effect of REM deprivation on the hcrt tone shows an increase in the hcrt levels during the rest period but not at the end of the active period[45] because of high spontaneous levels at that time of the day. Indeed, hcrt-1 release is under the control of the clock, since lesion of the suprachiasmatic nucleus eliminates the daily rhythm of hcrt-1 release in the rat. An effect mediated via the dorsomedial hypothalamic nucleus, a structure that receives direct and indirect SCN inputs and that send projections to the lateral hypothalamus.[51] These latter data should again be taken account for the interpretation of results depending on the time of the day the collection of samples is performed.

Another limitation in the studies examining hcrt level in patients and controls is, in contrary to patients who have narcolepsy with cataplexy, the absence of information on the human leucocyte antigen (HLA) subtype of the subjects. Indeed, the hypothesis of an association between the HLA-DR2 subtype and hcrt deficiency has been suggested as detailed before.

10. CONCLUSION

In summary, deficient hypocretin-1 transmission is highly specific and sensitive for sporadic narcolepsy with typical excessive daytime somnolence, cataplexy and HLA DQB1*0602 allele. Hypocretin ligand deficiency appears not to be the major cause for other hypersomnias but a pathophysiological continuum between narcolepsy without cataplexy and idiopathic hypersomnia could be hypothesized. Undetectable CSF hcrt-1 levels are highly sensitive and specific for narcolepsy and are a useful diagnostic tool for pediatric and atypical cases. A decreased level however should not systematically suggest the diagnosis of narcolepsy. Indeed, low levels can be observed in various neurological conditions: neuroimmune disorders such as Guillain Barré syndrome and Miller Fisher syndrome; neurological conditions associated with lesion or dysfunction of the lateral hypothalamus; advanced forms of Parkinson disease. Finally, exploring the hypocretins in neuropsychiatric disorders is a growing field that requires further investigation to understand the role of hypocretins in pathology. The evaluation methods of the hypocretinergic system needs also to be improved with other techniques in order to evaluate the functioning of hypocretin-1 or 2 receptors as well as the consequences of partial lesions of hypocretin cells without low CSF hypocretin-1. Measurement and interpretation of intermediate and high levels of hypocretin, as reported in restless legs syndrome for example, requires more study.

11. REFERENCES

1. L. Lin, M. Hungs, and E. Mignot, Narcolepsy and the HLA region, *J. Neuroimmunol.* **117**,9-20 (2001).
2. L. Lin, J. Faraco, R. Li, H. Kadotani, W. Rogers, X. Lin, X. Qiu, P. J. de Jong, S. Nishino, and E. Mignot, The sleep disorder canine narcolepsy is caused by a mutation in the hypocretin (orexin) receptor 2 gene, *Cell* **98**,365-376 (1999).
3. R. M. Chemelli, J. T. Willie, C. M. Sinton, J. K. Elmquist, T. Scammell, C. Lee, J. A. Richardson, S. C. Williams, Y. Xiong, Y. Kisanuki, T. E. Fitch, M. Nakazato, R. E. Hammer, C. B. Saper, and M. Yanagisawa, Narcolepsy in orexin knockout mice: molecular genetics of sleep regulation, *Cell* **98**,437-451 (1999).
4. C. Peyron, J. Faraco, W. Rogers, B. Ripley, S. Overeem, Y. Charnay, S. Nevsimalova, M. Aldrich, D. Reynolds, R. Albin, R. Li, M. Hungs, M. Pedrazzoli, M. Padigaru, M. Kucherlapati, J. Fan, R. Maki, G. J. Lammers, C. Bouras, R. Kucherlapati, S. Nishino, and E. Mignot, A mutation in a case of early onset narcolepsy and a generalized absence of hypocretin peptides in human narcoleptic brains, *Nat. Med.* **6**,991-997 (2000).
5. T. C. Thannickal, R. Y. Moore, R. Nienhuis, L. Ramanathan, S. Gulyani, M. Aldrich, M. Cornford, and J. M. Siegel, Reduced number of hypocretin neurons in human narcolepsy, *Neuron* **27**,469-474 (2000).
6. S. Nishino, B. Ripley, S. Overeem, G. J. Lammers, and E. Mignot, Hypocretin (orexin) deficiency in human narcolepsy, *Lancet* **355**,39-40 (2000).
7. B. Ripley, S. Overeem, N. Fujiki, S. Nevsimalova, M. Uchino, J. Yesavage, D. Di Monte, K. Dohi, A. Melberg, G. J. Lammers, Y. Nishida, F. W. Roelandse, M. Hungs, E. Mignot, and S. Nishino, CSF hypocretin/orexin levels in narcolepsy and other neurological conditions, *Neurology* **57**,2253-2258 (2001).
8. A. Melberg, B. Ripley, L. Lin, J. Hetta, E. Mignot, and S. Nishino, Hypocretin deficiency in familial symptomatic narcolepsy, *Ann. Neurol.* **49**,136-137 (2001).
9. G. Hartwig, J. Harsh, B. Ripley, S. Nishino, and E. Mignot, Low cerebrospinal fluid hypocretin levels found in familial narcolepsy, *Sleep Med.* **2**,451-453 (2001).
10. L. E. Krahn, V. S. Pankratz, L. Oliver, B. F. Boeve, and M. H. Silber, Hypocretin (orexin) levels in cerebrospinal fluid of patients with narcolepsy: relationship to cataplexy and HLA DQB1*0602 status, *Sleep* **25**,733-736 (2002).
11. E. Mignot, G. J. Lammers, B. Ripley, M. Okun, S. Nevsimalova, S. Overeem, J. Vankova, J. Black, J. Harsh, C. Bassetti, H. Schrader, and S. Nishino, The role of cerebrospinal fluid hypocretin measurement in the diagnosis of narcolepsy and other hypersomnias, *Arch. Neurol.* **59**,1553-1562 (2002).
12. T. Kanbayashi, Y. Inoue, S. Chiba, R. Aizawa, Y. Saito, H. Tsukamoto, Y. Fujii, S. Nishino, and T. Shimizu, CSF hypocretin-1 (orexin-A) concentrations in narcolepsy with and without cataplexy and idiopathic hypersomnia, *J Sleep Res* **11**,91-93 (2002).
13. C. Bassetti, M. Gugger, M. Bischof, J. Mathis, C. Sturzenegger, E. Werth, B. Radanov, B. Ripley, S. Nishino, and E. Mignot, The narcoleptic borderland: a multimodal diagnostic approach including cerebrospinal fluid levels of hypocretin-1 (orexin A), *Sleep Med.* **4**,7-12 (2003).
14. Y. Dauvilliers, C. R. Baumann, B. Carlander, M. Bischof, T. Blatter, M. Lecendreux, F. Maly, A. Besset, J. Touchon, M. Billiard, M. Tafti, and C. L. Bassetti, CSF hypocretin-1 levels in narcolepsy, Kleine-Levin syndrome, and other hypersomnias and neurological conditions, *J. Neurol. Neurosurg. Psychiatry* **74**,1667-1673 (2003).
15. Y. Dauvilliers, S. Maret, C. Bassetti, B. Carlander, M. Billiard, J. Touchon, and M. Tafti, A monozygotic twin pair discordant for narcolepsy and CSF hypocretin-1, *Neurology* **62**,2137-2138 (2004).
16. T. Kanbayashi, Y. Inoue, K. Kawanishi, H. Takasaki, R. Aizawa, K. Takahashi, Y. Ogawa, M. Abe, Y. Hishikawa, and T. Shimizu, CSF hypocretin measures in patients with obstructive sleep apnea, *J. Sleep Res.* **12**,339-341 (2003).
17. M. Lecendreux, S. Maret, C. Bassetti, M. C. Mouren, and M. Tafti, Clinical efficacy of high-dose intravenous immunoglobulins near the onset of narcolepsy in a 10-year-old boy, *J. Sleep Res.* **12**,347-348 (2003).
18. S. Nishino, T. Kanbayashi, N. Fujiki, M. Uchino, B. Ripley, M. Watanabe, G. J. Lammers, H. Ishiguro, S. Shoji, Y. Nishida, S. Overeem, I. Toyoshima, Y. Yoshida, T. Shimizu, S. Taheri, and E. Mignot, CSF

hypocretin levels in Guillain-Barre syndrome and other inflammatory neuropathies, *Neurology* **61**,823-825 (2003).
19. T. Kanbayashi, H. Ishiguro, R. Aizawa, Y. Saito, Y. Ogawa, M. Abe, K. Hirota, S. Nishino, and T. Shimizu, Hypocretin-1 (orexin-A) concentrations in cerebrospinal fluid are low in patients with Guillain-Barre syndrome, *Psychiatry Clin. Neurosci.* **56**,273-274 (2002).
20. C. R. Baumann, Y. Dauvilliers, E. Mignot, and C. L. Bassetti, Normal CSF Hypocretin-1 (Orexin A) Levels in Dementia with Lewy Bodies Associated with Excessive Daytime Sleepiness, *Eur. Neurol.* **52**,73-76 (2004).
21. J. Dalmau, F. Graus, A. Villarejo, J. B. Posner, D. Blumenthal, B. Thiessen, A. Saiz, P. Meneses, and M. R. Rosenfeld, Clinical analysis of anti-Ma2-associated encephalitis, *Brain* **127**,1831-1844 (2004).
22. S. Overeem, J. Dalmau, L. Bataller, S. Nishino, E. Mignot, J. Verschuuren, and G. J. Lammers, Hypocretin-1 CSF levels in anti-Ma2 associated encephalitis, *Neurology* **62**,138-140 (2004).
23. I. O. Ebrahim, Y. K. Semra, S. De Lacy, R. S. Howard, M. D. Kopelman, A. Williams, and M. K. Sharief, CSF hypocretin (Orexin) in neurological and psychiatric conditions, *J. Sleep Res.* **12**,83-84 (2003).
24. E. G. Celius, H. F. Harbo, T. Egeland, F. Vartdal, B. Vandvik, and A. Spurkiand, Sex and age at diagnosis are correlated with the HLA-DR2, DQ6 haplotype in multiple sclerosis, *J. Neurol. Sci.* **178**,132-135 (2000).
25. T. Kato, T. Kanbayashi, K. Yamamoto, T. Nakano, T. Shimizu, T. Hashimoto, and S. Ikeda, Hypersomnia and low CSF hypocretin-1 (orexin-A) concentration in a patient with multiple sclerosis showing bilateral hypothalamic lesions, *Intern. Med.* **42**,743-745 (2003).
26. H. Kubota, T. Kanbayashi, Y. Tanabe, J. Takanashi, and Y. Kohno, A case of acute disseminated encephalomyelitis presenting hypersomnia with decreased hypocretin level in cerebrospinal fluid, *J. Child Neurol.* **17**,537-539 (2002).
27. R. F. Gledhill, P. R. Bartel, Y. Yoshida, S. Nishino, and T. E. Scammell, Narcolepsy caused by acute disseminated encephalomyelitis, *Arch. Neurol.* **61**,758-760 (2004).
28. J. D. Katz, and A. H. Ropper, Familial Kleine-Levin syndrome: two siblings with unusually long hypersomnic spells, *Arch. Neurol.* **59**,1959-1961 (2002).
29. J. Vankova, I. Stepanova, R. Jech, M. Elleder, L. Ling, E. Mignot, S. Nishino, and S. Nevsimalova, Sleep disturbances and hypocretin deficiency in Niemann-Pick disease type C, *Sleep* **26**,427-430 (2003).
30. J. E. Martinez-Rodriguez, L. Lin, A. Iranzo, D. Genis, M. J. Marti, J. Santamaria, and E. Mignot, Decreased hypocretin-1 (Orexin-A) levels in the cerebrospinal fluid of patients with myotonic dystrophy and excessive daytime sleepiness, *Sleep* **26**,287-290 (2003).
31. J. Arii, T. Kanbayashi, Y. Tanabe, J. Ono, S. Nishino, and Y. Kohno, A hypersomnolent girl with decreased CSF hypocretin level after removal of a hypothalamic tumor, *Neurology* **56**,1775-1776 (2001).
32. T. E. Scammell, S. Nishino, E. Mignot, and C. B. Saper, Narcolepsy and low CSF orexin (hypocretin) concentration after a diencephalic stroke, *Neurology* **56**,1751-1753 (2001).
33. O. J. Dempsey, P. McGeoch, R. N. de Silva, and N. J. Douglas, Acquired narcolepsy in an acromegalic patient who underwent pituitary irradiation, *Neurology* **61**,537-540 (2003).
34. D. B. Rye, and J. Jankovic, Emerging views of dopamine in modulating sleep/wake state from an unlikely source: PD, *Neurology* **58**,341-346 (2002).
35. P. Bourgin, S. Huitron-Resendiz, A. D. Spier, V. Fabre, B. Morte, J. R. Criado, J. G. Sutcliffe, S. J. Henriksen, and L. de Lecea, Hypocretin-1 modulates rapid eye movement sleep through activation of locus coeruleus neurons, *J. Neurosci.* **20**,7760-7765 (2000).
36. T. Mochizuki, and T. E. Scammell, Orexin/hypocretin: wired for wakefulness, *Curr. Biol.* **13**,R563-564 (2003).
37. B. A. Baldo, R. A. Daniel, C. W. Berridge, and A. E. Kelley, Overlapping distributions of orexin/hypocretin- and dopamine-beta-hydroxylase immunoreactive fibers in rat brain regions mediating arousal, motivation, and stress, *J. Comp. Neurol.* **464**,220-237 (2003).
38. S. Overeem, J. J. van Hilten, B. Ripley, E. Mignot, S. Nishino, and G. J. Lammers, Normal hypocretin-1 levels in Parkinson's disease patients with excessive daytime sleepiness, *Neurology* **58**,498-499 (2002).
39. X. Drouot, S. Moutereau, J. P. Nguyen, J. P. Lefaucheur, A. Creange, P. Remy, F. Goldenberg, and M. P. d'Ortho, Low levels of ventricular CSF orexin/hypocretin in advanced PD, *Neurology* **61**,540-543 (2003).
40. I. Arnulf, E. Konofal, M. Merino-Andreu, J. L. Houeto, V. Mesnage, M. L. Welter, L. Lacomblez, J. L. Golmard, J. P. Derenne, and Y. Agid, Parkinson's disease and sleepiness: an integral part of PD, *Neurology* **58**,1019-1024 (2002).
41. R. P. Allen, E. Mignot, B. Ripley, S. Nishino, and C. J. Earley, Increased CSF hypocretin-1 (orexin-A) in restless legs syndrome, *Neurology* **59**,639-641 (2002).
42. K. Stiasny-Kolster, E. Mignot, L. Ling, J. C. Moller, W. Cassel, and W. H. Oertel, CSF hypocretin-1 levels in restless legs syndrome, *Neurology* **61**,1426-1429 (2003).

43. J. E. Martinez-Rodriguez, R. Sanchez-Valle, A. Saiz, L. Lin, A. Iranzo, E. Mignot, and J. Santamaria, Normal hypocretin-1 levels in the cerebrospinal fluid of patients with fatal familial insomnia, *Sleep* **26**,1068 (2003).
44. D. Georgescu, V. Zachariou, M. Barrot, M. Mieda, J. T. Willie, A. J. Eisch, M. Yanagisawa, E. J. Nestler, and R. J. DiLeone, Involvement of the lateral hypothalamic peptide orexin in morphine dependence and withdrawal, *J. Neurosci.* **23**,3106-3111 (2003).
45. M. Pedrazzoli, V. D'Almeida, P. J. Martins, R. B. Machado, L. Ling, S. Nishino, S. Tufik, and E. Mignot, Increased hypocretin-1 levels in cerebrospinal fluid after REM sleep deprivation, *Brain Res.* **995**,1-6 (2004).
46. C. Peyron, D. K. Tighe, A. N. van den Pol, L. de Lecea, H. C. Heller, J. G. Sutcliffe, and T. S. Kilduff, Neurons containing hypocretin (orexin) project to multiple neuronal systems, *J. Neurosci.* **18**,9996-10015 (1998).
47. K. S. Eriksson, O. Sergeeva, R. E. Brown, and H. L. Haas, Orexin/hypocretin excites the histaminergic neurons of the tuberomammillary nucleus, *J. Neurosci.* **21**,9273-9279 (2001).
48. S. Taheri, J. Gardiner, S. Hafizi, K. Murphy, C. Dakin, L. Seal, C. Small, M. Ghatei, and S. Bloom, Orexin A immunoreactivity and preproorexin mRNA in the brain of Zucker and WKY rats, *Neuroreport* **12**,459-464 (2001).
49. R. M. Salomon, B. Ripley, J. S. Kennedy, B. Johnson, D. Schmidt, J. M. Zeitzer, S. Nishino, and E. Mignot, Diurnal variation of cerebrospinal fluid hypocretin-1 (Orexin-A) levels in control and depressed subjects, *Biol. Psychiatry* **54**,96-104 (2003).
50. C. Cajochen, D. P. Brunner, K. Krauchi, P. Graw, and A. Wirz-Justice, EEG and subjective sleepiness during extended wakefulness in seasonal affective disorder: circadian and homeostatic influences, *Biol. Psychiatry* **47**,610-617 (2000).
51. T. C. Chou, T. E. Scammell, J. J. Gooley, S. E. Gaus, C. B. Saper, and J. Lu, Critical role of dorsomedial hypothalamic nucleus in a wide range of behavioral circadian rhythms, *J. Neurosci.* **23**,10691-10702 (2003).
52. T. Nakamura, K. Uramura, T. Nambu, T. Yada, K. Goto, M. Yanagisawa, and T. Sakurai, Orexin-induced hyperlocomotion and stereotypy are mediated by the dopaminergic system, *Brain Res.* **873**,181-187 (2000).
53. M. Kuru, Y. Ueta, R. Serino, M. Nakazato, Y. Yamamoto, I. Shibuya, and H. Yamashita, Centrally administered orexin/hypocretin activates HPA axis in rats, *Neuroreport* **11**,1977-1980 (2000).
54. M. Jaszberenyi, E. Bujdoso, I. Pataki, and G. Telegdy, Effects of orexins on the hypothalamic-pituitary-adrenal system, *J Neuroendocrinol* **12**,1174-1178 (2000).
55. S. Nishino, B. Ripley, E. Mignot, K. L. Benson, and V. P. Zarcone, CSF hypocretin-1 levels in schizophrenics and controls: relationship to sleep architecture, *Psychiatry Res* **110**,1-7 (2002).
56. M. A. Dalal, A. Schuld, and T. Pollmacher, Lower CSF orexin A (hypocretin-1) levels in patients with schizophrenia treated with haloperidol compared to unmedicated subjects, *Mol. Psychiatry* **8**,836-837 (2003).
57. T. Kanbayashi, T. Yano, H. Ishiguro, K. Kawanishi, S. Chiba, R. Aizawa, Y. Sawaishi, K. Hirota, S. Nishino, and T. Shimizu, Hypocretin-1 (orexin-A) levels in human lumbar CSF in different age groups: infants to elderly persons, *Sleep* **25**,337-339 (2002).
58. D. Gerashchenko, E. Murillo-Rodriguez, L. Lin, M. Xu, L. Hallett, S. Nishino, E. Mignot, and P. J. Shiromani, Relationship between CSF hypocretin levels and hypocretin neuronal loss, *Exp. Neurol.* **184**,1010-1016 (2003).

HYPOCRETIN/OREXIN AND SLEEP

Implications for the pathophysiology of human narcolepsy

Gert Jan Lammers and Sebastiaan Overeem*

1. INTRODUCTION

In 1877, Westphal described the first unequivocal case of narcolepsy, although that name was coined only later by Gélineau, in 1880. He used the name to describe a combination of sleep attacks and attacks of "astasia".[1,2] Ever since, the definition of narcolepsy has been discussed: is cataplexy required to make the diagnosis, and is it a disease entity, a *morbus sui generis*?

A later issue, generated by the discovery of REM-sleep abnormalities in narcolepsy, was whether narcolepsy is a disease of REM-sleep alone, or a more general "regulation problem" of wakefulness and sleep. The focus on REM-sleep was induced by the assumption that cataplexy, the only specific symptom of narcolepsy, represents the atonia that normally only occurs during REM-sleep.

In the late 80's of the last century the regulatory theory gained support, emphasizing that narcolepsy is more than a disorder of REM-sleep. Broughton incorporated it in a new pathophysiological concept, which held that narcolepsy is characterized by a loss of "state-boundary-control".[3] According to this view, two aspects of regulation are lost. The first is that it is not possible to sustain a given sleep/wake state for any length of time: when awake, patients fall asleep quickly, and when asleep they awaken quickly. Second, phenomena that normally occur together during a certain sleep stage can occur dissociated, i.e. on their own and out of their context. The view that cataplexy represents REM sleep atonia at an inappropriate moment is an example. Although the validity of this concept may be questioned, in particular for cataplexy,[4,5] it probably is still the best description of the pathophysiology of narcolepsy. It is important to realize that according to this theory narcolepsy is characterized by a disturbed *distribution* of sleep (phenomena), rather than by an increased amount of sleep over the 24 hours of a day.

* Gert Jan Lammers and Sebastiaan Overeem, Department of Neurology, Leiden University Medical Center, PO BOX 9600, Leiden, the Netherlands. Tel: +31-71-5262895. Fax: +31-71-5248253

Interestingly, this concept is supported by findings in the recently developed hypocretin-deficient animal models for narcolepsy.

In the current chapter we will discuss the role of hypocretin in the regulation of human sleep and particularly in the pathophysiology and etiology of narcolepsy. The discovery of the hypocretin system represents a major step in understanding the pathophysiology of narcolepsy, but will not silence the historical debate summarized above. We also discuss the possible causes of disturbances in hypocretin transmission that lead to human narcolepsy. Finally, we summarize metabolic and neuroendocrine findings in human narcolepsy and speculate how hypocretin deficiency may affect metabolic rate, neuroendocrine rhythms, and the autonomic nervous system.

2. HYPOCRETIN DEFICIENCY IN HUMAN NARCOLEPSY

In 1999, it was discovered that a genetically induced deficiency of hypocretin neurotransmission in dogs and mice leads to a complex of symptoms remarkably similar to that seen in human narcolepsy.[6,7] These findings focused attention on the role of hypocretin in the pathophysiology of human narcolepsy.

A mutation screening of the hypocretin genes was performed in 74 patients, including patients with and without a family history of narcolepsy and patients with and without HLA DQB1*0602.[8] Of these, only one patient carried a disease-causing mutation: a G-to-T transversion in the hypocretin signal peptide. The disease phenotype of this patient differed significantly from the classical pattern: cataplectic attacks with a frequency of five to 20 per day had started at the age of 6 months. This low mutation yield was not surprising from a clinical point of view, as typical narcolepsy does not run in families. More unexpectedly no mutations in the hypocretin genes were identified in the uncommon familial cases. Up till now there have been no reports about any other human case or family with an identified defect (including polymorphisms) in the pre-prohypocretin or the hypocretin receptor genes.[9,10] However, there are known associations with other genes, which may be considered as susceptibility genes. The strongest association known is with the HLA DQB1*0602 allele, present in over 90% of narcoleptic patients.[11] As about 20% of the population carries this allele and only 0.05% of the population develops narcolepsy, this association is not very specific. Interestingly, the uncommon familial cases show a much weaker association with this HLA type.[12] Associations with polymorphisms in the genes for monoamine-oxidase-A, tumor necrosis factor alpha, and catechol-O methyltransferase have also been reported.[13,14]

Despite the lack of an identified genetic cause, hypocretin deficiency as cause of human narcolepsy remained an attractive hypothesis. To further study this an assay for the measurement of hypocretin was developed and cerebrospinal fluid (CSF) hypocretin-1 measurements were performed blindly in a small series of patients and controls.[15] The astonishing finding was that 7 out of 9 patients had an undetectable low hypocretin-1 level, in contrast to the control group having levels about 7 times the detection limit, and all in a narrow range. Subsequently, a large extension study was performed to confirm these findings and establish the sensitivity and specificity.[16,17] Several hundreds of subjects were included: healthy controls, patients with typical narcolepsy, atypical narcolepsy, idiopathic hypersomnia, sleep apnea, restless legs, periodic limb movements disorder, insomnia patients as well as patients suffering from a wide variety of neurological disorders. From these data the sensitivity and specificity of hypocretin deficiency for narcolepsy turned out to be very high, 87% and 99% respectively. Typical cataplexy, characterized by symmetrical weakness particularly triggered by laughter, was the best predictor of hypocretin deficiency in these studies. Narcoleptic patients with atypical or no cataplectic attacks, HLA DQB1*0602 negative patients and those suffering from familial forms of narcolepsy often show normal hypocretin-1 levels. These findings suggest other possible causes, such as "downstream" defects in the hypocretin system, involvement of substances that modulate hypocretin cells, non-genetic acquired receptor defects, and even leave the possibility for causes other than hypocretin deficiency. The identification of these potential causes is hampered by the limited knowledge concerning the hypocretin system.

The only other relatively frequently occuring disorder in which undetectable hypocretin-1 levels were found is the Guillain-Barré syndrome.[18,19] The explanation is unclear, and the deficiency is probably transitory. Whether this deficiency is accompanied by clinical symptoms has yet to be established. The Guillain-Barré studies were retrospective in nature and concerned patients with severe paralysis requiring artificial ventilation, preventing a reliable assessment of narcoleptic symptoms.

The identification of hypocretin-1 deficiency as a specific marker for human narcolepsy without any overt genetic explanation raised the question about the most plausible cause. Two controlled post-mortem neuropathological studies shed some light on this crucial question. One study used in situ hybridization against hypocretin mRNA in frozen brain tissue.[8] In the two brains a complete disappearance of hypocretin mRNA was demonstrated. Melanin-concentrating hormone mRNA, a component of adjacent non-hypocretin cells, was preserved in these patients, just as in control subjects. To extend these findings, hypocretin-1 and 2 peptide levels were measured in brain tissue, cortex and pons, of eight control subjects and six patients.[8] Both hypocretin-1 and hypocretin-2 were absent in the brains of the narcoleptic patients. Furthermore, there were no signs of inflammation in two brains studied: tumor necrosis factor-alpha levels and HLA DR2 expression were normal. There were also no obvious signs of gliosis or global cellular loss in the region of interest. In the other study, immunohistochemistry after antigen retrieval in fixed tissue was used (four brains - one from a patient without cataplexy).[20] A dramatically reduced number of neurons expressing hypocretin 1 and 2, when compared with control brains, was found. Five to 15% of hypocretin neurons were preserved. Interestingly, the highest number of remaining neurons was found in the one patient without cataplexy. In this study an increased number of GFAP stained astrocytes in the narcoleptic brains was found, supporting a degeneration of neurons.

These findings suggest that normally functioning hypocretin cells are present in patients early in life, and that narcoleptic symptoms appear only after degeneration of the majority of cells. However, it must be kept in mind that these assumptions have not yet been proven. In particular it has not been proven whether the hypocretin cells degenerate or only have an impaired hypocretin synthesis or release. The main reason for this lack of knowledge is the lack of a hypocretin-independent marker of hypocretin cells. Structural studies using advanced MRI techniques have not settled this question; voxel based morphometry yielded contradictory results.[21-23] Additional factors impeding the identification of the possible cell loss are the low number of cells and their anatomical distribution: they are not clustered in a nucleus but are intermingled with other cells in quite a large area of the perifornical hypothalamus.

Although CSF hypocretin-1 deficiency in humans correlates with the presence of excessive daytime sleepiness (EDS), it is most specifically correlated with the presence of cataplexy,[17] similar to the findings in animal models. Unfortunately, we are not informed whether undetectable levels reflect a total absence of (biological active) hypocretin, although this seems likely.[24] The exact role of hypocretin deficiency, respectively hypocretin loss, in the pathophysiology of human narcolepsy has therefore not been established. The extent of the hypocretin deficiency may possibly determine whether patients only suffer from EDS or do also develop cataplexy. This is suggested by one of the post mortem brain studies, showing more residual hypocretin cells in the only patient without cataplexy.[20] However, firm conclusions cannot be drawn from one patient without information regarding CSF hypocretin-1 levels. Alternatively, the difference between narcolepsy with and without cataplexy may be explained by a differential disturbance in the function of hypocretin 1 and 2 or their respective receptors, as the type of lesion in the various animal models leads to differential behavioral phenotypes (see below).[25,26] However, a completely different etiology, independent of a disturbed hypocretin transmission may also be possible.

To return to the historical debate about narcolepsy and the question whether the presence of cataplexy is a prerequisite for the diagnosis, and whether this is a disease *sui generis*, we still have no definite answer. The few cases with isolated long lasting EDS, typical REM sleep abnormalities on polysomnographic testing and undetectable levels of hypocretin-1 challenge this view.[17] However, the very high sensitivity and specificity of undetectable levels of hypocretin for the presence of cataplexy underscores that cataplexy is a cardinal symptom.

3. ANIMAL MODELS

By now there is a variety of narcoleptic animal models. A naturally occurring narcolepsy variant in dogs concerns the "classical" autosomal recessive model harboring a hypocretin receptor 2 mutation,[6] as well as sporadic forms with undetectable low hypocretin-1 levels in the CSF.[27] In mice, several knockout models have been created: pre-pro-hypocretin, hypocretin receptor-1, hypocretin receptor-2, and the receptor-1 -and 2 double-knockout. Furthermore, there is the so-called ataxin-3 model.[7,25,26,28] In this last model, the hypocretin promotor is used to express a truncated Machado-Joseph disease gene product (ataxin-3), resulting in a selective degeneration of the hypocretin neurons in the first 12 weeks of life. Recently, the first narcoleptic rat model has been published, also based on the ataxin-3 transgenic approach. The ataxin-3 model appears to resemble the pathophysiology of human narcolepsy most closely; the number of hypocretin cells is normal at birth, and decreases gradually afterwards. Moreover, these mice become obese, although eating less, very similar to what happens in human narcolepsy (see below).[28-30]

All these models, with the exception of the receptor-1 knockout, exhibit a narcolepsy-like phenotype, including attacks that resemble cataplexy, sleep fragmentation and abnormal expression and distribution of REM sleep. Moreover, these animals show fragmented sleep rather than an increased amount of sleep over the 24-hour period. Interestingly, the receptor 2 knockout mice are less affected than the ligand- and combined receptor knockout mice.[25,26] This observation suggests that the hypocretin-1 receptor has additional effects on sleep/wakefulness regulation.

In summary, deficient hypocretin transmission leads to narcoleptic symptoms. Whether it has developed congenitally or post-natal, is caused by a deficiency of the ligand, or a receptor dysfunction, has a genetic or acquired etiology, occurs in mice, dogs, rats or humans; in all cases similar narcoleptic symptoms develop.

4. THE CAUSE OF HUMAN NARCOLEPSY

Multiple etiologies may cause narcolepsy. The vast majority of patients, estimated at 90% of all patients, have cataplectic attacks that are typically induced by laughter, are HLA DQB1*0602 positive, have no detectable hypocretin-1 in their CSF, and a disease onset between 10-30 years of age.[31] A selective autoimmune degeneration of hypocretin neurons is the most likely cause in these patients. This hypothesis is supported by the tight HLA association and the post-mortem findings as presented by Thannickal et al., but direct evidence for this theory is lacking as of yet.[20] For these patients the development of narcolepsy seems to involve environmental factors acting on a specific genetic (HLA) predisposition. This is supported by the approximately 30% concordance among monozygotic twins, and the higher risk for narcolepsy and EDS in first-degree family members of these patients.[12] First degree family members have a risk of approximately 2% for narcolepsy and 2-4% for atypical EDS.

A definite autoimmune cause, with undetectable CSF hypocretin-1, has been identified in an uncommon disorder: the anti-Ma paraneoplastic syndrome.[32] Patients with this disorder develop autoantibodies against Ma proteins and, consequently, encephalitis that predominates in the limbic system, hypothalamus and brainstem.[33] In some cases, the patients additionally show excessive daytime sleepiness and cataplexy due to hypocretin deficiency.[32] Importantly these patients always have additional neurological symptoms. Other evidence that an autoimmune process can lead to hypocretin deficiency comes from two patients with acute disseminated encephalomyelitis who showed a decrease in CSF hypocretin-1 during their disease.[34,35]

In some patients with Guillain-Barré syndrome, a (possibly autoimmune mediated) transient hypocretin deficiency has been described,[19] but it is currently not known whether this is accompanied with narcoleptic symptoms. Conversely, autoantibodies typically found in Guillain-Barré syndrome, are not found in idiopathic narcolepsy.[36]

Various causes has been identified or suggested for the small remaining number of patients. These patients often lack at least one of the four earlier mentioned characteristics of typical narcolepsy: typical cataplexy, the HLA DQB1*0602 allele, undetectable hypocretin-1 level in the CSF, disease onset > 10 years of age. As previously described in only one patient a genetic defect in the genes coding for pre-pro-hypocretin and its receptors has been established: a mutation in the signal peptide for hypocretin.[8] In several families, usually showing an autosomal dominant pattern of inheritance, genome scans revealed several susceptibility loci.[37,38] These regions may contain genes involved in the regulation of the hypocretin system, or the immune system. In genetic disorders such as Prader-Willi syndrome, Niemann Pick, Norrie disease, and the Coffin-Lowry syndrome, in whom the prevalence of narcolepsy/cataplexy seems to be unexpectedly high, hypocretin cells may be affected among other cells of the central nervous system.[39-42] Finally, local damage to the hypothalamus or adjacent lesions may induce narcolepsy as described in various case reports. An associated hypocretin deficiency has been reported in such cases.[43-45]

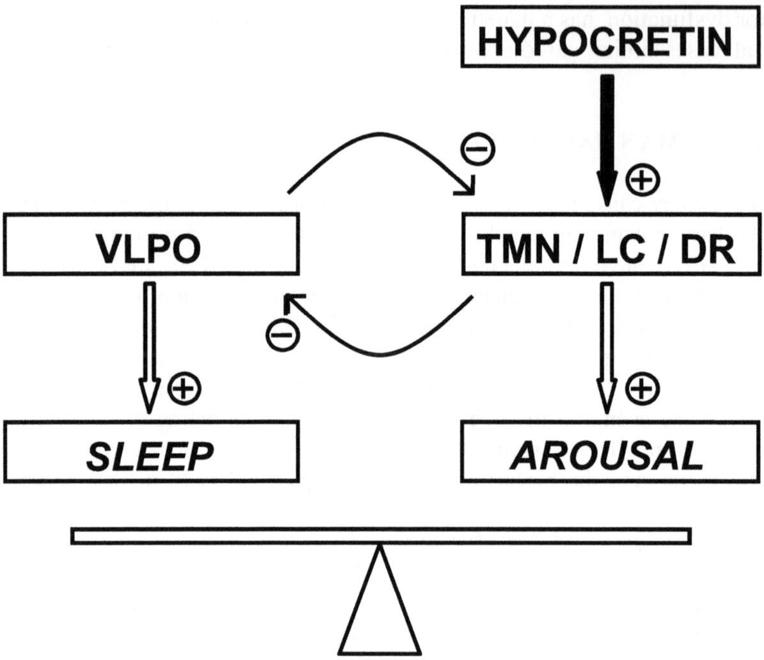

Figure 1. Simplified model, showing brain regions that promote wakefulness and sleep respectively, which mutually inhibited each other. This results in a bi-stable 'sleep switch'.[62] With the present knowledge, the hypocretin system seems to be responsible for stabilising the system, holding the switch in the wake position. VLPO: ventrolateral preoptic area, TMN: tuberomamillary nucleus, LC: locus coeruleus; DR: dorsal raphe.

All these so-called "symptomatic" cases of narcolepsy differ from the "idiopathic" form by the presence of additional neurological symptoms.

5. ROLE OF HYPOCRETIN IN SLEEP REGULATION

The hypocretin system is a crucial regulator of (REM-)sleep and wakefulness cycle, and seems to be particularly involved in sustained wakefulness since sleep deprivation is associated with an upregulated activity of the system.[46] It integrates the activity of nuclei involved in these states including the motor nuclei. Absence of hypocretin leads to attenuated daytime wakefulness and the earlier discussed loss of "state boundary control".[47] In this context it is important to realize that the hypocretin system is not required to generate wakefulness or other vigilance states, since these states still occur in almost the normal amounts, when there is no hypocretin transmission. Hypocretin merely stabilizes sleep and wakefulness states. This regulatory role is corroborated by neuro-anatomical findings. The hypocretin system comprises a small group of cells exclusively located in the lateral hypothalamus with efferent connections to virtually all parts of the brain.[48]

Afferent input is mainly from centers/nuclei that are not primarily involved in the generation of sleep and wakefulness.[49] This represents a pattern that may be expected from a "high order" regulatory system. Functionally, the activity of the system is entrained by the biological clock as shown in animal studies.[46,50]

Most likely, the promotion of wakefulness occurs through the excitatory effects of hypocretin on aminergic brain regions implicated in arousal such as the noradrenergic neurons of the locus coeruleus, the serotonergic neurons of the dorsal raphe, and the histaminergic neurons of the tuberomammillary nucleus.[51-55] Of these, the histaminergic neurons may play a prominent role because histamine concentrations are low in hypocretin receptor-2 mutated narcoleptic dogs, and mice lacking the H1 histamine receptor appear to have no waking response to hypocretin.[56,57] Hypocretin may also promote wakefulness by exciting cholinergic neurons.[58] Hypocretin neurons are active during wakefulness as indicated by the expression of Fos, and the extracellular concentration of hypocretin is high during wakefulness.[59-61,61] By increasing the activity of aminergic and pontine cholinergic brain regions, hypocretin may help an individual to maintain long episodes of wakefulness.

How to view the role of hypocretin in the current concept of the control of wakefulness and sleep stages? Briefly summarized, the most recent concept proposes that the sleep-promoting function of the ventrolateral preoptic area, and the wake-promoting function of several monoaminergic nuclei, including the tuberomamillary nucleus, form a reciprocal relationship with the properties of a flip-flop circuit (figure 1).[62] This essentially means that systems that promote wakening and sleep interact as an inherently unstable system with the tendency to avoid intermediate states, so that an organism is clearly waking or clearly sleeping with only brief times spent in transitions. The hypocretin system in this model forms a stabilizing "finger on the switch" or, in other words, acts as state boundary controller (figure 1).[47,62]

6. ENDOCRINE RHYTHMS, AUTONOMIC TONE AND OBESITY

Narcoleptic patients have a tendency to grow obese. This had already been observed more than 70 years ago.[63] Recent studies point to a specific role of hypocretin in the genesis of this obesity. A first hint came from the observation that narcoleptic patients with measurable hypocretin levels (a small minority of the total patient population) tended to be less obese than patients with undetectably low levels.[16] A subsequent study in a large group of typical narcoleptic patients, known to have undetectable hypocretin-1 in more than 90% of cases, and comparably sleepy patients with idiopathic hypersomnia (who have normal hypocretin levels), confirmed this impression and suggests that hypocretin deficiency *per se* promotes body weight gain.[29] Hypocretin-deficient narcoleptics were significantly heavier than age- and sex- matched controls, and the percentage of obese patients (BMI > $30 kg/m^2$) was significantly higher in the narcoleptic group. In addition, waist circumference and the percentage body fat were significantly increased as well in the narcoleptics (figure 2). The increased body weight is not explained by a higher caloric intake, as narcoleptics actually eat less during the day.[30]

These metabolic findings and the evidence of involvement of hypocretin in endocrine regulation in animal studies incited additional neuroendocrine studies in human narcolepsy. A study of diurnal rhythms of leptin, hypothalamo-pituitary-adrenal, somatotropic-, gonadal-, and thyroid hormones in hypocretin deficient male narcoleptic patients showed abnormalities in all secretion patterns, the most robust in leptin.

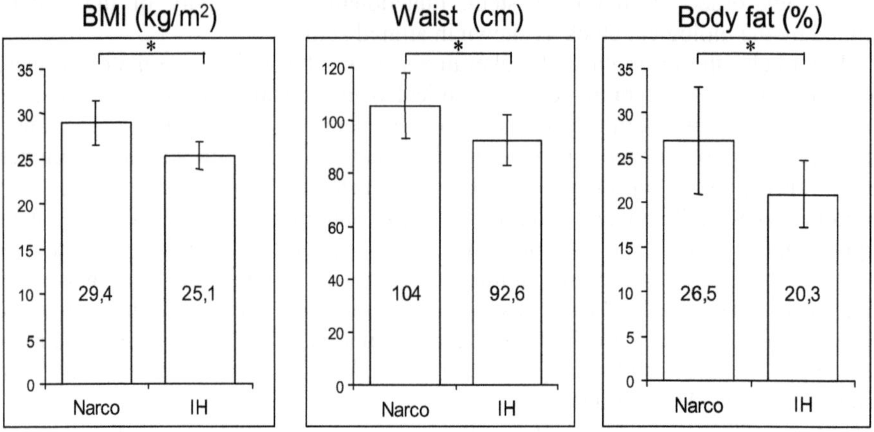

Figure 2. Body mass index (BMI), waist circumference and percentage body fat in a group of 64 male narcoleptic patients (mean age 45.6±12 years) and 21 male patients with idiopathic hypersomnia (42±8.7 years). Data based on Kok et al.[29]

Leptin levels in narcoleptic patients were about half of those in normal controls during the full 24 hour period, and the normal nocturnal rise in leptin concentrations was absent in narcoleptic subjects.[64]

The adrenal, gonadal, and thyroid axes showed an almost similar pattern of disturbance: a lowered secretion on the pituitary level with a preserved rhythm and normal levels of the end-hormones (with the exception of FSH).[65-67] For the somatotropic axis a shift towards daytime secretion of growth hormone (GH) was found.[68] The 24-hour mean serum concentration of GH was not significantly different in narcoleptic patients and controls, but patients secreted almost half of the total amount during the day, whereas controls produced only a quarter during daytime.

The findings prompt several new questions: how to explain these findings and are they clinically relevant? Are the changes in leptin a consequence of the sleep disturbance, or are they a direct consequence of the hypocretin deficiency itself? Are they mediated by a disruption of the diurnal sympathetic tone induced by sleep disruption or by the hypocretin deficiency? Furthermore, are the assessed abnormalities in the other neuroendocrine axes secondary to the hypoleptinemia or has hypocretin deficiency itself a direct effect on hypothalamic releasing hormones? In this context it is tempting to speculate about a central role for the autonomic nervous system, in particular because this may not only mediate the endocrine disturbances, but also the associated obesity.

Therefore, we hypothesize that the obesity and leptin disturbances are induced by changes in autonomic "tone", and consider the changes in the adrenal, gonadal, and thyroid axis to be secondary to the hypoleptinemia. It is not likely that the changes in the somatotropic axis are explained by hypoleptinemia, they are probably induced by diminished hypocretin mediated daytime inhibition of GH-releasing-hormone production. Future studies to verify these hypotheses may not only deepen the insight in the pathophysiology of narcolepsy, but will also provide important insight in human neuroendocrine regulation and the interaction between the endocrine axes and the autonomic nervous system, sleep and metabolic rate.

7. REFERENCES

1. C. Westphal, Eigenthümliche mit Einschlafen verbundene Anfälle, *Archiv Psychiat.Nervenkr.* **7**, 631-635 (1877).
2. J. B. Gélineau, De la narcolepsie, *Gaz.Hôp.(Paris)* **53**, 626-628 (1880).
3. R. Broughton, V. Valley, M. Aguirre, J. Roberts, W. Suwalski, and W. Dunham, Excessive daytime sleepiness and the pathophysiology of narcolepsy- cataplexy: a laboratory perspective, *Sleep* **9**, 205-215 (1986).
4. S. Nishino, J. Riehl, J. Hong, M. Kwan, M. S. Reid, and E. Mignot, Is narcolepsy a REM sleep disorder? Analysis of sleep abnormalities in narcoleptic Dobermans, *Neurosci.Res.* **38**, 437-446 (2000).
5. S. Overeem, G. J. Lammers, and J. G. van Dijk, Cataplexy: 'tonic immobility' rather than 'REM-sleep atonia'?, *Sleep Med* **3**, 471-477 (2002).
6. L. Lin, J. Faraco, R. Li, H. Kadotani, W. Rogers, X. Lin, X. Qiu, P. J. de Jong, S. Nishino, and E. Mignot, The sleep disorder canine narcolepsy is caused by a mutation in the hypocretin (orexin) receptor 2 gene, *Cell* **98**, 365-376 (1999).
7. R. M. Chemelli, J. T. Willie, C. M. Sinton, J. K. Elmquist, T. Scammell, C. Lee, J. A. Richardson, S. C. Williams, Y. Xiong, Y. Kisanuki, T. E. Fitch, M. Nakazato, R. E. Hammer, C. B. Saper, and M. Yanagisawa, Narcolepsy in orexin knockout mice: molecular genetics of sleep regulation, *Cell* **98**, 437-451 (1999).
8. C. Peyron, J. Faraco, W. Rogers, B. Ripley, S. Overeem, Y. Charnay, S. Nevsimalova, M. Aldrich, D. Reynolds, R. Albin, R. Li, M. Hungs, M. Pedrazzoli, M. Padigaru, M. Kucherlapati, J. Fan, R. Maki, G. J. Lammers, C. Bouras, R. Kucherlapati, S. Nishino, and E. Mignot, A mutation in a case of early onset narcolepsy and a generalized absence of hypocretin peptides in human narcoleptic brains, *Nat.Med* **6**, 991-997 (2000).
9. B. R. Olafsdottir, D. B. Rye, T. E. Scammell, J. K. Matheson, K. Stefansson, and J. R. Gulcher, Polymorphisms in hypocretin/orexin pathway genes and narcolepsy, *Neurology* **57**, 1896-1899 (2001).
10. M. Hungs, L. Lin, M. Okun, and E. Mignot, Polymorphisms in the vicinity of the hypocretin/orexin are not associated with human narcolepsy, *Neurology* **57**, 1893-1895 (2001).
11. E. Mignot, R. Hayduk, J. Black, F. C. Grumet, and C. Guilleminault, HLA DQB1*0602 is associated with cataplexy in 509 narcoleptic patients, *Sleep* **20**, 1012-1020 (1997).
12. E. Mignot, Genetic and familial aspects of narcolepsy, *Neurology* **50**, S16-S22 (1998).
13. Y. Dauvilliers, E. Neidhart, M. Lecendreux, M. Billiard, and M. Tafti, MAO-A and COMT polymorphisms and gene effects in narcolepsy, *Mol.Psychiatry* **6**, 367-372 (2001).
14. H. Hohjoh, T. Nakayama, J. Ohashi, T. Miyagawa, H. Tanaka, T. Akaza, Y. Honda, T. Juji, and K. Tokunaga, Significant association of a single nucleotide polymorphism in the tumor necrosis factor-alpha (TNF-alpha) gene promoter with human narcolepsy, *Tissue Antigens* **54**, 138-145 (1999).
15. S. Nishino, B. Ripley, S. Overeem, G. J. Lammers, and E. Mignot, Hypocretin (orexin) deficiency in human narcolepsy, *Lancet* **355**, 39-40 (2000).
16. S. Nishino, B. Ripley, S. Overeem, S. Nevsimalova, G. J. Lammers, J. Vankova, M. Okun, W. Rogers, S. Brooks, and E. Mignot, Low cerebrospinal fluid hypocretin (orexin) and altered energy homeostasis in human narcolepsy, *Ann.Neurol.* **50**, 381-388 (2001).
17. E. Mignot, G. J. Lammers, B. Ripley, M. Okun, S. Nevsimalova, S. Overeem, J. Vankova, J. Black, J. Harsh, C. Bassetti, H. Schrader, and S. Nishino, The role of cerebrospinal fluid hypocretin measurement in the diagnosis of narcolepsy and other hypersomnias, *Arch.Neurol.* **59**, 1553-1562 (2002).
18. B. Ripley, S. Overeem, N. Fujiki, S. Nevsimalova, M. Uchino, J. Yesavage, D. Di Monte, K. Dohi, A. Melberg, G. J. Lammers, Y. Nishida, F. W. Roelandse, M. Hungs, E. Mignot, and S. Nishino, CSF hypocretin/orexin levels in narcolepsy and other neurological conditions, *Neurology* **57**, 2253-2258 (2001).
19. S. Nishino, T. Kanbayashi, N. Fujiki, M. Uchino, B. Ripley, M. Watanabe, G. J. Lammers, H. Ishiguro, S. Shoji, Y. Nishida, S. Overeem, I. Toyoshima, Y. Yoshida, T. Shimizu, S. Taheri, and E. Mignot, CSF hypocretin levels in Guillain-Barre syndrome and other inflammatory neuropathies, *Neurology* **61**, 823-825 (2003).

20. T. C. Thannickal, R. Y. Moore, R. Nienhuis, L. Ramanathan, S. Gulyani, M. Aldrich, M. Cornford, and J. M. Siegel, Reduced number of hypocretin neurons in human narcolepsy, *Neuron* **27**, 469-474 (2000).
21. S. Overeem, S. C. Steens, C. D. Good, M. D. Ferrari, E. Mignot, R. S. Frackowiak, M. A. van Buchem, and G. J. Lammers, Voxel-based morphometry in hypocretin-deficient narcolepsy, *Sleep* **26**, 44-46 (2003).
22. B. Draganski, P. Geisler, G. Hajak, G. Schuierer, U. Bogdahn, J. Winkler, and A. May, Hypothalamic gray matter changes in narcoleptic patients, *Nat.Med.* **8**, 1186-1188 (2002).
23. C. Kaufmann, A. Schuld, T. Pollmacher, and D. P. Auer, Reduced cortical gray matter in narcolepsy: preliminary findings with voxel-based morphometry, *Neurology* **58**, 1852-1855 (2002).
24. D. Gerashchenko, E. Murillo-Rodriguez, L. Lin, M. Xu, L. Hallett, S. Nishino, E. Mignot, and P. J. Shiromani, Relationship between CSF hypocretin levels and hypocretin neuronal loss, *Exp.Neurol* **184**, 1010-1016 (2003).
25. Y. Kisanuki, R. M. Chemelli, S. Tokita, J. T. Willie, C. M. Sinton, and M. Yanagisawa, Behavioral and polysomnographic characterization of orexin-1 receptor and orexin-2 receptor double knockout mice (abstract), *Sleep* **24 (suppl)**, A22-A22 (2001).
26. J. T. Willie, R. M. Chemelli, C. M. Sinton, S. Tokita, S. C. Williams, Y. Y. Kisanuki, J. N. Marcus, C. Lee, J. K. Elmquist, K. A. Kohlmeier, C. S. Leonard, J. A. Richardson, R. E. Hammer, and M. Yanagisawa, Distinct narcolepsy syndromes in Orexin receptor-2 and Orexin null mice: molecular genetic dissection of Non-REM and REM sleep regulatory processes, *Neuron* **38**, 715-730 (2003).
27. B. Ripley, N. Fujiki, M. Okura, E. Mignot, and S. Nishino, Hypocretin levels in sporadic and familial cases of canine narcolepsy, *Neurobiol.Dis.* **8**, 525-534 (2001).
28. J. Hara, C. T. Beuckmann, T. Nambu, J. T. Willie, R. M. Chemelli, C. M. Sinton, F. Sugiyama, K. Yagami, K. Goto, M. Yanagisawa, and T. Sakurai, Genetic ablation of orexin neurons in mice results in narcolepsy, hypophagia, and obesity, *Neuron* **30**, 345-354 (2001).
29. S. W. Kok, S. Overeem, T. L. Visscher, G. J. Lammers, J. C. Seidell, H. Pijl, and A. E. Meinders, Hypocretin deficiency in narcoleptic humans is associated with abdominal obesity, *Obes.Res.* **11**, 1147-1154 (2003).
30. G. J. Lammers, H. Pijl, J. Iestra, J. A. Langius, G. Buunk, and A. E. Meinders, Spontaneous food choice in narcolepsy, *Sleep* **19**, 75-76 (1996).
31. S. Overeem, E. Mignot, J. G. van Dijk, and G. J. Lammers, Narcolepsy: clinical features, new pathophysiologic insights, and future perspectives, *J.Clin.Neurophysiol.* **18**, 78-105 (2001).
32. S. Overeem, J. Dalmau, L. Bataller, S. Nishino, E. Mignot, J. Verschuuren, and G. J. Lammers, Hypocretin-1 CSF levels in anti-Ma2 associated encephalitis, *Neurology* **62**, 138-140 (2004).
33. J. Dalmau, F. Graus, A. Villarejo, J. B. Posner, D. Blumenthal, B. Thiessen, A. Saiz, P. Meneses, and M. R. Rosenfeld, Clinical analysis of anti-Ma2-associated encephalitis, *Brain* (2004).
34. H. Kubota, T. Kanbayashi, Y. Tanabe, J. Takanashi, and Y. Kohno, A case of acute disseminated encephalomyelitis presenting hypersomnia with decreased hypocretin level in cerebrospinal fluid, *J.Child Neurol.* **17**, 537-539 (2002).
35. R. F. Gledhill, P. R. Bartel, Y. Yoshida, S. Nishino, and T. E. Scammell, Narcolepsy caused by acute disseminated encephalomyelitis, *Arch Neurol* **61**, 758-760 (2004).
36. S. Overeem, K. Geleijns, M. P. Garssen, B. C. Jacobs, P. A. Van Doorn, and G. J. Lammers, Screening for anti-ganglioside antibodies in hypocretin-deficient human narcolepsy, *Neurosci.Lett.* **341**, 13-16 (2003).
37. J. Nakayama, M. Miura, M. Honda, T. Miki, Y. Honda, and T. Arinami, Linkage of human narcolepsy with HLA association to chromosome 4p13-q21, *Genomics* **65**, 84-86 (2000).
38. Y. Dauvilliers, J. L. Blouin, E. Neidhart, B. Carlander, J. F. Eliaou, S. E. Antonarakis, M. Billiard, and M. Tafti, A narcolepsy susceptibility locus maps to a 5MB region of chromosome 21q, *Ann Neurol* (2004, in press).
39. Z. B. Helbing, H. A. Kamphuisen, and M. S. Mourtazaev, The origin of excessive daytime sleepiness in the Prader-Willi syndrome, *J.Intellect.Disabil.Res.* **37**, 533-541 (1993).
40. R. S. Kandt, R. G. Emerson, H. S. Singer, D. L. Valle, and H. W. Moser, Cataplexy in variant forms of Niemann-Pick disease, *Ann Neurol* **12**, 284-288 (1982).
41. D. G. Vossler, A. R. Wyler, R. J. Wilkus, G. Gardner-Walker, and B. W. Vlcek, Cataplexy and monoamine oxidase deficiency in Norrie disease, *Neurology* **46**, 1258-1261 (1996).
42. G. B. Nelson and J. S. Hahn, Stimulus-induced drop episodes in Coffin-Lowry syndrome, *Pediatrics* **111**, e197-e202 (2003).
43. T. E. Scammell, S. Nishino, E. Mignot, and C. B. Saper, Narcolepsy and low CSF orexin (hypocretin) concentration after a diencephalic stroke, *Neurology* **56**, 1751-1753 (2001).
44. J. Arii, T. Kanbayashi, Y. Tanabe, J. Ono, S. Nishino, and Y. Kohno, A hypersomnolent girl with decreased CSF hypocretin level after removal of a hypothalamic tumor, *Neurology* **56**, 1775-1776 (2001).

45. T. Kato, T. Kanbayashi, K. Yamamoto, T. Nakano, T. Shimizu, T. Hashimoto, and S. Ikeda, Hypersomnia and low CSF hypocretin-1 (orexin-A) concentration in a patient with multiple sclerosis showing bilateral hypothalamic lesions, *Intern.Med.* **42**, 743-745 (2003).
46. T. Deboer, S. Overeem, N. A. Visser, H. Duindam, M. Frolich, G. J. Lammers, and J. H. Meijer, Hypocretin-1 is under influence of circadian and homeostatic mechanisms (abstract), *Sleep* **27 (suppl)**, A1-A2 (2004).
47. S. Overeem, T. E. Scammell, and G. J. Lammers, Hypocretin/orexin and sleep: implications for the pathophysiology and diagnosis of narcolepsy, *Curr.Opin.Neurol.* **15**, 739-745 (2002).
48. C. Peyron, D. K. Tighe, A. N. Den Pol, L. de Lecea, H. C. Heller, J. G. Sutcliffe, and T. S. Kilduff, Neurons containing hypocretin (orexin) project to multiple neuronal systems, *J Neurosci* **18**, 9996-10015 (1998).
49. K. Yoshida, R. A. Espana, S. L. McCormack, A. J. Crocker, and T. E. Scammell, Afferents to the orexin neurons (abstract), *Sleep* **27 (suppl)**, A13-A13 (2004).
50. S. Zhang, J. M. Zeitzer, Y. Yoshida, J. P. Wisor, S. Nishino, D. M. Edgar, and E. Mignot, Lesions of the suprachiasmatic nucleus eliminate the daily rhythm of hypocretin-1 release, *Sleep* **27**, 619-627 (2004).
51. J. J. Hagan, R. A. Leslie, S. Patel, M. L. Evans, T. A. Wattam, S. Holmes, C. D. Benham, S. G. Taylor, C. Routledge, P. Hemmati, R. P. Munton, T. E. Ashmeade, A. S. Shah, J. P. Hatcher, P. D. Hatcher, D. N. Jones, M. I. Smith, D. C. Piper, A. J. Hunter, R. A. Porter, and N. Upton, Orexin A activates locus coeruleus cell firing and increases arousal in the rat, *Proc.Natl.Acad Sci U.S.A.* **96**, 10911-10916 (1999).
52. P. Bourgin, S. Huitron-Resendiz, A. D. Spier, V. Fabre, B. Morte, J. R. Criado, J. G. Sutcliffe, S. J. Henriksen, and L. de Lecea, Hypocretin-1 modulates rapid eye movement sleep through activation of locus coeruleus neurons, *J Neurosci* **20**, 7760-7765 (2000).
53. R. E. Brown, O. Sergeeva, K. S. Eriksson, and H. L. Haas, Orexin A excites serotonergic neurons in the dorsal raphe nucleus of the rat, *Neuropharmacology* **40**, 457-459 (2001).
54. L. Bayer, E. Eggermann, M. Serafin, B. Saint-Mleux, D. Machard, B. Jones, and M. Muhlethaler, Orexins (hypocretins) directly excite tuberomammillary neurons, *Eur.J.Neurosci.* **14**, 1571-1575 (2001).
55. K. S. Eriksson, O. Sergeeva, R. E. Brown, and H. L. Haas, Orexin/hypocretin excites the histaminergic neurons of the tuberomammillary nucleus, *J.Neurosci.* **21**, 9273-9279 (2001).
56. S. Nishino, N. Fujiki, B. Ripley, E. Sakurai, M. Kato, T. Watanabe, E. Mignot, and K. Yanai, Decreased brain histamine content in hypocretin/orexin receptor-2 mutated narcoleptic dogs, *Neurosci.Lett.* **313**, 125-128 (2001).
57. Z. L. Huang, W. M. Qu, W. D. Li, T. Mochizuki, N. Eguchi, T. Watanabe, Y. Urade, and O. Hayaishi, Arousal effect of orexin A depends on activation of the histaminergic system, *Proc.Natl.Acad.Sci.U.S.A.* **98**, 9965-9970 (2001).
58. S. Burlet, C. J. Tyler, and C. S. Leonard, Direct and indirect excitation of laterodorsal tegmental neurons by hypocretin/orexin peptides: implications for wakefulness and narcolepsy, *J.Neurosci.* **22**, 2862-2872 (2002).
59. I. Estabrooke, M. T. McCarthy, E. Ko, T. C. Chou, R. M. Chemelli, M. Yanagisawa, C. B. Saper, and T. E. Scammell, Fos expression in orexin neurons varies with behavioral state, *J.Neurosci.* **21**, 1656-1662 (2001).
60. Y. Yoshida, N. Fujiki, T. Nakajima, B. Ripley, H. Matsumura, H. Yoneda, E. Mignot, and S. Nishino, Fluctuation of extracellular hypocretin-1 (orexin A) levels in the rat in relation to the light-dark cycle and sleep-wake activities, *Eur.J.Neurosci.* **14**, 1075-1081 (2001).
61. L. I. Kiyashchenko, B. Y. Mileykovskiy, N. Maidment, H. A. Lam, M. F. Wu, J. John, J. Peever, and J. M. Siegel, Release of hypocretin (orexin) during waking and sleep states, *J.Neurosci.* **22**, 5282-5286 (2002).
62. C. B. Saper, T. C. Chou, and T. E. Scammell, The sleep switch: hypothalamic control of sleep and wakefulness, *Trends Neurosci.* **24**, 726-731 (2001).
63. L. E. Daniels, Narcolepsy, *Medicine* **13**, 1-122 (1934).
64. S. W. Kok, A. E. Meinders, S. Overeem, G. J. Lammers, F. Roelfsema, M. Frolich, and H. Pijl, Reduction of plasma leptin levels and loss of its circadian rhythmicity in hypocretin (orexin)-deficient narcoleptic humans, *J.Clin.Endocrinol.Metab* **87**, 805-809 (2002).
65. S. W. Kok, F. Roelfsema, S. Overeem, G. J. Lammers, R. L. Strijers, M. Frolich, A. E. Meinders, and H. Pijl, Dynamics of the pituitary-adrenal ensemble in hypocretin deficient narcoleptic humans: blunted basal ACTH release and evidence for normal time-keeping by the master pacemaker, *J.Clin.Endocrinol.Metab* **87,** 5085-5091 (2002).
66. S. W. Kok, F. Roelfsema, S. Overeem, G. J. Lammers, M. Frolich, A. E. Meinders, and H. Pijl, Pulsatile LH release is Diminished,while FSH Secretion is Normal in Hypocretin Deficient Narcoleptic Men, *Am J Physiol Endocrinol.Metab* (2004).
67. G. J. Lammers, S. W. Kok, F. Roelfsema, S. Overeem, M. Frolich, A. E. Meinders, and H. Pijl, The thyreotropic axis in hypocretin deficient human narcolepsy: decreased total and pulsatile secretion of TSH (abstract), *Sleep* **26 (suppl)**, A284-A284 (2003).

68. S. Overeem, S. W. Kok, G. J. Lammers, A. A. Vein, M. Frolich, A. E. Meinders, F. Roelfsema, and H. Pijl, Somatotropic axis in hypocretin-deficient narcoleptic humans: altered circadian distribution of GH-secretory events, *Am.J.Physiol Endocrinol.Metab* **284**, E641-E647 (2003).

MODULATION OF CORTICAL ACTIVITY AND SLEEP-WAKE STATES BY HYPOCRETIN/OREXIN

Barbara E. Jones and Michel Muhlethaler*

1. INTRODUCTION

As evidenced by the narcoleptic syndrome that occurs in mice following knock out of the gene for the peptide hypocretin/orexin (Hcrt/Orx),[1] in dogs following knock out of the gene for Hcrt/Orx receptors[2] and in humans in association with the loss of Hcrt/Orx peptide and neurons,[3,4] Hcrt/Orx appears to be essential for the maintenance of waking. This role may be fulfilled through the widespread projections and influence of the Hcrt/Orx neurons on multiple systems including the hypothalamo-pituitary-adrenal (HPA) axis, the sympathetic nervous system and central arousal systems, as reviewed in this volume. Indeed, Hcrt/Orx appears to have an excitatory influence upon all the brainstem arousal systems, including the noradrenergic locus coeruleus neurons,[5,6] the cholinergic pontomesencephalic neurons[7] and the histaminergic tuberomammillary neurons (Fig. 1).[8,9] As we will elaborate in the present chapter, Hcrt/Orx also activates the major subcortical relays from the brainstem reticular formation to the cerebral cortex, the diffuse thalamo-cortical projection system and the cholinergic basalo-cortical projection system. Lastly, it also acts directly upon a select group of cortical neurons that may in turn together with the latter systems stimulate widespread cortical activation that supports the waking state. Interestingly, as we will also mention, Hcrt/Orx does not directly inhibit presumed sleep-promoting neurons in the forebrain.

* B.E. Jones, Department of Neurology and Neurosurgery, McGill University, Montreal Neurological Institute, 3801 University Street, Montreal, Quebec, Canada H3A 2B4 M. Muhlethaler, Département de Neurosciences fondamentales, Centre Médical Universitaire, 1 rue Michel-Servet, 1211 Genève 4, Switzerland

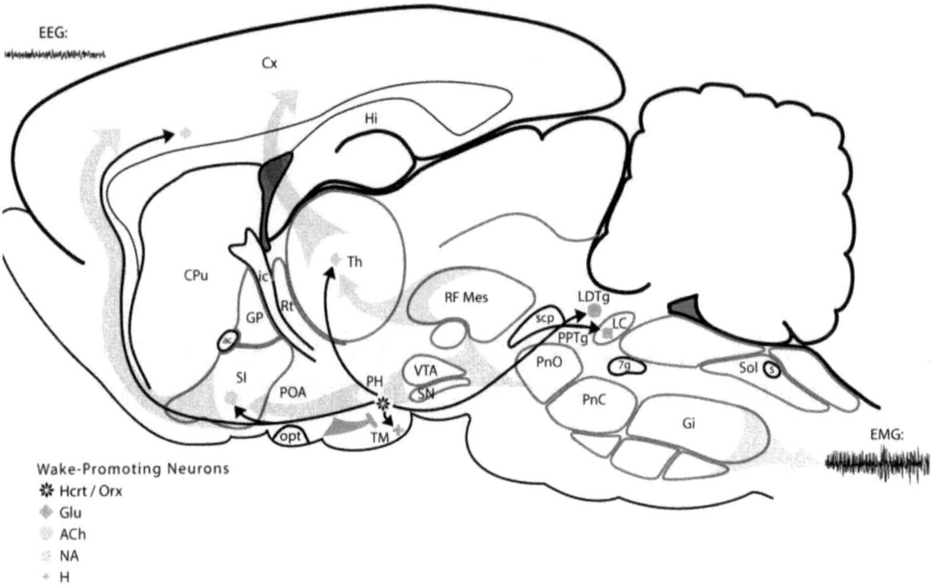

Figure 1. Hcrt/Orx neurons amid arousal systems. Schematic sagittal view of rat brain showing the major neuronal systems and major excitatory pathways (light gray arrows) involved in promoting the EEG fast activity (upper left) and EMG high muscle tone and activity (lower right) characteristic of the waking state. Hcrt/Orx neurons are located in the perifornical and lateral, tuberal and posterior hypothalamus (PH), where they are situated amongst ascending pathways from the brainstem arousal systems. The major ascending pathways emerge from the brainstem reticular formation (RF, most densely from the mesencephalic (RF Mes) and oral pontine (PnO) fields) to ascend along 1) a dorsal trajectory into the thalamus (Th) where they terminate upon (midline, medial and intralaminar) nuclei of the nonspecific thalamo-cortical projection system, which in turn projects in a widespread manner to the cerebral cortex (Cx) and 2) a ventral trajectory through the lateral hypothalamus up to the basal forebrain where they terminate upon magnocellular basal neurons, shown in the substantia innominata (SI), which also in turn project in a widespread manner to the cerebral cortex.[10] Descending projections collect from multiple levels of the reticular formation (though most densely from the caudal pontine, PnC, and medullary gigantocellular, Gi, fields) to form the reticulo-spinal pathways. The major transmitter systems that promote waking and contribute to these ascending and descending systems are represented by symbols where their cell bodies are located. Glutamatergic (Glu) neurons comprise the vast population of neurons of the reticular formation (not shown) and the diffuse thalamo-cortical projection system. Cholinergic neurons, containing acetylcholine (ACh), are located in the laterodorsal and pedunculopontine tegmental (LDTg and PPTg) nuclei in the brainstem from where they project along with other reticular neurons dorsally to the thalamus and ventrally to the posterior hypothalamus and basal forebrain, as well as to the brainstem reticular formation. Cholinergic neurons in the forebrain (SI) comprise the basalo-cortical projection. Noradrenergic (NA) neurons of the locus coeruleus (LC) send axons along the major ascending and descending pathways to project in a diffuse manner to the cortex, the subcortical relay stations, brainstem and spinal cord. Histaminergic (H) neurons of the tuberomammillary nucleus (TM) also project in a diffuse manner to the forebrain and cortex. The Hcrt/Orx neurons receive input from glutamatergic RF neurons, noradrenergic LC neurons and cholinergic LDTg/PPTg neurons of the brainstem. They also receive input from nearby histaminergic neurons. And they receive inhibitory input from GABAergic neurons of the basal forebrain (represented by a small dark gray arrow ending in a block to represent inhibition). Like other neurons of the arousal systems, Hcrt/Orx neurons project diffusely to spinal cord, the brainstem and the forebrain along the same major pathways utilized by other arousal systems. Through these projections (shown as black lines ending in arrowheads to indicate excitation), they excite many neurons, including the noradrenergic LC neurons and cholinergic LDTg/PPTg neurons in the brainstem and histaminergic neurons in the hypothalamus. In the forebrain, they directly excite the glutamatergic neurons of the nonspecific thalamo-cortical projection system (in Th) and the cholinergic neurons of the basalo-cortical projection system (in SI), which in turn stimulate widespread cortical activation. And they also directly excite cortical neurons within the deepest layer of the

cortex, which in turn influence other cortical neurons. *Anatomical abbreviations*: 7g, 7th nerve genu; ac, anterior commissure; CPu, caudate-putamen; Cx, cortex; Gi, gigantocellular reticular formation; GP, globus pallidus; Hi, hippocampus; ic, internal capsule; LC, locus coeruleus; LDTg, laterodorsal tegmental nucleus; opt, optic tract; PH, posterior hypothalamus; POA, preoptic area; PnC, pontis caudalis reticular formation; PnO, pontis oralis reticular formation; PPTg, pedunculopontine tegmental nucleus; RF Mes, mesencephalic reticular formation; Rt, reticularis nucleus of the thalamus; s, solitary tract; scp, superior cerebellar peduncle; SI, substantia innominata; Sol, solitary tract nucleus; SN, substantia nigra; Th, thalamus; TM, tuberomammillary nucleus; VTA, ventral tegmental area. *Other abbreviations*: ACh, acetylcholine; EEG, electroencephalogram; EMG, electromyogram; Glu, glutamate; H, histamine; Hcrt/Orx, hypocretin/orexin; NA, noradrenaline. Copied and modified with permission from Jones (2002) In *Biological Psychiatry* (D'Haenen, D., den Boer, J. A. & Willner, P., eds.), Vol. 2, pp. 1215-1228.[11]

2. MODULATION AND ACTIVITY OF HCRT/ORX NEURONS

The Hcrt/Orx neurons are situated within the lateral hypothalamus where fibers ascend within the medial forebrain bundle (MFB) from the brainstem arousal systems, including the reticular formation, the locus coeruleus noradrenergic neurons and the cholinergic pontomesencephalic neurons (Fig. 1).[10, 12, 13] They can thus be activated by these constituents of the ascending reticular and brainstem activating systems. Indeed like most neurons, they are excited by glutamate or its agonists[14] that would be released by neurons of the reticular formation.[10] They would also act in series with the noradrenergic and cholinergic neurons which discharge during waking,[15-17] if excited by these transmitters. Yet, surprisingly in electrophysiological studies performed in hypothalamic slices of transgenic mice expressing green fluorescent protein (GFP) through the Hcrt/Orx promoter, noradrenaline (NA) was found to inhibit and acetylcholine (ACh) to have no effect upon the GFP fluorescent neurons,[14, 18] In contrast, however, we recently found in hypothalamic slices of rats on immunohistochemically identified Hcrt/Orx neurons that both NA and ACh (or its agonist, carbachol) exerted a direct depolarizing and excitatory effect.[19] These results indicate that the Hcrt/Orx neurons would be excited in tandem with the major arousal systems of the brainstem. By their reciprocal projections and excitatory actions, they would in turn sustain the excitation of the brainstem arousal systems, while influencing in parallel the relay neurons in the thalamus and basal forebrain to promote cortical activation.

From studies examining release of Hcrt/Orx and c-Fos expression in association with waking and sleep in rats, cats or primates, the Hcrt/Orx neurons appear to be most active during the period of the day when the animals are most active and thus also during the waking state.[20-26] Moreover, they would appear to be most active during periods of maximal motor activity. Only by recording from identified Hcrt/Orx neurons in the hypothalamus will it be known whether they discharge or not during rapid eye movement sleep (REMS).[27] Given their excitation by NA and ACh (above), they would be excited and active during waking when both the noradrenergic and cholinergic brainstem neurons are active, and depending upon the balance of inputs could also be excited during REMS when the cholinergic neurons discharge[16,28] or alternatively they could be inhibited by GABAergic inputs (below). C-Fos expression is maximal during waking and minimal during sleep, including REMS, in Hcrt/Orx neurons, in contrast to that in surrounding melanin concentrating hormone (MCH) neurons.[25, 29]

From *in vitro* studies, we have discovered that Hcrt/Orx neurons are endowed with particular membrane properties that provide them with an intrinsic mechanism for sustained tonic discharge independent of synaptic input.[30] Identified by Neurobiotin labeling combined with immunohistochemical staining, Hcrt/Orx cells show a low threshold spike (LTS), after-depolarization potential (ADP) and inward rectification upon hyperpolarization (indicative of an h current, I_h, Fig. 2). They are spontaneously active with an average spike rate of ~3 Hz and tonically depolarized with a membrane potential of ~-46 mV in the presence of tetrodotoxin (TTX, which eliminates sodium action potentials). These properties are not found in adjacent MCH neurons that are not spontaneously active nor depolarized (having a resting membrane potential of ~-62 mV, Fig. 2). The relatively depolarized membrane potential of the Hcrt/Orx neurons was found to be independent of synaptic input, as evident under conditions of synaptic uncoupling (with TTX, block of glutamatergic and GABAergic receptors or high Mg^{2+}/low Ca^{2+}, Fig. 3). It was found to be maintained by a calcium-activated nonselective cation current (I_{CAN}) that was manifest by the ADP. The Hcrt/Orx cells could accordingly sustain tonic discharge and influence upon their target neurons during the waking state independent of other arousal systems. Indeed, given the innervation and excitation of the brainstem arousal systems by Hcrt/Orx, the Hcrt/Orx neurons may represent the central generator for eliciting and maintaining activity in the arousal systems of the brain during waking (Fig. 1).

In order to stop their spontaneous discharge, Hcrt/Orx neurons must be actively inhibited as can be achieved by the inhibitory neurotransmitter, GABA (Fig. 3).[30] It is thus presumed that GABAergic neurons that would become active during slow wave sleep (SWS) could turn off the Hcrt/Orx neurons. Since no neurons have been recorded in the lateral hypothalamus that are active during SWS,[31] such SWS-active GABAergic neurons are most likely located in the basal forebrain or preoptic area. Indeed, GABAergic neurons have been shown by retrograde and most recently anterograde transport to project from basal forebrain into the posterior lateral hypothalamus and directly onto the Hcrt/Orx neurons.[32-34] Electrophysiological and c-Fos studies have identified GABAergic SWS-active and promoting neurons in the basal forebrain[35-38] that could thus have the capacity to directly inhibit Hcrt/Orx neurons and remove an otherwise tonic excitatory influence upon all other brainstem arousal systems (Fig. 1).

3. EXCITATORY INFLUENCE OF HCRT/ORX UPON THE DIFFUSE THALAMO-CORTICAL PROJECTION SYSTEM

The Hcrt/Orx neurons project to the thalamus to provide a particularly dense innervation to the midline and intralaminar nuclei.[39] These thalamic neurons give rise in turn to widespread projections to the cerebral cortex[40] via which they stimulate widespread cortical activation.[41-43] This system thus also serves as the dorsal relay of the ascending reticular activating system for stimulating widespread and prolonged fast cortical activity.[44] Input to this system by the Hrct/Orx neurons thus provides an important route by which they may influence cortical activation (Fig. 1).

We have found that Hcrt/Orx depolarizes and excites neurons in the rhomboid (Rh) and centromedial (CM) nuclei of the thalamus, which give rise to very widespread projections to the cerebral cortex (Fig. 4).[45] This action is likely mediated by Orx_2 receptors, given the potent effect of Orx B relative to Orx A on these cells. Interestingly, the peptide has no effect upon neurons of the specific visual (dorsal lateral geniculate, DLG) or somatic sensory relay nuclei (ventral posterior lateral, VPL, Fig. 4). Hcrt/Orx would thus not alter transmission of specific sensory information but could prolong the cortical response to sensory input through the nonspecific projection system. The selective excitation of the nonspecific thalamo-cortical projection system would also maintain widespread fast cortical activity during the waking state.[43]

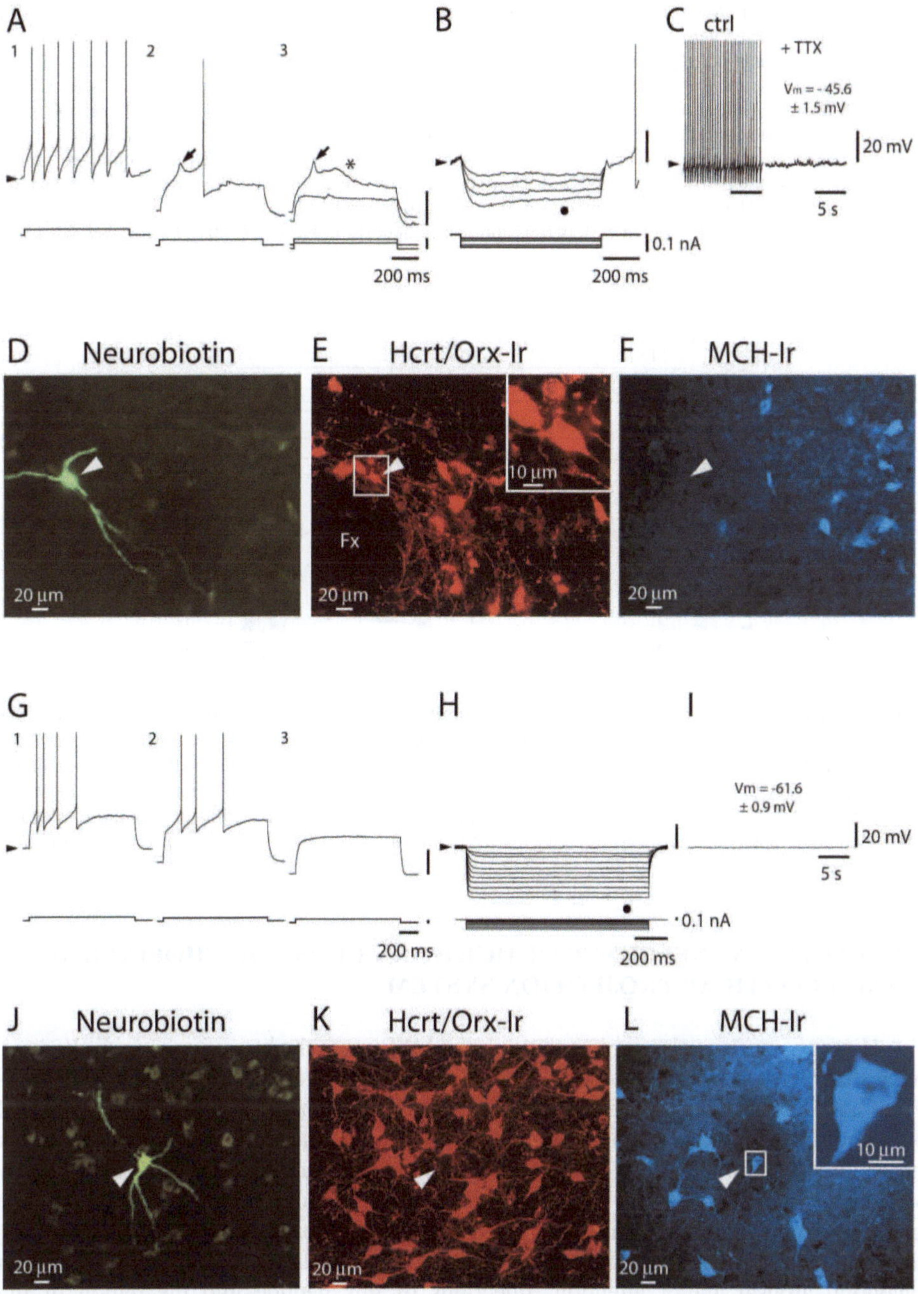

Figure 2. Characterization of neurons expressing Hcrt/Orx or MCH. (**A_1**) Tonic firing in response to a depolarizing current pulse delivered from the level of resting potential (arrowhead). (**$A_{2\text{-}3}$**) Low threshold spike (LTS, arrow) and after-depolarization (ADP, star) triggered by a depolarizing current pulse delivered from a hyperpolarized level. Further hyperpolarization eliminates the LTS and the ADP (lower trace in A_3). (**B**) Superimposed responses to hyperpolarizing pulses suggesting the presence of an h current (I_h, dot). Note that only the trace with the deepest hyperpolarization is shown in full. (**C**) Tonic firing at rest and its elimination by tetrodotoxin (TTX, 1.0 μM) to determine resting potentials. (**D - F**) Immunohistochemical identification of a Hcrt/Orx neuron injected with Neurobiotin (arrow in D) and expressing immunoreactivity (Ir) for Hcrt/Orx (E) but not for the melanin-concentrating hormone (MCH, F). (**G_1**) Firing with accommodation triggered by a depolarizing current pulse delivered from the resting potential level. (**$G_{2\text{-}3}$**) Absence of either LTS or ADP in response to depolarizing pulses applied from more hyperpolarized levels. (**H**) Responses to hyperpolarizing current pulses demonstrating the absence (dot) of any sag that could have indicated the presence of an I_h. (**I**) Absence of spontaneous firing in such neurons and their mean resting potential. (**J - L**) Immunohistochemical identification of an MCH neuron injected with Neurobiotin (arrow in J) and expressing immunoreactivity for MCH (L) and not for Hcrt/Orx (K). Membrane potentials (arrowheads): –47 mV (A), –44 mV (B), –48 mV (C), –61 mV (G), –61 mV (H, I). Copied with permission from Eggermann, E. et al. (2003) *J. Neurosci.* 23, 1557-1562.[30]

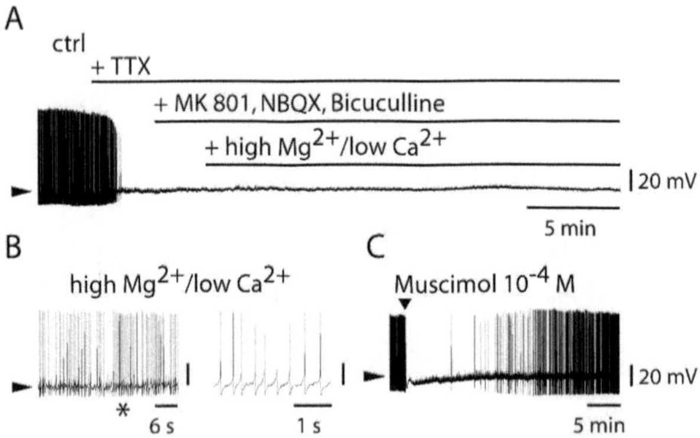

Figure 3. Spontaneous activity and GABAergic inhibition of Hcrt/Orx neurons. (**A**) Persistence of the membrane depolarization in TTX (1.0 μM), ionotropic blockers (MK801 at 20 μM, NBQX at 10 μM and bicuculline at 10 μM) and in a solution containing 10 mM Mg^{2+} and 0.1 mM Ca^{2+}. (**B**) Persistence of the spontaneous activity in presence of synaptic blockade (right panel showing an enlargement of the area identified by a star in the left panel). (**C**) Inhibition by a brief application of the $GABA_A$ agonist, muscimol (5 s at 100 μM). Membrane potentials (arrowheads): –42 mV (A), –43 mV (B), –44 mV (C). Copied with permission from Bayer et al. (2002) *J. Neurosci.* 22, 7835-7839.[30]

4. EXCITATORY INFLUENCE OF HCRT/ORX UPON THE CHOLINERGIC BASALO-CORTICAL PROJECTION SYSTEM

Hcrt/Orx neurons also project through the MFB up to the basal forebrain.[39] We have found that Hcrt/Orx directly depolarizes and excites cholinergic basal forebrain neurons (Fig. 5).[46] As in the thalamus, the effect of the peptide appears to be mediated in the basal forebrain through Orx_2 receptors. By direct widespread projections to the cerebral cortex and excitatory actions therein, the cholinergic basal forebrain neurons have the capacity to stimulate widespread and prolonged cortical activation, marked by fast gamma activity.[47] Hcrt/Orx could thus activate the cholinergic basalo-cortical system during waking for the elicitation and maintenance of cortical activation associated with behavioral arousal and/or attention. Injections of this peptide into the region of the cholinergic basal forebrain neurons have been shown to stimulate cortical activation along with a behaviorally active state.[48]

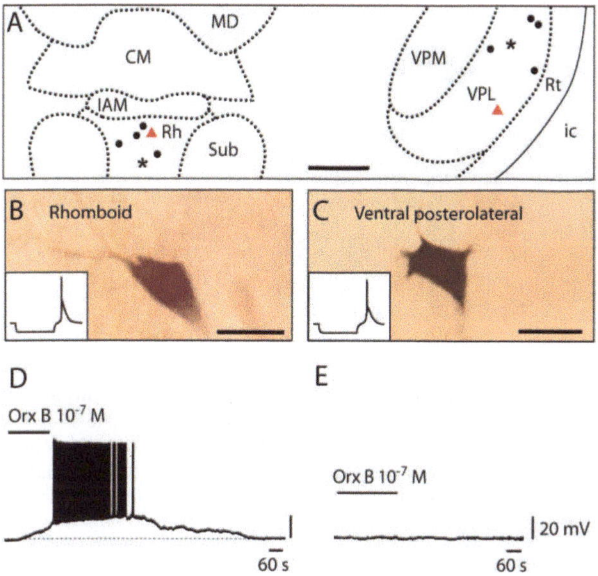

Figure 4. Actions of Hcrt/Orx on nonspecific and specific thalamo-cortical projection neurons. Nuclei of nonspecific systems (left, including CM, where units were also recorded, and Rh, as shown) and specific systems (right, somatosensory relay nuclei, VPM and VPL, where units were recorded as shown and) where neurons were tested for their response to the orexin (Orx) peptide and some labeled with Neurobiotin. (**A**) Red triangles correspond to injected cells shown in B and C. (**B, C**) Neurobiotin-filled neurons in the Rh and VPL nuclei (insets showing characteristic responses to hyperpolarizing pulses). (**D**) Depolarizing and excitatory effect of Orx B on Rh neurons. (**E**) Absence of effect of Orx B on VPL neurons. *Calibrations*: 500 µm in A and 20 µm in B and C. *Abbreviations*: IAM, interanteromedial thalamic nucleus; ic, internal capsule; CM, centromedial nucleus; MD, mediodorsal thalamic nucleus; Sub, submedius thalamic nucleus; VPL, ventral posterolateral thalamic nucleus; VPM, ventral posteromedial thalamic nucleus; Rh, rhomboid nucleus; Rt, reticular thalamic nucleus. Copied with permission from Bayer et al. (2002) *J. Neurosci.* 22, 7835-7839.[45]

5. EXCITATORY INFLUENCE OF HCRT/ORX UPON CORTICO-CORTICAL PROJECTION NEURONS

Like other diffuse projecting neurons of the arousal systems, Hrct/Orx neurons project directly to the cerebral cortex in addition to projecting onto subcortical relay neurons.[39] In the cortex, Hcrt/Orx fibers are distributed in a widespread manner and through all layers, although most densely in the deeper layers. In a systematic study of postsynaptic effects of Hcrt/Orx in the cortex, we recently discovered that the peptide has no direct effect upon cortical neurons in layers 1 through 6a, even though it has indirect effects through presynaptic terminals of afferent inputs to the cortical neurons (Fig. 6). Such presynaptic effects were demonstrated upon layer 5 pyramidal cells through actions upon terminals of the nonspecific thalamo-cortical afferents.[49] On the other hand, we found that Hrct/Orx does have a postsynaptic depolarizing action upon cortical neurons located exclusively in layer 6b (Fig. 6).[50] This action appears to be mediated by an Orx_2

receptor. Neurons in layer 6b give rise to widespread projections to layer 1 of surrounding cortical areas[51, 52] and could thus propagate widespread cortical activation.

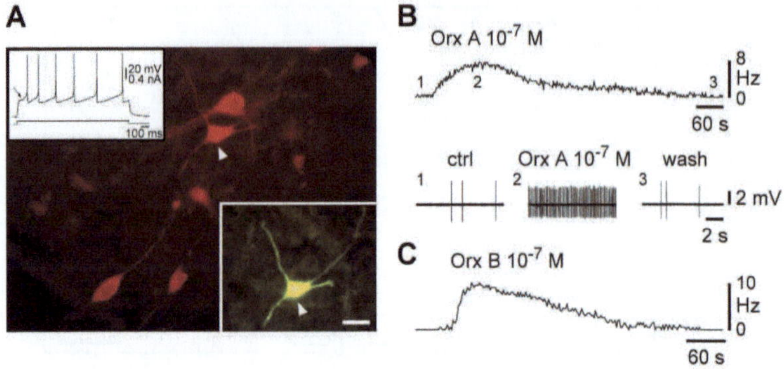

Figure 5. Excitatory action of Hcrt/Orx on cholinergic basal forebrain neurons. (**A**) Electrophysiological identification (upper left inset) and immunohistochemical confirmation (white arrowhead indicating a ChAT-immunopositive cell) of a basal forebrain cholinergic neuron labeled with Neurobiotin (lower right inset). *Calibration*: 25 µm. (**B, C**) Such cells were excited by both orexin (Orx) A and B. Extracellular action potentials (lower panel B) before (1) during (2) and after (3) Orx effects are demonstrated. Copied with permission from Eggermann et al. (2001) *Neuroscience* 108, 177-181.[46]

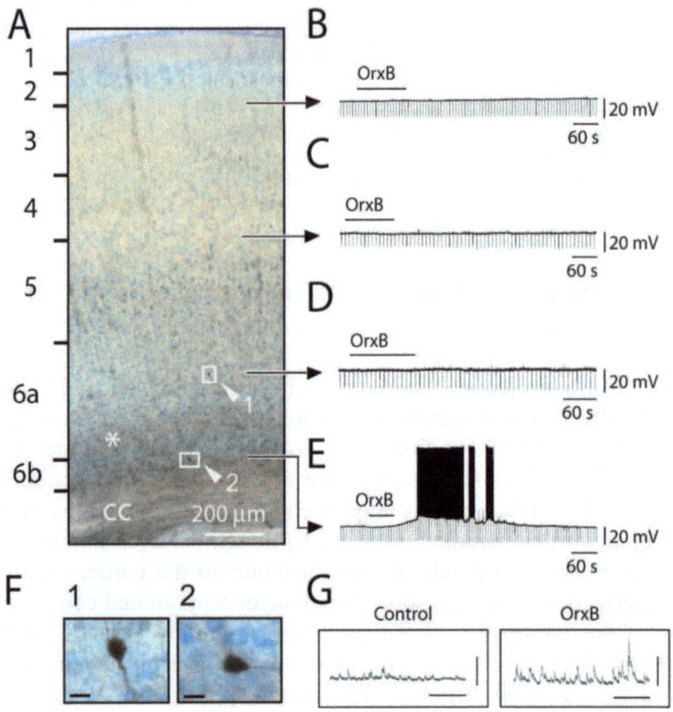

Figure 6. Exclusive action of Hcrt/Orx on cortical neurons of sublayer 6b in cortex. (**A**) Toluidine blue counter-stained cortical slice slab containing two recorded neurons (arrowheads) labeled with Neurobiotin in sublayers 6a and 6b (separated by a horizontal band of fibers, *) and whose responses to OrxB are shown in panels D and E. (**B**, **C**) Absence of responses to Orx B of neurons in layers 2/3 and 4/5. (**D**) Absence of response to OrxB of a neuron in layer 6a. (**E**) Depolarizing response to OrxB of a neuron in layer 6b. (**F_{1-2}**) Enlargement of neurons 1 and 2 from panel A. *Calibrations:* 15 μm. (**G**) Increase in post-synaptic potentials (PSPs) in a layer 5 neuron in the presence of OrxB, which is impeded in the presence of a high Mg^{2+}/low Ca^{2+} solution (not shown). *Calibrations*: 5 mV/2 sec. *Abbreviation*: cc, corpus callosum. Copied and modified with permission from Bayer (In press) *J. Neurosci.*[50]

6. INDIRECT INFLUENCE OF HCRT/ORX UPON SLEEP-PROMOTING NEURONS

Given the importance of Hcrt/Orx for maintaining a waking state, it was thought likely that the peptide could have a direct inhibitory influence upon sleep-promoting neurons. We tested this possibility on the presumed GABAergic sleep-promoting neurons of the ventrolateral preoptic area (VLPO), identified by their inhibitory response to NA.[53] Somewhat to our surprise, there was no evidence for a direct inhibitory effect of Hcrt/Orx upon these neurons (Fig. 7).[46] Yet, no direct inhibitory action of Hcrt/Orx has been found to date in the central nervous system. Although no indirect effect of Hcrt/Orx was evident in the slice upon the VLPO neurons, it is very possible that Hcrt/Orx could excite particular GABAergic neurons that would in turn inhibit sleep-promoting neurons. Without this, Hcrt/Orx would act indirectly to inhibit sleep-promoting neurons through its excitatory action upon the noradrenergic locus coeruleus neurons,[5, 6] since NA inhibits GABAergic sleep-promoting neurons in the preoptic region and basal forebrain (Fig. 7).[35,38,53]

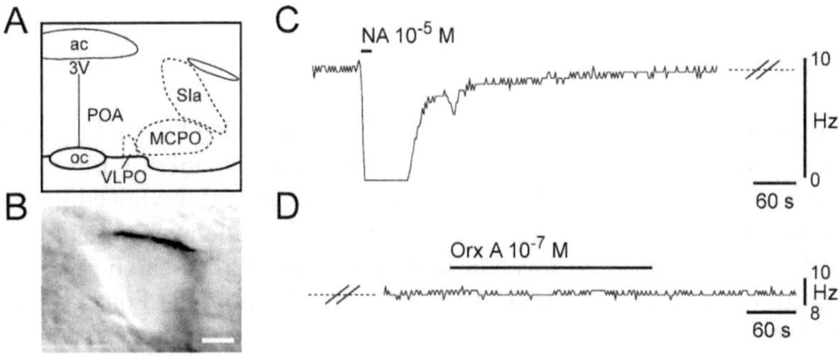

Figure 7. Absence of effect of Hcrt/Orx on ventrolateral preoptic area (VLPO) neurons. (**A**) The basal forebrain/preoptic area. (**B, C**) VLPO cells' identification by their shape (calibration, 5 μm) and their inhibition by noradrenaline (NA). (**D**) Such cells were unaffected by Orx A (or B, not shown). *Abbreviations:* ac, anterior commissure; oc, optic chiasm; MCPO, magnocellular preoptic nucleus; POA, preoptic area; SIa, substantia innominata pars anterior; VLPO, ventrolateral preoptic nucleus; 3V, third ventricle. Copied with permission from Eggermann et al. (2001) *Neuroscience* 108, 177-181.[46]

7. SUMMARY AND CONCLUSIONS

As evident in animal and human cases of narcolepsy, the hypocretin/orexin (Hcrt/Orx) peptides, receptors and neurons are necessary for the maintenance of the wake state. This critical role may derive from the diffuse projections and excitatory actions of Hcrt/Orx through the central nervous system. Lying within the medial forebrain bundle, Hcrt/Orx neurons receive input from and would be excited by other arousal systems, including noradrenergic and cholinergic brainstem neurons, and project back to excite these same systems to sustain their excitation. But they also have intrinsic properties that allow them to maintain tonic spontaneous discharge independent of afferent input. Accordingly, they may be tonically active during the wake state and serve as a central generator to sustain activity in other arousal systems. To be turned off during sleep, they can be inhibited by GABA, which is contained within afferent inputs from basal forebrain neurons that discharge during sleep. In addition to their innervation of the brainstem arousal systems, Hcrt/Orx neurons also project into the forebrain where they provide a diffuse innervation to subcortical relay stations and to the cerebral cortex. Hcrt/Orx peptides directly excite the midline and intralaminar nuclei of the nonspecific thalamo-cortical projection system and the cholinergic neurons of the basalo-cortical projection system. They can thus stimulate widespread cortical activation through these subcortical relay stations. In addition, Hcrt/Orx peptides exert a direct excitatory action upon cortical neurons located exclusively in layer 6b. These particular cortical neurons project to surrounding cortical areas and can thus also propagate widespread cortical activation that sustains the waking state. Finally, Hcrt/Orx peptides do not directly inhibit sleep-promoting neurons in the forebrain but would do so indirectly by exciting noradrenergic neurons, which inhibit the sleep-promoting cells. Given their intrinsic activity, diffuse projections and excitatory actions upon brainstem and forebrain arousal systems, the Hcrt/Orx neurons thus constitute a central generator for stimulating and maintaining the wake state.

8. ACKNOWLEDGMENTS

We thank Pablo Henny, Mandana Modirrousta, Maan Gee Lee and Lynda Mainville in Montreal and Mauro Serafin, Emmanuel Eggermann, Laurence Bayer and Daniel Machard in Geneva for their original contributions to the work reviewed in this chapter. We also thank James Galaty for assistance with the illustrations. Original research was supported by grants from the Canadian Institutes of Health Research (CIHR) and U.S. National Institutes of Health (NIH) to BEJ in Montreal and by the Fonds National Swiss to MM in Geneva.

9. REFERENCES

1. R. M. Chemelli, J. T. Willie, C. M. Sinton, J. K. Elmquist, T. Scammell, C. Lee, J. A. Richardson, S. C. Williams, Y. Xiong, Y. Kisanuki, T. E. Fitch, M. Nakazato, R. E. Hammer, C. B. Saper and M. Yanagisawa, Narcolepsy in orexin knockout mice: molecular genetics of sleep regulation, *Cell* **98**, 437-451 (1999).
2. L. Lin, J. Faraco, R. Li, H. Kadotani, W. Rogers, X. Lin, X. Qiu, P. J. de Jong, S. Nishino and E. Mignot, The sleep disorder canine narcolepsy is caused by a mutation in the hypocretin (orexin) receptor 2 gene, *Cell* **98**, 365-376 (1999).

3. T. C. Thannickal, R. Y. Moore, R. Nienhuis, L. Ramanathan, S. Gulyani, M. Aldrich, M. Cornford and J. M. Siegel, Reduced number of hypocretin neurons in human narcolepsy, *Neuron* **27**, 469-474. (2000).
4. C. Peyron, J. Faraco, W. Rogers, B. Ripley, S. Overeem, Y. Charnay, S. Nevsimalova, M. Aldrich, D. Reynolds, R. Albin, R. Li, M. Hungs, M. Pedrazzoli, M. Padigaru, M. Kucherlapati, J. Fan, R. Maki, G. J. Lammers, C. Bouras, R. Kucherlapati, S. Nishino and E. Mignot, A mutation in a case of early onset narcolepsy and a generalized absence of hypocretin peptides in human narcoleptic brains, *Nat Med.* **6**, 991-997 (2000).
5. T. L. Horvath, C. Peyron, S. Diano, A. Ivanov, G. Aston-Jones, T. S. Kilduff and A. N. van Den Pol, Hypocretin (orexin) activation and synaptic innervation of the locus coeruleus noradrenergic system, *J Comp Neurol.* **415**, 145-159 (1999).
6. P. Bourgin, S. Huitrón-Reséndiz, A. Spier, V. Fabre, B. Morte, J. Criado, J. G. Sutcliffe, S. Henriksen and L. de Lecea, Hypocretin-1 modulates REM sleep through activation of locus coeruleus neurons, *J. Neurosci.* **20**, 7760-7765 (2000).
7. S. Burlet, C. J. Tyler and C. S. Leonard, Direct and indirect excitation of laterodorsal tegmental neurons by Hypocretin/Orexin peptides: implications for wakefulness and narcolepsy, *J Neurosci.* **22**, 2862-2872 (2002).
8. K. S. Eriksson, O. Sergeeva, R. E. Brown and H. L. Haas, Orexin/hypocretin excites the histaminergic neurons of the tuberomammillary nucleus, *J Neurosci.* **21**, 9273-9279 (2001).
9. L. Bayer, E. Eggermann, M. Serafin, B. Saint-Mleux, D. Machard, B. Jones and M. Muhlethaler, Orexins (hypocretins) directly excite tuberomammillary neurons, *Eur J Neurosci.* **14**, 1571-1575 (2001).
10. B. E. Jones, Reticular formation. Cytoarchitecture, transmitters and projections. In *The Rat Nervous System* 2nd edit. (Paxinos, G., ed.), pp. 155-171, Academic Press Australia, Sydney (1995).
11. B. E Jones, Neurotransmitter systems regulating sleep-wake states. In *Biological Psychiatry* (D'Haenen, D., den Boer, J. A. & Willner, P., eds.), Vol. 2, pp. 1215-1228, John Wiley & Sons (2002).
12. B. E. Jones and T. Z. Yang, The efferent projections from the reticular formation and the locus coeruleus studied by anterograde and retrograde axonal transport in the rat, *J Comp Neurol.* **242**, 56-92 (1985).
13. B. Ford, C. J. Holmes, L. Mainville and B. E. Jones, GABAergic neurons in the rat pontomesencephalic tegmentum: codistribution with cholinergic and other tegmental neurons projecting to the posterior lateral hypothalamus, *J Comp Neurol.* **363**, 177-196 (1995).
14. Y. Li, X. B. Gao, T. Sakurai & A. N. van den Pol, Hypocretin/Orexin excites hypocretin neurons via a local glutamate neuron-A potential mechanism for orchestrating the hypothalamic arousal system, *Neuron.* **36**, 1169-1181 (2002).
15. G. Aston-Jones and F. E. Bloom, Activity of norepinephrine-containing locus coeruleus neurons in behaving rats anticipates fluctuations in the sleep-waking cycle, *J Neurosci.* **1**, 876-886 (1981).
16. M. el Mansari, K. Sakai and M. Jouvet, Unitary characteristics of presumptive cholinergic tegmental neurons during the sleep-waking cycle in freely moving cats, *Exp Brain Res.* **76**, 519-529 (1989).
17. M. Steriade, S. Datta, D. Pare, G. Oakson and R. C. Curro Dossi, Neuronal activities in brain-stem cholinergic nuclei related to tonic activation processes in thalamocortical systems. *J. Neurosci.* **10**, 2541-2559 (1990).
18. A. Yamanaka, Y. Muraki, N. Tsujino, K. Goto and T. Sakurai, Regulation of orexin neurons by the monoaminergic and cholinergic systems, *Biochem Biophys Res Commun.* **303**, 120-129 (2003).
19. L. Bayer, E. Eggermann, M. Serafin, J. Grivel, D. Machard, M. Muhlethaler and B. E. Jones, Opposite effects of noradrenaline and acetylcholine upon hypocretin/orexin versus melanin concentrating hormone neurons in rat hypothalamic slices (In preparation).
20. N. Fujiki, Y. Yoshida, B. Ripley, K. Honda, E. Mignot and S. Nishino, Changes in CSF hypocretin-1 (orexin A) levels in rats across 24 hours and in response to food deprivation, *Neuroreport* **12**, 993-997 (2001).
21. J. M. Zeitzer, C. L. Buckmaster, K. J. Parker, C. M. Hauck, D. M. Lyons and E. Mignot, Circadian and homeostatic regulation of hypocretin in a primate model: implications for the consolidation of wakefulness, *J Neurosci.* **23**, 3555-3560 (2003).
22. M. F. Wu, J. John, N. Maidment, H. A. Lam and J. M. Siegel, Hypocretin release in normal and narcoleptic dogs after food and sleep deprivation, eating, and movement, *Am J Physiol Regul Integr Comp Physiol.* **283**, R1079-86 (2002).
23. L. I. Kiyashchenko, B. Y. Mileykovskiy, N. Maidment, H. A. Lam, M. F. Wu, J. John, J. Peever and J. M. Siegel, Release of hypocretin (orexin) during waking and sleep states, *J Neurosci.* **22**, 5282-5286 (2002).
24. I. V. Estabrooke, M. T. McCarthy, E. Ko, T. C. Chou, R. M. Chemelli, M. Yanagisawa, C. B. Saper and T. E. Scammell, Fos expression in orexin neurons varies with behavioral state, *J Neurosci.* **21**, 1656-1662 (2001).

25. M. Modirrousta, L. Mainville and B. E. Jones, B. E, c-Fos expression in neurons of the posterior hypothalamus, including those containing orexin (orx) or melanin concentrating hormone (MCH), following sleep deprivation or recovery, *Sleep* **26**, A25 (2003).
26. Y. Yoshida, N. Fujiki, T. Nakajima, B. Ripley, H. Matsumura, H. Yoneda, E. Mignot and S. Nishino, Fluctuation of extracellular hypocretin-1 (orexin A) levels in the rat in relation to the light-dark cycle and sleep-wake activities, *Eur J Neurosci.* **14**, 1075-1081 (2001).
27. M. G. Lee and B. E. Jones, Discharge of identified orexin neurons across the sleep-waking cycle. *Soc. Neurosci. Abst. Online* (2004).
28. K. J. Maloney L. Mainville and B. E. Jones, Differential c-Fos expression in cholinergic, monoaminergic and GABAergic cell groups of the pontomesencephalic tegmentum after paradoxical sleep deprivation and recovery, *J. Neurosci.***19**, 3057-3072 (1999).
29. L. Verret, R. Goutagny, P. Fort, L. Cagnon, D. Salvert, L. Leger, R. Boissard, P. Salin, C. Peyron and P. H. Luppi, A role of melanin-concentrating hormone producing neurons in the central regulation of paradoxical sleep, *BMC Neurosci.* **4**, 19 (2003).
30. E. Eggermann, L. Bayer, M. Serafin, B. Saint-Mleux, L. Bernheim, D. Machard, B. E. Jones and M. Muhlethaler, The wake-promoting hypocretin-orexin neurons are in an intrinsic state of membrane depolarization. *J Neurosci.* **23**, 1557-1562 (2003).
31. T. L. Steininger, M. N. Alam, H. Gong, R. Szymusiak and D. McGinty, Sleep-waking discharge of neurons in the posterior lateral hypothalamus of the albino rat, *Brain Res.* **840**, 138-147 (1999).
32. P. Henny and B. E. Jones, Vesicular GABA (VGAT), glutamate (VGluT) or acetylcholine (VAChT) transporters in basal forebrain axon terminals innervating the lateral hypothalamus and orexin/hypocretin neurons: substrate for dual sleep-wake state regulation (In preparation).
33. P. Henny and B. E. Jones, Vesicular transporter proteins for glutamate (VgluT), GABA (VGAT) and acetylcholine (VAChT) in terminal varicosities of basal forebrain neurons projecting to the posterior lateral hypothalamus, *Sleep* **26**, A9 (2003).
34. I. Gritti, L. Mainville and B. E. Jones, Projections of GABAergic and cholinergic basal forebrain and GABAergic preoptic-anterior hypothalamic neurons to the posterior lateral hypothalamus of the rat. *J. Comp. Neurol.* **339**, 251-268 (1994).
35. I. D. Manns, M. G. Lee, M. Modirrousta, Y. P. Hou and B. E. Jones, Alpha 2 adrenergic receptors on GABAergic, putative sleep-promoting basal forebrain neurons, *Eur J Neurosci.* **18**, 723-727 (2003).
36. M. Modirrousta, L. Mainville and B. E. Jones, c-Fos expression in cholinergic and GABAergic neurons of the basal forebrain and/or preoptic area following sleep deprivation and recovery, *Soc. Neurosci. Abst. Online*, 341.312 (2003).
37. M. G. Lee, I. D. Manns, A. Alonso and B. E. Jones, Sleep-wake related discharge properties of basal forebrain neurons recorded with micropipettes in head-fixed rats, *J. Neurophysiol.* (In press).
38. M. Modirrousta, L. Mainville and B. E. Jones, GABAergic neurons with alpha2-adrenergic receptors in basal forebrain and preoptic area express c-Fos during sleep, *Neuroscience* (In press).
39. C. Peyron, D. K. Tighe, A. N. van den Pol, L. de Lecea, H. C. Heller, J. G. Sutcliffe and T. S. Kilduff, Neurons containing hypocretin (orexin) project to multiple neuronal systems, *J Neurosci.* **18**, 9996-10015 (1998).
40. M. Herkenham, New perspectives on the organization and evolution of nonspecific thalamocortical projections. In *Cerebral Cortex, Vol. 5* (Jones, E. G. & Peters, A., eds.), pp. 403-445. Plenum, New York (1986).
41. L. L. Glenn and M. Steriade, Discharge rate and excitability of cortically projecting intralaminar thalamic neurons during waking and sleep states, *J. Neurosci.* **2**, 1387-1404 (1982).
42. M. Steriade, Mechanisms underlying cortical activation: Neuronal organization and properties of the midbrain reticular core and intralaminar thalamic nuclei. In *Brain Mechanisms and Perceptual Awareness* (Pompeiano, O. & Ajmone Marsan, C., eds.), pp. 327-377, Raven Press, New York (1981).
43. M. Steriade, R. Curro Dossi and D. Contreras, Electrophysiological properties of intralaminar thalamocortical cells discharging rhythmic (approximately 40 HZ) spike-bursts at approximately 1000 HZ during waking and rapid eye movement sleep, *Neuroscience* **56**, 1-9 (1993).
44. B. E. Jones, Basic Mechanisms of Sleep-Wake States. In *Principles and Practice of Sleep Medicine* 3rd edit. (Kryger, M. H., Roth, T. & Dement, W. C., eds.), pp. 134-154, Saunders, Philadelphia (2000).
45. L. Bayer, E. Eggermann, B. Saint-Mleux, D. Machard, B. E. Jones, M. Muhlethaler and M. Serafin, Selective action of orexin (hypocretin) on nonspecific thalamocortical projection neurons, *J Neurosci.* **22**, 7835-7839 (2002).
46. E. Eggermann, M. Serafin, L. Bayer, D. Machard, B. Saint-Mleux, B. E. Jones and M. Muhlethaler, Orexins/hypocretins excite basal forebrain cholinergic neurones, *Neuroscience* **108**, 177-181 (2001).
47. B. E. Jones, Activity, modulation and role of basal forebrain cholinergic neurons innervating the cerebral cortex, *Progr. Brain Res.* **145**, 157-169 (2004).

48. R. A. Espana, B. A. Baldo, A. E. Kelley and C. W. Berridge, Wake-promoting and sleep-suppressing actions of hypocretin (orexin): basal forebrain sites of action, *Neuroscience* **106**, 699-715 (2001).
49. E. K. Lambe and G. K. Aghajanian, Hypocretin (orexin) induces calcium transients in single spines postsynaptic to identified thalamocortical boutons in prefrontal slice, *Neuron* **40**, 139-150 (2003).
50. L. Bayer, B. Saint-Mleux, M. Serafin, E. Eggermann, D. Machard, B. E. Jones and M. Muhlethaler, Hypocretin/orexin excites neurons exclusively in sublayer 6b of the cerebral cortex, *J. Neurosci.* (In press).
51. B. Clancy and L. J. Cauller, Widespread projections from subgriseal neurons (layer VII) to layer I in adult rat cortex, *J Comp Neurol.* **407**, 275-286 (1999).
52. R. L. Reep, Cortical layer VII and persistent subplate cells in mammalian brains. *Brain Behav Evol.* **56**, 212-234 (2000).
53. T. Gallopin, P. Fort, E. Eggermann, B. Cauli, P. H. Luppi, J. Rossier, E. Audinat, M. Muhlethaler and M. Serafin, Identification of sleep-promoting neurons *in vitro, Nature* **404**, 992-995 (2000).

THE HYPOCRETINS IN FEEDING AND ENERGY BALANCE

REGULATION OF HYPOCRETIN BY METABOLIC SIGNALS

Katherine E. Wortley and Sarah F. Leibowitz*

1. INTRODUCTION

Energy balance is maintained via a homeostatic system involving both the brain and the periphery. A key component of this system is the hypothalamus. Over the past two decades, major advances have been made in identifying an increasing number of neuropeptides within the hypothalamus that contribute to the process of energy homeostasis. Under stable conditions, equilibrium exists between anabolic peptides, which stimulate feeding behavior as well as decrease energy expenditure and lipid utilization in favor of fat storage, and catabolic peptides, which attenuate food intake while stimulating sympathetic nervous system activity and restricting fat deposition by increasing lipid metabolism. The equilibrium between these neuropeptides is dynamic in nature. It shifts across the day-night cycle and from day to day. It also responds to dietary challenges in addition to peripheral energy stores. These shifts occur in close relation to metabolic signals, e.g., circulating levels of the hormones, insulin, leptin, corticosterone and ghrelin, and also of the metabolites, glucose and lipids. These metabolic signals together with neural processes relay information regarding the availability of fuels needed for current cellular demand, in addition to the level of stored fuels needed for long-term use. Together, these signals have profound impact on the expression and production of neuropeptides in the hypothalamus that, in turn, initiate the appropriate anabolic or catabolic responses for restoring equilibrium.

There is accumulating evidence to suggest that hypocretin/orexin neurons, concentrated in the perifornical hypothalamus (PFH), are controlled by changes in diet and circulating metabolites that link short-term feeding to long-term body weight control. In this chapter, we summarize results showing the impact of various metabolic signals on these hypocretin/orexin neurons. We also compare this peptide system in the PFH to

* K. Wortley. Regeneron Pharmaceuticals, Tarrytown, NY, 10591; S. F. Leibowitz. The Rockefeller University, New York, NY 10021

other important peptides in the hypothalamus involved in energy balance. These are neuropeptide Y (NPY) and agouti-related protein (AgRP), which are expressed in the arcuate nucleus (ARC), and the peptides, galanin and enkephalin, which exist in neurons of the paraventricular nucleus (PVN) as well as the PFH and ARC.

2. REGULATION OF HYPOCRETIN/OREXIN SYSTEM BY METABOLIC SIGNALS RELATED TO NEGATIVE ENERGY BALANCE

Hypocretin/orexin neurons in the PFH are found to be stimulated by hypoglycemia.[1-4] This phenomenon can be seen in states of starvation or after insulin administration, although only in the absence of food. Conversely, these neurons are inhibited by glucose and by signals related to the ingestion of food.[1,3-6] Prior to the discovery of the hypocretin/orexin system, neurons in the PFH area had been identified as "glucose sensitive", i.e., stimulated by a decline in glucose.[7] More recent investigations of these glucose-sensitive neurons have shown that some, in fact, express hypocretin/orexin mRNA[2,8] and that some of these hypocretin/orexin neurons may be the same as those containing "prolactinlike-immunoreactivity",[9] which are similarly activated by hypoglycemia.[10] Thus, based on evidence linking the hypocretins/orexins to arousal and the sleep-wake cycle (see Chapter 6), these findings have led to the proposal that this peptide system in the PFH has a function in mediating hypoglycemic-associated arousal induced by fasting.[6]

Other metabolic signals that are affected by conditions of negative energy balance are corticosterone and ghrelin. These hormones rise during periods of food deprivation and glucoprivation, and they are restored by food consumption.[11,12] Recent studies demonstrate that hypocretin/orexin neurons in the PFH are stimulated by administration of exogenous glucocorticoids[13] or ghrelin.[6,14,15] Based on this evidence, it is suggested that these hormones mediate the stimulatory effect of food deprivation and hypoglycemia on this peptide system. Its activation by these hormones, as well as by metabolic signals related to negative energy balance, may function to ensure increased arousal and locomotor activity, which would facilitate food-seeking behavior.

The findings above highlight some similarities between the hypocretin/orexin system in the PFH and the orexigenic systems in the ARC. Similar to the former, the NPY/AgRP neurons are stimulated by food deprivation,[16] glucocorticoids,[17] and ghrelin.[18] Further still, a number of studies provide evidence for interactions between hypocretin/orexin neurons in the PFH and the NPY/AgRP neurons in the ARC (see Chapter 3). However, some differences are evident between these systems. For example, hypocretin/orexin gene expression is reduced during pregnancy and lactation,[19] states of increased energy demand that stimulate NPY and AgRP.[20,21] Since these peptide systems can be activated independently of one another, they may also perform different functions in specific physiological states.

3. REGULATION OF HYPOCRETIN/OREXIN SYSTEM BY METABOLIC SIGNALS RELATED TO POSITIVE ENERGY BALANCE

In addition to being stimulated by hypoglycemia, hypocretin/orexin neurons are also found to respond, in the opposite direction, to circulating lipids. This is suggested by a recent report showing hypocretin/orexin mRNA in the PFH to be stimulated by acute or chronic exposure to a high-fat diet as compared to a low-fat diet.[22] It also rises even further in association with obesity on a high-fat diet, which is characterized by hyperlipidemia. These findings liken the hypocretin/orexin peptide system to galanin and the opioid peptide, enkephalin, in the PVN and PFH, which are also stimulated in association with high-fat diet consumption.[23,24] In response to a brief, 4-hour period of consuming a high-fat vs low-fat diet, expression levels of the hypocretins/orexins as well as galanin and enkephalin, measured by real-time quantitative PCR, rise in close association with circulating levels of triglycerides and fatty acids, independently of changes in insulin and leptin (Table 1). These hypothalamic peptides in the PVN and PFH, but not the ARC, are strongly, positively correlated with the triglycerides. This has led to the proposal that these peptides, and perhaps others, form a class of "fat-stimulated" peptides, which are responsive specifically to lipids in the blood.[22,23]

	Low fat	High fat
Lipid metabolites		
TG (mg/dl)	137 ± 9	190 ± 29*
NEFA (mEq/l)	0.59 ± 0.1	1.53 ± 0.1*
Peptides		
Galanin mRNA/β-actin mRNA		
PVN	0.080 ± 0.001	0.091 ± 0.003*
PFH	0.021 ± 0.000	0.022 ± 0.000
ARC	0.058 ± 0.004	0.061 ± 0.005
Hcrt/ orexin mRNA/β-actin mRNA		
PFH	0.050 ± 0.002	0.067 ± 0.003*
Enkephalin mRNA/β-actin mRNA		
PVN	0.036 ± 0.002	0.041 ± 0.001*
PFH	0.041 ± 0.001	0.046 ± 0.001*
ARC	0.034 ± 0.002	0.035 ± 0.002

Table 1. Effects of a 4-hour high-fat diet compared to a low-fat diet. * $p<0.05$ for comparisons between low-fat and high-fat diet scores. Peptide expression is calculated as a ratio to β-actin mRNA. Abbreviations: TG, triglycerides; NEFA, non-esterified fatty acids; PVN, paraventricular nucleus; PFH, perifornical lateral hypothalamus; ARC, arcuate nucleus.

Similar to the high-fat diet, injection of Intralipid, a fat emulsion that raises triglyceride levels without affecting insulin and leptin, also stimulates hypocretin/orexin mRNA in the PFH, similar to galanin and enkephalin in the PVN (Figure 1).[22,23] These effects are similar to those seen in the natural feeding paradigms involving acute and chronic dietary fat, indicating that these peptides are responding to specific signals that

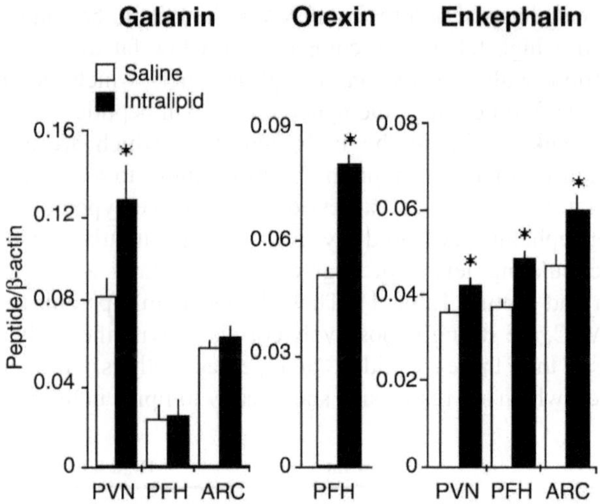

Figure 1: Galanin, hypocretin/ orexin and enkephalin mRNA (ratio to β-actin mRNA) in the paraventricular nucleus (PVN), perifornical lateral hypothalamus (PFH), and arcuate nucleus (ARC), 4 hrs after intraperitoneal administration of saline or Intralipid (5 ml, 20%). Given are mean ± sem. * $p<0.05$ for comparisons between saline and Intralipid scores.

are shared by Intralipid and a high-fat diet. These signals appear to be related specifically to the rise in circulating triglycerides, since Intralipid stimulates the expression of these peptides in the absence of changes in hormones as well as other factors related to body weight, adiposity, eating behavior, taste and palatability.[22,23]

Consistent with this relationship of the hypocretin/orexin system to circulating lipids is the circadian rhythm of the hypocretin/orexin peptide system. Like PVN galanin, this system in the PFH shows peak levels several hours after the onset of the natural feeding cycle,[25,26] when rats exhibit a natural rise in their preference for dietary fat.[27] The available evidence indicates that hypocretins/orexins, galanin and enkephalin are more closely related to the triglycerides than to levels of fatty acids.[22,23] This suggests that it is the process of triglyceride hydrolysis to fatty acids, perhaps via lipoprotein lipase in the hypothalamic capillary endothelium,[28,29] rather than circulating fatty acids from peripheral tissues, that may be involved in determining the level of peptide expression and synthesis. These hypothalamic systems are well positioned to provide the central mechanisms for a proposed role of triglyceride-fatty acid uptake in the control of feeding behavior.[30]

Results with double-labeling immunofluorescence further strengthen this idea of an effect of triglyceride-fatty acids on neurons expressing the hypocretins/orexins and other orexigenic peptides.[23] Administration of Intralipid increased c-Fos immunostaining in the hypothalamus, a marker of stimulus-induced neuronal activation. This effect was detected in the PVN, specifically in the anterior, parvocellular region, but not in the magnocellular region or the supraoptic nucleus, areas that should exhibit a change if Intralipid were affecting blood osmolality. It was also seen in the PFH, where a high-fat diet has been shown to increase c-Fos expression,[31] and fatty acid administration stimulates neuronal activity.[32] In addition, Intralipid increased c-Fos colocalization with the hypocretins/orexins in 20% of the peptide-synthesizing neurons in the PFH (Figure 2).[23]

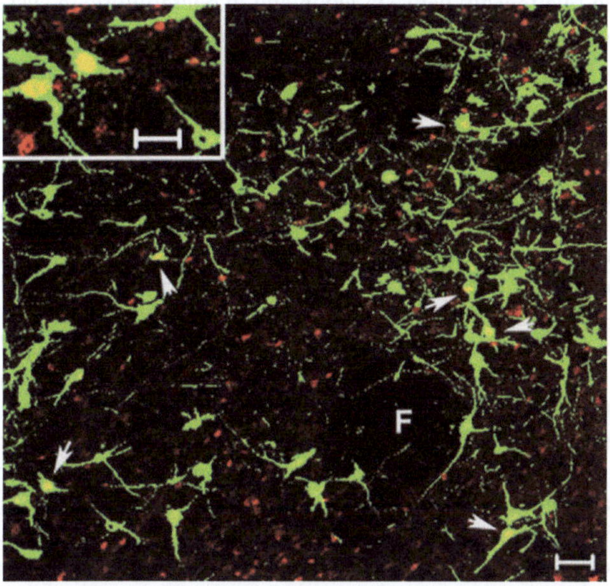

Figure 2: Double-labeled neurons (yellow), indicated by white arrows, in the PFH of Intralipid-injected rats, reflecting co-localization of c-Fos-ir (red) and orexin-ir (green). Saline-injected control rats showed no double-labeled neurons (not shown). Insert in upper left corner amplifies neurons in the lower left corner that co-express c-Fos and orexins. Abbreviations: F, fornix. Bar = 50 μm.

This provides support for an effect of lipids specifically on hypothalamic neurons containing the orexigenic peptides.

Taken together, these reports demonstrate that the hypocretin/orexin neurons are closely related to circulating lipids under conditions of positive energy balance. The role of this peptide system in such conditions may be to stimulate spontaneous physical activity. Activity level is known to increase with positive energy balance, in humans[33] and mice,[34] and this behavioral change is likely to have a function in mediating a resistance to weight gain. This idea receives support from the findings, first, that intra-hypothalamic injection of hypocretin/orexin 1 in rats induces a robust and dose-dependent increase in activity[35] and, second, that the increase in body fat mass of orexin-deficient mice is associated with a decrease in spontaneous activity level.[36]

Positive energy balance induces a rise in circulating leptin levels, which functions as an important signal for NPY/AgRP neurons in the ARC. Leptin receptors are found to exist on hypocretin neurons,[37,38] and exogenous leptin reduces hypocretin/orexin 1 levels[39] and the activity of hypocretin neurons.[6] However, there is little evidence for a direct interaction between leptin and the hypocretin/orexin peptides under physiological conditions. The hypocretins/orexins, in contrast to NPY and AgRP, remain stable or are actually increased in conditions or states with markedly elevated leptin, including overfeeding and dietary obesity.[4,22,40] Also, whereas NPY-induced feeding is completely inhibited by leptin or by anorectic peptides closely linked to leptin, the hypocretin/orexin feeding response is only partially suppressed by these compounds, suggesting the additional involvement of leptin-insensitive pathways.[41] A more detailed description of the relationship between leptin and hypocretin/orexin system can be found in other sections (see Chapter 7, The hypocretins, leptin and the regulation of metabolism).

4. CONCLUSION

These results indicate that hypocretin/orexin neurons in the PFH are highly responsive to changes in circulating and dietary nutrients. They have similar properties to NPY/AgRP-expressing neurons in the ARC, in being stimulated by negative energy balance, such as after food deprivation, when glucose levels are declining and corticosterone and ghrelin levels are elevated. They are also similar to galanin-expressing neurons in the PVN, in being stimulated by dietary fat and circulating lipids and in exhibiting a natural rise during the first half of the circadian feeding cycle when fat ingestion normally increases. In states of both energy deficiency and elevated fat consumption, the hypocretin/orexin peptides may provide an important link between energy homeostasis and brain mechanisms coordinating sleep/wakefulness and motivated behaviors. Specifically, they may function to increase arousal and locomotor activity that facilitate food-seeking behavior under conditions of food deprivation, and they may also facilitate activity-based energy expenditure and hyperphagia under conditions of excess dietary fat intake.

5. REFERENCES

1. B. Griffond, P. Y. Risold, C. Jacquemard, C. Colard and D. Fellmann, Insulin-induced hypoglycemia increases preprohypocretin (orexin) mRNA in the rat lateral hypothalamic area, *Neurosci Lett.* **262**, 77-80 (1999).
2. T. Moriguchi, T. Sakurai, T. Nambu, M. Yanagisawa and K. Goto, Neurons containing orexin in the lateral hypothalamic area of the adult rat brain are activated by insulin-induced acute hypoglycemia, *Neurosci Lett.* **264**, 101-104 (1999).
3. X. J. Cai, M. L. Evans, C. A. Lister, R. A. Leslie, J. R. Arch, S. Wilson and G. Williams, Hypoglycemia activates orexin neurons and selectively increases hypothalamic orexin-B levels: responses inhibited by feeding and possibly mediated by the nucleus of the solitary tract, *Diabetes* **50**, 105-112 (2001).
4. X. J. Cai, P. S. Widdowson, J. Harrold, S. Wilson, R. E. Buckingham, J. R. Arch, M. Tadayyon, J. C. Clapham, J. Wilding and G. Williams, Hypothalamic orexin expression: modulation by blood glucose and feeding, *Diabetes* **48**, 2132-2137 (1999).
5. K. P. Briski and P. W. Sylvester, Hypothalamic orexin-A-immunpositive neurons express Fos in response to central glucopenia, *Neuroreport* **12**, 531-534 (2001).
6. A. Yamanaka, C. T. Beuckmann, J. T. Willie, J. Hara, N. Tsujino, M. Mieda, M. Tominaga, K. Yagami, F. Sugiyama, K. Goto, M. Yanagisawa and T. Sakurai, Hypothalamic orexin neurons regulate arousal according to energy balance in mice, *Neuron* **38**, 701-713(2003).
7. L. L. Bernardis and L. L. Bellinger, The lateral hypothalamic area revisited: ingestive behavior, *Neurosci Biobehav Rev.* **20**, 189-287 (1996)
8. S. Muroya, K. Uramura, T. Sakurai, M. Takigawa and T. Yada, Lowering glucose concentrations increases cytosolic Ca(2+) in orexin neurons of the rat lateral hypothalamus, *Neurosci Lett.* **309**, 165-168 (2001).
9. P. Y. Risold, B. Griffond, T. S. Kilduff, J. G. Sutcliffe and D. Fellmann, Preprohypocretin (orexin) and prolactin-like immunoreactivity are coexpressed by neurons of the rat lateral hypothalamic area, *Neurosci Lett.* **259**, 153-156 (1999)

10. M. Bahjaoui-Bouhaddi, D. Fellmann and C. Bugnon, Induction of Fos-immunoreactivity in prolactin-like containing neurons of the rat lateral hypothalamus after insulin treatment, *Neurosci Lett.* **168**, 11-15 (1994).
11. D. L. Tempel and S. F. Leibowitz, Adrenal steroid receptors: interactions with brain neuropeptide systems in relation to nutrient intake and metabolism, *J Neuroendocrinol.* **6**, 479-501 (1994).
12. M. Tschop, D. L. Smiley and M. L. Heiman, Ghrelin induces adiposity in rodents, *Nature.* **407**, 908-913 (2000).
13. A. Stricker-Krongrad and B. Beck, Modulation of hypothalamic hypocretin/orexin mRNA expression by glucocorticoids, *Biochem Biophys Res Commun.* **296**, 129-133 (2002).
14. .K. Toshinai, Y. Date, N. Murakami, M. Shimada, M. S. Mondal, T. Shimbara, J. L. Guan, Q. P. Wang, H. Funahashi, T. Sakurai, S. Shioda, S. Matsukura, K. Kangawa and M. Nakazato, Ghrelin-induced food intake is mediated via the orexin pathway, *Endocrinology* **144**, 1506-1512 (2003).
15. P. K. Olszewski, D. Li, M. K. Grace, C. J. Billington, C. M. Kotz and A. S. Levine, Neural basis of orexigenic effects of ghrelin acting within lateral hypothalamus, *Peptides* **24**, 597-602 (2003).
16. T. M. Hahn, J. F. Breininger, D. G. Baskin and M. W. Schwartz, Coexpression of Agrp and NPY in fasting-activated hypothalamic neurons, *Nat Neurosci.* **1**, 271-272 (1998).
17. E. Savontaus, I. M. Conwell and S. L. Wardlaw, Effects of adrenalectomy on AGRP, POMC, NPY and CART gene expression in the basal hypothalamus of fed and fasted rats, *Brain Res.* **958**, 130-138 (2002).
18. M. Nakazato, N. Murakami, Y. Date, M. Kojima, H. Matsuo, K. Kangawa and S. Matsukura, A role for ghrelin in the central regulation of feeding, *Nature* **409**, 194-198 (2001).
19. M. C. Garcia, M. Lopez, O. Gualillo, L. M. Seoane, C. Dieguez and R. M. Senaris, Hypothalamic levels of NPY, MCH, and prepro-orexin mRNA during pregnancy and lactation in the rat: role of prolactin, *Faseb J.* **17**, 1392-1400 (2003).
20. P. Chen, C. Li, C. Haskell-Luevano, R. D. Cone and M. S. Smith, Altered expression of agouti-related protein and its colocalization with neuropeptide Y in the arcuate nucleus of the hypothalamus during lactation, *Endocrinology* **140**, 2645-2650 (1999).
21. M. Rocha, C. Bing, G. Williams and M. Puerta, Pregnancy-induced hyperphagia is associated with increased gene expression of hypothalamic agouti-related peptide in rats, *Regul Pept.* **114**, 159-165 (2003).
22. K. E. Wortley, G. Q. Chang, Z. Davydova and S. F. Leibowitz, Orexin gene expression is increased during states of hypertriglyceridemia, *Am J Physiol Regul Integr Comp Physiol* **284**, R1454-1465 (2003).
23. G. Q. Chang, O. Karatayev, Z. Davydova and S. F. Leibowitz, Circulating triglycerides impact on orexigenic peptides and neuronal activity in hypothalamus, *Endocrinology* **145**, 3904-3912 (2004).
24. M. Odorizzi, J. P. Max, P. Tankosic, C. Burlet and A. Burlet, Dietary preferences of Brattleboro rats correlated with an overexpression of galanin in the hypothalamus, *Eur J Neurosci.* **11**, 3005-3014 (1999).
25. G. S. Martinez, L. Smale and A. A. Nunez, Diurnal and nocturnal rodents show rhythms in orexinergic neurons, *Brain Res.* **955**, 1-7 (2002).
26. Y. Yoshida, N. Fujiki, T. Nakajima, B. Ripley, H. Matsumura, H. Yoneda, E. Mignot and S. Nishino, Fluctuation of extracellular hypocretin-1 (orexin A) levels in the rat in relation to the light-dark cycle and sleep-wake activities, *Eur J Neurosci.* **14**, 1075-1081 (2001).
27. G. Shor-Posner, C. Ian, G. Brennan, T. Cohn, H. Moy, A. Ning and S. F. Leibowitz, Self-selecting albino rats exhibit differential preferences for pure macronutrient diets: characterization of three subpopulations, *Physiol Behav.* **50**, 1187-1195 (1991).
28. D. H. Bessesen, C. L. Richards, J. Etienne, J. W. Goers and R. H. Eckel, Spinal cord of the rat contains more lipoprotein lipase than other brain regions, *J Lipid Res.* **34**, 229-238 (1993).
29. R. H. Eckel and R. J. Robbins, Lipoprotein lipase is produced, regulated, and functional in rat brain, *Proc Natl Acad Sci U S A.* **81**, 7604-7607 (1984).
30. M. Merkel, R. H. Eckel and I. J. Goldberg, Lipoprotein lipase: genetics, lipid uptake, and regulation, *J Lipid Res.* **43**, 1997-2006 (2002).
31. H. Wang, L. H. Storlien and X. F. Huang, Influence of dietary fats on c-Fos-like immunoreactivity in mouse hypothalamus, *Brain Res.* **843**, 184-192 (1999).
32. Y. Oomura, T. Nakamura, M. Sugimori and Y. Yamada, Effect of free fatty acid on the rat lateral hypothalamic neurons, *Physiol Behav.* **14**, 483-486 (1975).
33. J. A. Levine, N. L. Eberhardt and M. D. Jensen, Role of nonexercise activity thermogenesis in resistance to fat gain in humans, *Science.* **283**, 212-214 (1999).
34. A. A. Butler, D. L. Marks, W. Fan, C. M. Kuhn, M. Bartolome and R. D. Cone, Melanocortin-4 receptor is required for acute homeostatic responses to increased dietary fat, *Nat Neurosci.* **4**, 605-611 (2001).

35. K. Kiwaki, C. M. Kotz, C. Wang, L. Lanningham-Foster and J. A. Levine, Orexin A (hypocretin 1) injected into hypothalamic paraventricular nucleus and spontaneous physical activity in rats, *Am J Physiol Endocrinol Metab.* **286**, E551-559 (2004).
36. J. Hara, C. T. Beuckmann, T. Nambu, J. T. Willie, R. M. Chemelli, C. M. Sinton, F. Sugiyama, K. Yagami, K. Goto, M. Yanagisawa and T. Sakurai, Genetic ablation of orexin neurons in mice results in narcolepsy, hypophagia, and obesity, *Neuron* **30**, 345-354 (2001).
37. T. L. Horvath, S. Diano and A. N. van den Pol, Synaptic interaction between hypocretin (Orexin) and neuropeptide Y cells in the rodent and primate hypothalamus: A novel circuit implicated in metabolic and endocrine regulations, *J Neurosci.* **19**, 1072-1087 (1999).
38. M. Hakansson, L. De Lecea, J. G. Sutcliffe, M. Yanagisawa and B. Meister, Leptin receptor- and STAT3-immunoreactivities in Hypocretin/Orexin neurones of the lateral hypothalamus, *J Neuroendocrinol.* **11**, 653-663 (1999).
39. B. Beck and S. Richy, Hypothalamic hypocretin/orexin and neuropeptide Y: divergent interaction with energy depletion and leptin, *Biochem Biophys Res Commun.* **258**, 119-122 (1999).
40. S. Taheri, M. Mahmoodi, J. Opacka-Juffry, M. A. Ghatei and S. R. Bloom, Distribution and quantification of immunoreactive orexin A in rat tissues, *FEBS Lett.* **457**, 157-161 (1999).
41. Y. Zhu, A. Yamanaka, K. Kunii, N. Tsujino, K. Goto and T. Sakurai, Orexin-mediated feeding behavior involves both leptin-sensitive and -insensitive pathways, *Physiol Behav.* **77**, 251-257 (2002).

THE HYPOCRETINS IN ADDICTION AND HYPERAROUSAL

HYPOCRETIN AND BRAIN REWARD FUNCTION

Benjamin Boutrel, Paul J. Kenny, Athina Markou and George F. Koob*

1. INTRODUCTION

Hypocretin-containing neurons arise exclusively in the lateral hypothalamus (LH) and project widely throughout the brain,[1] with a dense innervation of anatomical sites involved in regulating arousal, motivation, and stress states.[2] Their interaction with autonomic, neuroendocrine and neuroregulatory systems[2-12] strongly suggests they act as neuromodulators in a wide variety of neural circuits. The hypocretins (Hcrt-1 and Hcrt-2), also known as orexins, have been implicated in the modulation of noradrenergic,[6,13-15] cholinergic,[16] serotonergic,[17,18] histaminergic[19] and dopaminergic systems,[20,21] and in the regulation of the hypothalamic-pituitary-adrenal (HPA) axis,[22-24] possibly via the release of neuropeptide Y.[25] Hence, the hypocretin/orexin system could constitute a sensitive key relay for mediating stress behaviors.[7,26]

Food restriction, known to induce a stress reaction in animals,[27] modulates lateral hypothalamic self-stimulation (LHSS),[27-31] reinforces drug intake[27,29,32,33] and can also reinstate drug-seeking behavior in laboratory animals.[34-38] However, little is known about the neuronal mechanisms underlying sensitization to both LHSS and drug reward, and reinstatement of drug seeking by acute food deprivation. Both insulin and leptin hyperpolarize hypocretin neurons,[9,39-41] and have been proposed to play a role in the modulation of brain reward function under fasting conditions.[30,31,42] In contrast, neuropeptide Y does not take part in the process whereby food restriction and leptin modulate reward circuitry activated by stimulating restriction-sensitive sites.[44] In this context, the Hcrt system, already described as a crucial link between energy balance and arousal,[41] could also be considered as the link between the brain reward function and the control of energy homeostasis. We suggest here that Hrct-1 may play a role in the regulation of motivated behaviors and relapse to drug seeking.

* B. Boutrel. Center for Psychiatric Neuroscience, Department of Child and Adolescent Psychiatry, Site de Cery, CH-1008 Prilly-Lausanne, Switzerland, P J. Kenny, A. Markou, G.F. Koob. Department of Neuropharmacology, The Scripps Research Institute, La Jolla, CA 92037.

2. LATERAL HYPOTHALAMIC SELF-STIMULATION AND THE HYPOCRETIN SYSTEM

2.1. Lateral hypothalamic self-stimulation paradigm

Olds and colleagues first reported that rats would work vigorously to electrically self-stimulate the LH.[45,46] LH lesions significantly reduce brain reward function in rats.[47,48] These observations suggest that the LH constitutes an important component of the brain's endogenous reward system. However, the molecular mechanisms by which the LH regulates brain reward function remain unclear.

LHSS, also called intracranial self-stimulation (ICSS), directly activates brain reward circuitries, and LHSS thresholds provide an accurate measure of brain reward function. The procedure used by our laboratory is a modification of the discrete-trial current-threshold procedure of Kornetsky and Esposito,[49,50] as previously described in detail.[51] Briefly, rats are trained to respond to a non-contingent electrical stimulus by turning a wheel in order to receive an identical electrical stimulus. The current intensity delivered is adjusted for each animal and typically ranges from 50 to 200 μA. Current levels are varied (per 5 μA steps) in alternating descending and ascending series. The threshold for each series is defined as the midpoint between two consecutive current intensities that yielded "positive scores" (animals responded for at least two of the three trials) and two consecutive current intensities that yielded "negative scores" (animals did not respond for two or more of the three trials). The overall threshold of the session is defined as the mean of the thresholds for the four individual series.

Hence, a lowering of ICSS thresholds reflects an increase in sensitivity to the rewarding properties of ICSS, which is thought to represent an operational measure of increased brain reward function. In contrast, an elevation of ICSS thresholds reflects a decreased brain reward function.

Most drugs of abuse have been shown to increase brain reward function, as measured by lowering of ICSS reward thresholds.[52-54] This drug-induced lowering of ICSS thresholds, considered as an accurate measure of euphorigenic drug action,[55] suggests that drugs of abuse synergize with the rewarding ICSS and that both these rewarding stimuli likely act on the same reward substrates in the brain. In contrast to this acute lowering action, chronic exposure to drugs of abuse generates adaptations in reward circuitries such that drug withdrawal significantly elevates ICSS thresholds.[56-60]

2.2. Lateral hypothalamic self-stimulation may activate hypocretin neurons

In the LHSS procedure described above, a stimulating electrode is usually placed in the lateral hypothalamus. Therefore, neurons in the vicinity of the stimulating electrode are expected to play an important role in the reward mechanisms associated with this paradigm. To test whether hypocretin neurons were stimulated upon LHSS, double immunocytochemistry to detect hypocretin and c-fos, a marker of neuronal activation, was performed one hour after rats performed ICSS. Preliminary results suggest a significant increase in hcrt/c-fos double labeled neurons in rats that self-stimulated themselves compared with rats that had the electrode and were placed in the same environment, but were not permitted to self-stimulate.[61]

However, electrical stimulation is not selective; thus, such a heightened expression of c-Fos in hypocretin neurons may not reflect activation a contribution of these neurons to brain reward activity, but only local activation of neurons in the vicinity of the stimulating electrode. Further studies are needed to confirm whether or not activation of hypocretinergic perikarya after ICSS reflects a specific activation of the brain reward circuitry, or just represents a response to strong local depolarization.

3. BRAIN REWARD CIRCUITRY AND HYPOCRETIN PROJECTIONS

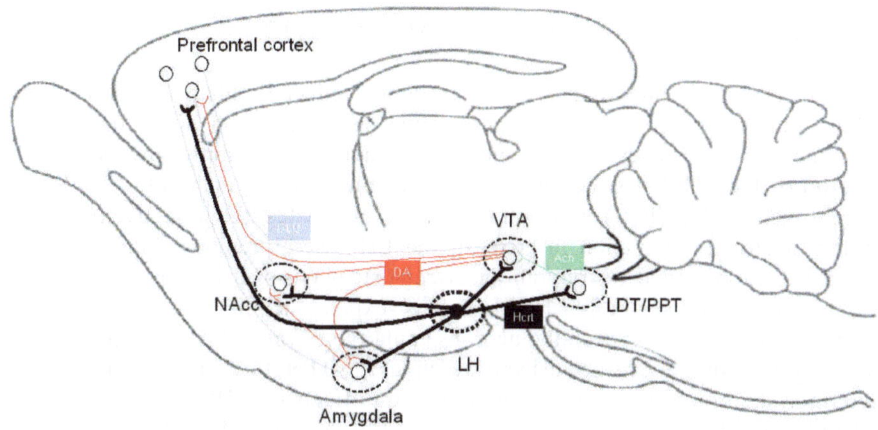

Figure 1: Schematic representation of interactions between the hypocretinergic system and the brain reward circuitry. Hypocretin (Hcrt) neurons project to several key structures involved in the regulation of motivated behaviors, namely the ventral tegmental area (VTA), the nucleus accumbens (NAcc), the amygdala and the laterodorsal tegmentum (LDT)/pedoculonpontine tegmentum (PPT). Abbreviations: Ach, acetylcholine; DA, dopamine; Glu, glutamate; LH, lateral hypothalamus

It is well accepted that the mesolimbic dopamine system is a key regulator of brain reward function. The ventral tegmental area (VTA) contains cell bodies of dopaminergic neurons projecting to the nucleus accumbens, the amygdala, the hippocampus and the prefrontal cortex. These neurons are critically implicated in mechanisms of reward, reinforcement and emotional arousal,[64] and their activity has been closely correlated to the availability of primary rewards such as food, water and sex.[65] The mesolimbic dopamine system receives glutamatergic input from cortical structures including the medial and occipital prefrontal cortex and the amygdala, it receives also GABAergic inputs from striatal sources and cholinergic inputs from the brainstem.[66] LHSS is thought to be rewarding in part because it activates cholinergic neurons in laterodorsal tegmental (LDT) and the pedoculopontine tegmental (PPT) nuclei that activates consequently dopaminergic neurons in the VTA.[67-70] However, the brain circuitry involved in the LDT/PPT activation is unknown. Interestingly, hypocretins act synergistically with glutamatergic afferents to depolarize cholinergic neurons in the LDT area, which is thought to coordinate the activation of the entire ascending reticular activating system,[16] and drive dopamine release in the nucleus accumbens by exciting dopaminergic neurons in the VTA.[67] Furthermore, Hcrt-1 and Hcrt-2 have been shown to depolarize cholinergic and non-cholinergic neurons in the LDT area.[16,71,72] It has been also shown that hypocretin neurons excite directly dopamine fibers in the VTA,[21,73,74] and that the VTA dopaminergic system is critically involved in hypocretin-induced hyperlocomotion and stereotypy.[20] Finally, hypocretin-immunoreactive fibers have been shown in the nucleus accumbens[1] and hcrt peptides hyperpolarize GABAergic neurons in this nucleus.[75] Hence, the hypocretin system could participate to the neuronal network involved in the regulation of the brain reward function.

4. HYPOCRETIN AND BRAIN REWARD MODULATION.

The mechanisms underlying food restriction-induced enhancement of central rewarding potency of both lateral hypothalamic self-stimulation and drugs of abuse remain unclear.[31,33] However, two hormones have been mainly involved in the modulation of the LHSS induced by food restriction, insulin[30] and leptin.[42,76] Notably, intracerebroventricular (ICV) infusion of leptin attenuates the effectiveness of the rewarding electrical stimulation in areas of the LH where previous stimulations were enhanced by chronic food restriction, an effect that persists as long as 4 days after a single injection.[42] Interestingly, Hcrt neurons express leptin receptors, and leptin regulates Hcrt neurons.[39,77] Secondly, neurons containing hypocretin/orexin in the LH are activated by insulin-induced acute hypoglycemia.[40,78] Therefore, it is possible that the hypocretin system mediates both insulin and leptin effects on LHSS, and thus could be considered as the link between the controls of energy homeostasis and the brain reward function. Interestingly, the hypocretin neurons have been proposed to regulate arousal according to energy balance in mice. Indeed, hypocretin/orexin neurons-ablated mice fail to exhibit fasting-induced arousal whereas fasting induces a significant increase in both wakefulness and locomotor activity in wild type mice.[41] Hence, evidence showing elevations in ICSS thresholds after Hcrt-1 administration reflects a transient decrease in brain reward function, which may ultimately reflect incentive motivation for food-seeking. All together, these observations suggest that the hypocretin system could integrate sensory stimuli (including metabolic needs), and relay the information to the brainstem nuclei, notably the LC, the LDT/PPT and the VTA. Activation of the HPA axis could also be involved given that the Hcrt system regulates CRF release,[22-24] and that CRF directly regulates brain reward function.[79] In this perspective, hypocretin-induced modulation of brain reward function (via the LDT/PPT) could act in concert with the activation of the LC and the HPA axis to increase the animal's arousal and motivation for food-seeking. The feeding-induced leptin release could then contribute to decrease the firing rate of hypocretin neurons,[41] leading ultimately to lowering of arousal threshold, and concomitant decreased motivation for food seeking.

Nevertheless, glucose has complex effects on Hcrt neurons;[80,81] indeed, whereas decreasing glucose concentrations increase the activity of these neurons (suggesting a role of the hypocretin/orexin system in the phenomenon of hypoglycemic awareness),[41] increasing glucose concentrations induce a cessation of firing in isolated Hcrt neurons. Hence, the feeding-induced leptin release provides a signal to the brain that energy stores are high, thus leptin-induced inhibition of Hcrt neurons contributes to a lowering of motivation for food-seeking as previously suggested. To consolidate this feed-back regulation, high concentrations of glucose also decrease Hcrt neurons firing.

To summarize, starvation activates the hypocretin/orexin system which relay these signals to the brainstem nuclei involved in the modulation of brain reward function and arousal. The concomitant brain reward function decrease, usually observed under drug withdrawal, can be considered here as a "food withdrawal" which will lead to food seeking and, ultimately, food taking. In this perspective, the hypocretin system could be considered as a key relay that promotes incentive motivation for food-seeking, and as such enhances wakefulness and locomotor activity.[82]

5. HYPOCRETIN AND RELAPSE FOR DRUG-SEEKING

In contrast to Hcrt-1-induced long-lasting elevations in ICSS thresholds, a single injection of Hcrt-1 was associated with a slight effect on cocaine self-administration, given that cocaine consumption remained nearly unchanged just after or 6 hours after the Hcrt-1 infusion.[61] Preliminary results obtained in our lab showed that Hcrt did not induce any significant increase on heroin intake in rats either. Furthermore, despite an effect on ICSS thresholds,[42] leptin does not modulate cocaine nor heroin self-administration in rats. Such a discrepancy between a marked elevation in ICSS thresholds and an absence of effect on both cocaine and heroin intake remains unclear. However, dissociation of neural systems subserving positive reinforcement and incentive motivation has been proposed previously.[83-85]

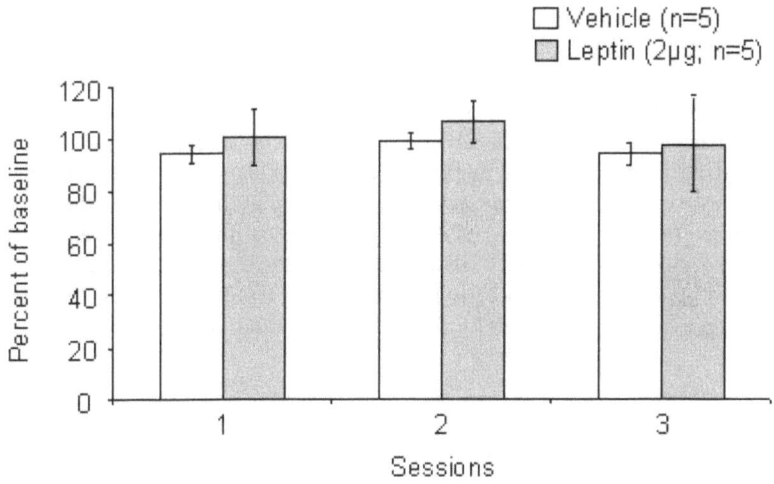

Figure 2: Effect of leptin (2ug ICV) on cocaine intake. Rats were injected with leptin 15 min prior to being allowed to self-administrate cocaine for 2 hours (c=0.25 mg/infusion). Leptin did not induce any significant increase in cocaine intake. Data are expressed as percent change (± SEM) from baseline cocaine intake during which rats received an ICV vehicle infusion.

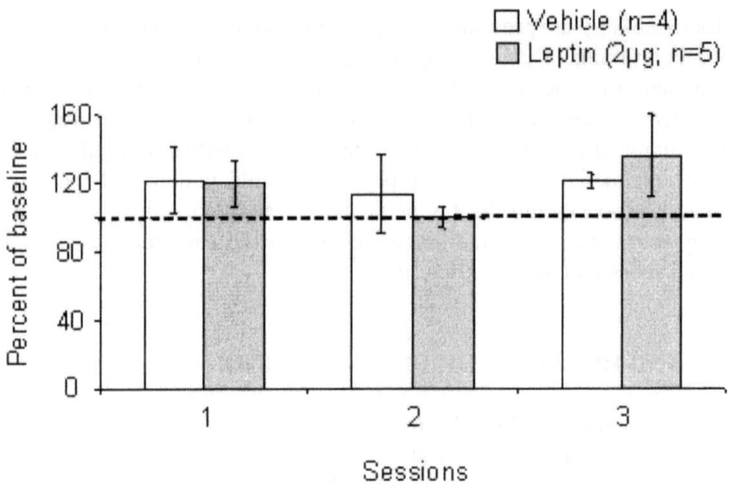

Figure 3: Effect of leptin (2ug ICV) on heroin intake. Rats were injected with leptin 15 min prior to being allowed to self-administrate cocaine for 2 hours (c=0.04 mg/infusion). Leptin did not induce any significant increase in heroin intake. Data are expressed as percent change (± SEM) from baseline cocaine intake during which rats received an ICV vehicle infusion.

Overall, the Hcrt system does not seem to drive directly drug consumption but, on the contrary, acts primarily as a strong modulator of brain reward function and drives appropriate and competing behaviors in order to maintain energy homeostasis.

In this perspective, the Hcrt system is an interesting candidate as a part of the circuitry that underlies vulnerability to relapse during prolonged abstinence from drugs of abuse. Indeed, activation of the hypocretin system has been hypothesized to play an important role during morphine intake and morphine withdrawal.[86,87] Hence, the role of the Hcrt system (receiving sensory stimuli and relaying them to brainstem nuclei and the HPA axis) could be interpreted as a marker of drug intoxication. At cessation of drug presentation the hypocretin system may act as an alarm signal that prepares the organism to face the danger of withdrawal for energy homeostasis. Interestingly, leptin which hyperpolarizes hypocretin neurons,[41] also attenuates fasting-induced heroin-seeking behavior.[37] In this perspective, it can be hypothesized that leptin may block activation of the Hcrt system (previously stimulated by food restriction) and therefore prevents relapse for drug seeking.

6. SUMMARY AND PERSPECTIVES

Fasting most likely leads to a particular state of metabolic needs. As a consequence, in a rat brain already exposed to drug, Hcrt could drive the most adapted behavior for a rat trapped in an operant box (e.g. drug seeking) to counter balance the stress induced by fasting. However, further studies are needed to attest the role of the Hcrt system in the circuitry that underlies the vulnerability to relapse during prolonged abstinence from drugs of abuse. Most likely, interactions between CRF and hypocretin systems should be of interest to delineate the mechanisms underlying fasting-induced relapse to drug seeking. However, Hcrt-1-induced drug seeking can not be limited to a simple stress like-reinstatement given that Hcrt-1 was shown to reinstate both drug- and food-seeking, whereas footshock stress, for example, was shown to reinstate cocaine-but not food-seeking behavior after extinction.[88]

In this perspective, studying the involvement of the hypocretin system in the persistence of a dysregulated reward system long after acute withdrawal remains a challenge for future studies, given the involvement of this system in the regulation of the extended amygdala,[2] a brain structure that was proposed to play a key role in protracted abstinence.[89]

7. REFERENCES

1. C. Peyron, D. K. Tighe, A. N. van den Pol, L. De Lecea, H. C. Heller, J. G. Sutcliffe and T. S. Kilduff, Neurons containing hypocretin (orexin) project to multiple neuronal systems, *J. Neurosci.* **18**, 9996-10015 (1998).
2. B. A. Baldo, R. A. Daniel, C. W. Berridge and A. E. Kelley, Overlapping distributions of orexin/hypocretin- and dopamine-beta-hydroxylase immunoreactive fibers in rat brain regions mediating arousal, motivation, and stress, *J. Comp Neurol.* **464**, 220-237 (2003).
3. R. Winsky-Sommerer, B. Boutrel and L. De Lecea, The role of the hypocretinergic system in the integration of networks that dictate the states of arousal, *Drug News Perspect.* **16**, 504-512 (2003).
4. C. T. Chen, L. L. Hwang, J. K. Chang and N. J. Dun, Pressor effects of orexins injected intracisternally and to rostral ventrolateral medulla of anesthetized rats, *Am. J. Physiol Regul. Integr. Comp Physiol* **278**, R692-R697 (2000).
5. Y. Date, M. S. Mondal, S. Matsukura, Y. Ueta, H. Yamashita, H. Kaiya, K. Kangawa and M. Nakazato, Distribution of orexin/hypocretin in the rat median eminence and pituitary, *Brain Res. Mol. Brain Res.* **76**, 1-6 (2000).
6. J. J. Hagan, R. A. Leslie, S. Patel, M. L. Evans, T. A. Wattam, S. Holmes, C. D. Benham, S. G. Taylor, C. Routledge, P. Hemmati, R. P. Munton, T. E. Ashmeade, A. S. Shah, J. P. Hatcher, P. D. Hatcher, D. N. Jones, M. I. Smith, D. C. Piper, A. J. Hunter, R. A. Porter and N. Upton, Orexin A activates locus coeruleus cell firing and increases arousal in the rat, *Proc. Natl. Acad. Sci. U. S. A* **96**, 10911-10916 (1999).
7. T. Ida, K. Nakahara, T. Murakami, R. Hanada, M. Nakazato and N. Murakami, Possible involvement of orexin in the stress reaction in rats, *Biochem. Biophys. Res. Commun.* **270**, 318-323 (2000).
8. L. K. Malendowicz, C. Tortorella and G. G. Nussdorfer, Orexins stimulate corticosterone secretion of rat adrenocortical cells, through the activation of the adenylate cyclase-dependent signaling cascade, *J. Steroid Biochem. Mol. Biol.* **70**, 185-188 (1999).
9. K. W. Nowak, P. Mackowiak, M. M. Switonska, M. Fabis and L. K. Malendowicz, Acute orexin effects on insulin secretion in the rat: in vivo and in vitro studies, *Life Sci.* **66**, 449-454 (2000).
10. W. K. Samson, B. Gosnell, J. K. Chang, Z. T. Resch and T. C. Murphy, Cardiovascular regulatory actions of the hypocretins in brain, *Brain Res.* **831**, 248-253 (1999).
11. T. Shirasaka, T. Kunitake, M. Takasaki and H. Kannan, Neuronal effects of orexins: relevant to sympathetic and cardiovascular functions, *Regul. Pept.* **104**, 91-95 (2002).
12. T. Shirasaka, M. Takasaki and H. Kannan, Cardiovascular effects of leptin and orexins, *Am. J. Physiol Regul. Integr. Comp Physiol* **284**, R639-R651 (2003).
13. P. Bourgin, S. Huitron-Resendiz, A. D. Spier, V. Fabre, B. Morte, J. R. Criado, J. G. Sutcliffe, S. J. Henriksen and L. De Lecea, Hypocretin-1 modulates rapid eye movement sleep through activation of locus coeruleus neurons, *J. Neurosci.* **20**, 7760-7765 (2000).
14. T. L. Horvath, C. Peyron, S. Diano, A. Ivanov, G. Aston-Jones, T. S. Kilduff and A. N. van den Pol, Hypocretin (orexin) activation and synaptic innervation of the locus coeruleus noradrenergic system, *J. Comp Neurol.* **415**, 145-159 (1999).
15. A. Ivanov and G. Aston-Jones, Hypocretin/orexin depolarizes and decreases potassium conductance in locus coeruleus neurons, *Neuroreport* **11**, 1755-1758 (2000).
16. S. Burlet, C. J. Tyler and C. S. Leonard, Direct and indirect excitation of laterodorsal tegmental neurons by Hypocretin/Orexin peptides: implications for wakefulness and narcolepsy, *J. Neurosci.* **22**, 2862-2872 (2002).

17. R. E. Brown, O. A. Sergeeva, K. S. Eriksson and H. L. Haas, Convergent excitation of dorsal raphe serotonin neurons by multiple arousal systems (orexin/hypocretin, histamine and noradrenaline), *J. Neurosci.* **22**, 8850-8859 (2002).
18. R. E. Brown, O. Sergeeva, K. S. Eriksson and H. L. Haas, Orexin A excites serotonergic neurons in the dorsal raphe nucleus of the rat, *Neuropharmacology* **40**, 457-459 (2001).
19. K. S. Eriksson, O. Sergeeva, R. E. Brown and H. L. Haas, Orexin/hypocretin excites the histaminergic neurons of the tuberomammillary nucleus, *J. Neurosci.* **21**, 9273-9279 (2001).
20. T. Nakamura, K. Uramura, T. Nambu, T. Yada, K. Goto, M. Yanagisawa and T. Sakurai, Orexin-induced hyperlocomotion and stereotypy are mediated by the dopaminergic system, *Brain Res.* **873**, 181-187 (2000).
21. T. M. Korotkova, O. A. Sergeeva, K. S. Eriksson, H. L. Haas and R. E. Brown, Excitation of ventral tegmental area dopaminergic and nondopaminergic neurons by orexins/hypocretins, *J. Neurosci.* **23**, 7-11 (2003).
22. M. Jaszberenyi, E. Bujdoso, I. Pataki and G. Telegdy, Effects of orexins on the hypothalamic-pituitary-adrenal system, *J. Neuroendocrinol.* **12**, 1174-1178 (2000).
23. A. Stricker-Krongrad and B. Beck, Modulation of hypothalamic hypocretin/orexin mRNA expression by glucocorticoids, *Biochem. Biophys. Res. Commun.* **296**, 129-133 (2002).
24. M. Kuru, Y. Ueta, R. Serino, M. Nakazato, Y. Yamamoto, I. Shibuya and H. Yamashita, Centrally administered orexin/hypocretin activates HPA axis in rats, *Neuroreport* **11**, 1977-1980 (2000).
25. S. H. Russell, C. J. Small, C. L. Dakin, C. R. Abbott, D. G. Morgan, M. A. Ghatei and S. R. Bloom, The central effects of orexin-A in the hypothalamic-pituitary-adrenal axis in vivo and in vitro in male rats, *J. Neuroendocrinol.* **13**, 561-566 (2001).
26. T. M. Reyes, J. R. Walker, C. DeCino, J. B. Hogenesch and P. E. Sawchenko, Categorically distinct acute stressors elicit dissimilar transcriptional profiles in the paraventricular nucleus of the hypothalamus, *J. Neurosci.* **23**, 5607-5616 (2003).
27. K. D. Carr , Feeding, drug abuse, and the sensitization of reward by metabolic need, *Neurochem. Res.* **21**, 1455-1467 (1996).
28. D. Cabeza, V, L. L. Krahne and K. D. Carr, A progressive ratio schedule of self-stimulation testing in rats reveals profound augmentation of d-amphetamine reward by food restriction but no effect of a "sensitizing" regimen of d-amphetamine, *Psychopharmacology (Berl)* (2004).
29. K. D. Carr , Augmentation of drug reward by chronic food restriction: behavioral evidence and underlying mechanisms, *Physiol Behav.* **76**, 353-364 (2002).
30. K. D. Carr, G. Kim and D. Cabeza, V, Hypoinsulinemia may mediate the lowering of self-stimulation thresholds by food restriction and streptozotocin-induced diabetes, *Brain Res.* **863**, 160-168 (2000).
31. D. Cabeza, V, S. Holiman and K. D. Carr, A search for the metabolic signal that sensitizes lateral hypothalamic self-stimulation in food-restricted rats, *Physiol Behav.* **64**, 251-260 (1998).
32. M. E. Carroll, C. P. France and R. A. Meisch, Food deprivation increases oral and intravenous drug intake in rats, *Science* **205**, 319-321 (1979).
33. D. Cabeza, V and K. D. Carr, Food restriction enhances the central rewarding effect of abused drugs, *J. Neurosci.* **18**, 7502-7510 (1998).
34. Y. Shaham, U. Shalev, L. Lu, H. De Wit and J. Stewart, The reinstatement model of drug relapse: history, methodology and major findings, *Psychopharmacology (Berl)* **168**, 3-20 (2003).
35. U. Shalev, M. Marinelli, M. H. Baumann, P. V. Piazza and Y. Shaham, The role of corticosterone in food deprivation-induced reinstatement of cocaine seeking in the rat, *Psychopharmacology (Berl)* **168**, 170-176 (2003).
36. U. Shalev, J. W. Grimm and Y. Shaham, Neurobiology of relapse to heroin and cocaine seeking: a review, *Pharmacol. Rev.* **54**, 1-42 (2002).
37. U. Shalev, J. Yap and Y. Shaham, Leptin attenuates acute food deprivation-induced relapse to heroin seeking, *J. Neurosci.* **21**, RC129 (2001).
38. D. A. Highfield, A. N. Mead, J. W. Grimm, B. A. Rocha and Y. Shaham, Reinstatement of cocaine seeking in 129X1/SvJ mice: effects of cocaine priming, cocaine cues and food deprivation, *Psychopharmacology (Berl)* **161**, 417-424 (2002).
39. M. Hakansson, L. De Lecea, J. G. Sutcliffe, M. Yanagisawa and B. Meister, Leptin receptor- and STAT3-immunoreactivities in hypocretin/orexin neurones of the lateral hypothalamus, *J. Neuroendocrinol.* **11**, 653-663 (1999).
40. T. Moriguchi, T. Sakurai, T. Nambu, M. Yanagisawa and K. Goto, Neurons containing orexin in the lateral hypothalamic area of the adult rat brain are activated by insulin-induced acute hypoglycemia, *Neurosci. Lett.* **264**, 101-104 (1999).

41. A. Yamanaka, C. T. Beuckmann, J. T. Willie, J. Hara, N. Tsujino, M. Mieda, M. Tominaga, K. Yagami, F. Sugiyama, K. Goto, M. Yanagisawa and T. Sakurai, Hypothalamic orexin neurons regulate arousal according to energy balance in mice, *Neuron* **38**, 701-713 (2003).
42. S. Fulton, B. Woodside and P. Shizgal, Modulation of brain reward circuitry by leptin, *Science* **287**, 125-128 (2000).
43. R. E. Campbell, M. S. Smith, S. E. Allen, B. E. Grayson, J. M. Ffrench-Mullen and K. L. Grove, Orexin neurons express a functional pancreatic polypeptide Y4 receptor, *J. Neurosci.* **23**, 1487-1497 (2003).
44. S. Fulton, B. Woodside and P. Shizgal, Does neuropeptide Y contribute to the modulation of brain stimulation reward by chronic food restriction?, *Behav. Brain Res.* **134**, 157-164 (2002).
45. J. Olds and P. Milner, Positive reinforcement produced by electrical stimulation of septal area and other regions of rat brain, *J. Comp Physiol Psychol.* **47**, 419-427 (1954).
46. J. Olds, Hypothalamic substrates of reward, *Physiol Rev.* **42**, 554-604 (1962).
47. I. Lestang, B. Cardo, M. T. Roy and L. Velley, Electrical self-stimulation deficits in the anterior and posterior parts of the medial forebrain bundle after ibotenic acid lesion of the middle lateral hypothalamus, *Neuroscience* **15**, 379-388 (1985).
48. B. Murray and P. Shizgal, Attenuation of medical forebrain bundle reward by anterior lateral hypothalamic lesions, *Behav. Brain Res.* **75**, 33-47 (1996).
49. C. Kornetsky and R. U. Esposito, Euphorigenic drugs: effects on the reward pathways of the brain, *Fed. Proc.* **38**, 2473-2476 (1979).
50. C. Kornetsky, R. U. Esposito, S. McLean and J. O. Jacobson, Intracranial self-stimulation thresholds: a model for the hedonic effects of drugs of abuse, *Arch. Gen. Psychiatry* **36**, 289-292 (1979).
51. A. Markou and G. F. Koob, Construct validity of a self-stimulation threshold paradigm: effects of reward and performance manipulations, *Physiol Behav.* **51**, 111-119 (1992).
52. R. A. Frank, P. Z. Manderscheid, S. Panicker, H. P. Williams and D. Kokoris, Cocaine euphoria, dysphoria, and tolerance assessed using drug-induced changes in brain-stimulation reward, *Pharmacol. Biochem. Behav.* **42**, 771-779 (1992).
53. A. A. Harrison, F. Gasparini and A. Markou, Nicotine potentiation of brain stimulation reward reversed by DH beta E and SCH 23390, but not by eticlopride, LY 314582 or MPEP in rats, *Psychopharmacology (Berl)* **160**, 56-66 (2002).
54. P. J. Kenny, I. Polis, G. F. Koob and A. Markou, Low dose cocaine self-administration transiently increases but high dose cocaine persistently decreases brain reward function in rats, *Eur. J. Neurosci.* **17**, 191-195 (2003).
55. C. Kornetsky , Brain-stimulation reward: a model for the neuronal bases for drug-induced euphoria, *NIDA Res. Monogr* **62**, 30-50 (1985).
56. L. Kokkinidis and B. D. McCarter, Postcocaine depression and sensitization of brain-stimulation reward: analysis of reinforcement and performance effects, *Pharmacol. Biochem. Behav.* **36**, 463-471 (1990).
57. A. Markou and G. F. Koob, Postcocaine anhedonia. An animal model of cocaine withdrawal, *Neuropsychopharmacology* **4**, 17-26 (1991).
58. M. P. Epping-Jordan, S. S. Watkins, G. F. Koob and A. Markou, Dramatic decreases in brain reward function during nicotine withdrawal, *Nature* **393**, 76-79 (1998).
59. D. Lin, G. F. Koob and A. Markou, Time-dependent alterations in ICSS thresholds associated with repeated amphetamine administrations, *Pharmacol. Biochem. Behav.* **65**, 407-417 (2000).
60. G. Schulteis, A. Markou, M. Cole and G. F. Koob, Decreased brain reward produced by ethanol withdrawal, *Proc. Natl. Acad. Sci. U. S. A* **92**, 5880-5884 (1995).
61. B.Boutrel, P.J.Kenny, R.Winsky, C.Wright, S.E.Specio, G.F.Koob, A.Markou, and L.de Lecea. Hypocretin regulates brain reward function and cocaine consumption in rats. Soc Neurosci Abstr 879, 7. 2003.
62. C. J. Langmead, J. C. Jerman, S. J. Brough, C. Scott, R. A. Porter and H. J. Herdon, Characterisation of the binding of [3H]-SB-674042, a novel nonpeptide antagonist, to the human orexin-1 receptor, *Br. J. Pharmacol.* **141**, 340-346 (2004).
63. C. T. Beuckmann, C. M. Sinton, S. C. Williams, J. A. Richardson, R. E. Hammer, T. Sakurai and M. Yanagisawa, Expression of a poly-glutamine-ataxin-3 transgene in orexin neurons induces narcolepsy-cataplexy in the rat, *J. Neurosci.* **24**, 4469-4477 (2004).
64. R. A. Wise and P. P. Rompre, Brain dopamine and reward, *Annu. Rev. Psychol.* **40**, 191-225 (1989).
65. W. Schultz , Predictive reward signal of dopamine neurons, *J. Neurophysiol.* **80**, 1-27 (1998).
66. R. A. Wise , Brain reward circuitry: insights from unsensed incentives, *Neuron* **36**, 229-240 (2002).
67. G. L. Forster and C. D. Blaha, Laterodorsal tegmental stimulation elicits dopamine efflux in the rat nucleus accumbens by activation of acetylcholine and glutamate receptors in the ventral tegmental area, *Eur. J. Neurosci.* **12**, 3596-3604 (2000).
68. J. Yeomans and M. Baptista, Both nicotinic and muscarinic receptors in ventral tegmental area contribute to brain-stimulation reward, *Pharmacol. Biochem. Behav.* **57**, 915-921 (1997).

69. J. S. Yeomans, A. Mathur and M. Tampakeras, Rewarding brain stimulation: role of tegmental cholinergic neurons that activate dopamine neurons, *Behav. Neurosci.* **107**, 1077-1087 (1993).
70. J. S. Yeomans, O. Kofman and V. McFarlane, Cholinergic involvement in lateral hypothalamic rewarding brain stimulation, *Brain Res.* **329**, 19-26 (1985).
71. M. A. Greco and P. J. Shiromani, Hypocretin receptor protein and mRNA expression in the dorsolateral pons of rats, *Brain Res. Mol. Brain Res.* **88**, 176-182 (2001).
72. K. Takahashi, Y. Koyama, Y. Kayama and M. Yamamoto, Effects of orexin on the laterodorsal tegmental neurones, *Psychiatry Clin. Neurosci.* **56**, 335-336 (2002).
73. K. Uramura, H. Funahashi, S. Muroya, S. Shioda, M. Takigawa and T. Yada, Orexin-a activates phospholipase C- and protein kinase C-mediated Ca2+ signaling in dopamine neurons of the ventral tegmental area, *Neuroreport* **12**, 1885-1889 (2001).
74. J. Fadel and A. Y. Deutch, Anatomical substrates of orexin-dopamine interactions: lateral hypothalamic projections to the ventral tegmental area, *Neuroscience* **111**, 379-387 (2002).
75. G. Martin, V. Fabre, G. R. Siggins and L. De Lecea, Interaction of the hypocretins with neurotransmitters in the nucleus accumbens, *Regul. Pept.* **104**, 111-117 (2002).
76. P. Shizgal, S. Fulton and B. Woodside, Brain reward circuitry and the regulation of energy balance, *Int. J. Obes. Relat Metab Disord.* **25 Suppl 5**, S17-S21 (2001).
77. H. Funahashi, T. Hori, Y. Shimoda, H. Mizushima, T. Ryushi, S. Katoh and S. Shioda, Morphological evidence for neural interactions between leptin and orexin in the hypothalamus, *Regul. Pept.* **92**, 31-35 (2000).
78. B. Griffond, P. Y. Risold, C. Jacquemard, C. Colard and D. Fellmann, Insulin-induced hypoglycemia increases preprohypocretin (orexin) mRNA in the rat lateral hypothalamic area, *Neurosci. Lett.* **262**, 77-80 (1999).
79. D. J. Macey, G. F. Koob and A. Markou, CRF and urocortin decreased brain stimulation reward in the rat: reversal by a CRF receptor antagonist, *Brain Res.* **866**, 82-91 (2000).
80. X. J. Cai, M. L. Evans, C. A. Lister, R. A. Leslie, J. R. Arch, S. Wilson and G. Williams, Hypoglycemia activates orexin neurons and selectively increases hypothalamic orexin-B levels: responses inhibited by feeding and possibly mediated by the nucleus of the solitary tract, *Diabetes* **50**, 105-112 (2001).
81. X. J. Cai, P. S. Widdowson, J. Harrold, S. Wilson, R. E. Buckingham, J. R. Arch, M. Tadayyon, J. C. Clapham, J. Wilding and G. Williams, Hypothalamic orexin expression: modulation by blood glucose and feeding, *Diabetes* **48**, 2132-2137 (1999).
82. C. B. Saper, T. C. Chou and J. K. Elmquist, The need to feed: homeostatic and hedonic control of eating, *Neuron* **36**, 199-211 (2002).
83. R. J. Hayes, S. R. Vorel, J. Spector, X. Liu and E. L. Gardner, Electrical and chemical stimulation of the basolateral complex of the amygdala reinstates cocaine-seeking behavior in the rat, *Psychopharmacology (Berl)* **168**, 75-83 (2003).
84. S. R. Vorel, X. Liu, R. J. Hayes, J. A. Spector and E. L. Gardner, Relapse to cocaine-seeking after hippocampal theta burst stimulation, *Science* **292**, 1175-1178 (2001).
85. J. W. Grimm and R. E. See, Dissociation of primary and secondary reward-relevant limbic nuclei in an animal model of relapse, *Neuropsychopharmacology* **22**, 473-479 (2000).
86. R. J. DiLeone, D. Georgescu and E. J. Nestler, Lateral hypothalamic neuropeptides in reward and drug addiction, *Life Sci.* **73**, 759-768 (2003).
87. D. Georgescu, V. Zachariou, M. Barrot, M. Mieda, J. T. Willie, A. J. Eisch, M. Yanagisawa, E. J. Nestler and R. J. DiLeone, Involvement of the lateral hypothalamic peptide orexin in morphine dependence and withdrawal, *J. Neurosci.* **23**, 3106-3111 (2003).
88. S. H. Ahmed and G. F. Koob, Cocaine- but not food-seeking behavior is reinstated by stress after extinction, *Psychopharmacology (Berl)* **132**, 289-295 (1997).
89. G. F. Koob , Neurobiology of addiction. Toward the development of new therapies, *Ann. N. Y. Acad. Sci.* **909**, 170-185 (2000).

OREXIN/HYPOCRETIN AND OPIOID DEPENDENCE

Ralph J. DiLeone*

1. INTRODUCTION

The ORX/HCRT/hypocretin (ORX/HCRT) neuropeptides are regulators of neuronal function and animal behavior. While their function in sleep and regulation of arousal is well studied, the broad projection patterns of the ORX/HCRT neurons suggest that the peptides have the potential to modulate diverse sets of behavior. Moreover, the expression of ORX/HCRT in the lateral hypothalamus (LH) is an indication that the neuropeptide may have an important role in a modulating a broad set of motivated behaviors. Many studies have suggested that the LH is a powerful brain center that has influence over an array of behavioral outputs. In this chapter, data is presented suggesting a role for ORX/HCRT in opioid dependence. These studies paint a more complete picture of the function of ORX/HCRT and may provide additional targets for therapeutic manipulation of drug dependence.

2. EARLY STUDIES OF THE LH

The hypothalamus has been studied for many years for its role in regulating the homeostatic state and physiology of animals. Early experiments demonstrated that decorticate sham rage is dependent on the hypothalamus,[1] and specific brain lesions have implicated the hypothalamus in thermoregulation, sleep, feeding, sex drive, and motivated behavior. Moreover, experiments performed as early as the 1940s implicated the hypothalamus as an important mediator of reward and motivated behavior. It was also demonstrated that different regions of the hypothalamus may antagonize each other: medial hypothalamus (MH) lesions led to increased feeding while lateral hypothalamus LH) lesions led to reduced feeding and reduced motivation. This led to the development of a "two-center" hypothesis where the LH mediates, and the MH inhibits, reward-related behavior.[2] The LH received further attention as a reward-mediating structure with the work of Olds and Milner, who showed that animals will robustly lever-press for electrical

* Ralph J. DiLeone, Department of Psychiatry, University of Texas Southwestern Medical Center, 5323 Harry Hines Blvd., Dallas, TX 75390

current (intracranial self-stimulation or ICSS) when the electrodes are placed within the LH.[3]

The early lesion and ICSS experiments attracted much attention, but were also criticized. The MH lesion experiments were reinterpreted as compensatory feeding in response to initial metabolic changes caused by the lesions.[4,5] The role of the LH neurons in ICSS was also brought into question, as the medial forebrain bundle, known to contain mesolimbic dopamine fibers, passes through the region.[6] However, lesions that spared the passing fibers were shown to produce some of the feeding symptoms,[7,8] and alter ICSS,[9,10] arguing for a role for the intrinsic LH cells in reward-related functions.

While the ability of the hypothalamus to broadly regulate the homeostatic state has been well appreciated and studied for many years, less is understood about the mechanisms of direct hypothalamic control over brain function. The many nuclei of the hypothalamus exhibit extensive neuronal projections connecting throughout the brain, and the hypothalamus is rich in neuropeptides, many of which have dramatic effects on neuronal function and behavior. The advent of molecular biology and gene cloning has helped to initiate mechanism-based study of hypothalamic function.

3. THE VIEW OF THE LH IN THE MODERN HYPOTHALAMUS

Genetic and molecular experiments support a role for the hypothalamus in the regulation of feeding and have begun to identify key molecules that control this function. It is believed that the leptin protein, generated in peripheral adipocytes, communicates with the brain in large part through the MH. Current models speculate that information from peripheral signaling molecules, such as leptin, travels from the MH to the paraventricular nucleus and LH in order to enact behavioral responses. Indeed, LH neurons express receptors for a number of neuropeptides made in the MH. However, leptin receptors are also expressed on the LH neurons,[11,12] raising the possibility that leptin might act directly on LH neurons to cause changes in gene expression[12,13] and behavior. Interestingly, leptin has been shown to modulate LH-mediated ICSS,[14] suggesting that peripheral signals, via direct or indirect pathways, interact with the LH. While the neurocircuitry details are not completely understood, these studies have helped to place our understanding of the hypothalamus on better molecular footing. However, until recently, the specific contribution that the LH neurons make to overall hypothalamic function has not been clear.

4. NEW LH PEPTIDES

Recent identification of two LH-specific peptides has revitalized interest in the LH, by allowing for direct studies of specific gene function. The ORX/HCRT and melanin-concentrating hormone (MCH) neuropeptides are localized to non-overlapping sets of neurons within the LH and have powerful effects on behavior. The ORX/HCRT gene product was independently isolated by two research groups and was found to exist in two forms which are the result of differential post-translational processing.[15,16] While the MCH knockout mice are lean and hypophagic,[17] the ORX/HCRT knockout mice display a narcolepsy-like phenotype,[18] suggesting an important role for ORX/HCRT in sleep control and arousal. However, experiments showing ORX/HCRT-mediated orexigenic activity,[16,19-21] together with evidence that activation of ORX/HCRT neurons correlates with orexigenic activity of antipsychotics,[22] suggest that ORX/HCRT neurons might have an important role in integrating arousal, feeding and metabolism reviewed in.[23,24]

The identification of these neuropeptides in the LH has allowed for direct test of function and clarification of the role of the LH in controlling animal behavior. The tools are available (knockout mice, pharmacological reagents) to begin to explore the role of ORX/HCRT in other behaviors that have been previously associated with the LH.

5. THE HYPOTHALAMUS AND ADDICTION

Many levels of hypothalamic function are likely to influence addiction. While most of the previous work on the role of the hypothalamus in addiction assessed effects of the HPA (hypothalamic-pituitary-adrenal) axis and various physiological states, such as stress, on addiction,[25-27] the LH-restricted expression of ORX/HCRT suggests a direct role for these peptides in controlling addiction-related behaviors. Indeed, the ORX/HCRT neurons project broadly to regions throughout the brain, including mesolimbic and brainstem nuclei known to be important in the development of drug addiction.[15,28-30]

5.1. The LH and Drugs of Abuse

The LH, in particular, has a history of being associated with drug addiction and drug abuse behavior. The relationship between the LH and both feeding and addiction may not be surprising since behavioral studies have shown cross-sensitization between 'natural' rewards and drugs of abuse. Both drugs of abuse and food restriction profoundly modulate LH-mediated ICSS and their effects can cross-sensitize.[31,32] Moreover, leptin acts to reduce heroin self-administration[33] and has can modulate LH-mediated ICSS. [14]

Furthermore, there are experiments that have more directly implicated the LH in the reinforcing properties of drugs. Both rats and mice will self-administer morphine and other opioid peptides directly into the lateral hypothalamus.[34-36] While some studies have found that LH lesions do not block opiate self-administration,[37] other groups have demonstrated that systemic self-administration of heroin can be blocked by opiate antagonist delivery directly to the LH.[38] Supporting work has established that direct administration of opioids to the LH region results in a conditioned place preference to the drug.[39] However, the molecular basis of this LH-mediated response to opiates has remained largely uncharacterized.

6. ORX/HCRT NEURONS RESPOND TO CHRONIC MORPHINE AND MORPHINE WITHDRAWAL

While it is clear that the LH has powerful effects on behavior, the specific neuronal cell types mediating these functions have not been identified. The LH-specific expression of both MCH and ORX/HCRT makes these neuropeptides prime candidates for mediating behaviors related to drug addiction. Toward this goal, the ORX/HCRT and MCH neurons were studied for their responses to chronic morphine and precipitated morphine withdrawal. The neurons were evaluated with two separate markers: (1) **c-Fos** immunoreactivity, and (2) **CRE-LacZ** reporter activity[40] were used to assess neural response and plasticity. Both c-Fos and the CRE-LacZ reporter are highly regulated by morphine and have been used extensively to map brain regions that demonstrate activation and neuronal plasticity in response to morphine and precipitated withdrawal.[41-43] Results from this work indicate that the ORX/HCRT neurons are highly responsive to morphine, with nearly 10X as many neurons showing CRE-LacZ, compared to control animals, after five days of morphine (Figure 1A) ($F_{(3.23)} = 49.93$; $p < 0.05$).

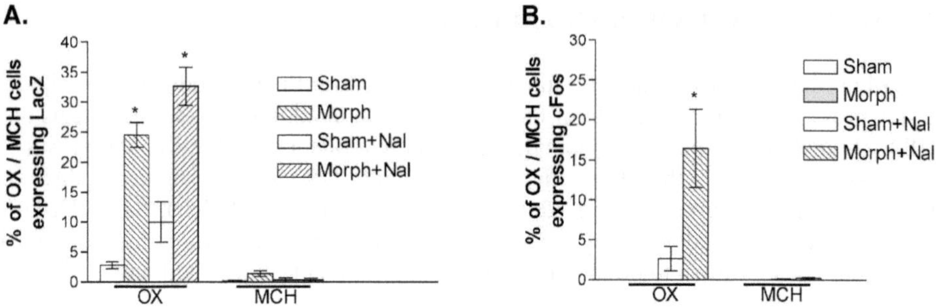

Figure 1. LH neurons expressing ORX/HCRT respond to morphine treatment while LH neurons expressing MCH do not. Mice were treated with sham or morphine pellets (2 x 25 mg) for five days followed by injection of saline or naltrexone (100 mg/kg, subcutaneous) four hours before being perfused and analyzed by immunohistochemistry. Neuronal responses to sham surgery (Sham) chronic morphine (Morph), sham surgery plus naltrexone (Sham+Nal), or chronic morphine plus naltrexone (Morph+Nal) were evaluated by use of a CRE-LacZ reporter mouse (A) or c-Fos detection (B). Figure modified from [44]. N=6 per group, *$p<0.05$.

In contrast, no c-Fos is seen in ORX/HCRT neurons (Figure 1B), consistent with previous data suggesting that the c-Fos response desensitizes over time. To further evaluate the ORX/HCRT neurons for their responsiveness, we precipitated withdrawal by administration of the μ-opioid receptor antagonist naltrexone after 5 days of morphine treatment. Under this scenario, dramatic c-Fos responses are seen ($F_{(3,20)} = 15.79$; $p < 0.05$) while a CRE-LacZ response is seen in over 30% of the neurons (Figure 1B).

The MCH neurons of the LH were also examined and did not show molecular responses to morphine or morphine + naltrexone (Figure 1A and 1B). Independent experiments have shown that the CRE-LacZ reporter is active in the MCH neurons as demonstrated by a large percentage of MCH neurons responding to starvation (data not shown). Thus, the data suggests a very specific response of ORX/HCRT neurons to both morphine exposure and precipitated withdrawal.

It should be noted that only a subset of ORX/HCRT neurons (25-35%) responded to morphine and precipitated withdrawal (Figure 1A and 1B) and nearly 50% of the cells showing CRE-LacZ response did not express ORX/HCRT. This suggests that the ORX/HCRT neurons are heterogeneous and also that there is a set of unidentified LH neurons that appears to be responsive to morphine.

7. THE ORX/HCRT GENE IS UPREGULATED AFTER PRECIPITATED MORPHINE WITHDRAWAL

Since the ORX/HCRT neurons are responsive to morphine, the expression of the ORX/HCRT gene was tested using paradigms described above. An orexin-tau-LacZ knock-in line was used to evaluate expression of the ORX/HCRT gene, as ß-galactosidase allows for better quantitative analysis of gene expression. Analysis of LH brain sections from reporter mice showed that, unlike c-Fos or the CRE-LacZ reporter gene, the Orx-tau-LacZ reporter is not upregulated after chronic morphine (Figure 2). Interestingly, however, the reporter gene is upregulated ~40% after naltrexone-induced withdrawal from morphine ($F_{(3,19)} = 5.78$; $p < 0.01$).

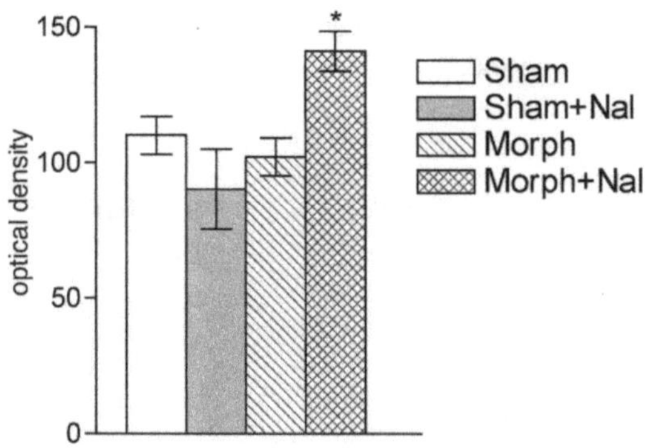

Figure 2. The ORX/HCRT gene is upregulated after chronic morphine and precipitation of withdrawal, but not after chronic morphine alone. Orx-tau-LacZ mice were treated with morphine as described in Figure 1. Brain sections were tested for ß-galactosidase activity and densitometry performed to assess levels of ORX/HCRT gene expression. Figure modified from [44]. N=5 per group, *$p<0.05$.

The data suggest that regulation of the ORX/HCRT gene may be occurring in response to the neuronal state. In particular, it may be that ORX/HCRT gene in upregulated in response to the release of the peptides during precipitated morphine withdrawal. However, future studies utilizing direct measurement of peptide release or receptor activation are needed to test this prediction.

8. ORX/HCRT NEURONS EXPRESS THE μ-OPIOID RECEPTOR

Previous in situ hybridization studies have indicated that the μ-opioid receptor is expressed in the LH region, but the identity of these neurons was not known. Triple immunostaining and confocal analysis has recently demonstrated that ORX/HCRT neurons express the μ-opioid receptor and may therefore be responding directly to morphine and naltrexone.[44] While not all ORX/HCRT neurons express μ-opioid receptor, all neurons that respond after precipitation of withdrawal (c-Fos or LacZ positive) appear to possess μ-opioid receptor. The expression of μ-opioid receptor on

ORX/HCRT neurons may explain why heroin is self-administered directly to the LH region and why opioid receptor antagonists administered to the LH can block this self-administration.[38]

9. ORX/HCRT KNOCKOUT MICE SHOW ATTENUATED MORPHINE WITHDRAWAL

To investigate the functional role of ORX/HCRT in response to drug administration, mice lacking the ORX/HCRT peptides[18] were analyzed for physical dependence to morphine. Following morphine exposure and precipitation of withdrawal, animals were evaluated for physical signs of withdrawal.[44] ORX/HCRT mutant mice exhibited dramatic attenuation of specific withdrawal symptoms (Figure 3). To control for general behavior abnormalities that might confound this data, the mutant mice were tested for normal open field activity and observed to confirm an absence of cataplexy or narcolepsy during the test period.[44]

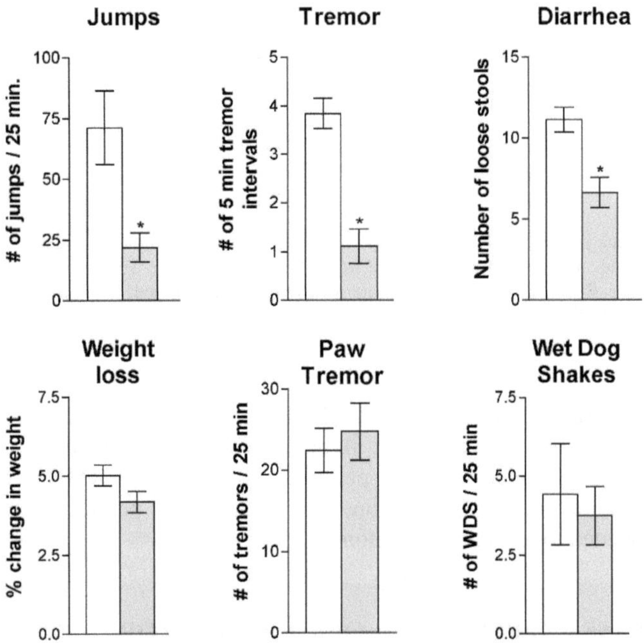

Figure 3. ORX/HCRT knockout mice (gray bars) show attenuated physical withdrawal from chronic morphine when compared to wildtype littermates (white bars). Two days after morphine pellet (25 mg) or sham surgery, mice were injected with naloxone (1 mg/kg subcutaneous) and monitored for 25 minutes for physical signs of withdrawal listed above each graph. ORX/HCRT mutant mice showed significant attenuation in a subset of physical withdrawal signs. Figure modified from [44]. N=8 per group, *$p<0.01$.

10. SUMMARY OF ORX/HCRT DATA

Taken together, the above data suggest that the ORX/HCRT neurons of the LH are responsive to opiates. Expression of the μ-opioid receptor on the ORX/HCRT neurons further implies a direct mode of action on the neurons. Finally, the attenuated withdrawal seen in mutant mice indicates that the ORX/HCRT neuropeptide is required for normal development of dependence and/or expression of withdrawal (see discussion below). Together, the data support a role for the ORX/HCRT neuropeptide system in molecular adaptations to morphine exposure.

11. INTERPRETATION AND IMPLICATIONS FOR ADDICTION BIOLOGY

The data discussed in this chapter is suggestive of a direct involvement of the ORX/HCRT system in response to morphine. However, a number of critical questions remain with regard to the role of ORX/HCRT neuropeptides in drug addiction. First, it should be emphasized that the above studies focus on morphine dependence and do not evaluate the role of ORX/HCRT in the reinforcing properties of drugs of abuse. Studies using conditioned place preference, or drug self-administration are needed to garner a more complete picture of the role of ORX/HCRT in addiction.

Nonetheless, the development of physical dependence is an important component of drug addiction. Koob and colleagues have proposed that negative physical and emotional states are the critical components that drive addiction.[45] Thus, it is essential to define the molecular adaptations that lead to dependence and the development of negative states since they represent potential targets for treatment of drug addiction. The development and application of ORX/HCRT receptor antagonists[46,47] will make it possible to test the effects of pharmacological blockage on morphine dependence and withdrawal.

11.1. Is ORX/HCRT Essential for the Development of Dependence and/or the Expression of Withdrawal?

Since the ORX/HCRT mutant mice lack gene function during both the chronic morphine exposure as well as the withdrawal phase, it is difficult to discern whether the neuropeptide is necessary for the development of drug dependence. It is possible that ORX/HCRT plays a role in both the development of dependence as well as the expression of physical withdrawal signs. In fact, our molecular experiments suggest that the ORX/HCRT neurons are modulated during both chronic morphine exposure (CRE-LacZ induction) as well as during withdrawal (c-Fos induction). However, it is not possible to conclude on the basis of this data that ORX/HCRT has a *function* in both the development of dependence and expression of withdrawal. Studies aimed at addressing this question require analysis of ORX/HCRT function during drug exposure or withdrawal only.

12. ORX/HCRT RECEPTORS AND NEURAL CIRCUITS RELEVANT TO ADDICTION

While other hypothalamic peptides have been studied for their role in addiction behavior and physiology (e.g. see refs 48,49-51), the ORX/HCRT neuropeptides exhibit several distinct characteristics. First, ORX/HCRT is expressed solely in the lateral hypothalamic region, an area implicated in reward and drug-related behaviors, but whose molecular effectors have remained unidentified. Second, neurons expressing ORX/HCRT have extensive projections to, and corresponding receptor expression within, the mesolimbic dopamine and noradrenergic (LC) pathways. This suggests multiple possible connections between the LH and regions well-studied regions for their role in drug addiction.

12.1. What Brain Regions are the Critical ORX/HCRT Targets?

The two ORX/HCRT receptors, OxR1 and OxR2, are expressed in distinct patterns in the adult brain,[52,53] and both are GPCRs; OxR1 appears to signal through Gq while OxR2 can couple to Gi/Go as well as Gq subunits.[16,54] ORX/HCRTs stimulate neuronal activity at target sites by raising cytoplasmic Ca^{2+} levels postsynaptically.[54] Interestingly, ORX/HCRT receptors are expressed at high levels in the ventral tegmental area (VTA) and the locus coeruleus (LC)[52,53] and ORX/HCRT can activate neurons in both regions.[55-59] The LC regulates alertness and vigilance, and is known to undergo dramatic molecular responses to chronic opiates and withdrawal.[60,61] In the LC, the neuronal responses are mediated largely via OxR1,[62] while the primary receptor remains untested at most other target sites. In addition, ORX/HCRTs have been shown to inhibit nucleus accumbens neurons,[63] indicating that the relationship between ORX/HCRT and neuronal response may be complex and site-specific.

There are, therefore, multiple potential connections between LH ORX/HCRT systems and drug-regulated neural circuitry. ORX/HCRT-OxR pathways modulate the activity of VTA and NAc neurons, potentially altering both ends of the mesolimbic reward pathway, while also playing a role in LC neuron excitability that may influence components of opiate withdrawal. These molecular connections provide a potential link between the classical substrates of brain stimulation reward and the well-studied neural and molecular circuitry of drug reward. Future studies are needed to better define the relevant neural circuits by which ORX/HCRT modulates morphine dependence and withdrawal.

13. REFERENCES

1. P. Bard, A diencephalic mechanism for the expression of rage with special reference to the sympathetic nervous system, *American Journal of Physiology.* **84**, 480-515 (1928).
2. E. Stellar, The physiology of motivation., *Psychol Rev.* **61**, 5-22 (1954).
3. J. Olds and P. Milner, Positive reinforcement produced by electrical stimulation of septal area and other regions of rat brain, *J Compar Physiological Psychol.* **47** (1954).
4. T. L. Powley and R. E. Keesey, Relationship of body weight to the lateral hypothalamic feeding syndrome, *J Comp Physiol Psychol.* **70**, 25-36 (1970).
5. T. L. Powley, The ventromedial hypothalamic syndrome, satiety, and a cephalic phase hypothesis, *Psychol Rev.* **84**, 89-126 (1977).
6. O. Millhouse, E., A Golgi study of the descending medial forebrain bundle, *Brain Res.* **12**, 341-61 (1969).
7. L. Velley, C. Chaminade, M. T. Roy, E. Kempf and B. Cardo, Intrinsic neurons are involved in lateral hypothalamic self-stimulation, *Brain Res.* **268**, 79-86 (1983).

8. P. Winn, A. Tarbuck and S. B. Dunnett, Ibotenic acid lesions of the lateral hypothalamus: comparison with the electrolytic lesion syndrome, *Neuroscience.* **12**, 225-40 (1984).
9. I. Lestang, B. Cardo, M. T. Roy and L. Velley, Electrical self-stimulation deficits in the anterior and posterior parts of the medial forebrain bundle after ibotenic acid lesion of the middle lateral hypothalamus, *Neuroscience.* **15**, 379-88 (1985).
10. S. Nassif, B. Cardo, F. Libersat and L. Velley, Comparison of deficits in electrical self-stimulation after ibotenic acid lesion of the lateral hypothalamus and the medial prefrontal cortex, *Brain Res.* **332**, 247-57 (1985).
11. M. L. Hakansson, H. Brown, N. Ghilardi, R. C. Skoda and B. Meister, Leptin receptor immunoreactivity in chemically defined target neurons of the hypothalamus, *J Neurosci.* **18**, 559-72 (1998).
12. M. Hakansson, L. de Lecea, J. G. Sutcliffe, M. Yanagisawa and B. Meister, Leptin receptor- and STAT3-immunoreactivities in hypocretin/orexin neurones of the lateral hypothalamus, *J Neuroendocrinol.* **11**, 653-63 (1999).
13. H. Funahashi, T. Yada, S. Muroya, M. Takigawa, T. Ryushi, S. Horie, Y. Nakai and S. Shioda, The effect of leptin on feeding-regulating neurons in the rat hypothalamus, *Neurosci Lett.* **264**, 117-20 (1999).
14. S. Fulton, B. Woodside and P. Shizgal, Modulation of brain reward circuitry by leptin, *Science.* **287**, 125-8 (2000).
15. L. de Lecea, T. S. Kilduff, C. Peyron, X. Gao, P. E. Foye, P. E. Danielson, C. Fukuhara, E. L. Battenberg, V. T. Gautvik, F. S. Bartlett, 2nd, W. N. Frankel, A. N. van den Pol, F. E. Bloom, K. M. Gautvik and J. G. Sutcliffe, The hypocretins: hypothalamus-specific peptides with neuroexcitatory activity, *Proc Natl Acad Sci U S A.* **95**, 322-7 (1998).
16. T. Sakurai, A. Amemiya, M. Ishii, I. Matsuzaki, R. M. Chemelli, H. Tanaka, S. C. Williams, J. A. Richardson, G. P. Kozlowski, S. Wilson, J. R. Arch, R. E. Buckingham, A. C. Haynes, S. A. Carr, R. S. Annan, D. E. McNulty, W. S. Liu, J. A. Terrett, N. A. Elshourbagy, D. J. Bergsma and M. Yanagisawa, Orexins and orexin receptors: a family of hypothalamic neuropeptides and G protein-coupled receptors that regulate feeding behavior, *Cell.* **92**, 573-85 (1998).
17. M. Shimada, N. A. Tritos, B. B. Lowell, J. S. Flier and E. Maratos-Flier, Mice lacking melanin-concentrating hormone are hypophagic and lean, *Nature.* **396**, 670-4. (1998).
18. R. M. Chemelli, J. T. Willie, C. M. Sinton, J. K. Elmquist, T. Scammell, C. Lee, J. A. Richardson, S. C. Williams, Y. Xiong, Y. Kisanuki, T. E. Fitch, M. Nakazato, R. E. Hammer, C. B. Saper and M. Yanagisawa, Narcolepsy in orexin knockout mice: molecular genetics of sleep regulation, *Cell.* **98**, 437-51 (1999).
19. C. M. Edwards, S. Abusnana, D. Sunter, K. G. Murphy, M. A. Ghatei and S. R. Bloom, The effect of the orexins on food intake: comparison with neuropeptide Y, melanin-concentrating hormone and galanin, *J Endocrinol.* **160**, R7-12 (1999).
20. A. C. Haynes, B. Jackson, P. Overend, R. E. Buckingham, S. Wilson, M. Tadayyon and J. R. Arch, Effects of single and chronic intracerebroventricular administration of the orexins on feeding in the rat, *Peptides.* **20**, 1099-105 (1999).
21. A. Yamanaka, T. Sakurai, T. Katsumoto, M. Yanagisawa and K. Goto, Chronic intracerebroventricular administration of orexin-A to rats increases food intake in daytime, but has no effect on body weight, *Brain Res.* **849**, 248-52 (1999).
22. J. Fadel, M. Bubser and A. Y. Deutch, Differential activation of orexin neurons by antipsychotic drugs associated with weight gain, *J Neurosci.* **22**, 6742-6 (2002).
23. J. T. Willie, R. M. Chemelli, C. M. Sinton and M. Yanagisawa, To eat or to sleep? Orexin in the regulation of feeding and wakefulness, *Annu Rev Neurosci.* **24**, 429-58 (2001).
24. J. G. Sutcliffe and L. De Lecea, The hypocretins: setting the arousal threshold, *Nat Rev Neurosci.* **3**, 339-49 (2002).
25. P. V. Piazza, S. Maccari, J. M. Deminiere, M. Le Moal, P. Mormede and H. Simon, Corticosterone levels determine individual vulnerability to amphetamine self-administration, *Proc Natl Acad Sci U S A.* **88**, 2088-92 (1991).
26. J. Ortiz, J. L. DeCaprio, T. A. Kosten and E. J. Nestler, Strain-selective effects of corticosterone on locomotor sensitization to cocaine and on levels of tyrosine hydroxylase and glucocorticoid receptor in the ventral tegmental area, *Neuroscience.* **67**, 383-97 (1995).
27. N. E. Goeders, The HPA axis and cocaine reinforcement, *Psychoneuroendocrinology.* **27**, 13-33 (2002).
28. C. Peyron, D. K. Tighe, A. N. van den Pol, L. de Lecea, H. C. Heller, J. G. Sutcliffe and T. S. Kilduff, Neurons containing hypocretin (orexin) project to multiple neuronal systems, *J Neurosci.* **18**, 9996-10015 (1998).
29. Y. Date, Y. Ueta, H. Yamashita, H. Yamaguchi, S. Matsukura, K. Kangawa, T. Sakurai, M. Yanagisawa and M. Nakazato, Orexins, orexigenic hypothalamic peptides, interact with autonomic, neuroendocrine and neuroregulatory systems, *Proc Natl Acad Sci U S A.* **96**, 748-53 (1999).

30. A. N. van den Pol, Hypothalamic hypocretin (orexin): robust innervation of the spinal cord, *J Neurosci.* **19**, 3171-82 (1999).
31. W. A. Carlezon, Jr. and R. A. Wise, Microinjections of phencyclidine (PCP) and related drugs into nucleus accumbens shell potentiate medial forebrain bundle brain stimulation reward, *Psychopharmacology (Berl).* **128**, 413-20 (1996).
32. S. Cabeza de Vaca and K. D. Carr, Food restriction enhances the central rewarding effect of abused drugs, *J Neurosci.* **18**, 7502-10 (1998).
33. U. Shalev, J. Yap and Y. Shaham, Leptin attenuates acute food deprivation-induced relapse to heroin seeking, *J Neurosci.* **21**, RC129 (2001).
34. M. E. Olds, Hypothalamic substrate for the positive reinforcing properties of morphine in the rat, *Brain Res.* **168**, 351-60 (1979).
35. M. E. Olds and K. N. Williams, Self-administration of D-ALA-MET-Enkephalinamide at hypothalamic self-stimulation sites, *Brain Res.* **194**, 155-170 (1980).
36. P. Cazala, C. Darracq and M. Saint-Marc, Self-administration of morphine into the lateral hypothalmus in the mouse, *Brain Res.* **416**, 283-8 (1987).
37. M. D. Britt and R. A. Wise, Opiate rewarding action: independence of the cells of the lateral hypothalamus, *Brain Res.* **222**, 213-17 (1981).
38. W. A. Corrigall, Heroin self-administration: effects of antagonist treatment in lateral hypothalamus, *Pharmacol Biochem Behav.* **27**, 693-700 (1987).
39. D. van der Kooy, R. F. Mucha, M. O'Shaughnessy and P. Bucenieks, Reinforcing effects of brain microinjections of morphine revealed by conditioned place preference, *Brain Res.* **243**, 107-17 (1982).
40. S. Impey, D. M. Smith, K. Obrietan, R. Donahue, C. Wade and D. R. Storm, Stimulation of cAMP response element (CRE)-mediated transcription during contextual learning, *Nat Neurosci.* **1**, 595-601. (1998).
41. M. D. Hayward, R. S. Duman and E. J. Nestler, Induction of the c-fos proto-oncogene during opiate withdrawal in the locus coeruleus and other regions of rat brain, *Brain Res.* **525**, 256-66 (1990).
42. B. Chieng, K. A. Keay and M. J. Christie, Increased fos-like immunoreactivity in the periaqueductal gray of anaesthetised rats during opiate withdrawal, *Neurosci Lett.* **183**, 79-82 (1995).
43. T. Z. Shaw-Lutchman, M. Barrot, T. Wallace, L. Gilden, V. Zachariou, S. Impey, R. S. Duman, D. Storm and E. J. Nestler, Regional and cellular mapping of cAMP response element-mediated transcription during naltrexone-precipitated morphine withdrawal, *J Neurosci.* **22**, 3663-72 (2002).
44. D. Georgescu, V. Zachariou, M. Barrot, M. Mieda, J. T. Willie, A. J. Eisch, M. Yanagisawa, E. J. Nestler and R. J. DiLeone, Involvement of the lateral hypothalamic peptide orexin in morphine dependence and withdrawal, *J Neurosci.* **23**, 3106-11 (2003).
45. G. F. Koob, S. H. Ahmed, B. Boutrel, S. A. Chen, P. J. Kenny, A. Markou, L. E. O'Dell, L. H. Parsons and P. P. Sanna, Neurobiological mechanisms in the transition from drug use to drug dependence, *Neurosci Biobehav Rev.* **27**, 739-49 (2004).
46. A. C. Haynes, B. Jackson, H. Chapman, M. Tadayyon, A. Johns, R. A. Porter and J. R. Arch, A selective orexin-1 receptor antagonist reduces food consumption in male and female rats, *Regul Pept.* **96**, 45-51 (2000).
47. A. C. Haynes, H. Chapman, C. Taylor, G. B. Moore, M. A. Cawthorne, M. Tadayyon, J. C. Clapham and J. R. Arch, Anorectic, thermogenic and anti-obesity activity of a selective orexin-1 receptor antagonist in ob/ob mice, *Regul Pept.* **104**, 153-9 (2002).
48. J. D. Alvaro, J. B. Tatro, J. M. Quillan, M. Fogliano, M. Eisenhard, M. R. Lerner, E. J. Nestler and R. S. Duman, Morphine down-regulates melanocortin-4 receptor expression in brain regions that mediate opiate addiction, *Mol Pharmacol.* **50**, 583-91 (1996).
49. J. D. Alvaro, J. B. Tatro and R. S. Duman, Melanocortins and opiate addiction, *Life Sci.* **61**, 1-9 (1997).
50. G. L. Kovacs, Z. Sarnyai and G. Szabo, Oxytocin and addiction: a review, *Psychoneuroendocrinology.* **23**, 945-62 (1998).
51. R. J. DiLeone, D. Georgescu and E. J. Nestler, Lateral hypothalamic neuropeptides in reward and drug addiction, *Life Sciences.* **73**, 759-68 (2003).
52. P. Trivedi, H. Yu, D. J. MacNeil, L. H. Van der Ploeg and X. M. Guan, Distribution of orexin receptor mRNA in the rat brain, *FEBS Lett.* **438**, 71-5 (1998).
53. J. N. Marcus, C. J. Aschkenasi, C. E. Lee, R. M. Chemelli, C. B. Saper, M. Yanagisawa and J. K. Elmquist, Differential expression of orexin receptors 1 and 2 in the rat brain, *J Comp Neurol.* **435**, 6-25 (2001).
54. A. N. van den Pol, X. B. Gao, K. Obrietan, T. S. Kilduff and A. B. Belousov, Presynaptic and postsynaptic actions and modulation of neuroendocrine neurons by a new hypothalamic peptide, hypocretin/orexin, *J Neurosci.* **18**, 7962-71 (1998).
55. T. Nakamura, K. Uramura, T. Nambu, T. Yada, K. Goto, M. Yanagisawa and T. Sakurai, Orexin-induced hyperlocomotion and stereotypy are mediated by the dopaminergic system, *Brain Res.* **873**, 181-7 (2000).

56. K. Uramura, H. Funahashi, S. Muroya, S. Shioda, M. Takigawa and T. Yada, Orexin-a activates phospholipase C- and protein kinase C-mediated Ca2+ signaling in dopamine neurons of the ventral tegmental area, *Neuroreport.* **12**, 1885-9 (2001).
57. J. J. Hagan, R. A. Leslie, S. Patel, M. L. Evans, T. A. Wattam, S. Holmes, C. D. Benham, S. G. Taylor, C. Routledge, P. Hemmati, R. P. Munton, T. E. Ashmeade, A. S. Shah, J. P. Hatcher, P. D. Hatcher, D. N. Jones, M. I. Smith, D. C. Piper, A. J. Hunter, R. A. Porter and N. Upton, Orexin A activates locus coeruleus cell firing and increases arousal in the rat, *Proc Natl Acad Sci U S A.* **96**, 10911-6 (1999).
58. T. L. Horvath, C. Peyron, S. Diano, A. Ivanov, G. Aston-Jones, T. S. Kilduff and A. N. van Den Pol, Hypocretin (orexin) activation and synaptic innervation of the locus coeruleus noradrenergic system, *J Comp Neurol.* **415**, 145-59 (1999).
59. P. Bourgin, S. Huitron-Resendiz, A. D. Spier, V. Fabre, B. Morte, J. R. Criado, J. G. Sutcliffe, S. J. Henriksen and L. de Lecea, Hypocretin-1 modulates rapid eye movement sleep through activation of locus coeruleus neurons, *J Neurosci.* **20**, 7760-5 (2000).
60. K. Rasmussen, D. B. Beitner-Johnson, J. H. Krystal, G. K. Aghajanian and E. J. Nestler, Opiate withdrawal and the rat locus coeruleus: behavioral, electrophysiological, and biochemical correlates, *J Neurosci.* **10**, 2308-17 (1990).
61. E. J. Nestler, M. Alreja and G. K. Aghajanian, Molecular control of locus coeruleus neurotransmission, *Biol Psychiatry.* **46**, 1131-9 (1999).
62. E. M. Soffin, M. L. Evans, C. H. Gill, M. H. Harries, C. D. Benham and C. H. Davies, SB-334867-A antagonises orexin mediated excitation in the locus coeruleus, *Neuropharmacology.* **42**, 127-33 (2002).
63. G. Martin, V. Fabre, G. R. Siggins and L. de Lecea, Interaction of the hypocretins with neurotransmitters in the nucleus accumbens, *Regul Pept.* **104**, 111-7 (2002).

DOPAMINE - HYPOCRETIN/OREXIN INTERACTIONS:

The Prefrontal Cortex And Schizophrenia

Ariel Y. Deutch, Jim Fadel, and Michael Bubser*

1. INTRODUCTION

The discovery of the hypocretins/orexins[1,2] has catalzyed research into sleep disorders, arousal, feeding and energy metabolism. The widespread projections of hypocretin/orexin (hereafter referred to as orexin for simplicity sake) cells almost insures that they will have correspondingly widespread functions. The excitement that accompanied the discovery of these hypothalamic peptides, and the finding that the loss of orexin cells in the lateral hypothalamus/perifornical area (LH/PFA) underlies narcolepsy, has not abated, and is increasingly being transferred to other fields of neuroscience.

Because of the functional role of the orexins in arousal and sleep considerable attention has focused on the regulation by orexin of cholinergic and noradrenergic neurons. The involvement of the orexins in feeding behavior has resulted in a focus on the elaborate interplay of peripheral and hypothalamic factors. Because it has been five years since the orexins were discovered, most transmitters have had their moment in the sun of orexin. Dopamine has remarkably diverse roles in such disorders as schizophrenia, Parkinson's's disease, attention deficit-hyperactivity disorder, and Tourette's syndrome. Because of the broad array of functions in which DA is in involved, our knowledge of how orexin impacts dopamine systems in both physiological and pathological conditions remains in its infancy. We have focused our efforts on the interactions between orexins and mesocorticolimbic dopamine system, with an eye toward exploring the possible involvement of the orexin in schizophrenia and the actions of antipsychotic drugs.

---

* Ariel Y. Deutch and Michael Bubser, Vanderbilt University Medical Center, Nashville, TN, 37212. Jim Fadel, University of South Carolina School of Medicine, Columbia, SC 29208

2. ANATOMICAL BASIS OF DOPAMINE-OREXIN INTERACTIONS

Despite the small number of orexin neurons, orexin-containing axons are distributed across virtually the entire neuraxis, from the frontal pole to the caudal spinal cord.[1,3,4] A reasonable assumption based on the discrepancy between the number of orexin neurons (estimated at ~3,000 in the rat) and the widespread distribution of orexin cells is that orexin neurons may collateralize extensively. The scattered distribution of orexin neurons in the LH/PFA presents difficulties in using anterograde tracing techniques to define distinct orexin-containing projections, but retrograde tract tracing studies do permit one to easily determine if all orexin cells or only a subset of orexin neurons innervate a particular region.

The dopaminergic neurons of the ventral mesencephalon are a major site of action of psychostimulants, such as amphetamine and cocaine. Lesions of the A10 DA neurons in the ventral tegmental area (VTA) or the dopaminergic axons of VTA neurons that innervate the nucleus accumbens block psychostimulant-induced hyperactivity.[5] The A10 DA neurons projection widely to corticolimbic forebrain sites, including the medial and suprarhinal prefrontal cortices, the nucleus accumbens, lateral septum, bed nucleus of the stria terminalis, and central and basolateral amygdala.

Because orexin-containing axons are found in the ventral mesencephalon, it follows that the orexin and dopamine systems may interact. The involvement of mesotelencephalic dopamine systems in such diverse functions as locomotor activity, cognition, and affect, and the alterations in these domains in neuropsychiatric disorders, suggests that orexin-dopamine interactions may prove to be a fruitful target for new approaches to schizophrenia, drug abuse, and other conditions.

2.1. Orexin Projections to Midbrain Dopamine Neurons

The VTA receives afferents from a large number of areas. Phillipson[6] provided what remains the most detailed analysis of afferents to the VTA, and noted that among the sources of inputs was a large projection from the lateral hypothalamus. However, subsequent studies of afferents that regulate the activity of VTA DA neurons focused extensively on forebrain sites that also received DA projections, and little more was published on hypothalamic projections to the VTA.

At the start of a study revisiting the hypothalamic projection to the ventral midbrain, we noticed that the distribution of hypothalamic cells that were retrogradely-labeled from the VTA overlapped with the distribution of orexin neurons. We soon determined that orexin neurons projected to the VTA, but very sparsely to the substantia nigra (SN), where the A9 DA nigrostriatal neurons are found.[7] Although hypothalamic cells that project to the VTA are scattered across the LH/PFA and dorsomedial nucleus of the hypothalamus, orexin cells projecting to the VTA are mainly situated in the lateral perifornical area.

The orexin innervation of the VTA is of moderate density, but represents a minority of the axons of the LH/PFA projection to the VTA. We have determined that a very small number of the LH/PFA projecting to the VTA express orphanin-FQ, but the transmitter identity of most of the hypothalamic cells that give rise to the VTA projection remains unknown. Orexin axons invest most areas of the VTA, with the most dense plexus of orexin fibers present in the caudal linear nucleus and dorsal extension of this nucleus into the dorsal raphe (the A10dc region of Hokfelt[8]).[7] The orexin axons in the

VTA have a beaded appearance, with varicosities that are frequently apposed to the dendrites of the DA neurons; occasional orexin fibers target the soma of DA neurons.

The innervation of the VTA by orexin neurons and the close association of orexin axons and A10 DA neurons suggest that orexin neurons can regulate the activity of DA neurons at the somatodendritic level. The overlap in orexin and dopamine inputs in forebrain sites raise the possibility of interactions at the terminal field level as well.

2.2. Overlap of Forebrain Hcrt/Orexin and DA Axons

Orexin fibers innervate all of the nuclei to which VTA DA neurons project, but within each nucleus there may be patterns of overlap or avoid. For example, the nucleus accumbens (NAS) receives a dense dopaminergic innervation, but in much of the NAS there is a very spare orexin innervation. However, in the mediodorsal shell and septal pole regions of the NAS[9] a low density orexin innervation is present.

Dopamine axons in the lateral septal nucleus form a distributed arc running from the ventrolateral to dorsomedial aspects; orexin axons can be seen in this region, but are usually in a complementary zone rather than intermingling with DA axons. In the central nucleus of the amygdala there is also a complementary pattern of distribution of orexin and dopamine axons; in contrast the dopamine and orexin fibers of the caudal basolateral nucleus are in register.

The general pattern across of mesolimbic DA terminal field regions, including the NAS, septum, and amygdala, is one in which orexin and dopamine neurons are both present but in usually in complementary zones instead of overlap. In contrast, in the medial prefrontal cortex (PFC), there is extensive intermingling of DA and orexin axons.

The PFC receives a moderately dense innervation from orexin neurons, with orexin axons found primarily in the deep layers of the infralimbic and prelimbic cortices, where they are intermixed with the DA axons. The density of the orexin innervation also seems to parallel the density of the DA innervation, decreasing more dorsally in the PFC in the shoulder (medial precentral) cortex. Orexin fibers are also seen in the suprarhinal cortex, admixed with the DA innervation.

It is worth noting that orexin neurons are considerably more dense in the PFC that in mesolimbic DA terminal fields, with the possible exception of the bed nucleus of the stria terminalis. The dopamine innervation of the PFC is derived from scattered cells across the parabrachial and paranigral nuclei of the VTA, but are most often encountered in the caudal linear and A10dc subnuclei.[10] These caudal VTA nuclei are the ones that receive the most dense orexin innervation. Because the PFC projects to both the LH/PFA and VTA,[11] the PFC may regulate both orexin and VTA DA neurons through a long-loop feedback. It will be interesting to determine if single PFC neurons collateralize to innervate both the orexin cells in the LH/PFA and the dopaminergic neurons of the VTA.

2.3. The dopaminergic innervation of the lateral hypothalamus and orexin

The dopamine projections from the midbrain to forebrain sites ascend in the median forebrain bundle (MFB), which runs at in the dorsolateral LH, forming a cap on the extreme dorsomedial aspect of the internal capsule. Although a catecholaminergic innervation of the LH has long been appreciated, most attention has focused on the noradrenergic innervations. However, fluorescent histochemistry studies have reported the presence of a dopaminergic innervation of the LH that is distinct from the ascending

DA tract in the median forebrain bundle.

In order to obtain a more complete picture of the dopaminergic innervation of the LH/PFA, we used immunohistochemistry with an antibody directed against dopamine. We observed a surprisingly dense dopaminergic innervation of the LH/PFA (see Fig. 1). Dopaminergic axons are distributed across the dorsal LH and extend medially to the dorsomedial nucleus of the hypothalamus; an especially dense plexus of coarse varicose DA axons is seen in the perifornical region. Relative fine caliber axons appear to enter the LH from the MFB, upon which they become very varicose, particularly in the perifornical region and a part of the LH ventrolateral to the fornix. In addition, fine caliber axons could be seen entering the perifornical area from the medial side of the fornix. This suggests that the dopaminergic innervation of the LH/PFA may derive from two sources: a midbrain source that ascends in the MFB, from which DA axons emanate to enter the lateral LH/PFA, and medially-situated hypothalamic DA neurons that may innervate the area medial to the fornix. Although the LH/PFA zone innervated by DA axons clearly overlaps with the general position of orexin cells, the antibody used to reveal DA requires that a fixative containing a high concentration of glutaraldehyde is used; this fixative is not compatible with available orexin antibodies, and thus it is not possible at the light microscopic level to determine if DA axons and orexin neurons are closely apposed. Nonetheless, the dopaminergic innervation of the LH/PFA suggests strongly that dopamine may directly regulate orexin neurons.

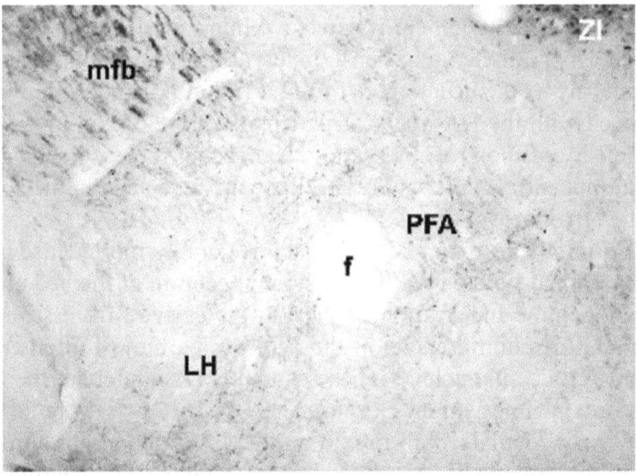

Figure 1. The distribution of dopamine-immunoreactive axons in the LH/PFA. Dopamine axons emanate from the median forebrain bundle (mfb) to invade the lateral hypothalamus and form a dense plexus around the fornix. The DA cells of the zona incerta (ZI) can be seen dorsally.

3. DOPAMINERGIC REGULATION OF OREXIN NEURONS

In view of the relatively dense dopaminergic innervation of the LH/PFA, we undertook an examination of the ability of indirect and direct DA agonists on the activity of orexin neurons, as reflected by induction of Fos, the protein product of the early-immediate gene c-*fos*. Because orexin neurons are a small population of LH/PFA cells that are scattered across the dorsal hypothalamus, reporters such as Fos are particularly useful in indicating the impact of a pharmacological treatment[12] on neurons identified by their transmitter phenotype; the recent introduction of a transgenic mouse with a GRP reporter under the prepro-orexin promotor[13,14] also offers a direct method for assessing the physiological changes in orexin cells.

3.1. Amphetamine and Apomorphine Effects on Fos expression in Orexin cells

We first examined the effects of the indirect DA agonist d-amphetamine on Fos expression in orexin neurons. Amphetamine markedly increases arousal and activity, and potently increases extracellular DA levels in mesocorticolimbic forebrain sties. Amphetamine challenge resulted in a marked increase in the percentage of double-labeled (Fos + orexin) cells, which were primarily seen medial to the fornix.[15]

Amphetamine results in a net increase in extracellular norepinephrine (NE) as well as DA, and hence one cannot know if the increase in Fos expression in orexin neurons is due to an increase in DA or NE. We therefore challenged animals with apomorphine, a mixed DA agonist at both D1 (D_1 and D_5)- and D2 (D_2, D_3, and D_4)-like receptors. Apomorphine markedly activated orexin neurons, with a particularly strong effect medial to the fornix (see Figure 2), indicating that activation of DA receptors can drive orexin neurons.

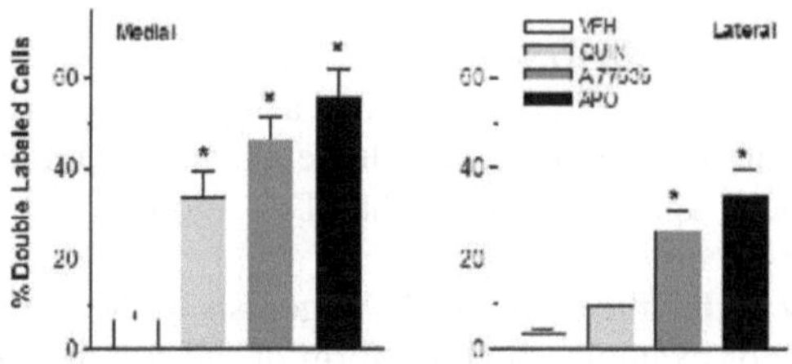

Figure 2. Dopamine agonist induced Fos expression in orexin cells in the medial and lateral LH/PFA. The D1 agonist A77636 and D2 agonist quinpirole (QUIN), as well as the mixed DA agonist apomorphine (APO), increased the percentage of orexin cells relative to vehicle (VEH) control-treated animals. Note the significantly greater effect of DA agonist in the medial sector of the LH/PFA. $^{*}p \leq .01$

3.2. Activation of Orexin Neurons by D1 and D2 Agonists

Accordingly, we then tested specific D1- and D2-like dopamine agonists. The full D1 agonist A77636 markedly increased Fos expression in both the medial and lateral LH/PFA, but had a greater effect on orexin neurons medial to a line vertically bisecting the fornix. The D2-like agonist quinpirole also activated orexin neurons, but to a lesser degree than observed in response to the D1 agonist (see Fig. 2).

The differential effect of DA agonists on orexin cells in the medial and lateral LH/PFA, and the greater effect elicited by the D1 agonist suggested that DA receptors may be distributed heterogeneously across the LH/PFA. However, we found only rare

LH/PFA cells that express D_2 mRNA, and no LH/PFA cells expressing D_1, D_3, D_4, or D_5 mRNAs. Thus despite the potent effect of DA agonists on orexin cells, particularly those in the medial LH/PFA, which is consistent with a dopaminergic innervation of the LH/PFA, DA receptors that would mediate a direct effect of the DA agonists are lacking.

3.3. Mechanisms of Dopamine Agonist-Induced Activation of Orexin Neurons

The striking effects of D1- and D2-like receptor agonists on Fos expression in orexin cells but the absence of DA receptors suggests several different possibilities that may account for the ability of DA agonists to drive Fos.

The first is that the agonists act on afferents to the orexin neurons. Among such sites could be neurons in the NAS, in which both D_1- and D_2-mRNA expressing cells are found, and where DA agonists exert potent effects. Acute pharmacological inactivation of the NAS resulted in an activation of orexin neurons in the lateral but not medial LH/PFA,[16] consistent with accumbal projections being directed to the LH and not invading the area medial to the fornix.[17] The PFC also contains D_1- and D_2-mRNA expressing cells, and an action of DA agonists on this cortical site might be expected to regulated orexin cells, but again in the lateral part of the LH/PFA. However, the fact that both the NAS and PFC primarily project to the area lateral to the fornix yet DA agonist act medial to the fornix suggests that neither accumbal nor prefrontal cortical neurons are the primary target in DA agonist-induced orexin activation. There are obviously neurons in many other sites that innervate the LH/PFA and express DA receptors, and indeed it is possible that D1-like agonists drive one afferent source and D2-like agonists a different set of afferents.

Another mechanism whereby DA agonists might activate orexin cells is by targeting a non-DA receptor. For example, dopamine has a high nanomolar affinity for the $\forall_{2C}$ receptor, which has been mapped to the LH. However, amphetamine does not bind to this receptor. Thus, it would appear unlikely that DA agonists act through non-cognate receptors to exert their effects.

A third possibility is that the ability of DA agonists to drive orexin cells is mediated by peripheral signals. For example, the long form of the receptor for leptin is expressed on orexin neurons.[18] DA agonists may trans-synaptically promote peripheral sources, such as adipocytes, to alter the release such neuroactive species as orexin, ghrelin, and adiponectin, which in turn would modulate the activity of orexin neurons.

4. ANTIPSYCHOTIC DRUGS AND OREXIN NEURONS

The ability of DA agonists to activate orexin cells, particularly those in the medial LH/PFA, suggests that drugs that block DA receptors may also modulate the activity of orexin neurons. We were particularly interested in antipsychotic drugs (APDs) because treatment of schizophrenics with many of these drugs causes marked weight gain,[19] leading to problems with patient compliance and inhibiting psychosocial integration, and because orexin potently produces feeding.[20-22] All APDs have some D_2 receptor affinity, although many of the APDs, particularly the newer generation (atypical) APDs, target many other receptors as well.[23, 24]

4.1. Effects Of Typical And Atypical Antipsychotic Drugs On Orexin Neurons.

We compared the effects of a panel of APDs, both typical and atypical, with different weight gain liabilities. Among these were several APDs with a high propensity for causing weight gain (e.g., clozapine), others that cause moderate weight gain (e.g., chlorpromazine), and some that do not elicit significant weight gain (e.g., haloperidol), as reported in a meta-analysis of weight change in schizophrenics six week after starting treatment.[19]

We found that those APDs that do not cause weight gain, including haloperidol, did not activate orexin neurons.[15] Those that do cause weight gain, such as clozapine, induced Fos in orexin neurons in both the medial and lateral orexin neurons of the LH/PFA (see Fig. 3). However, there was a direct correlation between the weight gain liability of a particular APD and the degree to which orexin neurons were activated in the lateral LH/PFA, while in the medial LH/PFA orexin cells the one-to-one correlation of weight gain liability and Fos induction was not observed.

Most of the APDs that activated orexin neurons are among the newer generation atypical APDs, including clozapine, olanzapine, and risperidone. However, the ability of APDs to activate orexin neurons is not restricted to atypical APDs: the typical APD chlorpromazine induces Fos in orexin neurons,[15] consistent with the moderate weight gain caused by this drug.[19] Conversely, the atypical APD ziprasidone, which does not appear to cause weight gain in patients,[19] did not activate orexin neurons.

4.2. How Do Both Anorexic And Orexigenic Drugs Both Active Orexin Cells?

We have seen that both amphetamine, a drug that inhibits feeding and causes weight loss with repeated administration, and certain APDs, which cause weight gain, are able to activate orexin cells. How can one resolve this paradox?

One clue was the differential distribution across the LH/PFA of orexin cells that were activated by amphetamine and clozapine. Amphetamine challenge, like apomorphine, induced Fos primarily in orexin cells medial to the fornix. Clozapine induced cells more broadly across the LH/PFA, although a significantly greater effect was observed in areas lateral to the fornix. This suggested that there are two different populations of orexin cells, and by extension raises the possibility that among the ways distinct groups of orexin cells might be differentiated is on the basis of projection target.

One concern about this hypothesis the paradox of the very small number of orexin cells that project so widely across the central nervous system. Because so many areas are innervated by orexin neurons, one possibility is that orexin neurons collateralize to innervate multiple targets. This would have evolutionary advantage over an axonal projection restricted to one or even two sites: the loss of even a very small number of target-specific orexin cells, such as by injury or as an age-related process, might result in dysfunction of the projection target.

The prefrontal cortex is strongly activated by both amphetamine and clozapine.[25-28] We therefore determined if orexin neurons that innervate the PFC also branch to innervate the dorsal raphe, in which the density of the orexin innervation is comparable to that in the PFC. We found that less than 5% of the orexic cells that were retrogradely labeled from the PFC were also retrogradely labeled from the dorsal raphe, suggesting the at least some orexin neurons project to specific targets.

We were particularly interested in the PFC because orexin neurons were activated by clozapine. Several studies have indicated that the weight gain in patients treated with clozapine is related to the therapeutic outcome.[29-31] The cognitive dysfunction in schizophrenia is thought to be related to a decrease in dopaminergic tone in the PFC,[32-34] and one action that differentiates clozapine from typical APDs, which are not very effective in targeting cognitive deficits, is the ability of the atypical drug to increase extracellular DA levels in the PFC.[35, 36]

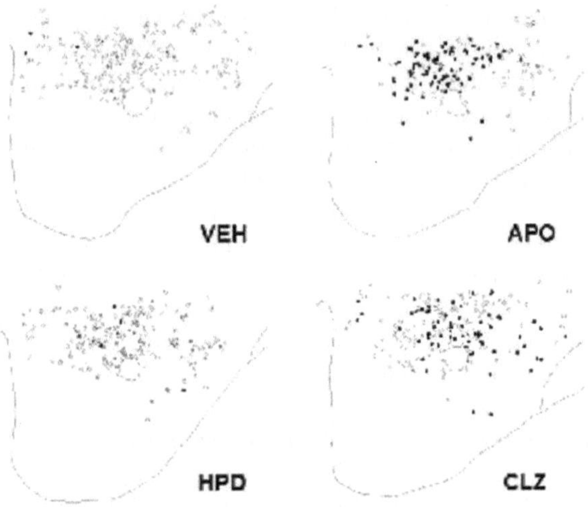

Figure 3. The distribution of orexin cells activated by apomorphine (APO), haloperidol (HPD), and clozapine (CLZ), as reflected by neurons expressing both Fos and orexin (filled circles). Single-labeled orexin cells that did not express Fos are depicted as open circles. Note that the DA agonist apomorphine strongly induced Fos in area just medial and dorsal to the fornix, whereas clozapine activated a more scattered population of orexin cells, of which the majority are in the lateral LH/PFA.

We therefore examined the effects of amphetamine and clozapine on orexin cells that project to the PFC. Clozapine increased Fos expression in the PFC-projecting orexin cells, while amphetamine challenge did not affect these hypothalamo-cortical orexin cells.[15] Moreover, the cells that innervated the PFC were concentrated lateral to the fornix, distinct from the DA agonist-driven cells.

These observations are consistent with clozapine and amphetamine regulating two distinct orexin systems: one comprised of orexin cells in the LH that project to the PFC and is activated by clozapine, and the other medially-situated group of orexin cells that project to an as yet unknown target that is activated by amphetamine. It is possible that amphetamine activates orexin cells involved in arousal, including those of medial septum-diagonal band complex, while clozapine targets orexin cells that are involved in feeding but also have a role in regulating the PFC.

5. DOPAMINE, OREXIN, AND SCHIZOPHRENIA

Several case reports have suggested that daytime hypnogogic hallucinations in narcolepsy may lead to an incorrect diagnosis of schizophrenia.[37-39] A small number of patients diagnosed with schizophrenia were subsequently shown to have narcolepsy and respond favorably to modafinil treatment.[39] On the basis of a structured interview of narcoleptic subjects, Vourdas[40] found that about 10% of the subjects had hallucinations, but suggested that the hallucinations were attributable to amphetamine treatment because they cleared when patients were switched to modafinil.

Among the current treatments for narcolepsy are psychostimulants, sodium oxybate, and modafinil;[41-44] the latter is gaining increasing favor.[43-45] Although the precise mode of action remains elusive, modafinil has been suggested to act through a primary action of orexin neurons, setting into play several downstream effectors, including histamine neurons of the tuberomammillary nucleus.[46, 47]

Most attention has focused on the wake-promoting effects of modafinil, particularly in narcolepsy. However, several studies have indicated the presence of cognitive dysfunction in narcolepsy, particularly in the realm of attention.[48-52] Moreover, recent case reports have suggested that modafinil may be useful in the treatment of clozapine-induced sedation.[53-55] Interestingly, Teitelman[55] noted that modafinil might also improve attention and reduce cognitive dysfunction. Although some investigators have suggested that in normal adult subjects modafinil does not affect cognition, but does affect mood,[56] another more extensive study concluded that modafinil enhances cognitive function, including spatial memory, delayed match-to-sample, and stop-signal reaction time, in normal adult subjects.[57] Moreover, an open trial has suggested that modafinil is an effective treatment for schizophrenia,[58] and a recent double-blind, placebo-controlled, cross-over study of modafinil concluded that modafinil improves cognitive function in schizophrenia.[59]

How might modafinil, a drug used extensively to treat narcolepsy, be beneficial in treating the cognitive deficits in schizophrenia? The dopaminergic innervation of the PFC is thought to be a major target of atypical APDs such as clozapine. Clozapine differs from typical APDs by targeting more effectively the cognitive deficits of schizophrenia.[60-63] The ability of clozapine to effectively treat some cognitive changes in schizophrenia has bee suggested to occur, at least in part, by increasing DA tone in the PFC.[24, 64-67]

A single report has suggested that systemic administration of modafinil increases extracellular DA and NE levels in the PFC of the rat.[68] More recently, we[69] and Berridge and colleagues[70] have found in pilot studies that orexin infusion into the PFC results in an increase in both extracellular dopamine and norepinephrine levels. This raises the possibility that orexin may be important in augmenting cortical catecholamine levels, and thereby improving cognitive function.

If orexin augments cortical DA levels and thereby restores to some degree dopaminergic tone in the PFC, do clozapine or other APDs act in part through promoting orexin release? As discussed previously, we found that clozapine and certain other atypical APDs activate orexin cells in the lateral part of the orexin cell cluster,[15] with many of these cells projecting to the PFC. In contrast, orexin cells innervating the PFC were not activated by amphetamine. We hypothesize that there are two separable systems involved in dopamine-orexin interactions. One involves PFC-projecting orexin cells and has a functional role in modulating dopaminergic tone and thereby enhancing cognition.

The other involves those orexin cells activated by DA agonists that function in arousal; although the target of such cells is not yet known, the medial septum-diagonal band cholinergic cells are a likely suspect.

Our data indicates that clozapine, olanzapine, and certain other APDs activate a distinct subpopulation of orexin neurons. However, a recent short communication reported that cerebrospinal fluid (CSF) levels of orexin-A are significantly lower in schizophrenic patients treated with haloperidol than unmedicated subjects, but that orexin levels do not differ between haloperidol-treated patients and another group of patients treated with either clozapine or olanzapine.[71] The authors suggested that the clozapine-induced activation of orexin neurons seen in the rat is not manifested in human schizophrenic subjects. However, the regional source(s) of orexin in the CSF is not known. We showed that clozapine specifically activated orexin cells projecting to the PFC.[15] It is not known if CSF orexin levels are primarily derived from the orexin innervation of the spinal cord, or if they correlate with orexin levels in the PFC or any other site. In other systems it has proven very difficult to define the source of CSF transmitters and their metabolites. For example, there is no correlation between CSF dopaminergic metabolites and striatal DA concentrations in probenecid-treated monkeys,[72] despite the fact striatal DA forms the largest single pool of the amine. Given the lack of information concerning the source of orexin in the CSF, and the fact that there are distinct orexin systems that project to different targets, the conclusion that atypical APDs do not activate orexin neurons in humans[71] is suspect.

6. CONCLUSIONS

The discovery of the hypocretins/orexins has allowed us to understand the etiology and proximate cause of narcolepsy, and continued advances in other neuropsychiatric disorders, including a role for the orexin in schizophrenia, seems likely. The advantages of having a small number of defined cells integrate a broad function, such as arousal, would seem intuitively attractive. However, continued data point to orexin cells forming several distinct functional systems, each with a correspondingly distinct anatomy, and each with its own set of pharmacological interactions. The elucidation of these systems and the roles they play will likely drive new means of treating several neuropsychiatric disorders.

7. ACKNOWLEDGMENTS

This work was supported in part by MH 45124 and MH 57995 (AYD), and by NARSAD Young Investigator Awards (JF and MB). We thank Tamara Geracci for assistance with figure preparation.

8. REFERENCES

1. L. de Lecea, T. S. Kilduff, C. Peyron, X. Gao, P. E. Foye, P. E. Danielson, C. Fukuhara, E. L. Battenberg, V. T. Gautvik, F. S. Bartlett, 2nd, W. N. Frankel, A. N. van den Pol, F. E. Bloom, K. M. Gautvik and J. G. Sutcliffe, The hypocretins: hypothalamus-specific peptides with neuroexcitatory activity, *Proc Natl*

Acad Sci U S A **95,** 322-327 (1998).
2. T. Sakurai, A. Amemiya, M. Ishii, I. Matsuzaki, R. M. Chemelli, H. Tanaka, S. C. Williams, J. A. Richarson, G. P. Kozlowski, S. Wilson, J. R. Arch, R. E. Buckingham, A. C. Haynes, S. A. Carr, R. S. Annan, D. E. McNulty, W. S. Liu, J. A. Terrett, N. A. Elshourbagy, D. J. Bergsma and M. Yanagisawa, Orexins and orexin receptors: a family of hypothalamic neuropeptides and G protein-coupled receptors that regulate feeding behavior, *Cell* **92,** 573-585 (1998).
3. C. Peyron, D. K. Tighe, A. N. van den Pol, L. de Lecea, H. C. Heller, J. G. Sutcliffe and T. S. Kilduff, Neurons containing hypocretin (orexin) project to multiple neuronal systems, *J Neurosci* **18,** 9996-10015 (1998).
4. T. Nambu, T. Sakurai, K. Mizukami, Y. Hosoya, M. Yanagisawa and K. Goto, Distribution of orexin neurons in the adult rat brain, *Brain Res* **827,** 243-260 (1999).
5. S. D. Iversen, in: *Behavioural effects of manipulation of basal ganglia neurotransmitters*, edited by D. Evered and M. O'Connor (Ciba Foundation, Pitman,1984), pp. 183-200.
6. O. T. Phillipson, Afferent projections to the ventral tegmental area of Tsai and interfascicular nucleus: a horseradish peroxidase study in the rat, *J Comp Neurol* **187,** 117-143 (1979).
7. J. Fadel and A. Y. Deutch, Anatomical substrates of orexin-dopamine interactions: lateral hypothalamic projections to the ventral tegmental area, *Neuroscience* **111,** 379-387 (2002).
8. T. Hokfelt, R. Martensson, A. Bjorklund, S. Kleinau and M. Goldstein, in: *Distributional maps of tyrosine hydroxylase-immunoreactive neurons in the rat brain*, edited by A. Bjorklund and T. Hokfelt (Elsevier, Amsterdam, 1984), pp. 277-379.
9. A. Y. Deutch, A. J. Bourdelais and D. S. Zahm, in: *The nucleus accumbens core and shell: accumbal compartments and their functional attributes*, edited by P. W. Kalivas and C. D. Barnes (CRC Press, Baton Rouge, LA, 1993), pp. 45-88.
10. M. Yoshida, M. Shirouzu, M. Tanaka, K. Semba and H. C. Fibiger, Dopaminergic neurons in the nucleus raphe dorsalis innervate the prefrontal cortex in the rat: a combined retrograde tracing and immunohistochemical study using anti-dopamine serum, *Brain Res* **496,** 373-376 (1989).
11. S. R. Sesack, A. Y. Deutch, R. H. Roth and B. S. Bunney, Topographical organization of the efferent projections of the medial prefrontal cortex in the rat: an anterograde tract-tracing study with Phaseolus vulgaris leucoagglutinin, *J Comp Neurol* **290,** 213-242 (1989).
12. A. Y. Deutch, in: *Sites and mechanisms of action of antipsychotic drugs as revealed by immediate-early gene expression*, edited by J. G. Csernansky (Springer, Berlin,1996), pp. 117-161.
13. A. N. van den Pol, P. K. Ghosh, R. J. Liu, Y. Li, G. K. Aghajanian and X. B. Gao, Hypocretin (orexin) enhances neuron activity and cell synchrony in developing mouse GFP-expressing locus coeruleus, *J Physiol* **541,** 169-185 (2002).
14. Y. Li, X. B. Gao, T. Sakurai and A. N. van den Pol, Hypocretin/Orexin excites hypocretin neurons via a local glutamate neuron-A potential mechanism for orchestrating the hypothalamic arousal system, *Neuron* **36,** 1169-1181 (2002).
15. J. Fadel, M. Bubser and A. Y. Deutch, Differential activation of orexin neurons by antipsychotic drugs associated with weight gain, *J Neurosci* **22,** 6742-6746 (2002).
16. B. A. Baldo, L. Gual-Bonilla, K. Sijapati, R. A. Daniel, C. F. Landry and A. E. Kelley, Activation of a subpopulation of orexin/hypocretin-containing hypothalamic neurons by GABAA receptor-mediated inhibition of the nucleus accumbens shell, but not by exposure to a novel environment, *Eur J Neurosci* **19,** 376-386 (2004).
17. D. S. Zahm, S. L. Jensen, E. S. Williams and J. R. Martin, 3rd, Direct comparison of projections from the central amygdaloid region and nucleus accumbens shell, *Eur J Neurosci* **11,** 1119-1126 (1999).
18. M. Hakansson, L. de Lecea, J. G. Sutcliffe, M. Yanagisawa and B. Meister, Leptin receptor- and STAT3-immunoreactivities in hypocretin/orexin neurones of the lateral hypothalamus, *J Neuroendocrinol* **11,** 653-663 (1999).
19. D. B. Allison, J. L. Mentore, M. Heo, L. P. Chandler, J. C. Cappelleri, M. C. Infante and P. J. Weiden, Antipsychotic-induced weight gain: a comprehensive research synthesis, *Am J Psychiatry* **156,** 1686-1696 (1999).
20. J. T. Willie, R. M. Chemelli, C. M. Sinton and M. Yanagisawa, To eat or to sleep? Orexin in the regulation of feeding and wakefulness, *Annu Rev Neurosci* **24,** 429-458 (2001).
21. D. Smart and J. Jerman, The physiology and pharmacology of the orexins, *Pharmacol Ther* **94,** 51-61 (2002).
22. Y. Date, Y. Ueta, H. Yamashita, H. Yamaguchi, S. Matsukura, K. Kangawa, T. Sakurai, M. Yanagisawa and M. Nakazato, Orexins, orexigenic hypothalamic peptides, interact with autonomic, neuroendocrine and neuroregulatory systems, *Proc Natl Acad Sci U S A* **96,** 748-753 (1999).
23. J. Arnt, Pharmacological differentiation of classical and novel antipsychotics, *Int Clin Psychopharmacol* **13** Suppl 3, S7-14 (1998).

24. A. Y. Deutch, B. Moghaddam, R. B. Innis, J. H. Krystal, G. K. Aghajanian, B. S. Bunney and D. S. Charney, Mechanisms of action of atypical antipsychotic drugs. Implications for novel therapeutic strategies for schizophrenia, *Schizophr Res* **4,** 121-156 (1991).
25. J. Q. Wang, A. J. Smith and J. F. McGinty, A single injection of amphetamine or methamphetamine induces dynamic alterations in c-fos, zif/268 and preprodynorphin messenger RNA expression in rat forebrain, *Neuroscience* **68,** 83-95 (1995).
26. B. Johansson, K. Lindstrom and B. B. Fredholm, Differences in the regional and cellular localization of c-fos messenger RNA induced by amphetamine, cocaine and caffeine in the rat, *Neuroscience* **59,** 837-849 (1994).
27. G. S. Robertson and H. C. Fibiger, Neuroleptics increase c-fos expression in the forebrain: contrasting effects of haloperidol and clozapine, *Neuroscience* **46,** 315-328 (1992).
28. A. Y. Deutch and R. S. Duman, The effects of antipsychotic drugs on Fos protein expression in the prefrontal cortex: cellular localization and pharmacological characterization, *Neuroscience* **70,** 377-389 (1996).
29. J. S. Lamberti, T. Bellnier and S. B. Schwarzkopf, Weight gain among schizophrenic patients treated with clozapine, *Am J Psychiatry* **149,** 689-690 (1992).
30. R. Leadbetter, M. Shutty, D. Pavalonis, V. Vieweg, P. Higgins and M. Downs, Clozapine-induced weight gain: prevalence and clinical relevance, *Am J Psychiatry* **149,** 68-72 (1992).
31. H. Y. Meltzer, E. Perry and K. Jayathilake, Clozapine-induced weight gain predicts improvement in psychopathology, *Schizophr Res* **59,** 19-27 (2003).
32. A. Abi-Dargham and H. Moore, Prefrontal DA transmission at D1 receptors and the pathology of schizophrenia, *Neuroscientist* **9,** 404-416 (2003).
33. T. S. Braver, D. M. Barch and J. D. Cohen, Cognition and control in schizophrenia: a computational model of dopamine and prefrontal function, *Biol Psychiatry* **46,** 312-328 (1999).
34. J. D. Jentsch, R. H. Roth and J. R. Taylor, Role for dopamine in the behavioral functions of the prefrontal corticostriatal system: implications for mental disorders and psychotropic drug action, *Prog Brain Res* **126,** 433-453 (2000).
35. K. D. Youngren, B. Moghaddam, B. S. Bunney and R. H. Roth, Preferential activation of dopamine overflow in prefrontal cortex produced by chronic clozapine treatment, *Neurosci Lett* **165,** 41-44 (1994).
36. B. Moghaddam and B. S. Bunney, Acute effects of typical and atypical antipsychotic drugs on the release of dopamine from prefrontal cortex, nucleus accumbens, and striatum of the rat: an in vivo microdialysis study, *J Neurochem* **54,** 1755-1760 (1990).
37. S. K. Bhat and R. Galang, Narcolepsy presenting as schizophrenia, *Am J Psychiatry* **159,** 1245 (2002).
38. A. B. Douglass, P. Hays, F. Pazderka and J. M. Russell, Florid refractory schizophrenias that turn out to be treatable variants of HLA-associated narcolepsy, *J Nerv Ment Dis* **179,** 12-17; discussion 18 (1991).
39. A. B. Douglass, J. E. Shipley, R. F. Haines, R. C. Scholten, E. Dudley and A. Tapp, Schizophrenia, narcolepsy, and HLA-DR15, DQ6, *Biol Psychiatry* **34,** 773-780 (1993).
40. A. Vourdas, J. M. Shneerson, C. A. Gregory, I. E. Smith, M. A. King, E. Morrish and P. J. McKenna, Narcolepsy and psychopathology: is there an association?, *Sleep Med* **3,** 353-360 (2002).
41. S. S. Chakravorty and D. B. Rye, Narcolepsy in the older adult: epidemiology, diagnosis and management, *Drugs Aging* **20,** 361-376 (2003).
42. G. J. Lammers and S. Overeem, Pharmacological management of narcolepsy, *Expert Opin Pharmacother* **4,** 1739-46 (2003).
43. M. M. Mitler and R. Hayduk, Benefits and risks of pharmacotherapy for narcolepsy, *Drug Saf* **25,** 791-809 (2002).
44. T. E. Scammell, The neurobiology, diagnosis, and treatment of narcolepsy, *Ann Neurol* **53,** 154-166 (2003).
45. S. Overeem, T. E. Scammell and G. J. Lammers, Hypocretin/orexin and sleep: implications for the pathophysiology and diagnosis of narcolepsy, *Curr Opin Neurol* **15,** 739-745 (2002).
46. Z. L. Huang, W. M. Qu, W. D. Li, T. Mochizuki, N. Eguchi, T. Watanabe, Y. Urade and O. Hayaishi, Arousal effect of orexin A depends on activation of the histaminergic system, *Proc Natl Acad Sci U S A* **98,** 9965-9970 (2001).
47. T. E. Scammell, I. V. Estabrooke, M. T. McCarthy, R. M. Chemelli, M. Yanagisawa, M. S. Miller and C. B. Saper, Hypothalamic arousal regions are activated during modafinil-induced wakefulness, *J Neurosci* **20,** 8620-8628 (2000).
48. S. Fulda and H. Schulz, Cognitive dysfunction in sleep disorders, *Sleep Med Rev* **5,** 423-445 (2001).
49. A. Naumann, J. Bierbrauer, H. Przuntek and I. Daum, Attentive and preattentive processing in narcolepsy as revealed by event-related potentials (ERPs), *Neuroreport* **12,** 2807-2811 (2001).
50. A. Naumann and I. Daum, Narcolepsy: pathophysiology and neuropsychological changes, *Behav Neurol* **14,** 89-98 (2003).
51. M. Rieger, G. Mayer and S. Gauggel, Attention deficits in patients with narcolepsy, *Sleep* **26,** 36-43 (2003).

52. H. Schulz and J. Wilde-Frenz, Symposium: Cognitive processes and sleep disturbances: The disturbance of cognitive processes in narcolepsy, *J Sleep Res* **4,** 10-14 (1995).
53. J. R. DeQuardo, Modafinil and antipsychotic-induced sedation, *J Clin Psychiatry* **65,** 278-9; author reply 279 (2004).
54. E. H. Makela, K. Miller and W. D. Cutlip, 2nd, Three case reports of modafinil use in treating sedation induced by antipsychotic medications, *J Clin Psychiatry* **64,** 485-486 (2003).
55. E. Teitelman, Modafinil for narcolepsy, *Am J Psychiatry* **158,** 970-971 (2001).
56. D. C. Randall, J. M. Shneerson, K. K. Plaha and S. E. File, Modafinil affects mood, but not cognitive function, in healthy young volunteers, *Hum Psychopharmacol* **18,** 163-173 (2003).
57. D. C. Turner, T. W. Robbins, L. Clark, A. R. Aron, J. Dowson and B. J. Sahakian, Cognitive enhancing effects of modafinil in healthy volunteers, *Psychopharmacology (Berl)* **165,** 260-269 (2003).
58. M. H. Rosenthal and S. L. Bryant, Benefits of adjunct modafinil in an open-label, pilot study in patients with schizophrenia, *Clin Neuropharmacol* **27,** 38-43 (2004).
59. D. C. Turner, L. Clark, E. Pomarol-Clotet, P. McKenna, T. W. Robbins and B. J. Sahakian, Modafinil improves cognition and attentional set shifting in patients with chronic schizophrenia, *Neuropsychopharmacology* **29,** 1363-1373 (2004).
60. R. M. Bilder, R. S. Goldman, J. Volavka, P. Czobor, M. Hoptman, B. Sheitman, J. P. Lindenmayer, L. Citrome, J. McEvoy, M. Kunz, M. Chakos, T. B. Cooper, T. L. Horowitz and J. A. Lieberman, Neurocognitive effects of clozapine, olanzapine, risperidone, and haloperidol in patients with chronic schizophrenia or schizoaffective disorder, *Am J Psychiatry* **159,** 1018-1028 (2002).
61. M. A. Lee, K. Jayathilake and H. Y. Meltzer, A comparison of the effect of clozapine with typical neuroleptics on cognitive function in neuroleptic-responsive schizophrenia, *Schizophr Res* **37,** 1-11 (1999).
62. H. Y. Meltzer and S. R. McGurk, The effects of clozapine, risperidone, and olanzapine on cognitive function in schizophrenia, *Schizophr Bull* **25,** 233-255 (1999).
63. C. A. Tamminga, Similarities and differences among antipsychotics, *J Clin Psychiatry* **64** Suppl 17, 7-10 (2003).
64. K. L. Davis, R. S. Kahn, G. Ko and M. Davidson, Dopamine in schizophrenia: a review and reconceptualization, *Am J Psychiatry* **148,** 1474-1486 (1991).
65. P. S. Goldman-Rakic, S. A. Castner, T. H. Svensson, L. J. Siever and G. V. Williams, Targeting the dopamine D(1) receptor in schizophrenia: insights for cognitive dysfunction, *Psychopharmacology (Berl)* **174,** 3-16 (2004).
66. D. A. Lewis and J. A. Lieberman, Catching up on schizophrenia: natural history and neurobiology, *Neuron* **28,** 325-334 (2000).
67. C. A. Tamminga and A. Carlsson, Partial dopamine agonists and dopaminergic stabilizers, in the treatment of psychosis, *Curr Drug Targets CNS Neurol Disord* **1,** 141-147 (2002).
68. Z. de Saint Hilaire, M. Orosco, C. Rouch, G. Blanc and S. Nicolaidis, Variations in extracellular monoamines in the prefrontal cortex and medial hypothalamus after modafinil administration: a microdialysis study in rats, *Neuroreport* **12,** 3533-353 (2001).
69. J. Fadel and A. Y. Deutch, Orexin regulation of catecholamine release in prefrontal cortex, *Soc Neurosci Abstr* 359.1 (2002).
70. N. M. Vittoz and C. W. Berridge, Hypocretin increases dopamine efflux in rat prefrontal cortex and not nucleus accumbens, *Soc Neurosci Abstr* 931.14 (2003).
71. M. A. Dalal, A. Schuld and T. Pollmacher, Lower CSF orexin A (hypocretin-1) levels in patients with schizophrenia treated with haloperidol compared to unmedicated subjects, *Mol Psychiatry* **8,** 836-837 (2003).
72. J. D. Elsworth, D. J. Leahy, R. H. Roth and D. E. Redmond, Jr., Homovanillic acid concentrations in brain, CSF and plasma as indicators of central dopamine function in primates, *J Neural Transm* **68,** 51-62 (1987).

HYPOCRETIN/OREXIN IN STRESS AND AROUSAL

Craig W. Berridge and Rodrigo A. España*

1. INTRODUCTION

Since the initial discovery of the hypocretin (HCRT)/orexin peptide system, substantial information has been gained regarding the behavioral and physiological actions of this putative neurotransmitter system. The presence of HCRT within a variety of arousal-related systems and the demonstration of dysregulation of HCRT neurotransmission in narcolepsy prompted intense interest in the wake-promoting actions of this peptide system. Beyond the regulation of waking *per se*, emerging evidence suggests prominent actions of HCRT under high-arousal conditions, including stress. This chapter will review the arousal-enhancing (wake-promoting) actions of HCRT and recent evidence suggesting a potential role of HCRT in stress and other high arousal states. Included in this, will be a review of evidence suggesting an interaction between HCRT and corticotrophin-releasing factor (CRF) in the modulation of certain physiological and behavioral processes. Prior to discussion of this topic, it is useful to review briefly the concept of stress and the neurotransmitter systems implicated in the regulation of physiology and behavior in stress.

2. THE NEUROBIOLOGY OF STRESS

The term stress typically refers to a behavioral state elicited by challenging or threatening events. This concept arises from nearly a century of research starting with the seminal work of Cannon[19] and Selye[86]. In these early studies, various physiological systems were similarly affected by disparate environmental events, which had in common a potential to disrupt homeostasis or threaten animal well-being. Initially, emphasis was placed primarily on stressor-induced activation of peripheral endocrine systems. This

* Craig W. Berridge, Psychology Department, University of Wisconsin, Madison, WI 53706. Rodrigo A. España, Department of Neurology, Beth Israel Deaconess Medical Center, Harvard Medical School, Boston, MA, 02115.

work identified the activation of both peripheral catecholamine systems and the pituitary-adrenal axis as hallmark features of the state of stress. The activation of these systems results in enhanced ability of the animal to physically contend with a challenging situation.

More recently, investigation has focused on the central neural systems involved in the affective, cognitive, and behavioral components of stress. This raises the longstanding issue regarding the psychologically defining features of stress. In contrast to the well-delineated physiological indices of stress, the affective and cognitive features of stress remain less clear. The terms stress and anxiety are frequently used interchangeably. However, the precise definition of these terms and the relationship between stress and anxiety are poorly understood. Typically, anxiety is viewed as having both anticipatory and affective components. In contrast, as defined above, stress is most commonly viewed as a response to a present challenge. Moreover, the extent to which stress has an affective component, which can or cannot be dissociated from anxiety is unclear. Regardless of the exact configuration of cognitive and affective responses associated with stress, it appears a heightened level of readiness for action is paramount to a state of stress. A prominent component of this preparatory state is an elevated level of arousal defined, for the purposes of this review, by a heightened awareness of, and sensitivity to, environmental stimuli. In fact, it can be argued that a defining feature of the state of stress is sustained high levels of arousal and the associated sympathetic activation. In addition to alterations in arousal level, stress is also associated with alterations in a variety of state-dependent processes, including attention, memory, and sensory information processing.[4,6,10]

Candidate neural systems that participate in stress-related alterations in arousal state and state-dependent behavioral processes include catecholamine neurotransmitter systems. For example, it has long-been known that stress is associated with a robust activation of cerebral noradrenergic and dopaminergic neurons resulting in increased release of norepinephrine (NE) and dopamine (DA) in a variety of terminal fields, particularly those terminal fields associated with higher cognitive and affective processes (e.g. prefrontal cortex, amygdala).[14,27,50,89,95] Relatively recent evidence suggests these neurotransmitter systems modulate behavioral and forebrain neuronal activity states as well as state-dependent cognitive (e.g. working memory, attention) and physiological processes.[5,10,16,79] Combined, these observations suggest a prominent role of central catecholaminergic systems in both adaptive and possibly maladaptive stressor-induced alterations in cognition and affect.

The peptide neurotransmitter, CRF, plays a pivotal role in stress-related activation of the HPA axis, involving the release of CRF from neurons located within the paraventricular nucleus of the hypothalamus (PVN) into the median eminence. Subsequent to the characterization of this peptide in 1981,[97] it was observed that CRF-containing neurons and fibers are found outside the hypothalamus, including within neocortex, limbic structures and autonomic nuclei.[70,85] Additional work demonstrated a variety of stress-like behavioral and physiological actions of CRF when injected centrally into animals. Among these stress-like actions of CRF are the activation of dopaminergic[28] and noradrenergic systems,[28,98] arousal-enhancing actions,[20,21] and elevation of behavioral indices of stress and/or anxiety.[29,54] These observations suggested a pivotal role of CRF in the coordination of a variety of behavioral and physiological responses in stress.[29] Consistent with this hypothesis, blockade of CRF neurotransmission via central administration of CRF antagonists reverses a variety of behavioral and physiological effects of stress.[29]

Combined, these observations suggest critical roles of catecholaminergic and CRF systems in regulating the constellation of behavioral and physiological processes associated with stress. For the purposes of this review, it is of interest that HCRT efferents target dopaminergic, noradrenergic and CRF systems, suggesting a possible role of HCRT in stress. Consistent with this hypothesis, recent evidence begins to suggest a role of HCRT in behavioral and physiological responding in stress and arousal.

3. HCRT AND WAKING

The hypocretins, also known as orexins, consist of two peptides (HCRT-1 and HCRT-2) that are synthesized solely in the lateral hypothalamus (LH) and adjacent regions. Despite their restricted origin, HCRT neurons extend a vast projection system that innervates virtually the entire neuroaxis. The widespread nature of the HCRT efferent projection system suggests multiple and varied actions of this neurotransmitter system in behavior. A comprehensive characterization of the actions of HCRT on cognition, affect/emotion and behavior remains to be elucidated. Nonetheless, a surprising amount of information exists regarding the behavioral actions of HCRT, given the short span of time since discovery of this peptide system.

Originally identified to enhance feeding[26,81,92] extensive evidence now suggests the HCRT system modulates arousal and arousal-related processes. For example, HCRT-containing fibers and receptors are located within a variety of brainstem and basal forebrain structures associated with the regulation of behavioral state.[25,65,72,77,81,93] Consistent with this, multiple observations indicate modulatory actions of the HCRT system on sleep and waking. For example, both clinical and animal studies demonstrate a dysregulation of HCRT neurotransmission is associated with the sleep/arousal disorder, narcolepsy.[22,6174,94] Moreover, when administered intracerebroventricularly (ICV), HCRT-1 and HCRT-2 increase time spent awake and decrease time spent in slow-wave and REM sleep (Figure 1).[11,35,36,42,48,78] HCRT-induced increases in time spent awake are associated with a variety of waking-related behaviors including eating, drinking, grooming, and locomotor activity.

A limited number of studies have begun to examine the neural circuitry associated with HCRT modulation of sleep-wake state. The locus coeruleus (LC) noradrenergic system exerts robust modulatory actions on behavioral state and state-dependent processes.[3,7,12,13,16,39,45,76] Thus, it was of interest that the LC receives a prominent HCRT innervation.[25,77] Moreover, HCRT application increases LC neuronal discharge *in vitro* and *in vivo*.[17,42,47] Consistent with these observations, infusion of HCRT into the LC region of the brainstem increases time spent awake.[17]

These observations raise the question of whether the LC is the primary site through which HCRT acts to impact behavioral state. To initially assess this issue, the wake-promoting effects of HCRT infused into the lateral ventricles were compared to those observed following HCRT infusions into the fourth ventricle, immediately adjacent to the LC. It was reasoned that if the LC were the primary site of action in HCRT-induced waking, more rapid and/or larger magnitude wake-promoting effects would be observed with fourth ventricular infusions. In contrast to this hypothesis, HCRT-1 infusion (0.07 nmol) into the fourth ventricle elicited less robust increases in time spent awake that

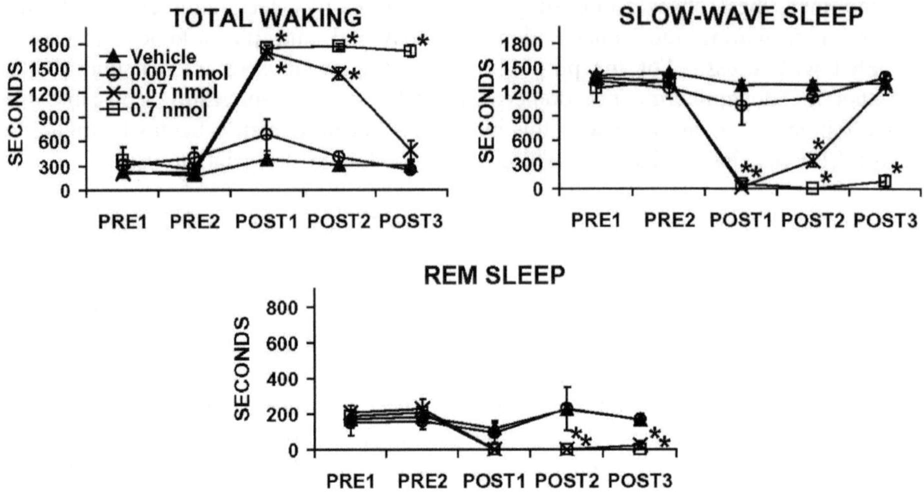

Figure 1. Effects of varying concentrations of HCRT-1 infused into the lateral ventricle on time spent awake (Total Waking) and in slow-wave sleep and REM sleep. Symbols represent mean (± SEM) time (sec) spent in the three different behavioral state categories per 30-min epoch. PRE1 and PRE2 represent pre-infusion epochs and POST1-POST3 represent post-infusion epochs. Vehicle-treated animals spent the majority of the testing period asleep. At 0.07 and 0.7 nmol, HCRT-1 significantly increased total waking and decreased slow-wave and REM sleep during POST-1 and POST-2. The 0.7-nmol dose significantly increased total waking and decreased slow-wave and REM sleep during POST-3. *P<0.01 significantly different from vehicle-treated animals. Modified from[35].

occurred with a longer latency than that observed following lateral ventricular infusions (Figure 2). These observations suggest that the LC is not the sole site wherein HCRT acts to elicit waking and that one or more sites anterior to LC participate in the behavioral state-modulatory actions of HCRT.

Various basal forebrain regions have been implicated in the regulation of behavioral state, including the general region of the medial septal area (MS), the general region of the medial preoptic area (MPOA), and the substantia innominata (SI).[13,15,18,39,45,57,64,71] Importantly, HCRT fibers and receptors are located within each of these structures.[8,24,25,41,44,65,72,77,81,91,93] Microinfusion of HCRT-1 (0.01 - 0.07 nmol) into MS, MPOA or SI elicited dose-dependent increases in time spent awake and decreases in slow-wave and REM sleep.[35] The largest increases in waking were observed following infusions within MPOA (see Figure 3). Infusions placed immediately outside these regions did not modulate sleep-wake state.

Two observations made in these studies are worth emphasizing. First, the latency to waking following ICV infusions of HCRT was extremely short (e.g. within 30 seconds of the start of the infusion). In contrast, for fourth ventricular and all intra-tissue infusions latency to waking was in the range of 3-8 minutes.[35] Second, a largely comparable dose range for HCRT-1-induced waking was observed with both ICV and intratissue infusion. Combined, these observations suggest that HCRT acts within multiple terminal fields to modulate behavioral state and/or state-dependent processes. Moreover, the extremely short latency to waking following lateral ventricular infusions of HCRT suggests that

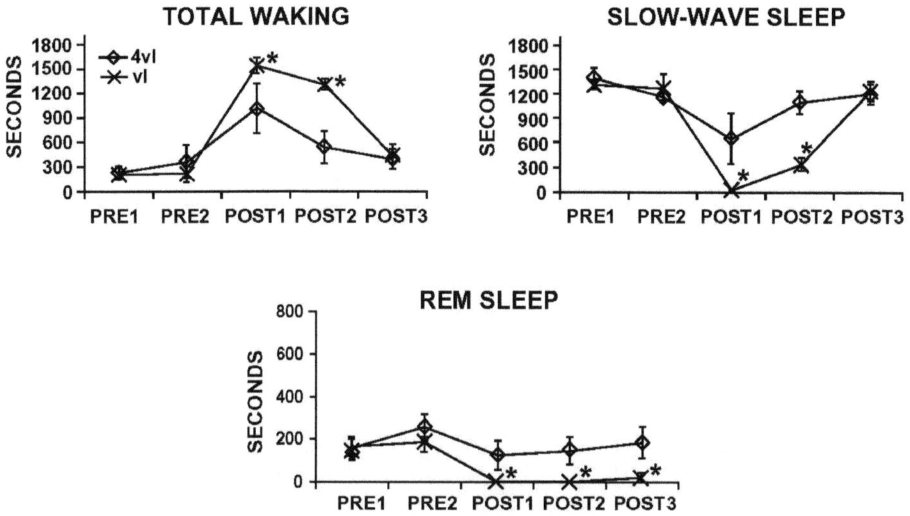

Figure 2. Effects of 0.07 nmol HCRT-1 infusions into the lateral and fourth ventricles on time spent awake (Total Waking), slow-wave sleep and REM sleep. Symbols represent mean (± SEM) time (sec) spent in the three different behavioral state categories per 30-min. PRE1 and PRE2 represent pre-infusion epochs and POST1-POST3 represent post-infusion epochs. HCRT-1 administered into the fourth ventricle increased total waking and decreased slow-wave and REM sleep. However, these effects were of a smaller magnitude than that observed following infusion into the lateral ventricles. Latency to waking following fourth ventricular infusion of HCRT (mean = 320 ± 40 sec from start of infusion; Range = 216-416 sec) was substantially longer than that observed following infusions into the lateral ventricles (mean = 191 ± 48 sec from the start of the infusion; Range = 123-375 sec; see results). *P<0.01 significantly different from fourth ventricle infusion values. Modified from.[35]

HCRT may act at a site within or extremely close to the ventricular wall. It has been posited previously select peptides may act within circumventricular sites (e.g. organum vasculosum) and that the ventricular system provides a mode of peptide transport to these sites.[33,46,60] Consistent with this idea, in immunohistochemical studies conducted in our laboratory, we observed occasional HCRT-ir fibers adjacent to, and in some cases, projecting directly into the lateral and third ventricles.

4. CIRCADIAN-INDEPENDENT ACTIONS OF HCRT

The wake-promoting actions of HCRT suggest a potential role of HCRT in circadian-dependent alterations in sleep and waking. In studies that have examined rates of HCRT release as assessed by *in vivo* microdialysis, HCRT levels increase slowly over the active period and decreases during the sleep period.[100] Similarly, when measured within cisternal cerebrospinal fluid, highest levels of HCRT-1 are observed during the latter third of the active period and lowest levels at the end of the sleep period.[102]

Although the temporal resolution of these measures is limited, these observations suggest that HCRT release is not initiating waking *per se*, but instead may be important in

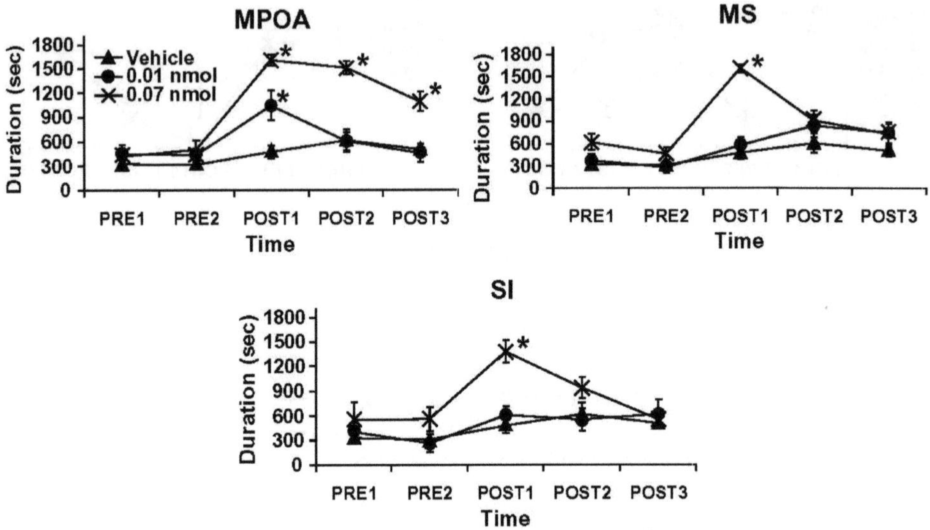

Figure 3. Effects of bilateral HCRT-1 infusions into MPOA, MS and SI on time spent awake (Total Waking). Symbols represent mean (± SEM) time (sec) spent awake per 30-min epoch. PRE1 and PRE2 represent pre-infusion epochs and POST1-POST3 represent post-infusion epochs. For MPOA, total time spent awake was significantly increased during POST1 in both the 0.07 nmol and 0.01 nmol groups and during POST2 and POST3 for the 0.07 nmol group. Less robust increases in waking were observed following infusions of HCRT within MS and SI. *P<0.01 significantly different from vehicle-treated animals. Modified from[35].

maintaining sustained waking, particularly toward the end of the activity portion of the circadian cycle (diurnal for monkeys, nocturnal for rats).

These observations are consistent with studies that have assessed neuronal/genomic activity within HCRT neurons using immunohistochemical measures of the protein product (Fos) of the immediate-early gene, *c-fos*, across various behavioral states. In these studies, waking per se was not associated with an increase in Fos immunoreactivity (Fos-ir)[37] within HCRT-synthesizing neurons (prepro-HCRT-ir neurons). Thus, as shown in Figure 4, low levels of Fos-ir are observed in HCRT-synthesizing neurons in rats that are spontaneously awake diurnally or nocturnally. Similar results were observed for HCRT-receptor-expressing (HCRT-r1) neurons within basal forebrain regions (MS, MPOA, SI) and LC (Figure 4).[37] Interestingly, correlation analyses indicated that the relatively small amount of Fos-ir observed in HCRT-synthesizing ne.[37]

Additional studies have examined the degree to which the behavioral actions of HCRT are circadian-dependent. In these studies, both circadian-dependent and circadian-independent behavioral actions of HCRT-1 have been observed. Thus, ICV HCRT-1 administration elicits comparable increases in time spent awake when administered during the sleep- and activity-periods.[35-37,42] During both nocturnal and diurnal periods, HCRT-1-induced waking was accompanied by significant increases in time spent grooming, chewing of inedible material (e.g., bedding) and locomotor activity as measured by quadrant entries and rears. In contrast to these largely circadian-independent actions of HCRT-1, when administered during the activity-period, HCRT-1

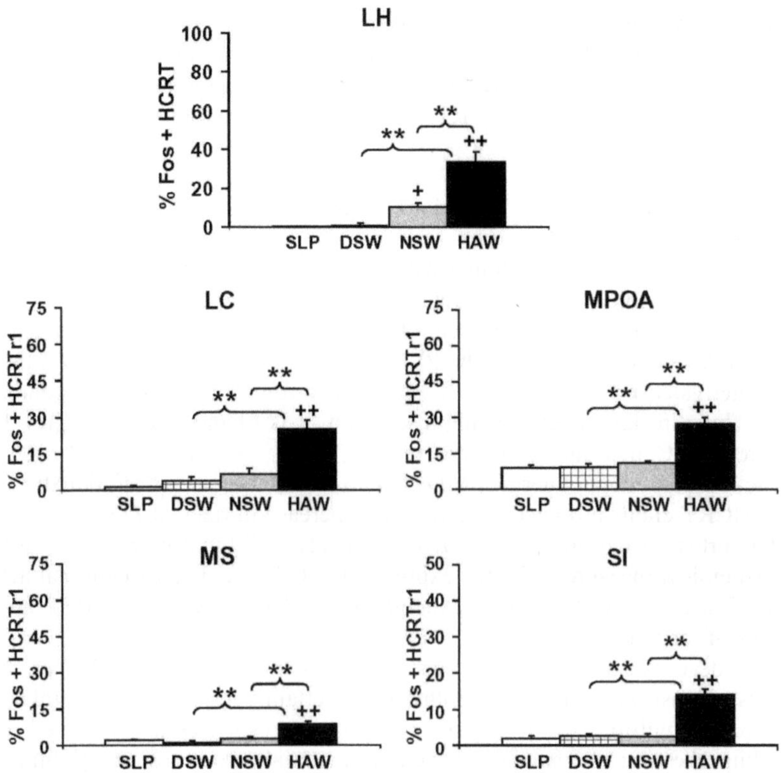

Figure 4. Effects of varying behavioral state/environmental condition on the percentage of Fos-ir nuclei within hypocretin-synthesizing (prepro-HCRT-ir) and hypocretin-1 receptor-expressing (HCRTr1-ir) neurons. Shown are diurnal sleeping (SLP), diurnal spontaneous waking (DSW), nocturnal spontaneous waking (NSW) and high-arousal waking (HAW). Neither diurnal sleeping nor diurnal spontaneous waking was associated with an increase in the percentage of Fos-ir within prepro-HCRT-ir neurons in LH. Relative to diurnal sleeping, nocturnal spontaneous waking was associated with a slight, yet significant, increase in the percentage of Fos-ir within prepro-HCRT-ir neurons. In contrast, high-arousal waking was associated with a significantly higher percentage of Fos-ir within prepro-HCRT-ir relative to diurnal sleeping, diurnal spontaneous waking and nocturnal spontaneous waking. Within HCRTr1-ir neurons across varying regions, only high-arousal waking was associated with increased levels of Fos. $^{+}P < 0.05$; $^{++}P < 0.01$ significantly different from diurnal sleeping. $^{**}P < 0.01$ significantly different from group indicated by brackets. Modified from[37].

had only minimal effects on eating and drinking.[37] These observations demonstrate a robust circadian dependence of HCRT-1-induced alterations in feeding and drinking but not chewing or locomotor activity.

5. HCRT AND STRESS

As reviewed above, HCRT increases arousal level and arousal-related behaviors in a circadian-independent manner. These actions are of particular relevance for the current discussion given the prominent role of arousal in stress. Combined, these observations suggest the hypothesis that HCRT may participate in the induction of a high-arousal, and possibly stress-like, behavioral state. In support of this hypothesis are a variety of observations indicating actions of HCRT on stress-related circuits and stress-related behavior, as reviewed below.

5.1. Stress-Like Physiological and Behavioral Actions of HCRT.

HCRT fibers and receptors are found within a variety of brain structures implicated in behavioral and physiological responding in stress. These include, the noradrenergic nucleus, LC; dopaminergic nuclei (e.g. ventral tegmental area, VTA); the bed nucleus of the stria terminalis, the amygdala, and the PVN.[25,37,38,65,77,81] As reviewed above, the LC-NE system is activated in stress and plays a prominent role in behavioral responding in stress.[3,10,73,90] Thus, in the context of this discussion, it is of interest that HCRT-1 and HCRT-2 increase LC neuronal discharge rates.[17,42,47,51,53] Similarly, HCRT exerts an excitatory action on dopaminergic VTA neurons in vitro.[56] Additionally, ICV administered HCRT elicits a robust and stress-like increase in rates of DA release within the prefrontal cortex, as measured by *in vivo* microdialysis.[99] Finally, in studies utilizing immunohistochemical measures of c-fos expression, HCRT modulates neuronal activity in a variety of stress-related structures, including the PVN, LC, VTA, BST and the central nucleus of the amygdala.[1,25,37,58,80]

The above-described observations suggest the possible action of HCRT on stress-related circuits. Consistent with this hypothesis are certain stress-like behavioral actions of HCRT. Across multiple species, various behavioral responses are observed during stress, including chewing or gnawing of inedible material, grooming and fighting, and motor activity (e.g. displacement behaviors).[2,10,32,43,55,59,68,96] These behaviors act to attenuate stressor-induced activation of a variety of central and peripheral physiological systems. Interestingly, ICV and basal forebrain administration of HCRT elicits a majority of these behaviors, including grooming,[48,49] chewing of inedible material,[37] and locomotor activity.[35-37,67]

In addition to the above-described actions of HCRT on neuronal activity and behavior, HCRT also promotes a variety of autonomic processes associated with stress and/or high levels of arousal. These include elevation of mean arterial blood pressure, heart rate, oxygen consumption, and body temperature.[23,62,83,88,101] Combined, these observations begin to suggest that HCRT simultaneously activates a variety of neural circuits that coordinate physiological and behavioral responding in stress and possibly other high-arousal states.

5.2. Effects of Stress On *c-Fos* Expression in HCRT-Synthesizing and HCRT-Receptive Neurons.

The above-described observations suggest the hypothesis that HCRT systems may be activated under high-arousal, stress-like conditions. To better assess this hypothesis, and as part of studies examining the circadian dependency of rates of HCRT neurotransmission described above, we examined Fos-ir levels in animals exposed to a

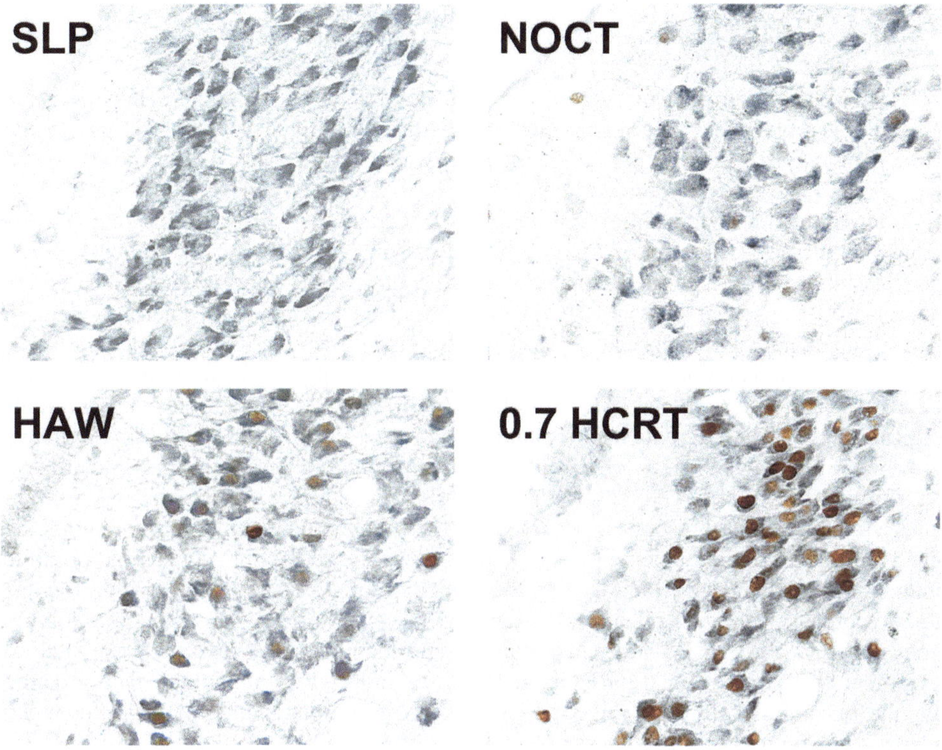

Figure 5 Photomicrographs depicting the effects of varying behavioral state/environmental condition on Fos-ir (brown) within prepro-HCRT-ir neurons (gray-blue) and non-prepro-HCRT-ir neurons within LH. Shown are diurnal sleeping (SLP), nocturnal spontaneous waking (NSW) and high-arousal waking (HAW). Neither diurnal sleeping nor diurnal spontaneous waking was associated with increased Fos-ir within LH. Relative to diurnal sleeping, nocturnal spontaneous waking was associated with a slight increase in Fos-ir within prepro-HCRT-ir neurons and non-prepro-HCRT-ir neurons. In contrast, high-arousal waking was associated with a substantial increase in Fos-ir nuclei within prepro-HCRT-ir neurons and non-prepro-HCRT-ir neurons relative to diurnal sleeping, diurnal spontaneous waking and nocturnal spontaneous waking. Modified from. [37]

stress-inducing, brightly lit novel environment.[14,43] It was observed that novelty-stress increased Fos-ir levels within HCRT-synthesizing neurons as well as HCRTr1-expressing neurons located within LC, MPOA, MS, and SI (Figure 4 and 5).[37] Interestingly, the magnitude of the novelty-stress-induced Fos-response within HCRT-receptive neurons was similar to that observed following ICV HCRT-administration. This latter observation further suggests the possibility that HCRT administration elicits at least a subset of stress-like actions within the brain. Similar activating effects on Fos expression within HCRT-synthesizing neurons have been observed following cold exposure, food deprivation, foot-shock, and immobilization stress.[49,80,82,103] Consistent with these observations, limited evidence indicates an increase in levels of HCRT mRNA within LH following immobilization stress.[49] Interestingly, although foot-shock increases Fos within

HCRT neurons, the conditioned stimulus that predicts the foot-shock, a stressor itself, does not alter Fos within these neurons.[103] Together, these observations indicate that presentation of some, but not all stressors, results in increased activity of HCRT neurons.

5.3. Activating Actions of HCRT on CRH Neurotransmission.

CRF plays a prominent role in coordinating the constellation of behavioral and physiological responses that define the state of stress.[30,54] Given this, it is of interest that HCRT fibers are located within close proximity of CRF neurons within the PVN and amygdala. Innervation of these structures by HCRT efferents suggests the possibility that HCRT may interact with central CRF systems to regulate the HPA axis and other stress-related processes. In support of this hypothesis, bath application of HCRT-1 elicits depolarization and increased spike frequency of magno- and parvocellular PVN neurons.[84,87] Consistent with this, HCRT-1 increases cortisol and corticosterone secretion from human and adrenocortical cells, respectively.[63,69] Similarly, *in vivo*, acute ICV HCRT-1 administration increases plasma levels of both corticosterone and adrenocorticotropin hormone (ACTH) release.[1,49,52,58,63,75,84] Together, these observations indicate that HCRT efferents exert excitatory actions on CRF PVN neurons. It remains to be determined whether HCRT modulates activity of extrahypothalamic CRF neurons.

Similar to that observed with HCRT, central administration of CRF and ACTH elicit a variety of behavioral responses observed in stress, including locomotion, grooming and chewing of inedible objects.[31,40] Combined, these observations suggest a possible interaction between HCRT and CRF in the regulation of stress-related behaviors. To test this hypothesis, Ida and colleagues[49] examined the effects of a CRF antagonist (alpha-helical $CRF_{9\text{-}41}$) on HCRT-induced grooming and locomotor activity. In these studies, treatment with alpha-helical CRF, reduced HCRT-induced grooming and face washing for the two hours following injection. Interestingly, the magnitude of HCRT-induced increases in locomotor activity was also reduced.

To further characterize the contribution of CRF systems to HCRT-induced arousal we have begun to examine the degree to which CRF participates in HCRT-induced waking and arousal-related behavioral activity. In these studies, animals were pretreated with vehicle or a CRF antagonist (alpha-helical $CRF_{12\text{-}41}$) 30-min prior to HCRT administration. All infusions were made via a remote-controlled infusion pump and thus the behavioral state of the animal was not influenced by the infusion procedures *per se*. Consistent with previous observations,[11,35,42,78] in animals pretreated with vehicle, HCRT-1 (0.7 nmol) increased EEG/EMG and behavioral indices of waking (Figure 6). In contrast, pretreatment with 15.0 nmol of the CRF antagonist attenuated HCRT-induced waking. Additionally, the CRF antagonist reduced HCRT-induced locomotion (rears and quadrant entries) and grooming (data not shown). Combined, evidence suggests the HCRT system may act, in part, to modulate behavioral and physiological processes under high arousal, stress-like, conditions and that at least a subset of these actions involves an interaction between HCRT and CRF systems.

6. APPETITIVE HIGH-AROUSAL STATES

The above discussion has focused on the potential role of HCRT in modulating behavioral and physiological processes under aversive and challenging situations (e.g.

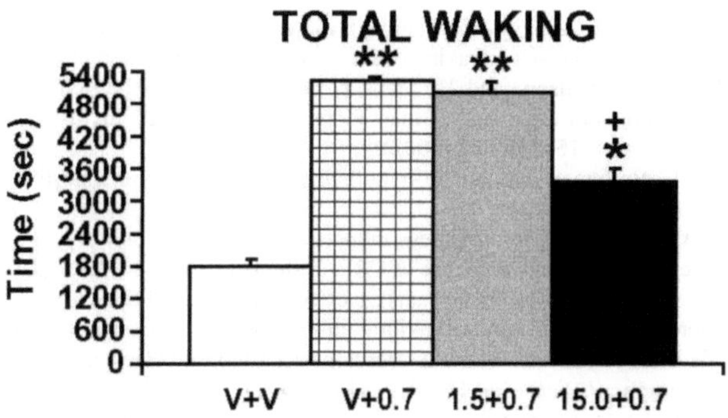

Figure 6. Effects of the CRF antagonist, alpha helical CRF_{12-41}, pretreatment on HCRT-induced waking. Shown are the effects of vehicle pretreatment followed by a vehicle infusion (V + V), vehicle pretreatment followed by a 0.7 nmol HCRT infusion (V + 0.7), 1.5 nmol alpha helical CRF pretreatment followed by a 0.7 nmol HCRT infusion (1.5 + 0.7), and 15.0 nmol alpha helical CRF pretreatment followed by a 0.7 nmol HCRT infusion (15.0 + 0.7). HCRT-1, preceded by vehicle pretreatment, significantly increased waking relative to the vehicle/vehicle condition. In contrast, pretreatment with alpha helical CRF, dose-dependently attenuated HCRT induced waking. *$P<0.05$; **$P<0.01$ significantly different from (V + V); $^{+}P < 0.01$ significantly different from (V + 0.7).

stress). However, it is important to note that elevated arousal levels occur under appetitive conditions and substantial evidence indicates an activation of the prototypical stress systems (hypothalamo-pituitary-adrenal axis as well as peripheral and central catecholaminergic systems) under both aversive and appetitive conditions.[66] This suggests the working hypothesis that at least a subset of the physiological indices of stress may be independent of affective valence (pleasant vs. unpleasant) and more closely aligned with arousal level, motivational state, and/or level of activity. Thus, HCRT may not only serve a critical function in behavior and physiology in stress, but also under high arousal, appetitive conditions. In this context, it is important to note that HCRT modulates appetitive behavior (e.g. feeding), and that LH has long been implicated in appetitive processes.[9,34] It remains for future work to delineate the degree to which the actions of HCRT are common to both aversive and appetitive high-arousal conditions.

7. REFERENCES

1. K. A. Al Barazanji, S. Wilson, J. Baker, D. S. Jessop, and M. S. Harbuz, Central Orexin-A Activates Hypothalamic-Pituitary-Adrenal Axis and Stimulates Hypothalamic Corticotropin Releasing Factor and Arginine Vasopressin Neurones in Conscious Rats, *J. Neuroendocrinol.* **13,** 421-424 (2001).
2. S. M. Antelman, H. Szechtman, P. Chin, and A. E. Fisher, Tail pinch-induced eating, gnawing and licking behavior in rats: dependence on the nigrostriatal dopamine system, *Brain Res* **99,** 319-337 (1975) .
3. A. F. Arnsten, The biology of being frazzled, *Science* **280,** 1711-1712 (1998).
4. A. F. Arnsten, Development of the cerebral cortex: XIV. Stress impairs prefrontal cortical function, *J. Am. Acad. Child Adolesc. Psychiatry* **38,** 220-222 (1999).

5. A. F. Arnsten, Stress impairs prefrontal cortical function in rats and monkeys: role of dopamine D1 and norepinephrine alpha-1 receptor mechanisms, *Prog. Brain Res.* **126,** 183-192 (2000).
6. A. F. Arnsten, C. Berridge, and D. S. Segal, Stress produces opioid-like effects on investigatory behavior, *Pharmacol. Biochem. Behav.* **22,** 803-809 (1985).
7. G. Aston-Jones and F. E. Bloom, Activity of norepinephrine-containing locus coeruleus neurons in behaving rats anticipates fluctuations in the sleep-waking cycle, *J. Neurosci.* **1,** 876-886 (1981).
8. M. Backberg, G. Hervieu, S. Wilson, and B. Meister, Orexin receptor-1 (OX-R1) immunoreactivity in chemically identified neurons of the hypothalamus: focus on orexin targets involved in control of food and water intake, *Eur. J. Neurosci.* **15,** 315-328 (2002).
9. L. L. Bernardis and L. L. Bellinger, The lateral hypothalamic area revisited: ingestive behavior, *Neurosci. Biobehav. Rev.* **20,** 189-287 (1996).
10. C. W. Berridge and A. J. Dunn, Restraint-stress-induced changes in exploratory behavior appear to be mediated by norepinephrine-stimulated release of CRF, *J. Neurosci.* **9,** 3513-3521 (1989).
11. C. W. Berridge and R. A. España, Synergistic sedative effects of noradrenergic α_1- and β- receptor blockade on forebrain electroencephalographic and behavioral indices, *Neuroscience* **99,** 495-505 (2000).
12. C. W. Berridge and S. L. Foote, Effects of locus coeruleus activation on electroencephalographic activity in neocortex and hippocampus, *J. Neurosci.* **11,** 3135-3145 (1991).
13. C. W. Berridge and S. L. Foote, Enhancement of behavioral and electroencephalographic indices of waking following stimulation of noradrenergic beta-receptors within the medial septal region of the basal forebrain, *J. Neurosci.* **16,** 6999-7009 (1996).
14. C. W. Berridge, E. Mitton, W. Clark, and R. H. Roth, Engagement in a non-escape (displacement) behavior elicits a selective and lateralized suppression of frontal cortical dopaminergic utilization in stress, *Synapse* **32,** 187-197 (1999).
15. C. W. Berridge and J. O'Neill, Differential sensitivity to the wake-promoting actions of norepinephrine within the medial preoptic area and the substantia innominata, *Behav. Neurosci.* **115,** 165-174 (2001).
16. C. W. Berridge and B. D. Waterhouse, The locus coeruleus-noradrenergic system: modulation of behavioral state and state-dependent cognitive processes, *Brain Res. Brain Res. Rev.* **42,** 33-84 (2003).
17. P. Bourgin, S. Huitron-Resendiz, A. D. Spier, V. Fabre, B. Morte, J. R. Criado, J. G. Sutcliffe, S. J. Henriksen, and L. de Lecea, Hypocretin-1 modulates rapid eye movement sleep through activation of locus coeruleus neurons, *J. Neurosci.* **20,** 7760-7765 (2000).
18. G. Buzsaki, R. G. Bickford, G. Ponomareff, L. J. Thal, R. Mandel, and F. H. Gage, Nucleus basalis and thalamic control of neocortical activity in the freely moving rat, *J. Neurosci.* **8,** 4007-4026 (1988) .
19. W. B. Cannon, The emergency function of the adrenal medulla in pain and the major emotions, *Am. J. Physiol* **33,** 356-372 (1914).
20. F. C. Chang and M. R. Opp, Corticotropin releasing hormone (CRF) as a regulator of waking, *Neurosci. Biobehav. Rev.* **25,** 445-453 (2001).
21. F. C. Chang and M. R. Opp, A corticotropin-releasing hormone antisense oligodeoxynucleotide reduces spontaneous waking in the rat, *Regul. Pept.* **117,** 43-52 (2004).
22. R. M. Chemelli, J. T. Willie, C. M. Sinton, J. K. Elmquist, T. Scammell, C. Lee, J. A. Richardson, S. C. Williams, Y. Xiong, Y. Kisanuki, T. E. Fitch, M. Nakazato, R. E. Hammer, C. B. Saper, and M. Yanagisawa, Narcolepsy in orexin knockout mice: molecular genetics of sleep regulation, *Cell* **98,** 437-451 (1999).
23. C. T. Chen, L. L. Hwang, J. K. Chang, and N. J. Dun, Pressor effects of orexins injected intracisternally and to rostral ventrolateral medulla of anesthetized rats, *Am. J. Physiol Regul. Integr. Comp Physiol* **278,** R692-R697 (2000).
24. J. E. Cluderay, D. C. Harrison, and G. J. Hervieu, Protein distribution of the orexin-2 receptor in the rat central nervous system, *Regul. Pept.* **104,** 131-144 (2002).
25. Y. Date, Y. Ueta, H. Yamashita, H. Yamaguchi, S. Matsukura, K. Kangawa, T. Sakurai, M. Yanagisawa, and M. Nakazato, Orexins, orexigenic hypothalamic peptides, interact with autonomic, neuroendocrine and neuroregulatory systems, *Proc. Natl. Acad. Sci.* **96,** 748-753 (1999).
26. M. G. Dube, S. P. Kalra, and P. S. Kalra, Food intake elicited by central administration of orexins/hypocretins: identification of hypothalamic sites of action, *Brain Res* **842,** 473-477 (1999).
27. A. J. Dunn, Stress-related activation of cerebral dopaminergic systems, *Ann. N. Y. Acad. Sci.* **537,** 188-205 (1988).
28. A. J. Dunn and C. W. Berridge, Corticotropin-releasing factor administration elicits a stress-like activation of cerebral catecholaminergic systems, *Pharmacol. Biochem. Behav.* **27,** 685-691 (1987).
29. A. J. Dunn and C. W. Berridge, Is corticotropin-releasing factor a mediator of stress responses?, *Ann. N. Y. Acad. Sci.* **579,** 183-191 (1990).

30. A. J. Dunn and C. W. Berridge, Physiological and behavioral responses to corticotropin-releasing factor administration: is CRF a mediator of anxiety or stress responses?, *Brain Res. Brain Res. Rev.* **15,** 71-100 (1990).
31. A. J. Dunn, C. W. Berridge, Y. I. Lai, and T. L. Yachabach, CRF-induced excessive grooming behavior in rats and mice, *Peptides* **8,** 841-844 (1987).
32. A. J. Dunn, A. L. Guild, N. R. Kramarcy, and M. D. Ware, Benzodiazepines decrease grooming in response to novelty but not ACTH or beta-endorphin, *Pharmacol. Biochem. Behav.* **15,** 605-608 (1981).
33. A. J. Dunn and R. W. Hurd, ACTH acts via an anterior ventral third ventricular site to elicit grooming behavior, *Peptides* **7,** 651-657 (1986).
34. J. K. Elmquist, C. F. Elias, and C. B. Saper, From lesions to leptin: hypothalamic control of food intake and body weight, *Neuron* **22,** 221-232 (1999).
35. R. A. España, B. A. Baldo, A. E. Kelley, and C. W. Berridge, Wake-promoting and sleep-suppressing actions of hypocretin (orexin): Basal forebrain sites of action. *Neuroscience* **106,** 699-715 (2001).
36. R. A. España, S. Plahn, and C. W. Berridge, Circadian-dependent and circadian-independent behavioral actions of hypocretin/orexin, *Brain Res.* **943,** 224-236 (2002).
37. R. A. España, R. J. Valentino, and C. W. Berridge, Fos immunoreactivity in hypocretin-synthesizing and hypocretin-1 receptor-expressing neurons: effects of diurnal and nocturnal spontaneous waking, stress and hypocretin-1 administration, *Neuroscience* **121,** 201-217 (2003).
38. J. Fadel and A. Y. Deutch, Anatomical substrates of orexin-dopamine interactions: lateral hypothalamic projections to the ventral tegmental area, *Neuroscience* **111,** 379-387 (2002).
39. S. L. Foote, G. Aston-Jones, and F. E. Bloom, Impulse activity of locus coeruleus neurons in awake rats and monkeys is a function of sensory stimulation and arousal, *Proc. Natl. Acad. Sci.* **77,** 3033-3037 (1980).
40. W. H. Gispen and R. L. Isaacson, ACTH-induced excessive grooming in the rat, *Pharmacol. Ther.* **12,** 209-246 (1981).
41. M. A. Greco and P. J. Shiromani, Hypocretin receptor protein and mRNA expression in the dorsolateral pons of rats, *Brain Res Mol. Brain Res* **88,** 176-182 (2001).
42. J. J. Hagan, R. A. Leslie, S. Patel, M. L. Evans, T. A. Wattam, S. Holmes, C. D. Benham, S. G. Taylor, C. Routledge, P. Hemmati, R. P. Munton, T. E. Ashmeade, A. S. Shah, J. P. Hatcher, P. D. Hatcher, D. N. Jones, M. I. Smith, D. C. Piper, A. J. Hunter, R. A. Porter, and N. Upton, Orexin A activates locus coeruleus cell firing and increases arousal in the rat, *Proc. Natl. Acad. Sci.* **96,** 10911-10916 (1999).
43. M. B. Hennessy and T. Foy, Nonedible material elicits chewing and reduces the plasma corticosterone response during novelty exposure in mice, *Behav. Neurosci.* **101,** 237-245 (1987).
44. G. J. Hervieu, J. E. Cluderay, D. C. Harrison, J. C. Roberts, and R. A. Leslie, Gene expression and protein distribution of the orexin-1 receptor in the rat brain and spinal cord, *Neuroscience* **103,** 777-797 (2001).
45. J. A. Hobson, R. W. McCarley, and P. W. Wyzinski, Sleep cycle oscillation: reciprocal discharge by two brainstem neuronal groups, *Science* **189,** 55-58 (1975).
46. W. E. Hoffman and M. I. Phillips, Regional study of cerebral ventricle sensitive sites to angiotensin II, *Brain Res.* **110,** 313-330 (1976).
47. T. L. Horvath, C. Peyron, S. Diano, A. Ivanov, G. Aston-Jones, T. S. Kilduff, and A. N. van Den Pol, Hypocretin (orexin) activation and synaptic innervation of the locus coeruleus noradrenergic system, *J. Comp Neurol.* **415,** 145-159 (1999).
48. T. Ida, K. Nakahara, T. Katayama, N. Murakami, and M. Nakazato, Effect of lateral cerebroventricular injection of the appetite- stimulating neuropeptide, orexin and neuropeptide Y, on the various behavioral activities of rats, *Brain Res.* **821,** 526-529 (1999).
49. T. Ida, K. Nakahara, T. Murakami, R. Hanada, M. Nakazato, and N. Murakami, Possible involvement of orexin in the stress reaction in rats, *Biochem. Biophys. Res. Commun.* **270,** 318-323 (2000).
50..P. M. Iuvone and A. J. Dunn, Tyrosine hydroxylase activation in mesocortical 3,4-dihydroxyphenylethylamine neurons following footshock, *J. Neurochem.* **47,** 837-844 (1986).
51. A. Ivanov and G. Aston-Jones, Hypocretin/orexin depolarizes and decreases potassium conductance in locus coeruleus neurons, *Neuroreport* **11,** 1755-1758 (2000).
52. M. Jászberényi, E. Bujdosó, I. Pataki, and G. Telegdy, Effects of orexins on the hypothalamic-pituitary-adrenal system, *J. Neuroendocrinol.* **12,** 1174-1178 (2000).
53. L. I. Kiyashchenko, B. Y. Mileykovskiy, Y. Y. Lai, and J. M. Siegel, Increased and decreased muscle tone with orexin (hypocretin) microinjections in the locus coeruleus and pontine inhibitory area, *J. Neurophysiol.* **85,** 2008-2016 (2001).
54. G. F. Koob and F. E. Bloom, Corticotropin-releasing factor and behavior, *Fed. Proc.* **44,** 259-263 (1985).
55. G. F. Koob, K. Thatcher-Briton, A. Tazi, and M. Le Moal, Behavioral pharmacology of stress: Focus on CNS corticotropin-releasing factor, *Adv. Exp. Med. Biol.* **245,** 25-34 (1988).

56. T. M. Korotkova, O. A. Sergeeva, K. S. Eriksson, H. L. Haas, and R. E. Brown, Excitation of ventral tegmental area dopaminergic and nondopaminergic neurons by orexins/hypocretins, *J. Neurosci.* **23,** 7-11 (2003).
57. V. M. Kumar, S. Datta, G. S. Chhina, and B. Singh, Alpha adrenergic system in medial preoptic area involved in sleep-wakefulness in rats, *Brain Res. Bull.* **16,** 463-468 (1986).
58. M. Kuru, Y. Ueta, R. Serino, M. Nakazato, Y. Yamamoto, I. Shibuya, and H. Yamashita, Centrally administered orexin/hypocretin activates HPA axis in rats, *Neuroreport* **11,** 1977-1980 (2000).
59. S. Levine, in: A definition of stress?/*Animal stress,* edited by G., Moberg (Waverly Press, Baltimore,1985), pp. 51-69.
60. R. E. Lewis and M. I. Phillips, Localization of the central pressor action of bradykinin to the cerebral third ventricle, *Am. J. Physiol* **247,** R63-R68 (1984).
61. L. Lin, J. Faraco, R. Li, H. Kadotani, W. Rogers, X. Lin, X. Qiu, P. J. de Jong, S. Nishino, and E. Mignot, The sleep disorder canine narcolepsy is caused by a mutation in the hypocretin (orexin) receptor 2 gene, *Cell* **98,** 365-376 (1999).
62. M. Lubkin and A. Stricker-Krongrad, Independent feeding and metabolic actions of orexins in mice 714, *Biochem. Biophys. Res. Commun.* **253,** 241-245 (1998).
63. L. K. Malendowicz, C. Tortorella, and G. G. Nussdorfer, Orexins stimulate corticosterone secretion of rat adrenocortical cells, through the activation of the adenylate cyclase-dependent signaling cascade, *J. Steroid Biochem. Mol. Biol.* **70,** 185-188 (1999).
64. B. N. Mallick and M. N. Alam, Different types of norepinephrinergic receptors are involved in preoptic area mediated independent modulation of sleep-wakefulness and body temperature, *Brain Res* **591,** 8-19 (1992).
65. J. N. Marcus, C. J. Aschkenasi, C. E. Lee, R. M. Chemelli, C. B. Saper, M. Yanagisawa, and J. K. Elmquist, Differential expression of orexin receptors 1 and 2 in the rat brain, *J. Comp Neurol.* **435,** 6-25 (2001).
66. M. Marinelli and P. V. Piazza, Interaction between glucocorticoid hormones, stress and psychostimulant drugs, *Eur. J. Neurosci.* **16,** 387-394 (2002).
67. P. J. Martins, V. D'Almeida, M. Pedrazzoli, L. Lin, E. Mignot, and S. Tufik, Increased hypocretin-1 (orexin-a) levels in cerebrospinal fluid of rats after short-term forced activity, *Regul. Pept.* **117,** 155-158 (2004).
68. J. W. Mason, The scope of psychoendocrine research, *Psychosom Med*565-575 (1968).
69. G. Mazzocchi, L. K. Malendowicz, L. Gottardo, F. Aragona, and G. G. Nussdorfer, Orexin A Stimulates Cortisol Secretion from Human Adrenocortical Cells through Activation of the Adenylate Cyclase-Dependent Signaling Cascade, *J. Clin. Endocrinol. Metab* **86,** 778-782 (2001).
70. I. Merchenthaler, Corticotropin Releasing-Factor (Crf)-Like Immunoreactivity in the Rat Central Nervous-System - Extrahypothalamic Distribution, *Peptides* **5,** 53-69 (1984).
71. R. Metherate, C. L. Cox, and J. H. Ashe, Cellular bases of neocortical activation: modulation of neural oscillations by the nucleus basalis and endogenous acetylcholine, *J. Neurosci.* **12,** 4701-4711 (1992).
72. T. Nambu, T. Sakurai, K. Mizukami, Y. Hosoya, M. Yanagisawa, and K. Goto, Distribution of orexin neurons in the adult rat brain, *Brain Res.* **827,** 243-260 (1999).
73. L. K. Nisenbaum, M. J. Zigmond, A. F. Sved, and E. D. Abercrombie, Prior exposure to chronic stress results in enhanced synthesis and release of hippocampal norepinephrine in response to a novel stressor, *J. Neurosci.* **11,** 1478-1484 (1991).
74. S. Nishino, B. Ripley, S. Overeem, G. J. Lammers, and E. Mignot, Hypocretin (orexin) deficiency in human narcolepsy, *Lancet* **355,** 39-40 (2000).
75. K. W. Nowak, P. Mackowiak, M. M. Switonska, M. Fabis, and L. K. Malendowicz, Acute orexin effects on insulin secretion in the rat: in vivo and in vitro studies, *Life Sci.* **66,** 449-454 (2000).
76. M. E. Page, C. W. Berridge, S. L. Foote, and R. J. Valentino, Corticotropin-releasing factor in the locus coeruleus mediates EEG activation associated with hypotensive stress, *Neurosci. Lett.* **164,** 81-84 (1993).
77. C. Peyron, D. K. Tighe, A. N. van Den Pol, L. de Lecea, H. C. Heller, J. G. Sutcliffe, and T. S. Kilduff, Neurons containing hypocretin (orexin) project to multiple neuronal systems, *J. Neurosci.* **18,** 9996-10015 (1998).
78. D. C. Piper, N. Upton, M. I. Smith, and A. J. Hunter, The novel brain neuropeptide, orexin-A, modulates the sleep-wake cycle of rats, *Eur. J. Neurosci.* **12,** 726-730 (2000).
79. T. W. Robbins, Cortical noradrenaline, attention and arousal, *Psychol. Med.* **14,** 13-21 (1984).
80. F. Sakamoto, S. Yamada, and Y. Ueta, Centrally administered orexin-A activates corticotropin-releasing factor-containing neurons in the hypothalamic paraventricular nucleus and central amygdaloid nucleus of rats: possible involvement of central orexins on stress-activated central CRF neurons, *Regul. Pept.* **118,** 183-191 (2004).
81. T. Sakurai, A. Amemiya, M. Ishii, I. Matsuzaki, R. M. Chemelli, H. Tanaka, S. C. Williams, J. A. Richarson, G. P. Kozlowski, S. Wilson, J. R. Arch, R. E. Buckingham, A. C. Haynes, S. A. Carr, R. S. Annan, D. E. McNulty, W. S. Liu, J. A. Terrett, N. A. Elshourbagy, D. J. Bergsma, and M. Yanagisawa,

Orexins and orexin receptors: a family of hypothalamic neuropeptides and G protein-coupled receptors that regulate feeding behavior, *Cell* **92,** 1 (1998).

82. T. Sakurai, T. Moriguchi, K. Furuya, N. Kajiwara, T. Nakamura, M. Yanagisawa, and K. Goto, Structure and function of human prepro-orexin gene, *J. Biol. Chem.* **274,** 17771-17776 (1999).
83. W. K. Samson, B. Gosnell, J. K. Chang, Z. T. Resch, and T. C. Murphy, Cardiovascular regulatory actions of the hypocretins in brain, *Brain Res* **831,** 248-253 (1999).
84. W. K. Samson, M. M. Taylor, M. Follwell, and A. V. Ferguson, Orexin actions in hypothalamic paraventricular nucleus: physiological consequences and cellular correlates, *Regul. Pept.* **104,** 97-103 (2002).
85. P. E. Sawchenko and L. W. Swanson, Localization, Colocalization, and Plasticity of Corticotropin-Releasing Factor Immunoreactivity in Rat-Brain, *Federation Proceedings* **44,** 221-227 (1985).
86. H. Selye, The general adaptation syndrome and the diseases of adaptation, *J. Clin. Endocrinol. Metab* **6,** 117-230 (1946).
87. T. Shirasaka, S. Miyahara, T. Kunitake, Q. H. Jin, K. Kato, M. Takasaki, and H. Kannan, Orexin depolarizes rat hypothalamic paraventricular nucleus neurons, *Am. J. Physiol Regul. Integr. Comp Physiol* **281,** R1114-R1118 (2001).
88. T. Shirasaka, M. Nakazato, S. Matsukura, M. Takasaki, and H. Kannan, Sympathetic and cardiovascular actions of orexins in conscious rats, *Am. J. Physiol* **277,** R1780-R1785 (1999).
89. E. A. Stone, Effect of stress on norepinephrine-stimulated cyclic AMP formation in brain slices, *Pharmacol. Biochem. Behav.* **8,** 583-591 (1978).
90. E. A. Stone and J. E. Platt, Brain adrenergic receptors and resistance to stress, *Brain Res.* **237,** 405-414 (1982).
91. R. Suzuki, H. Shimojima, H. Funahashi, S. Nakajo, S. Yamada, J. L. Guan, S. Tsurugano, K. Uehara, Y. Takeyama, S. Kikuyama, and S. Shioda, Orexin-1 receptor immunoreactivity in chemically identified target neurons in the rat hypothalamus, *Neurosci. Lett.* **324,** 5-8 (2002).
92. D. C. Sweet, A. S. Levine, C. J. Billington, and C. M. Kotz, Feeding response to central orexins, *Brain Res* **821,** 535-538 (1999).
93. S. Taheri, M. Mahmoodi, J. Opacka-Juffry, M. A. Ghatei, and S. R. Bloom, Distribution and quantification of immunoreactive orexin A in rat tissues, *FEBS Lett.* **457,** 157-161 (1999).
94. T. C. Thannickal, R. Y. Moore, R. Nienhuis, L. Ramanathan, S. Gulyani, M. Aldrich, M. Cornford, and J. M. Siegel, Reduced number of hypocretin neurons in human narcolepsy, *Neuron* **27,** 469-474 (2000).
95..A. M. Thierry, J. P. Tassin, G. Blanc, and J. Glowinski, Selective activation of mesocortical DA system by stress, *Nature* **263,** 242-244 (1976).
96. A. Tsuda, M. Tanaka, Y. Ida, I. Shirao, Y. Gondoh, M. Oguchi, and M. Yoshida, Expression of aggression attenuates stress-induced increases in rat brain noradrenaline turnover, *Brain Res* **474,** 174-180 (1988).
97. W. Vale, J. Spiess, C. Rivier, and J. Rivier, Characterization of A 41-Residue Ovine Hypothalamic Peptide That Stimulates Secretion of Corticotropin and Beta-Endorphin, *Science* **213,** 1394-1397 (1981).
98. R. J. Valentino, S. L. Foote, and G. Aston-Jones, Corticotropin-releasing factor activates noradrenergic neurons of the locus coeruleus, *Brain Res.* **270,** 363-367 (1983).
99. N. M. Vittoz and C. W. Berridge, Hypocretin increases dopamine efflux in rat prefrontal cortex and not in nucleus accumbens, *Society for Neuroscience Meeting* November 8-12 (New Orleans, LA), abstract 931.14, (2003).
100. Y. Yoshida, N. Fujiki, T. Nakajima, B. Ripley, H. Matsumura, H. Yoneda, E. Mignot, and S. Nishino, Fluctuation of extracellular hypocretin-1 (orexin A) levels in the rat in relation to the light-dark cycle and sleep-wake activities, *Eur. J. Neurosci.* **14,** 1075-1081 (2001).
101. G. Yoshimichi, H. Yoshimatsu, T. Masaki, and T. Sakata, Orexin-A regulates body temperature in coordination with arousal status, *Exp. Biol. Med.* **226,** 468-476 (2001).
102. J. M. Zeitzer, C. L. Buckmaster, K. J. Parker, C. M. Hauck, D. M. Lyons, and E. Mignot, Circadian and homeostatic regulation of hypocretin in a primate model: implications for the consolidation of wakefulness, *J. Neurosci.* **23,** 3555-3560 (2003).
103. L. Zhu, T. Onaka, T. Sakurai, and T. Yada, Activation of orexin neurones after noxious but not conditioned fear stimuli in rats, *Neuroreport* **13,** 1351-1353 (2002).

ROLE OF HYPOCRETINS ON PERIPHERAL SYSTEMS

THE HYPOCRETINS/OREXINS AND THE HYPOTHALAMO-PITUITARY-ADRENAL AXIS

Willis K. Samson, Meghan M. Taylor, and Alastair V. Ferguson*

1. INTRODUCTION

The presence of hypocretin/orexin receptors in brain sites known to be important in autonomic and neuroendocrine function in general, and the control of stress hormone secretion in particular, predicted a role for the peptides in the cardiovascular, behavioral and neuroendocrine responses to stress. In addition, these receptors are also present in the pituitary and adrenal glands, suggesting actions of the hypocretins/orexins at several levels of the hypothalamo-pituitary-adrenal (HPA) axis. Early in our studies of the pharmacologic effects of hypocretin/orexin in conscious animals, we observed a general increase in locomotor activity subsequent to intracerebroventricular (i.c.v.) administration of the peptides, and behaviors reminiscent of those observed during stressful stimuli. We focused therefore our attention on the cardiovascular and neuroendocrine effects of exogenously administered hypocretin/orexin and summarize here our findings, and those of many others, which demonstrate significant central actions of the peptides on cardiovascular function and the HPA axis.

2. THE ANATOMICAL FRAMEWORK

The paraventricular nucleus of the hypothalamus (PVN) is a paired cluster of neurons adjacent to the walls of the dorsal aspect of the third cerbroventricle. It is situated medial and adjacent to the hypocretin/orexin-producing cells of the lateral hypothalamus/perifornical area, some of which directly innervate the PVN.[1] Both hypocretin receptors (hcrtr1 and hcrtr2) are present in the PVN,[2,3] with the hcrtr2 most abundant.[4] The PVN is the central commander of the HPA axis and is the site of co-ordinated neural responses to stress.[5] It receives afferent information from

* Willis K. Samson and Meghan M. Taylor, Saint Louis University, Saint Louis, Missouri 63131. Alastair V. Ferguson, Queen's University, Kingston, Ontario, Canada K7L 3N6

circumventricular organs such as the subfornical organ (SFO) and organum vasculosum lamina terminalis (OVLT), sites lacking a blood brain barrier and responsive to changes in plasma sodium and osmolar content. Ascending fibers from brain stem centers responsive to changes in blood pressure and circulating carbon dioxide concentrations deliver baroreceptive and chemoreceptive information from peripheral sensors to the PVN where that information is assembled and interpreted. Circulating hormones that can cross the blood brain barrier influence neuronal activity in the PVN directly as well, as do relay neurons from a variety of brain sites including cerebral cortex, limbic lobe structures, and relays from retina and visual cortex. All this converging information must be translated by cells in the PVN into appropriate behavioral, endocrine, neuroendocrine and cardiovascular responses to changes in our internal and external environment. Thus the PVN plays a critical role in our adaptation to stress, both physical and emotional, and our ability to respond to metabolic, environmental and pshycological demands. It does this through well-organized and segregated subpopulations of neurons that contribute to the efferent projections of the nucleus. Typically two major subpopulations of PVN neurons have been identified, the magnocellular (larger) cells that project to the posterior lobe of the pituitary gland and the parvocellular (smaller) cells that project to other brain sites. Perhaps a more up-to-date categorization, however, is the functional classification of PVN neurons into four distinct cell types: neuroendocrine, endocrine, preautonomic and behavioral.[6-8]

The neuroendocrine cells of the PVN responsible for the control of adrenocorticotropin (ACTH) production and secretion from the anterior lobe of the pituitary gland express the 41 amino acid peptide, corticotropin releasing hormone (CRH). CRH-positive neurons project to a variety of brain sites, most important of which is the median eminence on the midline of the hypothalamus at the floor of the third cerebral ventricle. It is here that the superior hypophyseal arteries break up into a capillary bed characterized by fenestrated endothelial cells. These fenestrations permit access of peptides and other neural factors released in the median eminence to the portal vessels that connect this capillary bed with a second in the anterior lobe of the pituitary gland. Thus the hypophysal portal vessels communicate chemical messages released from hypothalamic neurons to the endocrine cells of the anterior pituitary gland.[9] It is via these vessels that CRH accesses the corticotrophs, cells that produce ACTH. Regulation of ACTH release therefore is dependent upon the activity of those parvocellular neurons in PVN, the neuroendocrine cells that produce CRH. They are inhibited by the end product of the HPA axis, corticosterone in rodents and cortisol in humans, and this is a major mechanism by which the HPA axis is regulated. However, equally important, indeed of primary importance in the acute regulation of the HPA axis, are the multiple neural afferents that innervate the PVN, among them the hypocretin producing cells of the adjacent lateral hypothalamus (vide infra).

Endocrine neurons of the PVN produce vasopressin (AVP) and oxytocin (OT). These magnocellular elements project primarily to the posterior lobe of the pituitary gland where they release their contents into the general circulation.[10] Vasopressin is released in response to stress, hypotension, hyperosmolar challenge, and inflammatory insults. Vasopressin acts in kidney to facilitate capture of free water from the tubule and thus help maintain adequate blood volume and circulatory pressures. It also acts as a vasoconstrictor in the vasculature, further facilitating the maintenance of adequate perifusion pressures for the vital organs of the body. The physiologic roles of oxytocin are more related to reproduction and lactation, but it too has been identified as a hormone released in response to stress. For the purposes of this chapter, it is important to identify what are often overlooked actions of AVP and OT. Receptors for both peptides are present in the anterior pituitary gland. Like CRH these peptides are delivered to cells of the anterior pituitary gland via the hypophysial portal vessels. Both peptides, at relatively high concentrations, are capable of stimulating ACTH release from corticotrophs in vitro.[11] More importantly, they both can significantly potentiate the ACTH releasing action of the primary regulator of corticotroph function, CRH. The physiologic significance of those potentiating effects of AVP and OT, observed in vitro, has been established in a variety of stress-related paradigms of HPA axis activation.[12] Thus whenever one considers the neural factors controlling the functional activity of the HPA axis, the effect of those neural factors on AVP and OT release must also be addressed.

Preautonomic neurons of the PVN project to brain stem and spinal cord sites important in the regulation of cardiovascular function.[7] These neurons are reponsive to a variety of inputs, many similar to those that activate the neighboring, parvocellular, neuroendocrine elements of the PVN. Importantly some of these preautonomic neurons also express CRH and it is clear that CRH producing neurons are important components of the central regulation of autonomic function. Thus factors that stimulate the release of CRH into the hypophyseal portal vessels may also stimulate the activity of CRH-producing preautonomic neurons assuring a co-ordinated neuroendocrine (i.e. ACTH) and autonomic (i.e. sympathetic activation) response to stress. It appears that the hypocretins act in this manner.

Finally, parvocellular neurons of the PVN that do not project to median eminence (neuroendocrine function) or brain stem cardiovascular centers (preautonomic function) project to a variety of other brain sites, including limbic centers and cerebral cortex where complex behavioral responses to stress are generated. Some of these cells also produce CRH and CRH is known to be a neurotransmitter involved in the stereotypic behavioral patterns of the stress response.[13] Since central administration of hypocretin elicits similar stress related behaviors, a role for these peptides in the behavioral response to stress is suggested and this too may be co-ordinated with the endocrine and autonomic responses observed.

3. HYPOCRETIN AND THE CENTRAL ARM OF THE HPA AXIS

The first hints that endogenous hypocretins/orexins might be involved in the central component of the HPA axis came from one of the initial descriptions of the behavioral effects of the peptides.[14] In addition to stimulating increased locomotor activity and grooming behaviors, responses similar to those observed following central administration of CRH, intracerebroventricular (i.c.v.) injection of orexin A resulted in increased corticosterone levels in conscious male rats. Growth hormone levels were reduced following orexin A administration. With the exception that prolactin levels fell to undetectable concentrations, these initial endocrine responses indicated an action of the peptide which, along with the behavioral responses, was related to the stress response. These observations were not lost on our group,[15] or for that matter on many other neuroendocrinologists.[16-21] The most likely explanation for the ability of orexin A to stimulate corticosterone secretion was an action of the peptide on the CRH-producing neurons of the PVN. Doses of orexin that were capable of stimulating feeding behavior and locomotor activity activated immediate early gene (*c-fos*) expression in PVN.[18,21,22]

In one recent study those FOS expressing neurons were identified to be CRH-positive.[21] However, the visualization of FOS-like immunoreactivity in CRH-positive cells does not confirm direct effects of hypocretin on those cells since the effect could have been mediated via an interneuron. More indirect evidence came from behavioral and neuroendocrine studies. Pretreatment with a CRH-antagonist blocked the behavioral (grooming and face-washing) effects of orexin.[17] Plasma corticosterone and ACTH levels were reported to be elevated following i.c.v. administration of hypocretin/orexin[14-21] and this central action of the peptides to stimulate the HPA axis was compromised by pretreatment with a CRH-antagonist.[15,16] These data strongly suggested that endogenous hypocretin/orexin could directly or indirectly activate CRH-producing neurons in the PVN, resulting in ACTH release and corticosterone secretion.

Hcrtr1-immunoreactivity has been localized to CRH-positive cells[23] and thus the effects of hypocretin observed in vivo on the HPA axis may have been the result of a direct, membrane effect of the peptide on those cells. Since AVP also, under certain conditions, can stimulate ACTH release it is possible that the ability of centrally administered hypocretin to stimulate ACTH secretion is due to an effect of the peptide on AVP-producing neurons in PVN. Hcrtr1 immunoreactivity has also been localized to AVP-positive cells,[23] and *c-fos* expression was observed in some magnocellular elements of the PVN[22]. However, it is not known if i.c.v. administration of hypocretin results in increased AVP levels in the hypophyseal portal vessels, the route the peptide must take to stimulate ACTH release by itself or in concert with CRH. There is indirect evidence both for and against this possibility. Russell and colleagues[20] were able to demonstrate CRH and NPY release from hypothalamic explants in response to orexin A; however, AVP release was not significantly affected. These explants were not the conventional hypothalamo-neurophypophyseal system (HNS) explants normally used to study AVP and OT release in vitro.[24] Thus it is not clear that the intact AVP-producing cell bodies and axon projections were included. If they were not, then the failure to observe any effect of orexin A on AVP release may have been a false negative result.

One group has demonstrated the ability of i.c.v. administration of orexin to increase plasma AVP levels in conscious rabbits;[25] however, the effect was reported at only one time point and then only following one dose of orexin A. Our studies (vide infra) have demonstrated direct membrane effects on magnocellular neurons in PVN, but those cells were not chemically identified to be AVP-producing neurons.[15,26] Similarly, parvocellular neurons in PVN, some of which may produce AVP and project to median eminence, respond to direct application of orexin A and the "peptide footprint" of those cells is unkown.[15,26] No study to date has presented evidence that antagonism of AVP's action in pituitary gland by either passive immunoneutralization or antagonist pretreatment can abrogate the ACTH release observed following i.c.v. administration of hypocretin. Thus the role of AVP in orexin's ability to activate the HPA axis remains unproven. Even less is known about the potential effect of hypocretin on OT release and the possible role of OT in the central action of hypocretin on the HPA axis.

Evidence for direct actions of hypocretin on identified PVN neurons comes from electrophysiologic studies. As mentioned above hypocretin receptors are present in PVN with the hcrtr2 most abundant.[4] Bath application of orexin A or orexin B resulted in depolarizations of the majority of PVN neurons in one study[27]. These effects, observed in whole cell patch clamp recording were not blocked by tetrodotoxin (TTX) suggesting postsynaptic effects. In our hands similar depolarizations, as well as increases in spike frequencies, were observed in whole cell patch recordings from PVN neurons in hypothalamic slices.[15,26] Magnocellular (i.e. AVP and OT producing) neurons in PVN slice preparations were depolarized by bath application of orexin A in a concentration-dependent, reversible manner and these effects were were abolished by TTX treatment[26]. Thus the effect of the hypocretins/orexins on the endocrine cells of PVN is in all likelihood indirect. Further evidence for this comes from our observations that the depolarizing effect of orexin A on magnocellular neurons of the PVN was abolished by kynurenic acid, demonstrating the role of glutaminergic interneurons in the action of orexin A.[26] The situation was quite the opposite for the effects of orexin A on parvocellular PVN neurons recorded in whole cell patch configuration. Concentration dependent and reversible depolarizations of these neurons were not blocked by TTX pretreatment. In voltage clamp configuration, these parvocellular PVN neurons were demonstrated to respond to orexin A with an increase in a non-selective cation conductance, something we have also observed in cultured area postrema neurons[28] and nucleus tractus solitarius neurons in slice preparations.[29]

While the exact identity of the parvocellular neurons in PVN that respond to orexin A administration is not known, it is clear that the peptide can stimulate the release of CRH into the portal vessels. The physiologic relevance of this action has not yet been established and with the development of selective hcrtr1 and hcrtr2 receptor antagonists, this may soon be determined. Similarly, the availability of genetic ablation models for the study of the role of endogenous hypocretin/orexin in the central component of the HPA axis promises new insight. This is particularly promising now that a post differentiation orexin depletion rat model has been developed.[30]

4. HYPOCRETIN ACTIONS AT THE LEVEL OF THE PITUITARY GLAND

Hypocretin-positive nerve terminals have been visualized in the external layer of the median eminence, adjacent to the fenestrated capillary endothelium of the portal vessels, and in the posterior lobe of the pituitary gland.[31] Thus the peptides may be delivered to the hormone secreting cells of the adenohypophysis by either the long portal vessels (from median eminence), or the short portal vessels that connect the neural lobe with the anterior lobe of the pituitary gland. Importantly, both hcrtr1 and hcrtr2 are expressed in anterior pituitary gland and thus hypocretin reaching the gland may exert trophic actions on hormone production or secretion. There appears to be some sexual dimorphism with regard to the receptor subtypes present in pituitary gland.[32] Hcrtr1 expression is much higher in male than in female pituitaries and ovariectomy resulted in 12-fold increases in hcrtr1 expression in female rats.[33] This increase in hcrtr1 expression was inhibited by estrogen replacement. Similarly, orchidectomy increased hcrtr1 expression in male pituitaries, but to a lesser extent, and this was reversed by testosterone treatment. In human pituitary tissues, hypocretin receptors co-localize with somatotrophs (growth hormone producing cells) and corticotrophs (ACTH producing cells).

Little is know about the effects of hypocretin on anterior pituitary hormone hormone production; however, one group has demonstrated the ability of orexin A to augment growth hormone (GH) secretion from cultured ovine somatotrophs.[34,35] The effects were apparently exerted via voltage-gated L-type calcium channels and protein kinase C signaling.[34] We were unable to detect any significant effect of a wide range of orexin A or orexin B concentrations on GH release from cultured rat pituitary cells, under basal secretion conditions or in co-incubation with effective, stimulatory concentrations of GH-releasing hormone.[36] The reason for this apparent species difference is unclear. However, important to this review, we have observed significant effects of the hypocretins on ACTH release from cultured, rat anterior pituitary cells.[36]

We had originally hypothesized that since the hypocretins acted centrally to activate the HPA axis, they might stimulate ACTH by a direct action in the pituitary gland. Certainly the presence of hypocretin receptors on human corticotrophs[37] and the fact that the receptors were identified to be members of the G-protein-coupled-receptor (GPCR) superfamily,[38] in particular GPCRs that coupled to G_q, suggested such an action. Furthermore, activation of these receptors had been demonstrated to stimulate phospholipase C activity[39] and increase intracellular calcium levels in cultured neurons,[40] suggesting that the peptides might stimulate hormone secretion directly in vitro. Much to our surprise, log molar concentrations of orexin A and orexin B, ranging from 1 picoMolar to 1microMolar, failed to significantly alter basal ACTH release in static incubations, whereas these cells responded with the expected, dose-dependent stimulation of hormone secretion in response to CRH. Instead, we observed that orexin A and orexin B exerted significant, dose-related inhibitory effects on CRH-induced ACTH release from cultured rat pituitary cells and AtT-20 cells, a corticotroph cell line derived from a mouse pituitary gland[36]. Orexin A was more potent than orexin B, suggesting that the inhibitory effect was mediated via the hcrtr1 receptor. The effect of orexin A was already significant at 1.0 nanoMolar concentrations,[36] a level of peptide similar to the affinity constant for the receptor.[38]

The observed inhibitory effect of orexin A was not blocked by pertussis toxin and the peptide did not inhibit the ability of CRH to increase cellular levels of cAMP.[36] This suggested to us that the effect of orexin A was not mediated via a G_i- signaling mechanism. Instead a G_q-mediated mechanism of action was suggested by the fact that inhibition of protein kinase C activity by inclusion of calphostin C in the incubation medium blocked the effect of orexin A. This became a bit of a conumdrum since the mechanism we were uncovering at the time suggested that orexin A should stimulate phospholipase C activity, resulting in generation of inositol trisphosphate (IP_3). If this were the case, then intracellular levels of free calcium should rise and hormone should be secreted. Thus it became important for us to examine calcium mobilization in response to orexin A and hormone secretion in the more acute setting. Our preliminary results, recently presented in abstract form and now being prepared for full publication, demonstrate that indeed orexin A can stimulate calcium entry into the cytosol and a transient stimulation of ACTH release. This transient elevation of intracellular calcium levels appears to be important for the longer lasting inhibitory effect of orexin on CRH-induced ACTH secretion. We hypothesize now that PLC activation by orexin results in formation of diacycl glycerol (DAG), which activates PKC. This activation of PKC results in opening of potassium channels and membrane hyperpolarization. Indeed, we can block the inhibitory action of orexin with the potassium channel blocker, glyburide, and we have observed hyperpolarizing currents in individual corticotrophs in response to orexin A application (manuscript in preparation). While our current hypothesis is that the inhibitory action of orexin A in corticotrophs is due to its hyperpolarizing actions, it is possible that the peptide acts downstream to cAMP formation in response to CRH and that mechanism is currently under investigation.

We need better methods with which to study the cellular actions of the hypocretins in corticotrophs. While AtT20 cells are derived from corticotrophs, they do not recapitulate the full secretory phenotype of primary ACTH producing cells.[41] Furthermore we are limited currently to recording from primary pituitary cells identified to be corticotrophs by their response (calcium imaging and depolarization) to CRH application. Our current focus is to develop POMC promotor-GFP constructs that will facilitate identification of ACTH producing cells for patch clamp studies and cell sorting.

5. HYPOCRETIN ACTIONS IN THE ADRENAL GLAND

Although it was originally thought that the sole locus of hypocretin production was in hypothalamus, some of the peptide may be produced in testis.[32,38] Whether the hypocretin apparently produced in testis, or peptide produced in brain, gains access to the general circulation in sufficient amounts to act as a true hormone is unknown. However, hypocretin receptors are present in a variety of peripheral tissues, including the adrenal gland.[33,42-45] In vitro studies have demonstrated significant actions of the hypocretins on adrenal steroid and catecholamine release. Cultured rat zona fasciculata and reticularis cells released increased amounts of corticosterone under basal, but not stimulated conditions.[46] Orexin A also increased cAMP formation in these cells and the effect on corticosterone release was blocked by a protein kinase A inhibitor. No significant effects of orexin on basal or agonist stimulated aldosterone secretion were observed in zona glomerulosa cell cultures. These authors also reported that s.c. injection of orexin A stimulated corticosterone, but not aldosterone secretion in conscious rats.[46] In another publication, these same authors reported that s.c. injection of orexin A stimulated ACTH, corticosterone and aldosterone secretion in rats.[47] They then followed those acute administration studies with a report that prolonged, systemic orexin administration (seven days) failed to alter adrenal weight or morphology but resulted in elevated concentrations of corticosterone and aldosterone, but not ACTH, in plasma.[48] Others have contributed similar observations.[49]

Using dispersed cell preparations of human adrenal tissues, Mazzochi and colleagues reported that orexin A but not orexin B stimulated basal cortisol secretion and enhanced the secretory response to angiotensin II, endothelin and ACTH.[42] No significant effects were observed on aldosterone or catecholamine release from adrenal slices. As was the case in their studies on rat tissues, both hypocretin receptors were detected in human adrenal gland extracts, in this case the mRNA for the hcrtr1 was most abundant. Similarly, as in rat tissues, the effect of orexin A appeared to be via activation of adenylyl cyclase. These same investigators also demonstrated the expression of the hcrtr2, but not the hcrtr1, gene in human pheochromocytomas and the ability of orexin A and orexin B to stimulate catecholamine release from these cells. The hcrtr2-mediated effect of the orexins was mediated not by adenylyl cyclase, but instead via activation of PLC and PKC.[42] Opposing results were reported in an immortalized, rat pheochromocytoma cell line, PC12 cells.[50] Here orexin A and B reduced PACAP-stimulated transcription of the gene for tyrosine hydroxylase, the rate limiting step of catecholamine synthesis, and inhibited PACAP-induced dopamine secretion, probably via an action on the hcrtr2 receptor. These effects were mediated at least in part via activation of adenylyl cyclase.

How does one reconcile these apparently disparate results? Could it be merely a species difference? Are the effects in vitro dependent on culture conditions? One thing is certain, when administered systemically in vivo, the hypocretins appear to increase glucocorticoid secretion. Is this pharmacology or do these results indicate a direct, physiologically relevant role for the peptides in adrenal function? Is there enough hypocretin produced in adrenal gland, or delivered via the general circulation, to exert autocrine or endocrine effects in the gland? Is there a relationship between these findings and the ability of glucocorticoids to increase hypocretin gene expression in hypothalamic neurons?[51] Are the secretory dynamics of mineralocorticoids and glucocorticoids altered in genetic models of orexin absence, and if so do those alterations support the stimulatory effects of the peptides advanced by the studies of Malendowicz and colleagues?

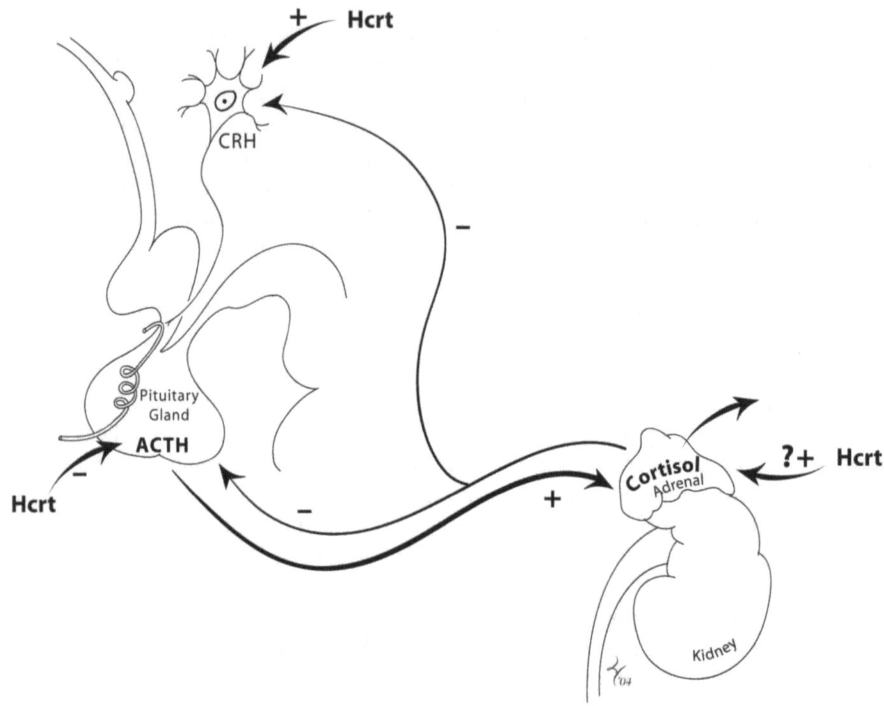

Figure 1. Schematic summary of the actions of hypocretin at the three levels of the hypothalamo-pituitary-adrenal axis. (-, inhibitory effect; +, stimulatory effect; ?, conflicting data exists).

6. AN INTEGRATIVE VIEW OF THE ACTIONS OF THE HYPOCRETINS IN THE HYPOTHALAMO-PITUITARY-ADRENAL AXIS (Figure 1)

There is consensus that the hypocretins act in brain to stimulate the HPA axis, elevate sympathetic tone, and elicit behaviors typical of the stress reaction. These effects are not unique to the hypocretins. Several other peptides exert similar effects in brain to activate the HPA axis and sympathetic tone, while stimulating ingestive and locomotor activities.[52-55] Is it possible that the ability of the hypocretins to stimulate the HPA axis by an action in brain is a reflection of, but not the primary response to, the generalized stress response?[56] Is the effect of orexin any more important than the response of the HPA axis to those other peptides? Genetic models lend some insight into this question.

If the direct, pharmacological effects of the hypocretins to stimulate glucocorticoid or mineralocorticoid secretion are physiologically relevant, then one would expect that knockout mice in which the gene was deleted might manifest altered electrolyte status and metabolism. However, these animals at 13-15 weeks of age displayed normal plasma electrolytes and blood glucose levels.[57] Hypocretin expression was already absent in the embryos, and thus endogenous hypocretin is not essential for development of the adrenal gland or for the ability to shift from a liquid diet to solid food at weaning. Interestingly, even though these animals have never "seen" orexin, their cataplexic phenotype can be rescued by ectopic hypocretin expression or acute peptide administration,[58] thus the hypocretin receptors remain functional.

These animals have been useful for the examination of the physiologic relevance of the sympathostimulatory and behavioral actions of the hypocretins. In a resident-intruder model of emotional stress, the orexin knockout mice displayed a significantly lower cardiovascular and behavioral response to the introduction of the intruder, or when the knockout mouse was the intruder, to the novelty of the paradigm.[59] These observations, taken together with the fact that basal blood pressures were lower in the knockout mice, strongly suggests that the effects of exogenous hypocretin on autonomic function[27,56,60] have a physiologic correlate. It would be very interesting to measure circulating corticosterone levels in those animals to determine, in a similar manner, the role of endogenous hypocretin in the control of HPA axis activity. We would predict that basal corticosterone levels in these animals would be elevated and that hormone secretion in response to emotional stress would be enhanced, reflecting the loss of hypocretin's inhibitory effect in the pituitary gland.[36,61]

The transgenic model that perhaps best mirrors human narcolepsy is the ataxin-3 expressing mouse[62] and now rat.[30] These animals express normal levels of the hypocretins until early adulthood, when expression of the truncated Machado-Joseph disease gene (driven by the orexin promotor) product causes lethality in neurons expressing the protein. The mice display as adults the behavioral phenotype most similar to human narcolepsy and in addition develop late-onset obesity, and what appears to be insulin resistance. Transgenic mice ate less food, but consistently gained more weight than wild type littermates. The decreased food intake may have been due to a loss of the orexigenic action of the hypocretins. These transgenic animals did display less spontaneous motor activity than controls during the dark phase, when the cataplexic attacks were most prevalent, and therefore a decreased metabolic rate may have contributed to the weight gain, although his was not directly assessed. This is suggested by the observation that in normal animals orexin stimulates oxygen consumption, an effect that reflects increased metabolic rate.[63]

The orexin/ataxin-3 mice present a potential model for the physical appearance of many human narcoleptics, in whom obesity and type 2 diabetes are common.[64,65] This combination of obesity and insulin resistant diabetes is also a common feature of Cushing's Syndrome patients, who suffer from excess production of cortisol, and display an obese body habitus and insulin resistance secondary to the elevated cortisol levels.[66] Could it be that in the absence of orexin's actions in pituitary gland to "buffer" the release of ACTH in response to endogenous CRH, more corticosterone is secreted in the knockout mice and, that in human narcoleptics some of the increased weight gain and insulin resistance are due to a similar loss of orexin action? This in fact is our working hypothesis and we are beginning a collaborative study in a human population of narcoleptics to examine this possibility.

7. CONCLUDING REMARKS

What then is the significance of the pharmacologic effects of exogenous hypocretin on the HPA axis? In a narrow sense, information gained on the central and possible peripheral actions of the peptides in several model systems has provided insight into the physiologic regulation of several homeostatic mechanisms and the pathologic basis of at least one disease. The original observations of increased alertness and activity certainly predicted a role for the peptides in the mechanisms of arousal and the regulation of autonomic function, which the gene deletion models now have established. Perhaps in a broader sense, these pharmacologic findings predict the potential for success for selective agonists in the treatment of narcolepsy, but warn also of potentially dangerous side effects of uncontrolled use of those compounds. Indeed, if the pharmacologic effect of hypocretin in pituitary gland is physiologically relevant, then caution must be exerted in the design and initial testing of hypocretin analogs in the clinical setting, in order to avoid the creation of an Addison's-like syndrome (secondary adrenal insufficiency)[66] that may replace the narcoleptic disease state.

8. REFERENCES

1. C. Peyron, D. K. Tighe, A. N. van den Pol, L. de Lecea, H. C. Heller, J. G, Sutcliffe and T. S. Kilduff, Neurons containing hypocretin (orexin) project to multiple neuronal systems, *J. Neurosci.* **18**(23), 9996-10015 (1998).
2. J. N. Marcus, C. J. Aschkenasi, C. E. Lee, R. M. Chemelli, C. B. Saper, M. Yanagisawa and J. K. Elmquist, Differential expression of orexin receptors 1 and 2 in the rat brain, *J. Comp. Neurol.* **435**, 6-25 (2001).
3. G. J. Hervieu, J. E. Cluderay, D. C. Harrison, J. C. Roberts and R. A. Leslie, Gene expression and protein distribution of the orexin-1 receptor in the rat brain and spinal cord, *Neuroscience* **103**, 777-797 (2001).
4. X. Y. Lu, D. Bagnol, S. Burke, H. Akil and S.J. Watson, Differential distribution and regulation of OX1 and OX2 orexin/hypocretin receptor messenger RNA in the brain upon fasting, *Horm. Behav.* **37**, 335-2344 (2000).
5. L. W. Swanson, in: *Handbook of Physiology, Section 1: The Nervous System,* edited by V. Mountcastle, F. E. Bloom, and S. R. Geiger (American Physiological Society, Bethesda, 1986), pp. 317-364.
6. J. G. Tasker and F. E. Dudek, Electrophysiological properties of neurones in the region of the paraventricular nucleus in slices of rat hypothalamus, *J. Physiol.* **434**, 271-293 (1991).
7. J. E. Stern, Electrophysiological and morphological properties of pre-autonomic neurons in the rat hypothalamic paraventricular nucleus, *J. Physiol.* **537**(1), 161-177 (2001).
8. J. A. Luther, S. S. Daftary, C. Boudaba, G. C. Gould, K. C. Halmos and J. G. Tasker, Neurosecretory and non-neurosecretory parvocellular neurons of the hypothalamic paraventricular nucleus express distinct electrophysiological properties.,*J. Neuroendocrinol.* **14**, 929-932 (2002).
9. R. D. Cone, M. J. Low, J. K. Elmquist and J. L. Cameron, in: *Williams Textbook of Endocrinology, 10th edition*, edited by P. R. Larsen, H. M. Kronenberg, S. Melmed, and K. S. Polonsky (Saunders, Philadelphia, 2003), pp. 81-176.
10. A. G. Robinson and J. G. Verbalis, in: *Williams Textbook of Endocrinology, 10th edition*, edited by P. R. Larsen, H. M. Kronenberg, S. Melmed, and K. S. Polonsky (Saunders, Philadelphia, 2003), pp. 281-330.
11. W. Vale, J. Vaughan, M. Smith, G. Yamamoto, J. Rivier and C. Rivier. Effects of synthetic ovine corticotropin-releasing factor, glucocorticoids, catecholamines, neurohypophyseal peptides, and other substances on cultured corticotropic cells, *Endocrinology* **113**, 1121-1131 (1983).
12. S. Melmed and D. L. Kleinberg, in: *Williams Textbook of Endocrinology, 10th edition*, edited by P. R. Larsen, H. M. Kronenberg, S. Melmed, and K. S. Polonsky (Saunders, Philadelphia, 2003), pp. 177-280.
13. L. W. Swanson, P. E. Sawchenko and R. W. Lind, Regulation of multiple peptides in CRF parvocellular neurosecretory neurons: implications for the stress response, *Progress Brain Res.* **68**, 169-90 (1986).

14. J. J. Hagan, R. A. Leslie, S. Patel, M. L. Evans, T. A. Wattam, S. Holmes, C. D. Benham, S. G. Taylor, C. Rutledge, P. Hemmati, R. P. Munton, T. E. Ashmeade, A. S. Shah, J. P. Hatcher, P. D. Hatcher, D. N. C. Jones, M. I. Smith, D. C. Piper, A. J. Hunter, R. D. Porter and N. Upton, Orexin A activates locus coeruleus cell firing and increases arousal in the rat, *Proc. Natl. Acad. Sci. USA* **96**, 10911-10916 (1999).
15. W. K. Samson, M. M. Taylor, M. J. Follwell and A. V. Ferguson, Orexin actions in hypothalamic paraventricular nucleus: physiological consequences and cellular correlates, *Regul. Pept.* **104**, 97-103 (2002).
16. M. Jaszberenyi, E. Bujdoso, I. Pataki and G. Telegdy, Effects of orexins on the hypothalamic-pituitary-adrenal system, *J. Neuroendocrinol.* 12, 1755-1758 (2000).
17. T. Ida, K. Nakahara, T. Murakami, R. Hanada, M. Nakazato and N. Murakami, Possible involvement of orexin in the stress reaction in rats, Biochem. *Biophys. Res. Commun.* **270**, 318-323 (2000).
18. M. Kuru, Y. Ueta, R. Serino, M. Nakazato, Y. Yamamoto, I. Shibuya and H. Yamashita, Centrally administered orexin/hypocretin activate the HPA axis in rats, *Neuroreport* **11**, 1977-1980 (2000).
19. K. A. Al-Barzanji, S. Wilson, J. Baker, D. S. Jessop and M. S. Harbuz, Central orexin-A activates the hypothalamic-pituitary-adrenal axis and stimulates hypothalamic corticotropin releasing factor and arginine vasopressin neurons in conscious rats, *J. Neuroendocrinol.* **13**, 421-424 (2001).
20. S. H. Russell, C. J. Small, C. L. Dakin, C. R. Abbott, D. G. A. Morgan, M. A. Ghatei and S. R. Bloom, The central effects of orexin-A in the hypothalamic-pituitary-adrenal axis in vivo and in vitro in male rats.,*J. Neuroendocrinol.* **13**, 561-566 (2001).
21. F. Sakamoto, S. Yamada and Y. Ueta, Centrally administered orexin-A activates corticotropin-releasing factor-containing neurons in the hypothalamic paraventricular nucleus and central amygdaloid nucleus of rats: possible involvement of central orexins on stress-activated central CRF neurons, *Regul. Pept.* **118**, 183-191 (2004).
22. Y. Date, Y. Ueta, H. Yamashita, H. Yamaguchi, S. Matsukura, K. Kangawa, T. Sakurai, M. Yanagisawa and M. Nakazato, Orexins, orexigenic hypothalamic peptides, interact with autonomic, neuroendocrine and neuroregulatory systems, *Proc. Natl. Acad. Sci. USA*. **96**, 748-753 (1999).
23. M. Backberg, G. Hervieu, S. Wilson and B. Meister, Orexin receptor-1(OX-R1) immunoreactivity in chemically identified neurons of the hypothalamus: focus on orexin targets involved in control of food and water intake, *Eur. J. Neurosci.* **15**, 315-328 (2002).
24. C. D. Sladek and J. R. Kapoor, Neurotransmitter/neuropeptide interactions in the regulation of neurohypophyseal hormone release, *Exp. Neurol.* **171**, 200-209 (2001).
25. K. Matsumura, T. Tsuchihashi and I. Abe, Central orexin-A augments sympathoadrenal outflow in conscious rabbits, *Hypertension* **37**, 1382-1387 (2001).
26. M. J. Follwell and A.V. Ferguson, Cellular mechanisms of orexin actions on paraventricular nucleus neurons in rat hypothalamus, *J. Physiol.* **545**, 855-867 (2002).
27. T. Shirasaka, S. Miyahara, T. Kunitake, Q. H. Lin, K. Kato, M. Takasaki and H. Kannan, Orexin depolarizes rat hypothalamic paraventricular nucleus neurons, Am. J. Physiol. **281**, R1114-R1118 (2001).
28. B. Yang and A.V. Ferguson, Orexin-A depolarizes dissociated rat area postrema neurons through activation of a nonselective cationic conductance, *J. Neurosci.* **22**, 6303-6308 (2002).
20. B. Yang and A.V. Ferguson, Orexin-A depolarizes nucleus tractus solitarius neurons through effects on nonselective cationic and K^+ conductances. *J. Neurophsyiol.* **89**, 2167-2175 (2003).
30. C. T. Beuckmann, C. M. Sinton, S. C. Williams, J. A. Richardson, R. E. Hammer, T. Sakurai and M. Yanagisawa, Expression of a poly-glutamine-ataxin-3 transgene in orexin neurons induced narcolepsy-cataplexy in the rat, *J. Neurosci.* **24**, 4469-4477 (2004).
31. Y. Date, M. S. Mondal, S. Matsukura, T. Ueta, H. Yamashita, H. Kaiya, K. Kangawa and M. Nakazato, Distribution of orexin/hypocretin in the rat median eminence and pituitary, Mol. Brain Res. **76**, 1-6 (2000).
32. O. Joehren, S. J. Neidert, M. Kummer, A. Dendorfer and P. Dominiak, Prepro-orexin and orexin receptor mRNA are differentially expressed in peripheral tissues of male and female rats, *Endocrinology* **142**, 3324-3331 (2001).
33. O. Joehren, N. Brueggemann, A. Dendorfer and P. Dominiak, Gonadal steroids differentially regulate the messenger ribonucleic acid expression of pituitary orexin type 1 receptors and adrenal orexin type 2 receptors, *Endocrinology* **144**, 1219-1225 (2003).
34. R. W. Xu, Q. L. Wang, M. Yan, M. Hernandez, C. H. Gong, W. C. Boon, Y. Murata, Y. Ueta and C. Chen, Orexin-A augments voltage-gated Ca^{2+} currents and synergistically increases growth hormone (GH) secretion with GH-releasing hormone in primary cultured ovine somatotropes, *Endocrinology* **143**, 4609-4619 (2002).
35. C. Chen and R. W. Xu, The in vitro regulation of growth hormone secretion by orexins, *Endocrine.* **22**: 57-66 (2003).

36. W. K. Samson and M. M. Taylor, Hypocretin/orexin suppresses corticotroph responsiveness in vitro, *Am. J. Physiol.* **281**, R1140-R1145 (2001).
37. M. Blanco, M. Lopez, T. Garcia-Caballero, R. Gallego, A. Vazquez-Boquette, G. Morel, R. Senaris, F. Casanueva, C. Dieguez and A. Beiras, Cellular localization of orexin receptors in human pituitary, *J. Clin. Endocrinol. Metab.* **86**, 1616-1619 (2001).
38. T. Sakurai, A. Amemiya, M. Ishii, I. Matsuzaki, R. M. Chemelli, H. Tanaka, S. C. Williams, J. A. Richardson, G. P. Kozlowski, S. Wilson, J. R. S. Arch, R. E. Buckingham, A. C. Haynes, S. A. Carr, R. S. Annan, D. E. McNulty, W/S. Liu, J. A. Terrett, N. A. Elshourbagy, D. J. Bergsma and M. Yanagisawa, Orexins and orexin receptors: a family of hypothalamic neuropeptides and G protein-coupled receptors that regulate feeding behavior, *Cell.* **92**, 573-585 (1998).
39. A. N. van den Pol, X. B. Gao, K. Obrietan, T. S. Kilduff and A. B. Belousov, Presynsptic and postsynaptic actions and modulation of neuroendocrine neurons by a new hypothalamic peptide, hypocretin/orexin, *J. Neurosci.* **18**, 7962-7971 (1998).
40. A. N. van den Pol, P. R. Patrylo, P. K. Ghosh and X. B. Gao, Lateral hypothalamus: early developmental expression and response to hypocretin (orexin*), J. Comp. Neurol*, **433**, 349-363 (2001).
41. F. A. Antoni, Calcium regulation of adenylyl cyclase: elevance for endocrine control, *Trends Endocrinol. Metab.* **8**, 7-14 (1997).
42. G. Mazzocchi, L. K. Malendowicz, L. Gottardo, F. Aragona and G. G. Nussdorfer, Orexin stimulates cortisol secretion from human adrenocortical cells through activation of the adenylate cyclase-dependent signaling cascade, *J. Clin. Endocrinol. Metab.* **86**, 778-782 (2001).
43. M. Lopez, R. Senaris, R. Gallego, T. Garcia-Cabellero, F. Lago, L. Seoane, F. Casanueva and C. Dieguez Orexin receptors are expressed in the adrenal medulla of the rat, *Endocrinology* **140**, 5991-5994 (1999).
44. H. S. Randeva, E. Karteris, D. Grammatopoulos and E. W. Hillhouse, Expression of orexin-A and functional orexin type 2 receptors in the human adult adrenals: implications for adrenal function and energy homeostasis, *J. Clin. Endocrinol. Metab.* **86**, 4808-4813 (2001).
45. L. K. Malendowicz, A. Hochol, A. Ziolkowska, M. Nowak, L. Gottardo and G. G. Nussdorfer, Prolonged orexin administration stimulates steroid-hormone secretion, acting directly on the rat adrenal gland, *Int. J. Mol. Med.* **7**, 401-404 (2001).
46. L. K. Malendowicz, C. Tortorella and G. G. Nussdorfer, Orexins stimulate corticosterone secretion of rat adrenocortical cells, through the activation of the adenylate cyclase-dependent signaling cascade, *J. Steroid Biochem. Mol. Biol.* **70**, 185-188 (1999).
47. L. K. Malendowicz, C. Tortorella and G. G. Nussdorfer, Acute effects of orexins A and B on the rat pituitary-adreocortical axis, *Biomed. Res.* 20, 301-304 (1999).
48. L. K. Malendowicz, N. Jedrzejczak, A. S. Belloni, M. Trejter, A. Hochol and G. G. Nussdorfer, Effects of orexins A and B on the secretory and proliferative activity of immature and regenerating rat adrenal glands, *Histol. Histopathol.* **16**, 713-717 (2001).
49. M. Nowak, A. Hochol, C. Tortorella, N. Jedrzejczak, A. Ziolkowska, G. G. Nussdorfer and L. K. Malendowicz, Modulatory effects of orexins on the function of rat pituitary-adrenocortical axis under basal and stressful conditions, *Biomed. Res.* **21**, 89-93 (2000).
50. T. Nanmoku, K. Isobe, T. Sakurai, A. Yamanaka, K. Takekoshi, Y. Kawakami, K. Ishii, K, Goto and T. Nakai, Orexins suppress catecholamine synthesis and secretion in cultured PC12 cells, *Biochem. Biophys. Res. Commun.* **274**, 310-315 (2000).
51. A. Stricker-Krongrad and B. Beck, Modulation of hypothalamic hypocretin/orexin mRNA expression by glucocorticoids, *Biochem. Biophys. Res. Commun.* **296**, 129-133 (2002).
52. W. K. Samson, T. C. Murphy and Z. T. Resch, Central mechanisms for the hypertensive effects of preproadrenomedullin-derived peptides in conscious rats, *Am. J. Physiol.* **274**, R1505-R1509 (1998).
53. M. M. Taylor and W. K. Samson, A possible mechanism for the action of adrenomedullin in brain to stimulate stress hormone secretion, *Endocrinology* in press (2004).
54. W. K. Samson, Z. T. Resch and T. C. Murphy, A novel action of the newly described prolactin-releasing peptides: cardiovascular regulation, *Brain Res.* **858**, 19-25 (2000).
55. W. K. Samson, C. Keown, C. K. Samson, H. W. Samson, B. Lane, J. R. Baker and M. M. Taylor, Prolactin-releasing peptide and its homolog RFRP-1 act in hypothalamus but not in anterior pituitary gland to stimulate stress hormone secretion, *Endocrine.* **20**, 59-66 (2003).
56. W. K. Samson, B. Gosnell, J. K. Chang, Z. T. Resch and T.C. Murphy, Cardiovascular regulatory actions of the hypocretins in brain, *Brain Res.* **831**, 248-253 (1999).
57. R. M. Chemelli, J. T. Willie, C. M. Sinton, J. K. Elmquist, T. Scammell, C. Lee, J. A. Richardson, S. C. Williams, Y. Xiong, Y. Kisanuki, T. E. Fitch, M. Nakazato, R. E. Hammer, C. B. Saper and M. Yanagisawa, Narcolepsy in orexin knockout mice: molecular genetics of sleep regulation, *Cell.* **98**, 437-451 (1999).

58. M. Mieda, J. T. Willie, J. Hara, C. M. Sinton, T. Sakurai and M. Yanagisawa, Orexin peptides prevent cataplexy and improve wakefulness in an orexin neuron-ablated model of narcolepsy in mice, *Proc. Natl. Acad. Sci. USA.* **101**,4649-4654 (2004).
59. Y. Kayaba, A. Nakamura, Y. Kasuya, T. Ohuchi, M. Yanagisawa, I. Komuro, Y. Fukuda and T. Kuwaki, Attenuated defense response and low basal blood pressure in orexin knockout mice, *Am. J. Physiol.* **285**, R581-R593 (2003).
60. A. V. Ferguson and W. K. Samson, The orexin/hypocretin system: a critical regulator of neuroendocrine and autonomic function, *Frontiers Neuroendocrinol.* **24**, 141-150 (2003).
61. M. M. Taylor and W. K. Samson, The other side of the orexins: endocrine and metabolic actions, *Am. J. Physiol.* **284**, E13-E17 (2003).
62. J. Hara, C. T. Beuckmann, T. Nambu, J. T. Willie, R. M. Chemelli, C. M. Sinton, F. Sugiyama, K. Yagami, K. Goto, M. Yanagisawa and T. Sakurai, Genetic ablation of orexin neurons in mice results in narcolepsy, hypophagia, and obesity, *Neuron* **30**, 345-354 (2001).
63. M. Lubkin and A. Stricker-Krongard, Independent feeding and metabolic actions of orexins in mice.,*Biochem. Biophys. Res. Commun.* **253**, 241-245 (1998).
64. Y. Honda, Y. Doi, R. Ninomiya and C. Ninomiya, Increased frequency of non-insulin-dependent diabetes mellitus among narcoleptic patients, *Sleep* **9**, 254-259 (1986).
65. A. Schuld, J. Hedebrand, F. Geller and T. Pollmaecher, Increased body mass index in patients with narcolepsy, *Lancet* **355**, 1274-1275 (2000).
66. P. M. Stewart, in: *Williams Textbook of Endocrinology, 10th edition*, edited by P. R. Larsen, H. M. Kronenberg, S. Melmed and K. S. Polonsky (Saunders, Philadelphia, 2003), pp.491-551.

OREXINS (HYPOCRETINS) IN THE GUT

Annette L Kirchgessner and Erik Näslund *

1. INTRODUCTION

The orexins (from the Greek for "appetite", orexin A (synonymous with hypocretin-1; 33 amino acids) and orexin B (hypocretin-2; 28 residues)) are derived from the prepro-hcrt/orexin (preprohypocretin) gene. These two peptides interact with two G-protein-coupled receptors, hcrt/orexin receptor-1 and hcrt/orexin receptor-2. The hcrt/orexin receptor-1 is selective for hcrt 1/orexin A and the hcrt/orexin receptor-2 exhibits similar affinity for hcrt1 and 2/orexin A and B.[1] (N.E. See note on nomenclature).

Hcrt/orexins were first localized to a specific population of neurons in the lateral hypothalamic area the classical "feeding center".[1] This prompted studies of the hcrt/orexin peptides' effect on food intake. Both hcrt/orexin 1 and 2 have been reported to stimulate food intake in rats when injected centrally.[1,2] Antibodies against hcrt/orexin 1 inhibit fasting-induced feeding[3] and the hyperphagic action of hcrt/orexin 1 appears to be mediated mainly by the hcrt/orexin receptor-1, as specific hcrt/orexin receptor-1 antagonists (SB-334867) inhibit spontaneous, fasting-induced and hcrt/orexin 1-induced feeding.[4] It is now generally recognized that hcrt/orexin 2 causes little or no stimulation of feeding.[5]

Hcrt/orexin neurons also respond to the availability of metabolic fuels. They are sensitive to glucose and are activated by hypoglycemia.[6-8] In addition, hcrt/orexin neurons project extensively throughout the CNS, allowing communication with other hypothalamic regions, the locus coeruleus and the nucleus of the solitary tract[9], a site that receives feeding-related signals, such as glucose availability and gastric distension, and relays this information to the hypothalamus.[7,8] These data support the involvement of hcrt/orexin neurons in feeding, perhaps enhancing hunger and initiating feeding in response to falling glucose levels when the gut is empty.

* A.Kirchgessner, NPS Pharmaceuticals Inc., 30 College Street, Suite 301, Toronto, Ontario, M5G 1K2 Canada and Eric Naslund, Division of Surgery, Karolinska Institutet, Danderyd Hosptial, SE-182 88 Stockholm, Sweden

2. DISTRIBUTION OF HCRT/OREXIN IN THE GASTROINTESTINAL TRACT

2.1. Rodent Gastrointestinal Tract

In 1999, Kirchgessner and Liu[10] reported that both hcrt/orexins and their receptors are expressed in the enteric nervous system (ENS; for review, see ref 11). These findings were subsequently confirmed.[11,12] Neurons in the submucosal and myenteric plexus and endocrine cells in the intestinal mucosa of the rat[10,12], mouse and guinea-pig [10] display hcrt/orexin 1 and hcrt/orexin receptor-1 immunoreactivity. In the rat myenteric plexus of the small intestine hcrt/orexin 1 and nitric oxide synthase (NOS) co-localize and there is hcrt/orexin receptor-1 immunoreactivity on NOS positive cell bodies.

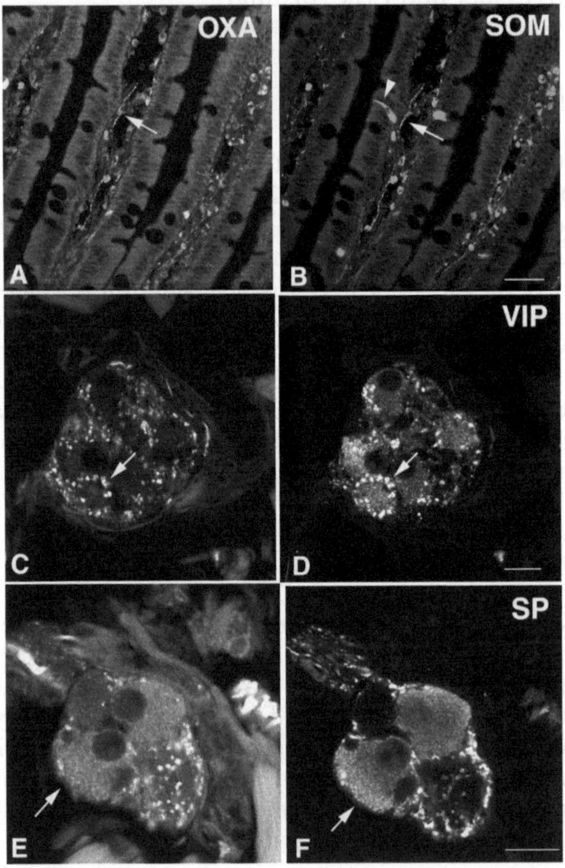

Figure 1. Hcrt/orexin A (hcrt1/OXA) immunoreactivity in the human duodenum. Varicose OXA nerve fibres (A; arrow) are abundant in the lamina propria of the mucosa. OXA nerve fibres contain somatostatin (SOM) (arrow; B) and substance P (SP) (not illustrated) and are close to SOM-containing enteroendocrine cells (arrowhead). OXA nerve fibres surround neurons in the submucosal plexus (C) and these nerve fibers co-express vasoactive intestinal peptide (VIP) (D). OXA cell bodies (E; arrow) in the submucosa contain SP (F; arrow) and SOM (not illustrated). Horizontal markers, 10 μm.

In addition hcrt/orexin 1-immunoreactive nerve fibers frequently encircle NOS-positive neurons.[13] This is also observed in the myenteric plexus of the rat gut (unpublished data). Furthermore, in the rat duodenum, hcrt/orexin 1 and 5-hydroxytryptamine (5-HT) are co-stored in enterochromaffin (EC) cells that express hcrt/orexin receptor-2 immunoreactivity suggesting modulation of hcrt/orexin 1/5-HT secretion. Hcrt/orexin 1-immunoreactive EC cells do not contain cholecystokinin (CCK) but some co-store substance P (SP). In both the submucosal and myenteric plexus, hcrt/orexin 1 immunoreactivity is found in cholinergic neurons, a subset of which contain vasoactive intestinal peptide (VIP).[12]

Hcrt/orexin-positive neurons in the gut, like those in the hypothalamus[14], are activated by fasting[10], indicating a functional role for hcrt/orexin 1 in the gut in connection with food intake. Indeed, plasma concentrations of hcrt/orexin 1 increase during fasting in both rats[15] and humans.[16] A likely source of plasma hcrt/orexin 1 would be the gut.

Another possible source of regulation of hcrt/orexin neurons involves ghrelin, a peptide predominantly produced in endocrine cells (X-cells) of the upper stomach[17], and the endogenous ligand for the GH secretagogue receptor.[18] Like hcrt/orexin 1, ghrelin stimulates feeding and blood glucose levels regulate its secretion. Thus, plasma levels of ghrelin also increase with fasting and fall promptly after eating.[19] Interestingly, ghrelin has been shown to activate hcrt/orexin neurons in the CNS.[20] The anatomical and functional relationship between ghrelin and hcrt/orexins in the gut has not yet been explored. Further investigation might provide novel insights into the regulation of feeding and energy homeostasis by gut peptides.

2.2. Human Gastrointestinal Tract

We recently examined the distribution of hcrt/orexins and hcrt/orexin receptors in the human ENS (unpublished data). Hcrt/orexin- and hcrt/orexin receptor-like immunoreactivity was observed in all regions of the human bowel. Hcrt/orexin 1 immunoreactivity was observed in neuronal cell bodies in the submucosal plexus and varicose nerve fibers in the mucosa, ganglia and circular muscle layer. Hcrt/orexin A-positive neurons contained somatostatin (SOM) and SP (Figure 1). Endocrine cells in the stomach and small intestine expressed hcrt/orexin A, and a subset of these cells contained gastrin (antrum), 5-HT or SP (small intestine). Hcrt/orexin receptor-1 immunoreactivity was found on parietal cells in the corpus, supporting a role for hcrt/orexin 1 in gastric acid secretion. Neuronal cell bodies and epithelial cells also expressed hcrt/orexin receptor-1. In contrast, enteroendocrine cells expressed hcrt/orexin receptor-2, as in the rat.

3. HCRT/OREXIN EFFECTS ON VAGAL AFFERENT SIGNALING

In the last 30 years a substantial amount of data has shown that CCK released from the small intestine acts directly on CCK-A receptors expressed by vagal afferent neurons innervating the stomach and jejunum.[21] This leads to inhibition of food intake, relaxation of the stomach and inhibition of gastric emptying. The idea emerging is that CCK effects on the vagus can be modulated by both potentiating and inhibitory factors, thus affecting food intake.

Vagal afferent neurons that express CCK-A receptors also express leptin receptors and there is functional evidence for potentiating interactions between leptin and CCK both in stimulating the discharge of vagal afferents and in inhibiting food intake.

Recently, it has been shown that the hcrt/orexin receptor-1 is also expressed by human and rat vagal afferent neurons and most importantly, that hcrt/orexin A inhibits the action of CCK on vagal afferent nerve discharge[22], suggesting a role in the modulation of gut to brain signaling. Inhibitory effects of orexigenic peptides on the vagus nerve are not limited to hcrt/orexin. Ghrelin receptors are also found on vagal afferents.[23]

Based on the localization of hcrt/orexin and hcrt/orexin receptors in the gut, the morphological data suggest a role for hcrt/orexin 1 in regulating gastrointestinal function. In addition, as reviewed in previous chapters, hcrt/orexin 1 neurons project to areas of the brainstem associated with gastrointestinal function, such as the dorsal motor nucleus of the vagus (DMN).[24] These areas of the CNS also express hcrt/orexin receptors.[25] Recent data also demonstrate that hcrt/orexin receptor-1 is present in the nodose ganglia in both rodents and humans[22], where humans also express hcrt/orexin receptor-2. This study also demonstrates that hcrt/orexin 1 by itself does not affect the resting discharge of afferent vagal nerve fibers but inhibits responses to CCK.[22] These data, taken together with the expression of hcrt/orexin A in enteric neurons and enteroendocrine cells suggest that hcrt/orexin A may have local effects in the gastrointestinal tract, including modulating responses to other gastrointestinal peptides (satiety signals) and signaling to the CNS. This review will now examine the effects of hcrt/orexin 1 on gastrointestinal motility and secretion.

4. FASTING SMALL BOWEL MOTILITY

Few studies have examined the effects of peripheral hcrt/orexin 1 on gastrointestinal function *in vivo. In vitro,* hcrt/orexin 1 has been demonstrated to increase colonic motility in guinea pigs.[10] In both rats and humans, fasting small bowel motility is characterized by the myoelectric motor complex (MMC). The MMC is characterized by three phases; in phase 1 there is motor quiescence, in phase 2 irregular spiking and in phase 3 a propagated motor pattern resulting in an activity front. In rats phase 3 of the MMC occurs approximately every 15 min and in humans 90-180 min. IV administration of both hcrt/orexin 1 and hcrt/orexin 2 to rats disrupts the MMC and lengthens the time between the phase 3 of the MMC.[12] The effect of hcrt/orexin A on the MMC is dose dependent (50 to 5000 pmol/kg/min), where the higher dose completely replaces the MMC with irregular spiking during the 60 min period studied. By employing the hcrt/orexin receptor-1 selective antagonist SB-334867-A[26], it was shown that the effect of hcrt/orexin 1 on fasting motility is mediated through the hcrt/orexin receptor-1.[13] Neither vagotomy nor pretreatment with guanethidine influenced the effect of hcrt/orexin 1, suggesting that the effect of the peptide on the MMC is independent of the CNS. The effect of hcrt/orexin 1 on the MMC is, however, partly mediated by nitric oxide (NO) as pretreatment with the NOS inhibitor, N -nitro-L-arginine (L-NNA), inhibited the effect of hcrt/orexin 1 on the MMC.[13] These results are consistent with the finding that hcrt/orexin 1-containing neurons in the myenteric plexus are in close proximity to NOS-immunoreactive nerve fibers. In addition, an IV bolus of SB-334867-A induces a premature activity front of the MMC and shortens the interval between activity fronts of the MMC.[13] These data taken together suggest that hcrt/orexin 1 is involved in regulating fasting small bowel motility in the rat.

5. GASTRIC EMPTYING

Central administration of hcrt/orexin 1 has been shown to stimulate gastric motility in rats and this effect is mediated by vagal excitatory motor neurons in the DMN.[27] Further studies have shown that neurons in the DMN demonstrate a viscerotopically organized sensitivity to hcrt/orexin 1. Seventy percent of neurons projecting to the fundus or corpus of the stomach increase their firing rate when hcrt/orexin A is applied, whereas only 3/13, 4/18 and 1/13 of neurons projecting to the antrum/pylorus, duodenum and caecum, respectively, increase their firing rate when hcrt/orexin 1 is applied.[28]

Preliminary data from our laboratory has demonstrated that IV infusion of hcrt/orexin 1 (500 pmol/kg/min) does not effect gastric emptying of a non-nutrient (radioactively-marked polyethylene glycol (PEG) 4000) or nutrient liquid (50% radioactively-marked PEG 4000 and 50% standard enteral feed (1 kcal/mL)) meal in rats after an overnight fast. However, the administration of the hcrt/orexin receptor-1 antagonist SB-334867-A (10 mg/kg bolus and 0.1 mg/kg/min infusion) results in increased gastric retention of the nutrient liquid (Figure 2), suggesting that endogenous hcrt/orexin A, but not exogenous hcrt/orexin A may influence gastric emptying in rats.

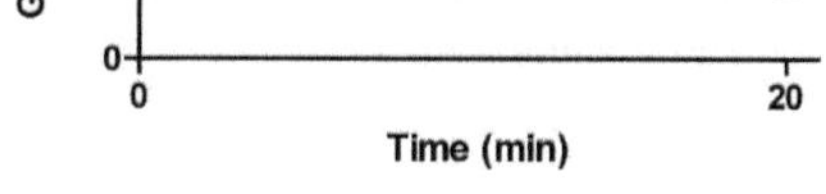

Figure 2. Effect of intravenous hcrt/orexin 1 and an hcrt/orexin receptor-1 antagonist (SB-334867-A) on gastric emptying of a non-nutrient liquid load (PEG 4000) or a liquid nutrient (50% standard enteral feed and 50% PEG 4000). Data shown percent remaining in the stomach at 0 and 20 min. (*) $p<0.05$ SB-334867-A enteral feed only vs other data, Tukey-Kramer HSD test (Saline + PEG 4000 (solid square), hcrt1/OXa+PEG 4000 (solid triangle), SB-334867-A+PEG 4000 (solid circle), saline+enteral feed (open square), hcrt1/OXA+enteral feed (open triangle), SB-334867-A+enteral feed (open circle).

We also have preliminary data regarding the effect of IV hcrt/orexin 1 on gastric emptying of a solid meal in humans. Healthy human volunteers were studied on two occasions with either saline or hcrt/orexin 1 IV (10 mg/kg/min) for 190 min. Ten min after the infusion of hcrt/orexin 1 began the subjects consumed a 310 kcal radioactively-marked omelet. Gastric emptying was followed for 120 min and visual analogue scales for hunger and satiety were recorded for 180 min. The lag phase and half-emptying time were not affected by hcrt/orexin 1, but there was a weak inhibitory effect on the rate of gastric emptying (%/min) by hcrt/orexin 1. VAS ratings for hunger, satiety, desire to eat and prospective consumption were unchanged by hcrt/orexin 1 compared to saline. Thus, in parallel to rats exogenous hcrt/orexin 1 in humans does not seem to significantly influence gastric emptying.

6. GASTRIC AND INTESTINAL SECRETIONS

In analogy with gastric emptying the central administration of hcrt/orexin 1 stimulates gastric acid secretion and this effect is mediated through vagal efferent nerve fibers and the hcrt/orexin receptor-1.[29,30] Again, our laboratory has preliminary data regarding the effect of peripherally administered and endogenous hcrt/orexin 1 on gastric acid secretion *in vivo* in rats. After an overnight fast, rats equipped with a chronic gastric fistula were administered either saline, hcrt/orexin 1 (500 pmol/kg/min) or SB-334867-A (10 mg/kg bolus, infusion 0.1 mg/kg/min) with and without pentagastrin stimulation (90 pmol/kg/min) for 30 min during which gastric secretions were collected, with the amount of acid secreted expressed as μmol/30 min. Similar to gastric emptying, IV hcrt/orexin 1 alone had no effect on gastric acid secretion while the hcrt/orexin receptor-1 antagonist SB-3348678-A significantly decreased gastric acid secretion (Fig 3). This effect is independent of gastrin as plasma gastrin concentrations were unchanged during hcrt/orexin A or SB-334867-A infusion. Again, these data suggest that exogenous hcrt/orexin 1 in contrast to endogenous hcrt/orexin 1 does not influence gastric acid secretion.

The effect of hcrt/orexin 1 on duodenal secretions has also been studied. Hcrt/orexin 1 (60-600 pmol/kg/min) caused a dose-dependent stimulation of duodenal secretion (HCO_3) in fed but not rats fasted overnight. This study also demonstrated a similar effect of an overnight fast on the effect of the muscarinic agonist bethanechol on duodenal secretion, where after the overnight fast an 100-fold increase of bethanechol was required to obtain the same secretory effect as seen in the fed animals.[31] These data demonstrate that IV hcrt/orexin 1 influences duodenal secretions and that this effect is markedly dependent on the nutritional state of the animal.

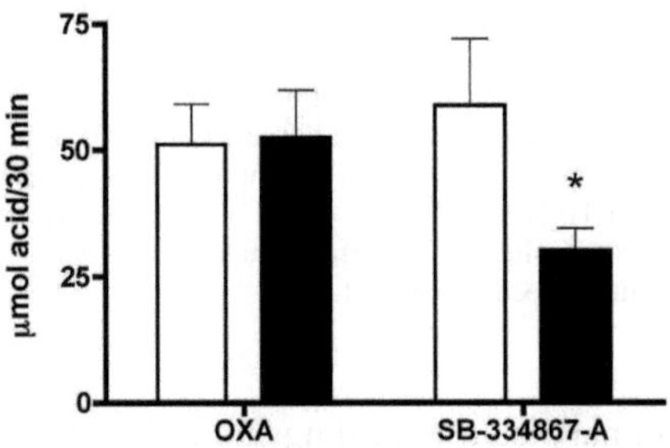

Figure 3. The effect of hcrt/orexin 1 (hcrt1/OXA) and hcrt/orexin receptor-1 antagonist SB-334867-A on gastric acid secretion in the rat (control, unfilled bar and OXA or SB-334867-A, filled bar) ($p<0.05$).

7. SUMMARY

Hcrt/orexin 1 and the hcrt/orexin receptors are abundant in the ENS and enteroendocrine cells of the mucosa of the gastrointestinal tract. Hcrt/orexin 1 and its receptors co-localize with several other neuropeptides and transmitters such as 5-HT, VIP, NOS and gastrin, to mention a few. Hcrt/orexin 1 (iv) seems to have a direct effect on fasting small bowel motility and duodenal secretions, but not gastric emptying and gastric acid secretion. In myenteric neurons of the guinea pig ileum, hcrt/orexin 1 increases the presynaptic release of acetylcholine, suggesting that hcrt/orexin 1 may influence gastrointestinal motility by presynaptic actions in the myenteric plexus.[32] Hcrt/orexin A has been reported to pass the blood brain barrier[33], yet other studies have found the central concentrations of hcrt/orexin 1 after peripheral administration to be too low to exert a biological effect.[34] Thus, we postulate that there are two pools of hcrt/orexin 1. One peripheral which seems to influence fasting motility and duodenal secretions and a central pool that influences gastric emptying and gastric acid secretion, demonstrated by the fact that there was no effect of peripherally administered hcrt/orexin 1 but an effect with the centrally-penetrant hcrt/orexin receptor-1 antagonist SB-334867-A. Clearly a peripheral effect of SB-334867-A cannot be discounted until studies with only peripherally acting hcrt/orexin receptor-1 antagonists are tested, but the results with SB-334867-A are consistent with the results of centrally administered hcrt/orexin 1 on gastric acid secretion and gastric emptying. A likely link between peripheral and central hcrt/orexin is the vagus nerve where hcrt/orexin modulates the activity in the vagus nerve induced by other gut peptides.

8. ACKNOWLEDGEMENTS

This work has in part been supported by grants from the Swedish Research Council, NIH grant NS27645 and The American Diabetes Association.

9. REFERENCES

1. T. Sakurai, A. Amemiya, M. Ishii, I. Matsuzaki, R. M. Chemelli, H. Tanaka, S. C. Williams, J. A. Richardson, G. P. Kozlowski, S. Wilson, J. R. Arch, R. E. Buckingham, A. C. Haynes, S. A. Carr, R. S. Annan, D. E. McNulty, W. S. Liu, J. A. Terrett, N. A. Elshourbagy and M. Yanagiswa, Orexin and orexin receptors: a family of hypothalamic neuropeptides and G protein-coupled receptors that regulate feeding behavior., *Cell.* **92**, 573-585 (1998).
2. C. M. Kotz, J. M. Teske, J. A. Levine and C. Wang, Feeding annd activity induced by orexin A in the lateral hypothalamus in rats., *Regul Pept.* **104**, 27-32 (2002).
3. H. Yamada, T. Okumura, W. Motomura, Y. Kobayashi and Y. Kohgo, Inhibition of food intake by central injection of anti-orexin antibody in fasted rats., *Biochem Biophys Res Commun.* **267**, 527-531 (2000).
4. R. J. Rodgers, J. C. Halford, R. L. Nunes de Souza, A. L. Canto de Souza, D. C. Pi, J. R. Arch, N. Upton, R. A. Porter, A. Johns and J. E. Blundell, SB-334867-A, a selective orexin-1 receptor antagonist, enhances behavioural satiety and blocks the hyperphagic effect of orexin A in rats., *Eur J Neurosci.* **13**, 1444-1452 (2001).
5. C. M. Edwards, S. Abusnana, D. Sunter, K. G. Murphy, M. A. Ghatei and S. R. Bloom, The effect of orexins on food intake: comparision with neuropeptide Y, melanin-concentrating hormone and galanin., *J Endocrinol.* **160**, R7-12 (1999).

6. X. J. Cai, X. H. Lui, M. L. Evans, J. C. Clapham, S. Wilson, J. R. Arch, R. Morris and G. Williams, Orexins and feeding: special occasions oreveryday occurrence., *Regul Pept.* **104**, 1-9 (2002).
7. X. J. Cai, P. S. Widdowson, J. Harrold, S. Wilson, R. E. Buckingham, J. R. Arch, M. Tadayyoon, J. C. Clapham, J. Wilding and G. Williams, Hypothalamic orexin supression: modulation by blood glucose and feeding., *Diabetes.* **48**, 2132-2137 (1999).
8. X. J. Cai, M. L. Evans, C. A. Lister, R. A. Leslie, J. R. Arch, S. Wilson and G. Williams, Hypoglycemia activates orexin neurons and selectively increases hypothalamic orexin-B levels: responses inhibited by feeding possibly mediated by the nucleus of the solitary tract., *Diabetes.* **50**, 105-112 (2001).
9. C. Peyron, D. K. Tighe, A. N. Van den Pool, L. De Lecea, H. C. Heller, J. G. Sutcliff and T. S. Kilduff, Neurons containing hypocretin (orexin) project to multiple neuronal systems., *J Neurosci.* **18**, 9996-10015 (1998).
10. A. L. Kirchgessner and M. Liu, Orexin synthesis and response in the gut, *Neuron.* **24**, 941-951 (1999).
11. M. J. de Miguel and M. A. Burrell, Immunocytochemical detection of orexin A in endocrine cells of the developing mouse gut., *J Histochem Cytochem.* **50**, 63-69 (2002).
12. E. Näslund, M. Ehrström, J. Ma, P. M. Hellström and A. L. Kirchgessner, Localization and effects of orexin on fasting motility in the rat duodenum., *Am J Physiol.* **282**, G470-G479 (2002).
13. M. Ehrström, E. Näslund, J. Ma, A. L. Kirchgessner and P. M. Hellström, Physiological regulation and NO-dependent inhibition of migrating myoelecctric complex in the rat small bowel by orexin A, *Am J Physiol.* **285**, G688-695 (2003).
14. M. S. Mondal, M. Nakazata, Y. Date, N. Murakami, M. Yanagisawa and S. Matsukura, Widespread distribution of orexin in rat brain and its regulation upon fasting., *Biochem Biophys Res Commun.* **256**, 495-499 (1999).
15. R. Ouedrago, E. Näslund and A. L. Kirchgessner, Glucose regulates the release of orexin-A from the endocrine pancreas., *Diabetes.* **52**, 111-117 (2003).
16. G. Komaki, Y. Matsumoto, H. Nishikata, K. Kawai, T. Nozaki, M. Takii, H. Sogawa and C. Kubo, Orexin-A and leptin change inversely in fasting non-obese subjects., *Eur J Endocrinol.* **144**, 645-651 (2001).
17. Y. Date, M. Kojima, H. Hosoda, A. Sawaguchi, M. S. Mondal, T. Suganuma, S. Matsukura, K. Kangawa and M. Nakazato, Ghrelin, a novel growth hormone-releasing acylated peptide, is synthesized in a distinct endocrine cell type in the gastrointestinal tracts of rats and humans., *Endocrinology.* **141**, 4255-4261 (2000).
18. M. Kojima, H. Hosada and Y. Date, Ghrelin is a growth-hormone releasing acylated peptide from stomach., *Nature.* **402**, 656-660 (1999).
19. D. E. Cummings, J. Q. Purnell, R. S. Frayo, K. Schmidova, B. E. Wisse and D. S. Weigle, A preprandial rise in plasma ghrelin levels suggests a role in meal initiation in humans., *Diabetes.* **50**, 1714-1719 (2001).
20. C. B. Lawrence, A. C. Snape, F. M. Baudoin and S. M. Luckman, Acute central ghrelin and GH secretagogues induce feeding and activate appetite centers., *Endocrinology.* **143**, 115-162 (2002).
21. R. D. Reidelberger, J. Hernandez, B. Fritzsch and M. Hulce, Abdominal vagal mediation of the satiety effects of CCK in rats., *Am J Physiol.* **286**, R1005-R1012 (2004).
22. G. Burdyga, S. Lal, D. Spiller, W. Jiang, D. Thompson, S. Attwood, S. Saeed, D. Grundy, A. Varro, R. Dimaline and G. J. Dockray, Localization of orexin-1 receptors to vagal afferent neurons in the rat and humans., *Gastroenterology.* **124**, 129-139 (2003).
23. I. Sakata, M. Yamazaki, K. Inoue, Y. Hayashi, K. Kangawa and T. Sakai, Growth hormone secretagogue receptor expression in the cells of the stomach-projected afferent nerve in the rat nodose ganglion., *Neurosci Lett.* **2003**, 183-186 (2003).
24. T. A. Harrison, C. T. Chen, N. J. Dun and J. K. Chang, Hypothalamic orexin-A-immunoreactive neurons project to the rat dorsal medulla., *Neurosci Lett.* **273**, 17-20 (1999).
25. G. J. Hervieu, J. E. Cluderay, D. C. Harrison, J. C. Roberts and R. A. Leslie, Gene expression and protein distribution of the orexin-1 receptor in the rat brain and spinal cord., *Neuroscience.* **103**, 777-797 (2001).
26. D. Smart, C. Sabido-David, S. J. Brough, F. Jewitt, A. Johns, R. A. Portor and J. C. Jerman, SB-334867-A: the first selective orexin-1 receptor antagonist., *Br J Pharmacol.* **132**, 1179-1182 (2001).
27. Z. K. Krowicki, H. R. Berthoud, M. A. Burmeister and P. J. Hornby, Orexins in rat dorsal motor nucleus of the vagus potently stimulate gastric motor function., *Am J Physiol.* **283**, G465-G472 (2002).
28. G. Grabauskas and H. C. Moises, Gastrointestinal projecting neurons in the dorsal motor nucleus of the vagus exhibit direct and viscerotopically organized sensitivity to orexin., *J Physiol.* **549**, 37-56 (2003).
29. N. Takahashi, T. Okumura, H. Yamada and Y. Kohgo, Stimulation of gastric acid secretion by centrally administered orexin A in conscious rats., *Biochem Biophys Res Commun.* **254**, 623-627 (1999).
30. T. Okumura, T. Takeuchi, W. Motomura, H. Yamada, S. Egashira, S. Asashi, A. Kanatani, M. Ihara and Y. Kohgo, Requirement of intact disulfide bonds in orexin-A-induced stimulation of gastric acid secretion which is mediated by OX1 receptor activation., *Biochem Biophys Res Commun.* **280**, 976-981 (2001).

31. G. Flemström, M. Sjöblom, G. Jedstedt and K. E. Åkerman, Short fasting dramatically decreases rat duodenal secretory responsiveness to orexin A but not to VIP or melatonin., *Am J Physiol.* **285**, G1091-1096 (2003).
32. Y. Katayama, T. Homma, K. Honda and K. Hirai, Actions of orexin-A in the myenteric plexus of the guinea-pig small intestine., *Neuroreport.* **114**, 1515-1518 (2003).
33. A. J. Kastin and V. Åkerström, Orexin A but not orexin B rapidly enters brain from blood by simple diffusion., *JPET.* **289**, 219-223 (1999).
34. N. Fujiki, Y. Yoshida, B. Ripley, E. Mignot and S. Nishino, Effects of IC and ICV hypocretin-1 (orexin A) in hypocretin receptor-2 gene mutated narcoleptic dogs and IV hypocretin-Q replacement therapy in a hypocretin-ligand-deficient narcoleptic dog., *Sleep.* **26**, 953-959 (2003).

HYPOCRETINS IN ENDOCRINE REGULATION

Miguel López, Manuel Tena-Sempere, Tomás García-Caballero, Rosa Señarís and Carlos Diéguez[*]

1. INTRODUCTION

The role of the hypothalamus in the regulation of endocrine and autonomic functions has been known for more than fifty years. These co-ordinate actions are responsible for the control of several processes such as feeding, drinking, reproduction, lactation, cardiovascular function, metabolic control, thermoregulation, sleep-wake cycle and hormone secretion.[1-3] The lateral hypothalamic area (LHA) is not part of the hypophysiotropic area (HTA) of the hypothalamus.[3] However, compelling evidence has pointed out its implication in neuroendocrine control and hormonal secretion from pituitary.[2,4,5]

In spite of all the data linking the LHA with the neuroendocrine regulation[2] the molecular mechanisms involved remained unknown. This situation changed at the beginning of 1998 when two different laboratories independently discovered of a new family of hypothalamic neuropeptides: the *hypocretins: hypocretin-1* (*Hcrtl*) *and hypocretin-2* (*Hcrt2*).[6,7] The hypocretins were originally believed to be primarily important in the regulation of feeding;[7-9] however, one of their major functions is the regulation of sleep, wakefulness and arousal.[10-12] Furthermore, they are important regulators of all the endocrine axes.[10, 3-17] Since hormonal secretion from pituitary and other peripheral endocrine glands is intimately linked to the sleep-wake cycle[18-29] and to the feeding state of the animal[30-33] it is likely that hypocretins link sleep, arousal and feeding to endocrine function.

* M. López, R. Señarís and C. Diéguez, Department of Physiology, School of Medicine, University of Santiago de Compostela, c/ S. Francisco s/n 15705, Santiago de Compostela, Spain. M. Tena-Sempere Department of Cell Biology, Physiology and Immunology, Section of Physiology, Faculty of Medicine, University of Cordoba, Avda. Menendez Pidal s/n, 14004 Cordoba, Spain. ¶ T. García-Caballero, Department of Morphological Sciences, School of Medicine-University Clinical Hospital, University of Santiago de Compostela, 15705 Santiago de Compostela, Spain

Taking in account all this evidence, the aim of this chapter is to describe and discuss the current knowledge regarding the endocrine aspects of the hypocretin system. The study and understanding of these neuropeptides could open new therapeutic possibilities in the treatment of sleep, body weight and endocrine pathologies in humans.

2. NEUROENDOCRINE ANATOMY OF THE HYPOCRETIN SYSTEM

In the rat brain there are about 1100-4000 (depending of the authors) immunoreactive cell bodies for hypocretins.[34] These neurons are mainly localized in the tuberal region of the hypothalamus within the perifornical and dorsomedial hypothalamic (DMN) nuclei and the dorsal and lateral (LHA) hypothalamic areas in rat,[6,7,35-39] mouse,[35,40,41] hamster,[42-44] guinea pig,[45] the rodent *Arvicanthis niloticus*,[46] cat,[47-49] sheep,[50] monkey,[39,51] human,[52-54] frog[55] and *Xenopus*.[56]

In contrast with the restricted localization of hypocretin cell bodies, hypocretin fibres are widely distributed throughout the brain and all levels of the spinal cord.[35-40,57] In the hypothalamus, hypocretin projections are found in the preoptic (PO), suprachiasmatic (SPN), supraoptic (SON), tuberomamillary, dorsomedial (DMN), paraventricular (PVN), periventricular (PeN), ventromedial (VMN) and arcuate (ARC) nuclei[35,36,38,39,57] all of them with major roles in the control of the different endocrine axes.

Hypocretin receptors (Hcrtrs) show widespread distribution in the rat brain and spinal cord and their pattern of expression is consistent with the map of hypocretin containing projections. In the rat hypothalamus neurons expressing hypocretin receptor 1 (Hcrtr1) were observed in the preoptic, suprachiasmatic, periventricular, paraventricular (magno- and parvocellular divisions), supraoptic, arcuate, ventromedial, dorsomedial and tuberomamillary nuclei and the lateral hypothalamic area.[58-61] In the rat, in all these hypothalamic nuclei Hcrtr1 co-localizes with another neuropeptide systems implicated in endocrine control. Thus, in magnocellular neurons of the paraventricular and supraoptic nuclei Hcrtr1 immunoreactivity is localized in arginine vasopressin (ADH) and oxytocin (OT) expressing neurons. Moreover, Hcrtr1 is present in vasopressin and vasoactive intestinal polypeptide (VIP) neurons in the suprachiasmatic nucleus, in somatostatin (SST) neurons in the periventricular nucleus, and in corticotropin-releasing hormone (CRH) neurons in the parvocellular part of the paraventricular nucleus.[61] In the arcuate nucleus, Hcrtr1 is detected in neuropeptide Y (NPY) neurons of the ventromedial part and in proopiomelanocortin (POMC) neurons of the ventrolateral part.[61] Moreover, Hcrtr1 is expressed in the septal and medial preoptic area of the rat brain[58,60,62] on LHRH neurons.[63] These results suggest that Hcrtr1 signalling may play a role in the control of several endocrine systems such as the adrenal, the somatotrope and the reproductive axes.[14,58,60,62]

The central expression pattern of hypocretin receptor 2 (Hcrtr2) has been extensively studied in the rat [58-60, 64] Therefore, in the rat hypothalamus, Hcrtr2 mRNA has been predominantly found in the medial parvocellular part of the paraventricular nucleus, in the arcuate nucleus, in the premammillary and tuberomamillary nuclei and in the lateral hypothalamic area. Moderate Hcrtr2 mRNA content is detected in the ventromedial, dorsomedial, periventricular hypothalamic nuclei and in the posterior and preoptic hypothalamic areas. Hcrtr2 mRNA levels are low in the supraoptic nucleus and in the dorsal and lateral parts of the paraventricular nucleus. No Hcrtr2 mRNA has been detected in the magnocellular neurons of the paraventricular nucleus.[58-60] So far, only one study has described in detail Hcrtr2 protein distribution in the rat hypothalamus.[64] However, there are no available results about the co-expression of this receptor with other neuropeptide systems implicated in endocrine regulation.

3. THE HYPOCRETIN SYSTEM IN PERIPHERAL ENDOCRINE TISSUES

The presence of the hypocretin system is not restricted to the central nervous system (CNS). In the periphery, hypocretin peptides and receptors are widely expressed, especially in endocrine tissues.

Hypocretins are expressed in the rat and human pituitary.[65-67] In human adenohypophysis Hcrt1 is present in lactotropes, thyrotropes, somatotropes, gonadotropes (FSH and LH) but not in corticotropes. In contrast, Hcrt2 was found in all corticotropes of the anterior pituitary.[67] These findings could suggest a paracrine role for hypocretins in human pituitary cell function. Finally, in addition to innervation of neuroendocrine centres in the hypothalamus, hypocretin-positive fibers are present in both the internal and external layers of the median eminence.[65] Hypocretin receptors have also been shown to be present in the rat and human pituitary.[65,68,69] Both Hcrtr1 and Hcrtr2 mRNAs are highly expressed in the intermediate lobe of rat pituitary, whereas in the anterior lobe, the Hcrtr1 is more markedly expressed than the Hcrtr2. The two receptors mRNAs are also found in the posterior lobe of rat pituitary.[65] In human pituitary Hcrtr1 is expressed by somatotrope cells and Hcrtr2 by corticotrope cells[68] (Fig. 1A and 1B). All these data suggest direct neuroendocrine actions of hypocretins on the pituitary gland, in addition to their possible neuromodulatory role in the hypothalamus.

The hypocretin system is highly expressed in the adrenal gland and shows a species-specific pattern of expression. Thus, while no expression of prepro-Hcrt has been detected in the rat adrenal gland,[69,70] immunoreactivity for prepro-Hcrt and Hcrt1 has been described in human adrenal gland.[71,72] Hypocretin receptors have been described in the adrenal cortex and/or in the adrenal medulla of rat,[66,70,73] pig[74] and human[71,75-77] (Fig. 1C and 1D).

Hypocretin and Hcrtr1 mRNA expression, but not Hcrtr2 mRNA, has been detected in the rat testis,[7,66,70,78] with presence of prepro-Hcrt signal in interstitial Leydig cells and seminiferous tubules, and predominant location of Hcrtr1 mRNA in the tubular compartment of the testis[78] (our unpublished data). In contrast, in humans prepro-Hcrt is only expressed in the epididymis and penis whereas both Hcrtrs are present in testis, epididymis, seminal vesicle and penis.[79]

In the gastrointestinal tract the hypocretin system is highly expressed in endocrine cells. Thus, Hcrt1 immunoreactivity is detected in numerous endocrine enterochromaffin (EC) cells of the gastric and intestinal mucosa as well as in the α and β cells of the pancreas.[72,80-83] Both Hcrtrs were also found in endocrine cells in the gut and in the pancreas, suggesting a role for hypocretins in the control of gastrointestinal and pancreatic function.[80,82]

Low levels of hypocretin and hypocretin receptors have been detected in other peripheral tissues. In humans prepro-Hcrt is expressed in the stomach, kidney, colon, colorectal epithelial cells and placenta.[72] In the rat, Hcrtr1 mRNA has been detected in the kidney, thyroid, ovary and placenta[66] (our unpublished data). In addition, Hcrtr2 mRNA has been found in rat lung, placenta[66] (our unpublished data) and pineal gland.[84]

Finally, it is interesting to note the presence of Hcrt1 in plasma in the human[85-92] and rat.[83,93] The exact source of these circulating hypocretins is unknown; although the hypothalamic hypocretin neurons project to the median eminence[65] their plasma levels are too high to be explained by a unique central origin. In addition, no differences in plasma hypocretin levels are detected between men and women suggesting that a testicular origin for plasma hypocretins is unlikely.[85] A gut origin has been proposed for circulating Hcrt.[45,80,85] However, the fact that Hcrt1 plasmatic levels are increased after fasting in both humans[85,86] and rats[83] and that glucose levels regulate its pancreatic secretion[83] suggest that Hcrt1 could be secreted by the pancreas.[83] In any case, independent of their origin, the presence of hypocretins in the circulation and the widespread distribution of hypocretin receptors in the periphery suggest the existence of direct endocrine actions (not mediated through the hypothalamus-pituitary unit) of these peptides, acting like true hormones.

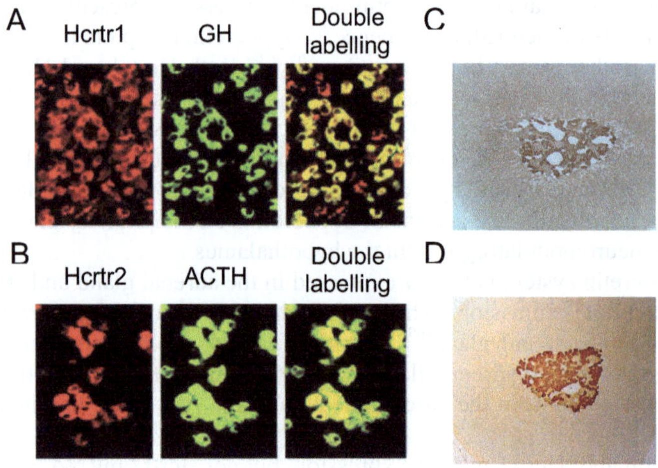

Figure 1. Hypocretin receptors immunoreactivity in endocrine tissues of human and rat. A. Colocalization of Hcrtr1 and GH in the human pituitary. Double immunofluorescence for Hcrtr1 and GH. Hcrtr1 expressing cells are identified by tetramethylrhodamine isothiocyanate (TRICT) immunofluorescence (red) and GH positive cells are identified by fluorescein isothiocyanate (FICT) immunofluorescence (green). Colocalization of both Hcrtr1 and GH in the same cells is illustrated by the yellow colour obtained using the double exposure method of photomicrography. (225X). B. Colocalization of Hcrtr2 and ACTH in the human pituitary. Double immunofluorescence for Hcrtr2 and ACTH. All Hcrtr2 displaying cells (Cy3 immunofluorescence, red) are also positive for ACTH (FICT immunofluorescence, green). Coexpression is confirmed by the double exposed microphotography (yellow). (250X). C. Hcrtr1 in the adrenal medulla of the rat (4X). D. Hcrtr2 in the adrenal medulla of the rat (4X).

4. ENDOCRINE ACTIONS OF HYPOCRETINS

The presence of abundant immunoreactive hypocretin fibres and hypocretin receptors in hypothalamic nuclei such as the arcuate nucleus (ARC), the paraventricular nucleus (PVN), the periventricular nucleus (PeN) and the preoptic area (PO) suggests a neuroendocrine regulatory role for hypocretins.[14] Moreover, the presence of both hypocretin receptors in rat and human pituitary[65,68] could indicate direct actions of hypocretins at this level. In this section we will review the endocrine and neuroendocrine actions of hypocretins.

4.1. Hypocretins and Adrenal Axis

Anatomical and physiological data demonstrate that the lateral hypothalamus (LHA) is involved in the regulation of the hypothalamic-pituitary-adrenal (HPA) axis and there is substantial evidence to support this contention. There are neuronal connections between the LHA and the corticotropin-releasing hormone (CRH) neurons in the parvocellular part of the paraventricular hypothalamic nucleus (PVN),[2] CRH is expressed in the LHA[94] and in rats with selective destruction of the LHA a disruption of the normal circadian corticosterone rhythm is observed.[95]

The HPA axis is the endocrine axis in which hypocretins play their most extensive role; in fact the rat adrenal gland was the first peripheral organ in which Hcrtrs were detected.[70] Additional papers have demonstrated the existence of hypocretin receptors at all levels of the HPA. Thus, immunohistochemical studies have revealed the anatomical contact between hypocretin terminals and corticotropin-releasing hormone (CRH) cells in the parvocellular part of the paraventricular hypothalamic nucleus,[38,57] which express Hcrtr1.[61] Moreover, in the human pituitary there is abundant expression of Hcrtr2 in corticotrope cells,[68] suggesting direct hypocretin actions, and in the rat adenohypophysis all the corticotrope cells present Hcrt2 immunoreactivity,[67] suggesting a paracrine role of hypocretins in these cells.

In addition to this anatomical evidence, physiological studies have revealed the existence of a functional link between hypocretins and the HPA axis. It has been shown that intracerebroventricular administration of both hypocretins results in the induction of *c-fos* activity and a depolarization of CRH-synthesizing parvocellular neurons of the paraventricular hypothalamic nucleus.[96-99] Furthermore, central administration of Hcrt1 increased mRNA levels of CRH and arginine vasopressin (AVP) in this neuronal population[100] (our unpublished data). Central administration of Hcrt1 also raised the circulating levels of adrenocorticotropin hormone (ACTH) and corticosterone.[98,101,102]

The hypothalamic interaction between hypocretins and the adrenal axis is mediated by neuropeptide Y (NPY). This is demonstrated by the blockade of central NPY function using antagonists or antibodies. Blocking NPY activity inhibits the hypocretin-induced corticosterone release.[103] In this sense, it is interesting to note that Hcrt1 stimulates the expression of NPY in the rat, both *in vitro*[104] and *in vivo*[9] and in the goldfish (*Carassius auratus*).[105] The participation of NPY arcuate neurons in the hypothalamic actions of hypocretins is not exclusive of the adrenal axis. In fact, the hypocretin-NPY neuronal circuit[39,61,106] plays a fundamental role in hypocretin actions in the gonadal axis,[107,108] prolactin release,[109] regulation of body temperature,[110,111] drinking behaviour,[106] orexigenic actions of Hcrt1 and Hcrt2[9,112-115] and probably the GH-axis (see below, *Hypocretins and Growth Hormone Axis*).

In addition to their hypothalamic and pituitary actions on the HPA axis, hypocretins also play an important role in the development and proliferation[116] and secretory activity of rat adrenal cortex, acting on mineralocorticoid and glucocorticoid secretion. In this sense, peripheral administration of Hcrt1 and Hcrt2 to rats enhances the blood levels of both aldosterone and corticosterone[73] and *in vitro* studies have reported stimulatory effects of both hypocretins on cortisol secretion.[75,117]

The physiological relevance of hypocretins in the HPA axis remains unclear. Some evidence suggests that some behavioural effects of hypocretins, such as grooming and face washing, are mediated through CRH and glucocorticoids.[118] Moreover, the ability of hypocretins to modulate the response to stress is supported by neuroendocrine[96,101,102,118,119] and cardiovascular studies.[120-122] In this sense, it is very interesting to note that during late pregnancy in the rat, the responsiveness of paraventricular CRH neurons to Hcrt1, and hence the pituitary-adrenal axis, is markedly reduced,[123] suggesting that hypocretin system plays a role in the attenuation of the HPA response to stressors observed in the pregnancy.[123]

The relevance of the hypocretin-HPA axis in humans remains unclear. Nevertheless, regulatory actions of hypocretins on pituitary-adrenal axis are supported by data obtained in narcoleptic humans. These patients exhibit normal cortisol levels despite the fact that ACTH secretion is markedly decreased, suggesting alterations in the interaction between the HPA hormones.[124,125] In this sense, a recent study in rat has reported that the expression of the enzymes involved in the steroid metabolism is altered during sleep.[126] Thus, it is tempting to speculate that changes in the hypocretin tone during sleep could act directly on adrenocortical metabolism, explaining then the lack of association between ACTH and cortisol levels.

On the other hand, it is possible that hypocretins play a role in the regulation of adrenal axis during the sleep cycle in normal healthy humans. In this sense, in humans deep sleep has an inhibitory effect on the HPA axis and sleep deprivation has a stimulatory effect on it.[127-129] Whether these changes are mediated by the hypocretinergic tone in humans merits further investigation.

Despite all the evidence linking hypocretins and the HPA axis, the actions of glucocorticoids on hypocretin expression are not very well established. It has been demonstrated that bilateral adrenalectomy induces a marked reduction in prepro-Hcrt mRNA levels in the LHA, which is reversed by peripheral dexamethasone treatment[130] but not reversed by corticosterone administration.[131] Additionally, treating normal rats with dexamethasone does not cause any change in prepro-Hcrt expression[11] (our unpublished data). Further work is required to understand the precise relation between hypocretins and glucocorticoids.

Finally, the role of hypocretins in the adrenal medulla is still unclear, although high expression of both hypocretin receptors has been detected.[70,75] *In vitro* studies have provided contradictory results. Using human adrenal slices, no effect of hypocretins on catecholamine release has been demonstrated.[75] However, in porcine adrenal cultures[74] and in the rat pheochromocytoma cell line PC12,[132] hypocretins decreased basal catecholamine secretion. *In vivo* studies, using conscious rabbits, have demonstrated that central administration of Hcrt1 increases plasma adrenaline.[133] These results suggest that Hcrt1 activates sympathoadrenal outflow and then it could influence cardiovascular physiology.[120-122,134,135]

4.2. Hypocretins and Growth Hormone Axis

Although the lateral hypothalamic area (LHA) does not belong to the hypophysiotropic area (HTA) of the hypothalamus, there are data to suggest that it is implicated in the regulation of the growth hormone (GH) axis. Both growth hormone-releasing hormone (GHRH)[136-139] and somatostatin (SST)[140] have been found in the LHA. Moreover, the LHA contains glucorreceptors that are implicated in the hypoglycaemia-induced GH secretion.[141] Finally, LHA destruction in rats results in linear growth reduction possibly related to GH deficiency.[5]

The relationship between hypocretins and the GH axis is well supported by anatomical and functional evidence. Hypocretin-containing neurons project into the periventricular (PeN), paraventricular (PVN) and arcuate (ARC) hypothalamic nuclei.[35,36] Hypocretin receptors are present in all these three nuclei.[60,62,64] Furthermore, it has been demonstrated that somatostatin neurons in the PeN express Hcrtr1[61] and that Hcrt1 stimulates SST release from hypothalamic explants.[142] Finally, Hcrt1 has been shown to decrease endogenous rat plasma GH levels[101] and to inhibit pulsatile GH secretion in the rat[143] (Fig. 2). Intriguingly, in primary culture of sheep somatotropes, it has been found that GH secretion is stimulated by Hcrt1[144] and Hcrt2.[145,146] These discrepancies, between *in vivo* and *in vitro* data, suggest that hypocretin actions on growth hormone release are mainly exerted at hypothalamic level.

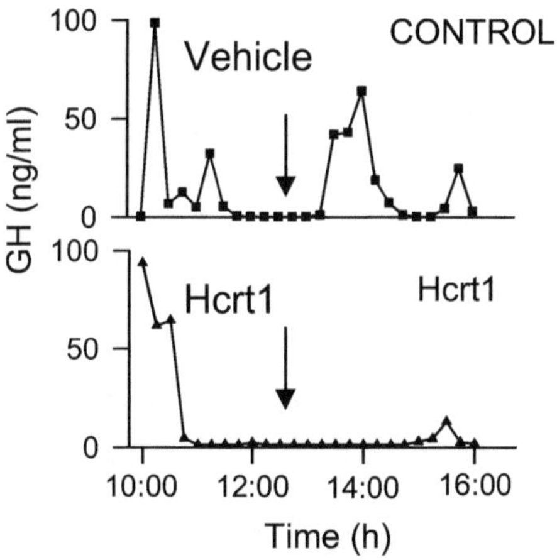

Figure 2. Effect of i.c.v. Hcrt1 (3 nmol) treatment on GH plasma levels. Representative plasma GH profiles in normally fed male rats that received either vehicle or Hcrt1.

Recent data suggest that Hcrt1 actions on GH secretion in the rat are mediated by a hypothalamic mechanism involving GHRH and SST neurons. First of all, Hcrt1 and Hcrt2 do not modify GH secretion in rat pituitary cultures[119,143] suggesting a lack of a direct effect of hypocretins on pituitary GH release. Secondly, in keeping with this, Hcrt1 failed to modify *in vivo* GH responses to GHRH, although it markedly blunted GH responses to ghrelin.[143] And thirdly, central Hcrt1 treatment induces a decrease in the GHRH mRNA levels in the parvocellular part of the PVN (Fig. 3A) and a GH-dependent stimulatory effect on somatostatin mRNA content in the PeN.[147] However, GHRH cells in the ARC, the classically implicated GHRH cell group controlling GH secretion,[148,149] is not regulated by Hcrt1 (Fig. 3B). The implication of PVN GHRH neurons in GH regulation is not well established; however it is supported by anatomical studies, showing that GHRH containing neurons in the PVN project their axons to the fenestrated capillaries of portal vessels in the median eminence.[150] Curiously, in spite of the abundant presence of Hcrtrs in the parvocellular part of the PVN and in the ventrolateral ARC, and the presence of GHRH neurons in the PVN and ARC, no study has so far addressed show whether Hcrtrs are express by GHRH neurons. However, these data suggest that Hcrtrs could be expressed in those cells.

NPY plays a role in the regulation of the somatotropic axis by inhibiting GH synthesis and secretion.[151,152] Since that Hcrt1 treatment increases NPY mRNA expression[9] and release[104] it is possible that NPY could mediate the inhibitory effects of hypocretins on the GH axis. In relation to this, it is interesting to note that in models with GH deficiency, namely hypophysectomized rats and dwarf rats, Hcrt1 does not affect the mRNA levels of NPY (our unpublished data) suggesting that GH tone could regulate this interaction. Additional work is required to address this issue.

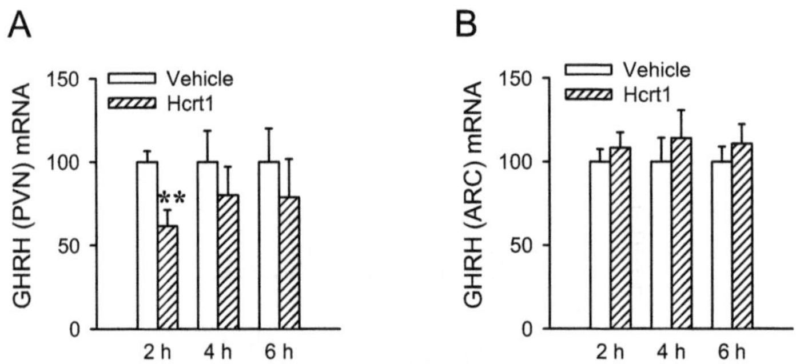

Figure 3. Effect of i.c.v. Hcrt1 (3 nmol) treatment on GHRH mRNA levels in the rat hypothalamus. A. mRNA levels in the paraventricular nucleus (PVN) of the hypothalamus of the described experimental groups. **: P<0.01 *vs.* vehicle 2 hours. B. GHRH mRNA levels in the arcuate nucleus of the hypothalamus (ARC) of the described experimental groups. The data (mean +/- SEM) are expressed as the percentage of change in relation to the control at each time (vehicle treated animals = 100%).

Although the physiological relevance of the hypocretin-GH axis interaction is still unclear, several options should be considered. Firstly, it is possible that hypocretins serve as a signal linking metabolic status and GH secretion. It is known that the levels of prepro-Hcrt are markedly influenced by nutritional status, being up-regulated upon fasting.[7,8,153] This is particularly interesting, since food-deprivation in the rat is associated with a marked decrease in spontaneous GH secretion.[30] Thus, it is possible that a fasting-induced increase in hypocretin-gene expression inhibits GH secretion, constituting an additional mechanism for adaptation of the hypothalamus to nutritional status.

Furthermore, the interaction between hypocretins and GH axis could be important as well in terms of sleep regulation. As it was described above, the hypocretins[11] and hypocretin receptors[154] are important regulators of the sleep-wake cycle. On the other hand, a growing body of evidence has linked GH secretion to sleep. It is well established that a major burst of GH secretion occurs during non-rapid eyes movement (NREM) sleep in humans and in several animal species.[29,30,149] Furthermore, functional alteration in the somatotropic axis can lead to marked changes in the regulation of sleep-wake activity. Thus, GH-deficiency in children is often associated with decreases in REMs.[29,155,156] Also, data obtained in genetic strains of GH-deficient rats, e.g. dwarf Lewis rats[157] or in transgenic models,[158] linked to an impairment in GH secretion, showed the presence of sleep alterations. Although the nature of the hypothalamic mechanisms through which the different components of GH-axis influences sleep is poorly understood, several possibilities have been put forward. Therefore, GH[155,157,159] SST[160,161] and GHRH[157,162] have been implicated in the regulation of the sleep-wake cycle in humans and rodents. Interestingly, GHRH neurons in the PVN play a major role in sleep control, stimulating REM sleep patterns.[163,164] The central inhibitory action of Hcrt1 on REM sleep[101,165] could be mediated, at least in part, by decreasing GHRH mRNA expression in the PVN.[147]

Data in narcoleptic humans link hypocretins and the somatotrope axis. GH secretion in narcolepsy shows alterations, such as abnormal circadian pattern and the decrease of plasma GH peak associated with slow wave sleep at the nocturnal sleep onset.[166-168] These alterations are likely mediated by changes in the expression and levels of GHRH in the hypothalamus.[147] In this sense, it has been recently demonstrated that hypocretin deficiency disrupts the circadian release of GHRH in narcoleptic patients causing abnormal circadian GH release and promoting sleep alterations.[168] All this evidence suggests that the relationship between hypocretins and the somatotropic axis is playing an important role in the regulation of sleep in physiological and pathological states.

Although the actions of hypocretins on GH and GHRH are well established, neither GH deficiency nor GHRH central treatment affect hypocretin gene expression.[147] These data indicate that the sleep alterations associated with GH deficiency do not appear to be mediated by changes in hypocretin gene expression and that the central effects of GHRH are mediated through a hypocretin-independent mechanism. Finally, prepro-Hcrt mRNA content in the LHA was unchanged after treatment with ghrelin[169] indicating that hypocretins are not involved in the hypothalamic action of this hormone on the GH axis.[143,169]

4.3. Hypocretins and Gonadal Axis

The role of the lateral hypothalamus (LHA) in the regulation of the reproductive axis was discovered in experiments involving LHA lesions. Animals with LHA lesions show behavioural and feeding patterns very close to those induced by estrogen.[170-174] Moreover, electrical stimulation of the LHA in rats affect ovarian steroidogenesis[175] and sexual behaviour.[176] On the other hand, the LHA has been implicated in the cholinergic blockade of both follicle-stimulating hormone (FSH) and luteinizing hormone (LH).[177]

Anatomical data suggest that hypocretins play important roles in the regulation of the hypothalamic-pituitary-gonadal (HPG) axis and sexual behaviour. The hypocretin neurons in the LHA project to CNS sites involved in the regulation of the HPG axis. In rat and sheep hypothalamus, Hcrt1 immunoreactive fibers project to the septal (SPO), medial preoptic area (mPO) and the arcuate nucleus-median eminence (ARC-ME) region.[38,50,57] Luteinizing hormone-releasing hormone (LHRH) neuronal bodies and their axons terminals are located in all three nuclei (SPO, MPO and ARC-ME).[178] Furthermore, hypocretin fibers make contact with LHRH neurons, and Hcrtr1 is expressed in the septal and medial preoptic area of the rat[58,60,62] on LHRH neurons.[63]

Curiously, LHRH nerve terminals in the median eminence do not express Hcrtr1.[63] Outside the hypothalamus Hcrt1 immunoreactive fibers project to the medulla, the midbrain dorsal raphe, the pontine locus coeruleus, the amygdala, the olfactory bulbs and the central gray matter, all of them important areas in the control of the HPG axis and sexual behaviour.[38,57] Moreover, Hcrtr1 is expressed at high levels in the monoaminergic locus coeruleus and dorsal/median raphe.[58-60]

These morphological data have been supported by functional studies showing that central administration of both Hcrt1 and Hcrt2 stimulate or suppress, depending on the presence or the absence of ovarian steroids respectively, the pulsatile secretion of luteinizing hormone (LH) in ovariectomized rats.[107,179-181] Such a bimodal mode of action is apparently linked with site-specific effects of Hcrt1 within different hypothalamic nuclei; Hcrt1 being stimulatory at the rostral preoptic area (rPO), and inhibitory at the medial PO or the arcuate nucleus-median eminence.[182] It has also been demonstrated that immunoneutralization of Hcrt1 in the brain abolished the LH surge,[180] whereas the central administration of the Hcrtr1 antagonist SB 334867-A attenuates this LH surge.[182] Moreover, the central administration of Hcrt1 leads to a dose-dependent recovery of fasting-suppressed LH surge.[180] These data are important in terms of food intake regulation because they support the idea that in fasting there is a depletion of Hcrt1/Hcrt2 in several hypothalamic areas[183] despite the fact that prepro-Hcrt mRNA levels are strongly increased after fasting.[7,8,153]

The effects of hypocretins on the gonadal axis appear to be entirely mediated through the stimulation of LHRH secretion[108,179,182] because either Hcrt1 or Hcrt2 failed to inhibit LH secretion from primary cultures of rat pituitary.[119] However, Hcrt1 inhibited LHRH-stimulated LH release in dispersed pituitaries from proestrous female (but not from male) rats.[108] The stimulatory action of hypocretins on LHRH is mediated by an NPY-dependent mechanism, as evidenced by the treatment with selective NPY Y1-receptor antagonist that abolishes the stimulatory effect of Hcrt1 on LHRH release *in vitro*.[108] On the other hand, recent evidence suggests that NPY regulates hypocretin neurons through the NPY Y4 receptor in a postsynaptic fashion.[63,106] All this evidence suggests that the NPY-hypocretin circuit plays a major role in the hypothalamic regulation of the gonadal axis.

Notably, hypocretins exert their endocrine action at testicular level as well. Both prepro-Hcrt and the Hcrtr1 are expressed in rat testis[7,66,70,78] and Hcrt1 stimulates testosterone secretion *in vitro* and *in vivo*, further suggesting a role for the hypocretin system in the control of the male reproductive axis[78] (Fig. 4). In addition, considering the predominant location of Hcrtr1 in the tubular compartment within the rat testis, we are currently investigating additional biological actions of Hcrt1 in the direct control of the seminiferous epithelium.

The physiological relevance of the hypocretin system in reproductive function could be related to the adaptation of the gonadal axis to changes in the metabolic state. It is well known that variations in the metabolical status of the animal are linked to changes in reproductive function.[178] Hypocretin neurons in the lateral hypothalamic area express the long form of the leptin receptor in rat[184,185] and sheep,[186] as well as the NPY Y4 receptor in the rat.[63,106] On the other hand, hypocretin neurons sense the nutritional status of the animal, being regulated by leptin,[8,187,188] glucose,[153,189-192] pancreatic polypeptide (PP)[106] and visceral signals,[153,192] Taken together, these data suggest that hypocretin neurons provide an important link between energy balance and reproductive function by integrating central (NPY) and peripheral signals (glucose, leptin, PP).

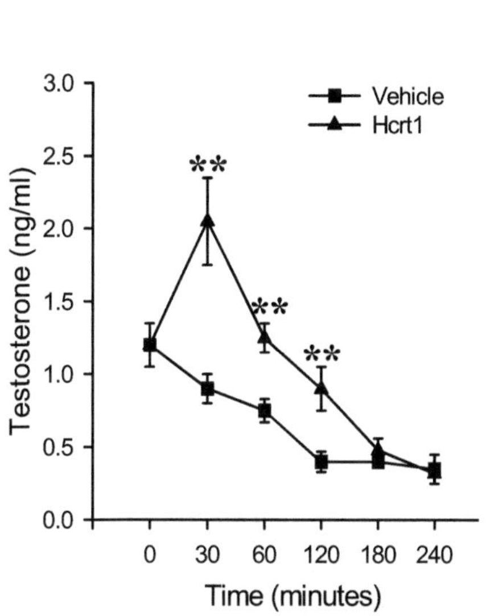

Figure 4. Regulation of testicular testosterone secretion by Hcrt1 *in vivo*. Bilateral intratesticular injections of Hcrt1 (10^{-6} M) were performed. Profiles of testosterone secretion along the study-period in the experimental groups are presented. **: P<0.01 vs. vehicle same time.

Besides its potential function as link between the metabolic state and reproductive function, it is tempting to speculate that hypocretins could connect reproduction with sleep cycle. In this sense, it is well known that the circadian secretion of testosterone and estradiol is associated with the sleep-wake cycle in human (women and men)[23,25,193,194] and it is markedly affected by sleep disturbances, such as sleep deprivation.[195-198] In spite of this, reproductive capacity is unaltered in human narcolepsy,[11] and no reproductive problems have been reported in prepro-hypocretin knockout mice.[199] Anyhow, whether hypocretin are regulating the sleep-dependent pattern of sexual hormones secretion, as well as the alterations observed in sleep disturbances, merits further investigation.

The hypocretin system shows a gender specific pattern of expression in the CNS and the periphery. In the rat, hypothalamic prepro-Hcrt mRNA[200] and Hcrt1 protein levels[201] are higher in the female than in the male. Moreover, prepro-Hcrt and hypocretin receptor mRNA levels in the hypothalamus and Hcrt1 content in several brain areas, such as the hypothalamus, thalamus, midbrain and medulla, are regulated by estrous cycle, pregnancy, parturition and lactation.[108,202,203] Hypocretin receptors also show a sexually dimorphic pattern of expression with higher Hcrtr1 mRNA content in the hypothalamus of female rats than in male rats and lower Hcrtr1 and Hcrtr2 mRNA levels in the pituitary and adrenal gland of female rats.[66] Intriguingly, however gonadectomy does not affect hypothalamic prepro-Hcrt and hypocretin receptor mRNA levels in rats (female and male)[69,108] or sheep.[204] These results indicate that hypothalamic expression of the hypocretin system is not under the control of gonadal steroids. However taking into account the total RNA approach used in some of these studies[69,108] we cannot exclude the possibility of hypothalamic nuclei-specific regulation of the hypocretin receptors. On the other hand, gonadal status regulates the mRNA levels of both hypocretin receptors in peripheral tissues (pituitary and adrenal) from male and female rats. Therefore, in female rats, pituitary Hcrtr1 mRNA levels and adrenal mRNA levels of Hcrtr2 are increased after ovarectomy and reduced by treatment with estradiol.[69] Alternatively, in male rats, orchidectomy increased the mRNA levels of pituitary Hcrtr1 and decreased the mRNA levels of adrenal Hcrtr2; testosterone treatment reversed the effect of orchidectomy in both cases.[69] Finally, follicle-stimulating hormone (FSH) stimulates Hcrtr1 mRNA levels in the rat testis.[78] Testicular expression of Hcrtr1 appears to be regulated also by its own ligand (Hcrt1).[78]

4.4. Hypocretins and Lactoprope Axis

Hypocretins actions on prolactin (PRL) are controversial and disagreement exists in the literature about the effect of central hypocretins on PRL release. Suppressive effects of Hcrt1 and Hcrt2 in male rats[101,142,205] and Hcrt1 in estrogen-primed ovariectomized female rats[109] have been described. This inhibitory effect is mediated through dopamine release from the tuberoinfundibular dopaminergic neurons (TIDA) in the arcuate hypothalamic nucleus.[109] This hypocretin action on dopamine release seems to be indirect, it being mediated by NPY acting on Y1 receptor on TIDA neurons.[109] Alternatively, Hcrt1 effects on PRL release could be mediated by NPY inhibitory actions on thyrotropin-releasing hormone (THR) (a potent PRL-releasing factor) neurons in the paraventricular nucleus, but at the moment there are no experimental data supporting this hypothesis.

In spite of all this evidence supporting the inhibitory actions of Hcrt1 on PRL secretion, hypocretins appear to play a fundamental role in the pre-ovulatory events involving PRL release in female rats. Thus, central Hcrt1 administration abolished the fasting inhibitory effects on PRL (and LH) surges.[180] Moreover, central administration of anti-Hcrt1 antibodies to normally fed rats completely blunted the pre-ovulatory PRL (and LH) surge.[180] Finally, although Hcrtrs are abundantly expressed in rat and human pituitary[65,68,69] the hypocretin actions on PRL release appear to be entirely conducted at the hypothalamus because both Hcrt1 and Hcrt2 failed to inhibit PRL secretion from primary cultures of rat pituitary.[119] It must be noted that in the human[67] and bullfrog pituitary[206] Hcrt1 immunoreactivity is detected in prolactin cells, suggesting autocrine or paracrine effects of hypocretins at this level. Additional work will be necessary to understand the interaction between hypocretins and PRL and its physiological implications. In any case, all this evidence suggests that the levels of gonadal steroids (male rats, ovariectomized and normal females) modulate these actions.

Finally, we can argue, that as is the case for other endocrine axes, hypocretins are linking PRL function to the sleep-wake cycle. It is well known that PRL secretion is intimately associated with sleep and arousal in humans.[23,26,207-210] In fact, PRL levels are lower in narcoleptic patients.[166,167] Further work is needed to assess this hypothesis.

Contrary to other neuroendocrine axes, PRL plays an important role in the regulation of the hypocretin system, particularly during pregnancy and lactation. It has been described that hyperprolactinemia induced by pituitary graft decreases prepro-Hcrt mRNA levels in the lateral hypothalamus of female rats.[202] On the other hand, in the hypothalamus of pregnant and lactating rats, there is a marked reduction of hypocretin expression, even after a fast of 72 hours.[202] These results suggest that the hyperprolactinemia observed in both states could act by decreasing hypocretin expression. The physiological relevance of this interaction could be related to the effects of PRL on the sleep alterations exhibited during pregnancy and lactation.[178,202,211] In addition, the ability of PRL to modulate central Hcrt expression might contribute to the well-known effects of PRL on the gonadotropin axis.[202,212]

4.5. Hypocretins and Thyroid Axis

Studies during the forties and fifties showed that stimulation of the lateral hypothalamus (LHA) induced morphological changes in the thyroid gland indicating hormone secretion.[2,213-215] Furthermore, rats with lesions in the LHA had lower T_3 and T_4 levels[214,216] and microinjection of thyrotropin-releasing hormone (TRH) into the LHA depressed food intake.[217]

A stimulatory effect of Hcrt1 on thyroid-stimulating hormone (TSH) levels has been reported.[205] Other studies, have demonstrated a functional relationship between hypocretins and hypothalamic-pituitary-thyroid axis (HPT). Peripheral Hcrt1 administration in rats inhibits TRH release from the hypothalamus,[218] resulting in a fall in the levels of TSH,[218,219] This effect is entirely mediated by the decrease in TRH levels because Hcrt1 and Hcrt2 both failed to inhibit TSH secretion from primary cultures from rat pituitary.[119,218]

Plasma thyroid hormone levels showed no changes after peripheral Hcrt1 administration.[218,219] It has been suggested that this lack of effect could be related to the slow metabolism of thyroid hormones[219] which usually show a great delay in their secretory responses.[220] Finally, hypocretins could exert direct effect of on the thyroid gland because the expression of Hcrtr1 has been detected at this level in the rat.[66]

The physiological relevance of the hypocretin-HPT axis interaction could be related to thermogenesis and metabolic control. Classically, the lateral hypothalamus has been implicated in the regulation of metabolism because animals with LHA lesions become hypercatabolic.[2,170] Several papers have demonstrated that hypocretins are implicated in the regulation of thermogenesis and central administration of Hcrt1 increases metabolic rate and oxygen consumption in mice.[221,222] When administered to rats, both Hcrt1 and Hcrt2 increase oxygen consumption and energy expenditure.[223] In this sense, mice lacking prepro-Hcrt expression[199] or having selective ablation of hypocretin neurons[224] are hypophagic but show normal body weight, suggesting hypometabolism. Interestingly, these alterations are present also in narcoleptic patients with hypocretin-deficiency, which show lower caloric intake[225,226] and a increase in body mass index,[227] indicating reduced metabolic rate or energy expenditure. Whether these metabolic alterations in the narcoleptic patients are related with hypocretin-deficient will need further investigations.

Similar to the somatotrope (GH) and adrenal (steroids) axes, thyroid hormone levels do not regulate the hypocretin system, either in the hypothalamus or in the periphery.[228,229] These data suggest that feeding and, more importantly, sleep alterations observed in patients with thyroid dysfunction are not mediated by the hypocretin system in the hypothalamus.

4.6. Hypocretins and Pancreatic Function

The lateral hypothalamic area has classically been implicated in the regulation of pancreatic endocrine function. Functional studies have shown that direct administration of noradrenaline in the LHA induces insulin secretion.[230, 231] This response is mediated by the parasympathetic nervous system, because atropine, given systemically, or vagotomy abolish it.[230] Moreover, lesions in the LHA are followed by hypoinsulinemia.[232-234] On the other hand, some reports have implicated the LHA with glucagon because electrical stimulation of the LHA increases plasma glucagon concentrations.[235]

Recent evidence has implicated hypocretin neurons in the regulation of pancreatic physiology. Morphological studies have demonstrated that melanin-concentrating hormone neurons and hypocretin neurons project from the lateral hypothalamus to first order centres implicated in the autonomic control of the pancreas, such as the dorsal motor nucleus of the vagus (DMV) and preganglionic spinal cord neurons.[236] In a very recent paper it has been reported that after central neuroglucopenia the increase in the hypocretin release from LHA neurons acts on DMV neurons to regulate pancreatic functions.[237]

Finally, it is interesting to note that the central hypocretin regulation of the pancreas is not only restricted to the endocrine portion, as central administration of Hcrt1 stimulates pancreatic exocrine secretion via the vagus nerve.[238]

Hypocretins and hypocretin receptors were also found in endocrine cells in the human and rat pancreas[72,80,83] and glucose levels regulate hypocretin secretion: Hcrt1 release is stimulated by a decrease in glycaemia and inhibited by a increase in glucose levels,[83] similar responses to those in the hypothalamus[153,189-192] and in the gut.[45,80] In this context, it is interesting to note that Hcrt1 plasma concentration falls after fasting in the human[86] and rat.[83] suggesting that the pancreas might be a physiological source for circulating Hcrt1, which may act as hormonal regulator of glucose homeostasis,[45,83,221] acting at peripheral (pancreas) or central levels. Moreover, these data suggest that peripheral Hcrt1 may act on central regulatory centres since Hcrt1 crosses the blood-brain-barrier when it is injected peripherally.[239]

Morphological studies have demonstrated that Hcrt1 immunoreactivity is displayed by insulin β cells[80,83] and in pancreatic glucagon α cells,[83] suggesting a potential role of these peptides in the regulation of insulin and glucagon secretion. In this sense, it has been demonstrated that Hcrt1 administration increases the glucagon secretion from isolated islets incubated in low glucose and that intravenous administration of Hcrt1 to fasted rats elevates plasma glucagon levels.[83]

There are contradictory reports about the actions of hypocretins on insulin release. An initial paper reported that subcutaneous administration (1 or 2 nmol) to rats of either Hcrt1 or Hcrt2 increased blood insulin; Hcrt1 administration also increased blood glucose. Moreover, in the same work, Hcrt1 and Hcrt2 stimulated insulin secretion *in vitro*.[240] In a more recent paper, however the opposite effect has been reported; using cultures of isolated islets, incubated with low glucose, Hcrt1 inhibited insulin release after glucose incubation; furthermore, intravenous administration of Hcrt1 (100 pmol) to fasted rats decreased insulin levels, in spite of a rise in glucose levels.[83] The reasons causing these discrepancies are not clear, although they could be related to the dose used. In any case, the hypocretin-insulin-glucagon interaction could be physiologically relevant in the regulation of glucose homeostasis and its adaptive response to feeding and fasting conditions. Thus, the increase in Hcrt1 levels during fasting[83,86] and its decrease after feeding[86] would modulate insulin and glucagon levels in order to get normoglycaemia.[83]

Alternatively, insulin is a potent regulator of NPY in the arcuate nucleus of the hypothalamus. Insulin acts to inhibit NPY gene expression.[241,242] It is temping to speculate that peripheral changes in Hcrt1 levels could indirectly regulate, by modulating insulin levels, the NPY orexigenic actions at central level. Further work will be required to test this hypothesis.

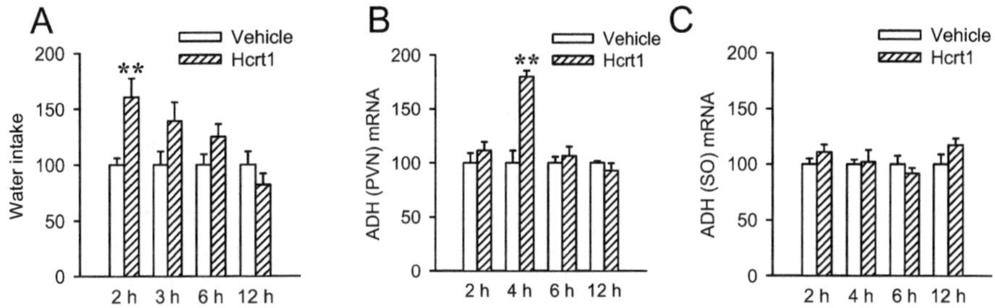

Figure 5. Effect of i.c.v. Hcrtl treatment (3 nmol) on water intake and ADH mRNA levels in the rat hypothalamus. A.Water intake was examined 2 hours, 3 hours, 6 hours and 12 hours after the treatment. Data (mean ± SEM) were expressed in relation to vehicle treated rats. **:P<0.01 *vs.* vehicle. B. ADH mRNA levels in the paraventricular nucleus of the hypothalamus (PVN) of the described experimental groups. **: P<0.01 *vs.* vehicle 2 hours. C. ADH mRNA levels in the supraoptic nucleus of the hypothalamus (SO) of the described experimental groups. The data (mean +/- SEM) are expressed as the percentage of change in relation to the control at each time (vehicle treated animals = 100%).

4.7. Endocrine Actions of Hypocretins in the Gastrointestinal Tract

Hypocretin fibres innervate the dorsal vagal complex, a key site in the control of gastrointestinal function[10,57,243] where Hcrtr1 is present.[244] Furthermore central administration of Hcrt1, but not Hcrt2, stimulates gastric fluid secretion.[245] On the other hand, hypocretin and hypocretin receptors expression is high in numerous endocrine cells of the gut mucosa[45,80-82] regulating gut motility and peristaltic reflex.[45,80,246,247]

All these evidences suggest that hypocretins are important regulators of gut physiology playing roles in the central control of digestion and gastrointestinal function,[45,245] as well as in the paracrine and endocrine regulation of the gut.[45,80,246,247] This aspect of Hcrt physiology, however, is beyond of the scope of our chapter, and will be extensively reviewed elsewhere in this book.

4.8. Hypocretins and Drinking Behaviour

The lateral hypothalamus has been historically implicated in the regulation of water intake.[248,249] LHA lesions cause adipsia and the animals only drink when they eat, just enough to enable mastication.[2,250,251] On the other hand, stimulation of the LHA, either chemically or electrically, increases water intake.[250,251]

It has been observed that hypocretins significantly increase water intake when administered centrally to rats[252] (our unpublished data) (Fig. 5A). Moreover, the effect of Hcrtl is greater than Hcrt2 suggesting the implication of Hcrtr1. Prepro-Hcrt mRNA levels are up regulated when rats are deprived of water.[252] This evidence indicates a physiological role for hypocretins as mediators of drinking behaviour.

The exact mechanism by which hypocretins induce drinking is not clear and several hypotheses have been proposed. Hypocretin-immunoreactive varicose axons are observed in the subfornical organ and area postrema, regions implicated in drinking behaviour.[252]

Moreover, hypocretin fibers innervate magnocellular neurons in the paraventricular (PVN) and supraoptic (SO) nuclei and Hcrtr1 immunoreactivity has been detected in vasopressin (ADH) and oxytocin (OT) neurons at these nuclei[61] indicating that hypocretins may influence the secretion of neurohypophysial hormones. In this sense, it has been reported that Hcrt1 treatment depolarizes magnocellular neurons *in vitro*.[97] However, although central Hcrt1 treatment increases vasopressin mRNA levels at the parvocellular part of the PVN[100] (our unpublished data)(Fig. 5B), no data are available on the effect of hypocretin on ADH gene expression in the magnocellular neurons of the PVN. Additional work is necessary to determinate whether hypocretins regulate the expression or secretion of ADH at this level. Concerning to the ADH neurons in the supraoptic nucleus (SO), unpublished data obtained by our group indicates that the stimulation of drinking after central administration of Hcrt1 is not related to changes in the expression of vasopressin in this nucleus (Fig. 5C).

Finally, a recent paper demonstrated that hypocretin neurons in the LHA express NPY Y4 receptor and the direct administration of NPY in this area increased water intake. These results indicate that the interaction between NPY-hypocretins could be implicated in the regulation of drinking behaviour.[106]

5. THE ROLE OF HCRT2

Most of the endocrine and neuroendocrine actions of hypocretins known so far have attributed to Hcrt1. To date, very limited data are available on the physiological relevance of Hcrt2; in fact, most of its actions are questioned.[10] This fact could be related with a linear structure, susceptible to peptidase actions.[10]

Here, we hypothesize that Hcrt2 could acts in the hypothalamus as paracrine factor regulating prepro-Hcrt expression. We based this idea on two pieces of evidence: 1) Hcrtr2, that binds Hcrt2, is very abundant in the lateral hypothalamic area, where hypocretin neurons are located[58-60,64] and 2) Hcrt1 does not regulate prepro-Hcrt mRNA expression in the LHA.[9] Further work is essential to test this hypothesis.

6 SUMMARY

This chapter has reviewed the endocrine aspects of the hypocretin system. Remarkable evidence obtained since the discovery of the hypocretins in 1998 has revealed that these peptides play a prominent role in the regulation of virtually all the neuroendocrine axes (Fig. 6). Moreover, hypocretins may serve as pivotal signals in the coordination of

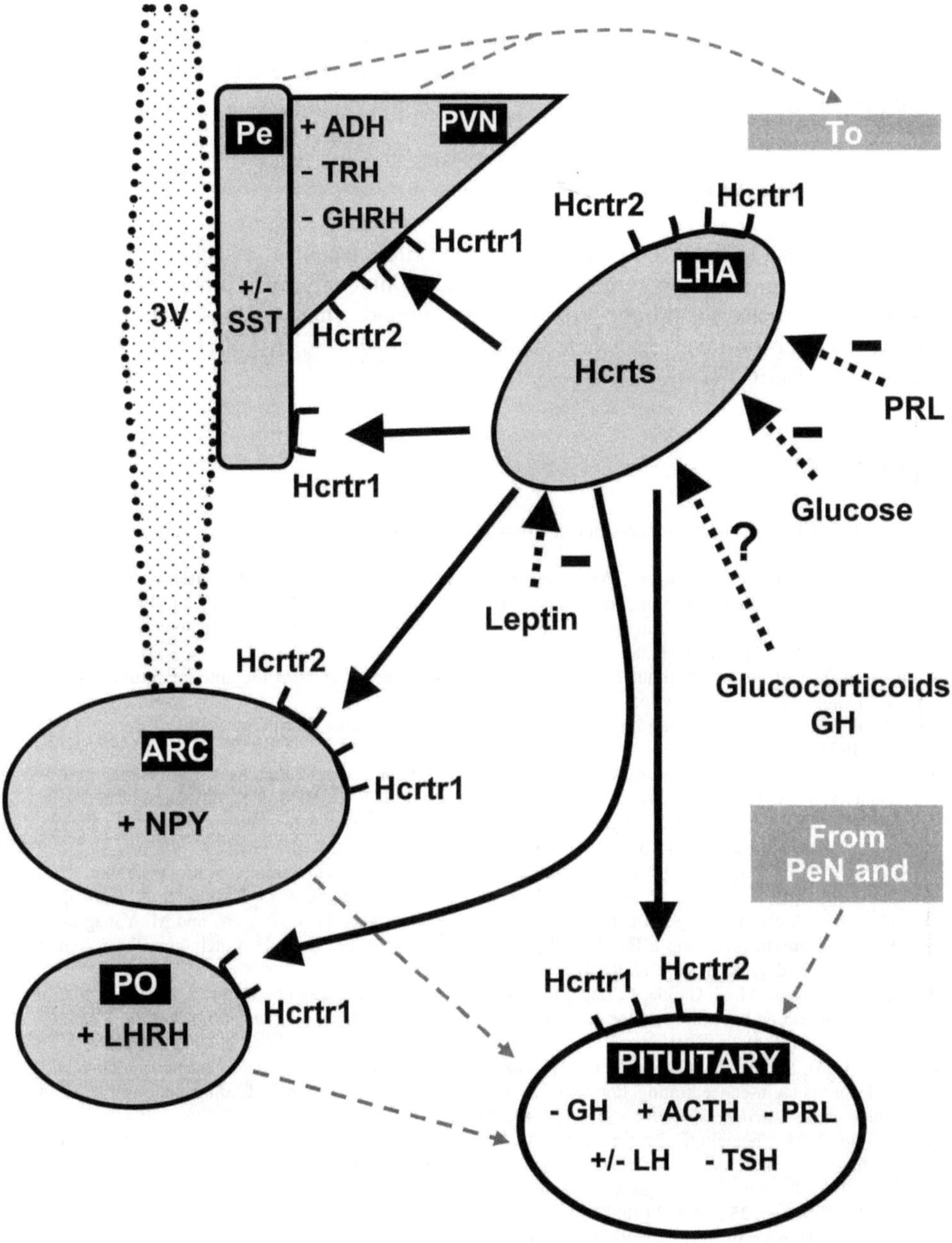

Figure 6. Neuroendocrine actions of hypocretins. The figure shows the actions of hypocretins on pituitary and hypothalamic nuclei regulating neuroendocrine function. In order to simplify the figure the relations between hypothalamic nuclei have been omitted. ARC: arcuate hypothalamic nucleus; LHA: lateral hypothalamic area; PO: preoptic area; PeN: periventricular hypothalamic nucleus; PVN: paraventricular hypothalamic nucleus; 3V: third ventricle. +: Stimulation; -: Inhibition.

endocrine responses with sleep, arousal and metabolic regulation. Additional work is fundamental to understand the interaction between hypocretins and other known (and unknown) systems implicated in these homeostatic events. These are the key questions that Neuroscience and Endocrinology must answer in the coming years.

7. ACKNOWLEDGEMENTS

We would like to thank Luz Casas for her excellent technical assistance in this last six years working on hypocretins, Sam Virtue (University of Cambridge) for his comments and criticisms and Irene Piñeiro for her support. This work has been funded by grants from Fondo de Investigaciones Sanitarias, Spanish Ministry of Health, Xunta de Galicia and the DGICYT.

8. REFERENCES

1. B. J. Everitt and T. Hokfelt. Neuroendocrine anatomy of the hypothalamus, *Acta Neurochir.Suppl (Wien.)* **47** 1-15 (1990).
2. L. L. Bernardis and L. L. Bellinger. The lateral hypothalamic area revisited: neuroanatomy, body weight regulation, neuroendocrinology and metabolism, *Neurosci.Biobehav.Rev.* **17**(2), 141-193 (1993).
3. L. L. Bernardis and L. L. Bellinger. The dorsomedial hypothalamic nucleus revisited: 1998 update, *Proc.Soc.Exp.Biol.Med.* **218**(4), 284-306 (1998).
4. M. Okada. Effect of electrical stimulation of the hypothalamus on the cells of the anterior pituitary gland, *Osaka Daigaku Igaku Zasshi* **3** 365-383 (1954).
5. C. R. Almli and G. T. Golden. Infant rats: effects of lateral hypothalamic destruction, *Physiol Behav.* **13**(1), 81-90 (1974).
6. L. de Lecea, T. S. Kilduff, C. Peyron, X. Gao, P. E. Foye, P. E. Danielson, C. Fukuhara, E. L. Battenberg, V. T. Gautvik, F. S. Bartlett, W. N. Frankel, A. N. van den Pol, F. E. Bloom, K. M. Gautvik and J. G. Sutcliffe. The hypocretins: hypothalamus-specific peptides with neuroexcitatory activity, *Proc.Natl.Acad.Sci.U.S.A* **95**(1), 322-327 (1998).
7. T. Sakurai, A. Amemiya, M. Ishii, I. Matsuzaki, R. M. Chemelli, H. Tanaka, S. C. Williams, J. A. Richardson, G. P. Kozlowski, S. Wilson, J. R. Arch, R. E. Buckingham, A. C. Haynes, S. A. Carr, R. S. Annan, D. E. McNulty, W. S. Liu, J. A. Terrett, N. A. Elshourbagy, D. J. Bergsma and M. Yanagisawa. Orexins and orexin receptors: a family of hypothalamic neuropeptides and G protein-coupled receptors that regulate feeding behavior, *Cell* **92**(4), 573-585 (1998).
8. M. López, L. Seoane, M. C. García, F. Lago, F. F. Casanueva, R. Senarís and C. Diéguez. Leptin regulation of prepro-orexin and orexin receptor mRNA levels in the hypothalamus, *Biochem.Biophys.Res.Commun.* **269** (1), 41-45 (2000).
9. M. López, L. M. Seoane, M. C. García, C. Diéguez and R. Señarís. Neuropeptide Y, but not agouti-related peptide or melanin-concentrating hormone, is a target Peptide for orexin-a feeding actions in the rat hypothalamus, *Neuroendocrinology* **75**(1), 34-44 (2002).
10. J. T. Willie, R. M. Chemelli, C. M. Sinton and M. Yanagisawa. To eat or to sleep? Orexin in the regulation of feeding and wakefulness, *Annu.Rev.Neurosci.* **24** 429-458 (2001).
11. S. Taheri, J. M. Zeitzer and E. Mignot. The role of hypocretins (orexins) in sleep regulation and narcolepsy, *Annu.Rev.Neurosci.* **25** 283-313 (2002).
12. J. M. Siegel. Hypocretin (orexin): Role in Normal Behavior and Neuropathology, *Annu.Rev.Psychol.* **55** 125-148 (2004).
13. J. G. Sutcliffe and L. de Lecea. The hypocretins: setting the arousal threshold, *Nat.Rev.Neurosci.* **3**(5), 339-349 (2002).
14. M. López, C. Diéguez, R. Señarís, in: *Recent Research in Endocrinology*, edited by S. G. Pandalai (Transworld Research Network, Kerala (India), 2001),pp.87-98
15. J. P. Kukkonen, T. Holmqvist, S. Ammoun and K. E. Akerman. Functions of the orexinergic/hypocretinergic system, *Am.J Physiol Cell Physiol* **283**(6), C1567-C1591 (2002).

16. M. M. Taylor and W. K. Samson. The other side of the orexins: endocrine and metabolic actions, *Am.J Physiol Endocrinol.Metab* **284**(1), E13-E17 (2003).
17. A. V. Ferguson and W. K. Samson. The orexin/hypocretin system: a critical regulator of neuroendocrine and autonomic function, *Front Neuroendocrinol.* **24**(3), 141-150 (2003).
18. B. J. Murawski and J. Crabbe. Effect of sleep deprivation on plasma 17-hydroxycorticosteroids, *J Appl.Physiol* **15** 280-282 (1960).
19. E. D. Weitzman, D. Fukushima, C. Nogeire, H. Roffwarg, T. F. Gallagher and L. Hellman. Twenty-four hour pattern of the episodic secretion of cortisol in normal subjects, *J Clin.Endocrinol.Metab* **33**(1), 14-22 (1971).
20. J. F. O'Connor, G. Y. Wu, T. F. Gallagher and L. Hellman. The 24-hour plasma thyroxin profile in normal man, *J Clin.Endocrinol.Metab* **39**(4), 765-771 (1974).
21. E. D. Weitzman, C. Nogeire, M. Perlow, D. Fukushima, J. Sassin, P. McGregor and L. Hellman. Effects of a prolonged 3-hour sleep-wake cycle on sleep stages, plasma cortisol, growth hormone and body temperature in man, *J Clin.Endocrinol.Metab* **38**(6), 1018-1030 (1974).
22. U. Beck, V. Brezinova, W. M. Hunter and I. Oswald. Plasma growth hormone and slow wave sleep increase after interruption of sheep, *J Clin.Endocrinol.Metab* **40**(5), 812-815 (1975).
23. A. Miyatake, Y. Morimoto, T. Oishi, N. Hanasaki, Y. Sugita, S. Iijima, Y. Teshima, Y. Hishikawa and Y. Yamamura. Circadian rhythm of serum testosterone and its relation to sleep: comparison with the variation in serum luteinizing hormone, prolactin, and cortisol in normal men, *J Clin.Endocrinol.Metab* **51**(6), 1365-1371 (1980).
24. T. Akerstedt, J. Palmblad, T. B. de la, R. Marana and M. Gillberg. Adrenocortical and gonadal steroids during sleep deprivation, *Sleep* **3**(1), 23-30 (1980).
25. A. Baumgartner, M. Dietzel, B. Saletu, R. Wolf, A. Campos-Barros, K. J. Graf, I. Kurten and U. Mannsmann. Influence of partial sleep deprivation on the secretion of thyrotropin, thyroid hormones, growth hormone, prolactin, luteinizing hormone, follicle stimulating hormone, and estradiol in healthy young women, *Psychiatry Res.* **48**(2), 153-178 (1993).
26. M. Sadamatsu, N. Kato, H. Iida, S. Takahashi, K. Sakaue, K. Takahashi, S. Hashida and E. Ishikawa. The 24-hour rhythms in plasma growth hormone, prolactin and thyroid stimulating hormone: effect of sleep deprivation, *J Neuroendocrinol.* **7**(8), 597-606 (1995).
27. F. Obal, Jr. and J. M. Krueger. The somatotropic axis and sleep, *Rev.Neurol.(Paris)* **157**(11 Pt 2), S12-S15 (2001).
28. C. D. Boethel. Sleep and the endocrine system: new associations to old diseases, *Curr.Opin.Pulm.Med.* **8**(6), 502-505 (2002).
29. E. Van Cauter, F. Latta, A. Nedeltcheva, K. Spiegel, R. Leproult, C. Vandenbril, R. Weiss, J. Mockel, J. J. Legros and G. Copinschi. Reciprocal interactions between the GH axis and sleep, *Growth Horm IGF.Res.* **14** (Suppl A) 10-17 (2004).
30. M. F. Scanlon, B. G. Issa and C. Diéguez. Regulation of growth hormone secretion, *Horm.Res.* **46**(4-5), 149-154 (1996).
31. F. F. Casanueva and C. Diéguez. Neuroendocrine regulation and actions of leptin, *Front Neuroendocrinol.* **20**(4), 317-363 (1999).
32. R. S. Ahima. Leptin and the neuroendocrinology of fasting, *Front Horm.Res.* **26** 42-56 (2000).
33. R. S. Ahima, C. B. Saper, J. S. Flier and J. K. Elmquist. Leptin regulation of neuroendocrine systems, *Front Neuroendocrinol.* **21**(3), 263-307 (2000).
34. J. G. Sutcliffe and L. de Lecea. The hypocretins: excitatory neuromodulatory peptides for multiple homeostatic systems, including sleep and feeding, *J.Neurosci.Res.* **62**(2), 161-168 (2000).
35. C. Broberger, L. de Lecea, J. G. Sutcliffe and T. Hokfelt. Hypocretin/orexin- and melanin-concentrating hormone-expressing cells form distinct populations in the rodent lateral hypothalamus: relationship to the neuropeptide Y and agouti gene-related protein systems, *J.Comp Neurol.* **402**(4), 460-474 (1998).
36. C. F. Elias, C. B. Saper, E. Maratos-Flier, N. A. Tritos, C. Lee, J. Kelly, J. B. Tatro, G. E. Hoffman, M. M. Ollmann, G. S. Barsh, T. Sakurai, M. Yanagisawa and J. K. Elmquist. Chemically defined projections linking the mediobasal hypothalamus and the lateral hypothalamic area, *J.Comp Neurol.* **402** (4), 442-459 (1998).
37. T. Nambu, T. Sakurai, K. Mizukami, Y. Hosoya, M. Yanagisawa and K. Goto. Distribution of orexin neurons in the adult rat brain, *Brain Res.* **827**(1-2), 243-260 (1999).
38. Y. Date, Y. Ueta, H. Yamashita, H. Yamaguchi, S. Matsukura, K. Kangawa, T. Sakurai, M. Yanagisawa and M. Nakazato. Orexins, orexigenic hypothalamic peptides, interact with autonomic, neuroendocrine and neuroregulatory systems, *Proc.Natl.Acad.Sci.U.S.A* **96**(2), 748-753 (1999).
39. T. L. Horvath, S. Diano and A. N. van den Pol. Synaptic interaction between hypocretin (orexin) and neuropeptide Y cells in the rodent and primate hypothalamus: a novel circuit implicated in metabolic and endocrine regulations, *J.Neurosci.* **19**(3), 1072-1087 (1999).

40. A. N. van den Pol. Hypothalamic hypocretin (orexin): robust innervation of the spinal cord, *J.Neurosci.* **19**(8), 3171-3182 (1999).
41. N. A. Tritos, J. W. Mastaitis, E. Kokkotou and E. Maratos-Flier. Characterization of melanin concentrating hormone and preproorexin expression in the murine hypothalamus, *Brain Res.* **895**(1-2), 160-166 (2001).
42. A. B. Reddy, A. S. Cronin, H. Ford and F. J. Ebling. Seasonal regulation of food intake and body weight in the male Siberian hamster: studies of hypothalamic orexin (hypocretin), neuropeptide Y (NPY) and pro-opiomelanocortin (POMC), *Eur.J.Neurosci.* **11**(9), 3255-3264 (1999).
43. P. A. McGranaghan and H. D. Piggins. Orexin A-like immunoreactivity in the hypothalamus and thalamus of the Syrian hamster (Mesocricetus auratus) and Siberian hamster (Phodopus sungorus), with special reference to circadian structures, *Brain Res.* **904**(2), 234-244 (2001).
44. E. M. Mintz, A. N. van den Pol, A. A. Casano and H. E. Albers. Distribution of hypocretin-(orexin) immunoreactivity in the central nervous system of Syrian hamsters (Mesocricetus auratus), *J.Chem.Neuroanat.* **21**(3), 225-238 (2001).
45. A. L. Kirchgessner. Orexins in the brain-gut axis, *Endocr.Rev.* **23**(1), 1-15 (2002).
46. C. M. Novak and H. E. Albers. Localization of hypocretin-like immunoreactivity in the brain of the diurnal rodent, Arvicanthis niloticus, *J.Chem.Neuroanat.* **23**(1), 49-58 (2002).
47. J. H. Zhang, S. Sampogna, F. R. Morales and M. H. Chase. Orexin (hypocretin)-like immunoreactivity in the cat hypothalamus: a light and electron microscopic study, *Sleep* **24**(1), 67-76 (2001).
48. S. J. Fung, J. Yamuy, S. Sampogna, F. R. Morales and M. H. Chase. Hypocretin (orexin) input to trigeminal and hypoglossal motoneurons in the cat: a double-labeling immunohistochemical study, *Brain Res.* **903**(1-2), 257-262 (2001).
49. J. H. Zhang, S. Sampogna, F. R. Morales and M. H. Chase. Co-localization of hypocretin-1 and hypocretin-2 in the cat hypothalamus and brainstem, *Peptides* **23**(8), 1479-1483 (2002).
50. J. Iqbal, S. Pompolo, T. Sakurai and I. J. Clarke. Evidence that Orexin-Containing Neurones Provide Direct Input to Gonadotropin-Releasing Hormone Neurones in the Ovine Hypothalamus, *J.Neuroendocrinol.* **13**(12), 1033-1041 (2001).
51. S. Diano, B. Horvath, H. F. Urbanski, P. Sotonyi and T. L. Horvath. Fasting activates the nonhuman primate hypocretin (orexin) system and its postsynaptic targets, *Endocrinology* **144**(9), 3774-3778 (2003).
52. C. Peyron, J. Faraco, W. Rogers, B. Ripley, S. Overeem, Y. Charnay, S. Nevsimalova, M. Aldrich, D. Reynolds, R. Albin, R. Li, M. Hungs, M. Pedrazzoli, M. Padigaru, M. Kucherlapati, J. Fan, R. Maki, G. J. Lammers, C. Bouras, R. Kucherlapati, S. Nishino and E. Mignot. A mutation in a case of early onset narcolepsy and a generalized absence of hypocretin peptides in human narcoleptic brains, *Nat.Med.* **6**(9), 991-997 (2000).
53. T. C. Thannickal, R. Y. Moore, R. Nienhuis, L. Ramanathan, S. Gulyani, M. Aldrich, M. Cornford and J. M. Siegel. Reduced number of hypocretin neurons in human narcolepsy, *Neuron* **27**(3), 469-474 (2000).
54. T. C. Thannickal, J. M. Siegel, R. Nienhuis and R. Y. Moore. Pattern of hypocretin (orexin) soma and axon loss, and gliosis, in human narcolepsy, *Brain Pathol.* **13**(3), 340-351 (2003).
55. L. Galas, H. Vaudry, B. Braun, A. N. van den Pol, L. de Lecea, J. G. Sutcliffe and N. Chartrel. Immunohistochemical localization and biochemical characterization of hypocretin/orexin-related peptides in the central nervous system of the frog Rana ridibunda, *J.Comp Neurol.* **429**(2), 242-252 (2001).
56. M. Shibahara, T. Sakurai, T. Nambu, T. Takenouchi, H. Iwaasa, S. I. Egashira, M. Ihara and K. Goto. Structure, tissue distribution, and pharmacological characterization of *Xenopus* orexins, *Peptides* **20**(10), 1169-1176 (1999).
57. C. Peyron, D. K. Tighe, A. N. van den Pol, L. de Lecea, H. C. Heller, J. G. Sutcliffe and T. S. Kilduff. Neurons containing hypocretin (orexin) project to multiple neuronal systems, *J.Neurosci.* **18**(23), 9996-10015 (1998).
58. P. Trivedi, H. Yu, D. J. MacNeil, L. H. Van Der Ploeg and X. M. Guan. Distribution of orexin receptor mRNA in the rat brain, *FEBS Lett.* **438**(1-2), 71-75 (1998).
59. X. Y. Lu, D. Bagnol, S. Burke, H. Akil and S. J. Watson. Differential distribution and regulation of OX1 and OX2 orexin/hypocretin receptor messenger RNA in the brain upon fasting, *Horm.Behav.* **37**(4), 335-344 (2000).
60. J. N. Marcus, C. J. Aschkenasi, C. E. Lee, R. M. Chemelli, C. B. Saper, M. Yanagisawa and J. K. Elmquist. Differential expression of orexin receptors 1 and 2 in the rat brain, *J.Comp Neurol.* **435**(1), 6-25 (2001).
61. M. Backberg, G. Hervieu, S. Wilson and B. Meister. Orexin receptor-1 (OX-R1) immunoreactivity in chemically identified neurons of the hypothalamus: focus on orexin targets involved in control of food and water intake, *Eur.J.Neurosci.* **15**(2), 315-328 (2002).
62. G. J. Hervieu, J. E. Cluderay, D. C. Harrison, J. C. Roberts and R. A. Leslie. Gene expression and protein distribution of the orexin-1 receptor in the rat brain and spinal cord, *Neuroscience* **103**(3), 777-797 (2001).

63. R. E. Campbell, K. L. Grove and M. S. Smith. Gonadotropin-releasing hormone neurons coexpress orexin 1 receptor immunoreactivity and receive direct contacts by orexin fibers, *Endocrinology* **144**(4), 1542-1548 (2003).
64. J. E. Cluderay, D. C. Harrison and G. J. Hervieu. Protein distribution of the orexin-2 receptor in the rat central nervous system, *Regul.Pept.* **104**(1-3), 131-144 (2002).
65. Y. Date, M. S. Mondal, S. Matsukura, Y. Ueta, H. Yamashita, H. Kaiya, K. Kangawa and M. Nakazato. Distribution of orexin/hypocretin in the rat median eminence and pituitary, *Brain Res.Mol.Brain Res.* **76**(1), 1-6 (2000).
66. O. Johren, S. J. Neidert, M. Kummer, A. Dendorfer and P. Dominiak. Prepro-orexin and orexin receptor mRNAs are differentially expressed in peripheral tissues of male and female rats, *Endocrinology* **142**(8), 3324-3331 (2001).
67. M. Blanco, R. Gallego, T. García-Caballero, C. Diéguez and A. Beiras. Cellular localization of orexins in human anterior pituitary, *Histochem.Cell Biol.* **120**(4), 259-264 (2003).
68. M. Blanco, M. López, T. García-Caballero, R. Gallego, A. Vázquez-Boquete, G. Morel, R. M. Señarís, F. Casanueva, C. Diéguez and A. Beiras. Cellular localization of orexin receptors in human pituitary, *J.Clin.Endocrinol.Metab* **86**(4), 1616-1619 (2001).
69. O. Johren, N. Bruggemann, A. Dendorfer and P. Dominiak. Gonadal steroids differentially regulate the messenger ribonucleic acid expression of pituitary orexin type 1 receptors and adrenal orexin type 2 receptors, *Endocrinology* **144**(4), 1219-1225 (2003).
70. M. López, R. Senarís, R. Gallego, T. García-Caballero, F. Lago, L. Seoane, F. Casanueva and C. Diéguez. Orexin receptors are expressed in the adrenal medulla of the rat, *Endocrinology* **140**(12), 5991-5994 (1999).
71. H. S. Randeva, E. Karteris, D. Grammatopoulos and E. W. Hillhouse. Expression of orexin-A and functional orexin type 2 receptors in the human adult adrenals: implications for adrenal function and energy homeostasis, *J.Clin.Endocrinol.Metab* **86**(10), 4808-4813 (2001).
72. M. Nakabayashi, T. Suzuki, K. Takahashi, K. Totsune, Y. Muramatsu, C. Kaneko, F. Date, J. Takeyama, A. D. Darnel, T. Moriya and H. Sasano. Orexin-A expression in human peripheral tissues, *Mol.Cell Endocrinol.* **205**(1-2), 43-50 (2003).
73. L. K. Malendowicz, A. Hochol, A. Ziolkowska, M. Nowak, L. Gottardo and G. G. Nussdorfer. Prolonged orexin administration stimulates steroid-hormone secretion, acting directly on the rat adrenal gland, *Int.J.Mol.Med.* **7**(4), 401-404 (2001).
74. T. Nanmoku, K. Isobe, T. Sakurai, A. Yamanaka, K. Takekoshi, Y. Kawakami, K. Goto and T. Nakai. Effects of orexin on cultured porcine adrenal medullary and cortex cells, *Regul.Pept.* **104**(1-3), 125-130 (2002).
75. G. Mazzocchi, L. K. Malendowicz, F. Aragona, P. Rebuffat, L. Gottardo and G. G. Nussdorfer. Human pheochromocytomas express orexin receptor type 2 gene and display an in vitro secretory response to orexins A and B, *J.Clin.Endocrinol.Metab* **86**(10), 4818-4821 (2001).
76. G. Mazzocchi, L. K. Malendowicz, L. Gottardo, F. Aragona and G. G. Nussdorfer. Orexin A stimulates cortisol secretion from human adrenocortical cells through activation of the adenylate cyclase-dependent signaling cascade, *J.Clin.Endocrinol.Metab* **86**(2), 778-782 (2001).
77. M. Blanco, T. García-Caballero, M. Fraga, R. Gallego, J. Cuevas, J. Forteza, A. Beiras and C. Diéguez. Cellular localization of orexin receptors in human adrenal gland, adrenocortical adenomas and pheochromocytomas, *Regul.Pept.* **104**(1-3), 161-165 (2002).
78. M. L. Barreiro, R. Pineda, V. M. Navarro, M. López, J. S. Suominen, L. Pinilla, R. Senarís, J. Toppari, E. Aguilar, C. Diéguez and M. Tena-Sempere. Orexin 1 receptor messenger ribonucleic Acid expression and stimulation of testosterone secretion by orexin-a in rat testis, *Endocrinology* **145**(5), 2297-2306 (2004).
79. E. Karteris, J. Chen and H. S. Randeva. Expression of human prepro-orexin and signaling characteristics of orexin receptors in the male reproductive system, *J Clin.Endocrinol.Metab* **89**(4), 1957-1962 (2004).
80. A. L. Kirchgessner and M. Liu. Orexin synthesis and response in the gut, *Neuron* **24**(4), 941-951 (1999).
81. M. J. de Miguel and M. A. Burrell. Immunocytochemical Detection of Orexin A in Endocrine Cells of the Developing Mouse Gut, *J.Histochem.Cytochem.* **50**(1), 63-70 (2002).
82. E. Naslund, M. Ehrstrom, J. Ma, P. M. Hellstrom and A. L. Kirchgessner. Localization and effects of orexin on fasting motility in the rat duodenum, *Am.J Physiol Gastrointest.Liver Physiol* **282**(3), G470-G479 (2002).
83. R. Ouedraogo, E. Naslund and A. L. Kirchgessner. Glucose regulates the release of orexin-a from the endocrine pancreas, *Diabetes* **52**(1), 111-117 (2003).
84. J. D. Mikkelsen, F. Hauser, L. deLecea, J. G. Sutcliffe, T. S. Kilduff, C. Calgari, P. Pevet and V. Simonneaux. Hypocretin (orexin) in the rat pineal gland: a central transmitter with effects on noradrenaline-induced release of melatonin, *Eur.J.Neurosci.* **14**(3), 419-425 (2001).

85. Z. Arihara, K. Takahashi, O. Murakami, K. Totsune, M. Sone, F. Satoh, S. Ito and T. Mouri. Immunoreactive orexin-A in human plasma, *Peptides* **22**(1), 139-142 (2001).
86. G. Komaki, Y. Matsumoto, H. Nishikata, K. Kawai, T. Nozaki, M. Takii, H. Sogawa and C. Kubo. Orexin-A and leptin change inversely in fasting non-obese subjects, *Eur.J.Endocrinol.* **144**(6), 645-651 (2001).
87. M. A. Dalal, A. Schuld, M. Haack, M. Uhr, P. Geisler, I. Eisensehr, S. Noachtar and T. Pollmacher. Normal plasma levels of orexin A (hypocretin-1) in narcoleptic patients, *Neurology* **56**(12), 1749-1751 (2001).
88. T. Matsumura, M. Nakayama, A. Nomura, A. Naito, K. Kamahara, K. Kadono, M. Inoue, T. Homma and K. Sekizawa. Age-related changes in plasma orexin-A concentrations, *Exp.Gerontol.* **37**(8-9), 1127 (2002).
89. J. A. Adam, P. P. Menheere, F. M. van Dielen, P. B. Soeters, W. A. Buurman and J. W. Greve. Decreased plasma orexin-A levels in obese individuals, *Int.J.Obes.Relat Metab Disord.* **26**(2), 274-276 (2002).
90. T. Sugimoto, Y. Nagake, S. Sugimoto, S. Akagi, H. Ichikawa, Y. Nakamura, N. Ogawa and H. Makino. Plasma orexin concentrations in patients on hemodialysis, *Nephron* **90**(4), 379-383 (2002).
91. S. Higuchi, A. Usui, M. Murasaki, S. Matsushita, N. Nishioka, A. Yoshino, T. Matsui, H. Muraoka, Y. Ishizuka, S. Kanba and T. Sakurai. Plasma orexin-A is lower in patients with narcolepsy, *Neurosci.Lett.* **318**(2), 61-64 (2002).
92. S. W. Kok, A. E. Meinders, S. Overeem, G. J. Lammers, F. Roelfsema, M. Frolich and H. Pijl. Reduction of plasma leptin levels and loss of its circadian rhythmicity in hypocretin (orexin)-deficient narcoleptic humans, *J.Clin.Endocrinol.Metab* **87**(2), 805-809 (2002).
93. J. L. Li, F. L. Zheng, H. B. Tan, S. Y. Yin, J. H. Yang, Y. Li and Y. F. Bu. [Orexin A and neuropeptide Y levels in plasma and hypothalamus of rats with chronic renal failure], *Zhonghua Yi.Xue.Za Zhi.* **83**(11), 992-995 (2003).
94. M. Palkovits, M. J. Brownstein and W. Vale . Distribution of corticotropin-releasing factor in rat brain, *Fed.Proc.* **44**(1 Pt 2), 215-219 (1985).
95. S. Welle and G. D. Coover. Meal-induced decreases in serum corticosterone in the lateral hypothalamic syndrome, *Physiol Behav.* **23**(3), 547-555 (1979).
96. M. Kuru, Y. Ueta, R. Serino, M. Nakazato, Y. Yamamoto, I. Shibuya and H. Yamashita. Centrally administered orexin/hypocretin activates HPA axis in rats, *Neuroreport* **11**(9), 1977-1980 (2000).
97. M. J. Follwell and A. V. Ferguson. Cellular mechanisms of orexin actions on paraventricular nucleus neurones in rat hypothalamus, *J.Physiol* **545**(Pt 3), 855-867 (2002).
98. W. K. Samson, M. M. Taylor, M. Follwell and A. V. Ferguson. Orexin actions in hypothalamic paraventricular nucleus: physiological consequences and cellular correlates, *Regul.Pept.* **104**(1-3), 97-103 (2002).
99. T. Shirasaka, S. Miyahara, T. Kunitake, Q. H. Jin, K. Kato, M. Takasaki and H. Kannan. Orexin depolarizes rat hypothalamic paraventricular nucleus neurons, *Am.J.Physiol Regul.Integr.Comp Physiol* **281**(4), R1114-R1118 (2001).
100. K. A. Al Barazanji, S. Wilson, J. Baker, D. S. Jessop and M. S. Harbuz. Central orexin-A activates hypothalamic-pituitary-adrenal axis and stimulates hypothalamic corticotropin releasing factor and arginine vasopressin neurones in conscious rats, *J.Neuroendocrinol.* **13**(5), 421-424 (2001).
101. J. J. Hagan, R. A. Leslie, S. Patel, M. L. Evans, T. A. Wattam, S. Holmes, C. D. Benham, S. G. Taylor, C. Routledge, P. Hemmati, R. P. Munton, T. E. Ashmeade, A. S. Shah, J. P. Hatcher, P. D. Hatcher, D. N. Jones, M. I. Smith, D. C. Piper, A. J. Hunter, R. A. Porter and N. Upton. Orexin A activates locus coeruleus cell firing and increases arousal in the rat, *Proc.Natl.Acad.Sci.U.S.A* **96**(19), 10911-10916 (1999).
102. M. Jaszberenyi, E. Bujdoso, I. Pataki and G. Telegdy. Effects of orexins on the hypothalamic-pituitary-adrenal system, *J.Neuroendocrinol.* **12**(12), 1174-1178 (2000).
103. M. Jaszberenyi, E. Bujdoso and G. Telegdy . The role of neuropeptide Y in orexin-induced hypothalamic-pituitary- adrenal activation, *J.Neuroendocrinol.* **13**(5), 438-441 (2001).
104. S. H. Russell, C. J. Small, C. L. Dakin, C. R. Abbott, D. G. Morgan, M. A. Ghatei and S. R. Bloom. The central effects of orexin-A in the hypothalamic-pituitary-adrenal axis in vivo and in vitro in male rats, *J.Neuroendocrinol.* **13**(6), 561-566 (2001).
105. H. Volkoff and R. E. Peter. Interactions between orexin A, NPY and galanin in the control of food intake of the goldfish, *Carassius auratus*, *Regul.Pept.* **101**(1-3), 59-72 (2001).
106. R. E. Campbell, M. S. Smith, S. E. Allen, B. E. Grayson, J. M. ffrench-Mullen and K. L. Grove. Orexin neurons express a functional pancreatic polypeptide Y4 receptor, *J Neurosci.* **23**(4), 1487-1497 (2003).
107. S. Pu, M. R. Jain, P. S. Kalra and S. P. Kalra. Orexins, a novel family of hypothalamic neuropeptides, modulate pituitary luteinizing hormone secretion in an ovarian steroid-dependent manner, *Regul.Pept.* **78**(1-3), 133-136 (1998).
108. S. H. Russell, C. J. Small, A. R. Kennedy, S. A. Stanley, A. Seth, K. G. Murphy, S. Taheri, M. A. Ghatei and S. R. Bloom. Orexin a interactions in the hypothalamo-pituitary gonadal axis, *Endocrinology* **142**(12), 5294-5302 (2001).

109. Y. C. Hsueh, S. M. Cheng and J. T. Pan. Fasting stimulates tuberoinfundibular dopaminergic neuronal activity and inhibits prolactin secretion in oestrogen-primed ovariectomized rats: involvement of orexin a and neuropeptide y, *J.Neuroendocrinol.* **14**(9), 745-752 (2002).
110. M. Jaszberenyi, E. Bujdoso, E. Kiss, I. Pataki and G. Telegdy. The role of NPY in the mediation of orexin-induced hypothermia, *Regul.Pept.* **104**(1-3), 55-59 (2002).
111. M. Szekely, E. Petervari, M. Balasko, I. Hernadi and B. Uzsoki. Effects of orexins on energy balance and thermoregulation, *Regul.Pept.* **104**(1-3), 47-53 (2002).
112. A. Yamanaka, K. Kunii, T. Nambu, N. Tsujino, A. Sakai, I. Matsuzaki, Y. Miwa, K. Goto and T. Sakurai. Orexin-induced food intake involves neuropeptide Y pathway, *Brain Res.* **859**(2), 404-409 (2000).
113. M. R. Jain, T. L. Horvath, P. S. Kalra and S. P. Kalra. Evidence that NPY Y1 receptors are involved in stimulation of feeding by orexins (hypocretins) in sated rats, *Regul.Pept.* **87**(1-3), 19-24 (2000).
114. T. Ida, K. Nakahara, T. Kuroiwa, K. Fukui, M. Nakazato, T. Murakami and N. Murakami. Both corticotropin releasing factor and neuropeptide Y are involved in the effect of orexin (hypocretin) on the food intake in rats, *Neurosci.Lett.* **293**(2), 119-122 (2000).
115. M. G. Dube, T. L. Horvath, P. S. Kalra and S. P. Kalra. Evidence of NPY Y5 receptor involvement in food intake elicited by orexin A in sated rats, *Peptides* **21**(10), 1557-1560 (2000).
116. L. K. Malendowicz, N. Jedrzejczak, A. S. Belloni, M. Trejter, A. Hochol and G. G. Nussdorfer. Effects of orexins A and B on the secretory and proliferative activity of immature and regenerating rat adrenal glands, *Histol.Histopathol.* **16**(3), 713-717 (2001).
117. L. K. Malendowicz, C. Tortorella and G. G. Nussdorfer. Orexins stimulate corticosterone secretion of rat adrenocortical cells, through the activation of the adenylate cyclase-dependent signaling cascade, *J.Steroid Biochem.Mol.Biol.* **70**(4-6), 185-188 (1999).
118. T. Ida, K. Nakahara, T. Murakami, R. Hanada, M. Nakazato and N. Murakami. Possible involvement of orexin in the stress reaction in rats, *Biochem.Biophys.Res.Commun.* **270**(1), 318-323 (2000).
119. W. K. Samson and M. M. Taylor. Hypocretin/orexin suppresses corticotroph responsiveness in vitro, *Am.J.Physiol Regul.Integr.Comp Physiol* **281**(4), R1140-R1145 (2001).
120. C. T. Chen, L. L. Hwang, J. K. Chang and N. J. Dun. Pressor effects of orexins injected intracisternally and to rostral ventrolateral medulla of anesthetized rats, *Am.J.Physiol Regul.Integr.Comp Physiol* **278**(3), R692-R697 (2000).
121. W. K. Samson, B. Gosnell, J. K. Chang, Z. T. Resch and T. C. Murphy. Cardiovascular regulatory actions of the hypocretins in brain, *Brain Res.* **831**(1-2), 248-253 (1999).
122. T. Shirasaka, M. Nakazato, S. Matsukura, M. Takasaki and H. Kannan. Sympathetic and cardiovascular actions of orexins in conscious rats, *Am.J.Physiol* **277**(6 Pt 2), R1780-R1785 (1999).
123. P. J. Brunton and J. A. Russell. Hypothalamic-pituitary-adrenal responses to centrally administered orexin-A are suppressed in pregnant rats, *J Neuroendocrinol.* **15**(7), 633-637 (2003).
124. B. B. Gallagher. Regulation of cortisol secretion in Parkinson's syndrome and narcolepsy, *J Clin.Endocrinol.Metab* **32**(6), 796-800 (1971).
125. S. W. Kok, F. Roelfsema, S. Overeem, G. J. Lammers, R. L. Strijers, M. Frolich, A. E. Meinders and H. Pijl. Dynamics of the pituitary-adrenal ensemble in hypocretin-deficient narcoleptic humans: blunted basal adrenocorticotropin release and evidence for normal time-keeping by the master pacemaker, *J Clin.Endocrinol.Metab* **87**(11), 5085-5091 (2002).
126. K. Morita, A. Kuwada, H. Fujihara, Y. Morita and H. Sei. Changes in the expression of steroid metabolism-related genes in rat adrenal glands during selective REM sleep deprivation, *Life Sci.* **72**(17), 1973-1982 (2003).
127. J. R. Davidson, H. Moldofsky and F. A. Lue. Growth hormone and cortisol secretion in relation to sleep and wakefulness, *J Psychiatry Neurosci.* **16**(2), 96-102 (1991).
128. A. N. Vgontzas, G. Mastorakos, E. O. Bixler, A. Kales, P. W. Gold and G. P. Chrousos. Sleep deprivation effects on the activity of the hypothalamic-pituitary-adrenal and growth axes: potential clinical implications, *Clin.Endocrinol.(Oxf)* **51**(2), 205-215 (1999).
129. F. Chapotot, A. Buguet, C. Gronfier and G. Brandenberger. Hypothalamo-pituitary-adrenal axis activity is related to the level of central arousal: effect of sleep deprivation on the association of high-frequency waking electroencephalogram with cortisol release, *Neuroendocrinology* **73**(5), 312-321 (2001).
130. A. Stricker-Krongrad and B. Beck. Modulation of hypothalamic hypocretin/orexin mRNA expression by glucocorticoids, *Biochem.Biophys.Res.Commun.* **296**(1), 129-133 (2002).
131. D. L. Drazen, L. M. Coolen, A. D. Strader, M. D. Wortman, S. C. Woods and R. J. Seeley. Differential effects of adrenalectomy on melanin-concentrating hormone and orexin A, *Endocrinology* (2004) (EPub 24/3/2004 as doi:10.1210/en2003-1760).
132. T. Nanmoku, K. Isobe, T. Sakurai, A. Yamanaka, K. Takekoshi, Y. Kawakami, K. Ishii, K. Goto and T. Nakai. Orexins suppress catecholamine synthesis and secretion in cultured PC12 cells, *Biochem.Biophys.Res.Commun.* **274** (2), 310-315 (2000).

133. K. Matsumura, T. Tsuchihashi and I. Abe. Central orexin-A augments sympathoadrenal outflow in conscious rabbits, *Hypertension* **37**(6), 1382-1387 (2001).
134. T. Shirasaka, T. Kunitake, M. Takasaki and H. Kannan. Neuronal effects of orexins: relevant to sympathetic and cardiovascular functions, *Regul.Pept.* **104**(1-3), 91-95 (2002).
135. P. Smith, B. Connolly and A. Ferguson. Microinjection of orexin into the rat nucleus tractus solitarius causes increases in blood pressure, *Brain Res.* **950**(1-2), 261 (2002).
136. P. E. Sawchenko, L. W. Swanson, J. Rivier and W. W. Vale. The distribution of growth-hormone-releasing factor (GRF) immunoreactivity in the central nervous system of the rat: an immunohistochemical study using antisera directed against rat hypothalamic GRF, *J.Comp Neurol.* **237**(1), 100-115 (1985).
137. D. Fellmann, C. Bugnon, J. Verstegen and G. N. Lavry. Coexpression of human growth hormone-releasing factor 1-37-like and alpha-melanotropin-like immunoreactivities in neurones of the rat lateral dorsal hypothalamus, *Neurosci.Lett.* **68**(1), 122-126 (1986).
138. S. Daikoku, H. Kawano, M. Noguchi, J. Nakanishi, M. Tokuzen, K. Chihara and I. Nagatsu. GRF neurons in the rat hypothalamus, *Brain Res.* **399**(2), 250-261 (1986).
139. J. L. Bresson, M. C. Clavequin, D. Fellmann and C. Bugnon. Human hypothalamic neuronal system revealed with a salmon melanin- concentrating hormone (MCH) antiserum, *Neurosci.Lett.* **102**(1), 39-43 (1989).
140. M. Brownstein, A. Arimura, H. Sato, A. V. Schally and J. S. Kizer. The regional distribution of somatostatin in the rat brain, *Endocrinology* **96**(6), 1456-1461 (1975).
141. R. L. Himsworth, P. W. Carmel and A. G. Frantz. The location of the chemoreceptor controlling growth hormone secretion during hypoglycemia in primates, *Endocrinology* **91**(1), 217-226 (1972).
142. S. H. Russell, M. S. Kim, C. J. Small, C. R. Abbott, D. G. Morgan, S. Taheri, K. G. Murphy, J. F. Todd, M. A. Ghatei and S. R. Bloom. Central administration of orexin A suppresses basal and domperidone stimulated plasma prolactin, *J.Neuroendocrinol.* **12**(12), 1213-1218 (2000).
143. L. M. Seoane, S. Tovar, D. Pérez, F. Mallo, M. López, R. Señarís, F. F. Casanueva and C. Diéguez. Orexin A suppresses in vivo GH secretion, *Eur.J Endocrinol.* **150**(5), 731-736 (2004).
144. R. Xu, Q. Wang, M. Yan, M. Hernandez, C. Gong, W. C. Boon, Y. Murata, Y. Ueta and C. Chen. Orexin-A augments voltage-gated Ca2+ currents and synergistically increases growth hormone (GH) secretion with GH-releasing hormone in primary cultured ovine somatotropes, *Endocrinology* **143**(12), 4609-4619 (2002).
145. R. Xu, S. G. Roh, C. Gong, M. Hernandez, Y. Ueta and C. Chen. Orexin-B augments voltage-gated L-type Ca(2+) current via protein kinase C-mediated signalling pathway in ovine somatotropes, *Neuroendocrinology* **77**(3), 141-152 (2003).
146. C. Chen and R. Xu. The in vitro regulation of growth hormone secretion by orexins, *Endocrine.* **22**(1), 57-66 (2003).
147. M. López, L. M. Seoane, S. Tovar, R. Nogueiras, C. Diéguez and R. Señarís. Orexin-A regulates growth hormone releasing hormone mRNA content in a nucleus specific manner and somatostatin mRNA content in a growth hormone-dependent fashion in the rat hypothalamus, *Eur.J.Neurosci.* **19**(8), 2080-2088 (2004).
148. F. Camanni, E. Ghigo and E. Arvat. Growth hormone-releasing peptides and their analogs, *Front Neuroendocrinol.* **19**(1), 47-72 (1998).
149. F. F. Casanueva and C. Diéguez. Growth Hormone Secretagogues: Physiological Role and Clinical Utility, *Trends Endocrinol.Metab* **10**(1), 30-38 (1999).
150. Z. Liposits. Ultrastructure of hypothalamic paraventricular neurons, *Crit Rev.Neurobiol.* **7**(2), 89-162 (1993).
151. V. Rettori, L. Milenkovic, M. C. Aguila and S. M. McCann. Physiologically significant effect of neuropeptide Y to suppress growth hormone release by stimulating somatostatin discharge, *Endocrinology* **126**(5), 2296-2301 (1990).
152. K. Okada, H. Sugihara, S. Minami and I. Wakabayashi. Effect of parenteral administration of selected nutrients and central injection of gamma-globulin from antiserum to neuropeptide Y on growth hormone secretory pattern in food-deprived rats, *Neuroendocrinology* **57**(4), 678-686 (1993).
153. X. J. Cai, P. S. Widdowson, J. Harrold, S. Wilson, R. E. Buckingham, J. R. Arch, M. Tadayyon, J. C. Clapham, J. Wilding and G. Williams. Hypothalamic orexin expression: modulation by blood glucose and feeding, *Diabetes* **48**(11), 2132-2137 (1999).
154. J. T. Willie, R. M. Chemelli, C. M. Sinton, S. Tokita, S. C. Williams, Y. Y. Kisanuki, J. N. Marcus, C. Lee, J. K. Elmquist, K. A. Kohlmeier, C. S. Leonard, J. A. Richardson, R. E. Hammer and M. Yanagisawa. Distinct narcolepsy syndromes in Orexin receptor-2 and Orexin null mice: molecular genetic dissection of Non-REM and REM sleep regulatory processes, *Neuron* **38**(5), 715-730 (2003).
155. M. Hayashi, M. Shimohira, S. Saisho, K. Shimozawa and Y. Iwakawa. Sleep disturbance in children with growth hormone deficiency, *Brain Dev.* **14**(3), 170-174 (1992).

156. R. H. Wu and M. J. Thorpy. Effect of growth hormone treatment on sleep EEGs in growth hormone-deficient children, *Sleep* **11**(5), 425-429 (1988).
157. F. Obal, Jr., J. Fang, P. Taishi, B. Kacsoh, J. Gardi and J. M. Krueger. Deficiency of growth hormone-releasing hormone signaling is associated with sleep alterations in the dwarf rat, *J.Neurosci.* **21**(8), 2912-2918 (2001).
158. J. Zhang, F. Obal, Jr., J. Fang, B. J. Collins and J. M. Krueger. Non-rapid eye movement sleep is suppressed in transgenic mice with a deficiency in the somatotropic system, *Neurosci.Lett.* **220**(2), 97-100 (1996).
159. E. Van Cauter, L. Plat and G. Copinschi. Interrelations between sleep and the somatotropic axis, *Sleep* **21**(6), 553-566 (1998).
160. J. Danguir. Intracerebroventricular infusion of somatostatin selectively increases paradoxical sleep in rats, *Brain Res.* **367**(1-2), 26-30 (1986).
161. J. Danguir and K. S. Saint-Hilaire. Reversal of desipramine-induced suppression of paradoxical sleep by a long-acting somatostatin analogue (octreotide) in rats, *Neurosci.Lett.* **98**(2), 154-158 (1989).
162. C. L. Ehlers, T. K. Reed and S. J. Henriksen. Effects of corticotropin-releasing factor and growth hormone-releasing factor on sleep and activity in rats, *Neuroendocrinology* **42**(6), 467-474 (1986).
163. J. Toppila, M. Asikainen, L. Alanko, F. W. Turek, D. Stenberg and T. Porkka-Heiskanen. The effect of REM sleep deprivation on somatostatin and growth hormone- releasing hormone gene expression in the rat hypothalamus, *J.Sleep Res.* **5**(2), 115-122 (1996).
164. J. Toppila, L. Alanko, M. Asikainen, I. Tobler, D. Stenberg and T. Porkka-Heiskanen. Sleep deprivation increases somatostatin and growth hormone-releasing hormone messenger RNA in the rat hypothalamus, *J.Sleep Res.* **6**(3), 171-178 (1997).
165. P. Bourgin, S. Huitron-Resendiz, A. D. Spier, V. Fabre, B. Morte, J. R. Criado, J. G. Sutcliffe, S. J. Henriksen and L. de Lecea. Hypocretin-1 modulates rapid eye movement sleep through activation of locus coeruleus neurons, *J.Neurosci.* **20**(20), 7760-7765 (2000).
166. R. W. Clark, H. S. Schmidt and W. B. Malarkey. Disordered growth hormone and prolactin secretion in primary disorders of sleep, *Neurology* **29**(6), 855-861 (1979).
167. T. Higuchi, Y. Takahashi, K. Takahashi, Y. Niimi and A. Miyasita. Twenty-four-hour secretory patterns of growth hormone, prolactin, and cortisol in narcolepsy, *J Clin.Endocrinol.Metab* **49**(2), 197-204 (1979).
168. S. Overeem, S. W. Kok, G. J. Lammers, A. A. Vein, M. Frolich, A. E. Meinders, F. Roelfsema and H. Pijl. Somatotropic axis in hypocretin-deficient narcoleptic humans: altered circadian distribution of GH-secretory events, *Am.J.Physiol Endocrinol.Metab* **284**(3), E641-E647 (2003).
169. L. M. Seoane, M. López, S. Tovar, F. Casanueva, R. Señarís and C. Diéguez. Agouti-related peptide, neuropeptide Y, and somatostatin-producing neurons are targets for ghrelin actions in the rat hypothalamus, *Endocrinology* **144** 544-551 (2003).
170. T. L. Powley and R. E. Keesey. Relationship of body weight to the lateral hypothalamic feeding syndrome, *J Comp Physiol Psychol.* **70**(1), 25-36 (1970).
171. R. F. Drewett. The meal patterns of the oestrous cycle and their motivational significance, *Q.J Exp.Psychol.* **26**(Pt3), 489-494 (1974).
172. J. D. Blaustein and G. N. Wade. Ovarian influences on the meal patterns of female rats, *Physiol Behav.* **17**(2), 201-208 (1976).
173. D. M. Nance and R. A. Gorski. Neurohormonal determinants of sex differences in the hypothalamic regulation of feeding behavior and body weight in the rat, *Pharmacol.Biochem.Behav.* **3**(Suppl 1), 155-162 (1975).
174. G. C. Sieck, D. M. Nance, J. A. Ramaley, A. N. Taylor and R. A. Gorski. Prepubertal cyclicity in feeding behavior and body weight regulation in the female rat, *Physiol Behav.* **18**(2), 299-305 (1977).
175. H. Saito, H. Kaba, T. Sato, M. Kondo, Y. Takeshima, N. Edashige, K. Seto and M. Kawakami. Influence of electrical stimulation of the hypothalamus on ovarian steroidogenesis in hypophysectomized and adrenalectomized rats, *Exp.Clin.Endocrinol.* **95**(2), 259-261 (1990).
176. S. Hansen, C. Kohler, M. Goldstein and H. V. Steinbusch. Effects of ibotenic acid-induced neuronal degeneration in the medial preoptic area and the lateral hypothalamic area on sexual behavior in the male rat, *Brain Res.* **239**(1), 213-232 (1982).
177. W. L. Benedetti, R. Lozdziejsky, M. A. Sala, J. M. Monti and E. Grino. Blockade of ovulation after atropine implants in the lateral hypothalamus of the rat, *Experientia* **25**(11), 1158-1159 (1969).
178. M. S. Smith and K. L. Grove. Integration of the regulation of reproductive function and energy balance: lactation as a model, *Front Neuroendocrinol.* **23**(3), 225-256 (2002).
179. T. Tamura, M. Irahara, M. Tezuka, M. Kiyokawa and T. Aono. Orexins, orexigenic hypothalamic neuropeptides, suppress the pulsatile secretion of luteinizing hormone in ovariectomized female rats, *Biochem.Biophys.Res.Commun.* **264**(3), 759-762 (1999).

180. A. Kohsaka, H. Watanobe, Y. Kakizaki, T. Suda and H. B. Schioth. A significant participation of orexin-A, a potent orexigenic peptide, in the preovulatory luteinizing hormone and prolactin surges in the rat, *Brain Res.* **898**(1), 166-170 (2001).
181. M. Furuta, T. Funabashi and F. Kimura. Suppressive action of orexin A on pulsatile luteinizing hormone secretion is potentiated by a low dose of estrogen in ovariectomized rats, *Neuroendocrinology* **75**(3), 151-157 (2002).
182. C. J. Small, M. L. Goubillon, J. F. Murray, A. Siddiqui, S. E. Grimshaw, H. Young, V. Sivanesan, T. Kalamatianos, A. R. Kennedy, C. W. Coen, S. R. Bloom and C. A. Wilson. Central orexin A has site-specific effects on luteinizing hormone release in female rats, *Endocrinology* **144**(7), 3225-3236 (2003).
183. M. S. Mondal, M. Nakazato, Y. Date, N. Murakami, M. Yanagisawa and S. Matsukura. Widespread distribution of orexin in rat brain and its regulation upon fasting, *Biochem.Biophys.Res.Commun.* **256**(3), 495-499 (1999).
184. M. L. Hakansson, H. Brown, N. Ghilardi, R. C. Skoda and B. Meister. Leptin receptor immunoreactivity in chemically defined target neurons of the hypothalamus, *J.Neurosci.* **18**(1), 559-572 (1998).
185. M. Hakansson, L. de Lecea, J. G. Sutcliffe, M. Yanagisawa and B. Meister. Leptin receptor- and STAT3-immunoreactivities in hypocretin/orexin neurones of the lateral hypothalamus, *J.Neuroendocrinol.* **11**(8), 653-663 (1999).
186. J. Iqbal, S. Pompolo, T. Murakami, E. Grouzmann, T. Sakurai, B. Meister and I. J. Clarke. Immunohistochemical characterization of localization of long-form leptin receptor (OB-Rb) in neurochemically defined cells in the ovine hypothalamus, *Brain Res.* **920**(1-2), 55-64 (2001).
187. B. Beck and S. Richy. Hypothalamic hypocretin/orexin and neuropeptide Y: divergent interaction with energy depletion and leptin, *Biochem.Biophys.Res.Commun.* **258**(1), 119-122 (1999).
188. A. Yamanaka, C. T. Beuckmann, J. T. Willie, J. Hara, N. Tsujino, M. Mieda, M. Tominaga, K. Yagami, F. Sugiyama, K. Goto, M. Yanagisawa and T. Sakurai. Hypothalamic orexin neurons regulate arousal according to energy balance in mice, *Neuron* **38**(5), 701-713 (2003).
189. T. Moriguchi, T. Sakurai, T. Nambu, M. Yanagisawa and K. Goto. Neurons containing orexin in the lateral hypothalamic area of the adult rat brain are activated by insulin-induced acute hypoglycemia, *Neurosci.Lett.* **264**(1-3), 101-104 (1999).
190. B. Griffond, P. Y. Risold, C. Jacquemard, C. Colard and D. Fellmann. Insulin-induced hypoglycemia increases preprohypocretin (orexin) mRNA in the rat lateral hypothalamic area, *Neurosci.Lett.* **262**(2), 77-80 (1999).
191. X. H. Liu, R. Morris, D. Spiller, M. White and G. Williams. Orexin a preferentially excites glucose-sensitive neurons in the lateral hypothalamus of the rat in vitro, *Diabetes* **50**(11), 2431-2437 (2001).
192. X. J. Cai, X. H. Liu, M. Evans, J. C. Clapham, S. Wilson, J. R. Arch, R. Morris and G. Williams. Orexins and feeding: special occasions or everyday occurrence?, *Regul.Pept.* **104**(1-3), 1-9 (2002).
193. R. M. Boyar, R. S. Rosenfeld, S. Kapen, J. W. Finkelstein, H. P. Roffwarg, E. D. Weitzman and L. Hellman. Human puberty. Simultaneous augmented secretion of luteinizing hormone and testosterone during sleep, *J Clin.Invest* **54**(3), 609-618 (1974).
194. P. K. Opstad and A. Aakvaag. Decreased serum levels of oestradiol, testosterone and prolactin during prolonged physical strain and sleep deprivation, and the influence of a high calorie diet, *Eur.J Appl.Physiol Occup.Physiol* **49**(3), 343-348 (1982).
195. V. Cortes-Gallegos, G. Castaneda, R. Alonso, I. Sojo, A. Carranco, C. Cervantes and A. Parra. Sleep deprivation reduces circulating androgens in healthy men, *Arch.Androl* **10**(1), 33-37 (1983).
196. Y. Touitou, O. Benoit, J. Foret, A. Aguirre, A. Bogdan, M. Clodore and C. Touitou. Effects of a two-hour early awakening and of bright light exposure on plasma patterns of cortisol, melatonin, prolactin and testosterone in man, *Acta Endocrinol.(Copenh)* **126**(3), 201-205 (1992).
197. K. Opstad. Circadian rhythm of hormones is extinguished during prolonged physical stress, sleep and energy deficiency in young men, *Eur.J Endocrinol.* **131**(1), 56-66 (1994).
198. F. Singer and B. Zumoff. Subnormal serum testosterone levels in male internal medicine residents, *Steroids* **57**(2), 86-89 (1992).
199. R. M. Chemelli, J. T. Willie, C. M. Sinton, J. K. Elmquist, T. Scammell, C. Lee, J. A. Richardson, S. C. Williams, Y. Xiong, Y. Kisanuki, T. E. Fitch, M. Nakazato, R. E. Hammer, C. B. Saper and M. Yanagisawa. Narcolepsy in orexin knockout mice: molecular genetics of sleep regulation, *Cell* **98**(4), 437-451 (1999).
200. O. Johren, S. J. Neidert, M. Kummer and P. Dominiak. Sexually dimorphic expression of prepro-orexin mRNA in the rat hypothalamus, *Peptides* **23**(6), 1177-1180 (2002).
201. S. Taheri, M. Mahmoodi, J. Opacka-Juffry, M. A. Ghatei and S. R. Bloom. Distribution and quantification of immunoreactive orexin A in rat tissues, *FEBS Lett.* **457**(1), 157-161 (1999).

202. M. C. García, M. López, O. Gualillo, L. Seoane, C. Diéguez and R. Señarís. Hypothalamic levels of NPY, MCH, and prepro-orexin mRNA during pregnancy and lactation in the rat: role of prolactin, *FASEB J.* **17**(11), 1392-1400 (2003).
203. J. B. Wang, T. Murata, K. Narita, K. Honda and T. Higuchi. Variation in the expression of orexin and orexin receptors in the rat hypothalamus during the estrous cycle, pregnancy, parturition, and lactation, *Endocrine.* **22**(2), 127-134 (2003).
204. Z. A. Archer, P. A. Findlay, S. M. Rhind, J. G. Mercer and C. L. Adam. Orexin gene expression and regulation by photoperiod in the sheep hypothalamus, *Regul.Pept.* **104**(1-3), 41-45 (2002).
205. D. N. Jones, J. Gartlon, F. Parker, S. G. Taylor, C. Routledge, P. Hemmati, R. P. Munton, T. E. Ashmeade, J. P. Hatcher, A. Johns, R. A. Porter, J. J. Hagan, A. J. Hunter and N. Upton. Effects of centrally administered orexin-B and orexin-A: a role for orexin-1 receptors in orexin-B-induced hyperactivity, *Psychopharmacology (Berl)* **153**(2), 210-218 (2001).
206. T. Yamamoto, H. Suzuki, H. Uemura, K. Yamamoto and S. Kikuyama. Localization of orexin-A-like immunoreactivity in prolactin cells in the bullfrog (*Rana catesbeiana*) pituitary, *Gen.Comp Endocrinol.* **135**(2), 186-192 (2004).
207. U. Beck and D. Marquetand. Effects of selective sleep deprivation on sleep-linked prolactin and growth hormone secretion, *Arch.Psychiatr.Nervenkr.* **223**(1), 35-44 (1976).
208. U. Beck. Hormonal secretion during sleep in man. Modification of growth hormone and prolactin secretion by interruption and selective deprivation of sleep, *Int.J Neurol.* **15**(1-2), 17-29 (1981).
209. A. Steiger. Sleep and endocrine regulation, *Front Biosci.* **8** s358-s376 (2003).
210. C. A. Everson and W. R. Crowley. Reductions in circulating anabolic hormones induced by sustained sleep deprivation in rats, *Am.J Physiol Endocrinol.Metab* (2004).
211. S. Q. Zhang, M. Kimura and S. Inoue. Effects of prolactin on sleep in cyclic rats, *Psychiatry Clin.Neurosci.* **53**(2), 101-103 (1999).
212. M. S. Smith. Site of action of prolactin in the suppression of gonadotropin secretion during lactation in the rat: effect on pituitary responsiveness to LHRH, *Biol.Reprod.* **24**(5), 967-976 (1981).
213. K. Fujita. Changes of the finer structures in the cells of the thyroid gland induced by electrical stimulation of the hypothalamus in rabbits, *Proc Doctor's Treatises, Osaka Univ* F-10 (1947).
214. J. R. Davis. Decreased metabolic rates contingent upon lateral hypothalamic lesion-induced body weight losses in male rats., *J Comp Physiol Psychol.* **91** 1019-1031 (1977).
215. T. Kurotsu. Changes of the finer structures of the different gland cells induced by electrical stimulation of the hypothalamus, *Med.J.Osaka Univ.* **5**(87), 104 (1954).
216. L. N. Kaufman, S. W. Corbett and R. E. Keesey. Relationship of thyroid hormones and norepinephrine to the lateral hypothalamic syndrome, *Metabolism* **35**(9), 847-851 (1986).
217. T. Suzuki, H. Kohno, T. Sakurada, T. Tadano and K. Kisara. Intracranial injection of thyrotropin releasing hormone (TRH) suppresses starvation-induced feeding and drinking in rats, *Pharmacol.Biochem.Behav.* **17**(2), 249-253 (1982).
218. T. Mitsuma, Y. Hirooka, Y. Mori, M. Kayama, K. Adachi, N. Rhue, J. Ping and T. Nogimori. Effects of orexin A on thyrotropin-releasing hormone and thyrotropin secretion in rats, *Horm.Metab Res.* **31**(11), 606-609 (1999).
219. S. H. Russell, C. J. Small, D. Sunter, I. Morgan, C. L. Dakin, M. A. Cohen and S. R. Bloom. Chronic intraparaventricular nuclear administration of orexin A in male rats does not alter thyroid axis or uncoupling protein-1 in brown adipose tissue, *Regul.Pept.* **104**(1-3), 61-68 (2002).
220. M. S. Kim, C. J. Small, S. A. Stanley, D. G. Morgan, L. J. Seal, W. M. Kong, C. M. Edwards, S. Abusnana, D. Sunter, M. A. Ghatei and S. R. Bloom. The central melanocortin system affects the hypothalamo-pituitary thyroid axis and may mediate the effect of leptin, *J.Clin.Invest* **105**(7), 1005-1011 (2000).
221. M. Lubkin and A. Stricker-Krongrad. Independent feeding and metabolic actions of orexins in mice, *Biochem.Biophys.Res.Commun.* **253**(2), 241-245 (1998).
222. A. Asakawa, A. Inui, K. Goto, H. Yuzuriha, Y. Takimoto, T. Inui, G. Katsuura, M. A. Fujino, M. M. Meguid and M. Kasuga. Effects of agouti-related protein, orexin and melanin-concentrating hormone on oxygen consumption in mice, *Int.J Mol.Med.* **10**(4), 523-525 (2002).
223. J. Wang, T. Osaka and S. Inoue. Energy expenditure by intracerebroventricular administration of orexin to anesthetized rats, *Neurosci.Lett.* **315**(1-2), 49-52 (2001).
224. J. Hara, C. T. Beuckmann, T. Nambu, J. T. Willie, R. M. Chemelli, C. M. Sinton, F. Sugiyama, K. Yagami, K. Goto, M. Yanagisawa and T. Sakurai. Genetic ablation of orexin neurons in mice results in narcolepsy, hypophagia, and obesity, *Neuron* **30**(2), 345-354 (2001).
225. G. J. Lammers, H. Pijl, J. Iestra, J. A. Langius, G. Buunk and A. E. Meinders. Spontaneous food choice in narcolepsy, *Sleep* **19**(1), 75-76 (1996).
226. T. Sakurai. Roles of orexins in the regulation of feeding and arousal, *Sleep Med.* **3** (Suppl 2) S3-S9 (2002).

227. A. Schuld, J. Hebebrand, F. Geller and T. Pollmacher. Increased body-mass index in patients with narcolepsy, *Lancet* **355**(9211), 1274-1275 (2000).
228. M. López, L. Seoane, R. Señarís and C. Diéguez. Prepro-orexin mRNA levels in the rat hypothalamus, and orexin receptors mRNA levels in the rat hypothalamus and adrenal gland are not influenced by the thyroid status, *Neurosci.Lett.* **300**(3), 171-175 (2001).
229. S. Ishii, J. Kamegai, H. Tamura, T. Shimizu, H. Sugihara and S. Oikawa. Hypothalamic neuropeptide Y/Y1 receptor pathway activated by a reduction in circulating leptin, but not by an increase in circulating ghrelin, contributes to hyperphagia associated with triiodothyronine-induced thyrotoxicosis, *Neuroendocrinology* **78**(6), 321-330 (2003).
230. A. B. Steffens. The modulatory effect of the hypothalamus on glucagon and insulin secretion in the rat, *Diabetologia* **20** (Suppl) 411-416 (1981).
231. J. H. Strubbe and A. B. Steffens. Rapid insulin release after ingestion of a meal in the unanesthetized rat, *Am.J Physiol* **229**(4), 1019-1022 (1975).
232. M. G. Tordoff, C. V. Grijalva, D. Novin, L. L. Butcher, J. H. Walsh, F. X. Pi-Sunyer and D. A. VanderWeele. Influence of sympathectomy on the lateral hypothalamic lesion syndrome, *Behav.Neurosci.* **98**(6), 1039-1059 (1984).
233. C. V. Grijalva, D. Novin and G. A. Bray. Alterations in blood glucose, insulin and free fatty acids following lateral hypothalamic lesions and parasagittal knife cuts, *Brain Res.Bull.* **5**(Suppl 4), 109-117 (1980).
234. L. L. Bernardis, L. L. Bellinger and A. Awad. Metabolic-endocrine correlates of the lateral hypothalamic syndrome: the first 48 hours, *Pharmacol.Biochem.Behav.* **37**(3), 393-398 (1990).
235. A. M. Helman, P. Giraud, S. Nicolaidis, C. Oliver and R. Assan. Glucagon release after stimulation of the lateral hypothalamic area in rats: predominant beta-adrenergic transmission and involvement of endorphin pathways, *Endocrinology* **113**(1), 1-6 (1983).
236. R. M. Buijs, S. J. Chun, A. Niijima, H. J. Romijn and K. Nagai. Parasympathetic and sympathetic control of the pancreas: a role for the suprachiasmatic nucleus and other hypothalamic centers that are involved in the regulation of food intake, *J.Comp Neurol.* **431**(4), 405-423 (2001).
237. X. Wu, J. Gao, J. Yan, C. Owyang and Y. Li. Hypothalamus-brainstem circuitry responsible for vagal efferent signaling to the pancreas evoked by hypoglycemia in rat, *J Neurophysiol.* **91**(4), 1734-1747 (2004).
238. K. Miyasaka, M. Masuda, S. Kanai, N. Sato, M. Kurosawa and A. Funakoshi. Central Orexin-A stimulates pancreatic exocrine secretion via the vagus, *Pancreas* **25**(4), 400-404 (2002).
239. A. J. Kastin and V. Akerstrom. Orexin A but not orexin B rapidly enters brain from blood by simple diffusion, *J.Pharmacol.Exp.Ther.* **289**(1), 219-223 (1999).
240. K. W. Nowak, P. Mackowiak, M. M. Switonska, M. Fabis and L. K. Malendowicz. Acute orexin effects on insulin secretion in the rat: in vivo and in vitro studies, *Life Sci.* **66**(5), 449-454 (2000).
241. D. G. Baskin, L. D. Figlewicz, R. J. Seeley, S. C. Woods, D. Porte, Jr. and M. W. Schwartz. Insulin and leptin: dual adiposity signals to the brain for the regulation of food intake and body weight, *Brain Res.* **848**(1-2), 114-123 (1999).
242. M. W. Schwartz, S. C. Woods, D. Porte, Jr., R. J. Seeley and D. G. Baskin. Central nervous system control of food intake, *Nature* **404**(6778), 661-671 (2000).
243. G. Grabauskas and H. C. Moises. Gastrointestinal-projecting neurones in the dorsal motor nucleus of the vagus exhibit direct and viscerotopically organized sensitivity to orexin, *J Physiol* **549**(Pt 1), 37-56 (2003).
244. G. Burdyga, S. Lal, D. Spiller, W. Jiang, D. Thompson, S. Attwood, S. Saeed, D. Grundy, A. Varro, R. Dimaline and G. J. Dockray. Localization of orexin-1 receptors to vagal afferent neurons in the rat and humans, *Gastroenterology* **124**(1), 129-139 (2003).
245. N. Takahashi, T. Okumura, H. Yamada and Y. Kohgo. Stimulation of gastric acid secretion by centrally administered orexin- A in conscious rats, *Biochem.Biophys.Res.Commun.* **254**(3), 623-627 (1999).
246. M. Kobashi, Y. Furudono, R. Matsuo and T. Yamamoto. Central orexin facilitates gastric relaxation and contractility in rats, *Neurosci.Lett.* **332**(3), 171-174 (2002).
247. K. Matsuo, M. Kaibara, Y. Uezono, H. Hayashi, K. Taniyama and Y. Nakane. Involvement of cholinergic neurons in orexin-induced contraction of guinea pig ileum, *Eur.J.Pharmacol.* **452**(1), 105 (2002).
248. M. A. Greer. Suggestive evidence of a primary drinking center in hypothalamus of the rat, *Proc.Soc.Exp.Biol.Med.* **89**(1), 59-62 (1955).
249. N. Rowland. Recovery of regulatory drinking following lateral hypothalamic lesions: nature of residual deficits analyzed by NaCl and water infusions, *Exp.Neurol.* **53**(2), 488-507 (1976).
250. J. F. Marshall and P. Teitelbaum. Further analysis of sensory inattention following lateral hypothalamic damage in rats, *J Comp Physiol Psychol.* **86**(3), 375-395 (1974).

251. G. A. Oltmans and J. A. Harvey. Lateral hypothalamic syndrome in rats: a comparison of the behavioral and neurochemical effects of lesions placed in the lateral hypothalamus and nigrostriatal bundle, *J Comp Physiol Psychol.* **90**(11), 1051-1062 (1976).
252. K. Kunii, A. Yamanaka, T. Nambu, I. Matsuzaki, K. Goto and T. Sakurai. Orexins/hypocretins regulate drinking behaviour, *Brain Res.* **842**(1), 256-261 (1999).

THE HYPOCRETINS IN CARDIOVASCULAR REGULATION

Tetsuro Shirasaka and Hiroshi Kannan*

1. INTRODUCTION

The hypothalamus plays a pivotal role in the integrated control of feeding and energy homeostasis and acts as major regulatory center for the autonomic and endocrine systems[1]. Hypocretin-1 and -2,[2] also known as orexin-A and –B,[3] respectively are two isolated hypothalamic peptides. Recent observations[4,5] implicate hypocretinergic neuronal activity in the sleep disorder narcolepsy, which is a debilitating neurological disease characterized by excessive daytime sleepiness, premature transitions to rapid eye movement (REM) sleep, and cataplexy. During arousal, hypocretin neurons become activated. This is demonstrated by Fos expression.[6] Increases in the sympathetic drive also occur during arousal and affects cardiovascular and renal function.[7] Hypocretin receptors[3] are expressed and hypocretinergic neurons[8] project throughout the central nervous system (CNS). In addition to controlling sleep, they are also responsible for the control of feeding, neuroendocrine homeostasis, and autonomic regulation. Within the hypothalamus, hypocretin nerve fibers[8] and hypocretin receptors (OX_1R and OX_2R),[3] especially OX_2R, are found extensively in the hypothalamic paraventricular nucleus (PVN). The PVN is thought to be involved in the control of the autonomic nervous system, cardiovascular function, and the neuroendocrine system.[9,10] Thus hypocretins may have a role in the regulation of the cardiovascular system. The aim of this chapter is to summarize our recent studies in which we used direct recording of sympathetic nerve activity in conscious rats[11] and an in vitro whole cell patch-clamp technique to examine the direct effect of hypocretins on PVN neurons using a hypothalamic slice.[12] These studies were performed to elucidate the central actions of hypocretins on cardiovascular function and the autonomic nervous system.

* T Shirasaka Dept. of Anesthesiology; H. Kannan, Dept of Physiology. Miyazaki Medical College, 5200 Kihara, Kiyotake, Miyazaki 889-1692, Japan

2. HYPOCRETINS AND CARDIOVASCULAR REGULATIONS *IN VIVO*

Hypocretin-containing neurons of the hypothalamus innervate multiple sites throughout the CNS, exerting widespread excitatory effects related to arousal and may also be involved in the fight-or-flight response. Immunohistochemical study[8] and the viral transneuronal labeling method[13] indicated that hypocretin systems are capable of modulating sympathetic outflow systems. We examined the effects of intracerebroventricularly (i.c.v)-administered hypocretins on mean arterial pressure (MAP), heart rate (HR), renal sympathetic nerve activity (RSNA), and plasma catecholamine (CA) in conscious rats.[11] I.c.v. administered hypocretin-1 dose-dependently increased MAP, HR and RSNA (Fig. 1A).

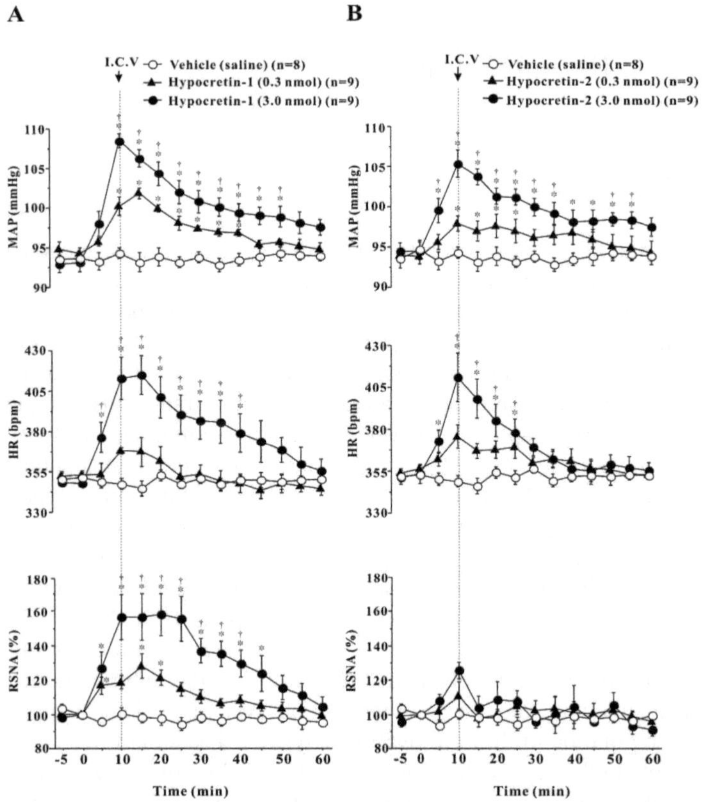

Figure 1. Time course of changes in mean arterial pressure (MAP), heart rate (HR), and renal sympathetic nerve activity (RSNA) during the 60 min following intracerebroventricular administration of hypocretin-1 (0.3, 3.0 nmol) (A) or -2 (0.3, 3.0 nmol) (B) in conscious rats. Vertical dotted line indicates time 0 min; bpm, beats per min. All data are mean ± SE; *n* is no. of animals. $P < 0.05$ vs. vehicle. $^{\dagger}P < 0.05$ vs. hypocretin-1 or -2 (0.3 nmol). Results are from ref. 11.

MAP and HR increased rapidly and reached peak value 10-15 min after hypocretin-1 administration. In also urethane or pentobarbital-anesthetized rats, central-administrations of hypocretin-1 or -2 dose-dependently increased MAP and HR.[14,15] On the other hand, i.v.-injection of the same dose of hypocretin-1 or -2 used in the i.c.v.-injection experiment failed to cause any cardiovascular and SNA change.[11,14] These findings suggest that pressor effects induced by hypocretin-1 or -2 were mediated via a CNS site of action. High dose of hypocretin-1 additionally produced a significant increase in RSNA 10 min after injection, which persisted for ~15 min. RSNA also increased transiently at a low dose (0.3 nmol) of hypocretin-1. There was a statistically significant correlation coefficient (r) between RSNA and MAP ($r = 0.69$ and $r = 0.83$, respectively; both P values < 0.001) or HR ($r = 0.76$ and $r = 0.89$, respectively; both P values < 0.001) at 0.3- and 3.0- nmol doses in hypocretin-1 injected group (Fig. 2). There was also a statistically significant correlation between MAP and plasma norepinephrine (NE) concentration when i.c.v.-administration of hypocretin-1 (5.0 nmol).[16]

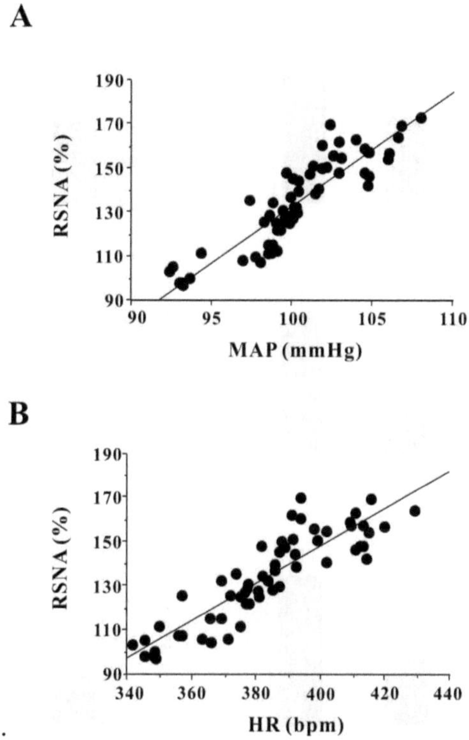

Figure 2. The relationship of renal sympathetic nerve activity (RSNA) and mean arterial pressure (MAP) (A) or heart rate (HR) (B). There was a statistically significant correlation between RSNA and MAP or HR ($r = 0.83$ and $r = 0.89$, respectively; both P values < 0.001) in the hypocretin-1 (3.0 nmol)-injected group. Results are from ref. 11.

I.c.v.-administered hypocretin-2 (3.0 nmol) also significantly increased MAP, this response pattern being similar to what was observed for hypocretin-1 administration (Fig. 1B). HR also rapidly increased and returned to control level within 30 min after hypocretin-2 (3.0 nmol) administration. In contrast to the results with hypocretin-1,

RSNA did not increase singnificantly at 0.3 and 3.0 nmol dose of hypocretin-2. However, at 30 nmol dose of hypocretin-2 resulted in ~40 % (p <0.001) increase in RSNA 20 min after injection (unpublished data). For each dose, the maximum changes from control values during recording time (60 min) were compared for hypocretin-1 and -2 (Fig. 3A).

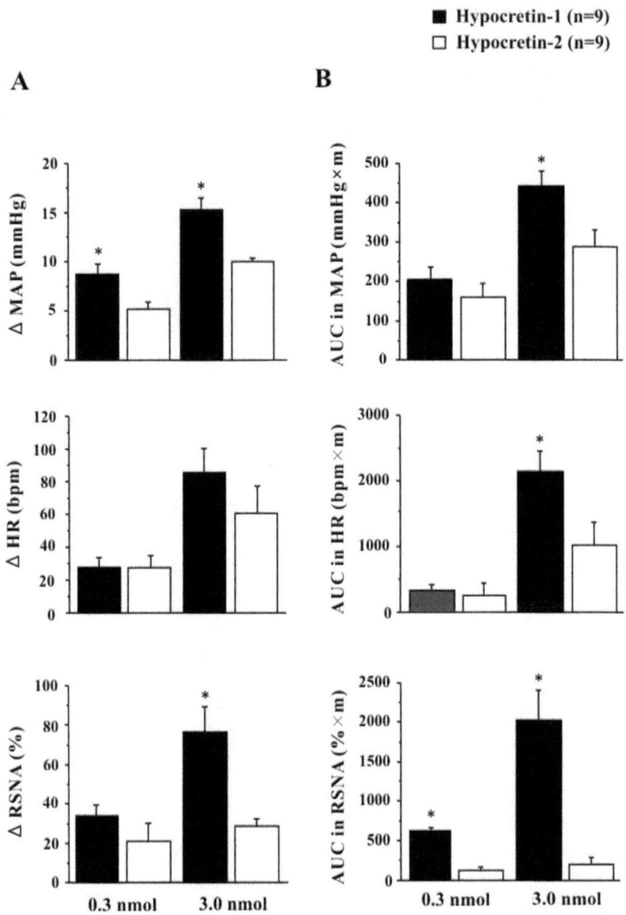

Figure 3. Bar graph showing maximal changes from control values (A) and the area under the curve (AUC; B) for MAP, HR, and RSNA during the 60 min following intracerebroventricular administration of hypocretin-1 and -2 (0.3, 3.0 nmol, respectively) in conscious rats; bpm, beats per min. All data are mean ± SE; n is no. of animals. $P < 0.05$ vs. hypocretin-2 for each dose. Results are from ref. 11.

An increase in MAP induced by central hypocretin-1 was 1.5-fold larger than that of hypocretin-2 for both doses, but significant differences were not observed in HR. Our results that the pressor response induced by hypocretin-1 is larger than -2, are agreement with other study.[15] The increase in RSNA produced by i.c.v.-administered hypocretin-1 was larger than that of -2 at 3.0 nmol. In addition, the area under the curve (AUC) in MAP and HR was significantly larger in hypocretin-1 than -2 at only 3.0 nmol (Fig. 3B). The AUC in RSNA was larger in hypocretin-1 than -2 at each dose. In almost all hypocretin i.c.v.-administered rats, increases in locomotor activities, such as chewing and grooming, which is known to be related with a stress response[17] were observed. Stress, wakefulness, muscle exercise and postural change are well known to induce the activation of sympathetic outflow.[18,19] To exclude the effects of stress, wakefulness, and/or locomotion on these parameters, hypocretin-1 and -2 (3.0 nmol) were administered centrally in rats anesthetized with pentobarbital (50 mg/kg i.p.). I.c.v.-administered hypocretin-1 significantly increased MAP, HR, and RSNA, and hypocretin-2 increased MAP and HR in anesthetized rats, indicating that the increases in these parameters were not due to the rats' activated locomotion and/or stress.

Noradrenergic neurons are active especially during waking,[20] and norepinephrine (NE) inhibits the activity of sleep promoting neurons in the ventrolateral preoptic nucleus, an area critical in the regulation of sleep.[21] The existence of regional differences in sympathetic outflow has been demonstrated.[22] Thus, to examine systemic sympathetic

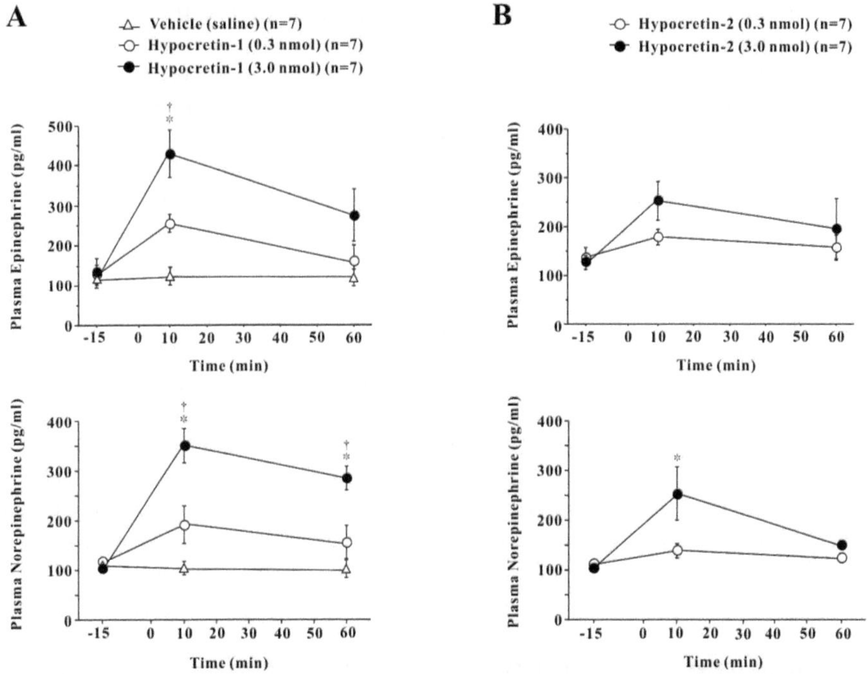

Figure 4. Effect of i.c.v. administration of vehicle (saline), or hypocretin-1 (0.3, 3.0 nmol) (A) and hypocretin-2 (0.3, 3.0 nmol) (B) on plasma concentration of epinephrine (Epi) and norepinephrine (NE) in conscious rats. 0 min, time of administration. All data are mean ± SE; *n* is no. of animals. $P < 0.05$ vs. pre-administration values. $^{\dagger}P < 0.05$ vs. hypocretin-1 (0.3 nmol). Results are from ref. 11.

outflow induced by central hypocretins, plasma CA was measured under similar conditions to record nerve activity (Fig. 4). High doses of hypocretin-1 and -2 increased plasma NE, the effect being larger and lasting longer with hypocretin-1. Therefore, it is likely that the hypocretin-induced increase in sympathetic nerve outflow leads to the increase in plasma NE, which produces cardiovascular responses.

I.c.v.-administered hypocretin-1 also significantly increased plasma epinephrine (Epi) level 10 min after injection. Rapid increases in plasma concentrations of ACTH and corticosterone, and the mRNA levels of CRF and AVP induced by hypocretin-1 were demonstrated in the parvocellular neurons of the PVN.[23] These results suggests that hypocretin-1 acts centrally to activate the hypothalamic-pituitary-adrenal (HPA) axis involving stimulation of both CRF and AVP expression. And also, central hypocretin-1 has been reported to increase plasma Epi, glucose and AVP levels in conscious rabbits.[14] The elevated circulating levels of Epi as well as NE after injections of the high dose of hypocretin-1 suggests the activation of the sympatho-adrenomedullary system (SA system). In contrast to hypocretin-1, central hypocretin-2 did not produce an increase in plasma Epi. The large pressor response induced by central hypocretin-1, compared to that induced by hypocretin-2, may be due to activation of the SA system in addition to sympathetic outflow. These results indicate that i.c.v.-administered hypocretin-1 and -2 produce cardiovascular responses via different central mechanisms.

Hypocretin-1 and -2 are endogenous neuropeptide agonists for hypocretin receptor-1 (OX_1R) and hypocretin receptor-2 (OX_2R).[3] Hypocretin-1 has equal affinity for OX_1R and OX_2R, while hypocretin-2 displays higher affinity for OX_2R.[3] The novel OX_1R antagonist (SB-334867), which alone did not significantly change baseline hemodynamic variables and plasma CA levels, markedly attenuated increases in MAP, HR and plasma NE concentration induced by hypocretin-1.[16] The results suggests that hypocretin-1 may regulate sympathetic and cardiovascular activity through mainly OX_1R. On the other hand, hypocretin-containing neurons in the hypothalamus including PVN neurons project to intermediolateral (IML) cell column of the thoracolumbar spinal cord,[8,24] which is the site of sympathetic preganglionic motor neurons involved in the regulation of HR and blood pressure (BP). The increases in MAP induced by central hypocretin-1 was abolished by a ganglion-blocking agent pentolinium.[14] The α_1-adrenergic receptor antagonist prazosin or the β-adrenergic receptor antagonist proranolol markedly diminished, respectively, the hypocretin-1 induced increase of MAP and HR.[25] These results support the contension that the pressor and tachycardic response induced by hypocretin-1 is mediated by the activation of sympathetic nervous system. Moreover, intrinsic hypocretin participates in BP maintenance at basal conditions probably through activation of the sympathetic vasoconstrictor outflow were demonstrated using hypocretin knockout mice.[26] The results indicate endogenous hypocretins play important role in cardiovascular and sympathetic regulations in the central nervous system.

3. HYPOCRETINS AND CARDIOVASCULAR REGULATIONS *IN VITRO*

The mRNA of two known orexin receptors (OX_1R and OX_2R) belongs to the G protein-coupled receptor superfamily and has a proposed seven-transmembrane topology that is observed exclusively in the rat brain.[3] The OX_1R and OX_2R mRNAs are differentially distributed, with OX_1R mRNA being most abundant in the ventromedial

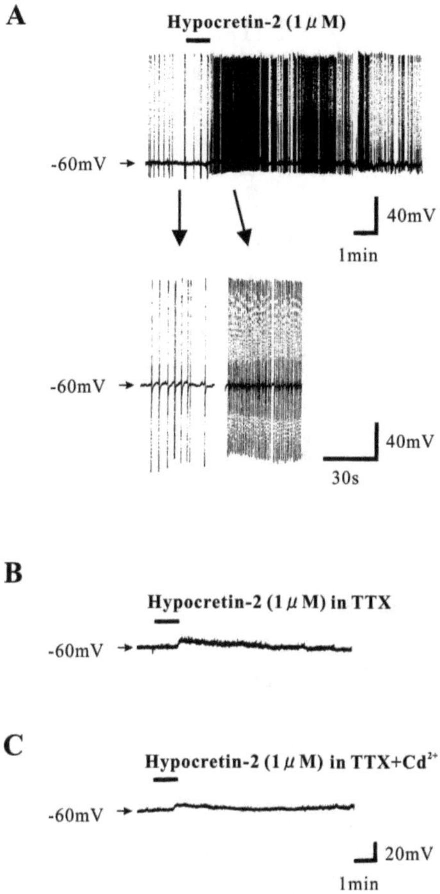

Figure 5. The effects of hypocretin-2 on membrane potential in hypothalamic paraventricular nucleus (PVN) neurons. Horizontal bars indicate the peptide application time (1 min). Arrows indicate the resting membrane potential (RMP). (A) (Upper) The application of hypocretin-2 (1 μM) induced a transient depolarization in magnocellular neuron in the PVN in normal artificial cerebrospinal fluid (ACSF). (Lower; left) Baseline firing at the RMP. (Lower; right) Increased action potential frequency at the peak of the response to 1 μM hypocretin-2. (B) In the presence of TTX (1 μM), hypocretin-2 (1 μM) depolarized a parvocellular neuron in the PVN. (C) Addition of Cd^{2+} (1 mM) in ACSF containing TTX (1 μM) significantly reduced the depolarization induced by hypocretin-2 (1 μM) in same parvocellular neurons. Results are from ref. 12.

hypothalamus (VMH), which plays an important role in the homeostatic regulation of body metabolism mediated through sympathetic nerves,[27] and OX_2R mRNA being predominantly expressed in the hypothalamic PVN, which is a heterogeneous structure comprised of neuronal populations that are grouped generally into magnocellular (type 1) and parvocellular (type 2) neurons.[9,10] The pre-autonomic parvocellular neurons of PVN send long descending projections to several areas within the CNS that are known to be important in cardiovascular function.[28] These regions include the NTS, where baroreceptor and chemorecepor afferents terminate, and the vagal complex present in the dorsomedial medulla, the (rostral ventrolateral medulla) RVLM, which is probably the

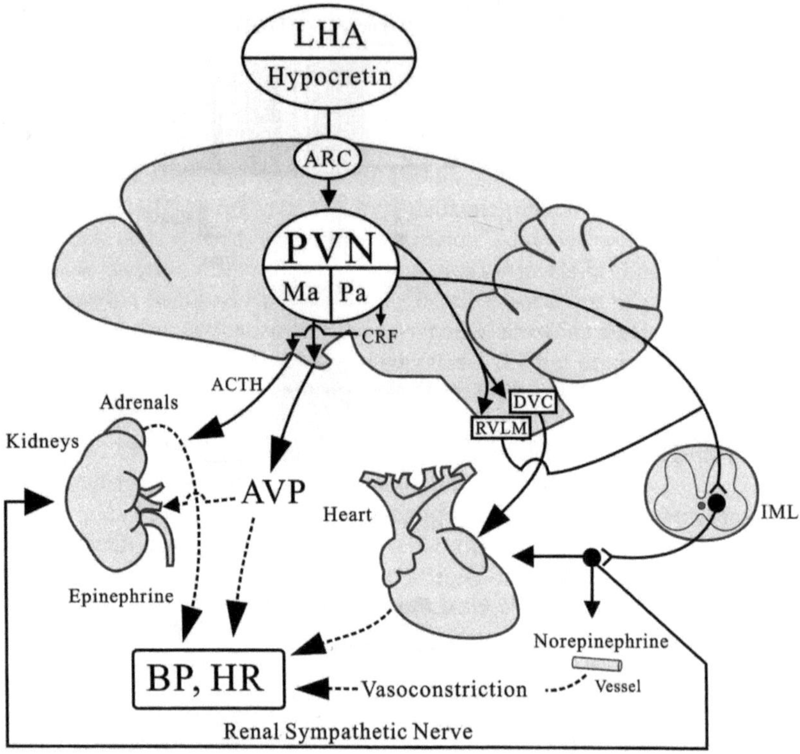

Figure. 6. Schematic diagram of possible mechanisms for the action of central hypocretin in cardiovascular, neuroendocrine and sympathetic outflows. Hypocretin bind to their receptors of the magnocellular or parvocellular neurons of the hypothalamic paraventricular nucleus, or arcuate nucleus neurons, causing their depolarization. Excitation of magnocellular neurons induces secretion of arginine vasopressin (AVP) from the posterior pituitary, anti-diuresis, and vasoconstriction. Conversely, parvocellular neurons activate autonomic centers in the brainstem and spinal cord, increasing the heart rate and blood pressure, or causing the release of corticotropin-releasing factor (CRF). Secretion of adrenocorticotropic hormone (ACTH) from the anterior pituitary is controlled by CRF and AVP, synthesized by the parvocellular neuron. The right lower vessel indicates that norepinephrine released from the sympathetic nerve ending induces vasoconstriction. Activation of the renal sympathetic nerve induces anti-diuresis and secretion of renin. The solid lines indicate a neural or humoral pathway. The dotted lines indicate a functional influence. ARC, arcuate nucleus; DVC, dorsal vagal complex; IML, inter mediolateral cell column; LHA, lateral hypothalamus; PVN, paraventricular nucleus; Ma, magnocellular neuron; Pa, parvocellular neuron; RVLM, rostral ventrolateral medulla. (From ref. 31.)

major site for generation of sympathetic tone for the vasculature, and the intermediolateral (IML) cell column of the thoraco-lumbar spinal cord, which is the site of sympathetic pregangliоic neurons (SPN) involved in the regulation of HR and BP. Hypothalamic PVN is also reported to be involved in arousal mechanisms induced by hypocretins.[29] To determine the effect of hypocretins on PVN physiology, whole cell patch clamp recordings were obtained from PVN neurons in rat brain slices.[12,30] About 70 - 80 % of magnocellular and parvocellular neurons in the PVN responded to hypocretin-1 or -2. Hypocretin-1 produced a depolarization accompanied by an increase in action potential firing in a dose-dependent manner[12] and also produced increases in the excitatory postsynaptic current (EPSC) frequency and amplitude in magnocellular neurons.[30] The depolarizing effects of hypocretin-1 on magnocellular neurons were mediated by the activation of glutamatergic transmission.[30] Depolarization of parvocellular neurons by hypocretin-1 were mediated by the direct activation of hypocretin receptors on parvocellular neurons.[12] A voltage-clamp study revealed that hypocretin-1 induced a non-selective cationic conductance with a reversal potential of -40 mV in parvocellular neurons.[30] Hypocretin-2 also depolarizes both types of PVN neurons in a concentration-dependent manner[12] (Fig. 5). The depolarizing responses were greater in parvocellular than in magnocellular neurons. The effects of hypocretin-2 persisted in the presence of (tetrodotoxin) TTX, indicating them to be a direct effect (Fig. 5B). Addition of Cd^{2+} (1 mM) to ACSF containing TTX significantly reduced the depolarizing effect in parvocellular neurons (Fig. 5C).

The excitation of magnocellular neurons induces the secretion of arginine vasopressin (AVP) from the posterior pituitary, antidiuresis, and vasoconstriction (Fig. 6).[31] On the other hand, parvocellular neurons activate autonomic centers in the brain stem and spinal cord, increasing the HR and BP or causing the release of CRF. Secretion of ACTH from the posterior pituitary is controlled by CRF and AVP, synthesized by the parvocellular neuron. These studies suggest that hypocretin-1 and -2 depolarize PVN neurons and increase the firing rate, leading to modulation of autonomic, cardiovascular, and neuroendocrine functions. In addition, dense orexinergic projections innervate the IML of the spinal cord. Hypocretin-1 or -2 directly and reversibly depolarized sympathetic preganglionic neurons (SPN) in spinal cord slices.[25] These findings indicate that hypocretins also excite sympathetic activity at the spinal level and modulate cardiovascular functions.

4. CONCLUSIONS

The hypocretin system is involved in the integration of hypothalamic functions with sleep-wakefulness, and diverse autonomic functions, including cardiovascular control, hormone secretion, and energy metabolism. Hypocretins have a direct excitatory postsynaptic effect on PVN neurons leading to diverse pathophysiological consequences, including autonomic, neuroendocrine and cardiovascular function (Fig. 6).[31] The functional significance of the activation of cardiovascular and sympathetic function has not been established; several lines of evidence suggest that hypocretins may be activated under stress conditions[26] and cause an increase in BP, HR, and SNA as an autonomic adaptive response. Hypocretins are crucial in the maintenance of wakefulness.[32] Activation of PVN also induces cortical arousal[29] during non-REM sleep, eliciting bursts of SNA and increases in BP.[33] These results suggest that hypocretin arousal stimuli may, at least in part, be involved in increases in BP, HR and SNA.

Besides increasing feeding behavior, central injection of hypocretin-1 increases oxygen consumption indicating an increased metabolic rate.[34] Hypocretins also increase sympathetic outflow. These effects generally result in increased energy consumption. Hypothalamic prepro-hypocretin mRNA expression is reduced in grossly obese animals with a defective leptin system[35] and is increased by a high-fat diet, especially in those rats that become most obese.[36] Metabolic abnormalities in food intake and/or energy expenditure may exist in human narcolepsy. This is evidenced by increased frequencies of obesity and non-insulin dependent (type2) diabetes in narcoleptic patients compared with a control group.[37,38] Orexin/ataxin-3 mice showed late-onset obesity, consistent with human cases.[5] These results are consistent with the hypocretin system participating in a counter-regulatory response to obesity. Although the pathophysiological role of the sympathoexcitatory effects of hypocretins is not clear, the close relationship between obesity, hypertension, and altered cardiovascular responses has been documented in a number of studies.[39] Therefore, hypocretins may be the chemical mediators in the brain responsible for the generation and maintenance of hypertension that is associated with conditions of energy imbalance, such as obesity. The discovery of hypocretins marks an important milestone in our understanding of narcolepsy-cataplexy physiology and will stimulate further research into effector mechanisms in the brain and other organs involved in energy homeostasis.

5. REFERENCES

1. L. L. Bernardis, and L. L. Bellinger. The lateral hypothalamic area revisited: neuroanatomy, body weight regulation, neuroendocrinology and metabolism, *Neurosci. Biobehav. Rev.* **17**, 141-193 (1993).
2. L. de Lecea, T. S. Kilduff, C. Peyron, X. B. Gao, P. E. Foye, P. E. Danielson, C. Fukuhara, E. L. F. Battenberg, V. T. Gautvik, F. S. Bartlett, W. N. Frankel, A. N. van den Pol, F. E. Bloom, K. M. Gautvik, and J. G. Sutcliffe. The hypocretins: Hypothalamus-specific peptides with neuroexcitatory activity, *Proc. Natl. Acad. Sci. USA* **95**, 322-327 (1998).
3. T. Sakurai, A. Amemiya, M. Ishii, I. Matsuzaki, R. M. Chemelli, H. Tanaka, S. C. Williams, J. A. Richardson, G. P. Kozlowski, S. Wilson, J. R. S. Arch, R. E. Buckingham, A. C. Haynes, S. E. Carr, R. S. Annan, D. E. McNulty, W. S. Liu, J. A. Terrett, N. A. Elshourbagy, D. J. Bergsma, and M. Yanagisawa. Orexins and orexin receptors: a family of hypothalamic neuropeptides and G protein-coupled receptors that regulate feeding behavior, *Cell* **92**, 573-585 (1998).
4. M. Mieda, J. T. Willie, J. Hara, C. M. Sinton, T. Sakurai, and M. Yanagisawa. Orexin peptides prevent cataplexy and improve wakefulness in an orexin neuron-ablated model of narcolepsy in mice, *Proc. Natl. Acad. Sci. USA* **101**, 4649-4654 (2004).
5. J. Hara, C. T. Beuckmann, T. Nambu, J. T. Willie, R. M. Chemelli, C. M. Sinton, F. Sugiyama, K. Yagami, K. Goto, M. Yanagisawa, and T. Sakurai. Genetic ablation of orexin neurons in mice results in narcolepsy, hypophagia, and obesity, *Neuron* **30**, 345-354 (2001).
6. I. V. Estabrooke, M. T. McCarthy, E, Ko, T. C. Chou, R. M. Chemelli, M. Yanagisawa, C. B. Saper, and T. E. Scammell. Fos expression in orexin neurons varies with behavioral state, *J. Neurosci.* **21**, 1656-1662 (2001).
7. H. A. Futuro-Neto, and J. H. Coote. Changes in sympathetic activity to heart and blood vessels during desynchronized sleep, *Brain Res.* **252**, 259-268 (1982).
8. Y. Date, Y. Ueta, H. Yamashita, Y. Yamaguchi, S. Matsukura, K. Kangawa, T. Sakurai, M. Yanagisawa, and M. Nakazato. Orexins, novel orexigenic hypothalamic peptides, interact with autonomic, neuroendocrine and neuroregulatory systems, *Proc. Natl. Acad. Sci. USA* **96**, 748-753 (1999).
9. L. W. Swanson, and H. G. Kuypers. The paraventricular nucleus of the hypothalamus: cytoarchitectonic subdivisions and organization of projections to the pituitary, dorsal vagal complex, and spinal cord as demonstrated by retrograde fluorescence double-labeling methods, *J. Comp. Neurol.* **194**, 555-570 (1980).
10. L. W. Swanson, and P. E. Sawchenko. Hypothalamic integration: organization of the paraventricular and supraoptic nuclei, *Annu. Rev. Neurosci.* **6**, 269-324 (1983).
11. T. Shirasaka, M. Nakazato, S. Matsukura, M. Takasaki, and H. Kannan. Sympathetic and cardiovascular actions of orexins in conscious rats, *Am. J. Physiol.* **277**, R1780-R1785 (1999).
12. T. Shirasaka, S, Miyahara, T. Kunitake, Q. H. Jin, K. Kato, M. Takasaki, and H. Kannan. Orexin depolarizes rat hypothalamic paraventricular nucleus neurons, *Am. J. Physiol.* **281**, R1114-R1118 (2001).
13. J. C. Geerling, T. C. Meetenleiter, and A. D. Loewy. Orexin neurons project to diverse sympathetic outflow systems, *Neuroscience* **122,** 541-550 (2003).
14. K. Matsumura, T. Tsuchihashi, and I. Abe. Central orexin-A augments sympathoadrenal outflow in conscious rabbits, *Hypertention* **37**, 1382-1387 (2001).
15. C. T. Chen, L. L. Hwang, J. K. Chang, and N. J. Dun. Pressor effects of orexins injected intracisternally and to rostral ventrolateral medulla of anesthetized rats. *Am. J. Physiol.* **278**, R692-R697 (2000).
16. K. Hirota,Y. Kushikata, M. Kudo, T. Kudo, D. Smart, and A. Matsuki. Effects of central hypocretin-1 administration on hemodynamic responses in young-adult and middle-aged rats, *Brain Res.* **981**, 143-150 (2003).

17. W. H. Gispen, and R. L. Isaacson. ACTH-induced excessive grooming in the rat, *Pharmacol. Ther.* **12**, 209-246 (1981).
18. K. Matsukawa, J. H. Mitchell, P. T. Wall, and L. B. Wilson. The effect of static exercise on renal sympathetic nerve activity in conscious cats, *J. Physiol. (Lond)* **434**, 453-467 (1991).
19. H. A. Futuro-Neto, and J. H. Coote. Changes in sympathetic activity to heart and blood vessels during desynchronized sleep, *Brain Res.* **252**, 259-268 (1982).
20. Y. Kayama, and Y. Koyama. Brainstem neural mechanisms of sleep and wakefulness, *Eur. Urol.* **3**, 12-15 (1998).
21. T. Gallopin, P. Fort, E. Eggermann, B. Cauli, P. H. Luppi, J. Rossier, E. Audinat, M. Muhlethaler, and M. Serafin. Identification of sleep-promoting neurons in vitro, *Nature* **404**, 992-995 (2000).
22. O. E. Walther, M, Iriki, and E, Simon. Antagonistic changes of blood flow and sympathetic activity in different vascular beds following central thermal stimulation. . Cutaneous and visceral sympathetic activity during spinal cord heating and cooling in anesthetized rabbits and cats, *Pflügers Arch* 319, 162-184 (1970).
23. K. A. Al-Barazanji, S. Wilson, J. Baker, D. S. Jessop, and M. S. Harbuz. Central orexin-A activates hypothalamic corticotropin releasing factor and arginine vasopressin neurons in conscious rats, *J. Neuroendocrinol.* **13**, 421-424 (2001).
24. A. N. Van den Pol. Hypothalamic hypocretin (orexin): robust innervation of the spinal cord, *J. Neurosci.* **19**, 3171-3182 (1999).
25. V. R. Antunes, G. C. Brailoiu, E. H. Kwok, P. Scruggs, and N. J. Dun. Orexin/hypocretin excite rat sympathetic preganglionic neurons in vivo and in vitro, *Am. J. Physiol.* **281**, R1801-R1807 (2001).
26. Y. Kayaba, A. Nakamura, Y. Kasuya, T. Ohuchi, M. Yanagisawa, I. Komuro, Y. Fukuda, and T. Kuwaki. Attenuated defense response and low basal blood pressure in orexin knockout mice, *Am. J. Physiol.* **285**, R581-R593 (2003).
27. T. Shimazu. Nervous control of peripheral metabolism, *Acta. Physiol. Pol.* **30**, 1-18 (1979).
28. A. D. Shafton, A. Ryan, and E. Badoer. Neurons in the hypothalamic paraventricular nucleus send collaterals to the spinal cord and to the rostral ventrolateral medulla in the rat, *Brain Res.* **801**, 239-243 (1998).
29. I. Sato-Suzuki, I. Kita, Y. Seki, M. Oguri, and H. Arita. Cortical arousal induced by microinjection of orexins into the paraventricular nucleus of the rat, *Behav. Brain Res.* **128**, 169-177 (2002).
30. M. J. Follwell, and A. V. Ferguson. Cellular mechanisms of orexin actions on paraventricular nucleus neurones in rat hypothalamus, *J. Physiol.* **545**, 855-867 (2002).
31. T. Shirasaka, M. Takasaki , and H. Kannan. Cardiovascular effects of leptin and orexins, *Am. J. Physiol.* **284**, R639-R651 (2003).
32. C. B. Saper, T. C. Chou, and T. E. Scammel. The sleep switch: hypothalamic control of sleep and wakefulness, *Trends. Neurosci.* **24**, 726-731 (2001).
33. V. K. Somers, M. E. Dyken, A. L. Mark, and F. M. Abboud. Sympathetic nerve activity during sleep in normal subjects, *N. Engl. J. Med.* **328**, 303-307 (1993).
34. M. Lubkin, and A. Stricker-Krongrad. Independent feeding and metabolic actions of orexins in mice, Biochem. Biophys. Res. Commun**.** **253**, 241-245 (1998).
35. Y. Yamamoto, Y. Ueta, Y. Date, M. Nakazato, Y. Hara, R. Serino, M. Nomura, I. Shibuya, S. Matsukura, and H. Yamashita. Down regulation of the *prepro-orexin* gene expression in genetically obese mice, *Mol. Brain Res.* **65**, 14-22 (1999).
36. K. E. Wortley, G. Q. Chang, Z. Davydova, and S. F. Leibowitz. Peptides that regulate food intake: orexin gene expression is increased during states of hypertriglyceridemia, *Am. J. Physiol.* **284**, R1454-R1465 (2003).
37. A. Schuld, J. Hebebrand, F. Geller, and T. Pollmacher. Increased body-mass index in patients with narcolepsy, *Lancet* **355**, 1274-1275 (2000).
38. Y. Honda, Y. Doi, R. Ninomiya, and C. Ninomiya. Increased frequency of non-insulin-dependent diabetes mellitus among narcoleptic patients, *Sleep* **9**, 254-259 (1986).
39. N. M. Kaplan. The deadly quartet: upper-body obesity, glucose intolerance, hyperglycemia, and hypertension, *Arch. Intern. Med.* **149**, 1514-1520 (1989).

INDEX

Acetylcholine, 46, 69, 109, 127, 153, 180
ACTH, 360, 370, 397, 426
Addiction, 317-321, 327-331
Adenosine A1 receptor, 63, 172
Adenylyl cyclase, 224, 375
Adrenal gland, 375, 395, 397
Afterdepolarization, 126, 139, 178, 292
Agonists, 204
Agouti-related peptide, 69, 81, 106, 306
Amino acid sequence, 6, 14
Amphetamines, 33, 49, 178, 246, 338, 341
Amygdala, 66, 317, 352, 358, 360
Antagonist, (see also *SB-334867*) 206
Antidepressants, 47, 176, 246
Antinociception, 211
Antisera to hypocretin, 7
Anxiety, 176
Arcuate nucleus, 65, 81, 89, 104, 306, 394
Arginine-vasopressin, 69, 370, 394, 397, 426
Arousal, 95, 132, 137, 150, 203, 315, 351, 393
Arthritis, 214
Atomoxetine, 246
Attention-deficit hyperactivity disorder, 176, 337
Autonomic functions, 393, 423

Barbiturate anaesthesia, 212
Basal forebrain, 48, 180, 210, 291, 294, 353
Basal ganglia, 110
Bed nucleus of the stria terminalis, 66, 338, 358
Behavioural satiety sequence, 207
Blood pressure, 212, 358, 370, 377, 424
Body temperature, 172, 358, 397
Brain reward function, 316

Caffeine, 33, 172
Calcium-activated non-selective cation current, 126, 133, 139, 160, 219, 292
Calcium imaging, 163, 198
canarc-1, 40, 43, 44
Canine narcolepsy, 39, 45, 237, 254
Carbachol, 48, 110, 128, 291
Cat, 64
Cataplexy, 31, 40, 235, 257, 277
Centromedial nucleus, 292
c-Fos, 106, 108, 110, 132, 172, 283, 291, 308, 316, 328, 340, 358, 371, 397, 423
Chemical structure, 13, 14
Chicken, 64
Cholecystokinin, 385
Cholinergic, 46, 69, 109, 127, 153, 169, 180, 291, 317
Circadian rhythm, 308, 354, 356
Circular permutation, 6, 7
Clozapine, 343
Cocaine- and amphetamine-related transcript, 62, 98, 100
Colocalization, 62, 97
Comparative neuroanatomy, 63

Cortex, 66, 111, 180, 191, 289, 292, 317, 339, 352
Corticosterone, 360, 370, 398
Corticotrophin releasing hormone, 81, 85, 271, 318, 351, 370, 394, 426
C-terminal amidation, 6, 22

Dementia, 267
Demonstration of excitatory activity, 8
Depression, 176, 270
Detection at synapses, 8, 62, 66, 89, 126
Dopamine beta-hydroxylase, 66, 109, 126
Dopamine, 46, 49, 169, 178, 210, 241, 317, 352
Dorsomedial hypothalamic nucleus, 108, 138, 147, 394
Drinking, 407
Drugs of abuse (see also *Addiction*)
  withdrawal, 327, 328
  relapse, 319
Duodenum, 385
Dynorphin, 63, 98, 124, 172

EEG recordings, 28, 40, 154, 209, 360
Encephalitis, 264
Endocrine effects, 283, 351, 369, 393, 423
Energy balance, 16, 36, 95, 104, 132, 203, 207, 242, 256, 305, 317, 377, 405
Enkephalin, 306
Enteric nervous system, 212, 384
Estrogen replacement, 373
Excitatory postsynaptic currents, 128, 130, 157, 195, 430
Extrahypothalamic expression, 65, 375, 385, 395
Follicle-stimulating hormone, 404
Food deprivation, 255, 306, 317, 385, 400
Food intake, 36, 106, 107, 207, 306, 377, 383
Frog, 64

G protein-coupled receptors, 16, 219, 374
G proteins, 220-224
GABA, 69, 82, 99, 100, 126, 142, 169, 170, 292
$GABA_A$, 63
Galanin, 63, 99, 108, 172, 305
Gamma-hydroxybutyrate, 246
Gastric emptying, 385, 387, 395, 407
Gastrin, 385, 388
Gene mutation in narcolepsy, 44, 279
Gene promoter, 15
Gene structure, 15
Genetic mapping, 5, 15
Ghrelin, 82, 132, 305, 385
Glucagon-like peptide 1, 132
Glucocorticoid, 82, 306, 375, 397
Glucose, 82, 132, 305, 319, 383, 402, 426
Glutamate, 63, 81, 100, 124, 126, 128, 157, 169, 170, 378
Glutamatergic synapses, 82, 103, 126
Gonadal axis, 401
Green fluorescent protein, 123, 172, 291, 375
Grooming, 210, 356, 398, 427
Growth hormone, 371, 398
Guillain-Barré syndrome, 263, 279, 281

Hamster, 64
Hcrtr-2, 16, 43
Heart rate, 172, 212, 358, 423
Hippocampus, 66, 169, 180, 317
Histamine, 51, 110, 128, 169, 172, 241, 283
Homology to secretin, 6, 7
Human, 64
Hyperlocomotion, 210
Hypersomnia, 262
Hypocretin
  afferents, 69, 81, 103
  autoregulation, 103
  cell number, 61, 102
  efferents, 65, 87
  in CSF, 45, 209, 233, 240, 254, 261, 278, 346, 354
  in development, 63
  in plasma, 396
  neuroanatomy, 61, 77, 95, 142
  neuronal morphology, 61, 77-79, 97, 123

neuronal recordings, 124-128
orexin equivalence, 15
replacement therapy, 53, 214, 247
Hypoglycemia, 306, 383
Hypophagia, 36
Hypothalamus-enriched mRNAs, 3
Hypothalamo-pituitary-adrenal axis, 289, 369-378, 397,

Idiopathic hypersomnia, 262
I_h current, 126, 292
Immune polyneuropathies, 263
Immunohistochemistry, 7, 61, 155, 340
In situ hybridization, 4, 5, 61, 97
Insomnia, 214
Insulin, 82, 104, 132, 305, 315, 317, 406
Interneurons, 81, 373
Intralaminar thalamic neurons, 191, 292
Intralipid, 308

Kleine-Levin syndrome, 265
Knockout mice, 17, 27, 207, 280, 289, 331, 377

Lactotrophs, 404
Lateral hypothalamic area, 61, 95, 315, 325, 340, 393
Lateral hypothalamic syndrome, 104
Lateral hypothalamic self-stimulation, 315, 326
Laterodorsal tegmental nucleus, 66, 153, 190, 317
Layer 6b neurons, 295
Layer V pyramidal neurons, 195, 295
Leptin receptor, 63, 68, 402
Leptin, 82, 104, 132, 173, 283, 305, 315, 317, 326, 386, 402
Locus coeruleus, 66, 89, 126, 132, 137, 180, 209, 221, 254, 283, 289, 332, 353, 383
Long-term potentiation, 182
Luteinizing hormone, 401

Maturation of preprohypocretin, 6, 17
Medial forebrain bundle, 291, 294, 326, 339
Medial preoptic area, 354, 401
Medial septal area, 354
Medullary reticular formation, 66
Melanin-concentrating hormone, 62, 81, 85, 96, 102, 125, 176, 208, 326
Membrane potential, 125, 139, 155, 292, 429
Memory, 169, 345, 352
Mesopontine cholinergic neurons, 153-159
Metabotropic glutamate receptor, 128, 130
MHC II linkage of narcolepsy, 43, 237, 239, 261, 278
Midline thalamus, 133, 191, 292
Migraine, 176
Miller-Fisher syndrome, 263
Modafinil, 33, 50, 178, 214, 246, 345
Models of narcolepsy, 27, 280
Monkey, 64
Morphine withdrawal, 124, 320, 328
Motivation, 315, 325
Multiple sclerosis, 245
Multiple sleep latency test, 233
Myenteric plexus, 384
Myoelectric motor complex, 386
Myotonic dystrophy, 265

Narcolepsy, 200, 208, 233, 253, 261, 277, 289
Nematosomes, 87
Neurodegeneration, 265, 281-283
Neuronal pentraxin (Narp), 63
Neuropeptide Y, 66, 68, 81, 89, 104, 208, 306, 315, 372, 394, 400, 406
Neurotensin, 99
Niemann-Pick disease, 265, 281
Nitric oxide synthase, 63, 100, 153, 155, 384, 386
Nomenclature, 9, 10, 13, 61
Norepinephrine, 46, 69, 82, 109, 126, 137, 169, 180, 212, 291, 345, 352, 425
Nucleus accumbens, 69, 110, 317, 332, 339
Nucleus of the solitary tract, 66, 383
Number of hypocretin neurons, 61, 102

Obesity, 36, 109, 132, 176, 208, 256, 283, 310, 377, 429
Obstructive sleep apnea syndrome, 263
Olanzapine, 343
Olfactory bulb, 66
Opioid dependence, 325
Opioid receptors, 124, 195, 328
Orchidectomy, 373, 404
orexin/ataxin-3 mice, 31, 35, 132, 207, 256, 280, 317, 377
orexin:EGFP mice 123-127
Organum vasculosum lamina terminalis, 370
Ovariectomy, 373, 402, 404
Oxygen consumption, 358, 377, 405
Oxytocin, 370, 394

Pain disorders, 269
Pain, 211
Pancreas, 395, 405
Pancreatic polypeptide Y4 receptor, 63
Paraventricular hypothalamic nucleus, 65, 81, 87, 306, 352, 358, 360, 369, 394, 423
Parkinson's disease, 173, 245, 266, 337
Parvalbumin, 66
Pedunculopontine nucleus, 66, 153, 317
Periaqueductal gray, 66, 178, 211
Parabrachial pontine region, 66
Perifornical hypothalamus, 61, 78, 105, 123, 142, 305, 340, 369, 394
Peripheral expression, 65
Pertussis toxin , 221
Phospholipase C, 224
Pig, 64
Pineal, 395
Pituitary, 373, 395
Plasticity, 169, 180 (see also *Synapses*)
Pontine reticular formation, 48, 153
Prader Willi syndrome, 265, 281
Precursor-protein convertase, 63
Prion disorders, 268
Prolactin, 371, 397, 404
Prolactin-like immunoreactive neurons, 63, 100, 306
Propopiomelanocortin, 81, 104, 106, 394

Raphe, 66, 126, 132, 169, 175, 283
Reboxetine, 246
Receptor binding parameters, 16, 204
Receptor coupling, 17, 18, 169, 221, 374
Receptor distribution, 20, 67, 126, 206, 223, 369, 395, 429
REM sleep, 28, 40, 109, 150, 154, 164, 171, 175, 208, 233, 270, 277, 291, 353
Replacement therapy, 214, 247
Rescue of narcoleptic mice, 35
Restless legs syndrome, 267
Reticular activating system, 153, 317
Reticular formation, 291
Reverse pharmacology, 13
Rhomboid nucleus, 292
Risperidone, 343

SB-334867, 205, 383, 386, 402, 426
SB-408124, 205
Schizophrenia, 271, 337
Secondary Narcolepsy, 245
Secretogranin II, 63, 99
Septum, 66
Serotonin, 46, 69, 82, 109, 126, 169, 175, 210
Sheep, 64
Sleep deprivation, 254
Sleep-off, 254
Small bowel motility, 386
Small bowel motor control, 212
Sodium-calcium exchanger, 170
Somatostatin, 385, 394, 399
Spinal cord, 66, 132, 211
Spontaneous spike frequency, 124
STAT-3, 63
Stress, 315, 351, 369, 398, 427
Subfornical organ, 370
Substance P, 99, 385
Substantia innominata, 354
Substantia nigra, 66, 169, 178, 221, 266
Subtractive hybridization, 3, 97
Suprachiasmatic nucleus, 69, 87, 107, 147, 394
Supramammillary nucleus, 66, 87
Supraoptic nucleus, 87, 394

Synapses on hypocretin neurons, 82, 126, 169

Testis, 375, 395
Testosterone treatment, 373
Thalamus
paraventricular and reuniens nuclei, 66
midline intralaminar nuclei, 193-200
Thalamocortical activating system, 191, 291
Thyroid hormone, 82, 405
Thyrotropin-releasing hormone, 99, 404
Tractus solitarius, 221
Transgenic mice, 27, 31, 35, 123-127 132, 207, 256, 280, 317, 377
Transgenic rat, 31
Tuberomammillary nucleus, 51, 110, 133, 169, 172, 211, 283, 289, 345, 394

Vagus, 386, 405, 429
Vasoactive intestinal peptide, 69, 385, 394
Venlafaxine, 246
Ventral tegmental area, 66, 154, 169, 178, 266, 317, 332, 338, 358
Ventrolateral preoptic area, 108, 172, 283, 297
Ventromedial hypothalamic nucleus, 87, 108
Vesicular glutamate transporter, 81, 84, 124, 126
Vote on nomenclature, 9

Weight gain, 343
Whole-cell recordings, 128-132, 158-161

Zebrafish, 64

MIX
Papier aus verantwortungsvollen Quellen
Paper from responsible sources
FSC® C105338

If you have any concerns about our products,
you can contact us on
ProductSafety@springernature.com

In case Publisher is established outside the EU,
the EU authorized representative is:
Springer Nature Customer Service Center GmbH
Europaplatz 3, 69115 Heidelberg, Germany

Printed by Libri Plureos GmbH
in Hamburg, Germany